Comprehensive Electrocardiology

Peter W. Macfarlane · A. van Oosterom · Olle Pahlm ·
Paul Kligfield · Michiel Janse · John Camm (Eds.)

Comprehensive Electrocardiology

Second Edition

Volume 4

With 698 Figures and 122 Tables

Springer

Editors
Peter W. Macfarlane
University of Glasgow
Glasgow
UK

A. van Oosterom
Radboud University Nijmegen
Nijmegen
The Netherlands

Olle Pahlm
Lund University
Lund
Sweden

Paul Kligfield
Weill Cornell Medical College
New York, NY
USA

Michiel Janse
University of Amsterdam
Amsterdam
The Netherlands

John Camm
St. George's, University of London
London
UK

ISBN 978-1-84882-045-6 ISBN 978-1-84882-046-3 (eBook)
Print and electronic bundle ISBN 978-1-84882-047-0
DOI 10.1007/978-1-84882-046-3

Library of Congress Control Number: 2010937436

Printed on acid-free paper

Springer is part of Springer Science+Business Media (www.springer.com) SPIN: 10981870 2109SPi–543210

Editors-in-Chief

Peter W. Macfarlane
University of Glasgow
Glasgow
UK

A. van Oosterom
Radboud University Nijmegen
Nijmegen
The Netherlands

Olle Pahlm
Lund University
Lund
Sweden

Paul Kligfield
Weill Cornell Medical College
New York, NY
USA

Michiel Janse
University of Amsterdam
Amsterdam
The Netherlands

John Camm
St. George's, University of London
London
UK

Preface

The first edition of *Comprehensive Electrocardiology* was published in 1989, when e-mail was still in its infancy (!!), and it was never envisaged at that time that a new edition would be prepared. It is probably fair to say that the majority of physicians would have regarded electrocardiography in particular as having reached its maximum usefulness with little additional information to be obtained therefrom. The intervening 20 years have shown how untrue this was.

An update to the book is long overdue. Sadly, some of the former contributors have died since the first edition was published and it is with regret that I note the passing of Philippe Coumel, Rudolph van Dam, David Detweiler, Karel den Dulk, Ramesh Gulrajani, Kenici Harumi, John Milliken, Jos Willems and Christoph Zywietz. Where relevant, their contributions continue to be acknowledged but in some cases, chapters have been completely rewritten by new contributors. On the other hand, eight completely new chapters have been added and the appendices restructured.

In some ways, it is inconceivable what has taken place in the field of electrocardiology since the first edition. New ECG patterns have been recognised and linked with sudden death, new prognostic indices have been developed and evaluated, the ECG has assumed a pivotal role in the treatment of an acute coronary syndrome and among many other things, automated ECG interpretation is now commonplace. Significant advances have been made in the field of mathematical modelling and a solution to the inverse problem is now applied in routine clinical use. Electrophysiological studies have taken giant steps over the past 20 years and biventricular pacing is a relatively recent innovation. Electrocardiology has certainly not stood still in the last 20 years. Of course there have been parallel advances in imaging techniques but the ECG still retains a unique position in the armamentarium of the physician, let alone the cardiologist.

For this edition, my previous co editor, Professor T D Veitch Lawrie, decided to step aside and I wish to congratulate him on reaching his 90th birthday in September 2010. However, I am pleased that other very eminent individuals agreed to assist with the editing of the book, namely Adriaan van Oosterom, Olle Pahlm, Paul Kligfield, Michiel Janse and John Camm. In the nature of things, some of these co-editors undertook much more work than others. I particularly have to acknowledge the support of Adriaan van Oosterom with preparation of volume 1 where he has now authored 4 chapters and edited 4 others. Olle Pahlm read copious numbers of chapters on which he commented while Paul Kligfield similarly helped review and edit many chapters. I am also grateful to Michiel Janse and John Camm for their assistance with relevant chapters in their field. Without the support of all, this edition would not have been possible.

Locally, I am very much indebted to my secretary Pamela Armstrong for a huge contribution in checking and subediting every chapter which went out from my office to the publisher. This was a Herculean task carried out with great aplomb. I would also like to thank Ms. Julie Kennedy for her contribution to a variety of tasks associated with preparing selected chapters, including enhancements to the English presentation on occasions.

I also wish to thank Springer for their considerable support throughout. Grant Weston initially commissioned the book and I am grateful to him for his confidence in supporting the preparation of a new edition. Jennifer Carlson in New York also assisted very significantly, as did the team in Pondicherry, India under the able supervision of Sivakumar Kunchithapatham.

I also must thank my long suffering wife Irene who has had to fight to gain access to our home PC almost every night over these past few years!

This 2nd Edition of *Comprehensive Electrocardiology* aims to bring together truly comprehensive information about the field. A book can never be completely up to date given the speed of publication of research findings over the internet these days but hopefully this publication will continue to be of significant use to readers for many years to come. This is particularly true with the many reference values to be found in the appendices, some of which are published for the first time, particularly data relating to the neonatal ECG collected in my own lab.

Now that this huge effort has been completed and the book is available electronically, it should be much easier to produce the next edition.......!!

Peter Macfarlane
Glasgow
Summer 2010

Table of Contents

Volume 4

List of Contributors

K. Martijn Akkerhuis
Erasmus University Medical Centre
Rotterdam
The Netherlands

Luigi De Ambroggi
University of Milan
Milan
Italy

Frits W. Bär
University of Maastricht
Maastricht
The Netherlands

R.C. Barr
Duke University
Durham, NC
USA

Dana Brooks
Northeastern University
Boston, MA
USA

Martin Buist
National University of Singapore
Singapore
Singapore

John Camm
St. George's, University of London
London
UK

Leo K. Cheng
The University of Auckland
Auckland
New Zealand

Alexandru D. Corlan
University Emergency Hospital of Bucharest
Bucharest
Romania

Alain Coulomb
Hopital Pitie-Salpetriere
Paris
France

Iwona Cygankiewicz
Medical University of Lodz
Lodz
Poland

D. Wyn Davies
Imperial College London
London
UK

David K. Detweiler
University of Pennsylvania
Philadelphia, PA
USA

Parvin C. Dorostkar
University of Minnesota
Minneapolis, MN
USA

Gordon E. Dower
Loma Linda University Medical Centre
Loma Linda, CA
USA

Karel Den Dulk
University of Maastricht
Maastricht
The Netherlands

Thomas Fåhraeus
University Hospital
Lund
Sweden

Guy Fontaine
Hopital Pitie-Salpetriere
Paris
France

Robert Frank
Hopital Pitie-Salpetriere
Paris
France

Alireza Ghodrati
Draeger Medical
Andover, MA
USA

Anton P.M. Gorgels
University of Maastricht
Maastricht
The Netherlands

Helena Hänninen
Helsinki University Central Hospital
Helsinki
Finland

Johan Herlitz
Sahlgrenska University Hospital
Gothenburg
Sweden

Gerard Van Herpen
Erasmus University Medical Centre
Rotterdam
The Netherlands

V. Hombach
University Hospital of Ulm
Ulm
Germany

B. Milan Horáček
Dalhousie University
Halifax, NS
Canada

Michiel J. Janse
University of Amsterdam
Amsterdam
The Netherlands

Julian Jarman
Imperial College London
London
UK

Per Johansson
Sahlgrenska University Hospital
Gothenburg
Sweden

Raija Jurkko
Helsinki University Central Hospital
Helsinki
Finland

Demosthenes G. Katritsis
Athens Euroclinic
Athens
Greece

Paul Khairy
Montreal Heart Institute
Montreal, QC
Canada

Michael Koa-Wing
Imperial College London
London
UK

Petri Korhonen
Helsinki University Central Hospital
Helsinki
Finland

Jan A. Kors
Erasmus University Medical Centre
Rotterdam
The Netherlands

Jèrôme Lacotte
Hopital Pitie-Salpetriere
Paris
France

T.D. Veitch Lawrie
University Of Glasgow
Glasgow
UK

Jerome Liebman
Case Western Reserve University School of Medicine
Cleveland, OH
USA

S.M. Lobodzinski
California State University
Long Beach, CA
USA

Mário G. Lopes
Institute of Preventive Cardiology
Lisbon
Portugal

Robert L. Lux
University of Utah
Salt Lake City, UT
USA

Peter W. Macfarlane
University of Glasgow
Glasgow
UK

Rob MacLeod
University of Utah
Salt Lake City, UT
USA

J.E. Madias
Mount Sinai School of Medicine
New York, NY
USA

Markku Mäkijärvi
Helsinki University Central Hospital
Helsinki
Finland

Ariane J. Marelli
McGill University Health Centre
Montreal, QC
Canada

Vias Markides
Imperial College London
London
UK

Nuno Marques
Institute of Preventive Cardiology
Lisbon
Portugal

Andrew D. Mcgavigan
Royal Melbourne Hospital
Melbourne, VIC
Australia

John A. Milliken
Queens University
Kingston, ON
Canada

N. Sydney Moïse
Cornell University
Ithaca, NY
USA

Martyn P. Nash
The University of Auckland
Auckland
New Zealand

Peter Okin
Weill Cornell Medical College
New York, NY
USA

A. van Oosterom
Radboud University Nijmegen
Nijmegen
The Netherlands

Tobias Opthof
University Medical Centre Utrecht
Utrecht
The Netherlands

Fernando de Pádua
Institute of Preventive Cardiology
Lisbon
Portugal

Olle Pahlm
Lund University
Lund
Sweden

Armando Pereirinha
Institute of Preventive Cardiology
Lisbon
Portugal

Nicholas Peters
Imperial College London
London
UK

R. Plonsey
Duke University
Durham, NC
USA

Andrew J. Pullan
The University of Auckland
Auckland
New Zealand

F. Russell Quinn
Glasgow Royal Infirmary
Glasgow
UK

Alan P. Rae
Royal Infirmary Glasgow
Glasgow
UK

Andrew C. Rankin
University of Glasgow
Glasgow
UK

Pentti M. Rautaharju
Wake Forest University School of Medicine
Winston-Salem, NC
USA

Michael R. Rosen
Columbia University
New York, NY
USA

Oliver R. Segal
University College London
London
UK

Ronald H. Selvester
University of Southern Carolina
Columbia, SC
USA

Pentti Siltanen
Helsinki University Central Hospital
Helsinki
Finland

Maarten L. Simoons
Erasmus University Medical Centre
Rotterdam
The Netherlands

Michael B. Simson
Hospital of the University of Pennsylvania
Philadelphia, PA
USA

Leif Sörnmo
Lund University
Lund
Sweden

Maciej Sosnowski
Medical University of Silesia
Katowski
Poland

David G. Strauss
Lund University
Lund
Sweden

Borys Surawicz
The Care Group at St Vincent Hospital
Indianapolis, IN
USA

Leif Svensson
Stockholm Prehospital Center
Stockholm
Sweden

Elin Trägårdh-Johansson
Lund University
Lund
Sweden

Heikki Väänänen
Aalto University School of Science and Technology
Espoo
Finland

Galen S. Wagner
Duke University
Durham, NC
USA

Hein J.J. Wellens
University of Maastricht
Maastricht
The Netherlands

Ronald Wilders
University of Amsterdam
Amsterdam
The Netherlands

Andrew L. Wit
Columbia University
New York, NY
USA

Wojciech Zareba
University of Rochester Medical Centre
Rochester, NY
USA

Antonio Zaza
Università di Milano-Bicocca
Milano
Italy

Section 9

Specialized Aspects of Electrocardiography Part 2

36 Exercise Electrocardiography and Exercise Testing

K. Martijn Akkerhuis · Maarten L. Simoons

36.1	***Introduction***	***1679***
36.2	***Safety of Exercise Testing, Precautions, and Contraindications***	***1679***
36.3	***Exercise Protocols***	***1680***
36.4	***Exercise Endpoints***	***1683***
36.5	***Recording and Computer Processing of the Electrocardiogram***	***1684***
36.5.1	Recording of the Electrocardiogram	1684
36.5.2	Computer Processing of the Electrocardiogram	1685
36.5.3	Lead Systems for Exercise Electrocardiography	1687
36.5.3.1	Right-Sided Chest Leads	1687
36.6	***Interpretation of the Electrocardiogram***	***1687***
36.6.1	Changes of the ECG During Exercise in Normal Subjects	1687
36.6.2	Changes of the ECG During Exercise in Patients with Coronary Artery Disease	1688
36.6.2.1	ST Segment Depression	1688
36.6.2.2	ST Segment Elevation	1696
36.6.2.3	Changes of the QRS Complex	1697
36.6.2.4	T-Wave Changes	1699
36.7	***Exercise Testing to Diagnose Coronary Artery Disease***	***1699***
36.7.1	Pretest Probability	1700
36.7.2	Diagnostic Characteristics and Test Performance	1701
36.7.3	Probability Analysis	1702
36.7.4	Diagnostic Accuracy of the Standard Exercise Test	1703
36.7.5	Electrocardiographic Factors Influencing Sensitivity and Specificity	1703
36.7.6	Exercise Electrocardiography Versus Noninvasive Stress Imaging Studies	1704
36.8	***Exercise Testing in Patients with Coronary Artery Disease***	***1704***
36.8.1	Risk Stratification and Assessment of Prognosis with Exercise Testing	1705
36.8.2	Exercise Testing to Guide Patient Treatment	1706
36.9	***Exercise Testing after Acute Myocardial Infarction***	***1706***
36.9.1	Exercise Testing in Patient Management	1706
36.9.2	Risk Stratification and Prognostic Assessment	1707
36.10	***Exercise Testing after Revascularization***	***1707***
36.10.1	Exercise Testing after Coronary Artery Bypass Grafting	1707
36.10.2	Exercise Testing after Percutaneous Coronary Intervention	1708

P. W. Macfarlane, A. van Oosterom, O. Pahlm, P. Kligfield, M. Janse, J. Camm (eds.), *Comprehensive Electrocardiology*, DOI 10.1007/978-1-84882-046-3_36,

36.1 Introduction

Exercise testing is a widely used method for diagnosis in patients with suspected ischemic heart disease and for functional evaluation of patients with known heart disease. Throughout the years, most attention has been given to the information obtained from the electrocardiogram during exercise. However, other information that can be obtained during the test is of equal importance. Such information, as listed in ➋ Table 36.1, can be obtained by observation of the patient, measurement of the heart rate and blood pressure responses, measurement or estimation of total body oxygen consumption, and from other noninvasive investigations such as myocardial-perfusion scintigraphy and evaluation of left ventricular function by radionuclide angiography or echocardiography. This chapter focuses on exercise electrocardiography.

Compared with the noninvasive stress imaging tests such as stress echocardiography and myocardial perfusion scintigraphy, exercise testing can be performed at a much lower cost. However, the noninvasive imaging modalities have been increasingly more used in the last decade and have been shown to outperform the standard exercise test in terms of diagnostic accuracy. This does not mean, however, that exercise electrocardiography should be replaced by these imaging modalities. The standard exercise test remains the initial test of choice in many clinical circumstances. However, new technical improvements in the field of the stress imaging tests are expected to further increase their performance and clinical use. Furthermore, new techniques are being developed that permit direct, noninvasive imaging of the coronary arteries (e.g., multislice computed tomography). These developments will require a continuous reassessment of the relative role of all commonly used diagnostic tests and procedures.

The report of an exercise test should contain a summary of the previous history of the patient, his present symptoms, medication, and an interpretation of the ECG at rest. The exercise protocol should be described with the expected performance of a normal subject of the same age, sex, and body size. The reasons for termination of the test should be stated and symptoms which occur during the test should be described, including the workload at which symptoms began, the type of symptoms, their severity, and the duration of the symptoms after exercise has ceased. Heart rate and blood pressure should be reported at rest, at the onset of symptoms, at peak exercise, and after approximately 6 min into the recovery period. A description of the ECG should contain any arrhythmias which may have occurred and changes in the QRS complex and ST segment. Finally, a conclusion should be drawn which answers the clinical questions that were posed before the test. The report should not be limited to words like "positive" or "negative."

36.2 Safety of Exercise Testing, Precautions, and Contraindications

Exercise testing is a well-established procedure that has been in widespread clinical use for many decades. Although exercise testing is generally a safe procedure, both myocardial infarction and death have been reported. A large survey reported

Table 36.1
Information obtained by exercise testing in patients with (suspected) heart disease

Information	Measurement
• Exercise tolerance	• Maximum workload • Heart rate response • Maximum oxygen consumption
• Limiting symptoms	• Symptoms during test
• Myocardial ischemia	• Angina pectoris • ST segment changes • Myocardial perfusion scintigraphy
• Left ventricular function	• Maximum workload • Blood pressure response • Echocardiography (Radionuclide angiography) (Left ventricular filling pressure)
• Arrhythmias	• Electrocardiogram

17 deaths among 712,285 patients tested in three German-speaking countries [1]. In addition, nonfatal ventricular fibrillation occurred at a rate of 1/7,000 tests and nonfatal myocardial infarction at a rate of 1/70,000 tests. Since these risks are not negligible, all stress testing should be done in a setting where emergencies can be treated efficiently and expeditiously. A defibrillator and an emergency kit of appropriate drugs should be immediately available, and the staff of the exercise laboratory should be trained in cardiopulmonary resuscitation [2]. In addition, exercise testing should be supervised by an appropriately trained physician, who should be in the immediate vicinity and available for emergencies [3]. However, the most important safety factors in stress testing are patient selection and good clinical judgment, deciding which patient should undergo exercise testing and knowing when not to start and when to stop a test are essential in reducing the risk of stress testing. Absolute and relative contraindications to exercise stress testing are listed in ➋ Table 36.2 [4].

36.3 Exercise Protocols

Exercise should be performed on a treadmill or a calibrated bicycle ergometer [5, 6]. Other modalities may be used in special situations; for example, in sports medicine a rowing ergometer or a canoe ergometer might be used. The various forms of step tests are outdated, since these do not permit quantification of the work performed. The choice between a bicycle ergometer and a treadmill can be made on the basis of personal preferences or local custom. In general, normal subjects can reach higher heart rates and a higher level of oxygen consumption on a treadmill [7]. Furthermore, virtually all subjects can be exercised on a treadmill, while riding a bicycle requires some skill. On the other hand, the body position is more stable on a bicycle ergometer, resulting in less motion artifacts on the ECG. For studies of left ventricular function during exercise or studies with indwelling catheters, only bicycles can be used, usually with the patient in the supine position. Guidelines for the design of a clinical exercise testing laboratory have been published by the American Heart Association [2].

Exercise should start at a low workload and increment stepwise or continuously, according to a fixed protocol. A large number of treadmill protocols has been described (➋ Fig. 36.1). Probably the most widely used protocol has been described by Bruce [8]. In this protocol, both the grade (slope) of the treadmill and its speed are altered at the end of

Table 36.2

Contraindications to exercise testing (Adapted from ACC/AHA 2002 Guideline Update for Exercise Testing [4])

Absolute
• Acute myocardial infarction (within 2 days)
• High-risk unstable angina
• Symptomatic severe aortic stenosis
• Uncontrolled symptomatic heart failure
• Acute pulmonary embolus or pulmonary infarction
• Uncontrolled cardiac arrhythmias causing symptoms or hemodynamic compromise
• Acute myocarditis or pericarditis
• Acute aortic dissection
Relative
• Left main coronary stenosis
• Moderate stenotic valvular heart disease
• Electrolyte abnormalities
• Severe arterial hypertension[a]
• Tachyarrhythmias or bradyarrhythmias
• Hypertrophic (obstructive) cardiomyopathy and other forms of outflow tract obstruction
• Mental or physical impairment leading to inability to exercise adequately
• High-degree atrioventricular block

[a]Systolic blood pressure of >200 mmHg and/or diastolic blood pressure of >110 mmHg, as suggested by the ACC/AHA Committee on Exercise Testing

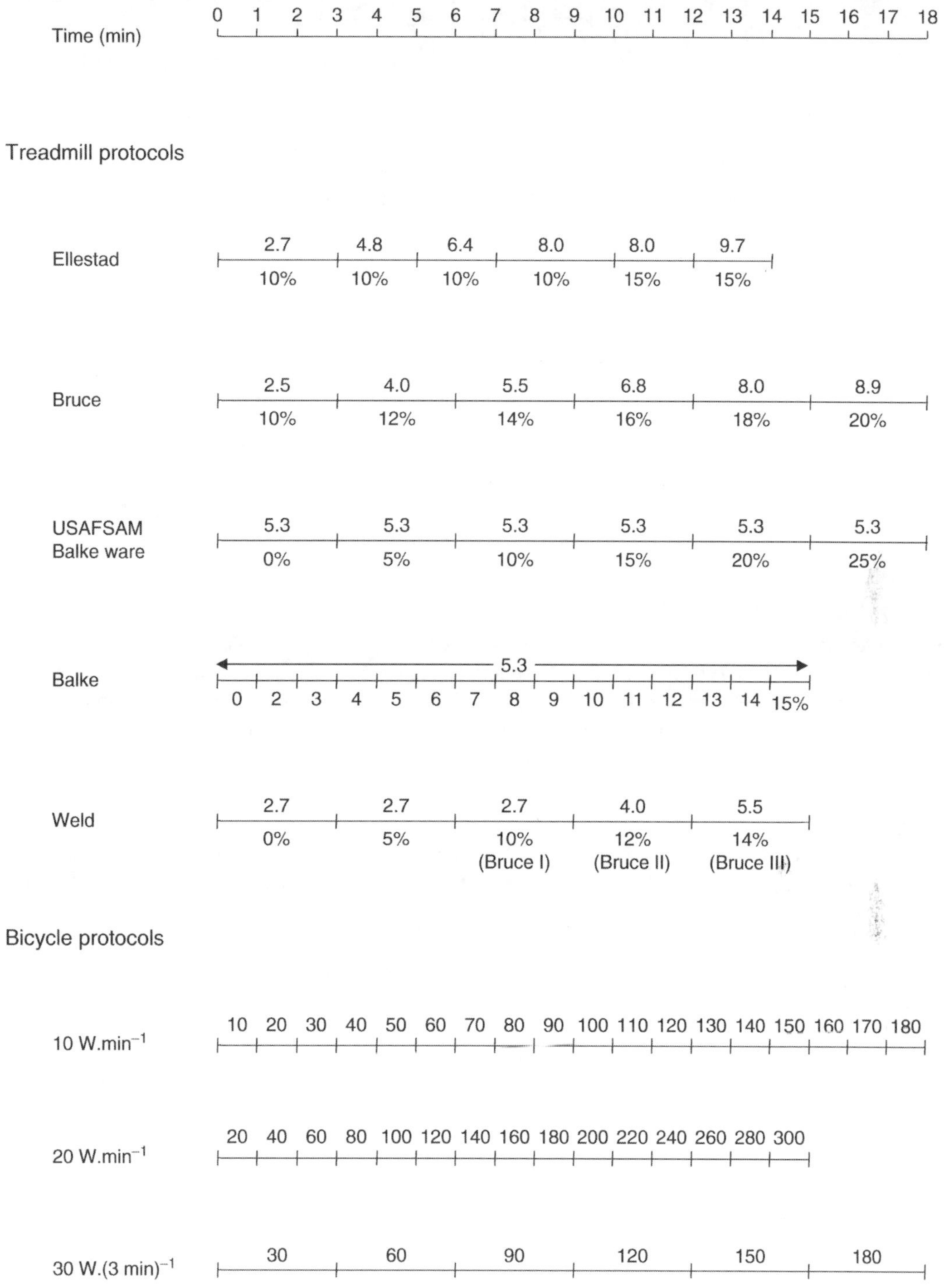

Figure 36.1

Summary of five treadmill protocols and three bicycle protocols. The figures above the lines of the treadmill protocols represent treadmill speed in kilometers per hour, and the figures below the lines represent the slope. Some treadmill protocols use steps of 3 min each (Bruce, USAFSAM, Weld), while others use steps of 1 min (Balke) or alternating steps of 3 and 2 min (Ellestad). The figures above the lines of the bicycle protocols correspond to workload in Watts. Either 1-min steps or steps of 3 min are used. The value of the divisions marked on these lines is indicated on the *left*.

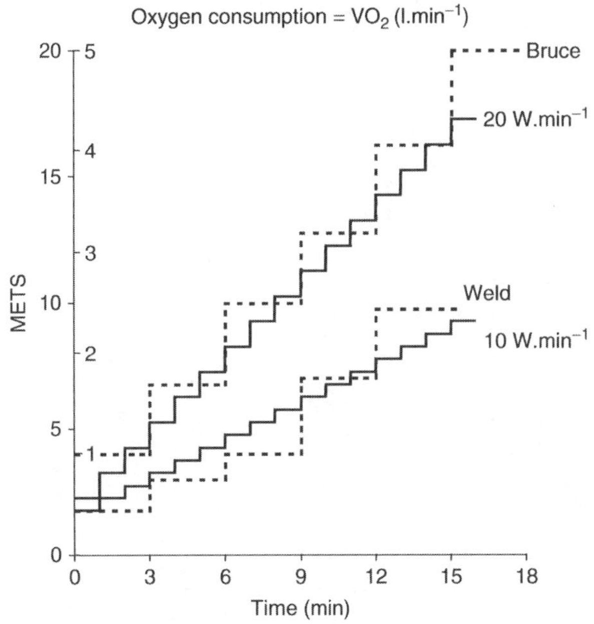

Figure 36.2

Comparison of oxygen requirements over time for two bicycle protocols (steps of 20 W/min or 10 W/min) and two treadmill protocols (Bruce and Weld) in a 75 kg male subject. The horizontal axis represents time in minutes and the vertical axis represents average oxygen uptake (l/min), which can be equated to metabolic equivalents (METS). It should be noted that the actual oxygen consumption of a given patient at a certain level of exercise varies widely and depends on their level of physical condition. Furthermore, oxygen consumption on the treadmill is dependent on body weight.

each 3-min stage. In the Ellestad protocol [9], the slope is constant during the first four stages, while the speed increases. At stage five, speed is kept constant while the slope is altered; for stage six, speed is again increased. The Balke protocol and its derivatives maintain a constant speed in the range of 4.8–6.4 km/h with changes of treadmill slope every minute or every 3 min [10]. Modifications of these protocols have been developed for the evaluation of patients prior to discharge from hospital after myocardial infarction. For example, the protocol used by Weld employs two stages with a low treadmill slope prior to the first stage of the Bruce protocol [11].

Workload on a treadmill is dependent on body weight. Accordingly, the workload performed using various protocols can be expressed as ml/kg/min oxygen consumption. In the literature, workload has frequently been expressed in metabolic equivalents (METS) [12, 13]. One MET equals 3.5 ml/kg/min oxygen consumption, which corresponds to the average resting oxygen consumption. In ➋ Fig. 36.2, mean oxygen consumption is given for various protocols and the corresponding METS have been indicated. Oxygen consumption rises more rapidly with the Bruce protocol or the Ellestad protocol than with the USAFSAM modification of the Balke protocol, which uses a speed of 5.3 km/h [14, 15]. The work on a bicycle ergometer is independent of body weight. Frequently used protocols on a bicycle ergometer in the sitting position increase the workload by 10 W/min or 20 W/min. For adult male subjects with average body weight, the metabolic requirement (total body oxygen consumption) of a protocol with 20 W/min workload increments is comparable to the Bruce protocol (➋ Fig. 36.2). Similarly, a protocol with 10 W/min steps is comparable to the Weld protocol for submaximal predischarge tests after myocardial infarction. Predicted normal values for the peak workload for a protocol with steps of 20 W/min are presented in ➋ Table 36.3. In the supine position, lower workload increments are normally used; for example, 20 or 30 W every 3–5 min.

It should be realized that the hemodynamic response to exercise in the supine position differs considerably from the response in the sitting position. In patients with coronary artery disease (CAD), the peak level of exercise in the supine position is approximately 70% of the peak level in the sitting position. Peak heart rate and peak systolic blood pressure are

Table 36.3
Predicted normal exercise tolerance (in Watts) using a bicycle ergometer protocol for exercise testing

	Exercise tolerance in *women*					Exercise tolerance in *men*				
	Height (cm)					Height (cm)				
Age (years)	160	170	180	190	200	160	170	180	190	200
20	176	192	208	225	241	220	240	261	281	301
25	167	183	199	215	232	209	229	249	269	290
30	158	174	190	206	222	197	217	238	258	278
35	149	165	181	197	213	186	206	226	246	267
40	139	156	172	188	204	174	195	215	235	255
45	130	146	163	179	195	163	183	203	223	244
50	121	137	153	170	186	151	172	192	212	232
55	112	128	144	160	177	140	160	180	201	221
60	103	119	135	151	167	129	149	169	189	209
65	94	110	126	142	158	117	137	157	178	198
70	84	101	117	133	149	106	126	146	166	186

Predicted normal exercise tolerance (in Watts) for women (left) and men (right). These normal values are applicable when a protocol with steps of 20 W/min is used on a bicycle ergometer. The normal range is between 85% and 115% of these values. For example, a 45-year old man with a height of 190 cm has a predicted normal exercise tolerance of 223 W, ranging from 190 to 256 W.

approximately 10% lower. On the other hand, pulmonary capillary wedge pressure during exercise is higher in the supine position owing to the greater venous return [16–18]. In order to obtain reproducible results, each hospital and laboratory should select one or two protocols which can be applied to all subjects. Reference values for such protocols are readily available as shown in Fig. 36.3 and Table 36.3. Nevertheless, each laboratory should verify whether these values are indeed applicable to the local population. Measurement of oxygen consumption is not very useful in a laboratory for exercise electrocardiography. If necessary, oxygen consumption can be estimated from the workload as shown in Fig. 36.2. The standard deviation of this estimation is between 5% and 10%.

36.4 Exercise Endpoints

For most purposes, exercise can be continued until symptoms occur. It is a fallacy to terminate exercise at an arbitrary percentage (70 or 85%) of the age-adjusted maximum predicted heart rate for a number of reasons. First, the heart-rate response in normal subjects is highly variable [13]. The mean value at each age may be predicted as 220 minus age, or 200 minus age/2. These mean values apply both to men and women up to 70 years of age. However, the standard deviation is large, approximately 10 beats/min. Thus, the 95% confidence interval is between 20 beats below and 20 beats above the predicted mean value [12]. Second, the maximum heart rate reached by patients with severe disease is considerably lower than the heart rate reached by normal subjects. Thus, when heart rate is used as an endpoint for terminating exercise, patients with severe disease may be stressed beyond their limits, while other patients will be unnecessarily prevented from exercising to their true capacity. Finally, the target heart rate approach has additional limitations in patients with heart rate impairment, those with excessive heart rate response, and those receiving medication such as beta-blockers, some calcium antagonists, and specific sinus-node inhibitors [20–23].

Therefore, other endpoints for terminating exercise testing, as summarized in Table 36.4, are strongly preferred. It should be noted that ST segment elevation developing in leads (other than aVR or V_1) without diagnostic Q-waves of a previous myocardial infarction usually indicates transmural ischemia rather than just a subendocardial problem. It is almost always associated with a high-grade obstruction proximal in the respective coronary artery. If exercise continues, myocardial infarction may be imminent, so that it is necessary to terminate the test and direct the patient to immediate follow-up care including coronary angiography.

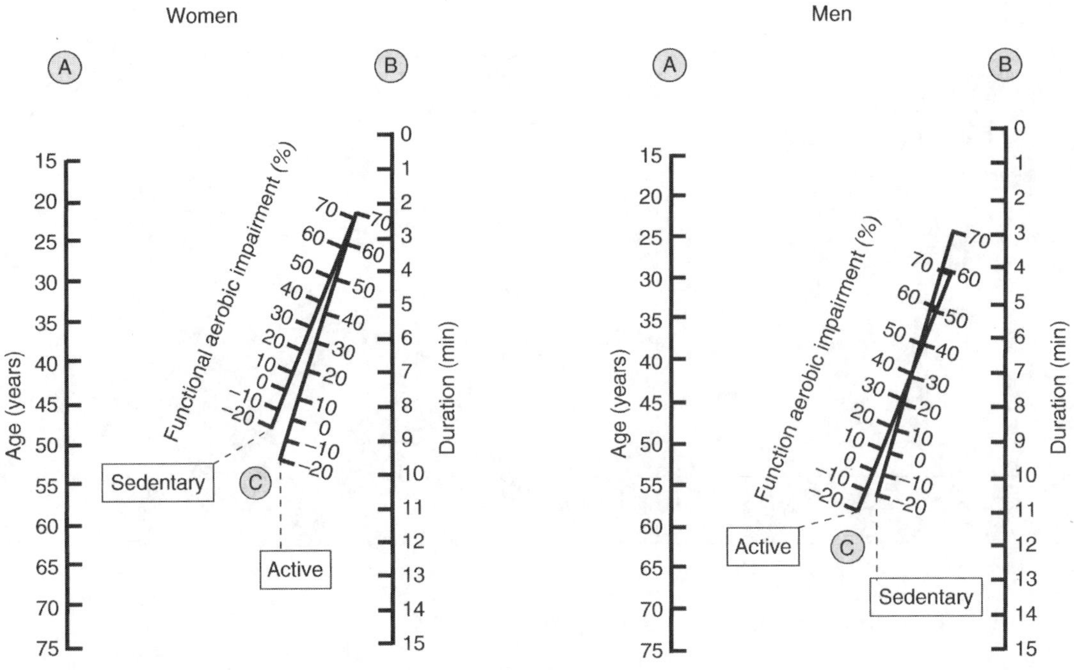

Figure 36.3

Nomogram for assessment of functional aerobic impairment of men and women according to age, duration of exercise by the Bruce multistage procedure, and habitual physical activity status. In order to find functional aerobic impairment (*C*), apply a straight-edge to age (*A*) and duration (*B*) and read intercepts of diagonal (Adapted from Bruce [19]).

It is advisable to express symptoms which occur during exercise on a semiquantitative scale, such as the Borg scale (➋ Table 36.5) [24]. The use of rating of perceived exertion scales is often useful in assessment of patient fatigue. Symptom-limited testing with the Borg scale as an aid is very important when the test is used to assess functional capacity. It also permits a comparison of the degree of a patient's symptoms over time.

It has previously been recommended that all medication be stopped one or more days before an exercise test is done. Although this may be useful for some scientific applications, this principle is not generally followed in clinical practice. In patients with stable angina, a period of instability may develop if beta-blockers, nitrates, or calcium antagonists are withdrawn. If the interpretation of the test is significantly hampered by medication, for example, a normal response is obtained in a patient with angina using beta-blockers, the medication may be gradually withdrawn and the test repeated. However, most patients with angina will develop symptoms and ischemic ECG changes in spite of the use of antianginal drugs, albeit at a higher workload than without such drugs [21, 22, 25, 26]. Despite the effect of beta-blockers on maximal exercise heart rate, no differences in test performance were found in a consecutive group of men being evaluated for possible CAD when they were subgrouped according to beta-blocker administration initiated by their referring physician [4, 27]. Therefore, for routine exercise testing, it appears unnecessary for physicians to accept the risk of stopping beta-blockers before testing when a patient exhibits possible symptoms of myocardial ischemia.

36.5 Recording and Computer Processing of the Electrocardiogram

36.5.1 Recording of the Electrocardiogram

It is essential to record a high-quality ECG. Proper ECG quality can be achieved in virtually all patients if skin preparation is meticulous. The skin should be abraded with sandpaper or a special, commercially available, drill. Special electrodes

Table 36.4
Indications for terminating exercise testing (Adapted from ACC/AHA 2002 Guideline Update for Exercise Testing [4])

Sensations felt by *patient*	Observations made by *physician*
Absolute indications	
• Moderate to severe (progressive) angina • Increasing nervous system symptoms (e.g., ataxia, dizziness, or near-syncope) • Subject's desire to stop	• ST-elevation (≥1.0 mm) in leads without diagnostic Q-waves (other than V_1 or aVR) • Technical difficulties in monitoring ECG or systolic blood pressure • Sustained ventricular tachycardia • Drop in systolic BP of >10 mmHg from baseline despite an increase in workload, when accompanied by other evidence of ischemia • Signs of poor perfusion (cyanosis or pallor)
Relative indications	
• Fatigue, shortness of breath, wheezing, leg cramps, or claudication • Increasing chest pain	• Drop in systolic BP of >10 mmHg from baseline despite an increase in workload, in the absence of other evidence of ischemia • ST- or QRS-changes such as excessive ST-depression (>2 mm of horizontal or down-sloping ST segment depression, especially when the magnitude of the ST-depression is increasing rapidly at low workloads) or marked axis shift • Arrhythmias other than sustained ventricular tachycardia, including multifocal PVCs, triplets of PVCs, supraventricular tachycardia, heart block, or bradyarrhythmias • Development of bundle-branch block or IVCD that cannot be distinguished from ventricular tachycardia • Hypertensive response (systolic blood pressure of >250 mmHg and/or diastolic blood pressure of >115 mmHg)[a]

BP, blood pressure; ECG, electrocardiogram; PVCs, premature ventricular contractions; and IVCD, intraventricular conduction delay
[a] As suggested by the ACC/AHA Committee on Exercise Testing

should be used to prevent motion artifacts. Furthermore, the ECG cables should be of special design to prevent artifacts owing to cable motion. As stated, cycle ergometers produce less motion of the upper body resulting in less motion artifacts on the ECG. Finally, ECG amplifiers which meet the American Heart Association standards, and have high input impedance, should be used [2]. If a proper combination of skin preparation, electrodes, cables, and amplifiers is used, a stable baseline can be achieved throughout the test in most subjects.

36.5.2 Computer Processing of the Electrocardiogram

Computerized exercise stress test systems have become the method of choice in most stress testing laboratories. The principles of computer-assisted interpretation of exercise ECGs have been described [28–32] and a review of methods for computer-based resting electrocardiography is also presented elsewhere in this textbook (➋ Chap. 37). A summary of methods for analysis of exercise ECGs is presented here in brief.

Table 36.5
Borg scale

Grade	Symptoms
0	Nothing at all
0.5	Extremely weak (just noticeable)
1	Very weak
2	Weak (light)
3	Moderate
4	Somewhat strong
5	Strong (heavy)
6	
7	Very strong
8	
9	Extremely strong (almost maximal)
10	Maximal

The Borg scale for rating of perceptual intensities constructed as a category scale with ratio properties [24]. This can be used for quantitative evaluation of symptoms, for example chest pain

Computer systems for exercise testing regulate the workload of the bicycle or treadmill ergometer according to one of several predefined protocols. The computer systems also maintain a record of time, heart rate, and workload, and display these data on a computer screen together with baseline ECG waveforms [33]. The continuous stream of ECG waveforms, data, and computer measurements are stored and processed online in the system. ECG waveforms are acquired by a process termed analog–digital conversion. Most current digital processing ECG and exercise test systems sample the analog waveforms at the rate of 250 samples/s. A representation of the original analog signal can be obtained by digital–analog reconstruction of the digitized waveform. The computer system subsequently processes the ECG waveforms online to minimize noise and artifactual effects. Especially, at the higher workloads and heart rates, the recorded basic ECG tracings may show so much noise and artifacts from exercising skeletal muscles and respiratory variability that measurements from the original ECG waveforms are unreliable.

The processing involves the steps of QRS-detection, temporal alignment, and signal averaging. The QRS complexes are detected with the aid of a combination of the derivatives of multiple ECG leads. Thus, the characteristic feature used for detection of the QRS complex is the large voltage changes that occur in all leads simultaneously during ventricular activation. The QRS complexes are then classified as normal or abnormal. Abnormal beats may be a result of premature ventricular or supraventricular complexes, or they may be normal beats distorted by excessive noise or baseline drift. The normal beats are then combined into a single representative complex by computation of an average (mean) or median beat. Signal averaging can be performed at preselected intervals, for example, during 20 s of each minute, or continuously. The latter is the method of choice because it permits continuous display of the updated ECG waveform. Since the averaging procedure might be subject to errors in some of the patients, the user should compare the shape of the averaged signal with the original ECG tracings.

From the representative complexes, measurements can be obtained. Since the noise level during exercise is rather high in comparison with resting ECGs, it is not appropriate to take measurements from individual beats as is the case in some resting ECG programs. In commercially available systems, the baseline and ST segment are defined at fixed intervals before and after a single fiducial point in the QRS complex. Therefore, it is necessary to define the proper onset and end of the QRS complex. Such precise definition of QRS-onset and end is only possible if a combination of multiple leads is used [29, 30]. Computerized algorithms for QRS-detection and time alignment are, therefore, more reliable and robust in multichannel recording. Additionally, algorithms are applied that reduce the effect of baseline drift during exercise caused by temperature changes, respiration, and body motion, especially at higher workloads and if there is poor skin–electrode contact.

The processed, online cleaned data allow more accurate measurements of the exercise ECG, especially the low-amplitude signals such as the P-wave, the PR-segment, the junction of the QRS complex and ST segment (J-point), and the ST segment itself [33]. The digital measurements are generally more accurate and reproducible than manual measurements. However, careful overreading to provide quality control for automated choice of onset–offset waveform fiducial markers on the signal-processed data is required to avoid error in interpretation resulting in false-positive test results [4, 33]. Therefore, ECG recordings of the raw, unprocessed ECG data should be available at each stage of the exercise protocol for comparison with the averages that the exercise test monitor generates. Recommendations and standards for the degree of filtering and processing of data have been published [2, 4].

36.5.3 Lead Systems for Exercise Electrocardiography

Various types of lead systems have been developed, studied, and used in the last decades. Bipolar precordial lead systems have been used for a number of years and have produced satisfactory results in patients with suspected CAD and a normal ECG at rest. The two optimal leads for detection of ST segment depression during exercise are the bipolar lead from the right infraclavicular region to the V5 position (CS5 bipolar lead system) and that from the manubrium of the sternum to the V5 position (CM5 bipolar lead system). They have been reported to detect up to 90% of all ST segment depression identified by multiple-lead systems [34–36]. In patients with a previous myocardial infarction, a single-lead system is inadequate and such patients should be tested with a multiple-lead system. Multiple-lead systems that have been evaluated include the pseudo-orthogonal lead system, the corrected orthogonal lead system (with computer-processed Frank leads), the precordial map, the conventional 12-lead ECG system, and systems using a combination of bipolar leads and the standard 12 leads [33–35, 38–41]. The arrival of powerful high-speed microcomputers has allowed the development of systems for continuous online recording and analysis of the standard 12 ECG leads, which has become the norm in most testing facilities today [33]. It should, however, be noted that lead V5 remains by far the most sensitive electrode position that outperforms the inferior leads and the combination of lead V5 with II, because the latter has a relatively high false-positive rate. Therefore, in patients without prior myocardial infarction and with normal resting ECGs, the precordial leads alone are a reliable marker for CAD, and monitoring of inferior limb leads adds little additional diagnostic information. In these patients, exercise-induced ST segment depression confined to the inferior leads is of little value for detection of CAD [4, 42].

36.5.3.1 Right-Sided Chest Leads

In a study of 245 patients, it was shown that the diagnostic accuracy of exercise testing was increased when right ventricular leads were added to the standard 12 ECG leads [43]. By using right-sided chest leads, the sensitivity for the detection of CAD by angiography was comparable to that of myocardial perfusion scintigraphy (92 versus 93%). However, it is important to realize that these data were obtained in a population with a very high prevalence of CAD and might well be less reliable if the prevalence of CAD were to be substantially lower. Therefore, routine clinical use of right-sided chest leads awaits confirmation of these results in differing populations [4].

36.6 Interpretation of the Electrocardiogram

36.6.1 Changes of the ECG During Exercise in Normal Subjects

Normal ECG changes during exercise have been studied in standard chest leads [44], in corrected orthogonal leads [45] and by body-surface mapping [46, 47]. It has been shown that such changes occur gradually in relation to heart-rate changes during stress.

The amplitude of the P-wave increases, without major changes in the waveform. At heart rates over approximately 120 beats/min, this is enhanced by the superposition of the T-wave and the P-wave. At peak exercise with heart rates over 160 beats/min, the P-wave amplitude is on average twice, but in some patients up to five times the amplitude at rest.

Obviously, the PQ-interval shortens, owing to increased sympathetic activity during stress. With the P-wave becoming taller and the Ta-wave (atrial repolarization) increasing, there is a downward displacement of the PQ-junction below the isoelectric line in the resting ECG. This point is considered to be the baseline for terms of measuring ST segment change [33]. Exaggerated atrial repolarization waves during exercise may extend into the ST segment and T-wave and can cause downsloping ST-depression in the absence of ischemia at high peak exercise heart rates [48, 49]. Patients with false-positive exercise tests based on this finding have a high peak exercise heart rate, absence of exercised-induced angina, and markedly downsloping PQ-segments in the inferior leads [4, 48, 49].

QRS-duration does not change significantly during exercise. In one specially designed study, a systematic reduction of QRS-duration of only a few milliseconds was observed [50]. Prolongation of QRS-duration, however, is certainly abnormal. Changes in QRS-morphology have been discussed extensively. At peak exercise, QRS-forces shift toward the right and superiorly. Accordingly, Q-waves in the left precordial leads deepen, R-waves in the left precordial leads decrease, and S-waves become wider and deeper. However, intra-individual variability is large. In standard precordial leads, R-wave amplitude increases frequently at intermediate workloads, with heart rates between 120 and 160 beats/min [44]. This is precisely the heart rate reached by most patients with CAD. At higher heart rates, R-wave amplitude decreases. The situation is even more complex when body-surface maps are analyzed [46, 51]. The position of the maximum amplitude on the chest wall during ventricular depolarization remains located in the left anterior region, but the precise location of the maximum shifted one electrode position or more in seven out of 25 normal subjects studied by Block [46]. No consistent pattern was observed in the changes of the maximum QRS-forces in body-surface maps in normal subjects exercised up to a heart rate of 150 beats/min during maximum exercise. The mechanism of the QRS-changes during exercise is still unclear. It has been claimed that these are related to reduction of end-diastolic volume, which would diminish the effect of the radially oriented depolarization fronts in the left ventricular wall traveling from endocardium to epicardium. However, no reduction in end-diastolic volume was seen in patients studied in the supine position with radionuclide ventriculography [52]. Furthermore, R-wave amplitude variations appeared not to be related to changes in left ventricular volume during experimental myocardial ischemia [53].

The QT-interval shortens during exercise, although it cannot be measured accurately when the T-wave and P-wave overlap at higher heart rates. Usually, there is depression of the J-point followed by an upsloping ST segment. However, the ST segment slope may vary considerably. The T-wave amplitude varies greatly during exercise but increases considerably in most normal subjects immediately after exercise. Generally, the ST segment amplitude 80 ms after the end of the QRS complex remains positive in normal subjects at heart rates around 140 beats/min in the left precordial area, while a minimum develops below the right clavicle. During exercise, depression of the J-point and ST segment are prominent in leads with a lateral and vertical orientation [54]. These changes are far less significant in anterior–posterior and transverse leads. This explains in part the relatively high incidence of so-called false-positive ST segment depressions in these leads. ECG changes in healthy female volunteers are similar to those in males, with the exception of ST segment measurements. Flat or even slightly negative ST segments are present in some of the females at rest. This contributes in part to the false-positive ST segment changes in females [55]. Accordingly, separate criteria were developed for quantitative interpretation of the ECG changes during exercise in females [56].

36.6.2 Changes of the ECG During Exercise in Patients with Coronary Artery Disease

The most prominent electrocardiographic findings associated with CAD are a depression or elevation of the ST segment in the left precordial leads.

36.6.2.1 ST Segment Depression

The probability that a given subject has CAD is higher if the ST segment depression adheres to the following pattern: it develops early during exercise at a low heart rate, is deeper and has a more horizontal, or even downsloping pattern. However, it should be appreciated that, in subjects with a normal ECG at rest, there remains a considerable overlap between ST segment measurements in normal subjects and in patients with CAD. Also, as mentioned, ST segment waveforms in females during exercise differ from males and show a greater tendency towards ST segment depression, particularly in

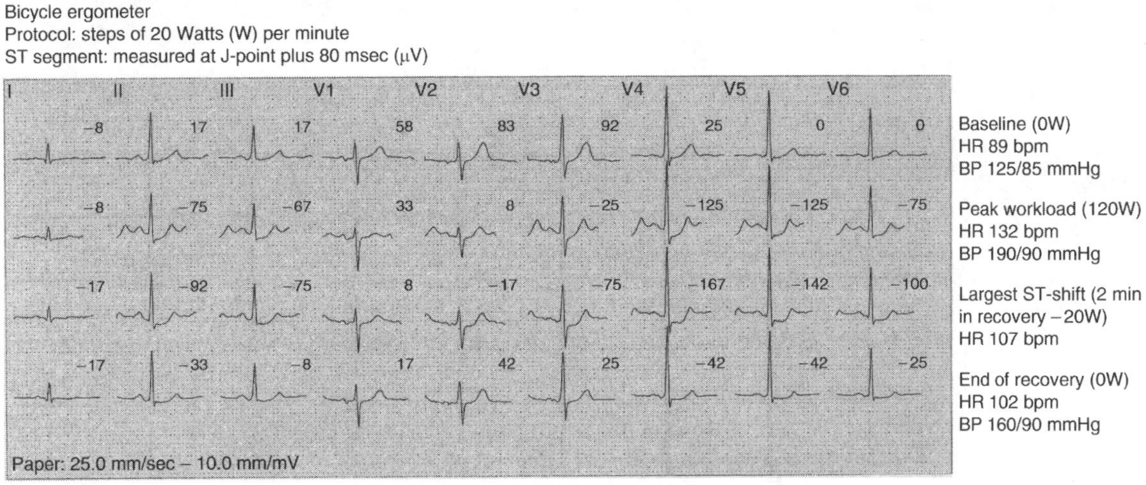

Figure 36.4

Bicycle exercise stress test in 56-year old male who underwent percutaneous coronary intervention of the left anterior descending artery 10 years earlier for stable angina. He was referred for screening before entering a fitness program. Upon request, he admitted that he had chest pain on exertion. The degree of functional impairment was difficult to assess because of an inactive lifestyle. The patient reached 120 W, whereas the predicted normal value for the peak workload was 174 W. He discontinued the test because of shortness of breath and chest pain. With increasing workload, there is a slow-upsloping ST segment depression up to 1 mm (0.1 mV) in the inferior and left precordial leads, evolving to a downsloping pattern in the recovery phase. The patient was referred for coronary angiography, which revealed a significant stenosis in the left circumflex artery that was subsequently treated by percutaneous coronary intervention.

the vertically-oriented leads [54]. In combination with the lower incidence of CAD in females, this results in an increased fraction of false-positive exercise test results in women, as reported in the literature [55, 57–59]. Nevertheless, exercise testing can be used in females when these factors are taken into account [57, 58]. Accordingly, the likelihood of CAD in females is lower than in males for most combinations of age, history, and ST segment depression, as will be discussed later.

Upsloping ST Segment Depression

The significance of upsloping ST segment depression has been studied extensively in the past decades and has generated much controversy. Clearly, downsloping ST segment depression is a stronger predictor of CAD than horizontal depression, and both are more predictive than upsloping depression. However, slowly-upsloping ST segment depression, for example, when the slope is less than 1 mV/s or when the degree of depression at 60–80 ms from the J-point is 1.5 mm (0.15 mV) or more below the baseline level of the PQ-junction, is associated with an increased probability of CAD (➲ Figs. 36.4 and ➲ 36.5) [33, 60–62]. In a study of 70 patients with these ST segment changes, 57% had either two- or three-vessel CAD [33, 61]. On the other hand, junctional depression with a very steep upsloping ST segment is probably not pathological. From this it follows that, if slowly-upsloping ST-depression is considered an abnormal finding, sensitivity of exercise testing is increased, albeit at the cost of decreased specificity resulting in more false-positive test results. In the most recent ACC/AHA guidelines on exercise testing, the use of the more commonly-used definition for a positive test of ≥1 mm (0.1 mV) of horizontal or downsloping ST segment depression is favored [4].

Horizontal and Downsloping ST Segment Depression

The most commonly-used definition for visual interpretation of a positive exercise test result from an electrocardiographic standpoint is ≥1 mm (0.1 mV) of horizontal or downsloping ST segment depression for at least 60–80 ms after the end of the QRS complex (➲ Fig. 36.6 (Continued)) [4]. As stated, downsloping ST segment depression is a stronger predictor of CAD than horizontal depression.

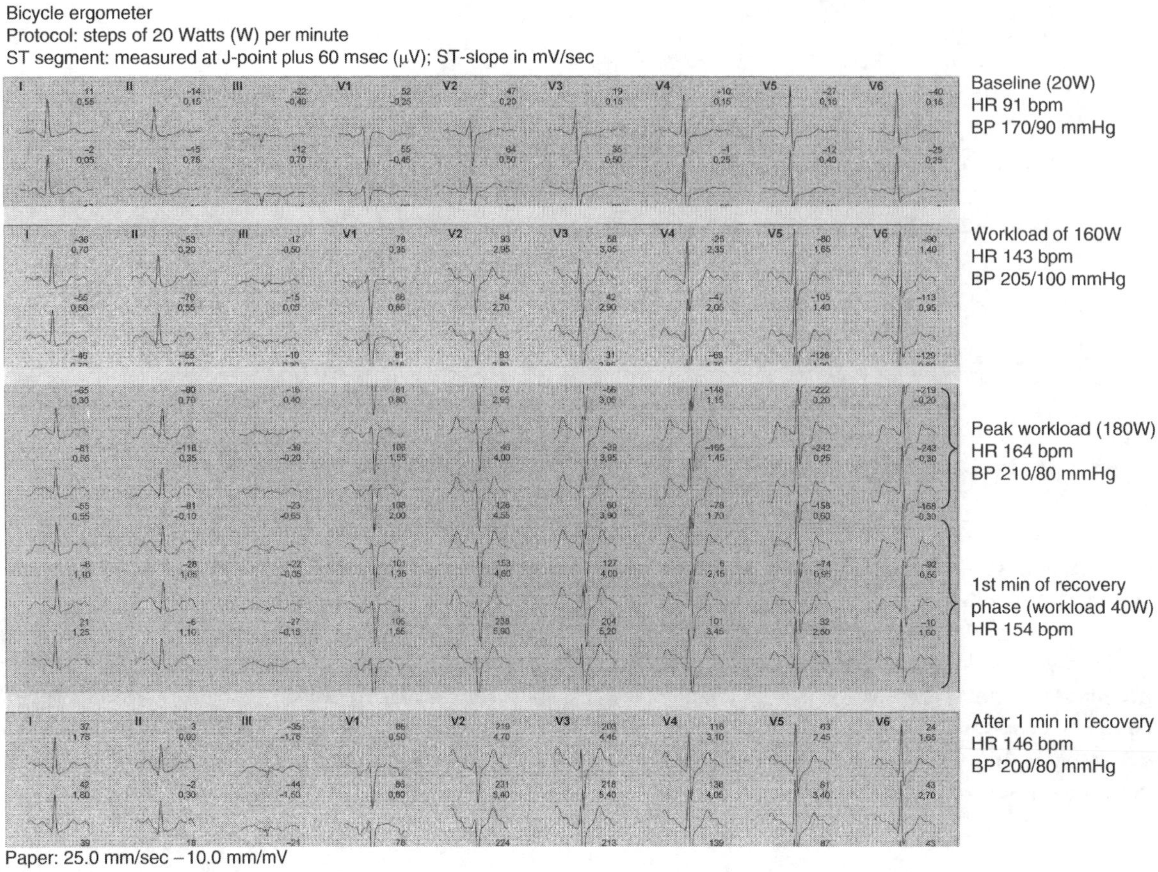

Figure 36.5

A 60-year old male treated for hypercholesterolemia by his general practitioner underwent exercise testing at the request of the specialist in allergy and immunology. He was suspected to have had a severe allergic reaction to the (rare) combination of wheat and exercise. Before planning an exercise test after ingestion of wheat to assess whether an allergic reaction could be provoked, a standard bicycle exercise test was performed as baseline. In daily life, the patient had a good exercise tolerance without any angina-like complaints. He reached 180 W (predicted peak workload 167 W) and discontinued because of fatigue. There were no angina-like symptoms or dyspnea. The baseline ECG was normal. Up to 160 W, there were no significant ST segment changes indicative of myocardial ischemia. From 160 W onward, minimal slow-upsloping ST segment depression developed in the left precordial leads, which got a more horizontal pattern and increased up to 2 mm (0.2 mV) at peak workload. The ST segment depression resolved directly following the discontinuation of exercise. This ECG pattern in which the ST segment becomes abnormal only at high workloads and returns to baseline in the immediate recovery phase may indicate a false-positive test in an asymptomatic subject. A nuclear perfusion scintigraphy using dipyridamole was subsequently performed which showed no signs of myocardial ischemia.

Magnitude of ST Segment Depression and Probability of Coronary Artery Disease

Table 36.6 summarizes data from various studies that related the magnitude of ST segment depression at peak exercise to the probability of CAD at coronary angiography [64–67]. It is evident that, for example, horizontal ST segment depression of ≥2 mm (0.2 mV) is highly specific (if only such results were considered abnormal, the specificity would be 99%), but occurs in the minority of patients with CAD (sensitivity would be only 28%). On the other hand, ST segment depression of 0.5 mm (0.05 mV) is nonspecific (sensitivity is 68%, but specificity only 80%), and should, therefore, not be used as a criterion for abnormality. A superior approach is to compute, from the data in Table 36.6, the ratio of

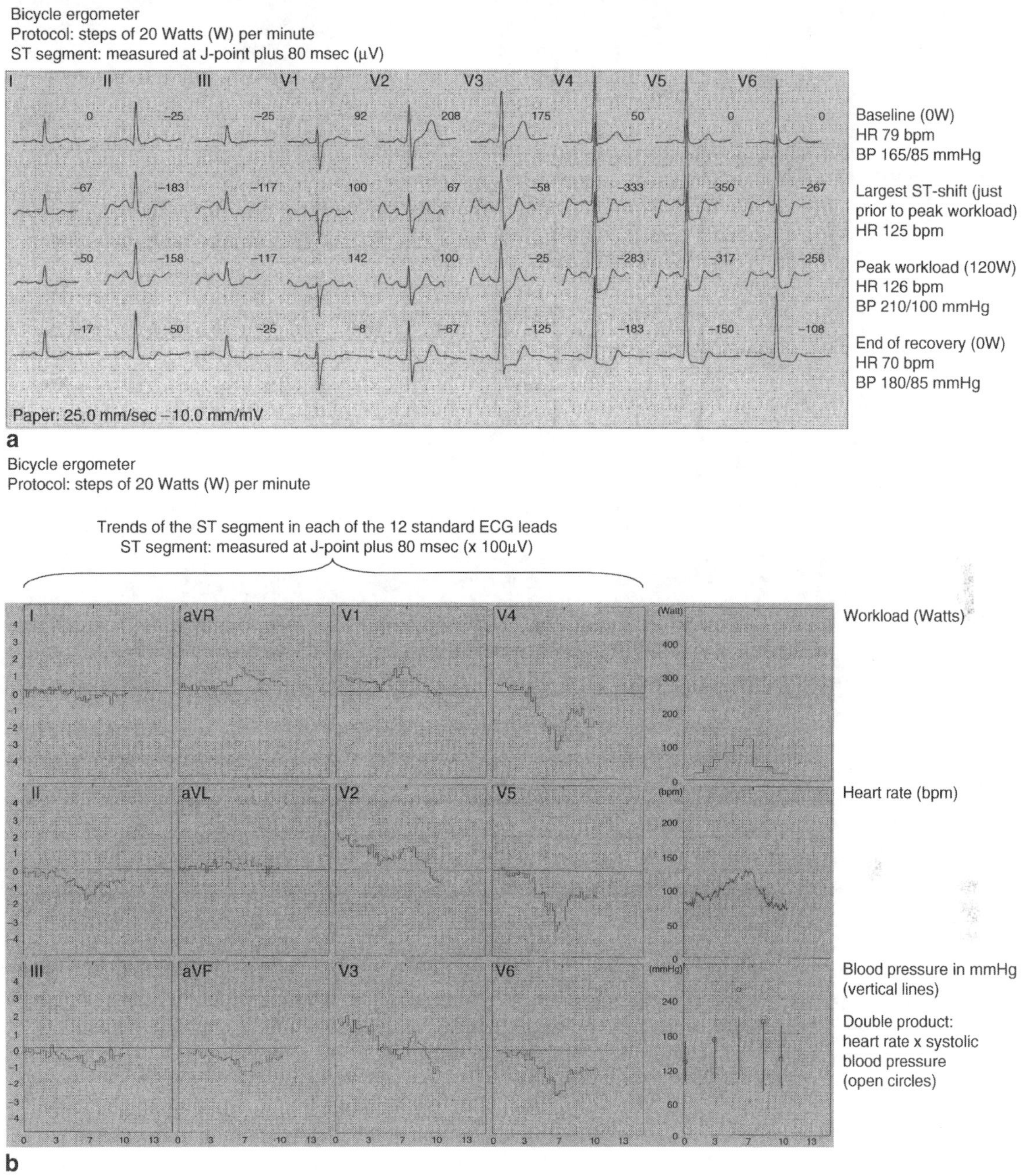

Figure 36.6

(a) Bicycle exercise stress test in 54-year old male with no history of cardiovascular disease who visited the outpatient clinic for angina-like chest pain complaints, class III according to the Canadian Cardiovascular Society [63]. The predicted maximum workload was 175 W. The patient discontinued the test at 120 W because of severe chest pain. Slow-upsloping to horizontal ST segment depression already developed at low workload (80 W) and increased to a maximum of 3.5 mm (0.35 mV) in the left precordial lead V_5 at maximal exertion. The ST segment depression evolved to a downsloping pattern in the recovery phase. Coronary angiography revealed a severe stenosis at the bifurcation of the left anterior descending artery and first diagonal branch, which was subsequently treated by percutaneous coronary intervention with bifurcation stenting according to the culotte technique. (b) Overview of ST segment trends and hemodynamic data of the same exercise test, showing the development of progressive ST segment depression in the inferior and precordial leads with increasing workload.

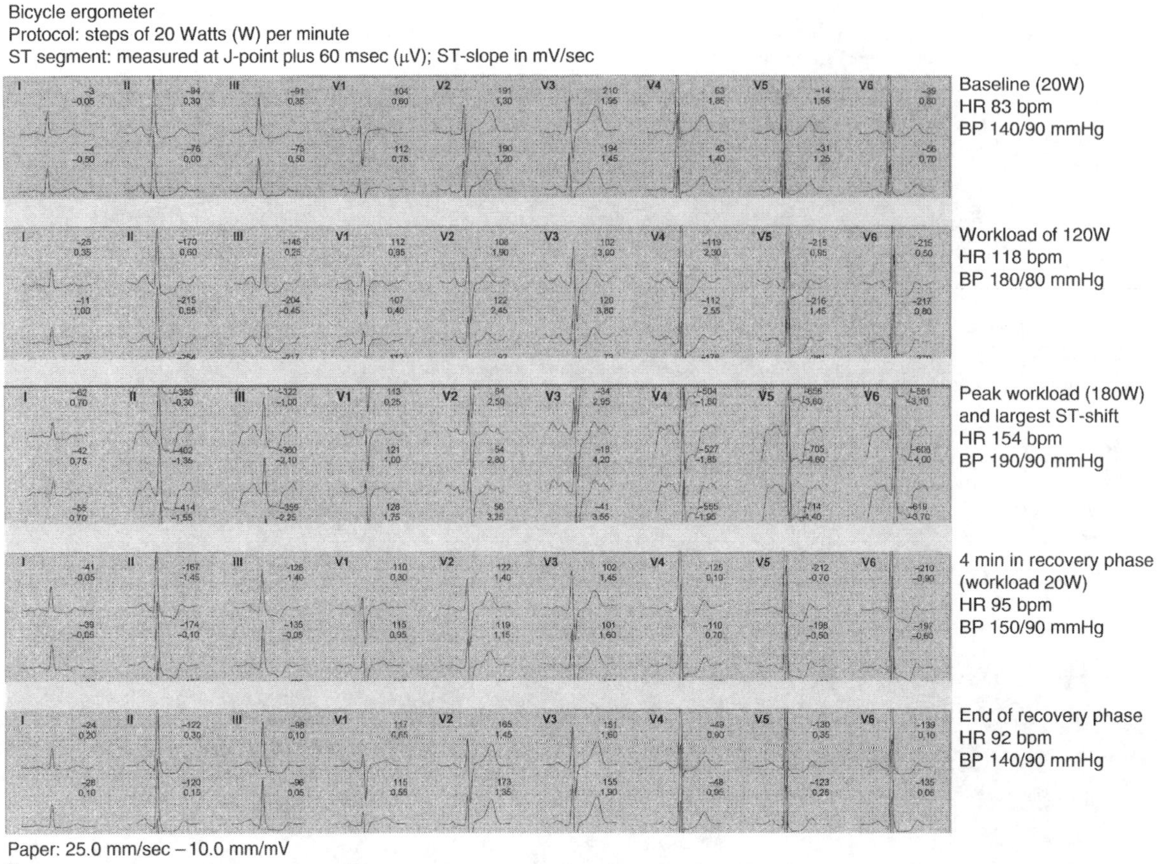

c

Figure 36.6 (Continued)

(c) The same patient visited the outpatient clinic approximately 1 year after the percutaneous coronary intervention. He indicated that he suffered again from chest pain similar to the complaints he had prior to the revascularization procedure. The only difference was that the chest pain now occurred with more strenuous exertion as compared to the complaints 1 year earlier (functional class II according to the Canadian Cardiovascular Society [63]). The patient reached the predicted workload of 171 W and discontinued the test because of tiredness and shortness of breath. The exercise electrocardiogram again showed marked ST segment depression up to 6 mm (0.6 mV) with a horizontal to downsloping pattern at maximal exertion. Coronary angiography revealed severe in-stent restenosis in the treated segments in the left anterior descending artery and first diagonal. The patient was subsequently successfully treated by repeat percutaneous coronary intervention.

the likelihood that a given range of ST segment depression is related to the presence or absence of CAD. This likelihood ratio is 0.4 for ST segment depression up to 0.9 mm (0.09 mV). Only ST segment depression greater than 1 mm (0.1 mV) is clearly associated with CAD. It is apparent that this likelihood increases when the ST segment depression is more pronounced.

Magnitude of ST Segment Depression and the Degree of Anatomical Coronary Artery Disease

Many investigators have attempted to correlate the magnitude of ST segment depression with the degree of anatomical CAD. Initial studies reported that an increased magnitude of ST-depression was associated with an increased degree of ischemia [33, 68]. Later studies have been unable to correlate the ischemia estimated from the magnitude of the ST segment depression in any lead or from the sum of the ST-changes in all leads with either the number of diseased coronary

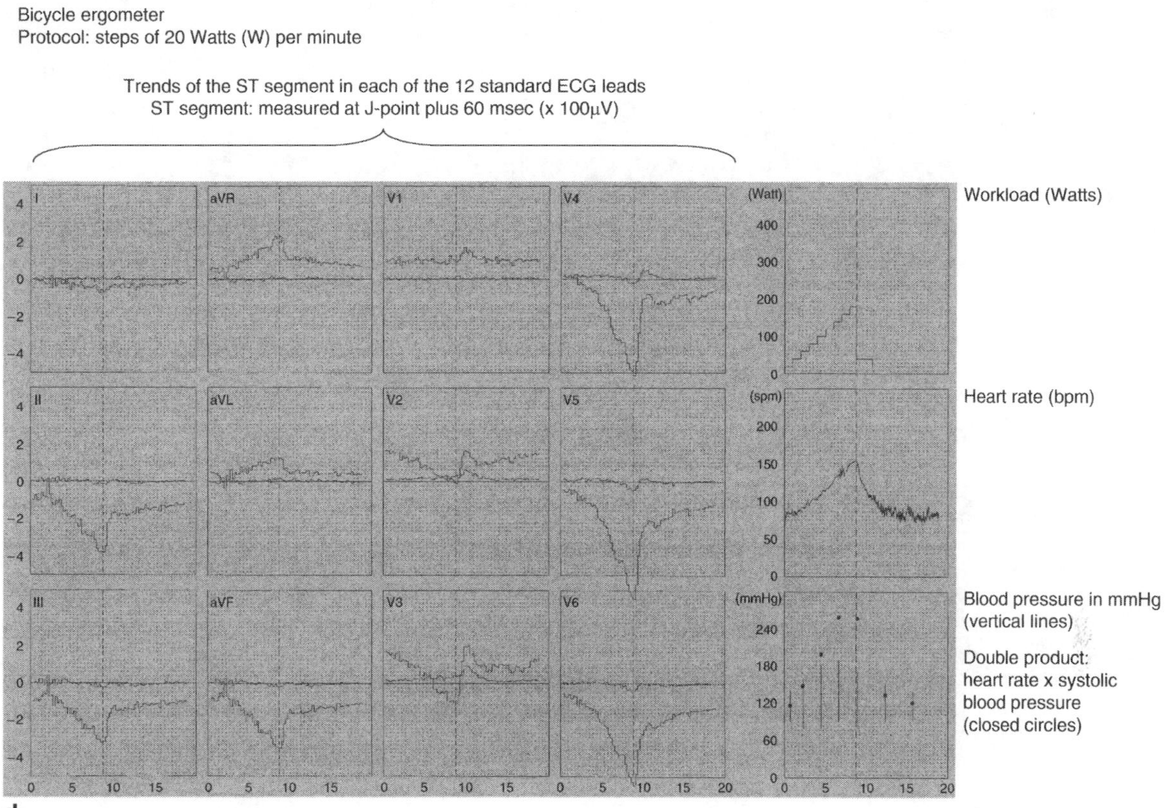

d

Figure 36.6 (Continued)

(d) The respective ST segment trends illustrate the marked ST segment depression, predominantly occurring in the inferior and left precordial leads.

Table 36.6

Probability of CAD in relation to degree of ST segment depression during exercise testing

ST-depression (mV)	Normal subjects (*n*=225)	Subjects with CAD (*n*=381)	Specificity (%)	Sensitivity (%)	Likelihood ratio for CAD
< 0.05	181	122			0.4
0.05–0.09	30	19	80	68	0.4
0.1–0.19	11	133	94	63	7.3
> 0.2	3	107	99	28	21.6

Pooled data from four studies [64–67] that provided information on the presence of coronary artery disease (CAD) in relation to the degree of ST segment depression during exercise stress testing. Seventeen patients with ST segment elevation have been excluded from reference 66.

arteries or the size of the area of reversible ischemia observed on myocardial perfusion scintigraphy [33, 69, 70]. As the development of ST segment depression as an electrophysiological phenomenon results from many influences, including those caused by electrolytes, hormones, and hemodynamic, metabolic, as well as anatomical changes, it is unlikely that the magnitude of ST segment depression would correlate well with the coronary anatomy in patients with CAD. Furthermore, the degree of ST-depression at higher workloads also depends on what is used as an indication to terminate exercise. For example, if patients with single vessel disease are encouraged to exercise strenuously or if they have some degree of left

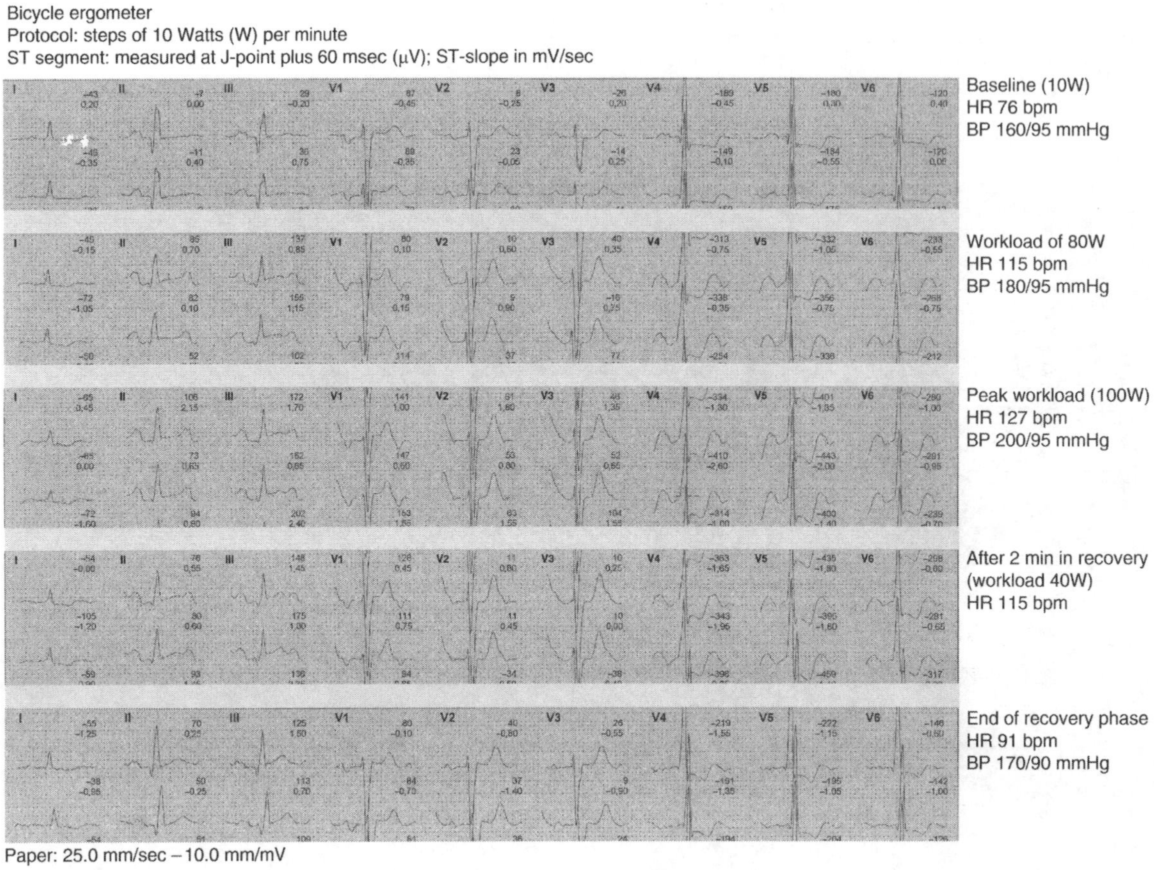

Figure 36.7

Bicycle exercise test (10 W/min protocol) in a 77-year old male with angina pectoris complaints, classified as II–III according to the Canadian Cardiovascular Society [63]. His medical history was unremarkable except for poorly controlled arterial hypertension. The resting ECG showed signs of an old inferior myocardial infarction (Q-waves in leads II, III, and aVF), as well as left ventricular hypertrophy with repolarization abnormalities in the left precordial leads (downsloping ST segment with inverted T-wave). With increasing workload, the preexistent ST segment depression increased progressively up to 4 mm (0.4 mV), persisting in the recovery period. Coronary angiography revealed three-vessel coronary artery disease and ventriculography showed a moderately impaired left ventricular systolic function. The patient subsequently underwent coronary artery bypass grafting.

ventricular hypertrophy, they may have severe ST-depression (see also Figs. 36.6 (Continued) and 36.7). Therefore, the magnitude of ST segment depression at maximum workload may not indicate the severity of CAD.

Time Course of ST Segment Depression

ST segment depression that comes on early during exercise at low workloads is associated with more severe disease and impaired prognosis [33, 68, 71–73]. In a study on the predictive value of the time course of ST segment depression during exercise testing in patients referred for coronary angiography, a direct relationship between the time of onset and offset of ST-depression and the number of diseased coronary arteries was demonstrated [33, 73]. Severe three-vessel CAD was found in 33% of patients with early onset and late offset of the ST-depression, while this increased up to 50% in those with resting ST-depression that increased with exercise [73]. Even ST segment depression occurring at high workloads and with rapid resolution during recovery may be associated with significant CAD, although it may also indicate a false-positive response in an asymptomatic subject (➋ Fig. 36.5).

Furthermore, although ST segment depression during exercise often persists into the recovery phase, it sometimes may not develop until exercise has been terminated (➋ Fig. 36.4). Whereas this phenomenon is still not completely understood, several studies have demonstrated that ST segment depression occurring only during recovery has the same significance as exercise-induced ST-depression in predicting CAD [33, 74, 75].

Adjustment of ST Segment Depression

Subjects with tall R-waves exhibit a greater amount of ST segment depression than those with smaller R-wave amplitude [76]. Accordingly, when the R-wave in the lateral precordial leads is less than 10 mm, the sensitivity of ST-depression for the detection of CAD is low if 1 mm (0.1 mV) of ST-depression is required for a positive test result [77]. Correcting ST-depression for R-wave amplitude has been proposed by dividing the amount of ST segment depression by the R-wave height (both expressed in mm) [78]. The ST/R-ratio may then be used as an alternative to the amount of ST segment depression per se, with 0.1 as a cutoff value for an abnormal test result [33]. In studies, however, adjustment of the amount of ST segment depression by the R-wave height has not been shown to consistently improve the diagnostic value of exercise-induced ST-depression [4].

Several methods of heart rate adjustment have been proposed to increase the diagnostic accuracy of the exercise ECG. The first technique is to derive the maximal slope of the ST segment relative to heart rate (ST/HR slope). The second method, termed the ST/HR index, divides the difference between ST-depression at peak exercise by the delta heart rate (difference between resting and maximum heart rate) [4, 33, 79–82]. The value of ST/HR adjustment in improving diagnostic accuracy has been evaluated in several studies [4, 83–87]. Most studies, however, were limited by work-up bias and enrollment of relatively healthy patients, which presents a limited challenge to the ST/HR index [4, 85–87]. In a large multicenter study without these limitations, the ST/HR slope or index was not found to be more accurate than simple measurement of the ST segment depression [88]. The most recent guidelines on exercise testing take the perspective that the ST/HR adjustment approach in symptomatic patients has at least equivalent accuracy to the standard approach [4]. Although the ST/HR approach might be useful in assessing certain borderline or equivocal ST segment responses during exercise (e.g., ST segment depression associated with a very high exercise heart rate), further validation is required.

ST Segment Depression and the Location of Coronary Artery Disease

In several studies, an attempt has been made to relate the location of ST segment depression during exercise to coronary anatomy. As stated, extensive CAD (three-vessel disease) is more likely to be present in a patient who develops major ST segment depression in multiple leads at low workloads. However, even in patients with single-vessel disease, there is no clear separation between the leads in which ST segment depression develops when either the right coronary artery or the left anterior descending coronary artery is involved [67]. Studies using body-surface mapping indicate that the ST segment depression is most prominent in the precordium. The actual position where the largest ST segment depression occurs varies widely, from the level of the third intercostal space to the tenth intercostal space [46]. However, even analysis of precordial surface maps has not permitted an accurate prediction of the location of CAD [40, 89], although this has previously been claimed [40]. Furthermore, in another study, no relation was found between the spatial orientation of the ST segment vectors during exercise and the location of myocardial ischemia as detected by myocardial perfusion scintigraphy [90].

ST Segment Depression and/or Exercise-Induced Angina

The occurrence of chest pain during stress testing is of equal diagnostic value as the development of ST segment depression [91–93]. Symptoms that cannot be clearly understood during an outpatient clinic visit may be clarified if they occur during the actual exercise procedure and disappear rapidly in the recovery phase. In order to permit optimal comparison between the history of the patient and the observations during the stress test, it is recommended that either the test is performed by the patient's own physician, or that the physician supervising the test reports in detail all the complaints that occur during the test and notes whether these are similar to the complaints that led the patient to the outpatient clinic. In ➋ Fig. 36.8, the incidence of CAD and the number of diseased vessels are presented for ST segment depression with and without angina [92].

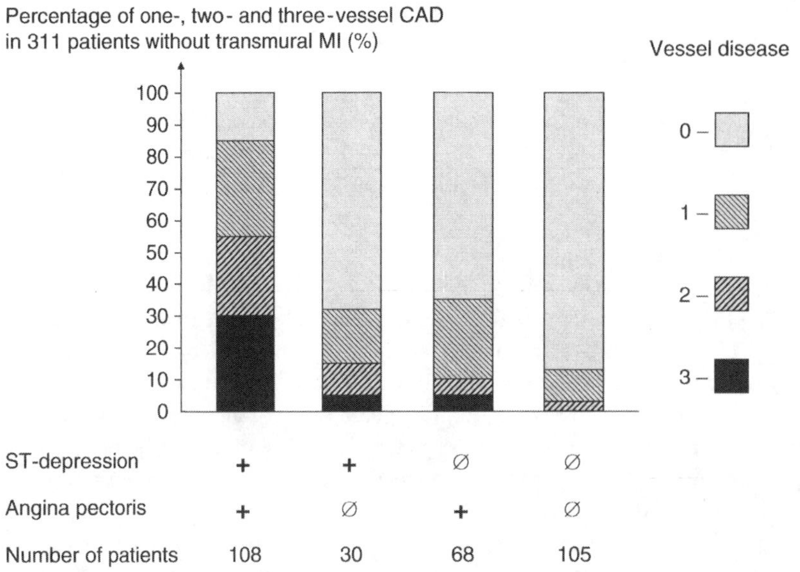

Figure 36.8

Incidence of normal coronary arteries (zero-) and one-, two-, or three-vessel disease in 311 patients without a history of transmural myocardial infarction. It should be noted that the incidence of CAD was similar in 30 patients with ST segment depression without symptoms (30%) and in 68 patients with exercise-induced angina without significant ECG changes (36%). ST-depression represents 0.1 mV or greater of ST segment depression in at least one precordial lead (Adapted from Roskamm [92]).

36.6.2.2 ST Segment Elevation

ST segment elevation during exercise is observed most frequently in leads with Q-waves in patients with a history of myocardial infarction (➋ Fig. 36.9). The significance of ST segment elevation in areas of Q-waves of an old myocardial infarction is controversial. Some studies have suggested that this ST segment elevation is caused by an area with abnormal wall motion (ventricular akinesia or dyskinesia) [64, 94–97]. However, other studies have shown that in some patients such ST-elevation disappears after coronary bypass surgery, which suggests that the ST-elevation was a result of myocardial ischemia [98]. More recent studies using myocardial perfusion scintigraphy have found such ST-elevation to be a marker of residual viability in the infarcted area [99–101]. Therefore, it seems advisable to consider additional tests such as myocardial perfusion scintigraphy in patients with chest pain complaints after myocardial infarction and ST segment elevation during exercise stress testing.

Exercise-induced ST segment elevation in leads without Q-waves on a normal ECG (other than in aVR or V_1) is very rare and represents severe transmural myocardial ischemia (whereas ST segment depression represents subendocardial ischemia) [94, 102–109]. It is caused by a coronary artery spasm that completely obliterates antegrade flow through the epicardial artery, or a high-grade proximal stenosis in a coronary artery (➋ Fig. 36.10). In patients with variant or Prinzmetal's angina, coronary spasm is most commonly seen at rest, but very occasionally it occurs with exercise, probably indicating the presence of hemodynamically significant coronary atherosclerosis [33, 110, 111]. In contrast to ST segment depression, exercise-induced ST segment elevation is very arrhythmogenic and localizes the ischemic area. When ST segment elevation occurs in leads V_2 through V_4, there is severe anterior wall ischemia with a high-grade stenosis in the proximal left anterior descending coronary artery (➋ Fig. 36.10); in leads II, III, and aVF, there is severe inferior wall ischemia with involvement of a proximal stenosis in a large right coronary artery; and in the lateral leads, the left circumflex or diagonals are involved [4]. As stated previously, this rare finding warrants immediate discontinuation of the stress test and prompt referral for coronary angiography.

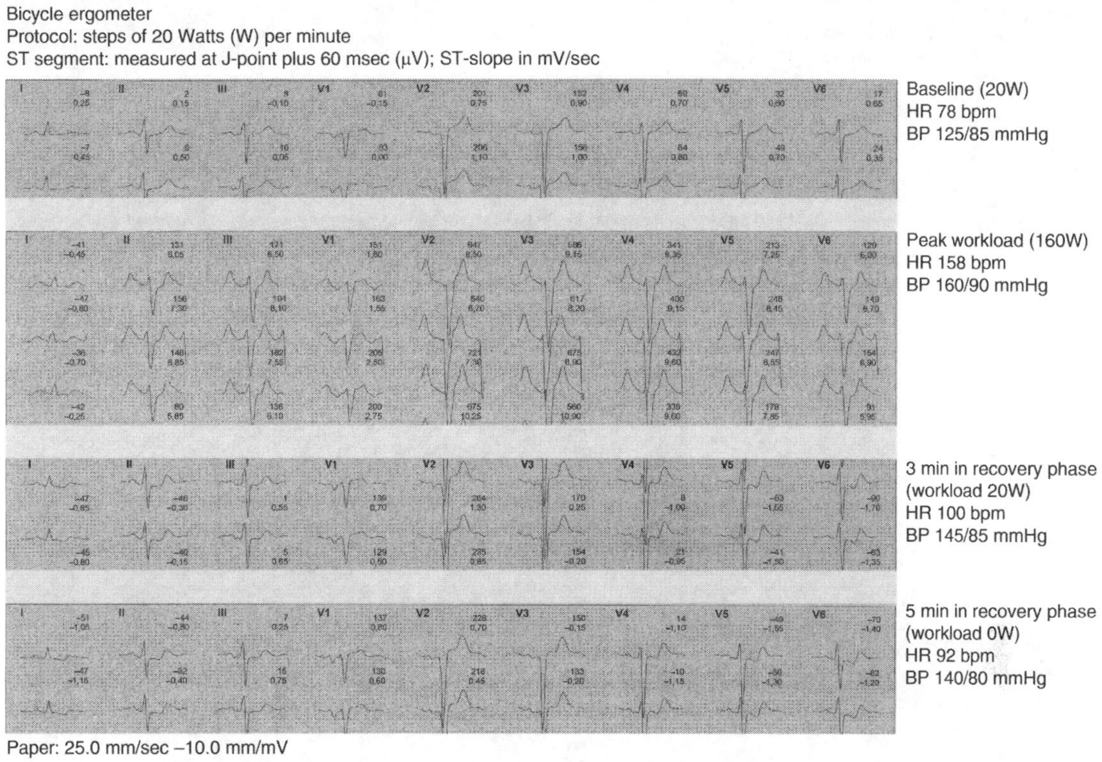

Figure 36.9

Bicycle exercise stress test in a 62-year-old male approximately 1 year after he underwent a primary coronary intervention with recanalization of the occluded left anterior descending coronary artery for an acute anterior myocardial infarction. The result of the intervention was suboptimal because there was no reflow in the infarct-related artery. No intervention was performed on a 50–70% stenosis in the left circumflex coronary artery. Echocardiography showed an impaired systolic left ventricular function with akinesia of the interventricular septum and hypokinesia of the anterior wall and apex. The baseline ECG showed sinus rhythm, a left anterior fascicular block, and a QRS-pattern compatible with an old antero-septal-apical myocardial infarction with QS-complexes in V_1 and V_2, a small R-wave in V_3, as well as a Q-wave in aVL. The patient reached 160 W and discontinued the test because of fatigue. The patient did not experience any anginal complaints or dyspnea during the test. With an increase in workload, there is an increasing level of ST segment elevation in the right precordial leads (V_1–V_3), representing either wall motion abnormalities or residual viability in the infarcted area. Furthermore, the duration of the QRS complex increased. In the recovery phase, minimal 0.5 mm (0.05 mV) downsloping ST segment depression developed in the inferior and left precordial leads.

36.6.2.3 Changes of the QRS Complex

R-Wave Amplitude

It has previously been reported that an increase in R-wave amplitude during exercise is a useful indicator of CAD [112]. The first reports on this phenomenon were based on an unusual series of patients with known false-positive or false-negative ST-changes during exercise. Although some studies supported these observations, others found that the R-wave amplitude changes were variable both in patients with CAD and in normal subjects [52–54, 113]. It is possible that the observed differences in R-waves between patients with CAD and normal subjects result in part from differences in heart rate, since most patients stop exercising at heart rates between 120 and 150 beats/min which is the rate at which normal subjects frequently exhibit increased R-wave amplitudes [44]. The mechanisms of these changes also remain unclear. In particular, the hypothesis that the increase in R-wave amplitude would be related to an increase in ventricular volume owing to

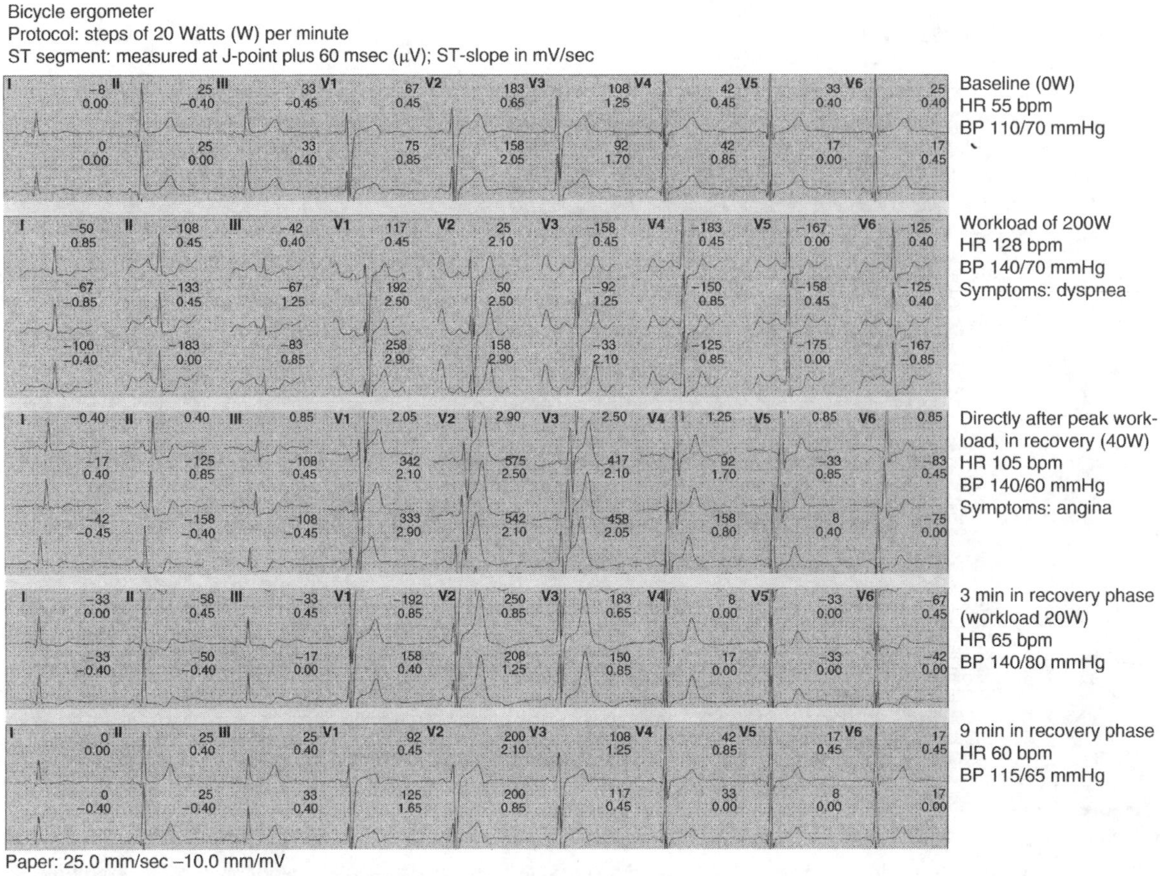

Figure 36.10

Bicycle exercise test in a 59-year-old male with no history of cardiovascular disease who underwent stress testing because of complaints of chest pain on exertion suspected for angina pectoris. The predicted maximum workload was 163 W. The test was discontinued at 200 W by the attending physician because of shortness of breath, as well as failure of the blood pressure to rise in the presence of ST segment depression. Starting at 180 W, up to 1.0–1.5 mm (0.1–0.15 mV) of slow-upsloping to horizontal ST segment depression developed in the inferior and left precordial leads, indicating subendocardial ischemia. Directly following discontinuation of exercise, the patient experienced angina-like chest pain. This was attended on the ECG by ST segment elevation and a marked increase in T-wave amplitude in the right precordial leads, suggesting transmural ischemia due to a high-grade stenosis in the proximal left anterior descending artery. Later on in the recovery phase, biphasic T-waves developed in the right precordial leads. At the end of the recovery period, the ECG had returned to normal. The patient was subsequently admitted to hospital. Coronary angiography was performed the same day and revealed, as expected, a severe stenosis in the proximal left anterior descending artery which was treated with stent implantation.

myocardial ischemia [114] was not supported by studies in which the volume changes were actually measured [52, 53, 115]. As exercise-induced changes in R-wave amplitude have provided very little, if any, discrimination for myocardial ischemia, the R-wave measurements are not routinely applied in clinical practice.

QRS-Duration

As stated, the duration of the QRS complex during exercise does not increase but may be reduced slightly because conduction velocity is increased by catecholamine release [50, 116]. Although a number of studies have shown that conduction velocity is decreased by myocardial ischemia [61, 117], QRS-duration measurements are not routinely used for the diagnosis of CAD as studies on this phenomenon have had equivocal results [61]. Prolongation of the QRS-duration during exercise was found to be associated with CAD in 330 patients who underwent exercise testing and coronary angiography [118]. The greatest prolongation was found in patients with three-vessel disease. However, the observations could not be used in individual patients to determine the presence or absence of myocardial ischemia as the variations around the mean value were too large [118].

Bundle-Branch Block

Rate-dependent bundle-branch block may become apparent during exercise and the interpretation of such a finding is the same as in rate-dependent bundle-branch block occurring under other circumstances. Rate-dependent left bundle-branch block, right bundle-branch block or left anterior fascicular block may be related to CAD [33, 119]. However, they have also been observed in healthy subjects without myocardial or coronary pathology [33]. The isolated finding of a rate-dependent bundle-branch block should, therefore, not be used as a proof of significant heart disease. On the other hand, the occurrence of a bundle-branch block during exercise should be viewed in light of the other clinical findings in each patient in order to assess the probability that CAD is present. Leftward rotation of the frontal axis with development of a left anterior fascicular block during exercise in a patient *with* CAD usually signifies involvement of the proximal left anterior descending coronary artery or three-vessel CAD [120, 121]. The axis shift associated with left anterior fascicular block may mask the ischemic ST segment changes in the frontal plane and to a lesser extent in the precordial leads.

The significance of ST segment depression as a predictor of ischemia in the presence of right bundle-branch block has been the subject of debate. Exercise-induced ST-depression usually occurs with right bundle-branch block in the anterior chest leads V_1 through V_3, probably because of secondary repolarization changes, and is not associated with ischemia [4, 122]. However, ST segment depression in the left precordial leads (V_5 and V_6) or inferior leads (II and aVF) has the same significance in predicting CAD as exercise-induced ST-depression in a normal resting ECG [123, 124].

Exercise-induced ST-depression usually occurs with left bundle-branch block and has no association with ischemia [125]. Even up to 10 mm of ST-depression occurs in healthy subjects. There is no level of ST-depression that confers diagnostic significance in left bundle-branch block. Therefore, there is a consensus, as stated in the guidelines, that exercise-induced ischemia cannot be diagnosed from the ECG in patients with left bundle-branch block [4].

36.6.2.4 T-Wave Changes

As previously mentioned, the T-wave amplitude varies greatly during exercise but increases considerably in most normal subjects immediately after exercise as a result of an increased stroke volume, which makes up for the lingering metabolic debt after the heart rate has dropped very rapidly. Many patients with flat or inverted T-waves at rest will manifest upright T-waves at the time of exercise. This is particularly true in women. This phenomenon of normalization of inverted T-waves or pseudonormalization of the T-waves is not considered to indicate ischemia. In earlier studies, T-wave normalization was accompanied by significant ST segment depression in 90% of patients with CAD, but by a negative test result based on ST segment criteria in all patients without ischemic heart disease [33, 126]. Similarly, exercise-induced deep T-wave inversion is almost always accompanied by significant ST segment depression and is then associated with a more severe degree of CAD.

36.7 Exercise Testing to Diagnose Coronary Artery Disease

The vast majority of exercise testing is performed in adults with symptoms of known or suspected ischemic heart disease. There has been considerable discussion on the value and use of exercise testing for the diagnosis of CAD and the literature on this subject is extensive. Knowledge of terminology used in describing diagnostic test characteristics and test performance is required for understanding the exercise test literature (➋ Table 36.7).

Table 36.7
Diagnostic test characteristics

True-positive test result (TP)	Abnormal test result in subject *with* disease
False-positive test result (FP)	Abnormal test result in subject *without* disease
True-negative test result (TN)	Normal test result in subject *without* disease
False-negative test result (FN)	Normal test result in subject *with* disease
Specificity	Percentage of subjects *without* disease who have a normal test result = TN/(TN + FP)
Sensitivity	Percentage of subjects *with* disease who have an abnormal test result = TP/(TP + FN)
Test accuracy	Percentage of true test results = (TP + TN)/total number of tests performed
Predictive value of a positive test result	Percentage of subjects with abnormal test who have the disease = TP/(TP + FP)
Predictive value of a negative test result	Percentage of subjects with normal test and without disease = TN/(TN + FN)
Formula for calculation of posttest probability (Bayes' theorem):	
Posttest odds (disease present) = pretest odds (disease present) × likelihood ratio (LR) where, odds = probability/(1 – probability) LR = sensitivity/(1 – specificity) in case of an abnormal test result = (1 – sensitivity)/specificity in case of a normal test result	
Posttest odds (disease absent) = pretest odds (disease absent) × likelihood ratio (LR) where, odds = probability/(1 – probability) LR = (1 – specificity)/sensitivity in case of an abnormal test result = specificity/(1 – sensitivity) in case of a normal test result	
The better (very high or very low) the likelihood ratio of the test (determined by sensitivity and specificity), the more discriminant the test is.	

Table 36.8
Pretest probability (%) of CAD in patients by age, sex, and chest pain characteristics (Adapted from Diamond and Forrester [127])

Age (years)	Sex	Non-anginal chest pain	Atypical angina	Typical angina
30–39	M	5.2 ± 0.8	21.8 ± 2.4	69.7 ± 3.2
	F	0.8 ± 0.3	4.2 ± 1.3	25.8 ± 6.6
40–49	M	14.1 ± 1.3	46.1 ± 1.8	87.3 ± 1.0
	F	2.8 ± 0.7	13.3 ± 2.9	55.2 ± 6.5
50–59	M	21.5 ± 1.7	58.9 ± 1.5	92.0 ± 0.6
	F	8.4 ± 1.2	32.4 ± 3.0	79.4 ± 2.4
60–69	M	28.1 ± 1.9	67.1 ± 1.3	94.3 ± 0.4
	F	18.6 ± 1.9	54.4 ± 2.4	90.6 ± 1.0

CAD denotes coronary artery disease

36.7.1 Pretest Probability

The pretest probability of obstructive CAD can be estimated from factors such as age, gender, risk factors, and chest pain characteristics [127–129]. The pretest probabilities described in this way by Diamond and Forrester in a series of over 60,000 patients are shown in Table 36.8 [127]. From this table, it is apparent that exercise testing is not very useful for establishing the diagnosis CAD in a 64-year-old man with typical or definite angina. The pretest probability of CAD

Table 36.9
Posttest probability (%) of coronary artery disease based on age, sex, symptom classification, and exercise test-induced electrocardiographic ST segment depression

		Typical angina		Atypical angina		Non-anginal chest pain		Asymptomatic	
Age	ST-depression (mV)	M	F	M	F	M	F	M	F
30–39	0.00–0.04	25	7	6	1	1	<1	<1	<1
	0.05–0.09	68	24	21	4	5	1	2	<1
	0.10–0.14	83	42	38	9	10	2	4	<1
	0.15–0.19	91	59	55	15	19	3	7	1
	0.20–0.24	96	79	76	33	39	8	18	3
	>0.25	99	93	92	63	68	24	43	11
40–49	0.00–0.04	61	22	16	3	4	1	1	<1
	0.05–0.09	86	53	44	12	13	3	5	1
	0.10–0.14	94	72	64	25	26	6	11	2
	0.15–0.19	97	84	78	39	41	11	20	4
	0.20–0.24	99	93	91	63	65	24	39	10
	>0.25	>99	98	97	86	87	53	69	28
50–59	0.00–0.04	73	47	25	10	6	2	2	1
	0.05–0.09	91	78	57	31	20	8	9	3
	0.10–0.14	96	89	75	50	37	16	19	7
	0.15–0.19	98	94	86	67	53	28	31	12
	0.20–0.24	99	98	94	84	75	50	54	27
	>0.25	>99	99	98	95	91	78	81	56
60–69	0.00–0.04	79	69	32	21	8	5	3	2
	0.05–0.09	94	90	65	52	26	17	11	7
	0.10–0.14	97	95	81	72	45	33	23	15
	0.15–0.19	99	98	89	83	62	49	37	25
	0.20–0.24	99	99	96	93	81	72	61	47
	>0.25	>99	>99	99	98	94	90	85	76

is so high that the test result does not substantially change this probability. As shown in Table 36.9, the likelihood of CAD is 79% if no ST segment depression occurs during the test, while it would be 99% if 2 mm (0.2 mV) of ST segment depression developed. However, the test may still be used to determine the functional impairment of that subject, to measure the blood-pressure and heart rate response, or to estimate the prognosis. Similarly, the diagnostic value of exercise electrocardiography is low in asymptomatic men and women (see also Fig. 36.11). The greatest diagnostic value of exercise testing is obtained in patients with an intermediate pretest probability of CAD, for example, between 20% and 80%, because the test result has the largest potential effect on diagnostic outcome. If the posttest likelihood is intermediate, another noninvasive (e.g., myocardial perfusion scintigraphy) or invasive test may be applied.

36.7.2 Diagnostic Characteristics and Test Performance

One of the problems with diagnostic tests for CAD is that there is a considerable overlap in the range of measurements for the normal population and those with CAD. Since the depth of exercise-induced ST segment depression and the extent of the myocardial ischemic response can be considered as continuous variables, a certain cutpoint or discriminant value (e.g., 1 mm (0.1 mV) of ST segment depression) cannot completely discriminate patients with CAD from those without

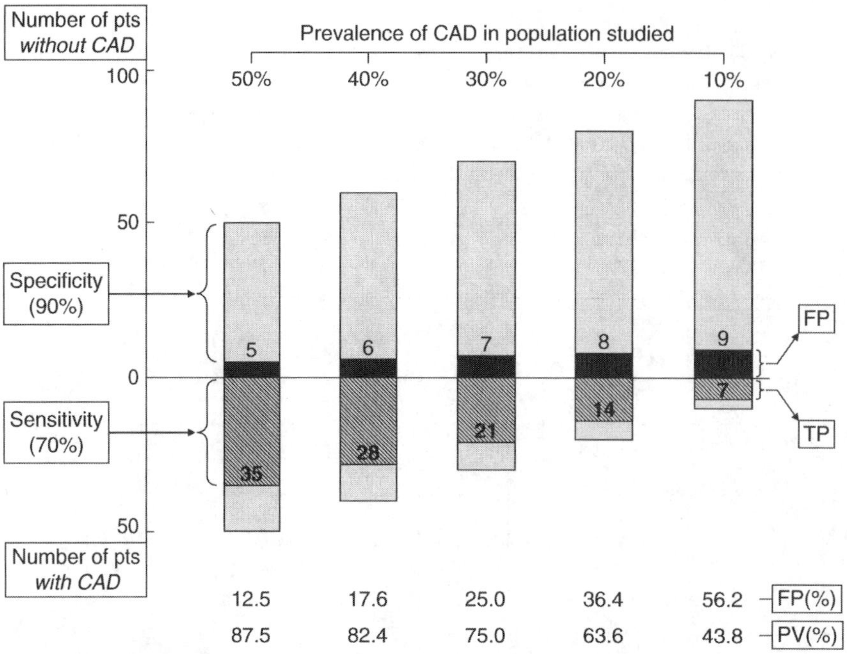

Figure 36.11

Illustration of the impact of the prevalence of CAD in the population studied on the fraction of false-positive test results and the predictive value of the exercise test. Each bar represents a population of 100 patients. In the left-most bar, 50 patients have CAD, while 50 do not have CAD. In the second bar, 40% has CAD, in the third 30%, in the fourth 20%, and in the right-most bar only 10%. In each theoretical example, the sensitivity for detection of CAD is 70%. Thus, 70% of patients with CAD do indeed have ischemic ST segment depression. On the other hand, the specificity is 90%. Thus, 10% of the patients without CAD exhibit false-positive ST segment depression (black area). In the population with a 50% prevalence of CAD (left-most bar), 40 abnormal tests are found, five of which are false-positive (12.5%). In the population with only a 10% prevalence of CAD (right-most bar), nine false-positive tests occur out of a total of 16 tests with ST segment depression (56.2%). In this latter population, which reflects the prevalence of CAD in many screening conditions, the number of false-positives is higher than the number of correct or true-positive test results. CAD, coronary artery disease; FP, false-positive; PV, predictive value; Pts, patients; and TP, true-positive.

disease. A higher discriminant value for ST segment depression improves specificity, but reduces the test's sensitivity. Therefore, sensitivity and specificity are inversely related and determined by the choice of a cutpoint or discriminant value. The positive predictive value (PPV) and negative predictive value (NPV) are also affected by the population tested. The PPV will be higher in a population with a higher prevalence of CAD. Accordingly, PPV is higher in patients with three-vessel CAD and lower in patients with one-vessel CAD. The NPV can be decreased if the test is used in patients in whom false-positive results are more likely such as those with ST segment depression on the resting ECG or left ventricular hypertrophy. ➋ Figure 36.11 illustrates these points.

36.7.3 Probability Analysis

The use of Bayes' theorem of conditional probability can assist in the interpretation of a test result and can also provide a meaningful estimate of the posttest probability in the individual patient. According to this theorem, posttest probability is a function of pretest probability, and the sensitivity and specificity (likelihood ratio) of the test [130].

Under the assumption of the independence of the exercise test result from the clinical (pretest) data, posttest probability can be calculated according to the formula listed in ➋ Table 36.7. Although this calculation is often made intuitively by the clinician, mathematical equations or scores have been developed from multivariate analysis of clinical and exercise test variables (including heart rate at peak exercise, ST segment response, the presence or absence of angina during the test, peak workload, and ST segment slope) that can provide a more accurate estimate of the probability that CAD is present, when compared with use of the ST segment measurements alone [4, 131–135]. However, the use of these statistical models remains limited. Nevertheless, they underline the importance of taking into account all relevant variables when estimating the probability of CAD in a given subject [136–138].

36.7.4 Diagnostic Accuracy of the Standard Exercise Test

A meta-analysis was performed on the diagnostic accuracy of the exercise test based on 147 consecutive published reports involving 24,074 patients who underwent both coronary angiography and exercise testing [139, 140]. There was a wide variability in the reported diagnostic accuracy of the standard exercise test among the studies (➋ Table 36.10). The mean sensitivity was 68% (range 23–100%) and mean specificity was 77% (range 17–100%). This large variability results from the fact that most studies do not fulfill two major criteria that are important when evaluating diagnostic tests [4]. The first concerns the fact that the population studied does not represent the diagnostic dilemma group in clinical practice. In particular, inclusion of patients who most certainly have the disease (e.g., post-myocardial infarction patients) in the test group presents a limited challenge to the diagnostic test. The second cause concerns the presence of work-up bias which refers to the fact that most reported studies were affected to some degree by clinical practice wherein the results of the exercise test were used to decide who would undergo coronary angiography and be included in the study [4]. The reported meta-analysis provides, however, the best estimate of the diagnostic accuracy of the exercise test. When only studies that excluded patients with a previous myocardial infarction were considered, mean sensitivity was 67% and mean specificity 72% (➋ Table 36.10). Only three studies avoided work-up bias and provide an estimate of the diagnostic accuracy of exercise testing in a more general population of patients presenting with chest pain [141–143]. The mean sensitivity and specificity in these studies were 50% and 90%, respectively (➋ Table 36.10).

36.7.5 Electrocardiographic Factors Influencing Sensitivity and Specificity

The influence of left ventricular hypertrophy, resting ST segment depression and use of digoxin on the exercise test characteristics are summarized in ➋ Table 36.10. Left ventricular hypertrophy with repolarization abnormalities is associated

Table 36.10
Meta-analysis of diagnostic performance of exercise test (Adapted from ACC/AHA Practice Guidelines on Exercise Testing [4] and from [139, 140])

(Sub-)Groups	Number of studies	Total number of patients	Sensitivity (%)	Specificity (%)	Predictive accuracy (%)
Meta-analysis of standard exercise test	147	24, 047	68	77	73
Meta-analysis *without* prior MI	58	11, 691	67	72	69
Meta-analysis *without* work-up bias	3	>1, 000	50	90	69
Meta-analysis *with* ST-depression	22	9, 153	69	70	69
Meta-analysis *without* ST-depression	3	840	67	84	75
Meta-analysis *with* digoxin	15	6, 338	68	74	71
Meta-analysis *without* digoxin	9	3, 548	72	69	70
Meta-analysis *with* LVH	15	8, 016	68	69	68
Meta-analysis *without* LVH	10	1, 977	72	77	74

LVH, left ventricular hypertrophy; MI, myocardial infarction with a decreased specificity, but sensitivity is unaffected. If the test result is negative, the probability of CAD is substantially reduced, but additional tests are indicated in patients with an abnormal test result (➋ Fig. 36.7). Although not apparent from ➋ Table 36.10, studies have suggested that digoxin also lowers specificity by producing an abnormal ST segment response to exercise, which occurs in up to 40% of healthy individuals [144, 145]. Resting ST segment depression is associated with a higher prevalence of severe CAD and a higher incidence of adverse cardiac events [4, 146–151]. As shown in ➋ Table 36.10, resting ST-depression lowers specificity of the exercise test. The most recent guidelines on exercise testing state that, if the resting ST segment depression is less than 1 mm (0.1 mV), the standard exercise test may still be the first test because sensitivity is increased [4]. In a retrospective study of male patients with resting ST segment depression, but without prior myocardial infarction, undergoing exercise testing, 2 mm (0.2 mV) of additional exercise-induced ST segment depression, or downsloping ST segment depression of 1 mm (0.1 mV) or more in the recovery phase were found to be important markers for the diagnosis of CAD, with a sensitivity of 67%, a specificity of 80%, and a likelihood ratio of 3.4 [4, 151]. In patients with less than 1 mm (0.1 mV) of resting ST segment depression who are also taking digoxin or have left ventricular hypertrophy, as well as in those with more than 1 mm (0.1 mV) of resting ST segment depression (➋ Fig. 36.7), the diagnostic merits of the standard exercise test may be insufficient and imaging tests such as myocardial perfusion scintigraphy or stress echocardiography are preferred [4].

Finally, as stated, specificity was lowered and sensitivity increased when upsloping ST segment depression was classified as abnormal. Furthermore, the presence of right bundle-branch block does not appear to reduce the diagnostic accuracy of the exercise test for the diagnosis of CAD. On the contrary, exercise testing should not be used for the diagnosis of CAD in patients with complete left bundle-branch block, or in those with pre-excitation (Wolff-Parkinson-White) syndrome, or electronically paced ventricular rhythm [4].

36.7.6 Exercise Electrocardiography Versus Noninvasive Stress Imaging Studies

Although it is beyond the scope of this chapter to include a complete discussion on the comparison of exercise electrocardiography with the noninvasive stress imaging tests for the diagnosis of CAD, some features are presented here in brief. Studies that compared the diagnostic performance between exercise electrocardiography and the noninvasive imaging modalities using pharmacological stress (e.g., stress echocardiography or stress single-photon emission computed tomography (SPECT) myocardial perfusion scintigraphy) indicate that the latter are both more sensitive and more specific for the detection of CAD [152–155]. Furthermore, myocardial perfusion scintigraphy and stress echocardiography permit, to a certain extent, separation of patients with single and multiple vessel CAD and detection of the location of significant coronary artery stenosis [152–156]. This does not mean that exercise electrocardiography should be replaced by these imaging modalities. In many patients, CAD can be effectively ruled out, or diagnosed, by conventional exercise testing. Noninvasive imaging tests may provide additional information if the diagnosis is uncertain after conventional exercise testing, for example, if there is a posttest likelihood of intermediate probability. This likelihood is then considered as the pretest likelihood for the stress imaging test, as shown in ➋ Table 36.11 for myocardial perfusion scintigraphy; a 50% likelihood will reduce to 15% if the scintigram is normal, or increase to 93% if a reversible perfusion defect occurs. Myocardial perfusion scintigraphy and stress echocardiography using pharmacological stress agents can also be a useful alternative to exercise stress protocols in patients who are unable to exercise because of neurological, orthopedic, or peripheral vascular disease, as well as in those patients in whom interpretation of the ECG is hampered [152–158]. For a more complete and detailed overview of the noninvasive imaging modalities and their usefulness in the evaluation of CAD, the reader is referred to the relevant literature and guidelines [152–158].

36.8 Exercise Testing in Patients with Coronary Artery Disease

The value of exercise testing for risk or prognostic stratification must be considered in the light of what is added to that which is already known about the patient's risk status. Most studies on risk assessment in patients with CAD using exercise testing have focused on the relation between test parameters and future survival. The strongest predictor of survival in patients with CAD is the function of the left ventricle. Other important prognostic factors include the anatomic extent

Table 36.11
Calculation of posttest probability of CAD after myocardial perfusion scintigraphy. The pretest probability (*after* exercise testing) can be obtained from the appropriate data in Table 36.9

	Posttest probability (after nuclear perfusion scintigraphy in %)		
Pretest probability (%)	No defect	Nonreversible defect	Reversible defect
10	2	14	56
20	4	26	74
30	7	38	83
40	10	49	89
50	15	59	92
60	21	68	95
70	29	77	96
80	41	85	98
90	61	93	99

CAD denotes coronary artery disease

and severity of CAD, evidence of a recent acute coronary syndrome resulting from a coronary plaque rupture, and the propensity for the development of ventricular arrhythmias.

36.8.1 Risk Stratification and Assessment of Prognosis with Exercise Testing

In patients with suspected or known CAD and symptoms suggestive of myocardial ischemia, exercise testing is the standard initial test in those with a normal ECG for identification of ischemia and risk assessment [4, 159–161]. In patients with a non-interpretable ECG, exercise testing may still provide useful prognostic information, but cannot be used to identify ischemia. Studies on the prognostic value of the exercise test in symptomatic patients with non-acute CAD identify the maximum exercise capacity to be the strongest and most consistent prognostic factor [4, 162–168]. Maximum exercise capacity can be expressed as maximum exercise duration, maximum workload or MET level achieved, maximum blood pressure, or double (rate-pressure) product and represents at least in part left ventricular function. Markers of exercise-induced ischemia (electrocardiographic and/or clinical) represent the second group of variables that bear adverse prognostic information. In particular, electrocardiographic evidence of myocardial ischemia in patients with a low maximum exercise capacity represents a high-risk population [4, 162, 169].

Using data of 2,842 patients with known or suspected CAD, without prior revascularization or recent myocardial infarction, who underwent exercise testing before coronary angiography, the Duke treadmill score was created [163, 170]. This score is calculated using multiple variables of prognostic importance from the exercise test and can subsequently be converted into an average annual mortality rate [163, 170]. Based on the individual scores, patients can be classified in a high-risk group with a high average annual cardiovascular mortality, an intermediate risk group, or a low-risk group. The Duke treadmill score was shown to independently add prognostic information to the standard clinical data plus the data resulting from cardiac catheterization. The score can also be applied in women, although women have a lower overall risk for any score value than men [171]. Comparable prognostic scores have been developed by other groups [164].

Recently, several studies have identified other parameters from the exercise test to bear important prognostic information. These include chronotropic incompetence, abnormal heart rate recovery, and delayed blood pressure response [172–179]. In one study of almost 10,000, mostly, asymptomatic patients, it was demonstrated that abnormal heart rate recovery and the Duke treadmill score were independent predictors of mortality [4, 177].

Exercise testing is a much stronger predictor of cardiovascular mortality than of nonfatal myocardial infarction. This may, in part, be a result of the fact that myocardial infarctions are mostly caused by rupture of relatively small, vulnerable atherosclerotic plaques that are difficult to detect by exercise testing because of their non-obstructive character, whereas exercise test results are correlated with the presence and severity of obstructive CAD [4, 163, 180].

36.8.2 Exercise Testing to Guide Patient Treatment

The results of exercise testing may be used to guide patient treatment. Patients with a low risk exercise test result and a low predicted average annual mortality rate can be treated medically, whereas patients at higher risk should be referred for additional testing or cardiac catheterization, especially in case of left ventricular dysfunction.

Patients with acute coronary syndromes without persistent ST segment elevation are stratified as low, intermediate, or high risk based on history, physical examination, 12-lead ECG, and cardiac markers of myocardial necrosis [181, 182]. High-risk patients will be scheduled to undergo coronary angiography and subsequent revascularization. In low or intermediate risk patients with unstable angina, exercise or pharmacologic stress testing plays an important part in risk stratification and identification of obstructive CAD. Exercise electrocardiography should be the standard mode of stress testing in patients with a normal resting ECG [4]. In general, stress testing can be performed as soon as the patient has stabilized clinically. Furthermore, studies have shown that exercise testing is safe when used in emergency department chest pain centers to provide risk stratification for chest pain patients believed to possibly have acute coronary disease [4, 183–185]. However, exercise testing in this setting should only be used in low, and intermediate risk patients on the basis of history, physical examination, 12-lead ECG, and markers of myocardial necrosis.

36.9 Exercise Testing after Acute Myocardial Infarction

Exercise testing after acute myocardial infarction can be used for patient management, risk stratification, and prognostic assessment. Treatment strategies for acute myocardial infarction have changed substantially over the past decades. In particular, the advent of reperfusion therapy involving the use of fibrinolytic agents or, more recently, direct or primary percutaneous coronary intervention has led to a marked improvement in the prognosis of patients after myocardial infarction. Contemporary medical treatment with beta-adrenergic blocking agents and angiotensin converting enzyme inhibitors has further improved prognosis. The patient population currently undergoing exercise testing after acute myocardial infarction is, therefore, far different from historical populations before the reperfusion era. The goals and basic principles of exercise testing have, however, not changed dramatically. Therefore, the role of exercise testing must be viewed in the context of the patients who present for exercise testing.

In patients following acute myocardial infarction, exercise testing is frequently performed before hospital discharge to establish the hemodynamic response and functional capacity for exercise prescriptions and cardiac rehabilitation, to detect serious ventricular arrhythmia, and to identify patients with inducible myocardial ischemia [186–191]. Furthermore, it is helpful in reestablishing patients' confidence in their ability to conduct their activities following discharge. Predischarge exercise testing in patients after acute myocardial infarction appears to be safe provided that the proper contraindications are observed [188, 192, 193]. Major contraindications in this patient population include manifest congestive heart failure and postinfarction angina. Predischarge exercise testing has historically been performed between 5 and 26 days following myocardial infarction [188, 192, 194, 196], although studies have suggested that exercise testing can also be performed safely within 3–4 days in patients with an uncomplicated myocardial infarction [196, 197]. Predischarge exercise testing following acute myocardial infarction has traditionally utilized a submaximal protocol that requires the patient to exercise until a target, predetermined workload (e.g., achievement of 5–6 METs or 70–80% of age-predicted maximum) has been reached [198]. However, it has been proposed that symptom-limited exercise testing prior to discharge may be safely performed in patients with an uncomplicated postinfarction course. As opposed to submaximal testing, performance of a symptom-limited test provides a better estimate of peak functional capacity and is associated with an increased detection rate of ischemic ST segment changes and angina pectoris [188, 193, 199].

36.9.1 Exercise Testing in Patient Management

Both in patients treated with thrombolysis and in those who have not received reperfusion therapy, a predischarge standard exercise test remains the test of choice to identify myocardial ischemia and select patients who might benefit from coronary angiography and revascularization. Imaging studies may be helpful for risk stratification and detection of myocardial ischemia in patients who have physical limitations that prevent them from exercising to an adequate workload, or in those with ECG abnormalities that preclude an accurate interpretation of ST segment changes. As expected, patients treated with thrombolytic therapy exhibit exercise-induced angina and ST segment depression less frequently than patients who have not received reperfusion therapy. In patients treated with direct or primary percutaneous coronary intervention, the coronary anatomy is known. If, besides the infarct-related artery, one or more of the other coronary arteries also shows a significant and important coronary obstruction at the time of angiography, additional coronary revascularization may be warranted. If the other coronary lesions found at the time of angiography are of intermediate or equivocal severity and significance, exercise testing or noninvasive stress imaging studies can be used to provoke residual ischemia and select patients who might benefit from additional revascularization, as well as those who can be managed conservatively.

36.9.2 Risk Stratification and Prognostic Assessment

As stated, the prognosis among patients after myocardial infarction has improved significantly, particularly in those who have received thrombolytic therapy or revascularization during hospitalization. Consequently, the low subsequent cardiac event rate associated with this improved treatment and survival substantially reduces the predictive accuracy of early exercise testing. Parameters derived from the exercise test following acute myocardial infarction that are associated with an increased risk of future death or recurrent nonfatal myocardial infarction include inability to perform the submaximal predischarge exercise test, poor exercise capacity, inability to increase - or a decrease in - systolic blood pressure, the development and magnitude of ST segment depression, especially at low workloads, and the development of angina [186, 194, 200–206].

Patients who have not undergone coronary revascularization and are unable to undergo exercise testing have the highest cardiac event rate. This was demonstrated both in trials in the thrombolytic era and in earlier studies in patients not receiving thrombolytic agents [194, 200–204]. In patients who are able to perform the test, exercise capacity is an important predictor of adverse cardiac events [189, 194, 195, 200, 203, 206–209]. Similarly, the hemodynamic response during the exercise test is of prognostic importance. Failure to increase systolic blood pressure by 10–30 mmHg or a decrease in blood pressure during exercise have been shown to be independent predictors of adverse outcome in patients after myocardial infarction [191, 195, 200, 203, 206, 209–211]. In a study in the prethrombolytic era, the degree of blood pressure rise during the exercise test was reported to be the single best predictive measurement [210]. The exercise capacity and the change in systolic blood pressure are, in fact, measures of left ventricular function which is the most important prognostic determinant of mortality following acute myocardial infarction. However, it was demonstrated that the maximum exercise capacity achieved during exercise testing provided an incremental prognostic value in patients with a low (less than 35%) left ventricular ejection fraction by gated radionuclide scintigraphy 1 month after acute myocardial infarction [212].

Exercise-induced ischemic ST segment depression in patients after myocardial infarction is an independent predictor of death or nonfatal myocardial infarction, particularly if the ST segment depression is accompanied by angina, occurs at a low level of exercise or in patients with controlled congestive heart failure [189, 203, 205, 206, 213, 214]. The predictive value of exercise-induced ischemia for adverse outcome is, however, limited by the fact that many patients who have an abnormal test result undergo coronary revascularization, which may alter the natural history of the disease process [193, 194, 201, 208, 215].

Finally, exercise testing can be used for activity counseling after hospital discharge and is an important tool in exercise training as part of a comprehensive cardiac rehabilitation program [4, 198].

36.10 Exercise Testing after Revascularization

36.10.1 Exercise Testing after Coronary Artery Bypass Grafting

The conversion of a positive exercise test result performed before coronary artery bypass grafting (CABG) to a negative postoperative test result is associated with successful revascularization [4, 216]. In patients with recurrent chest pain after

CABG, exercise testing may be used to demonstrate myocardial ischemia, although the exercise ECG is limited in this group of patients by the relatively high frequency of resting ECG abnormalities and the inability to document the site and extent of ischemia, as compared to stress imaging tests [4, 217, 218].

Exercise testing may be used for guiding exercise training as part of cardiac rehabilitation. It has been demonstrated that exercise testing in an asymptomatic individual, who has undergone successful CABG, is not predictive of outcome when the test is performed within the first few years after CABG [219, 220]. Therefore, routine periodic monitoring of asymptomatic patients after CABG is not indicated [4].

36.10.2 Exercise Testing after Percutaneous Coronary Intervention

Several studies that evaluated the diagnostic accuracy of exercise testing for identification of restenosis after percutaneous coronary intervention have shown that the exercise ECG is an insensitive predictor of restenosis [221–226], especially in asymptomatic patients, with sensitivities ranging from 40% to 50%, significantly less than those obtained by stress imaging tests [227, 228]. The insensitivity may be caused by the failure of moderate one-vessel stenoses to lead to significant ischemia on the exercise ECG. Routine, periodic exercise testing of asymptomatic patients after percutaneous coronary intervention without specific indications is therefore not recommended, especially since the prognostic benefit of controlling silent ischemia needs to be proved [4].

36.11 Exercise Testing and Heart Rhythm Disorders

Heart rhythm disorders occur frequently with exercise. The prevalence increases steadily with age, both in patients with heart disease and in normal individuals [229–231]. Increased sympathetic tone with withdrawal of much of the vagal tone, as well as myocardial ischemia may all play a role in the development of cardiac arrhythmias.

36.11.1 Sinus Node Dysfunction

Exercise testing may distinguish subjects with resting bradycardia with a normal increase in heart rate with exercise from those with sinus node dysfunction with a low resting heart rate that fails to accelerate normally with exercise, which is also labeled as chronotropic incompetence [4]. The definition of chronotropic incompetence has varied, the most common definition being failure to achieve 85% of (i.e., more than two standard deviations below) age-predicted maximum heart rate [4, 232]. The mechanisms involved are complex and not yet completely understood [233]. However, studies have confirmed the adverse prognostic implications of chronotropic incompetence [172, 233, 234]. Furthermore, a normal exercise test result does not negate the possibility of sinus node dysfunction.

36.11.2 Supraventricular Arrhythmias

The incidence of any supraventricular arrhythmia during exercise testing varies from 4% to 18% and increases with age. Atrial premature beats regularly occur at lower workloads, disappear as exercise increases and may return in the recovery period. They are considered to be of little clinical significance.

The majority of patients with atrial fibrillation demonstrate an abnormal heart rate response to exercise, which comprises an initial reduction of heart rate followed by delayed acceleration at lower workloads and a subsequent exaggerated increase in heart rate with tachycardia often persisting for a long period of time in the recovery phase [4, 235]. In patients with atrial fibrillation, exercise testing may help to evaluate the efficacy of drug regimens aimed at ventricular rate control. The ST segment changes associated with myocardial ischemia in atrial fibrillation are similar to those observed with sinus rhythm. However, in patients with atrial fibrillation and a very high ventricular response, the very short diastolic intervals may produce subendocardial ischemia because of the inadequate perfusion time in the absence of CAD. Atrial fibrillation initiated by exercise can be associated with CAD, rheumatic heart disease, or cardiomyopathy. However, it is also seen in subjects with no apparent cardiac abnormalities, in whom it may be a prelude to the development of sustained atrial fibrillation at a later stage [33].

Patients with Wolff-Parkinson-White syndrome may exhibit ST segment depression during exercise testing in the absence of CAD. In patients with Wolff-Parkinson-White syndrome, exercise testing may be used to help evaluate the risk of developing rapid ventricular response during atrial arrhythmias. When the pre-excitation disappears during exercise, the antegrade refractory period in the accessory pathway is longer than that in the atrioventricular node and it is unlikely that a rapid ventricular response will occur [4].

36.11.3 Ventricular Arrhythmias

In normal subjects, there is an increase in the incidence of resting premature ventricular contractions (PVCs) of 2–15% with age [236]. It has been demonstrated that, in the absence of overt heart disease, these resting PVCs are usually benign [237]. PVCs are often induced by exercise and the incidence in clinically normal, middle-aged or older subjects during maximum stress testing is about 35%–45%, usually at high workloads [238–240]. In general, exercise-induced ventricular arrhythmias in a large group of subjects without symptoms followed for 5–10 years were found to have no influence on subsequent morbidity and mortality, and appear to be benign [239, 241]. These data are generally believed to apply to all asymptomatic individuals although one study on a large cohort of asymptomatic men suggests that exercise-induced PVCs may have more adverse prognostic implications than previously reported [242].

In patients with CAD, the reported incidence of ventricular arrhythmias during exercise ranges from 40% to 65% [238, 243, 244]. In general, patients with CAD manifest arrhythmias at a lower heart rate than normal subjects. Despite the fact that PVCs are more easily evoked in patients with CAD than in normal subjects, there is too much overlap between those with and without ischemia to allow such arrhythmias to have diagnostic value. Accordingly, the appearance of PVCs, including multiform or repetitive PVCs, should not be interpreted as a sign of myocardial ischemia in diagnostic stress testing [245].

The appearance of frequent, multiform, or repetitive PVCs during exercise is associated with an increased risk of mortality in patients with a previous myocardial infarction, especially apparent in patients with an impaired left ventricular function [246–252]. Most studies suggest that exercise-induced PVCs are also associated with an impaired survival in patients with CAD without a previous myocardial infarction, especially in cases of multiform, repetitive PVCs or (non-sustained) ventricular tachycardia [238, 244, 253–256]. Some reports have disputed such an association, however, at least in low-risk patients with demonstrable stable CAD [245, 257, 258]. Significant multivessel disease is likely to be present in patients with angina and exercise-induced ventricular arrhythmias, especially if ischemic ST segment changes are also present [238, 244, 245, 255, 256]. Although the induction of PVCs by exercise is well recognized, ventricular ectopic activity may also be abolished by exercise in patients with CAD, just as it may be in normal subjects. Therefore, this finding does not exclude the presence of CAD [33, 238, 255].

In general, more arrhythmias are seen on recovery than during exercise. In the recovery period, the imbalance between oxygen supply and demand induced during exercise may be augmented in patients with CAD; peripheral dilatation induced by exercise combined with a reduced venous return caused by cessation of muscular activity may result in reduced cardiac output and coronary flow at a time when myocardial oxygen demand is still high owing to tachycardia. Furthermore, catecholamine levels are considerably elevated [259]. These changes can be minimized by a gradual cool down.

36.11.3.1 Exercise-Induced Sustained Ventricular Tachycardia

Sustained ventricular tachycardia during exercise testing is relatively rare and occurs most frequently in the group of patients with ventricular tachycardia or ventricular fibrillation as their primary complaint. As stated previously, because ventricular tachycardia can deteriorate into ventricular fibrillation, immediate termination of exercise is warranted. Sustained ventricular tachycardia or long runs of non-sustained ventricular tachycardia usually portray serious underlying diseases; either CAD with ischemia or some type of cardiomyopathy should be suspected. Ventricular tachycardia caused by ischemia almost never has a left bundle-branch block pattern.

36.11.3.2 Exercise Testing to Evaluate Spontaneous Ventricular Tachycardia

Exercise testing can play an important role in the workup of patients who survived sudden death, as well as in those with syncope and sustained ventricular tachycardia [4]. The usefulness of exercise-testing in patients with ventricular tachycardia is variable, according to the cause of the tachycardia. In patients with idiopathic right ventricular outflow tract tachycardia, the ventricular tachycardia can be reproducibly induced during stress testing as it is commonly provoked by exercise. During exercise testing, the patients also exhibit many PVCs and coupled PVCs. The ventricular tachycardia has a left bundle-branch block morphology and is more likely to be non-sustained [260].

In adrenergic-dependent ventricular tachycardia, including monomorphic ventricular tachycardia and polymorphic ventricular tachycardia related to long-QT syndromes, exercise testing may supply the circumstances necessary for induction of the ventricular tachycardia and is, therefore, a useful prelude to an electrophysiological study [4]. Furthermore, the occurrence and nature of exercise-induced ventricular ectopy is of prognostic value in these patients [4, 254, 261]. Even in patients at risk of ventricular tachycardia, maximal exercise testing can be conducted safely with the appropriate precautions [4, 262]. The main limitation of exercise testing in patients with ventricular arrhythmias is related to its limited reproducibility so that other testing modalities are also required in the evaluation of these patients [4, 263].

Exercise testing may also be used to unmask pro-arrhythmic responses with development of sustained ventricular tachycardia during exercise in patients receiving anti-arrhythmic therapy.

36.12 Exercise Testing in Valvular Heart Disease

The primary value of exercise testing in valvular heart disease is to objectively assess atypical symptoms, exercise tolerance, and extent of disability to guide decision making with regard to surgical treatment. This is particularly of importance when a patient is thought to be asymptomatic because of inactivity (e.g., as in the elderly) or when a discrepancy exists between the patient's symptom status and the echocardiographic severity of the valvular stenosis or regurgitation. Furthermore, exercise testing can be used in follow-up of asymptomatic patients with valvular heart disease to detect a reduction in exercise capacity over time [4]. Details regarding the uses of exercise testing in patients with valvular heart disease have also been described in the respective guideline for the management of patients with valvular heart disease [264].

36.12.1 Mitral Valve Stenosis and Regurgitation

Exercise testing in mitral valve stenosis is of most value when the patient's symptom status and mitral valve area show discrepancy. When exercise induces excessive heart rate responses to a relatively low level of exercise or hypotension as a sign of a reduction in cardiac output, a more aggressive therapeutic approach aimed at earlier surgery might be considered [264]. In a rare case, exercise may precipitate atrial fibrillation in a patient with mitral valve stenosis.

In patients with mitral valve regurgitation, exercise testing objectively determines the functional capacity of the patient. Patients with severe mitral valve regurgitation commonly demonstrate a reduction in exercise capacity and are usually limited by the development of dyspnea. In patients with moderately-severe mitral valve regurgitation, a combination of exercise testing and assessment of left ventricular function may be useful in documenting occult left ventricular dysfunction and provoking earlier surgery [4, 264, 265]. Furthermore, exercise testing can be used to monitor exercise tolerance over time in these patients.

36.12.2 Aortic Valve Stenosis and Regurgitation

Severe, symptomatic aortic valve stenosis is a contraindication to exercise testing. As aortic valve replacement is not indicated in asymptomatic patients [264], it is important to distinguish those who are truly asymptomatic from patients who are asymptomatic because they are inactive or have adjusted to their functional impairment. In these patients, exercise testing can be used to select a subpopulation of patients who are hemodynamically compromised by aortic valve stenosis and in whom surgery should be considered. Studies in adults with moderate to severe aortic valve stenosis have demonstrated that, with the appropriate precautions, exercise testing can be safely performed in these patients [4, 264, 266, 267]. Adverse hemodynamic responses that advocate aortic valve replacement include profound functional limitation, hypotension during exercise or failure to augment systolic blood pressure with exercise, and a rapid increase in heart rate indicating a fixed stroke volume. In this way, exercise testing can be combined with echocardiography in the follow-up of patients with aortic valve stenosis to help in determining the time at which aortic valve replacement should be performed.

Exercise tolerance is preserved until late in the course of aortic valve regurgitation. Exercise testing is not routinely required to guide treatment as the decision to proceed to valve surgery in chronic aortic regurgitation is primarily based on symptom status, left ventricular systolic (dys)function, and left ventricular size [4, 268].

References

1. Kaltenbach, M., D. Scherer, and S. Dowinsky, Complications of exercise testing. A survey in three German-speaking countries. *Eur. Heart J.*, 1982;**3**: 199–202.
2. Pina, I.L., G.J. Balady, P. Hanson, et al., Guidelines for clinical exercise testing laboratories: A statement for healthcare professionals from the Committee on Exercise and Cardiac Rehabilitation, American Heart Association. *Circulation*, 1995;**91**: 912–921.
3. Schlant, R.C., G.C. Friesinger II, and J.J. Leonard, Clinical competence in exercise testing: A statement for physicians from the ACP/ACC/AHA Task Force on Clinical Privileges in Cardiology. *J. Am. Coll. Cardiol.*, 1990;**16**: 1061–1065.
4. Gibbons, R.J., G.J. Balady, J.T. Bricker, B.R. Chaitman, G.F. Fletcher, V.F. Froelicher, D.B. Mark, B.D. McCallister, A.N. Mooss, M.G. O'Reilly, and W.L. Winters Jr, ACC/AHA 2002 guideline update for exercise testing: A report of the American College of Cardiology/American Heart Association Task Force on Practice Guidelines (Committee on Exercise Testing). 2002.
5. Taylor, H.L., W. Haskell, S.M. Fox, et al., Exercise tests: A summary of procedures and concepts of stress testing for cardiovascular diagnosis and function evaluation, in *Measurement in Exercise Electrocardiography*, H. Blackburn, Editor. Springfield, IL: Thomas, 1969.
6. Sheffield, L.T. and D. Roitman, Stress testing methodology. *Prog. Cardiovasc. Dis.*, 1976;**19**: 33–49.
7. Niederberger, M., R.A. Bruce, F. Kusumi, et al., Disparities in ventilatory and circulatory responses to bicycle and treadmill exercise. *Br. Heart J.*, 1974;**36**: 377–382.
8. Bruce, R.A., Methods of exercise testing. Step test, bicycle, treadmill, isometrics. *Am. J. Cardiol.*, 1974;**33**: 715–720.
9. Ellestad, M.H., Memorial hospital protocol, in *Stress Testing. Principles and Practice*, M.H. Ellestad, Editor. Philadelphia, PA: Davis, 1975, pp. 67–84.
10. Balke, B. and R.W. Ware, An experimental study of physical fitness of Air Force personnel. *US Armed Forces Med. J.*, 1959;**10**: 675–688.
11. Weld, F.M., K.-L. Chu, J.T. Bigger Jr, et al., Risk stratification with low-level exercise testing 2 weeks after acute myocardial infarction. *Circulation*, 1981;**64**: 306–314.
12. Ellestad, M.H., C.G. Blomqvist, and J.P. Naughton, Standards for adult exercise testing laboratories. *Circulation*, 1979;**58**: 421A–430A.
13. Fox, S.M., III, J.P. Naughton, and W.L. Haskell, Physical activity and the prevention of coronary heart disease. *Ann. Clin. Res.*, 1971;**3**: 404–432.
14. Froelicher, V.F., Jr, A.J. Thompson Jr, I. Noguera, et al., Prediction of maximal oxygen consumption. Comparison of the Bruce and Balke treadmill protocols. *Chest*, 1975;**68**: 331–336.
15. Pollock, M.L., R.L. Bohannon, K.H. Cooper, et al., A comparative analysis of four protocols for maximal treadmill stress testing. *Am. Heart J.*, 1976;**92**: 39–46.
16. Lecerof, H., Influence of body position on exercise tolerance, heart rate, blood pressure, and respiration rate in coronary insufficiency. *Br. Heart J.*, 1971;**33**: 78–83.
17. Thadani, U., R.O. West, T.M. Mathew, et al., Hemodynamics at rest and during supine and sitting bicycle exercise in patients with coronary artery disease. *Am. J. Cardiol.*, 1977;**39**: 776–783.
18. Kramer, B., B. Massie, and N. Topic, Hemodynamic differences between supine and upright exercise in patients with congestive heart failure. *Circulation*, 1982;**66**: 820–825.
19. Bruce, R.A., Exercise testing of patients with coronary heart disease: Principles and normal standards for evaluation. *Ann. Clin. Res.*, 1971;**3**: 323–332.
20. Erikssen, J., E. Thaulow, R. Mundal, et al., Comparison of beta-adrenoceptor blockers under maximal exercise (Pindolol v Metoprolol v Atenolol). *Br. J. Clin. Pharmacol.*, 1982;**13**: 201S–209S.
21. Samek, L. and H. Roskamm, Antianginal and antiarrhythmic effects of pindolol in post-infarct patients. *Br. J. Clin. Pharmacol.*, 1982;**13**: 297S.
22. Simoons, M.L., M. Taams, J. Lubsen, et al., Treatment of stable angina pectoris with verapamil hydrochloride: A double blind cross-over study. *Eur. Heart J.*, 1980;**1**: 269–274.
23. Harron, D.W.G., J.G. Riddell, and R.G. Shanks. Alinidine reduces heart rate without blockade of beta-adrenoceptors. *Lancet*, 1981;**1**: 351–353.
24. Borg, G., A. Holmgren, and I. Lindblad, Quantitative evaluation of chest pain. *Acta Med. Scand.*, 1981;Suppl. 644: 43–45.
25. Simoons, M.L., J. Tumraers, H. van Meurs-van Woezik, et al., Alinidine, a new agent which lowers heart rate in patients with angina pectoris. *Eur. Heart J.*, 1982;**3**: 542–545.
26. Simoons, M.L. and K. Balakumaran, The effects of drugs on the exercise electrocardiogram. *Cardiology*, 1981;**68**(Suppl. 2): 124–132.

27. Herbert, W.G., P. Dubach, K.G. Lehmann, et al., Effect of beta-blockade on the interpretation of the exercise ECG: ST level versus delta ST/HR index. *Am. Heart J.*, 1991;**122**: 993–1000.
28. Simoons, M.L., P.G. Hugenholtz, C.A. Ascoop, et al., Quantitation of exercise electrocardiography. *Circulation*, 1981;**63**: 471–475.
29. Wolf, H.K., P.J. MacInnis, S. Stock, et al., Computer analysis of rest and exercise electrocardiograms. *Comput. Biomed. Res.*, 1972;**5**: 329–346.
30. Simoons, M.L., H.B.K. Boom, and E. Smallenburg, On-line processing of orthogonal exercise electrocardiograms. *Comput. Biomed. Res.*, 1975;**8**: 105–117.
31. Bhargava, V., K. Watanabe, and V.F. Froelicher, Progress in computer analysis of the exercise electrocardiogram. *Am. J. Cardiol.*, 1981;**47**: 1143–1151.
32. Pahlm, O. and L. Sornmo, Data processing of exercise ECGs. *IEEE Trans. Biomed. Eng.*, 1987;**34**: 158–165.
33. Ellestad, M.H., *Stress Testing: Principles and Practice*, 5th edn. New York, NY: Oxford University Press.
34. Simoons, M.L. and P. Block, Toward the optimal lead system and optimal criteria for exercise electrocardiography. *Am. J. Cardiol.*, 1981;**47**: 1366–1374.
35. Phibbs, B.P. and L.J. Buckets, Comparative yield of ECG leads in multistage stress testing. *Am. Heart J.*, 1975;**90**: 275–276.
36. Miller, T.D., K.B. Desser, and M. Lawson, How many electrocardiographic leads are required for exercise treadmill tests? *J. Electrocardiol.*, 1987;**20**: 131–137.
37. Simoons, M.L., Optimal measurements for detection of coronary artery disease by exercise electrocardiography. *Comput. Biomed. Res.*, 1977;**10**: 483–499.
38. Mason, R.E., I. Likar, R.O. Biern, et al., Multiple-lead exercise electrocardiography. Experience in 107 normal subjects and 67 patients with angina pectoris, and comparison with coronary cinearteriography in 84 patients. *Circulation*, 1967;**36**: 517–525.
39. Chaitman, B.R., M.G. Bourassa, P. Wagniart, et al., Improved efficiency of treadmill exercise testing using a multiple lead ECG system and basic hemodynamic exercise response. *Circulation*, 1978;**57**: 71–79.
40. Fox, K., A. Selwyn, and J. Shillingford, Precordial electrocardiographic mapping after exercise in the diagnosis of coronary artery disease. *Am. J. Cardiol.*, 1979;**43**: 541–546.
41. Fox, K.M., J. Deanfield, P. Ribero, et al., Projection of ST segment changes on the front of the chest. *Br. Heart J.*, 1982;**48**: 555–559.
42. Miranda, C.P., J. Liu, A. Kadar, et al., Usefulness of exercise-induced ST segment depression in the inferior leads during exercise-testing as a marker for coronary artery disease. *Am. J. Cardiol.*, 1992;**69**: 303–307.
43. Michaelides, A.P., Z.D. Psomadaki, P.E. Dilaveris, et al., Improved detection of coronary artery disease by exercise electrocardiography with the use of right precordial leads. *N. Engl. J. Med.*, 1999;**340**: 340–345.
44. Wolthuis, R.A., V.F. Froelicher, A. Hopkirk, et al., Normal electrocardiographic waveform characteristics during treadmill exercise testing. *Circulation*, 1979;**60**: 1028–1035.
45. Simoons, M.L. and P.G. Hugenholtz, Estimation of the probability of exercise-induced ischemia by quantitative ECG analysis. *Circulation*, 1977;**56**: 552–559.
46. Block, P., J. Tiberghien, I. Raadschelders, et al., Diagnostic value of surface mapping recordings registered at rest and during exercise, in *Computers in Cardiology 1977*, H.G. Ostrow and K.L. Ripley, Editors. New York, NY: IEEE, 1977.
47. Mirvis, D.M., F.W. Keller Jr, J.W. Cox Jr, et al., Left precordial isopotential mapping during supine exercise. *Circulation*, 1977;**56**: 245–252.
48. Sapin, P.M., M.B. Blauwet, G.G. Koch, et al., Exaggerated atrial repolarization waves as a predictor of false positive exercise tests in an unselected population. *J. Electrocardiol.*, 1995;**28**: 313–321.
49. Sapin, P.M., G. Koch, M.B. Blauwet, et al., Identification of false positive exercise tests with use of electrocardiographic criteria: A possible role for atrial repolarization waves. *J. Am. Coll. Cardiol.*, 1991;**18**: 127–135.
50. Bhargava, V. and A.L. Goldberger, New method for measuring QRS duration using high-frequency electrocardiography. *Am. J. Physiol.*, 1982;**242**: H507–H511.
51. Sketch, M.H., S.M. Mohiuddin, C.K. Nair, et al., Automated and nomographic analysis of exercise tests. *Am. Med. Assoc.*, 1980;**243**: 1052–1055.
52. Battler, A., V.F. Froelicher, R. Slutsky, et al., Relationship of QRS amplitude changes during exercise to left ventricular function and volumes and the diagnosis of coronary artery disease. *Circulation*, 1979;**60**: 1004–1013.
53. David, D., M. Naito, C.C. Chen, et al., R-wave amplitude variations during acute experimental myocardial ischemia: An inadequate index for changes in intracardiac volume. *Circulation*, 1981;**63**: 1364–1371.
54. Deckers, J.W., R.V. Vinke, J.R. Vos, et al., Changes in the electrocardiographic response to exercise in healthy women. *Br. Heart J.*, 1990;**64**: 376–380.
55. Cumming, G.R., C. Dufresne, L. Kich, et al., Exercise electrocardiogram patterns in normal women. *Br. Heart J.*, 1973;**35**: 1055–1061.
56. Deckers, J.W., B.J. Rensing, M.L. Simoons, et al., Diagnostic merits of exercise testing in females. *Eur. Heart J.*, 1989;**10**: 543–550.
57. Detry, J.M., B.M. Capita, J. Cosyns, et al., Diagnostic value of history and maximal exercise electrocardiography in men and women suspected of coronary heart disease. *Circulation*, 1977;**56**: 756–761.
58. Val, P.G., B.R. Chaitman, D.D. Waters, et al., Diagnostic accuracy of exercise ECG lead systems in clinical subsets of women. *Circulation*, 1982;**65**: 1465–1474.
59. Barolsky, S.M., C.A. Gilbert, A. Faruqui, et al., Differences in electrocardiographic response to exercise of women and men: A non-Bayesian factor. *Circulation*, 1979;**60**: 1021–1027.
60. Rijneke, R.D., C.A. Ascoop, and J.L. Talmon, Clinical significance of upsloping ST segments in exercise electrocardiography. *Circulation*, 1980;**61**: 671–678.
61. Stuart, R.J. and M.H. Ellestad, Upsloping ST segments in exercise stress testing: Six-year follow-up study of 438 patients and correlation with 248 angiograms. *Am. J. Cardiol.*, 1976;**37**: 19–22.
62. Kurita, A., B.R. Chaitman, and M.G. Bourassa, Significance of exercise-induced ST depression in evaluation of coronary artery disease. *Am. J. Cardiol.*, 1977;**40**: 492–497.
63. Campeau, L., Grading of angina pectoris. *Circulation*, 1975;**54**: 522–523.
64. Simoons, M.L., M. van den Brand, and P.G. Hugenholtz, Quantitative analysis of exercise electrocardiograms and left ventricular angiocardiograms in patients with abnormal QRS complexes at rest. *Circulation*, 1977;**55**: 55–60.
65. Piessens, J., W. van Mieghem, H. Kesteloot, et al., Diagnostic value of clinical history, exercise testing and atrial pacing in patients with chest pain. *Am. J. Cardiol.*, 1974;**33**: 351–356.

66. Bartel, A.G., V.S. Behar, R.H. Peter, et al., Graded exercise stress tests in angiographically documented coronary artery disease. *Circulation*, 1974;**49**: 348–356.
67. Ascoop, C.A., M.L. Simoons, W.G. Egmond, et al., Exercise test, history, and serum lipid levels in patients with chest pain and normal electrocardiograms at rest: Comparison to findings at coronary arteriography. *Am. Heart J.*, 1971;**82**: 609–617.
68. Bogaty, P., et al., Does more ST segment depression on the 12-lead exercise electrocardiogram signify more severe ischemic heart disease. *Circulation*, 1993;**88**(Suppl. 2): 1.
69. Taylor, A.J. and G.A. Beller, Patients with greater than 2 mm of ST depression do not have a greater ischemic burden by thallium-201 scintigraphy. *Circulation*, 1992;**86**(Suppl. II): 138.
70. Husted, R., et al., The failure of multilead ST depression to predict severity of ischemia. *Am. J. Noninvas. Cardiol.*, 1994;**8**: 386.
71. Barlow, J.B., The "false positive" exercise electrocardiogram: Value of time course patterns in assessment of depressed ST segments and inverted T waves. *Am. Heart J.*, 1985;**110**: 1328–1336.
72. Miranda, C.P., K.G. Lehmann, and V.F. Froelicher, Correlation between resting ST segment changes, exercise testing, coronary angiography, and long term prognosis. *Am. Heart J.*, 1991;**122**: 1617–1628.
73. Ellestad, M.H., L. Thomas, R. Ong, et al., The predictive value of the time course of ST depression during exercise testing in patients referred for angiograms. *Am. Heart J.*, 1992;**123**: 904–908.
74. Lachterman, B., K.G. Lehmann, R. Detrano, et al., Comparison of ST segment/heart rate index to standard ST criteria for analysis of exercise electrocardiogram. *Circulation*, 1990;**82**: 44–50.
75. Rywik, T.M., R.C. Zink, N.S. Gittings, et al., Independent prognostic significance of ischemic ST segment response limited to recovery from treadmill exercise in asymptomatic subjects. *Circulation*, 1998;**97**: 2117–2222.
76. Hollenberg, M., M. Go Jr, B.M. Massie, et al., Influence of R-wave amplitude on exercise induced ST depression: Need for a "gain factor" correction when interpreting stress electrocardiograms. *Am. J. Cardiol.*, 1985;**56**: 13–17.
77. Hakki, A.H., A.S. Iskandria, S. Kutalek, et al., R-wave amplitude: A new determinant of failure of patients with coronary heart disease to manifest ST depression during exercise. *J. Am. Coll. Cardiol.*, 1984;**3**: 1155–1160.
78. Ellestad, M.H., R. Crump, and M. Surber, The significance of lead strength on ST changes during treadmill stress tests. *J. Electrocardiol.*, 1992;**25**(Suppl.): 31–34.
79. Elamin, M.S., R. Boyle, M.M. Kardash, et al., Accurate detection of coronary heart disease by new exercise test. *Br. Heart J.*, 1982;**48**: 311–320.
80. Okin, P.M. and P. Kligfield, Computer-based implementation of the ST-segment/heart rate slope. *Am. J. Cardiol.*, 1989;**64**: 926–930.
81. Detrano, R., E. Salcedo, M. Passalacqua, et al., Exercise electrocardiographic variables: A critical appraisal. *J. Am. Coll. Cardiol.*, 1986;**8**: 836–847.
82. Kligfield, P., O. Ameisen, and P.M. Okin, Heart rate adjustment of ST segment depression for improved detection of coronary artery disease. *Circulation*, 1989;**79**: 245–255.
83. Okin, P.M. and P. Kligfield, Heart rate adjustment of ST segment depression and performance of the exercise electrocardiogram: A critical evaluation. *J. Am. Coll. Cardiol.*, 1995;**25**: 1726–1735.
84. Fletcher, G.F., T.R. Flipse, P. Kligfield, et al., Current status of ECG stress testing. *Curr. Probl. Cardiol.*, 1998;**23**: 353–423.
85. Morise, A.P., Accuracy of heart rate-adjusted ST segments in populations with and without posttest referral bias. *Am. Heart J.*, 1997;**134**: 647–655.
86. Okin, P.M., M.J. Roman, J.E. Schwartz, et al., Relation of exercise-induced myocardial ischemia to cardiac and carotid structure. *Hypertension*, 1997;**30**: 1382–1388.
87. Viik, J., R. Lehtinen, and J. Malmivuo, Detection of coronary artery disease using maximum value of ST/HR hysteresis over different number of leads. *J. Electrocardiol.*, 1999;**32**(Suppl.): 70–75.
88. Froelicher, V.F., K.G. Lehmann, R. Thomas, et al., The electrocardiographic exercise test in a population with reduced workup bias: Diagnostic performance, computerized interpretation, and multivariable prediction. Veterans Affairs Cooperative Study in Health Services #016 (QUEXTA) Study Group. Quantitative Exercise Testing and Angiography. *Ann. Intern. Med.*, 1998;**128**: 965–974.
89. Macfarlane, P.W., A. Tweddel, D. Macfariane, et al., Body surface ECG mapping on exercise. *Eur. Heart J.*, 1984;**5**(Suppl. 1): 279.
90. Simoons, M.L., A. Withagen, R. Vinke, et al., ST-vector orientation and location of myocardial perfusion defects during exercise. *Nuklearmedizin*, 1978;**17**: 154–156.
91. Samek, L., H. Roskamm, P. Rentrop, et al., Belastungsprufungen und koronarangiogramm im chronischen infarktstadium. *Z. Kardiol.*, 1975;**64**: 809–814.
92. Roskamm, H., L. Samek, K. Zweigle, et al., Die beziehungen zwischen den befunden der koronarangiographie und des belastungs-ekg bei patienten ohne transmuralen myokardinfarkt. *Z. Kardiol.*, 1977;**66**: 273–280.
93. Fisher, L.D., J.W. Kennedy, B.R. Chaitman, et al., Diagnostic quantification of CASS (Coronary Artery Surgery Study) clinical and exercise test results in determining presence and extent of coronary artery disease: A multivariate approach. *Circulation*, 1981;**63**: 987–1000.
94. De Feyter, P.J., P.A. Majid, M.J. van Eenige, et al., Clinical significance of exercise-induced ST segment elevation. Correlative angiographic study in patients with ischaemic heart disease. *Br. Heart J.*, 1981;**46**: 84–92.
95. Alijarde, M., J. Soler-Soler, J. Perez-Jabaloyes, et al., Significance of treadmill stress testing in transmural myocardial infarction. Correlation with coronary angiography. *Eur. Heart J.*, 1982;**3**: 353–361.
96. Manvi, K.N. and M.H. Ellestad, Elevated ST-segments with exercise in ventricular aneurysm. *J. Electrocardiol.*, 1972;**5**: 317–323.
97. Haines, D.E., G.A. Beller, D.D. Watson, et al., Exercise-induced ST segment elevation 2 weeks after uncomplicated myocardial infarction: Contributing factors and prognostic significance. *J. Am. Coll. Cardiol.*, 1987;**9**: 996–1003.
98. Fox, K.M., A. Jonathan, and A. Selwyn, Significance of exercise induced ST segment elevation in patients with previous myocardial infarction. *Br. Heart J.*, 1983;**49**: 15–19.
99. Margonato, A., C. Ballarotto, F. Bonetti, et al., Assessment of residual tissue viability by exercise testing in recent myocardial infarction: Comparison of the electrocardiogram and myocardial perfusion scintigraphy. *J. Am. Coll. Cardiol.*, 1992;**19**: 948–952.
100. Margonato, A., S.L. Chierchia, R.G. Xuereb, et al., Specificity and sensitivity of exercise-induced ST segment elevation for detection of residual viability: Comparison with fluorodeoxyglucose and positron emission tomography. *J. Am. Coll. Cardiol.*, 1995;**25**: 1032–1038.

101. Lombardo, A., F. Loperfido, F. Pennestri, et al., Significance of transient ST-T segment changes during dobutamine testing in Q wave myocardial infarction. *J. Am. Coll. Cardiol.*, 1996;**27**: 599–605.
102. Dunn, R.F., B. Freedman, D.T. Kelly, et al., Exercise-induced ST segment elevation in leads V_1 or aVL. A predictor of anterior myocardial ischemia and left anterior descending coronary artery disease. *Circulation*, 1981;**63**: 1357–1363.
103. Dunn, R.F., B. Freedman, I.K. Bailey, et al., Localization of coronary artery disease with exercise electrocardiography: Correlation with thallium-201 myocardial perfusion scanning. *Am. J. Cardiol.*, 1981;**48**: 837–843.
104. Waters, D.D., B.R. Chaitman, M.G. Bourassa, et al., Clinical and angiographic correlates of exercise-induced ST segment elevation. Increased detection with multiple ECG leads. *Circulation*, 1980;**61**: 286–296.
105. Waters, D.D., J. Szlachcic, M.G. Bourassa, et al., Exercise testing in patients with variant angina: Results, correlation with clinical and angiographic features and prognostic significance. *Circulation*, 1982;**65**: 265–274.
106. Hegge, F.N., N. Tuna, and H.B. Burchell, Coronary arteriographic findings in patients with axis shifts or ST segment elevations on exercise stress testing. *Am. Heart J.*, 1973;**86**: 603–615.
107. Chahine, R.A., A.E. Raizner, and T. Ishimori, The clinical significance of exercise induced ST segment elevation. *Circulation*, 1976;**54**: 209–213.
108. Longhurst, J.C. and W.L. Kraus, Exercise-induced ST elevation in patients without myocardial infarction. *Circulation*, 1979;**60**: 616–629.
109. Mark, D.B., M.A. Hlatky, K.L. Lee, et al., Localizing coronary artery obstructions with the exercise treadmill test. *Ann. Intern. Med.*, 1987;**106**: 53–55.
110. Prinzmetal, M., A. Ekmekci, R. Kennamer, et al., Variant form of angina pectoris, previously undelineated syndrome. *JAMA*, 1960;**174**: 1794–1800.
111. Detry, J.M., P. Mengeot, M.F. Rousseau, et al., Maximal exercise testing in patients with spontaneous angina pectoris associated with transient ST segment elevation. Risks and electrocardiographic findings. *Br. Heart J.*, 1975;**37**: 897–903.
112. Bonoris, P.E., P.S. Greenberg, M.J. Castellanet, et al., Significance of changes in R wave amplitude during treadmill stress testing: Angiographic correlation. *Am. J. Cardiol.*, 1978;**41**: 846–851.
113. Wagner, S., J.C. Cohn, and A. Selzer, Unreliability of exercise induced R wave changes as indexes of coronary artery disease. *Am. J. Cardiol.*, 1979;**44**: 1241–1246.
114. Brody, D.A., A theoretical analysis of intracavitary blood mass influence on the heart-lead relationship. *Circ. Res.*, 1956;**4**: 731–738.
115. Greenberg, P.S., M.H. Ellestad, R. Berge, et al., Radionuclide angiographic correlation of the R-wave, ejection fraction, and volume responses to upright bicycle exercise. *Chest*, 1981;**80**: 459–464.
116. Froelicher, V.F., Jr, R. Wolthuis, N. Keiser, et al., A comparison of two bipolar exercise ECG leads to V5. *Chest*, 1976;**70**: 611–616.
117. Froelicher, V.F. and J.N. Myers, *Exercise and the Heart*, 4th edn. Philadelphia, PA: W.B. Saunders.
118. Michaelides, A.P., H. Boudoulas, H. Antonakoudis, et al., Effect of number of coronary arteries significantly narrowed and status of intraventricular conduction on exercise-induced QRS prolongation in coronary artery disease. *Am. J. Cardiol.*, 1992;**70**: 1487–1489.
119. Wayne, V.S., et al., Exercise-induced bundle branch block. *Am. J. Cardiol.*, 1983;**52**: 283.
120. Levy, S., R. Gerard, A. Castellanos, et al., Transient left anterior hemiblock during angina pectoris: Coronarographic aspects and clinical significance. *Eur. J. Cardiol.*, 1979;**9**: 215–225.
121. Hegge, F.N., N. Tuna, and H.B. Burchell, Coronary arteriographic findings in patients with axis shifts or ST segment elevations on exercise stress testing. *Am. Heart J.*, 1973;**86**: 603–615.
122. Whinnery, J.E., V.F. Froelicher Jr, M.R. Longo Jr, et al., The electrocardiographic response to maximal treadmill exercise in asymptomatic men with right branch bundle block. *Chest*, 1977;**71**: 335–340.
123. Kattus, A.A., Exercise electrocardiography. Recognition of the ischemic response: False positive and negative patterns. *Am. J. Cardiol.*, 1974;**33**: 721–731.
124. Tanaka, T., M.J. Friedman, R.D. Okada, et al., Diagnostic value of exercise-induced ST segment depression in patients with RBBB. *Am. J. Cardiol.*, 1978;**41**: 670–673.
125. Whinnery, J.E., V.F. Froelicher, and A.J. Stuart, The electrocardiographic response to maximal treadmill exercise in asymptomatic men with left bundle branch block. *Am. Heart J.*, 1997;**94**: 316–324.
126. Aravindakshan, V., W. Surawicz, and R.D. Allen, Electrocardiographic exercise test in patients with abnormal T waves at rest. *Am. Heart J.*, 1977;**93**: 706–714.
127. Diamond, G.A. and J.S. Forrester, Analysis of probability as an aid in the clinical diagnosis of coronary artery disease. *N. Engl. J. Med.*, 1979;**300**: 1350–1358.
128. Fisher, L.D. J.W. Kennedy, B.R. Chaitman, et al., Diagnostic quantification of CASS (Coronary Artery Surgery Study) clinical and exercise test results in determining presence and extent of coronary artery disease. A multivariate approach. *Circulation*, 1981;**63**: 987–1000.
129. Morise, A.P., W.J. Haddad, and D. Beckner, Development and validation of a clinical score to estimate the probability of coronary artery disease in men and women presenting with suspected coronary disease. *Am. J. Med.*, 1997;**102**: 350–356.
130. Altman, X., *Practical Statistics for Medical Research*. London: Chapman and Hall, 1991, pp. 414–417.
131. Detry, J.M., A. Robert, R.R.J. Luwaert, et al., Diagnostic value of computerized exercise testing in men without previous myocardial infarction. A multivariate, compartmental and probabilistic approach. *Eur. Heart J.*, 1985;**6**: 227–238.
132. Pryor, D.B., F.E. Harrell Jr, K.L. Lee, et al., Estimating the likelihood of significant coronary artery disease. *Am. J. Med.*, 1983;**75**: 771–780.
133. Ellestad, M.H., S. Savitz, D. Bergdall, et al., The false positive stress test: Multivariate analysis of 215 subjects with hemodynamic, angiographic and clinical data. *Am. J. Cardiol.*, 1977;**40**: 681–685.
134. Yamada, H., D. Do, A. Morise, et al., Review of studies using multivariable analysis of clinical and exercise test data to predict angiographic coronary artery disease. *Prog. Cardiovasc. Dis.*, 1997;**39**: 457–481.
135. Shaw, L.J., E.D. Peterson, L.K. Shaw, et al., Use of a prognostic treadmill score in identifying diagnostic coronary disease subgroups. *Circulation*, 1998;**98**: 1622–1630.

136. Deckers, J.W., B.J. Rensing, J.G. Tijssen, et al., A comparison of methods of analysing exercise tests for diagnosis of coronary artery disease. *Br. Heart J.*, 1989;**62**: 438–444.
137. Lee, K.L., D.B. Pryor, F.E. Harrell Jr, et al., Predicting outcome in coronary disease: Statistical models versus expert clinicians. *Am. J. Med.*, 1986;**80**: 553–560.
138. Detrano, R., M. Bobbio, H. Olson, et al., Computer probability estimates of angiographic coronary artery disease: Transportability and comparison with cardiologists' estimates. *Comput. Biomed. Res.*, 1992;**25**: 468–485.
139. Gianrossi, R., R. Detrano, D. Mulvihill, et al., Exercise induced ST depression in the diagnosis of coronary artery disease: A meta analysis. *Circulation*, 1989;**80**: 87–98.
140. Detrano, R., R. Gianrossi, and V. Froelicher, The diagnostic accuracy of the exercise electrocardiogram: A meta analysis of 22 years of research. *Prog. Cardiovasc. Dis.*, 1989;**32**: 173–206.
141. Morise, A.P. and G.A. Diamond, Comparison of the sensitivity and specificity of exercise electrocardiography in biased and unbiased populations of men and women. *Am. Heart J.*, 1995;**130**: 741–747.
142. DelCampo, J., D. Do, T. Umann, et al., Comparison of computerized and standard visual criteria of exercise ECG for diagnosis of coronary artery disease. *Ann. Noninvasive Electrocardiogr.*, 1996;**1**: 430–442.
143. Froelicher, V.F., K.G. Lehmann, R. Thomas, et al., The electrocardiographic exercise test in a population with reduced workup bias: Diagnostic performance, computerized interpretation, and multivariable prediction. Veterans Affairs Cooperative Study in Health Services #016 (QUEXTA) Study Group. Quantitative Exercise Testing and Angiography. *Ann. Intern. Med.*, 1998;**128**: 965–974.
144. Sketch, M.H., A.N. Mooss, M.L. Butler, et al., Digoxin-induced positive exercise tests: Their clinical and prognostic significance. *Am. J. Cardiol.*, 1981;**48**: 655–659.
145. LeWinter, M.M., M.H. Crawford, R.A. O'Rourke, et al., The effects of oral propranolol, digoxin and combination therapy on the resting and exercise electrocardiogram. *Am. Heart J.*, 1977;**93**: 202–209.
146. Blackburn, H., Canadian colloquium on computer assisted interpretation of electrocardiograms, VI: importance of the electrocardiogram in populations outside the hospital. *Can. Med. Assoc. J.*, 1973;**108**: 1262–1265.
147. Cullen, K., N.S. Stenhouse, K.L. Wearne, et al., Electrocardiograms and 13-year cardiovascular mortality in Busselton study. *Br. Heart J.*, 1982;**47**: 209–212.
148. Aronow, W.S., Correlation of ischemic ST segment depression on the resting electrocardiogram with new cardiac events in 1,106 patients over 62 years of age. *Am. J. Cardiol.*, 1989;**64**: 232–233.
149. Califf, R.M., D.B. Mark, F.E. Harrell Jr, et al., Importance of clinical measures of ischemia in the prognosis of patients with documented coronary artery disease. *J. Am. Coll. Cardiol.*, 1988;**11**: 20–26.
150. Harris, P.J., F.E. Harrell Jr, K.L. Lee, et al., Survival in medically treated coronary artery disease. *Circulation*, 1979;**60**: 1259–1269.
151. Miranda, C.P., K.G. Lehmann, and V.F. Froelicher, Correlation between resting ST segment depression, exercise testing, coronary angiography, and long term prognosis. *Am. Heart J.*, 1991;**122**: 1617–1628.
152. Geleijnse, M.L., P.M. Fioretti, and J.R.T.C. Roelandt, Methodology, feasibility, safety and diagnostic accuracy of dobutamine stress echocardiography. *J. Am. Coll. Cardiol.*, 1997;**30**: 595–606.
153. Geleijnse, M.L., A. Salustri, T.H. Marwick, et al., Should the diagnosis of coronary artery disease be based on the evaluation of myocardial function or perfusion? *Eur. Heart J.*, 1997;**18**(Suppl. D): D68–D77.
154. Schinkel, A.F.L., J.J. Bax, M.L. Geleijnse, et al., Noninvasive evaluation of ischemic heart disease: Myocardial perfusion imaging or stress echocardiography? *Eur. Heart J.*, 2003;**24**: 789–800.
155. Geleijnse, M.L., A. Elhendy, and R.T. Van Domburg, Cardiac imaging for risk stratification with dobutamine-atropine stress testing in patients with chest pain. Echocardiography, perfusion scintigraphy, or both? *Circulation*, 1997;**96**: 137–147.
156. Rigo, P., I.K. Bailey, L.C. Griffith, et al., Value and limitations of segmental analysis of stress thallium myocardial imaging for localization of coronary artery disease. *Circulation*, 1980;**61**: 973–981.
157. Ritchie, J.L., T.M. Bateman, R.O. Bonow, et al., Guidelines for clinical use of cardiac radionuclide imaging: Report of the American College of Cardiology/American Heart Association Task Force on Assessment of Diagnostic and Therapeutic Cardiovascular Procedures (Committee on Radionuclide Imaging), developed in collaboration with the American Society of Nuclear Cardiology. *J. Am. Coll. Cardiol.*, 1995;**25**: 521–547.
158. Cheitlin, M.D., J.S. Alpert, W.F. Armstrong, et al., ACC/AHA guidelines for the clinical application of echocardiography: A report of the American College of Cardiology/American Heart Association Task Force on Practice Guidelines (Committee on Clinical Application of Echocardiography). Developed in collaboration with the American Society of Echocardiography. *Circulation*, 1997;**95**: 1686–1744.
159. Christian, T.F., T.D. Miller, K.R. Bailey, et al., Exercise tomographic thallium 201 imaging in patients with severe coronary artery disease and normal electrocardiograms. *Ann. Intern. Med.*, 1994;**121**: 825–832.
160. Gibbons, R.J., A.R. Zinsmeister, T.D. Miller, et al., Supine exercise electrocardiography compared with exercise radionuclide angiography in noninvasive identification of severe coronary artery disease. *Ann. Intern. Med.*, 1990;**112**: 743–749.
161. Ladenheim, M.L., T.S. Kotler, B.H. Pollock, et al., Incremental prognostic power of clinical history, exercise electrocardiography and myocardial perfusion scintigraphy in suspected coronary artery disease. *Am. J. Cardiol.*, 1987;**59**: 270–277.
162. Weiner, D.A., T.J. Ryan, C.H. McCabe, et al., Prognostic importance of a clinical profile and exercise test in medically treated patients with coronary artery disease. *J. Am. Coll. Cardiol.*, 1984;**3**: 772–779.
163. Mark, D.B., M.A. Hlatky, F.E. Harrell Jr, et al., Exercise treadmill score for predicting prognosis in coronary artery disease. *Ann. Intern. Med.*, 1987;**106**: 793–800.
164. Morrow, K., C.K. Morris, V.F. Froelicher, et al., Prediction of cardiovascular death in men undergoing noninvasive evaluation for coronary artery disease. *Ann. Intern. Med.*, 1993;**118**: 689–695.
165. Brunelli, C., R. Cristofani, and A. L'Abbate, Long term survival in medically treated patients with ischaemic heart disease and prognostic importance of clinical and electrocardiographic data (the Italian CNR Multicentre Prospective Study OD1). *Eur. Heart J.*, 1989;**10**: 292–303.
166. Luwaert, R.J., J.A. Melin, C.R. Brohet, et al., Non invasive data provide independent prognostic information in patients

with chest pain without previous myocardial infarction: Findings in male patients who have had cardiac catheterization. *Eur. Heart J.*, 1988;**9**: 418–426.
167. Gohlke, H., L. Samek, P. Betz, et al., Exercise testing provides additional prognostic information in angiographically defined subgroups of patients with coronary artery disease. *Circulation*, 1983;**68**: 979–985.
168. Hammermeister, K.E., T.A. DeRouen, and H.T. Dodge, Variables predictive of survival in patients with coronary disease: Selection by univariate and multivariate analyses from the clinical, electrocardiographic, exercise, arteriographic, and quantitative angiographic evaluations. *Circulation*, 1979;**59**: 421–430.
169. McNeer, J.F., J.R. Margolis, K.L. Lee, et al., The role of the exercise test in the evaluation of patients for ischemic heart disease. *Circulation*, 1978;**57**: 64–70.
170. Mark, D.B., L. Shaw, F.E. Harrell Jr, et al., Prognostic value of a treadmill exercise score in outpatients with suspected coronary artery disease. *N. Engl. J. Med.*, 1991;**325**: 849–853.
171. Alexander, K.P., L.J. Shaw, L.K. Shaw, et al., Value of exercise treadmill testing in women [published erratum appears in J Am Coll Cardiol 1999;33:289]. *J. Am. Coll. Cardiol.*, 1998;**32**: 1657–1664.
172. Lauer, M.S., G.S. Francis, P.M. Okin, et al., Impaired chronotropic response to exercise stress testing as a predictor of mortality. *JAMA*, 1999;**281**: 524–529.
173. Cole, C.R., E.H. Blackstone, F.J. Pashkow, et al., Heart-rate recovery immediately after exercise as a predictor of mortality. *N. Engl. J. Med.*, 1999;**341**: 1351–1357.
174. Cole, C.R., J.M. Foody, E.H. Blackstone, et al., Heart rate recovery after submaximal exercise testing as a predictor of mortality in a cardiovascularly healthy cohort. *Ann. Intern. Med.*, 2000;**132**: 552–555.
175. Diaz, L.A., R.C. Brunken, E.H. Blackstone, et al., Independent contribution of myocardial perfusion defects to exercise capacity and heart rate recovery for prediction of all-cause mortality in patients with known or suspected coronary heart disease. *J. Am. Coll. Cardiol.*, 2001;**37**: 1558–1564.
176. Watanabe, J., M. Thamilarasan, E.H. Blackstone, et al., Heart rate recovery immediately after treadmill exercise and left ventricular systolic dysfunction as predictors of mortality: The case of stress echocardiography. *Circulation*, 2001;**104**: 1911–1916.
177. Nishime, E.O., C.R. Cole, E.H. Blackstone, et al., Heart rate recovery and treadmill exercise score as predictors of mortality in patients referred for exercise ECG. *JAMA*, 2000;**284**: 1392–1398.
178. Shetler, K., R. Marcus, V.F. Froelicher, et al., Heart rate recovery: Validation and methodologic issues. *J. Am. Coll. Cardiol.*, 2001;**38**: 1980–1987.
179. McHam, S.A., T.H. Marwick, F.J. Pashkow, et al., Delayed systolic blood pressure recovery after graded exercise: An independent correlate of angiographic coronary disease. *J. Am. Coll. Cardiol.*, 1999;**34**: 754–759.
180. Bogaty, P., G.R. Dagenais, B. Cantin, et al., Prognosis in patients with a strongly positive exercise electrocardiogram. *Am. J. Cardiol.*, 1989;**64**: 1284–1288.
181. Bertrand, M.E., M.L. Simoons, K.A. Fox, et al.; Task Force on the Management of Acute Coronary Syndromes of the European Society of Cardiology, Management of acute coronary syndromes in patients presenting without persistent ST segment elevation. *Eur. Heart J.*, 2002;**23**: 1809–1840. Erratum in: *Eur. Heart J.*, 2003;**24**: 1174–1175. *Eur. Heart J.*, 2003;**24**: 485.
182. Boersma, E., K.S. Pieper, E.W. Steyerberg, et al., Predictors of outcome in patients with acute coronary syndromes without persistent ST segment elevation. Results from an international trial of 9461 patients. The PURSUIT Investigators. *Circulation*, 2000;**101**: 2557–2567.
183. Stein, R.A., B.R. Chaitman, G.J. Balady, et al., Safety and utility of exercise testing in emergency room chest pain centers: An advisory from the Committee on Exercise, Rehabilitation, and Prevention, Council on Clinical Cardiology, American Heart Association. *Circulation*, 2000;**102**: 1463–1467.
184. Gibler, W.B., J.P. Runyon, R.C. Levy, et al., A rapid diagnostic and treatment center for patients with chest pain in the emergency department. *Ann. Emerg. Med.*, 1995;**25**: 1–8.
185. Farkouh, M.E., P.A. Smars, G.S. Reeder, et al., A clinical trial of a chest-pain observation unit for patients with unstable angina. Chest Pain Evaluation in the Emergency Room (CHEER) Investigators. *N. Engl. J. Med.*, 1998;**339**: 1882–1888.
186. Froelicher, E.S., Usefulness of exercise testing shortly after acute myocardial infarction for predicting 10-year mortality. *Am. J. Cardiol.*, 1994;**74**: 318–323.
187. Mark, D.B. and V.F. Froelicher, Exercise treadmill testing and ambulatory monitoring, in *Acute Coronary Care*, R.M. Califf, D.B. Mark, and G.S. Wagner, Editors. St. Louis, Mosby-Year Book, 1995.
188. Juneau, M., P. Colles, P. Théroux, et al., Symptom limited versus low level exercise testing before hospital discharge after myocardial infarction. *J. Am. Coll. Cardiol.*, 1992;**20**: 927–933.
189. Stevenson, R., V. Umachandran, K. Ranjadayalan, et al., Reassessment of treadmill stress testing for risk stratification in patients with acute myocardial infarction treated by thrombolysis. *Br. Heart J.*, 1993;**70**: 415–420.
190. Moss, A.J., R.E. Goldstein, W.J. Hall, et al., Detection and significance of myocardial ischemia in stable patients after recovery from an acute coronary event: Multicenter Myocardial Ischemia Research Group. *JAMA*, 1993;**269**: 2379–2385.
191. Arnold, A.E., M.L. Simoons, J.M. Detry, et al., Prediction of mortality following hospital discharge after thrombolysis for acute myocardial infarction: Is there a need for coronary angiography? *Eur. Heart J.*, 1993;**14**: 306–315.
192. Hamm, L.F., R.S. Crow, G.A. Stull, et al., Safety and characteristics of exercise testing early after acute myocardial infarction. *Am. J. Cardiol.*, 1989;**63**: 1193–1197.
193. Jain, A., G.H. Myers, P.M. Sapin, et al., Comparison of symptom limited and low level exercise tolerance tests early after myocardial infarction. *J. Am. Coll. Cardiol.*, 1993;**22**: 1816–1820.
194. Krone, R.J., J.A. Gillespie, F.M. Weld, et al., Low level exercise testing after myocardial infarction: Usefulness in enhancing clinical risk stratification. *Circulation*, 1985;**71**: 80–89.
195. Nielsen, J.R., H. Mickley, E.M. Damsgaard, et al., Predischarge maximal exercise test identifies risk for cardiac death in patients with acute myocardial infarction. *Am. J. Cardiol.*, 1990;**65**: 149–153.
196. Topol, E.J., K. Burek, W.W. O'Neill, et al., A randomized controlled trial of hospital discharge three days after myocardial infarction in the era of reperfusion. *N. Engl. J. Med.*, 1988;**318**: 1083–1088.
197. Senaratne, M.P., G. Smith, and S.S. Gulamhusein, Feasibility and safety of early exercise testing using the Bruce protocol after acute myocardial infarction. *J. Am. Coll. Cardiol.*, 2000;**35**: 1212–1220.

198. Fletcher, G.F., G.J. Balady, E.A. Amsterdam, et al., Exercise standards for testing and training: A statement for healthcare professionals from the American Heart Association. *Circulation*, 2001;**104**: 1694–1740.
199. Vanhees, L., D. Schepers, and R. Fagard, Comparison of maximum versus submaximum exercise testing in providing prognostic information after acute myocardial infarction and/or coronary artery bypass grafting. *Am. J. Cardiol.*, 1997;**80**: 257–262.
200. Villella, A., A.P. Maggioni, M. Villella, et al., Prognostic significance of maximal exercise testing after myocardial infarction treated with thrombolytic agents: The GISSI 2 data base. Gruppo Italiano per lo Studio della Sopravvivenza Nell'Infarto. *Lancet*, 1995;**346**: 523–529.
201. Chaitman, B.R., R.P. McMahon, M. Terrin, et al., Impact of treatment strategy on predischarge exercise test in the Thrombolysis in Myocardial Infarction (TIMI) II Trial. *Am. J. Cardiol.*, 1993;**71**: 131–138.
202. Newby, L.K., R.M. Califf, A. Guerci, et al., Early discharge in the thrombolytic era: An analysis of criteria for uncomplicated infarction from the Global Utilization of Streptokinase and tPA for Occluded Coronary Arteries (GUSTO) trial. *J. Am. Coll. Cardiol.*, 1996;**27**: 625–632.
203. Froelicher, V.F., S. Perdue, W. Pewen, et al., Application of meta analysis using an electronic spread sheet for exercise testing in patients after myocardial infarction. *Am. J. Med.*, 1987;**83**: 1045–1054.
204. Maggioni, A.P., F.M. Turazza, and L. Tavazzi, Risk evaluation using exercise testing in elderly patients after acute myocardial infarction. *Cardiol. Elder.*, 1995;**3**: 88–93.
205. Théroux, P., D.D. Waters, C. Halphen, et al., Prognostic value of exercise testing soon after myocardial infarction. *N. Engl. J. Med.*, 1979;**301**: 341–345.
206. Shaw, L.J., E.D. Peterson, K. Kesler, et al., A meta analysis of predischarge risk stratification after acute myocardial infarction with stress electrocardiographic, myocardial perfusion, and ventricular function imaging. *Am. J. Cardiol.*, 1996;**78**: 1327–1337.
207. Volpi, A., C. de Vita, M.G. Franzosi, et al., Predictors of nonfatal reinfarction in survivors of myocardial infarction after thrombolysis: Results of the Gruppo Italiano per lo Studio della Sopravvivenza nell'Infarto Miocardico (GISSI 2) Data Base. *J. Am. Coll. Cardiol.*, 1994;**24**: 608–615.
208. Ciaroni, S., J. Delonca, and A. Righetti, Early exercise testing after acute myocardial infarction in the elderly: Clinical evaluation and prognostic significance. *Am. Heart J.*, 1993;**126**: 304–311.
209. Stone, P.H., Z.G. Turi, J.E. Muller, et al., Prognostic significance of the treadmill exercise test performance 6 months after myocardial infarction. *J. Am. Coll. Cardiol.*, 1986;**8**: 1007–1017.
210. Fioretti, P., R.W. Brower, M.L. Simoons, et al., Prediction of mortality in hospital survivors of myocardial infarction: Comparison of predischarge exercise testing and radionuclide ventriculography at rest. *Br. Heart J.*, 1984;**52**: 292–298.
211. Fioretti, P., R.W. Brower, M.L. Simoons, et al., Relative value of clinical variables, bicycle ergometry, rest radionuclide ventriculography and 24 hour ambulatory electrocardiographic monitoring at discharge to predict 1 year survival after myocardial infarction. *J. Am. Coll. Cardiol.*, 1986;**8**: 40–49.
212. Pilote, L., J. Silberberg, R. Lisbona, et al., Prognosis in patients with low left ventricular ejection fraction after myocardial infarction. *Circulation*, 1989;**80**: 1636–1641.
213. Krone, R.J., E.M. Dwyer, H. Greenberg, et al., Risk stratification in patients with first non Q wave infarction: Limited value of the early low level exercise test after uncomplicated infarcts: The Multicenter Post Infarction Research Group. *J. Am. Coll. Cardiol.*, 1989;**14**: 31–37.
214. DeBusk, R.F. and W. Haskell, Symptom limited vs heart rate limited exercise testing soon after myocardial infarction. *Circulation*, 1980;**61**: 738–743.
215. Abboud, L., J. Hir, I. Eisen, and W. Markiewicz, Angina pectoris and ST segment depression during exercise testing early following acute myocardial infarction. *Cardiology*, 1994;**84**: 268–273.
216. McConahay, D.R., M. Valdes, B.D. McCallister, et al., Accuracy of treadmill testing in assessment of direct myocardial revascularization. *Circulation*, 1977;**56**: 548–552.
217. Visser, F.C., L. van Campen, and P.J. de Feyter, Value and limitations of exercise stress testing to predict the functional results of coronary artery bypass grafting. *Int. J. Card. Imaging*, 1993;**9**(Suppl. 1): 41–47.
218. Kafka, H., A.J. Leach, and G.M. Fitzgibbon, Exercise echocardiography after coronary artery bypass surgery: Correlation with coronary angiography. *J. Am. Coll. Cardiol.*, 1995;**25**: 1019–1023.
219. Yli Mayry, S., H.V. Huikuri, K.E. Airaksinen, et al., Usefulness of a postoperative exercise test for predicting cardiac events after coronary artery bypass grafting. *Am. J. Cardiol.*, 1992;**70**: 56–59.
220. Krone, R.J., R.M. Hardison, B.R. Chaitman, et al., Risk stratification after successful coronary revascularization: The lack of a role for routine exercise testing. *J. Am. Coll. Cardiol.*, 2001;**38**: 136–142.
221. Kadel, C., T. Strecker, M. Kaltenbach, et al., Recognition of restenosis: Can patients be defined in whom the exercise ECG result makes angiographic restudy unnecessary? *Eur. Heart J.*, 1989;**10**(Suppl. G): 22–26.
222. Honan, M.B., J.R. Bengtson, D.B. Pryor, et al., Exercise treadmill testing is a poor predictor of anatomic restenosis after angioplasty for acute myocardial infarction. *Circulation*, 1989;**80**: 1585–1594.
223. Schroeder, E., B. Marchandise, P. DeCoster, et al., Detection of restenosis after coronary angioplasty for single vessel disease: How reliable are exercise electrocardiography and scintigraphy in asymptomatic patients? *Eur. Heart J.*, 1989;**10**: 18–21.
224. Laarman, G., H.E. Luijten, L.G. van Zeyl, et al., Assessment of silent restenosis and long-term follow-up after successful angioplasty in single vessel coronary artery disease: The value of quantitative exercise electrocardiography and quantitative coronary angiography. *J. Am. Coll. Cardiol.*, 1990;**16**: 578–585.
225. Desmet, W., I. De Scheerder, and J. Piessens, Limited value of exercise testing in the detection of silent restenosis after successful coronary angioplasty. *Am. Heart J.*, 1995;**129**: 452–459.
226. Vlay, S.C., J. Chernilas, W.E. Lawson, et al., Restenosis after angioplasty: Don't rely on the exercise test. *Am. Heart J.*, 1989;**117**: 980–986.
227. Echt, H.S., R.E. Shaw, H.L. Chin, et al., Silent ischemia after coronary angioplasty: Evaluation of restenosis and extent of ischemia in asymptomatic patients by tomographic thallium 201 exercise imaging and comparison with symptomatic patients. *J. Am. Coll. Cardiol.*, 1991;**17**: 670–677.
228. Hecht, H.S., L. DeBord, R. Shaw, et al., Usefulness of supine bicycle stress echocardiography for detection of restenosis after percutaneous transluminal coronary angioplasty. *Am. J. Cardiol.*, 1993;**71**: 293–296.

229. Bigger, J.T., Jr, F.J. Dresdale, R.H. Heissenbuttel, et al., Ventricular arrhythmias in ischemic heart disease: Mechanism, prevalence, significance and management. *Prog. Cardiovasc. Dis.*, 1977;**19**: 255–300.
230. Faris, J.V., P.L. McHenry, J.W. Jordan, et al., Prevalence and reproducibility of exercise-induced ventricular arrhythmias during maximal exercise testing in normal men. *Am. J. Cardiol.*, 1976;**37**: 617–622.
231. Busby, M.J., E.A. Shefrin, and J.L. Fleg, Prevalence and long-term significance of exercise-induced frequent or repetitive ventricular ectopic beats in apparently healthy volunteers. *J. Am. Coll. Cardiol.*, 1989;**14**: 1659–1665.
232. Ellestad, M.H. and M.K. Wan, Predictive implications of stress testing. Follow up of 2700 subjects after maximum treadmill stress testing. *Circulation*, 1975;**51**: 363–369.
233. Ellestad, M.H., Chronotropic incompetence: The implications of heart rate response to exercise (compensatory parasympathetic hyperactivity?). *Circulation*, 1996;**93**: 1485–1487.
234. Lauer, M.S., P.M. Okin, M.G. Larson, et al., Impaired heart rate response to graded exercise: Prognostic implications of chronotropic incompetence in the Framingham Heart Study. *Circulation*, 1996;**93**: 1520–1526.
235. Corbelli, R., M. Masterson, and B.L. Wilkoff, Chronotropic response to exercise in patients with atrial fibrillation. *Pacing Clin. Electrophysiol.*, 1990;**13**: 179–187.
236. Fisher, F.D. and H.A. Tyroler, Relationship between ventricular premature contractions in routine electrocardiograms and subsequent sudden death from coronary heart disease. *Circulation*, 1973;**47**: 712–719.
237. Buckingham, T.A., The clinical significance of ventricular arrhythmias in apparently healthy subjects. *Pract. Cardiol.*, 1983;**9**: 37.
238. McHenry, P.L., S.N. Morris, M. Kavalier, et al., Comparative study of exercise-induced ventricular arrhythmias in normal subjects and patients with documented coronary artery disease. *Am. J. Cardiol.*, 1976;**37**: 609–616.
239. Froelicher, V.F., M.M. Thomas, C. Pillow, et al., Epidemiologic study of asymptomatic men screened by maximal treadmill testing for latent coronary artery disease. *Am. J. Cardiol.*, 1974;**34**: 770–776.
240. Blackburn, H., H.L. Taylor, B. Hamrell, et al., Premature ventricular complexes induced by stress testing. Their frequency and response to physical conditioning. *Am. J. Cardiol.*, 1973;**31**: 441–449.
241. McHenry, P.L., S.N. Morris, and M. Kavalier, Exercise-induced arrhythmia: Recognition, classification and clinical significance. *Cardiovasc. Clin.*, 1974;**6**: 245–254.
242. Jouven, X., M. Zureik, M. Desnos, et al., Long-term outcome in asymptomatic men with exercise-induced premature ventricular depolarizations. *N. Engl. J. Med.*, 2000;**343**: 826–833.
243. Jelinek, M.V. and B. Lown, Exercise stress testing for exposure of cardiac arrhythmia. *Prog. Cardiovasc. Dis.*, 1974;**16**: 497–522.
244. Goldschlager, N., D. Cake, and K. Cohn, Exercise-induced ventricular arrhythmias in patients with coronary artery disease: Their relationship to angiographic findings. *Am. J. Cardiol.*, 1973;**31**: 434–440.
245. Surawicz, B. and T.K. Knilans, *Chou's Electrocardiography in Clinical Practice: Adult and Pediatric*, 5th edn. Philadelphia, PA: W.B. Saunders.
246. Kotler, M.N., B. Tabatznik, M.M. Mower, et al., Prognostic significance of ventricular ectopic beats with respect to sudden death in the late postinfarction period. *Circulation*, 1973;**47**: 959–966.
247. Chiang, B.N., L.V. Perlman, L.D. Ostrander Jr, et al., Relationship of premature systoles to coronary heart disease and sudden death in the Tecumseh epidemiologic study. *Ann. Intern. Med.*, 1969;**70**: 1159–1166.
248. Coronary Drug Project Research Group, Prognostic importance of premature beats following myocardial infarction. Experience in the Coronary Drug Project. *JAMA*, 1973;**223**: 1116–1124.
249. Henry, R.L., G.T. Kennedy, and M.H. Crawford, Prognostic value of exercise-induced ventricular ectopic activity for mortality after acute myocardial infarction. *Am. J. Cardiol.*, 1987;**59**: 1251–1215.
250. Krone, R.J., J.A. Gillespie, F.M. Weld, et al., Low-level exercise testing after myocardial infarction: Usefulness in enhancing clinical risk stratification. *Circulation*, 1985;**71**: 80–89.
251. Waters, D.D., X. Bosch, A. Bouchard, et al., Comparison of clinical variables and variables derived from a limited predischarge exercise test as predictors of early and late mortality after myocardial infarction. *J. Am. Coll. Cardiol.*, 1985;**5**: 1–8.
252. Margonato, A., A. Mailhac, F. Bonetti, et al., Exercise-induced ischemic arrhythmias in patients with previous myocardial infarction: Role of perfusion and tissue viability. *J. Am. Coll. Cardiol.*, 1996;**27**: 593–598.
253. Marieb, M.A., G.A. Beller, R.S. Gibson, et al., Clinical relevance of exercise-induced ventricular arrhythmias in suspected coronary artery disease. *Am. J. Cardiol.*, 1990;**66**: 172–178.
254. Califf, R.M., R.A. McKinnis, J.F. McNeer, et al., Prognostic value of ventricular arrhythmias associated with treadmill exercise testing in patients studied with cardiac catheterization for suspected ischemic heart disease. *J. Am. Coll. Cardiol.*, 1983;**2**: 1060–1067.
255. Helfant, R.H., R. Pine, V. Kabde, et al., Exercise-related ventricular premature complexes in coronary heart disease: Correlations with ischemia and angiographic severity. *Ann. Intern. Med.*, 1974;**80**: 589–592.
256. Udall, J.A. and M.H. Ellestad, Predictive implications of ventricular premature contractions associated with treadmill stress testing. *Circulation*, 1977;**56**: 985–989.
257. Sami, M., B. Chaitman, L. Fisher, et al., Significance of exercise-induced ventricular arrhythmia in stable coronary artery disease: A Coronary Artery Surgery Study project. *Am. J. Cardiol.*, 1984;**54**: 1182–1188.
258. Schweikert, R.A., F.J. Pashkow, C.E. Snader, et al., Association of exercise-induced ventricular ectopic activity with thallium myocardial perfusion and angiographic coronary artery disease in stable, low-risk populations. *Am. J. Cardiol.*, 1999;**83**: 530–534.
259. Dimsdale, J.E., W. Ruberman, R.A. Carleton, V. DeQuattro, E. Eaker, R.S. Eliot, C.D. Furberg, C.W. Irvin Jr, A.P. Shapiro, et al., Sudden cardiac death. Stress and cardiac arrhythmias. *Circulation*, 1987;**76**(1 Pt. 2): I198–I201.
260. Pinski, S.L., The right ventricular tachycardias. *J. Electrocardiol.*, 2000;**33**: 103–114.
261. Podrid, P.J. and T.B. Graboys, Exercise stress testing in the management of cardiac rhythm disorders. *Med. Clin. North Am.*, 1984;**68**: 1139–1152.
262. Young, D.Z., S. Lampert, T.B. Graboys, et al., Safety of maximal exercise testing in patients at high risk for ventricular arrhythmia. *Circulation*, 1984;**70**: 184–191.
263. Saini, V., T.B. Graboys, V. Towne, et al., Reproducibility of exercise-induced ventricular arrhythmia in patients undergoing

evaluation for malignant ventricular arrhythmia. *Am. J. Cardiol.*, 1989;**63**: 697–701.

264. ACC/AHA guidelines for the management of patients with valvular heart disease: A report of the American College of Cardiology/American Heart Association Task Force on Practice Guidelines (Committee on Management of Patients with Valvular Heart Disease). *J. Am. Coll. Cardiol.*, 1998;**32**: 1486–1588.
265. Hochreiter, C. and J.S. Borer, Exercise testing in patients with aortic and mitral valve disease: Current applications. *Cardiovasc. Clin.*, 1983;**13**: 291–300.
266. Chandramouli, B., D.A. Ehmke, and R.M. Lauer, Exercise induced electrocardiographic changes in children with congenital aortic stenosis. *J. Pediatr.*, 1975;**87**: 725–730.
267. James, F.W., D.C. Schwartz, S. Kaplan, et al., Exercise electrocardiogram, blood pressure, and working capacity in young patients with valvular or discrete subvalvular aortic stenosis. *Am. J. Cardiol.*, 1982;**50**: 769–775.
268. Bonow, R.O., Management of chronic aortic regurgitation. *N. Engl. J. Med.*, 1994;**331**: 736–737.

37 Computer Analysis of the Electrocardiogram

Jan A. Kors · Gerard van Herpen

P. W. Macfarlane, A. van Oosterom, O. Pahlm, P. Kligfield, M. Janse, J. Camm (eds.), *Comprehensive Electrocardiology*, DOI 10.1007/978-1-84882-046-3_37,

This chapter deals with the use of digital computers for the handling of the resting ECG. The reader is referred to other chapters of this book for a review of computer-assisted interpretation of the exercise electrocardiogram, for ambulatory monitoring, on-line arrhythmia detection in the coronary care unit, and for surface mapping and modeling applications.

37.1 Some History

It is a deep rooted fantasy of man, told in many stories, to be able to handicraft a creature in his own image and endow it with the breath of life: a "homunculus." More often than not, the endeavor is initially successful but ends in disenchantment. The advent of the computer seemed to make it possible to equip such a creation with human intelligence. Electrocardiography appeared to be a field where the computer could be expected to deploy its intelligent capabilities to the best advantage.

The ECG was an attractive object for computerization for a number of reasons: (1) it is an electrical signal, easily recorded and easily digestible for a computer; (2) it is a rather simple, orderly and repetitive signal; (3) it carries important information about which a vast corpus of knowledge has been amassed to prime the computer program; (4) ECGs are produced in enormous quantities all over the world which makes it worthwhile to let a computer reduce human workload.

ECG diagnosis by computer would be an instance of artificial intelligence which might ultimately challenge the specific human faculty of diagnosis. When is artificial intelligence really intelligent? Turing devised a thought experiment to answer this question [1]. Suppose that a computer and a human being are sitting in separate closed chambers and that an outside investigator is questioning them. If the investigator cannot decide from the answers whether they are given by the computer or by man, then the computer has attained full human intelligence. This implies that the computer's diagnosis of the ECG must exactly copy that of man in order to satisfy this requirement. Indeed, the agreement between computer classification and man-made diagnosis of the ECG has been extensively studied. But how intelligent is man? Disagreement between computer and man might even be due to a computer's superior diagnostic power! A second approach in testing a diagnostic ECG program, therefore, had to be to gauge it against reality, the clinical diagnosis, and to compare its performance with the skills of the human reader in this respect. The outcomes of both approaches will be discussed later.

The first attempts to automate ECG analysis by computer were made as early as the late 1950s, but it took considerably more time to develop operational computer programs than had originally been anticipated. At that time, the analog electric computer dominated the field. Analog computers were being used in basic electrocardiographic research [2], but their impact on clinical electrocardiography remained insignificant and they soon disappeared from the scene [3]. To have an analog signal processed by a digital computer, it must be broken down into digital samples. However, analog-to-digital (A-D) conversion systems (see ➋ Chap. 12) were not available as they are today and a special purpose A-D conversion system for ECGs had to be developed first [4].

A very crude computer program for the separation of normal from abnormal records became operational in 1959 [5]. It was based on angles and magnitude of spatial ventricular gradients, calculated from the time integrals over one PQRST cycle in the three orthogonal leads. Cycle recognition was still obtained through analog circuitry. The beginning and end of P, QRS, and T were not identified in this simple screening program. The first automatic wave-recognition program became available in 1961, opening the way for more detailed analysis of ECG records [6]. This original program, as well as its successors developed by Pipberger and coworkers [7, 8], was based on the Frank XYZ leads for the main reason that three leads laid less claim to computer capacity, still very limited both in speed and in memory space, than 12 leads. Through the pioneering work of Caceres et al. [9] the first program for conventional 12-lead ECG analysis became available in 1962. It was supported by the US Public Health Service and intended to make expert ECG analysis immediately available to all who needed it, using the telephone system. The 12 ECG leads were transmitted sequentially and, therefore, had to be analyzed individually. The loss of context between the leads, together with inherent cable noise, caused considerable waveform recognition errors and the program came to an untimely end.

User satisfaction, indeed, was not won by these early accomplishments and the initial state of excitement was to give way to one of dogged perseverance. Fresh investigators were, according to their state of awareness, attracted by the seeming easiness or the proven difficulties of the subject. Commercial suppliers started to be active in this market [10]. Gradually, a number of operational systems came into being in North America, Europe, and Japan. In the bigger hospitals, main frame computers were installed, which received the ECG signals on tape from mobile carts and could store them in databases.

For the less well-equipped customers, ECG diagnosis via telephone was offered by all kinds of entrepreneurial services, which did a thriving business in the 1970s and 1980s, sometimes handling millions of ECGs a year [11, 12]. The scene changed completely with the advent of the microprocessor [13]. Microprocessor-based systems have made it possible to distribute computerized ECG analysis outside hospital and laboratory to even the smallest hospitals, to screening clinics and to practitioners of every kind. While it took minutes to process a single ECG on a main frame computer in the early years, a present-day simple PC takes a fraction of a second to do the analysis. Most of these electrocardiographs are "stand-alone," that is, have a computer and printer on board. Some manufacturers offer separate front-end equipment (amplifiers and A-D converter) that can be interfaced with the computer standing in the doctor's office.

With a widening choice of systems on the market, the consumer may wish to know how intelligent the system being considered for purchase is. But where can the consumer turn for an objective intelligence test? It is clear that objective assessment of the quality of computerized ECG systems is needed. We will return to this subject in later sections of this chapter.

37.2 ECG-Processing Computer Systems

37.2.1 Advantages

Although diagnostic accuracy and reliability are greatly desirable characteristics of an interpretation system, the program's utility in a clinical environment is also determined by other factors. Reduction of the labor spent by the cardiological and clerical staff has been a key determinant in moving toward computer-aided processing systems in hospitals [11, 14]; improvement of overall process quality has been another. Rigorous operational control, as imposed by a computer system, can substantially add to the recording quality and thus the reliability of interpretation. The quality of reporting is also improved through the use of standard terminology and formats.

Computer analysis abolishes the well-known subjective differences arising in visual interpretation, and through a quantitative approach may enhance correct classification. Pipberger et al. [15] have, since the early 1960s, pointed to improved accuracy of interpretation as the primary objective of computer ECG processing. Automatic storage and retrieval, with the possibility of comparing the new ECG with its predecessors (serial analysis) is another asset of computerized ECG processing.

Computerized analysis of the resting ECG has also increased the feasibility of large-scale cardiovascular screening and epidemiological studies. "Minnesota coding" of the ECG can nowadays be performed by computer, quickly and reliably (see ➋ Chap. 40).

37.2.2 ECG Management Systems

An ECG management system can come in several forms and undertake multiple functions. In its simplest form, it effectively acts as a database for storage of ECGs and offers straightforward retrieval of ECGs for display or printing. It could be in the form of a small PC, to which an ECG machine might be directly attached as might be found in a small family medical practice.

At the other extreme, a large ECG management system could be used to centrally store all ECGs recorded within a large hospital or group of hospitals. Such systems generally have the capability to store multiple ECGs for a single patient and thereby facilitate serial comparison either through visual display of several ECGs on the same monitor screen or through automated techniques (see ➋ Sect. 37.16). The large systems offer the user the ability to measure wave amplitudes, durations, and standard time intervals on an averaged ECG waveform, generally using digital calipers.

One of the major benefits of the larger systems is to allow editing, sometimes known as over-reading, of the automated interpretations. Generally, manufacturers provide facilities to assist with editing such as through acronyms to represent standard interpretative statements. In many environments, a physician will handwrite comments onto an automated report and a member of the administrative staff with appropriate knowledge can edit the changes on the ECG management system. A problem then arises if serial comparison is available on the system. In such a case, the original electronic copy of the automated interpretation also has to be edited so that any subsequent serial comparison is made with the corrected report. This is actually quite a complex issue and one which can inhibit automated serial comparison.

Clearly, the ECG management system offers the facility for the provision of statistics on throughput of ECGs, number of edits made by each cardiologist, and so on. Nowadays, there is a trend toward manufacturers providing multiple databases for ECGs, echocardiograms, x-rays, etc., on a single server, but flexibility of editing and retrieving ECGs easily on an ECG management system still remains of significant importance in many healthcare environments.

37.3 Information Content of Lead Systems; Lead Transformations

In the analysis of biological signals in general, an enormous flow of information is channeled through a limited number of transducers. The often still redundant transducer output stream is then reduced and transformed to a manageable number of parameters. These in turn are harnessed into decision machinery, which finally summarizes the input information in a few standardized "diagnostic" terms that carry significance for the user and are the basis for further action. The ECG is such a biological signal.

The electrical activity of the heart gives rise to a three-dimensional time-varying distribution of currents and potentials. To obtain a full picture of this process, it is necessary to measure the complete course of potential distributions in and on the body. Only the body surface signals are readily available for measurement and even then the amount of information generated by the heart is so large that it is necessary to curtail the information stream for practical purposes. For the recording of a body surface potential map (BSPM), grids of 64–192 electrodes are being used (see ➋ Chap. 31). In this way, in a single BSPM all ECG information of that moment may be considered to be present as to localization. However, in a sequence of BSPMs, it is not feasible to follow all these individual electrocardiograms in their time course. For practical reasons, global parameters like "isointegral maps" or "isochrone maps" have to be used, and for diagnostic purposes, "difference maps" and "departure maps" have also been used. All this manipulation of data has only become possible, thanks to the computer, but so far clinical acceptance of BSPM diagnostic systems has been poor.

The universally used standard 12-lead ECG system needs nine electrodes, the three original Einthoven electrodes and the six thoracic electrodes adapted from Wilson. With the use of nine electrodes, eight independent leads, as they are called, can be obtained. The four extra extremity leads, which complete the 12-lead system, are not independent as they are arithmetically derived from any two other extremity leads. Nine electrodes are a drastic reduction of the number of sampling sites used to create a BSPM; the diagnostic power retained in this reduced lead set is indeed amazing. The explanation lies in the large degree of redundancy present in the surface information. Even among the eight leads of the standard ECG there is much redundancy and the question arises how many leads are actually required to render all relevant information.

In the vectorcardiogram (VCG), all information is contained in only three orthogonal leads, the X, Y, and Z components of a single dipole vector that changes with time in direction and strength, but is assumed to be stationary in position. Clearly, the three leads of the VCG (composed from a set of selected primary leads by linear combination) carry less information than eight leads, but, in return, the vector depicts the temporal relation between leads, providing phase information that is largely neglected in the ECG. Vectorcardiography could properly be called "phase electrocardiography." This feature makes up for the restriction on information input so that the diagnostic performances of the ECG and VCG are comparable, as will be discussed later on. Thus, each choice of data input, whether it be BSPM, standard ECG, or VCG, incurs the sacrifice of some information aspect: time resolution, phase coherence, or localization.

Presently, all ECG computer programs which are in clinical use will handle the conventional 12-lead system. Systems processing only the VCG have been driven from the market. The advantage that the VCG offers a data reduction of 8:3 has become insignificant considering the power of present-day computers. The clinical acceptance of the VCG has been hampered by the variety of different lead systems (from which the Frank lead system [16] emerged as the most commonly embraced), by a lack of understanding by clinicians of the physical principles, by the requirement of dorsal electrodes found to be cumbersome to apply, and finally by the absence, at the time, of proper recording equipment [17].

The question remains: how many leads are actually required to render all relevant information? It is of more than anecdotal value that Dower [18] let his ECG service only take vectorcardiograms. The ECGs that were then delivered to the hospital were derived from these VCGs by means of a mathematical transformation – apparently for many years to everybody's satisfaction. The inverse transformation yields the VCG from the standard ECG [19, 20], or, for better adjustment, from the standard ECG fitted out with some extra electrodes [21]. Kornreich [22, 23] demonstrated that the ECG and VCG leave some clinically useful spaces on the surface map not covered by their electrodes. With a set of electrodes also serving these areas, nine leads were determined that yielded better diagnostic classification (using multivariate analysis) than either the VCG or ECG [24, 25]. Kors et al. [26] proposed a system containing the standard electrodes of which two chest electrodes were moved to positions higher up and lower down on the chest. From the seven not-displaced electrodes (4 chest + 3 limb) the 12 leads could be reconstructed by linear transformation in very good approximation. In fact, it was possible to compose any set of ECG leads with electrode positions chosen for local information density and ease of placement, standard or nonstandard, and from these leads derive the standard ECG by a mathematical transformation [27]. Computer analysis of such a "derived" ECG should give all but the same result as that of its directly recorded counterpart. The nonstandard extra recorded leads might then be exploited for additional information – with the difficulty that no standard diagnostic criteria are available for such leads. Even an entire BSPM can be simulated from a limited number of measured ECG leads.

The EASI lead system designed by Dower uses only four electrodes to produce the 12 standard ECG leads by linear transformation [28]. Four electrodes will yield three independent leads; four was also the number of electrodes used originally by Burger to derive the three orthogonal vector components [29]. Linear transformations were also used to reduce signal noise [30] and to identify interchanged leads [31]. The first to use a linear transformation in electrocardiography were Burger et al. [32] to transform VCGs obtained from different lead systems into those from each other.

For practical purposes, this review will be further restricted to computer analysis of the resting standard 12-lead ECG with some reference to the orthogonal 3-lead ECG. Some ECG processing systems have an option to display and print vector loops, either directly recorded from the regular (Frank) electrode positions or computed through linear transformation from the ECG leads. The reader is referred to ➋ Chap. 11 in this book for a review of other lead systems.

37.4 Computer Processing of the ECG

In teaching electrocardiography, and in order to start the novice reader on a career in electrocardiography, it suffices to say: "This is a QRS complex, this little wiggle in front is a P, and the hump following QRS is a T." Some more detailed descriptions are absorbed with equal ease and the final touch is added by the recommendations and guidelines for nomenclature and measurements as issued by various committees and task forces [33–36]. Reading the electrocardiogram is in the first place a matter of visual observation, if necessary, aided by calipers and magnifying glass, and relies on the amazing deftness in pattern recognition of the human brain. The subsequent electrocardiographic diagnosis is a logic built on correct pattern recognition. The computer must do without human "eye balling" skills and needs to be instructed punctiliously about every detail of measurement down to the microvolt and millisecond level. Rules and definitions for visual ECG analysis may then appear imprecise and not consistent enough for computer application, and may even show gaps in their logic. New computer-compatible prescriptions for ECG signal measurements should, however, adhere as much as possible to accepted methods of visual measurement [34]. In the sections on waveform recognition (37.12) and parameter computation (37.13) we will go further into the matter.

As in visual ECG analysis, in an ECG data-processing system, the first of the two main parts that can be distinguished is that which contains program components to perform measurements. The second part includes the program components that derive the clinical significance of these measurements. This results in a final classification, or interpretation, of the ECG in terms familiar to the physician. The words "classification" or "interpretation" are preferred by some over "diagnosis," which is then the term reserved for the human interpretative activity.

The principal components of the ECG measurement section are as follows:

1. Data acquisition
2. Preprocessing and signal conditioning
3. Detection of QRS complexes and of P waves
4. Typing of QRS complexes
5. Forming a representative (P)-QRS-T complex
6. Boundary recognition, that is, detection of points of onset and offset of waves
7. Parameter extraction

The principal components of the ECG interpretative section are:

1. Rhythm analysis
2. Diagnostic classification
3. Serial comparison

These separate program components will be discussed in detail in the following paragraphs. Many ECG processing systems implement these different tasks in separate modules, each of which has well-defined objectives [37]. The advantage of such a structured setup is its easy implementation, evaluation, and maintenance [38]. In addition, systems for clinical use will have to offer an over-reading facility, which allows manual correction of the computer output. A fixed set of codes, most commonly acronyms, takes care of producing the required text elements, thereby ensuring consistency of terminology. In addition, entries in free text are admissible.

To inform the consumer about the quality of a computerized ECG system, it is necessary to have access to reference standards for objective evaluation. A paragraph on reference databases will therefore precede the discussion of the various parts of computer processing mentioned above, since it will often be necessary to refer to these standards.

37.5 Reference Databases for Program Evaluation

"How to test the intelligence of the computer" was the question raised in the introduction of this chapter. The simple precept is to compare the computer statements with the objective "truth" embodied in the ECG readings of the human expert or the clinical findings. How to carry this out will be shown to be no simple matter.

Two qualities are to be considered separately in the operations of the computer program, namely, waveform recognition skill and diagnostic competence, with correct waveform recognition being a sine qua non for reliable interpretation. As to the first quality, large measurement differences may become apparent between programs while analyzing identical ECGs [39, 40]. As to the second aspect, where is it possible to obtain the diagnostic truth? Initially, a number of program designers compared the computer output with their own ECG interpretation [41]. Diagnostic criteria were identical for both. As could be expected, agreement between computer and human readers was excellent. It is obvious that not diagnostic accuracy but reliability of the wave-recognition and measurement parts of the programs was tested in these studies. For example, in the study by Crevasse and Ariet [42] a computer accuracy rate of 98% was reported for LVH. Both the computer and the human readers used the Romhilt-Estes point score. When this criterion was tested by the originators of this scoring system in 150 autopsy cases with LVH, they found a true accuracy of only 60% [43]. This is therefore the closest to "truth" that the computer could ever get on the strength of this criterion [8]. In other studies, computer results were compared to interpretations of human readers who used their own, individual diagnostic rules [44–47]. As had to be expected, percentages of disagreement rose sharply, reaching almost 50% in one early study [48].

Comparisons with ECG-independent clinical evidence were also carried out in due course [8, 49–53]. These investigations were mostly limited to one or two programs only. Moreover, they are difficult to compare since they utilized different databases, most of them collected within a single institute. More and more the need was felt for common, objective, impartial reference standards to test and improve different ECG computer programs [14, 54–58]. For this purpose, well-established test libraries are indispensable, together with well-defined measurement rules and evaluation procedures. On several occasions, this necessity was stressed, as at the first IFIP Working Conference on Computerized ECG Analysis [54] or in the editorial of Pipberger and Cornfield [15]. Detailed guidelines for program testing were presented at the 1977 Bethesda Conference on Optimal Electrocardiography [14].

In 1979, a concerted European action was started to develop "Common Standards for Quantitative Electrocardiography" (abbreviated: CSE). The CSE Working Party consisted of investigators from 25 institutes in Europe with the participation of investigators from six North-American and one Japanese center.

The CSE project was divided into two 5-year periods, the first dealing with measurements, the second with diagnostic interpretation [59]. The main objectives of the first part of the CSE project were, firstly, to establish standards for computer-derived ECG measurements, which implied agreement on definitions of waves and amplitude reference levels, and, secondly, to compare the results of measurements from different programs.

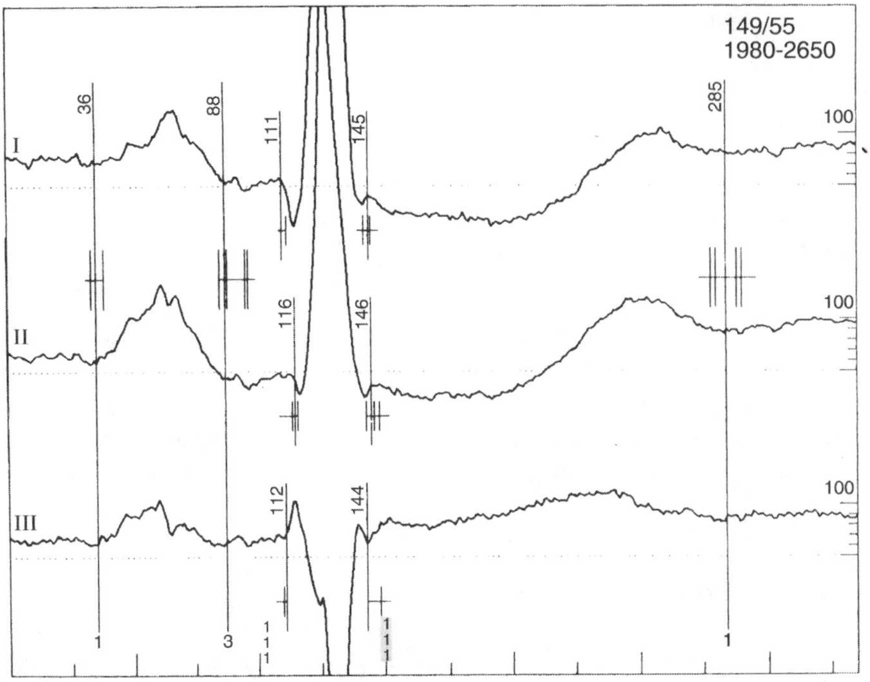

Figure 37.1

Example of wave boundary determination by five cardiologists (from Willems et al. [62]). The *short vertical lines* depict the individual, sometimes coinciding, referee estimates, the long ones the median results. The adjacent values denote the sample point locations relative to the onset of the recording window. The figures at the *bottom* of the *vertical lines* indicate the final reviewing round in which the estimates were obtained. Note that the onset of QRS in leads I and II is at a lower level than the onset of P, due to the presence of the atrial repolarization wave. In lead III the projection of this wave is isoelectric. Also, in lead III the T wave ends in an isoelectric segment. The observers disregarded this when they established T offset over the three leads simultaneously.

Thanks to this first CSE study, two reference databases came into being [60, 61]. They were composed of normal and a variety of abnormal cases. The first database [60] comprised 250 original and 310 so-called artificial ECGs. The recordings contain four lead groups of three simultaneously recorded leads each, as was common for most electrocardiographs at the time. From the original ECGs, companion artificial ECGs were constructed by concatenating one selected beat per lead group into strings of identical complexes. This artifice makes it possible to compare programs with different approaches for deriving a representative beat. To verify the effect of beat-to-beat variability, out of 30 original ECGs, two additional beats were selected to form an extra 60 artificial records. A group of five cardiologists determined the points of onset and offset of the various ECG waves (P, QRS, and T) on much enlarged tracings in an iterative, four-round, Delphi review process (➋ Fig. 37.1).

The second database [61] comprised 250 original and 250 artificial ECGs in which all 15 leads – the 12 standard ECG leads plus the Frank XYZ leads – were recorded simultaneously. In view of the work load involved, the visual analysis strategy could not be repeated. Instead, the median wave-recognition results of 11 ECG and six VCG programs participating in the CSE study were taken as the reference. The cardiologists reviewed a random set of about 20% of the cases in a two-round process, and it was shown that the median of the program results was almost identical to the median results of the referees [61].

Both for the three-lead and multilead database, the original and corresponding artificial ECGs were divided over two sets. The waveform reference results of one set have been released [63, 64]; the results of the other set remain under lock and key at the CSE coordinating center for independent testing.

Differing mathematical algorithms may lead to similar solutions in pattern recognition [34]. For that reason, it was not seen as fitting to propose any specific algorithm as the exclusive standard for ECG wave recognition. At the same time, the CSE reference databases are strongly recommended as a bench mark for ECG measurement programs. Comparison with the standard involves two quality measures: a program should approach the reference as closely as possible, and the standard deviation of the differences of its results with respect to the reference should not exceed certain limits. These limits have been described in detail in a paper with recommendations for measurement standards in quantitative electrocardiography [34].

The second part of the CSE project aimed at the assessment of the diagnostic performance of ECG computer programs. This study commenced in 1985 and finished in 1990. A database consisting of 1,220 multilead recordings was collected, comprising seven diagnostic groups: normal (n = 382); left (n = 183), right (n = 55), and biventricular (n = 53) hypertrophy; and anterior (n = 170), inferior (n = 273), and combined (n = 73) myocardial infarction. Also some cases with both infarction and hypertrophy (n = 31) were included, but ECGs showing major intraventricular conduction defects were not. The clinical diagnosis was documented by ECG-independent evidence, such as cardiac catheterization, coronary arteriography, echocardiography, cardiac enzymes, and patient history [65]. The ECGs and VCGs were analyzed by 15 computer programs and by nine cardiologists, from seven European countries [66]. Evaluation results will be discussed later.

The CSE reference libraries have become an international standard for the evaluation and improvement of ECG and VCG computer programs. A comprehensive overview of the CSE project appeared in 1990 [59].

37.6 Data Acquisition

Before the ECG signals are transmitted to a computer system for analysis, the operator has to enter patient identification data. The patient's name, sex, and date of birth together with an identification code are essential for proper automatic handling of the ECG data. Data like weight, height, blood pressure, and medication might be useful – if they are not based on guess-work by a lackadaisical technician! The system will also store the time and date of the recording, as well as, mostly, a technician and location code. For programs that perform serial ECG analysis, the unique patient identification code is indispensable for record linking.

In order to achieve optimal performance of an ECG-processing system it is essential that the data used in the analysis are of good quality. In practice, ECG records can be disturbed by power line interference, baseline wander caused by electrode polarization, electromagnetic muscle artifacts, spikes, sudden baseline shifts due to electrode contact interruption, and amplitude saturation. Although automated systems can achieve a great deal of signal conditioning (see the following paragraphs), it devolves upon the operator to correctly apply the electrodes and to detect and remedy signal errors and disturbances before entering them in the system, according to the maxim "garbage in, garbage out." Power-line interference can be prevented or reduced by proper shielding and grounding, by appropriate skin preparation and electrode application. Muscle tremor can be reduced or removed by having the patient comfortable and relaxed, while respiratory baseline wander can be minimized by breath holding.

The front-end of an ECG processing system consists of analog amplifiers. Present-day technology even allows them to be incorporated in the electrodes and to transmit the signals by a local wireless system to the processing unit. The amplifiers must be of adequate bandwidth for a faithful rendering of the signals. For further technical details, the reader is referred to ➋ Chap. 12.

A digital computer can handle data only in numerical form, and therefore, the original continuous (or analog) voltage variations of ECG signals need to be converted into a series of numbers, corresponding to the voltage levels of the leads at any moment of time. General principles concerning analog-to-digital (A-D) conversion of the ECG have been reviewed by Berson et al. [67, 68]. The time and amplitude resolution of the A-D conversion is dependent on the sampling rate, the word size of the converter and the dynamic range of the analog input.

Sampling rates for ECG data in various programs originally varied from 250 to 500 samples per second. The American Heart Association [69] and the CSE Working Party [34] have recommended for clinical application a high-frequency cutoff of the analog amplifiers of 150 Hz and a sampling rate of 500 Hz for the A-D conversion. In the pediatric ECG, the waves, especially the QRS, are of shorter duration at often elevated voltage which results in high rates of acceleration and deceleration of the signal with corresponding high frequency content. This requires an extension of the minimal high-frequency cutoff to 250 Hz [70].

The more bits per word in an A-D converter, the greater the possible resolution. In addition, the dynamic range of the amplifiers must be wide enough to accept the largest possible excursion of the signal amplitude. Modern electrocardiographs have 16 or 18 bit converters, with quantization levels of 1.22 μV–4.88 μV for the least significant bit.

Until the introduction of "stand-alone," microprocessor-equipped systems for ECG analysis, ECG records needed to be transmitted to a central computing facility. This was possible by transferring magnetic tapes on which ECGs were recorded, or by sending the signals on-line over ordinary telephone lines. This involved frequency modulation and demodulation which noticeably contributed to recorder noise. In the present digital era, the total analysis of the ECG can be done on a stand-alone electrocardiograph. The central ECG management system is required for storage and retrieval, a necessary utility for serial analysis; as well as for providing a means of editing interpretations. Interchange with the central facility may occur through a local area network, possibly by wireless transmission, or the ECGs are conveyed on some digital storage medium.

In the acquisition device, all 12 leads are acquired simultaneously so that time coherence between the signals is preserved. In practice, eight independent leads are recorded (I, II, VI-V6) and the other four are reconstructed from leads I and II using the classical relations such as lead III = II – I. In some systems, additional leads may be entered simultaneously for the VCG or for other purposes. The signals are stored in memory and can be displayed for a visual quality check. The computer may interact with the operator and warn of excessive noise in a lead or indicate electrode reversal. Reversal between right and left arm is detected by most programs. Other forms of electrode interchange require more sophisticated approaches [31, 71]. After approval by the operator, the signals are processed.

While the electrodes may have been placed in the correct order, without interchange, they still may be ill-positioned, for example, one interspace too high or too low on the chest. These errors are almost refractory to detection and correction [72, 73].

Conventionally, 12-lead ECGs were presented on multichannel writers in four lead groups of 2.5 s, often with a 10.0 s rhythm strip, on one page. Sometimes 5.0 s was taken per lead group, requiring two pages, or 6-channel writers were used. With the entire ECG present in memory and the availability of a thermal writer or laser printer, a modern device will allow considerable flexibility of output format. The signals, in addition, are preprocessed by digital techniques in such a way that mains interference and baseline drift have been reduced to a minimum to produce a neat paper output.

37.7 Signal Preprocessing and Conditioning

ECG records may be disturbed by different types of artifacts as enumerated previously, namely, power-line interference, baseline wander, muscle noise, spikes, and amplitude saturation. The aim of the preprocessing stage is to detect and correct these artifacts as far as possible. In hopeless cases, rejection of a part or the whole of the recording may be inevitable. The problems of signal improvement have been a challenge to many capable technical minds and have resulted in an abundant literature on the subject. Some of the algorithms require that the locations of the QRS complexes are known, and thus are not properly part of the preprocessing stage. This will be mentioned when the case occurs. The evaluation of the performance of these algorithms applied to real ECGs is not straightforward since the undisturbed signals are unknown. This may in particular be an issue in treating baseline wander and muscle noise. A common approach to studying the problem is to add simulated noise to a clean, "noise free," ECG signal. This approach can be carried one step further by simulating both noise and signal. As with all simulations, the validity of the evaluation results will depend on how well the simulated signals mimic those that occur in practice.

37.7.1 Power-Line Interference

Power-line interference is a common problem and is characterized by its periodicity of 50 or 60 Hz (higher harmonics may also be observed). Different digital filters have been proposed for the removal of power-line interference [74, 75]. They can be categorized in three types:

1. Notch filters, which attenuate frequencies in a narrow frequency band around the interference frequency [76, 77].
2. Global filters, which make a single estimate of the interfering noise over the total duration of the signal and then subtract this estimate from the signal. Cramer et al. [75] describe two such global approaches. One is based on a least-squares error fit of the interference. The other approach requires the sampling rate to be an integer multiple of the power-line frequency, say n, and calculates an average amplitude for each of the n phase points in one period of the interference. Both approaches only perform well if the line frequency is stable within 0.02 Hz of the nominal frequency [75], which in practice may often not be the case. Also, these methods, by definition, cannot adjust to amplitude changes of the interference. Levkov et al. [78] proposed estimating the interference from each isoelectric or slowly changing part of the ECG signal, which allows tracking of interference amplitude changes. To ensure that the sampling rate was an integer multiple of the power-line frequency, they used a hardware-synchronized A-D converter. More recently, a software solution to adjust for interference frequency variation was proposed, involving estimation of the interference period and resampling of the ECG signal [79]. An extensive review of the method is given by Levkov et al. [80].
3. Adaptive filters, which use an auxiliary reference signal containing the interference alone [74, 81–83]. This reference is adaptively filtered to match the interfering sinusoid as closely as possible, and is then subtracted from the primary signal. An interesting variant of the classical adaptive filter approach is the "incremental estimation" filter proposed by Mortara [84] and further investigated by others [74, 85–87]. The filter generates a prediction of the power-line interference $v(n)$ contained in the signal $x(n)$, based on the noise estimate $w(n)$ at the previous two samples:

$$v(n) = 2\cos(2\pi f/f_s)\, w(n-1) - w(n-2),$$

with f the nominal frequency of the interference and f_s the sampling rate. If $v(n)$ is a good prediction, then the difference $x(n) - v(n)$ should be zero apart from a possible constant offset. Assuming a slowly changing signal, the difference $x(n-1) - w(n-1)$ is taken as an estimate of this offset, and the error

$$e(n) = (x(n) - v(n)) - (x(n-1) - w(n-1))$$

indicates how well $v(n)$ predicts the interference amplitude. A final estimate is then produced by incrementing or decrementing the value of $v(n)$ with a fixed amount δ, depending on the sign of $e(n)$:

$$w(n) = v(n) + \delta \operatorname{sgn}(e(n)).$$

Finally, the output $y(n)$ of the filter is

$$y(n) = x(n) - w(n).$$

The value of the increment δ is chosen heuristically. A value which is too small may cause sluggish adaptation to the interference amplitude and poor tracking of its changes. A value which is too large, on the other hand, may cause the estimate $w(n)$ to jitter around the power-line interference, introducing extra noise. Also, since the assumption of a slowly changing signal does not hold true for the QRS complex, large increments will result in ringing artifacts. An incremental value of 1.25 μV was proposed by Mortara [84]. Using a few simplifying assumptions, Talmon [85] analyzed the relationship between the amplitude of the sinusoidal disturbance, the increment, and the bandwidth of the filter. Glover [88] showed that the filter reduces to a standard notch filter if the nonlinear sign function in the update equation is replaced by a linear increment function.

A number of different filters, representative of the above filter types, were compared by McManus et al. [74]. They applied the filters both to artificial test signals simulating various forms of interference (including second and third harmonics of the nominal frequency), and to a small subset of real ECGs from one of the CSE databases. Tested against a list of 14 desiderata for an ideal interference filter, no single filter consistently performed better than the others for all requirements. Remarkably, the incremental estimation filter and the global filter were the only ones that did not produce a ringing effect at the end of the QRS as is the usual accompaniment of large-amplitude QRS complexes when filtered, as illustrated by ➋ Fig. 37.2. In another study [87], the incremental estimation filter was compared to a nonadaptive second order notch filter. The better transient behavior of the adaptive filter produced less distortion in the ST segment and removed the interference more effectively.

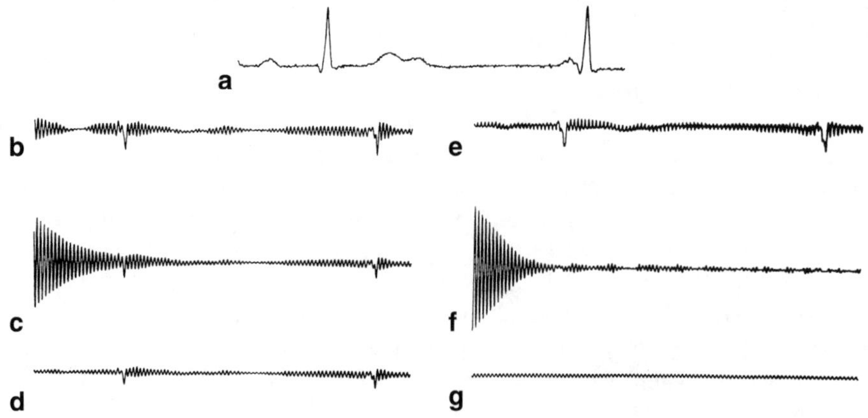

■ Figure 37.2

Effects of different interference removal filters on an ECG, which has no interference: (a) original ECG, and tenfold magnified differences between filter-input and output of two notch filter methods (b and c), three adaptive methods (d, e, and f; f is the incremental estimation filter), and a global method (g). Several filters can be seen to generate sizable differences during the QRS complex (From McManus et al. [74]. © Elsevier. Reproduced with permission).

37.7.2 Baseline Wander

Baseline wander is another annoyance. The source of low-frequency fluctuations of the baseline may be sought in changing electrode impedances, such as caused by respiratory movement. More abrupt changes may result from the patient being restless. The frequency content of baseline wander is typically less than 0.5 Hz. Baseline wander may severely disturb ECG beat morphology. A variety of techniques for estimating and removing baseline wander has, therefore, been developed.

In the 1975 recommendations of the American Heart Association (AHA) [89], for baseline wander removal a standard 0.05 Hz single pole high-pass filter was advised. While such a filter largely avoids the problem of phase nonlinearities, it does very little to suppress the baseline wander that can typically be observed in ECGs. In their 1990 recommendations, the AHA stipulated that a baseline removal filter should have a flat amplitude response within 0.5 dB from 1 to 30 Hz, with a −3 dB cutoff frequency of less than 0.67 Hz, and should adhere to certain test criteria based on triangular and rectangular wave impulse responses [69, 90].

A general problem in the evaluation of baseline correction methods is the difficulty in discriminating between baseline wander and the genuine ECG [85, 91]. Most studies only provide a qualitative assessment, showing ECGs before and after application of a correction algorithm. In a few studies, artificially generated baseline wander is added to "clean" ECGs constructed by concatenating identical beats [92, 93]. Since baseline correction may introduce new distortions, several algorithms try to identify periods with minimal or no baseline wander and skip these periods in the filtering [85, 93, 94].

Two main remedial approaches have been followed:

1. Interpolation.

Linear interpolation approximates the baseline by straight-line segments between isoelectric levels, usually estimated from the intervals preceding QRS onset [85, 94]. The estimated baseline is subsequently subtracted from the ECG signal. Boucheham et al. [95] proposed a piecewise linear correction based on "dominant" points as detected by a curve simplification algorithm [96]. An interesting feature of their algorithm is its capability to correct for sudden baseline shifts.

A more elaborate interpolation method estimates the baseline by a third-order polynomial, or cubic spline [97]. Each PR segment provides a "knot" through which the cubic spline must pass. Meyer [97] described an elegant and fast state-space approach for the computation of the cubic splines.

Talmon [85] compared the linear interpolation and cubic spline methods on a set of real ECGs, and concluded that both approaches performed similarly. Linear baseline correction, however, is to be favored because cubic spline correction is more difficult to apply in the presence of sudden baseline shifts.

Since all interpolation methods assume that reference points or knots can accurately be determined, they may break down if this assumption is not met, for example, in the case of a premature beat merging with the preceding T wave [98–100]. Performance may also degrade at lower heart rates as the knots become more separated.

2. High-pass filtering.

Infinite impulse response (IIR) filters are generally unacceptable due to their nonlinear phase response, which may induce distortions in particular in areas of the ECG where amplitudes change abruptly [90]. However, in off-line situations, or if some time delay between acquisition and processing of the signal is accepted, forward-backward or bidirectional filtering can be applied [100, 101], yielding overall linear-phase response [102]. Alternatively, De Pinto [103] proposed a linear-phase high-pass filter. He subtracted the output of a linear phase low-pass IIR filter from the original signal with a delay equal to the pass-band group delay of the low-pass filter. The filter was shown to adhere to the 1990 AHA recommendations.

Linear phase filtering is easily accomplished with finite impulse response (FIR) filters. However, these filters typically have very long impulse responses, resulting in many multiplications and long time delay [91, 104], which may be unacceptable for short-term resting ECG recordings. Van Alste et al. [104] proposed an FIR filter that combines removal of baseline wander with that of power line interference, and greatly saves on the number of computations. QRS complexes may heavily influence the baseline estimate. This may cause a shift of the assumed baseline with respect to the true one, resulting in measuring errors especially in diagnostically sensitive low-frequency segments such as the ST segment. Sörnmo et al. [92] described an approach in which the QRS complexes are removed prior to filtering the signal by one of a bank of linear low-pass filters with variable cut-off frequencies. The method was tested on ECGs with different types of simulated baseline wander and showed superior performance as compared to standard high-pass filtering or cubic spline interpolation (➋ Fig. 37.3), especially when the baseline wander contained frequencies >0.5 Hz. It must be noted that the method requires that beat classification is performed prior to correction, to minimize the effects of ectopic beats.

A two-step method combining interpolation and filtering techniques was proposed by Shusterman [93]. First, the magnitude of baseline wander is determined and classified as small or large. If large baseline drift is present, the signals are filtered with a bidirectional high-pass filter with a cut-off that depends on the estimated frequency content of the baseline wander. In a second step, any residual, small baseline wander is removed by simple linear interpolation between PQ and TP segments.

A two-stage cascade filter was described by Jane et al. [105]. First, a high-pass notch filter is applied with a cut-off frequency at 0.3 Hz [83]. Any remaining baseline contamination at higher frequencies is then removed with an adaptive impulse correlated filter. The reference input consists of a sequence of unit impulses correlated with the QRS complexes. This filter was shown to be equivalent to an exponentially weighted average [106] and requires a QRS detector. It also removes other disturbances not correlated with the QRS complex, such as muscle noise or line interference. When tested on a few records of the MIT-BIH database [107], the filter was shown to perform better than cubic spline correction.

Park et al. [108] described a wavelet adaptive filter. The filter consists of two parts. A wavelet transform decomposes the ECG signal into seven frequency bands. The signal of the lowest frequency band (0–1.4 Hz) is then fed into an adaptive filter. The wavelet adaptive filter was compared with a "commercial standard filter" with a cutoff of 0.5 Hz and with a general adaptive filter. Using test data from the MIT-BIH and the European ST-T [109] databases, the wavelet filter was shown to perform better than the other two filters, especially with respect to distortion of the ST segment. When tested on a triangular wave, as recommended by the AHA, the wavelet filter showed negligible distortion of the ST segment, whereas the standard filter and the adaptive filter produced severe distortions.

Finally, nonlinear filtering methods for baseline correction have been proposed [110]. Sun et al. [111], building on earlier work of Chu and Delp [112], described an approach using morphological filtering, a technique widely used in the field of image processing [113]. Testing on simulated ECG signals, they compared their approach with the wavelet adaptive filter [108] and found morphological filtering to produce better results.

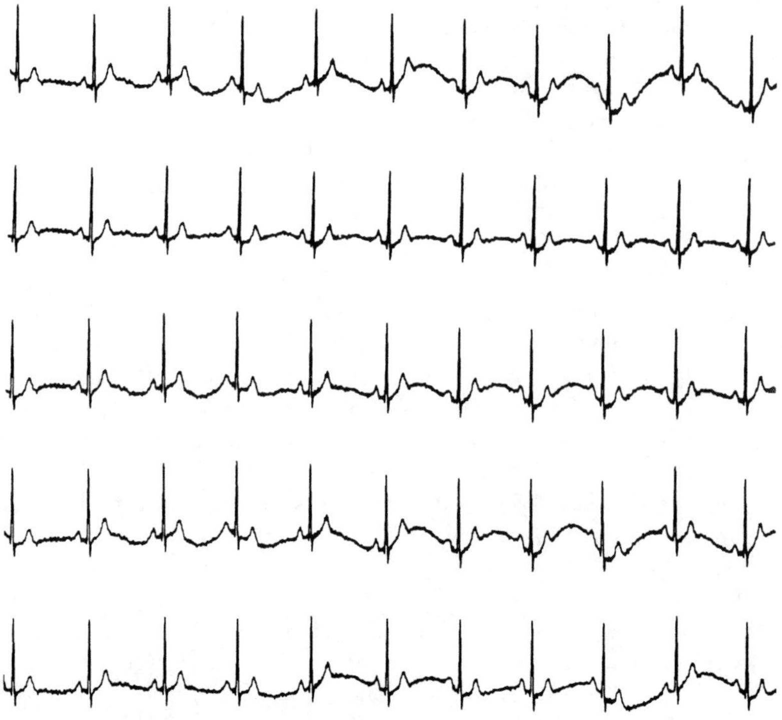

Figure 37.3
Example of baseline wander removal. (a) Original ECG, and resulting signal after baseline wander removal using (b) time-varying filter with beat subtraction, (c) time-varying filter without beat substraction, (d) time-invariant filter, and (e) cubic spline interpolation (from Sörnmo [92]. © Springer).

37.7.3 Muscle Noise

Muscle noise, another signal deformity, is caused by the electrical discharges of skeletal muscles. Common causes are patient restlessness, nervousness, cold shivers, and Parkinson tremor. Reduction of muscle noise is often effected by one of the other tasks in ECG signal processing as, for example, when the ECG signal is band-pass filtered for the purpose of QRS detection. Since most of the energy of the QRS complex is contained in the frequency band from 10 to 25 Hz, noise components outside the pass band can effectively be suppressed. For P-wave detection, even stronger noise reduction is possible because the frequency content of the P wave is lower. In these applications it is unnecessary to keep the original signal undistorted. For accurate amplitude and duration measurements, however, the requirement is to improve signal-to-noise ratio (SNR) without loss of signal information.

A common method to reduce muscle noise is coherent averaging. ECG complexes of one family are summed and the sum is divided by their number, giving an averaged ECG complex while uncorrelated noise averages out and disappears. Coherent averaging is one way of improving the SNR, as will be discussed later.

A number of other noise reduction techniques have been proposed. Adaptive filtering for muscle noise suppression was described by Thakor and Zhu [83]. To cancel the noise in a particular lead, they proposed employing, as a reference signal, another lead perpendicular to the first, in order to ensure that the noise in the two leads is uncorrelated.

Talmon et al. [114] describe an adaptive Gaussian filter, based on earlier work of Hodson [115]. The frequency characteristics of the filter are dependent on the estimated curvature of the signal to be filtered. A curvature estimate is obtained by fitting a polynomial function. The width of the filter is then adjusted according to the curvature, where low curvature (e.g., in the PR interval) implies a low-frequency cutoff, and a high curvature (e.g., in the QRS complex) a cutoff at higher frequencies. A related approach is proposed by De Pinto [103], who describes a low-pass filter with time-varying bandwidth. The bandwidth is maximal during the QRS complex, and decreases in the interval between QRS complexes. The bandwidth is varied in increments by selecting one of six sets of coefficients, controlled by an estimate of the slope of the ECG signal.

Another approach, termed noise consistency filtering, was initially proposed by Mortara [30] and later investigated by Wei et al. [116]. This filter method requires the availability of multiple, simultaneously recorded leads, and exploits the redundancy in the ECG signals. Reconstruction coefficients are determined to synthesize each of the eight independent ECG leads from the remaining seven leads. The filter output is the original lead signal multiplied by a time-varying coherence measure with a value between 0 and 1, dependent on the correspondence between the predicted and the original signal. The lower the coherence, the stronger is the filtering, and vice versa. The filter was reported to reduce noise by a factor of 10, provided the noise in the leads is uncorrelated [30].

Wei et al. [116] proposed two modifications to the filter as it was described by Mortara. Firstly, to avoid problems with baseline wander, the signal is split into a low-frequency and a high-frequency part, and only the latter is filtered according to the source consistency method. Secondly, a modified coherence measure is used, to reduce unwanted filtering effects that were observed by the authors. The modified filter performance was verified on simulated and real ECG signals, and compared with the original source consistency filter and conventional low-pass filters. Results show more effective noise suppression and less distortion of the QRS complex with the modified filter.

More elaborate filtering methods, involving the discrete cosine transform and singular-value decomposition, have been described by several investigators [117–119]. These approaches are computationally demanding and were mainly developed for the suppression of excessive muscle noise during exercise testing.

Finally, several studies used morphological filtering for muscle noise suppression [110–112]. Morphological filtering was also employed to detect muscle noise, without attempting to suppress it [120].

37.7.4 Spikes

Spikes are sudden pulses of short duration and high amplitude. They may be due to an interfering electrical source in the environment or to an implanted artificial pacemaker.

Accurate pacemaker spike detection has become increasingly important with the growing population of patients with implanted pacemakers. Because of the small spike width, in the order of 0.2–0.5 ms, many electrocardiographs use high-bandwidth front end amplifiers and analog circuitry to detect pacemaker spikes before A-D conversion. To reduce the number of false detections that this approach may induce, it has been combined with software algorithms that must confirm the presence of the spikes in the diagnostic bandwidth (0.05–150 Hz) signals. Alternatively, signals have been sampled at very high frequency, of the order of several kHz, and then software is used to detect the spikes.

Only a few algorithms have been described that attempt to detect spikes in diagnostic bandwidth signals. As these signals have been sampled at no more than 500 Hz; one may wonder how it is possible to detect pulses that have a duration well below the sampling interval. The reason is that the anti-aliasing low-pass filtering broadens the spikes. However, this filter also greatly reduces the amplitude of the spikes, which makes detection more difficult. The presence of narrow QRS complexes mimicking spikes, high-frequency noise and other artifacts further complicates the detection task.

The Louvain VCG analysis program [121] used a simple spike detector: if the spatial velocity exceeds a fixed threshold, a spike is assumed to be present [122]. In the AVA program [123] a number of tests based on slope differences between four consecutive points (8 ms) in a single lead were performed to detect spikes or discontinuities in the input signals [122]. No evaluation of these methods has been given.

Talmon [85] described a spike detector that operates in two stages. First, signals are filtered with a parabolic filter and the root-mean-square (RMS) of the residuals is computed. If the residual at a certain time instant exceeds three times the RMS value in that signal, a potential spike is assumed to be present. In a second stage, an additional number of criteria, structured in a decision tree, are tested to verify whether a spike has truly been detected. The algorithm was tested on an independent test set of 1,908 ECGs and VCGs, showing a sensitivity of 90.9% and a positive predictive value of 95.4%.

Helfenbein et al. [124] proposed another pacemaker spike detector. The algorithm detects a spike if a steep slope exceeding a threshold is followed by an opposite polarity slope within a short time window. The threshold is adaptive and computed as a function of the maximum slope in a window preceding the spike. When tested on a set of 1,108 adult ECGs containing a variety of pacemaker types and modes, excellent performance (sensitivity 99.7%, positive predictive value 99.5%) was obtained. On another set of 1,382 non-paced pediatric ECGs, only four false-positive QRS complexes were reported.

37.7.5 Amplitude Saturation and Sudden Baseline Shifts

No substantial literature is available dealing with the detection of amplitude saturation and sudden baseline shifts. In descriptions of various ECG processing systems, a statement is made that these artifacts are searched for, but the algorithms are not described.

37.8 Detection of QRS Complexes

The detection of QRS complexes is probably the most extensively studied problem in ECG signal analysis [86, 125–131]. A host of different algorithms has been proposed, most of them originating from applications in the fields of coronary care monitoring and Holter recording. Consequently, many of these QRS-detection algorithms were designed to operate on a single lead. In computerized resting ECG and VCG interpretation, three simultaneously recorded leads were used from early times. Now, multiple simultaneously recorded leads have also become increasingly common in the non-resting ECG.

Basically, a QRS detector consists of two stages [132]: a preprocessing stage, in which the signal is filtered and signal features are determined, and an identification stage, in which a decision is made about the presence and location of a QRS complex. Algorithms are commonly distinguished with respect to their preprocessing stages.

An extensive survey of single-lead QRS-detection algorithms is given by Kohler et al. [130]. The most common approaches are based on high-pass and band-pass digital filters, but many other approaches have been proposed, based on wavelet transforms [133, 134], artificial neural networks [127, 135], and genetic algorithms [136]. Some algorithms [137, 138] employ techniques from syntactic pattern recognition, but it has been difficult to demonstrate the practical utility of this approach [139].

In the multiple-lead algorithms, the simultaneous leads are transformed to a detection function. The transformation brings out the QRS complexes amongst the other parts of the signal, in order to increase the QRS detection rate. One of the most commonly used transformations is the computation of the spatial velocity of the VCG or of a similar derived function for the 12-lead ECG. The spatial velocity (SV) is defined as:

$$SV(n) = \sqrt{\sum_{k=1}^{3} (d_k(n))^2}$$

where $d_k(n)$ denotes the first derivatives of the VCG leads X, Y, and Z. Various difference equations have been proposed to approximate the derivatives [130]:

$$d(n) = x(n+1) - x(n-1);$$
$$d(n) = 2x(n+2) - x(n+1) - x(n-1) - 2x(n-2);$$
$$d(n) = x(n) - x(n-1).$$

In the case of the 12-lead ECG, the VCG leads can be reconstructed from the ECG leads by linear transformation, or approximated by a quasi-orthogonal set of ECG leads [140]. Alternatively, a pseudo-SV has been computed by combining the derivatives of all ECG leads.

The detection signal is then gauged against a threshold to detect the occurrence of a QRS complex. The threshold may be fixed, but more commonly is adaptive, changing with varying signal characteristics [140–142]. Some algorithms also require that the detected QRS complexes fulfill certain amplitude constraints. Other algorithms compute a second derivative,

$$d^2(n) = x(n+2) - 2x(n) + x(n-2),$$

and use a combination of first and second derivatives as the detection function. Balda et al. [143] used the sum of the absolute first and second derivatives over three simultaneous leads. This detection function was called the "waveform boundary indicator" as it also provided an estimate of the onset and the end of QRS complexes. A single-lead implementation of the method was given by Ahlstrom and Tompkins [86].

Once a potential QRS complex is detected, most of the algorithms apply further heuristic criteria to exclude false-positive detections, for example, by requiring a minimum time lag between adjacent QRS locations [140]. Laguna et al. [144] apply a single-lead detection algorithm to each of a set of simultaneously recorded leads, and then enter into a decision process comparing the detection positions over all leads to decide which detections are true or not.

An inventory of different methods used in seven VCG and eight ECG programs can be found in the second progress report of the CSE project [122]. However, only few developers of these ECG computer programs have published detailed evaluation results of their detection algorithms. On a set of 2,889 QRS complexes from the CSE multilead library, Kors et al. [140] found no false positive or false negative detections at all. Contrasting with the parsimonious communications on multilead QRS-detection algorithms, performance results for single-lead algorithms have been reported in fair abundance (see [130] for an overview). Many of these algorithms were evaluated on (part of) the MIT-BIH arrhythmia database, and achieved excellent results (>99% sensitivity and positive predictive value) [130]. A comparative study on noise sensitivity of nine single-lead QRS-detection algorithms for five different types of noise also indicated very good detection performance of most algorithms for all but the highest levels of noise [126]. In general, considering that the multiple leads of the standard 12-lead ECG offer redundancy of information and that noise levels in the resting situation are typically less troublesome than under monitoring conditions, it may be concluded that near-perfect QRS detection in the resting ECG is feasible.

37.9 Detection of P Waves

Detection of the P waves remains one of the most difficult tasks in automated ECG analysis. Failure of P-wave detection will jeopardize the rhythm interpretation program. Problems may arise owing to low amplitude, variable morphology and diverse timing of P waves. A P wave superimposed on a T wave, and even more so, a P wave that coincides with a QRS complex, are difficult to distinguish in surface ECGs (➲ Fig. 37.4). Their probable location can only be extricated by considering the sequence of the preceding and following P waves and by recognizing small irregularities in the expected contour of QRS or T. This requires long and continuous recordings. Even then, problems remain in deciphering morphology and polarity. While P-wave detection on surface ECG leads may already pose problems to the eye of the cardiologist,

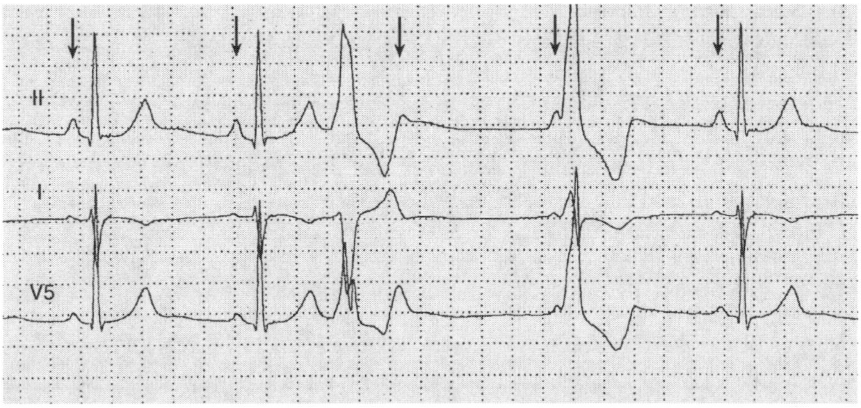

Figure 37.4
ECG showing an AV dissociation, with one P wave superimposed on a T wave and another P wave merging into a QRS complex.

more problems await the computer programmer working in this field [145]. The human ability to recognize even low-voltage P waves hidden in other waves and amidst noise and artifacts is indeed still far superior to the performance of all presently available ECG wave-recognition programs.

Different approaches with respect to automatic detection of P waves have been described in the literature. Stallman and Pipberger [6] applied a threshold approach to the smoothed spatial velocity curve. McManus [146] has worked further on this method. Bonner and Schwetman [147] used a piecewise approximation of the ECG, followed by tests on level crossings and slope changes. A two-stage detection method was introduced by Hengeveld and Van Bemmel [148], and later on refined by Talmon [85]. First, QRS-linked P waves are searched for, based on histograms of local signal extrema in the intervals preceding the QRS complexes. If they are not found, non-QRS-linked P waves are sought by cross-correlation of signal amplitudes with an empirical P-wave template. Schnyders and Jordan [149] applied an energy correlation technique on the 12-lead ECG with apparently good results. Martinez et al. [131] proposed a wavelet-based P-wave detector. Once the location of the QRS complex is found, local maxima in the wavelet transform of the higher scales (i.e., in the lower frequency bands) are sought in an RR-dependent search interval. If at least two local extrema exceed a certain threshold, a P wave is considered present. Gritzali et al. [150] used as a detection function the length transformation of a signal, which essentially is the length of the signal curve within a time window. The length transformation can be defined for more than one lead by adding the curve lengths of the individual leads. Simple thresholding of the transformation is used to detect the P waves, as well as their onsets and ends.

It can be very difficult to distinguish P waves from flutter waves. In the presence of 2:1 AV block it may be hard, even for a human observer, to choose between flutter and sinus tachycardia. In various programs, separate routines are applied to determine whether atrial flutter waves are present [37, 141]. Talmon et al. [151] described a method that detects the periodic components characteristic of flutter waves (➋ Fig. 37.5). After appropriate filtering and removal of the QRS complexes, the resulting signal is autocorrelated. Flutter waves are assumed to be present if the autocorrelation function has a sufficiently large local maximum at a lag between 150 and 300 ms (corresponding with flutter rates between 200 and 400/min) and a second local maximum at twice the lag of the first maximum. A sensitivity of 86% and a specificity of 99.9% were reported. Taha et al. [152] employed a QRST subtraction technique and spectral methods to distinguish between atrial flutter, atrial fibrillation, and other rhythms. They obtained a sensitivity for flutter of 80%, at a specificity of 98.7%. Giraldo et al. [153] tried to detect the boundaries of atrial flutter and fibrillation waves, and used the coefficient of variation of the wave amplitudes to distinguish between flutter and fibrillation. They tested their approach on 40 short signal segments with flutter or fibrillation from the MIT-BIH database, but did not include segments with other heart rhythms in their analysis. Christov et al. [142] propose an atrial flutter-fibrillation parameter that is the mean value of the filtered and rectified signal, after subtraction of QRS-T complexes. On a test set, they obtained 76% sensitivity for flutter and fibrillation combined, at 97.9% specificity.

The detection rate of P waves varies according to the different algorithms that are applied. McManus [146] reported failures from 0–3%. For the 250 ECGs in the multilead CSE library [61], Kors et al. [140] obtained a sensitivity of 98.1% and a positive predictive value of 99.9%. Oversensitive methods cause difficulties in cases with atrial fibrillation. In the AVA program [123], as well as in several other programs, a P-wave search is only performed in an interval between the end of the T wave and the onset of QRS. In addition, only one P wave can be detected per QRS complex, which practically excludes the diagnosis of AV block or AV dissociation with an atrial rate higher than the ventricular rate. Programs that do attempt to detect more than one P wave per cycle have not reported detailed results [37, 141, 154], but judging from the results of arrhythmia detection (see section on rhythm interpretation), the success rate of these algorithms must be rather unsatisfying. A confounding factor is that P waves may alter their appearance in one recording, suddenly, from one beat to the other, or gradually, and that precisely this behavior is an element in rhythm diagnosis. To our knowledge, no work has been done on P-wave typing, that is, the distinction between different P-wave morphologies in one and the same recording.

37.10 QRS Typing

QRS typing is essentially a clustering task followed by a classification task, as the case requires. The clustering attempts to distinguish between different types, or families, of QRS complexes (if more than one). Within one family the complexes are similar in QRS morphology. If more than one type has been detected, the classification task is to determine which one

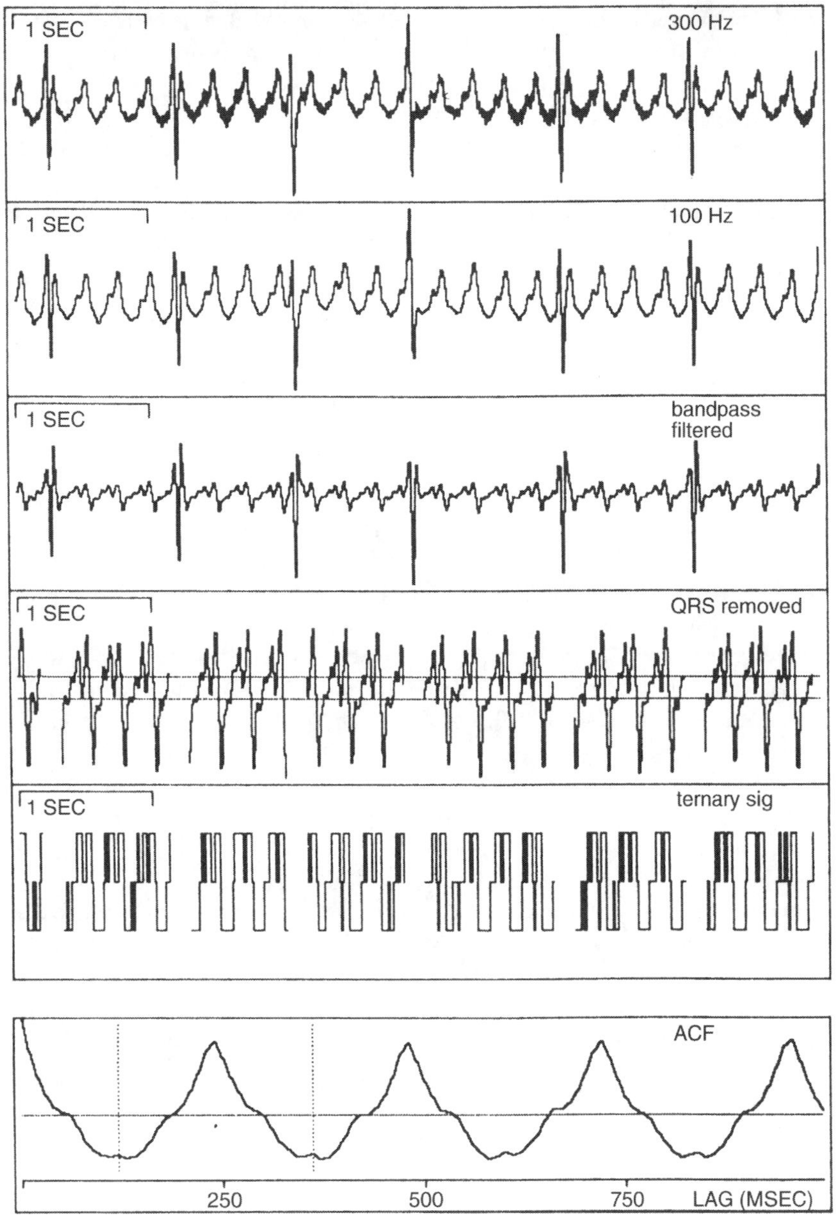

Figure 37.5

Example of the processing steps in a flutter wave detection algorithm (From Talmon et al. [151]). From *top* to *bottom*: original signal; signal after sampling rate reduction; signal after bandpass filtering; bandpass filtered signal after removal of QRS complexes; ternary signal; autocorrelation function.

is the dominant type. The term "dominant" is not synonymous with "most numerous." It may well be that in a recording the "dominant" sinus complexes are outnumbered by ectopics, for example. We define the term dominant as the indication for the family of complexes to be used for the morphological (or contour) diagnosis. The nondominant complexes may be further divided into several types, such as premature ventricular or supraventricular complexes (possibly from different foci) and escape beats. This makes it clear that correct QRS typing is indispensable for a reliable rhythm interpretation. Typing, however, is a less straightforward endeavor than one would wish. Commonly, the dominant complexes are sinus beats, but they might also be of atrial or nodal, or even ventricular origin. What if the pacemaker site changes during the recording? And, quite annoyingly, complexes from the same origin may show sudden variation in QRS morphology, for example, in intermittent right bundle branch block. Which one will be labeled as the dominant morphology? Differentiating aberrantly conducted complexes from ectopic complexes is the next problem. Also, similar QRS complexes are not always followed by similar ST-T segments, a complication that inspired only a few investigators to take a closer look at ST-T typing [85, 155].

A variety of clustering methods and features for QRS typing has been proposed. All methods compute a similarity measure between a newly presenting complex and one or more of the complexes in each of the already existing clusters. The new complex is then assigned to the group with the complexes that are most similar to it, unless the similarity is below a certain threshold, in which case a new group is formed. Different similarity measures have been proposed, such as the area between the normalized spatial velocity curves of the two complexes [156], the difference in so-called arc lengths of the complexes [94], or the Mahalanobis distance using four morphological QRS features [157]. To determine whether two complexes were of similar type, Talmon [85] used a decision tree that involved two similarity measures, reflecting similarity in shape and in power. Lagerholm et al. [158] decomposed the QRS complexes into five orthogonal Hermite functions and used the resulting coefficients, together with RR-interval parameters, as features to cluster the complexes into 25 groups by means of a self-organizing map.

When more than one type of QRS complex has been found, it must be decided which is the dominant type. As mentioned above, this may not be so self-evident since the most frequent type is not necessarily the dominant one. More elaborate decision logic has been developed [85, 141], involving QRS duration, RR interval length preceding the complex, and the number of beats of each type.

Few quantitative results on QRS typing in the resting ECG have been published. Using the ECGs from the CSE multilead library, Kors et al. [140] found an error rate of 0.3% for their multilead ECG program in classifying four types of complexes. Lagerholm et al. [158], using the MIT-BIH database with 16 different beat types, report a chance of 1.5% that a beat ends up in the wrong cluster.

Several investigators have approached QRS typing primarily as a beat classification task [127, 159–162], rather than as a clustering task, possibly followed by a classification stage. They trained one or more classifiers based on a labeled set of beats of different morphology. For example, Hu et al. [127] applied artificial neural networks to distinguish between normal and abnormal beats (involving 12 different beat morphologies), using raw QRS samples from the MIT-BIH arrhythmia database as the input. They obtained a total accuracy of 90.6%, which compared favorably with the 73.0% accuracy of a simple nearest neighbor classifier, but this level of accuracy would not seem high enough for clinical application. This may not come as a surprise considering the large interindividual variability of QRS morphology. Only when the classifier was trained on patient-specific ECG data, were more acceptable results obtained [159]. Christov et al. [162] applied a nearest neighbor rule to 26 parameters derived from two ECG leads. Again, when tested on the MIT-BIH arrhythmia database, a general classifier performed poorly, but patient-specific classifiers showed good performance. However, the patient-specific approach to QRS typing requires initial human labeling of beats and, therefore, is of little value for computer programs that should analyze the short-duration resting ECG fully automatically.

37.11 Forming a Representative Complex

Most diagnostic ECG computer programs operate on a single "representative" P-QRS-T complex from the "dominant" family [37, 121, 122, 141, 163, 164]. From this representative complex, the measurements needed for the diagnostic interpretation stage are derived. Mostly, the representative complex (sometimes called a "template" complex) is obtained by the computation of an averaged beat or a median beat from as many complexes in the record as may qualify for the purpose. Such a procedure presupposes complexes of similar morphology, and thus requires that QRS typing has been performed. The following techniques are in use:

1. Coherent or ensemble averaging is a simple technique in which the average of the time-aligned complexes is calculated at each sample point. The technique has been shown to be equivalent to low-pass comb filtering and to yield optimal noise reduction for Gaussian distributed noise [165, 166]. Under this condition, the SNR improvement is equal to the

square root of the number of complexes being averaged. Since the average is vulnerable for outliers, it is imperative that complexes affected by sudden baseline shifts or other major disturbances are excluded from averaging.

2. Another approach is finding the median value for each sample point of the time-aligned complexes. The median beat is less sensitive to baseline shifts, but low-frequency baseline wander may introduce discontinuities in the median beat and result in reduced noise suppression [85, 167].
3. To reduce the effect of averaging widely differing amplitudes, Macfarlane et al. [94] compute a so-called modal beat. They assign weights to each sample of each QRS complex, with similar amplitudes across complexes having a high weight, the others having lower weights. The approach performed well [85], but was computationally expensive. In later versions of the Glasgow program, the technique was, therefore, abandoned in favor of simple averaging or, if any individual beat significantly deviated from the average, computation of a median cycle [141].
4. A hybrid approach, trying to combine the advantages of the averaging and median techniques, was proposed by Mertens and Mortara [167]. They first split the QRS complexes in three equally-sized groups and determine the averaged complex for each group. These three averaged complexes are then partitioned in their low- and high-frequency components with a simple moving average filter, and for both signals the median complex is determined. Finally, the two median complexes are added to form the representative complex.

For all these methods precise beat alignment is essential. In several programs [122, 163, 164], synchronization is performed by a cross-correlation method. In the MEANS program [37, 85], alignment is based on a reference point within the QRS complex, the position of an extremum in the band-pass filtered signal, whereas the Louvain program [121] uses the onset of individual QRS complexes. If complexes are not accurately aligned, beat-to-beat variations in the timing between peaks in the complexes may reduce the peak amplitudes [85]. To avoid this problem, some programs select only one "typical" complex for further analysis [168]. While this approach obviously does not need beat alignment, it does not help to improve the SNR.

The single complex approach is being taken a step further when wave recognition and measurements are done on each individual complex followed by computing an average or median of the measurements across the complexes [123, 143, 154, 169, 170].

There has been some discussion on the merits of signal averaging versus measurement averaging. Talmon [85] was not able to show any significant difference in QRS duration if wave recognition were performed on single complexes or on an averaged complex, after proper beat alignment [85]. The averaging method, however, offers the advantage of increasing the SNR. Based on results of extensive noise tests performed in the CSE project, a measurement strategy that uses selective averaging has been recommended for diagnostic ECG computer programs [171, 172].

37.12 Waveform Recognition

The goal of waveform or boundary recognition is to determine the inflectional points P onset, P end, QRS onset, QRS end, and T end, as much as possible in conformity with their visually determined counterparts (➋ Fig. 37.6). No attempt is made in any ECG computer program to determine the end of the U wave, the orphan wave of electrocardiology. The CSE Working Party [34] stated that the inclusion or exclusion of isoelectric segments in the initial or terminal parts of a wave may lead to differences in wave duration of more than 10 ms between leads, and, therefore, urges that "the true onsets and offsets of P and QRS as well as the end of T should be determined using at least three simultaneously recorded leads." (cf. ➋ Fig. 37.1) The end of T deserves special attention in this respect. In 1990, Campbell and his group introduced the concept of QT dispersion (QTd) [173], defined as the difference between the longest and the shortest QT in any of the 12 surface ECG leads. QTd was supposed to reflect the dispersion of repolarization within the myocardium. The idea caused a tidal wave of papers but in due course became severely criticized [174–177]. Differences in QT duration will arise from differences in the projection of the spatial T vector onto the different leads, which in some leads may result in terminal isoelectric segments of various lengths and hence in shortened QT [174]. Also in QTd measurements, the U wave was regarded as a nuisance, obscuring the end of T. If it were too prominent, the lead was simply excluded from QTd determination. The U wave, however, might deserve more attention than this. Viewed in simultaneous leads, it may be seen in some as a separate wave, detached from the T; in other leads it may encroach on the T wave, and in still other leads it may blend in with the T wave. In fact, searching for the common end of the T wave is not pertinent if T and

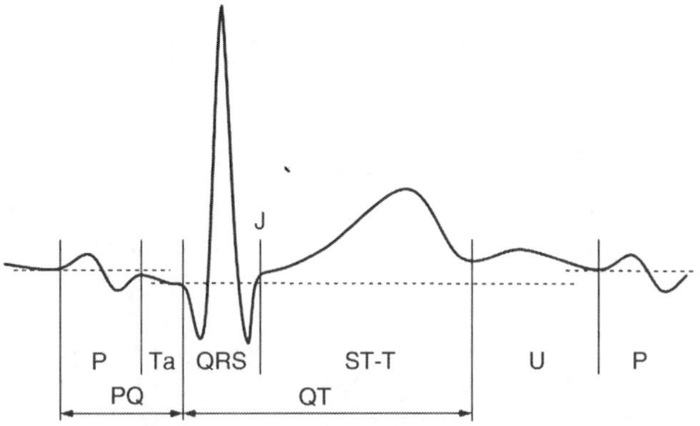

Figure 37.6
In the waveform recognition, onset and offset of P, QRS, and T are determined, as indicated by the *vertical bars*. QRS offset and J are identical points. A true 0-level cannot be identified: the onset of QRS is superimposed on the atrial repolarization wave (Ta), at the J point ventricular repolarization is already underway, and the end of T coalesces with the U wave, the end of which is hidden in the next P. As the operational 0-level for QRS and T the *horizontal line* through QRS onset is recommended, likewise for the P the horizontal through P onset. Amplitudes are measured with respect to these 0-levels.

U form a continuum as has been proposed [178], in the same way as it is impossible to ask for the common end of a Q wave in the QRS. Whatever the ultimate truth, it seems best for the time being and for practical purposes, for example, for QT duration measurement in drug safety testing, to let the computer determine T offset over multiple simultaneous leads.

As pointed out above, single lead, "one-dimensional" measurements systematically produce shorter measurements of wave duration than when, "multidimensionally," the first onset and latest offset in any lead are taken. Most ECG computer programs, therefore, use the spatial velocity or a similar multidimensional function as the detection function for waveform recognition (see Sect. 37.8). For single-lead recordings, other detection functions, such as the envelope of the ECG signal [179] or wavelet transforms [131, 133], have been applied. Basically, three different approaches of boundary recognition can be distinguished [180]. Each has to be trained and tested by human observers.

1. Thresholding

The most straightforward method is to apply a simple threshold to the detection function. For the onset, usually the maximum of the detection function is localized from where it is traced backward until where it first becomes smaller than an absolute or relative threshold. This point is considered to be the wave onset. The end of a wave is determined similarly.

According to an inventory of waveform-recognition methods used in various ECG computer programs in the CSE project [122], most programs applied threshold detectors. A number of different thresholding algorithms to detect the end of the T wave in single leads were described and compared by McLaughlin et al. [181, 182]. Using a threshold detector, Vila et al. [183] attempted to determine not only the end of the T but also the end of the U wave. They mathematically modeled the TU complex and used the modeled signal for detection. Laguna et al. [144] applied a single-lead threshold detector to each of the leads of multilead ECGs and then combined the wave boundaries in the individual leads to find overall onsets and offsets of P, QRS, and T.

2. Signal matching

A second approach is to search for the point where the weighted least-square difference between a reference waveform and the detection function is minimal [184]. The reference waveform is obtained from a learning set of detection functions with known wave boundaries, as indicated by one or more human observers. Rubel and Ayad [184] compared the performance of a signal matching algorithm for different detection functions using the CSE database. They obtained the best results for QRS onset and offset detection with an unfiltered spatial area detection function.

3. Template matching

The template method takes into account information on the time-amplitude distribution of the detection function in a window around the inflectional point [185]. A template is constructed from a series of detection functions in which the wave boundaries have visually been assessed by a human observer. The boundary point in a new ECG is then located at that point where the cross-correlation between the template and the new detection function is maximal. Details of this approach and an efficient way of implementation have been described by Talmon [85].

A somewhat different approach, obviating the need for thresholding or matching, was proposed by Zhang et al. [186]. As a detection function, they took the output of a signal integrator in a window sliding over the T wave. With recourse to some simple assumptions, the maximum of the detection function was shown to coincide with the end of the T wave. A model-based approach to delineate single-lead QRS onset and offset was described by Sörnmo [187]. He statistically modeled the low-frequency segments (containing P and T waves) and high-frequency segments (QRS complex), and applied a maximum-likelihood procedure to estimate the points where a change occurred between low- and high-frequency segments.

Using the multilead CSE database, Willems et al. [61] evaluated the measurement performance of 11 ECG and 6 VCG computer programs (Fig. 37.7), incorporating a variety of wave boundary detection algorithms [122]. Onset of QRS showed overall the smallest deviations from the reference, with the narrowest confidence intervals. End of T showed the largest scatter. The median of the waveform recognition results of all programs coincided best with the median results of referee cardiologists, corroborating findings of the former 3-lead CSE study [60]. However, individual program results

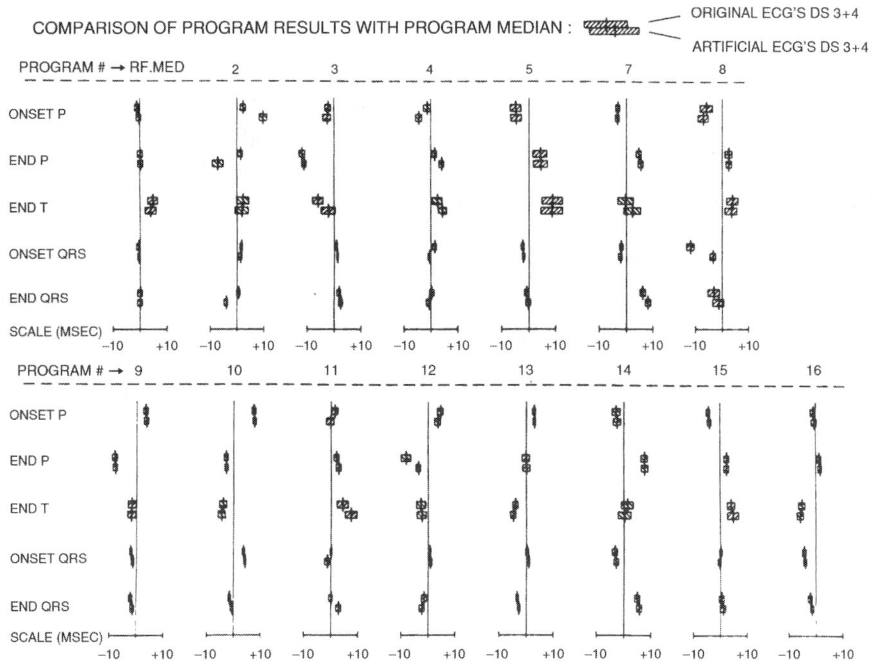

Figure 37.7

Comparison of individual program results (numbered 2–16) and median referee results (RF MED) with the median of all programs in the CSE multilead library. Mean differences and 99% confidence intervals are indicated by *small vertical lines* and *horizontal bars*, respectively (From Willems et al. [61]. © American College of Cardiology. Reproduced with permission).

were widely divergent. As a consequence, P, PR, QRS, and QT interval measurements also varied widely among the various programs.

Willems et al. [171] assessed the influence of noise on wave boundary recognition of eight ECG and six VCG programs. Seven different types of high- and low-frequency noise were added to each of ten recordings. Mains interference and baseline wander had no significant effect on boundary detection for the majority of programs, but increasing levels of high-frequency noise shifted the onsets and offsets of most programs outward. Programs which apply beat averaging techniques had more stable results than programs analyzing single beats, but it was noted that these results mainly occurred at high noise levels that reflect poor operational conditions and can be avoided by proper quality control [171]. In a second related study [172], the effect of noise on amplitude measurements was examined. Programs that showed the least variability in waveform onsets and offsets also exhibited the highest stability in waveform detection and amplitude measurements.

Laguna et al. [144] compared the waveform boundary results and interval measurements obtained by their own algorithm with the median program estimates and median referee estimates of the multilead CSE database. The standard deviations of the differences were shown to be within acceptable tolerance limits [34] for most of the measurements. Martinez et al. [131] used the same data set to evaluate their wavelet transform-based approach, but they obtained less favorable results for QRS end and T end.

An interesting method to assess and compare the results of various waveform recognition algorithms was proposed by Morlet et al. [188]. They use scatter graphs that picture the standard deviation of the differences between program results and reference when the largest differences are progressively removed. This information allows us to distinguish between the reliability of an algorithm, that is, its capacity to provide wave boundary estimates without flagrant errors, and its precision, that is, the standard deviation of the differences between its estimates and the references.

37.13 Parameter Computation

Once the onsets and offsets of the various ECG waveforms have been identified, parameter or feature extraction is the next step. Time intervals and wave durations follow directly from the established time points. The ambiguities around the QT interval have been pointed out in the previous section. For the amplitudes of the various deflections, a 0-line, or baseline, must be defined. It should be understood that there is no true 0-level anywhere in the ECG (➲ Fig. 37.6). The T-P interval which has been recommended as such contains the U wave. The CSE Working Party defined baseline as "a horizontal reference line computed from a single base level" [34] and declared itself against the use of non-horizontal lines. For QRS and T combined, it recommended strongly the uniform use of a baseline through the onset of QRS in combination with a limited number of preceding sample points. For the P wave, the base level may be chosen at the onset of P. Voltage amplitudes are then simply determined with respect to the chosen baselines. A wave is defined as a discernible deviation from the baseline where at least two opposite slopes can be identified and where discernible means that both the amplitude and the duration of the deviation exceed certain minimum values [34]. Various amplitude ratios, such as Q/R and R/S, and also integrals and angles can be derived from these measurements. When simultaneously recorded orthogonal leads are available or are reconstructed from the 12-lead ECG [20, 189], spatial or planar vector magnitudes and directions, gradients and polar vectors can be obtained. The relative areas of the spatial QRS loop in each octant of the three-dimensional space are also measured in some systems [168].

Pipberger et al. [7, 8] applied the technique of time normalization of the QRS complex and the ST-T segment. Each was divided into eight equal time segments, regardless of their duration. In this way, measurements from QRS complexes with different durations can be compared one with another.

Advanced mathematical techniques, like Fourier analysis, polynomial fitting, or Karhunen–Loeve expansion, have been applied by some early investigators [190, 191]. It has been pointed out by Van Bemmel [180] that parameters derived from these techniques may allow an accurate reconstruction of the overall shape of the ECG, but are not necessarily the best discriminating features in ECG classification.

Computer processing systems frequently make 250–300 measurements per ECG. Ultimately, only a limited subset is used in the diagnostic classification stage and a still smaller number is printed out in the final computer report, which is submitted to the requesting physician. Most ECG processing systems will optionally display the averaged or median beat, if desired with markings for wave onsets and offsets as determined by the program. The ECG reader, therewith, has a check on the correctness of the measurements underlying the computer diagnosis.

37.14 Diagnostic ECG Interpretation

37.14.1 Strategies for Diagnostic Classification

After the initial waveform-detection, pattern-recognition, and measurement algorithms have been applied, the diagnostic stage is entered in which the ECG is classified into one or more of the various possible diagnostic categories. Two basically different approaches to diagnostic classification have been developed since the early days of computer assisted electrocardiography [192]. In the first, the deterministic or heuristic approach, the cardiologist's method of interpreting ECGs is simulated. The majority of existing ECG computer programs follows this approach. In the second, a statistical or probabilistic approach is adopted whereby an attempt is made by mathematical means to establish the probability that a given ECG belongs to a particular diagnostic category.

More recently, other approaches to diagnostic ECG classification have been tried. Some investigators have applied fuzzy set theory [170, 193, 194], others the "expert system" approach [195, 196] and, finally, neural-network techniques have been used [197–201]. However, clinical application of the latter techniques on a larger scale is lacking.

37.14.2 Deterministic ECG Computer Programs

The analysis follows a logical path of questions regarding the presence or absence of certain predetermined criteria for every diagnostic category, to be answered with yes or no. The questions and answers can be arranged in a decision table or in a decision tree resulting in diagnostic statements [85]. Such decision schemes, often quite elaborate, are implemented in all programs using the deterministic approach. The likelihood of the resulting diagnostic statements may be indicated by modifiers like "definite," "probable," and "possible."

Deterministic programs have the advantage that the diagnostic criteria used are familiar to the cardiologist, and that the logic is easy to follow and comprehend [192]. Much experience with conventional ECG interpretation present in the brain of the developer or published in the literature can, theoretically at least, be incorporated in these programs. Decision-tree programs are organic and flexible in structure and remain open for modification. With advancing insight and experience diagnostic pathways can be improved. Criteria selection can be guided by deductive reasoning based on knowledge of electrophysiological processes, a design that is inherently excluded in the statistical diagnostic approach. Also, new diagnostic categories can be added quite easily without the need for recruiting new statistical data.

Emulation of the human reader brings advantages but has also its limitations. The criteria selection might tend to be arbitrary and based on impressions rather than founded on solid quantitative information. In some systems, criteria selection was the work of a single expert, in others the effort of a group [154]. The appreciation of ECG criteria may vary considerably between cardiologists, as demonstrated by many studies on intra- and inter-observer variability of ECG interpretation [202–204]. The set of criteria proposed by one cardiologist may be met with another cardiologist's disapproval. This lack of agreement on criteria and the dearth of reliable quantitative data on their sensitivity and specificity have resulted in a diversity of commercial systems, even in one where each user could incorporate his own preferences, if so desired [205].

Many studies on ECG criteria have demonstrated a considerable overlap between normal and abnormal populations with respect to almost all ECG parameters, notably of Q-wave durations and voltage measurements. This necessarily gives rise to a proportion of false-positive and false-negative statements. The addition of a new, independent criterion to an existing decision tree increases sensitivity, but is almost invariably accompanied by a loss in specificity. This practice led to intolerably high false-positive rates in some of the early programs that, as a result, were dismissed as unsuitable for clinical use [45, 206]. A weighty argument of the adherents of the probabilistic approach was that conventional ECG programs can never surpass an accuracy level of 54–60%, the level reached by eminent experts in Simonson's study [203], who read the ECGs and VCGs of patients whose diagnoses were known and documented by ECG-independent means. In the CSE study, later to be discussed, the level of accuracy against the clinical "truth" was much higher, ranging between 73% and 81% [207].

Nowadays, however, decision-logic programs may also employ statistical tools. Computer programs are available that automatically generate decision-tree logic based on a given choice of features [208, 209]. Sensitivity and specificity of thresholds used in different criteria can be statistically optimized on population samples with well-defined disease entities. In a way, conventional ECG programs thereby move toward the statistical programs and, as we will see, vice versa.

37.14.3 Statistical ECG Computer Programs

The availability of the computer to perform voluminous mathematical computations for statistical purposes would seem to make the statistical approach the easy solution for diagnostic classification, but it will be shown that there are no easy solutions. For each diagnostic category, a population of ECGs has to be collected. The presence of normality or of the various types of abnormality (infarction, LVH, etc.) has preferably been established from clinical data, although ECG information does not have to be excluded. Next, the probabilities of occurrence of the various ECG measurements are determined for each diagnostic category. Having done this in a learning stage, any new ECG can be assigned to its presumed appropriate population in a test stage to evaluate the validity of the classification method. The measurements going into the statistical procedure are in principle just the amplitude values of samples in each lead of an ECG, in addition to certain duration measurements, not the values of the customary diagnostic ECG criteria, although, again in principle, there could be nothing against using them.

The main claim to preeminence of statistical over conventional programs is that diagnostic performance should be better and that mathematical objectivity takes the place of the personal idiosyncrasies of conventional program developers [192, 210]. Also, in the neighborhood of critical thresholds, minute changes in measurements may alter the interpretation in heuristic programs, but less so in statistical programs and certainly not when the probabilities for the diagnoses are above 70% [211, 212].

There are, however, distinct disadvantages [192]. First of all, large databases are needed for developing and testing. For each diagnostic category, a separate population of clinically documented records has to be collected. Case collection carries with it the problem of selection. The AVA program [123] has been developed almost exclusively on a database collected in a male and war veteran population. Statistical parameters may differ significantly in other populations that are more heterogeneous with respect to sex or to age and more homogeneous, on the other hand, with respect to race (see the Appendix on normal limits).

Furthermore, the diagnostic categories are mutually exclusive. A case may have a probability of 60% of belonging to the single category of, say, left ventricular hypertrophy (LVH), 30% of belonging to anterior myocardial infarction (AMI), and 10% of being normal. The individual probabilities add up to 100%. With this approach, it is not possible to have 100% probability of LVH and at the same time 100% probability of AMI. For this to be possible a separate collection of cases with LVH + AMI would have to be created and every other combination of diagnoses would again necessitate a different data base. In this way, the required number of cases soon increases out of control. Also, when one ECG presents an abnormality which does not belong to any of the existing categories, the probabilities will be spread over the existing categories [7]. To handle this problem, some diagnostic-tree type decisions were added in the AVA program in order to deal with some abnormalities not listed by the main AVA classification [123].

The common denominator of all the proposed statistical techniques is multivariate analysis. The first program using such methods dates back to 1961 and was described by Cady et al. [213]. They separated 23 normal tracings from 19 records with a typical pattern of LVH. Bayes' theorem was first applied to ECG diagnosis by Kimura et al. [214]. There were more early attempts to use multivariate analysis for ECG classification [215–218], but particularly Pipberger and coworkers have for many years persistently promoted the statistical method at numerous conferences and in many articles [7, 8, 41, 123, 219–222]. Following this lead, other investigators have applied statistical techniques for computer processing mainly to the VCG [163, 223]. Multivariate analysis has less frequently been applied to the standard 12-lead electrocardiogram [53, 163, 224].

The Bayes' formula is given by

$$P(i|\mathbf{x}) = \frac{f(\mathbf{x}|i) \cdot P(i)}{\sum_{j=1}^{m} f(\mathbf{x}|j) \cdot P(j)}$$

where $P(i|\mathbf{x})$ is the posterior or conditional probability that an ECG belongs to diagnostic category i (out of m possible categories) given the measurement vector $\mathbf{x}$; $f(\mathbf{x}|i)$ represents the conditional probability that the measurement vector $\mathbf{x}$ is produced in the cases belonging to category i; and $P(i)$ is the unconditional or prior probability for an ECG to belong to category i [7].

An essential feature of the classification strategy in the AVA program developed by Pipberger et al. [7, 8], as well as by some other investigators [223, 224], is the application of prior probabilities as expressed in the Bayes' formula. For this purpose, the relative frequencies of occurrence of various clinical diagnostic categories in the environment in question must be estimated. Thus, in Pipberger's approach, different prior probabilities were assigned to patients of different provenance (cardiological out-patient, general medicine, pulmonary disease) even to the extent of introducing "individual prior codes" [8]. This concept was argued to emulate the thought process of a physician who, when interpreting an ECG, weighs his diagnosis in the light of age, sex, and clinical condition of the patient, and the likelihood of occurrence of the disease in the specific population to which he happens to belong. The "proper prior code" for an individual is indicated on the ECG request form as one of several broad clinical categories, according to the tentative clinical diagnosis [8]. When entered with the patient's vital statistics, the prior probabilities are automatically set for computation of the posterior probabilities. The sensitivity and specificity for certain diagnostic categories of the AVA program can easily be adapted through manipulation of prior probabilities [8].

By including the prior probabilities of the various diagnostic categories, non-electrocardiographic information is introduced in a formalized way. In an extreme example, when a prior probability is set to zero or to one for a certain group, no case or every case will respectively be diagnosed in this disease category, no matter what the ECG appearance may be. These are the mathematical facts of this classification procedure [192]. Proponents of the use of prior probabilities argue that in a non-formalized way, prior knowledge and bias are applied by the clinician as well in routine ECG reading, in a justifiable and natural way. The opponents state that, by using clinical information, the role of the ECG is reduced to futility. They claim that evaluation of the ECG, in the larger framework of a final clinical diagnosis, should independently contribute to the diagnosis. In all these considerations, the element of "cost" cannot be neglected. The cost of missing the diagnosis of a single case of bubonic plague in a population hitherto thought to be free of the disease may be huge, as is the cost of overdiagnosing cervical cancer in a population where the disease is not uncommon. Prior probabilities should not be applied without discernment!

37.14.4 Methodology of ECG Computer Program Evaluation

The following paragraphs mainly address the topics of diagnostic accuracy and reliability of ECG computer programs. For other aspects such as utility, efficacy, and cost-benefit analysis, the reader is referred to some specific publications [12, 225–229].

It was pointed out earlier that comparison between computer and truth is a more cumbersome and problematic task than one would anticipate. To list the main complications:

1. There are ECG statements for which there is no clinical confirmation. Three types of statements can be made by human readers or computer programs [14], as follows:
   (a) Type A statements refer to an anatomic lesion or pathophysiologic state, such as myocardial infarction or hypertrophy, the presence of which can be confirmed by non-electrocardiographic evidence like cardiac catheterization, serum enzyme levels, ventriculograms, echocardiograms, scintigrams, and autopsy findings.
   (b) Type B statements apply to conditions which belong intrinsically to the realm of electrocardiography itself. They refer to electrophysiological processes such as arrhythmias and atrioventricular and intraventricular conduction disturbances. Sometimes these type B statements can be corroborated by the results of other diagnostic methods, such as scintigraphy in the case of a statement of ischemia based on a typical ST-T change. Most often, the physician's interpretation is the reference.
   (c) Type C statements refer to purely descriptive ECG features for which no substrate can be demonstrated by other means. Under this category, statements are subsumed like "nonspecific ST-T changes" or "axis deviation."

At the tenth Bethesda conference, Task force III [14] strongly recommended evaluation of type A statements by means of independent, clinical evidence. For type B and C statements, the human observer is the reference. A distinction was made between a constrained and a free observer, the constraint being a given set of measurements or criteria agreed upon before the evaluation. In fact, this again boils down to a test on the waveform recognition proficiency of the program. The free observer applies his own individual rules, or the majority opinion of a group of free observers may be obtained, but if these rules are formalized they can be put into computer logic and the observer is again constrained.

2. Collecting a type A database is a laborious and tedious task. In the early years, investigators were quite satisfied when they could single out one disease condition, say D, from a population composed of only D and normals. A simple 2 × 2 contingency table suffices to describe the diagnostic results, with sensitivity, specificity, positive predictive value, etc., as the indices of performance. This simple situation is already spoilt when non-D can be anything, not just normal. Non-normal conditions other than D may resemble D in some features and lead to false positive diagnoses of D and so degrade performance. In the real-life situation, the population contains all types of abnormality and the program is expected to classify each one correctly. This results in as many sensitivities and specificities as there are diagnostic conditions, and total performance cannot be expressed in a single, representative figure. The same "total accuracy" could be attained by one program scoring low on infarct recognition but high on LVH and a second one doing the reverse. Whatever the problems of assessing the agreement between computer and reference, it is mandatory that the database includes a spectrum of disease states, with different grades of severity and a representative number in each category, as well as combinations of different diseases. A database without any bias, however, is an illusion. Mostly, noisy recordings are excluded from the collection although noise is a fact of life (for testing purposes noise could be added afterward). Also, arrhythmias and conduction defects are sometimes excluded. Finally, databases differ in the populations they were drawn from, with respect to race, age, sex, social status, etc.
3. Output statements are not standardized and their meanings must be established first. What is to be understood by septal myocardial infarction? Does it mean the same thing as antero-septal myocardial infarction? What are the clinical grounds for diagnosing right ventricular hypertrophy (RVH)? Is RVH caused by pulmonary valve stenosis the same as RVH caused by an atrial septal defect? Such questions were addressed in the CSE study [65]. These considerations not only apply to type A statements, but to B and C as well. What is a nonspecific ST-T change? How will RBBB or LBBB be defined? Regarding the latter, a set of criteria for conduction disturbances and pre-excitation have previously been recommended [35].
4. Probability statements constitute another problem. A human observer may make a diagnosis of "probable inferior infarct." Does this mean "70% probability of IMI," but still a chance of 30% normal? Do the probabilities add up to 100%? In the case of "probable anterior infarct, possible LVH" does this mean 70% probability of AMI, 30% of LVH, and 0% normal, or must we understand that the case is probably abnormal (70%) with a preference for AMI (50%), a residual chance of 20% for LVH and leaving a 30% likelihood of normality? The problem is compounded by the tendency of ECG readers to cover all contingencies by statements like "posterior wall infarct not entirely excluded." It is, therefore, necessary that the reporting follows strictly prescribed rules. As an example, in the CSE study the number of categories was limited to seven. For each category, four certainty qualifiers were allowed: definite, probable, possible, and absent with numerical values of 3, 2, 1, and 0, respectively. The computer statements had to be transcribed into the same format by the program developers. As per diagnostic category, the scores were averaged over all readers or programs, and rounded. The category with highest score was taken to be the diagnosis of the "combined cardiologist" or "combined program." Additional rules were defined to handle some more complex situations [230].
5. Instability of interpretation is not restricted to man. Intra-observer variability is recognized to be sometimes considerable, but computer programs are generally viewed as robust against the diagnostically insignificant short-term variations that may occur between two successive ECGs of the same person. This is not entirely the case [231–233]. Even one and the same ECG may give rise to two different outcomes, as demonstrated by Bailey et al. [44, 45, 234, 235]. They processed two sets of digital samples, 1 ms apart, from the same recording, and, surprisingly, the separate results were by no means identical. The reproducibility of some of the older programs tested in this way proved to be unacceptably low. Spodick and Bishop [236] assessed the variability in the interpretations by an unspecified computer program of 92 unselected pairs of ECGs recorded 1 min apart. In 36 (39%) pairs, they found "grossly" different interpretations. The clinical significance of part of the differences, however, is debatable.

37.14.5 Comparison of Computer Interpretations with Physician's Interpretations

The 1,220 ECGs and VCGs from the CSE diagnostic library were read by nine cardiologists of whom four read ECGs and VCGs, four read only the ECGs and one read only the VCGs. For each recording, a combined result was derived in the above described manner. The results of nine ECG and six VCG computer programs were then compared with the "combined cardiologist." Different misclassification matrices, as well as sensitivity, specificity, predictive values of positive and negative test results, total accuracy, information content, and other measures of performance, were calculated [66, 207, 237]. The specificity of the ECG programs (median 87.7%, range 73.4–96.8%) was higher than that of the VCG program (median 75.5%, range 69.1–84.8%). Median sensitivities for LVH, RVH, AMI, and inferior myocardial infarction (IMI) were 71.6%, 58.6%, 81.0%, and 78.1%, respectively, but among the programs, sensitivities varied widely (➋ Table 37.1). Total agreement between an individual program and the combined cardiologist varied between 68.1% and 80.3% for the ECG programs and between 69.2% and 78.1% for the VCG programs. The combined program agreed with the combined cardiologist in 87.9% of the cases, which was significantly higher than for any individual program. The figure may be compared with the intra-observer variability of the cardiologists, who interpreted 125 randomly selected cases a second time without their knowing. Reproducibility was then seen to vary between a low of 73.6% and 90.8% (median 82.4%) [66].

It should be noted that the cardiologists in the CSE study were experienced, highly-motivated electrocardiographers. In a study by Jakobsson et al. [238], the interpretations of 69 physicians (only four of them being cardiologists) were compared with those of a computer-based ECG recorder in routine clinical practice. The authors took their own judgment as the reference. The study population consisted of 474 routine ECGs taken in a general practice, a medical emergency department, and an out-patient department. No single physician interpreted more than 7% of the ECGs. A written physician's interpretation was lacking in 11% of the ECGs. The diagnostic sensitivity for myocardial infarction of the doctors was significantly lower than that of the program. The overall quality of the computer interpretations was judged as being satisfactory in 82% of the ECGs, whereas this was only 64% for the physicians' written interpretations. Most computer misinterpretations could be attributed to poor signal quality, and it was suggested that the nonexpert physician should team up with the computer to approach the performance of an experienced ECG reader.

The CSE study only dealt with type A diagnostic statements. Computer–physician comparison is the usual method for assessing type B and type C interpretative statements, for which the ECG itself is the reference. Although this may seem to be less essential than diagnostic type A accuracy, such comparison is important since agreement between computer and physician is a convincing argument for the clinical acceptability of an ECG computer program [46, 52, 239]. Further, physician review is mandatory for legal and other reasons [240].

37.14.6 Comparison of Computer Results with Clinical "Truth"

The results of the nine ECG and six VCG computer programs obtained in the same 1,220 clinically validated ECGs and VCGs in the CSE diagnostic library, as well as those from the nine readers, were now compared with the "clinical truth" [66, 207]. The classification accuracies of the different programs (➋ Table 37.2), and to a lesser extent of the cardiologists

Table 37.1

Percentage agreement of nine ECG and six VCG computer programs and the combined program results with the combined cardiologists' interpretations on the 1,220 cases of the CSE diagnostic library (Adapted from Willems [66], p. 183)

Program	Normal	LVH	RVH	Anterior MI	Inferior MI	Total accuracy
ECG	87.7 (73.4–96.8)[a]	73.4 (63.2–90.7)	58.3 (25.0–66.7)	83.6 (65.1–89.3)	71.1 (44.6–87.5)	76.6 (68.1–80.3)
VCG	75.5 (69.1–84.8)	71.3 (47.9–77.8)	63.3 (21.7–85.0)	72.3 (50.9–82.1)	85.1 (74.4–91.9)	71.1 (69.2–78.1)
Combined program results						
ECG	95.6	88.2	66.7	87.7	81.0	85.4
VCG	85.3	80.3	78.3	83.0	93.4	82.1
ECG+VCG	95.2	87.9	75.0	88.7	91.0	87.9

[a]Median (range).

Table 37.2

Percentage correct classifications of nine ECG and six VCG computer programs and the combined program results against the clinical "truth" on the 1,220 cases of the CSE diagnostic library (Adapted from Willems [66], p. 181)

Program	Normal	LVH	RVH	Anterior MI	Inferior MI	Total accuracy
ECG	91.3 (86.3–97.1)[a]	56.6 (50.3–76.2)	31.8 (14.5–52.8)	77.1 (58.8–81.5)	58.8 (38.7–82.8)	69.7 (62.0–77.3)
VCG	80.9 (71.4–86.6)	55.4 (47.2–76.2)	35.9 (28.2–64.5)	68.0 (54.5–74.1)	74.1 (63.6–86.5)	68.3 (64.3–76.2)
Combined program results						
ECG	96.7	67.9	40.6	79.6	68.8	76.3
VCG	87.1	67.3	53.0	77.3	82.2	77.0
ECG+VCG	95.5	69.0	45.8	80.0	76.7	78.5

[a]Median (range).

Table 37.3

Percentage correct classifications of eight ECG readers and five VCG readers and the combined cardiologist results against the clinical "truth" on the 1,220 cases of the CSE diagnostic library (Adapted from Willems [66], p. 180)

Cardiologist	Normal	LVH	RVH	Anterior MI	Inferior MI	Total accuracy
ECG	96.1 (92.7–97.6)[a]	63.9 (54.8–69.3)	46.6 (40.6–51.5)	84.9 (79.0–87.5)	71.7 (59.2–84.1)	76.3 (72.6–81.0)
VCG	80.6 (73.8–87.2)	63.4 (48.4–68.5)	50.0 (45.5–56.4)	65.2 (62.4–79.4)	75.1 (69.8–82.1)	70.3 (67.5–74.4)
Combined cardiologist results						
ECG	97.1	65.8	47.6	87.3	75.1	79.2
VCG	84.7	64.7	53.0	74.1	78.6	74.8
ECG+VCG	96.5	68.2	50.3	85.7	78.3	80.3

[a]Median (range).

(➲ Table 37.3), vary widely. The median specificity of the ECG programs was 91.3% (range 86.3–97.1%) against 80.9% (range 71.4–86.6%) for the VCG programs. Corresponding values for the cardiologists were 96.1% (range 92.7–97.6%) for the ECG and 80.6% (range 73.8–97.2%) for the VCG. The median sensitivities for LVH, RVH, AMI, and IMI were 55.7%, 32.7%, 74.1%, and 65.1%, respectively, for ECG and VCG computer programs together, versus 63.4%, 48.5%, 82.9%, and 73.3%, respectively, for all cardiologists. Total accuracy varied between 62.0% and 77.3% (median 69.7%) for the ECG programs, and between 64.3% and 76.2% (median 68.3%) for the VCG programs. Total accuracy of the cardiologists varied between 72.6% and 81.0% (median 76.3%) for the ECG, and between 67.5% and 74.4% (median 70.3%) for the VCG. However, the programs with the best performance reached almost equal levels as the best cardiologists.

The cardiologists obtained more accurate results for the ECG than for the VCG. The VCG readers had a lower specificity and, somewhat surprisingly, a lower sensitivity for anterior infarction, while sensitivities for RVH, inferior infarction, and combined infarction were slightly higher. These results largely corroborate those of other investigators demonstrating that the VCG is superior for most diagnostic categories as far as sensitivity is concerned, at the expense of specificity [241]. In another study [53], using 3,266 cases and the same statistical classification technique, the ECG and VCG were shown to have identical diagnostic information. One reason for the lower performance of the VCG readers in the CSE study may be, for some readers, an unusual display of the vector loops. It should be noted that the total accuracies of the combined ECG program (76.3%) and combined VCG program (77.0%) were almost identical.

Total accuracy of the four statistical programs was significantly higher (median 76.0%) than for the heuristic programs (median 67.4%). This may, at least in part, be explained by the fact that 87% of the database consisted of single-disease cases. The statistical programs were designed to classify a case into one of seven or eight disease categories, which is not the case for the open-ended heuristic programs. Thus, the composition of the database may have resulted in a bias favoring the statistical programs. Moreover, results of the AVA program may have been inflated thanks to its being fed with clinical information to adjust the prior probabilities on a case-by-case basis.

Combined results of programs and cardiologists were in almost all cases more accurate than those of the individual programs and cardiologists. This confirms previous findings on a subset of the CSE library [204].

The advantageous effect of combining programs was used in a study by Kors et al. [242] to improve the performance of their MEANS program. MEANS comprises two different classification programs, one for the ECG and one for the VCG. The interpretations of both programs in analyzing the ECGs and VCGs of the CSE diagnostic library were combined. The total accuracy of the combined program against clinical evidence was 74.2% (➋ Fig. 37.8), significantly better than the total accuracy of each program separately (69.8% for the ECG, 70.2% for the VCG). Interestingly, to avoid the necessity of recording a VCG, in addition to the ECG, the VCGs were also reconstructed from the ECGs and then interpreted by the VCG classification program. The combined ECG and reconstructed VCG results (73.6%) were almost the same as those of the combined ECG and original VCG. Thus, the performance of an ECG computer program was improved by incorporating both ECG and VCG classificatory knowledge, drawing only on the ECG itself.

Another strategy, which takes into account possible beat-to-beat variations in the ECG was proposed by the same group [243]. In the first step, all individual complexes of the dominant type are analyzed and a classification is derived for each. In a second round, the individual classifications are combined in one final interpretation. When tested on the CSE diagnostic library, the total accuracy of MEANS against the clinical evidence significantly increased from 69.8% for the interpretations of the averaged complexes to 71.2% for the combined interpretations of the individual complexes.

Rather than combining the diagnostic outputs of an ECG and a VCG program, Andresen et al. [244, 245] integrated ECG and VCG criteria for acute and prior MI in one algorithm. VCG leads were synthesized using the "inverse Dower" transformation [19]. Using a database containing normals and patients with infarctions validated by ECG-independent means, the algorithm achieved sensitivity equal to that of three cardiologists and three primary care physicians, but had

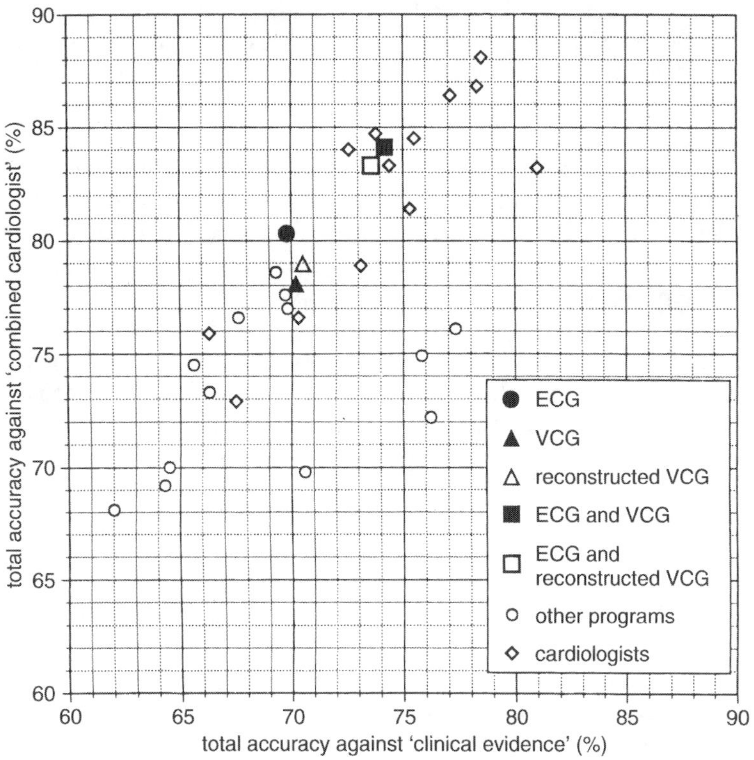

Figure 37.8

Total accuracies for the MEANS interpretation of the ECG, VCG, reconstructed VCG, and the combined interpretations. Also, the total accuracies of the other programs and of the cardiologists participating in the CSE study are shown (From Kors et al. [242]. © Elsevier. Reproduced with permission).

much higher specificity than the primary care physicians [246]. An improved version of the algorithm even outperformed the human readers [244], but these results might tend to be enhanced because the test database did not contain abnormalities other than infarctions.

The results of the CSE study showed that some ECG computer programs perform almost as well as the best cardiologists in classifying seven main diagnostic entities. However, it also became clear that some other programs were in need of considerable overhaul to meet reasonable standards.

37.14.7 Computer-Aided Physician's Interpretation of the ECG

Several studies evaluated the effect of computerized ECG interpretation on physicians' readings of ECGs. Milliken et al. [247] collected 180 ECGs from patients with various cardiac disorders known from ECG-independent data. Nine readers first read the ECG twice at an interval of several months without computer printouts. Months later, they reread the same ECGs, this time with the computer printout being available. The interpretations with several months intervening, but without computer output, resulted in an average change of 13.8% of the statements with 7.3% becoming correct and 6.5% becoming incorrect. When they reread the ECGs together with a computer output, the changes from incorrect to correct averaged 12% and from correct to incorrect 3%. From this study, it is apparent that the effect of priming the reader with computer results was in the direction of greater accuracy.

Hillson et al. [248] performed a randomized controlled trial to examine the effects of computer-assisted ECG interpretation on ECG reading time and agreement with the clinical diagnosis. Forty family physicians and general internists evaluated ten clinical vignettes accompanied by ECGs. Half of the physicians received the ECGs with computer-generated reports, the other half without. Those receiving the reports spent, on average, 25% less time in reading the ECGs. The first-listed diagnosis of those physicians who did not receive computer support agreed with the clinical diagnosis in only 15.3% of the cases, while the score rose to 30.1% in those who received support ($p = 0.004$). This effect could mainly be attributed to two cases with somewhat uncommon diagnoses (Wolff–Parkinson–White syndrome and pericarditis), which were correctly identified by the computer. However, in one of three cases that had erroneous computer reports, physicians who received the misleading reports were likely to adopt the diagnostic error. Another study [249] involved 22 cardiologists who each interpreted 80 ECGs, half of them with a computer report. Computer-assisted ECG interpretation gave an average reduction in reading time of 28% and significantly improved concordance of the cardiologists' interpretations with a gold standard established by a panel of five expert electrocardiographers.

In still another study, ten senior house officers were recruited in an emergency department [250]. They interpreted 50 ECGs and five of these junior doctors had access to the computer report. Their interpretations and the computer reports were compared with the consensus interpretation of two experienced clinicians. The computer made only two major errors. Access to the computer report improved the physicians' error rate (22.4% without report versus 18.4% with), but not significantly.

Tsai et al. [251] examined the effects of correct and incorrect computer advice. They performed a randomized controlled trial in which 30 internal medicine residents each interpreted 23 ECGs with a total of 54 findings. The gold standard was established by two cardiologists. Computer interpretations were correct in almost 60% of the findings. Overall, without computer report, the physicians' interpretations were correct in 48.9% of the findings. With the reports, they interpreted 55.4% correctly ($p < 0.0001$). For the subset in which the computer findings agreed with the gold standard, physicians without the computer report interpreted 53.1% correctly; when the computer report was included, accuracy increased to 68.1% ($p < 0.0001$). When computer advice that did not agree with the gold standard was not given to the physicians, accuracy was 56.7%. Accuracy dropped to 48.3% when the incorrect computer report was provided ($p = 0.13$). In the subset of findings in which the computer interpretation was incorrect, physicians agreed with the incorrect interpretation twice as often when they were prompted by the computer than when they were not (67.7% versus 34.6%, $p < 0.0001$).

Most studies have shown a clear improvement in ECG reading of physicians when they are provided with a computer-generated report. "Computerized electrocardiography – an adjunct to the physician" is the title of an editorial by Laks and Selvester [252] hailing the CSE report. Indeed, but there is a danger that nonexpert physicians let themselves be persuaded by an incorrect computer interpretation. This underlines the need for well-validated ECG computer programs that should perform at the level of expert electrocardiographers.

37.15 Rhythm Analysis Programs

Automatic identification of various arrhythmias poses major problems in the routine resting ECG. The rhythm interpretation logic in all clinically used programs follows a deterministic approach, using measurements (RR intervals and PR intervals, results of QRS wave typing, morphology of P waves, etc.) derived by the measurement program. The quality of the rhythm section rests largely on the quality of the measurement algorithms. While QRS complexes can be detected quite reliably, serious problems persist in P-wave recognition, especially in noisy recordings. For this reason, in many programs the logic has been constructed so as to allow for a degree of wave-recognition failure and measurement error [145, 253]. The record length poses another major problem. With the current multichannel recorders, usually 10 s of data are analyzed. This is too short for reliable detection of parasystole, for example, or other complicated arrhythmias. Providentially, the shortcomings of routine rhythm analysis programs are generally not too apparent, thanks to the low incidence of complex arrhythmias in the general hospital environment [254, 255].

The first extensive program for rhythm analysis was developed by Bonner and Schwetman in 1968 [256]. In a second program [257], five sets of three simultaneously recorded leads were used as input. The basic rhythm was derived from each 5 s record and a "combining" program made a decision as to the final rhythm statement. The total number of rhythm statements was forty.

These statements were also implemented in the AVA program [145]. The arrhythmia logic was divided into four main sections dealing, respectively, with regular and irregular rhythms and single and multiple aberrant beats. In the regular rhythm section, three major branchings were made depending on the ratio of P waves to the number of QRS complexes found (P:QRS). If P:QRS ≤ 0.25, entry into the AV-junctional section was made. If P:QRS ≥ 0.75, sinus rhythm was considered, as was AV-junctional rhythm depending on the polarity of the P wave. If $0.25 <$ P:QRS < 0.75, further tests were made and in the case of inconclusive results, the statement "regular rhythm" was printed. In the irregular-rhythm section, P:QRS was also used as a major branching point.

The same ratio (P:QRS) was also used in the more elaborate rhythm program developed by Plokker [253]. He considered 13 different types of arrhythmias, each represented by and programmed according to a detailed flowchart. Instead of decision trees, Wartak et al. [258] used decision tables. The whole arrhythmia logic is subdivided into a set of tables, which are linked to each other.

A comparison of results published by different investigators is difficult and sometimes delusive. For rhythm program evaluation, there are no approved databases comparable to that of CSE. The ideal case collection would contain sufficient numbers of records in every rhythm category and evaluation should provide figures for sensitivity and specificity for each category. However, arrhythmia cases tend to be in short supply. Evaluation is then often restricted to the sensitivity and specificity of sinus rhythm diagnosis, in which specificity is calculated with respect to all non-sinus cases. The results are thus heavily influenced by the diagnostic difficulty of the cases in the non-sinus category.

The results of a number of programs were evaluated by several authors. Bailey et al. [44] analyzed results of the IBM-Bonner program, and reported a sensitivity for arrhythmias of 87.2% while specificity was 98.2%. Of the 31 false-negative arrhythmias, the statement "undetermined rhythm" was made in 18 cases, all found to be atrial fibrillation. Sensitivity for atrial fibrillation was 85.5%. Of eight cases with second-degree AV block, the program correctly identified two. The prevalence of complex arrhythmias was too low to derive any meaningful conclusions.

Similar detection rates of cardiac arrhythmias have been reported for other processing systems [52, 239, 253, 259]. For example, Bernard et al. [239] reported an overall correct detection rate of 86% in 240 arrhythmias by the Telemed program. They did not specify whether the missed arrhythmias were incorrectly called sinus rhythm or any other arrhythmia.

A comparative study of five computer programs in the diagnosis of various types of AV block was undertaken by Shirataka et al. [260], using an ECG signal generator. Although all systems correctly detected normal sinus rhythm and first-degree AV block, only one system recognized second-degree AV block with classic Wenckebach periodicity, and no system was able to classify atypical Wenckebach periods. Most systems performed reasonably well for Mobitz II AV block, third-degree AV block, and ventricular bigeminy and trigeminy.

Plokker [253] tested the rhythm analysis of an early version of the MEANS program. On a set of 2,769 ECGs, he found a sensitivity for sinus rhythm of 96.6% and a specificity of 98.0%. Using the HP program, Thomson et al. [261] reported a sensitivity for sinus rhythm of 96.6% and a specificity of 97.0% on a set of 5,110 ECGs.

Several reports [255, 262, 263] were devoted to the rhythm analysis program of GE Health Technologies (formerly Marquette). Farrell et al. [263] used 70,000 physician-confirmed ECGs from four teaching hospitals. Patients with pacemakers were excluded. Primary rhythm statements of a new and a previous program version were compared with the confirmed interpretations. Overall disagreement decreased from 6.9% for the older version to 4.1% for the new version. Increased sensitivities were observed for sinus rhythm (98.2%), atrial fibrillation (89.0%), and AV-junctional rhythms (63.1%), while specificity and positive predictive value improved for all arrhythmias. However, specificity for sinus rhythm was 85.5%, which means that as much as 14.5% of abnormal rhythms were called sinus. This would disqualify such a program for arrhythmia case-finding purposes.

Using a nearly identical software version as in Farrell's study, Poon et al. [255] assessed the performance of the GE program in 4,297 consecutive ECGs recorded in a university teaching hospital. Over-reading was performed by either one of two cardiologists. Overall, 13.1% of the ECGs required revision of the computer's rhythm interpretation, but about half of these ECGs involved patients with pacemakers. In the unpaced population, sensitivity for sinus rhythm was 98.7%, but specificity was again rather low at 90.1%. For atrial fibrillation, sensitivity was 90.8% and specificity 98.9%. Sensitivity for atrial flutter was 61.0%, whereas sensitivity for atrial tachycardia was only 2.8%. It was concluded that physician over-reading remains mandatory, in particular to confirm a computer statement of normal sinus rhythm, or rhythm statements in patients with pacemakers.

In another study involving the GE 12SL program, 2,072 ECGs collected in a tertiary care hospital were processed by the computer and then over-read by two cardiologists [264]. In 9.9% of the ECGs there were significant disagreements between the computer and the cardiologists; 86% of these related to arrhythmias, conduction disorders, and electronic pacemakers. Sensitivities for atrial fibrillation and atrial flutter were 76.1% and 65.9%, respectively, at specificities of more than 99.5%.

Sinus rhythm appears uniformly well recognized, but other results depend quite heavily on the material analyzed, not only with respect to the mix of various arrhythmias, but also to the noise content of the database.

37.16 Serial Comparison Programs

Comparison with previous records is a part of normal routine in ECG reading. It must, therefore, also be seen as a necessary adjunct to computerized ECG analysis. A prerequisite is an efficient database management system for storage and retrieval of results. In the past, restricted storage capacity limited the number of records and the type of data that could be kept on line. Pryor et al. [265] developed the first serial ECG comparison program but the comparison was essentially limited to the final diagnostic statements. Others compared measurements as well [266]. In some systems, raw data from one representative cycle for up to three records were stored [267], or the entire record could potentially be regenerated in case of arrhythmias [143, 268].

Nowadays, with vastly expanded storage capacity and transmission and processing speed boosted to previously unimagined heights, technical limitations hardly seem to be a consideration. Nevertheless, an operational difficulty remains. The modern stand-alone microprocessor-based electrocardiograph is able to perform a complete ECG analysis but does not contain the information of the central management system. If one wants serial analysis, a way of communication with the management system must be chosen from several options, the most advanced being instantaneous wireless transmission of data [269, 270].

The more fundamental problem of serial comparison is that of semantics: which meaning must be attached to a difference in measurements or in diagnostic statements between successive interpretations? As is well known, there is a certain variability in recordings from 1 day to the next or from 1 year to the next [271, 272]. This variability is caused by differences in electrode placement, variations in depth of respiration, alterations in posture, changes in body fat, and by other factors [273]. Measurement changes from one tracing to the next should be given attention only if they exceed the natural variability. As mentioned earlier, in deterministic programs, it is quite possible that small, insubstantial measurement fluctuations around a threshold value cause a material difference in diagnostic statements [234]. But which of the statements in this borderline situation is the correct one? Each serial analysis program deals with this problem in its own way.

Different checks are made to ascertain whether or not changes are the result of a borderline crossover of decision criteria. As an option, a graphic display of trends in measurements and diagnostic statements has been implemented in some processing systems [274–276]. Also, serial comparison summary statements, such as "no significant change" or "descriptive differences only," have been proposed as a means of taking into account normal ECG fluctuations [277]. Alternative methods for improving the consistency of serial ECG analysis are the use of smooth decision functions rather than binary thresholds [278, 279], or VCG loop alignment (if not available as such, the VCGs may be derived from the two ECGs to be compared by mathematical transformation) [280, 281].

On the other hand, as long as the measurements are essentially identical, a good program will produce the same diagnostic result which may mean persisting in making the same error. Meanwhile, the initial statement may have been corrected by a reviewer [282]. Only if the corresponding changes are made in the central ECG management system may a serial analysis program take them into account in the comparison with a subsequent ECG from the same patient [277].

Most programs for serial ECG analysis use decision rules to arrive at their diagnostic interpretation [141, 163, 267, 283]. Two studies investigated the use of artificial neural networks. Sunemark et al. [281] described a neural network to classify serial changes indicative of newly developed infarcts, taking the consensus opinion of three interpreters as the reference. The input data consisted of measurements from the ECG and the reconstructed VCG. At 90% specificity, the use of only ECG or VCG measurements gave a sensitivity of 63% and 60%, respectively, and increased to 69% when measurements were combined. Ohlsson et al. [284] also used neural networks to detect acute MI based on either the current ECG only, or on the combination of the previous and the current ECGs. Acute MI was diagnosed according to characteristic chest pain, elevated enzyme levels, or characteristic ECG changes. There was a small but significant improvement in the neural network performance when a previous ECG was used as an additional input. On the same set of ECGs, the neural network appeared to perform better than two physicians, an experienced cardiologist and an intern.

No independent evaluation studies of different systems for serial ECG analysis have been published.

37.17 Computer Analysis of Pediatric Electrocardiograms

The ECG in children differs considerably in signal characteristics from that of the adult. Small children usually have high heart rates, and the recordings are often much noisier and show more baseline wander than in adults. There is a fast and profound evolution of electrocardiographic patterns especially in the first days and weeks after birth, but changes continue up to adolescence [285, 286]. Consequently, normal limits of ECG and VCG parameters are heavily dependent on age, and for diagnostic criteria, especially in young children, it is necessary to rely on extensive tables of values for amplitudes, durations, and angles [287–292]. Here, obviously, the computer can be of assistance. Also, congenital heart diseases can be quite complex and can produce a variety of electrocardiographic patterns. In addition, qualified readers of pediatric ECGs are rare and mostly located in university centers [293]. Inter- and intra-observer variability in reading pediatric ECGs was shown to be substantial [294]. These were the main incentives for the development of pediatric ECG programs.

In the past, the Mayo computer system routinely processed pediatric VCGs for some time [295], and in the VA Research Center for Cardiovascular Data, a large cooperative study has been undertaken in this field [296]. Other programs have been developed for pediatric VCG [163, 223] and 12-lead ECG [297–300]. The combination of ECG and synthesized VCG measurements to discriminate between mild RVH with terminal conduction delay and partial RBBB in children was described by Zhou et al. [301].

Relatively few evaluation studies of pediatric ECG programs have been performed. In a study of 248 pediatric ECGs that were diagnosed by ECG-independent means, the HP pediatric program had a 70% sensitivity and 82% specificity for RVH and a 38% sensitivity and 93% specificity for LVH [302].

Based on a test set of 642 ECGs diagnosed by two pediatric cardiologists, Rijnbeek et al. [300] assessed the performance of the pediatric version of the MEANS program (➔ Table 37.4). Sensitivities for RVH, LVH, RBBB, and LBBB were 74%, 79%, 84%, and 75%, respectively, at specificities of at least 95%. The program employs continuous age-dependent normal limits that in a separate study were shown to be considerably different from those commonly used in children [292].

In a study by Hamilton et al. [303], the diagnoses of RVH and LVH by the Glasgow program was compared with those of two pediatric cardiologists. When the cardiologists were not provided with clinical information, sensitivity of

Table 37.4
Performance of the computer program PEDMEANS on a training set (n = 1,076) and test set (n = 642) of pediatric ECGs (from Rijnbeek et al. [300])

	Training set		Test set	
Abnormality	Sensitivity (%)	Specificity (%)	Sensitivity (%)	Specificity (%)
LVH	71.7	97.1	74.3	96.5
RVH	80.0	95.0	79.4	95.6
LBBB	86.4	97.9	84.0	95.3
RBBB	62.5	99.3	75.0	99.7

the program for RVH was 73% at a specificity of 97%, but sensitivity for LVH was only 25% at 96% specificity though the sensitivity for LVH increased to 44% at the same specificity when the clinical information was provided. Interestingly, if the cardiologists had disagreed initially with each other, their consensus opinion was twice as likely to be in agreement with the program.

Finally, pediatric ECG interpretations by the Marquette 12 SL program were compared with those of emergency department physicians in a 12-month prospective study, taking the interpretation by a pediatric electrophysiologist as the reference [304]. The computer proved to be more accurate than the physicians for interpretations considered to be of minimal or indeterminate clinical significance, but both performed poorly in interpreting the few cases of definite clinical significance (prolonged QTc, acute MI, supraventricular tachycardia, and atrial fibrillation).

37.18 Conclusion

In 1989, the late Jos Willems, the author of this chapter on computerized electrocardiography in the first edition of this handbook, wrote: "Nowadays computerized ECG analysis is being utilized widely in many medical institutions" and "Microprocessor-equipped electrocardiographs are proliferating and are on the verge of widespread application in smaller hospitals, general practitioners' offices and the health-screening environment." Now, in 2010, it will be hard to find a non-microprocessor-based electrocardiograph outside a museum for medical instruments, where it may stand next to an Einthoven string-galvanometer-based electrocardiograph invented in 1902. Willems also wrote: "Since 1982, definite progress has been made in the development of reference standards aimed at the evaluation of ECG measurement programs. This has largely been the result of the cooperative study "Common Standards for Quantitative Electrocardiography" (CSE project). Much work still needs to be done in the objective assessment of the diagnostic performance of ECG-analysis computer programs. Many challenges are still ahead . . .."

In almost 20 years since the closure of the CSE project, certainly much work has been done, but many challenges remain. The volume per annum of publications on computerized ECG analysis has steadily decreased. It looks as if a certain amount of saturation has set in. The providers of "intelligent" ECG equipment seem to be less keen on improving their diagnostic programs: has not the CSE study shown that programs rival the cardiologists' intelligence in diagnosis? But the time has not arrived to sit back complacently. It must be remembered that the CSE study applies to an idealized breed of ECGs. In real life, ECGs are not selected for good appearance and straightforward character. They may be of unshapely physiognomy due to combinations of abnormalities and be disfigured by noise, arrhythmias, and conduction defects. Here, the computer's painstakingly taught tricks for detecting waves and inflectional points are eclipsed by man's innate talents for pattern recognition. Especially P-wave detection and, connected with it, rhythm analysis need improvement and it would help if electrocardiographs would routinely acquire records of at least 30 s for this purpose. On the other hand, man is unreliable and, except where waveform recognition is suddenly led astray – a weakness that still should be amended – computer measurements are much more consistent. Presently, computer measurement of the QT interval is widely used for drug safety testing. In general, computer diagnosis is very specific, that is, a statement of "normal" can be relied on almost blindly and computerized ECG analysis is, therefore, an adequate screening instrument. Also, population screening by Minnesota coding through computer is much more reliable, faster, and cheaper than by hand. For clinical purposes, an abnormal computer classification still requires over-reading by an expert. The confrontation of the reader with the computer report was shown to improve quality and consistency of diagnosis. The huge advantages offered by a computerized system for automatic reporting, filing, and retrieval do not have to be emphasized. Finally, on the list-to-do in computerized electrocardiography, we should put the further development of serial analysis and perfection of the diagnosis of acute coronary syndromes. For this latter purpose, well-validated data bases are necessary.

The electrocardiograph with diagnostic facilities is often called "intelligent." Shannon once exclaimed: "Man is a machine, man can think, therefore some machines can think." The reader may speculate whether the third term of this syllogism will at some point apply to the computerized ECG machine.

References

1. Turing, A.M., Computing machinery and intelligence. *Mind*, 1950;**59**: 433–460.
2. Rijlant, P.B.L., L'analyse par un calculateur analogique des electrocardiogrammes scalaires et vectoriels. *Bull. Acad. Royal Med. Belgique*, 1962;**2**: 363.
3. Rautaharju, P.M., The impact of computers on electrocardiography. *Eur. J. Cardiol.*, 1978;**8**: 237–248.
4. Taback, L., E. Marden, H.L. Mason, and H. Pipberger, Digital recording of electrocardiographic data for analysis by means of a digital electronic computer. *IRE Trans. Med. Electron.*, 1959;**6**: 167–171.
5. Pipberger, H.V., R.J. Arms, and F.W. Stallmann, Automatic screening of normal and abnormal electrocardiograms by means of a digital electronic computer. *Proc. Soc. Exp. Biol. Med.*, 1961;**106**: 130–132.
6. Stallmann, F.W. and H.V. Pipberger, Automatic recognition of electrocardiographic waves by digital computer. *Circ. Res.*, 1961;**9**: 1138–1143.
7. Cornfield, J., R.A. Dunn, C.D. Batchlor, and H.V. Pipberger, Multigroup diagnosis of electrocardiograms. *Comput. Biomed. Res.*, 1973;**6**: 97–120.
8. Pipberger, H.V., D. McCaughan, D. Littmann, H.A. Pipberger, J. Cornfield, R.A. Dunn, et al., Clinical application of a second generation electrocardiographic computer program. *Am. J. Cardiol.*, 1975;**35**: 597–608.
9. Caceres, C.A., C.A. Steinberg, S. Abraham, J. CW, J.M. McBride, and W.E. Tolles, et al., Computer extraction of electrocardiographic parameters. *Circulation*, 1962;**25**: 356–362.
10. Rautaharju, P.M., The current state of computer ECG analysis: a critique, in *Trends in Computer-Processed Electrocardiograms*, J.H. van Bemmel and J.L. Willems, Editors. Amsterdam: North-Holland, 1976, pp. 117–124.
11. Drazen, E.L., Use of computer-assisted ECG interpretation in the United States, in *Computers in Cardiology 1979*, K.L. Ripley and H.G. Ostrow, Editors. Long Beach: IEEE Comp Soc. 1979, pp. 83–85.
12. Drazen, E.L., N. Mann, R. Borun, M. Laks, and A. Bersen, Survey of computer-assisted electrocardiography in the United States. *J. Electrocardiol.*, 1988;**21**(Suppl): S98–104.
13. Macfarlane, P.W., A brief history of computer-assisted electrocardiography. *Methods Inf. Med.*, 1990;**29**: 272–281.
14. Rautaharju, P.M., M. Ariet, T.A. Pryor, R.C. Arzbaecher, J.J. Bailey, R. Bonner, et al., The quest for optimal electrocardiography. Task force III: computers in diagnostic electrocardiography. *Am. J. Cardiol.*, 1978;**41**: 158–170.
15. Pipberger, H.V., J. Cornfield, What ECG computer program to choose for clinical application. The need for consumer protection. *Circulation*, 1973;**47**: 918–920.
16. Frank, E., An accurate, clinically practical system for spatial vectorcardiography. *Circulation*, 1956;**13**: 737–749.
17. Rautaharju, P.M., H.W. Blackburn, H.K. Wolf, and M. Horacek, Computers in clinical electrocardiology. Is vectorcardiography becoming obsolete? *Adv. Cardiol.*, 1976;**16**: 143–156.
18. Dower, G.E., H.B. Machado, and J.A. Osborne, On deriving the electrocardiogram from vectoradiographic leads. *Clin. Cardiol.*, 1980;**3**: 87–95.
19. Edenbrandt, L. and O. Pahlm, Vectorcardiogram synthesized from a 12-lead ECG: superiority of the inverse Dower matrix. *J. Electrocardiol.*, 1988;**21**: 361–367.
20. Rubel, P., I. Benhadid, and J. Fayn, Quantitative assessment of eight different methods for synthesizing Frank VCGs from simultaneously recorded standard ECG leads. *J. Electrocardiol.*, 1992;**24**(Suppl): 197–202.
21. Macfarlane, P.W., M.P. Watts, and T.D.V. Lawrie, Hybrid electrocardiography, in *Optimization of Computer ECG Processing*, H.K. Wolf and P.W. Macfarlane, Editors. Amsterdam: North-Holland, 1980, pp. 57–61.
22. Kornreich, F. and P.M. Rautaharju, The missing waveform and diagnostic information in the standard 12 lead electrocardiogram. *J. Electrocardiol.*, 1981;**14**: 341–350.
23. Kornreich, F., The missing waveform information in the orthogonal electrocardiogram (Frank leads). I. Where and how can this missing waveform information be retrieved? *Circulation*, 1973;**48**: 984–995.
24. Kornreich, F., P. Smets, and J. Kornreich, About challenging the uniqueness of a new, so-called "optimal", "total" or "maximal" 9-lead system, in *Trends in Computer-Processed Electrocardiograms*, J.H. van Bemmel and J.L. Willems, Editors. Amsterdam: North-Holland, 1977, pp. 293–301.
25. Kornreich, F., R.L. Lux, and R.S. MacLeod, Map representation and diagnostic performance of the standard 12-lead ECG. *J. Electrocardiol.*, 1995;**28**(Suppl): 121–123.
26. Kors, J.A. and G. van Herpen, How many electrodes and where? A "poldermodel" for electrocardiography. *J. Electrocardiol.*, 2002;**35**(Suppl): 7–12.
27. Nelwan, S.P., J.A. Kors, S.H. Meij, J.H. van Bemmel, and M.L. Simoons, Reconstruction of the 12-lead electrocardiogram from reduced lead sets. *J. Electrocardiol.*, 2004;**37**: 11–18.
28. Dower, G.E., A. Yakush, S.B. Nazzal, R.V. Jutzy, and C.E. Ruiz, Deriving the 12-lead electrocardiogram from four (EASI) electrodes. *J. Electrocardiol.*, 1988;**21**(Suppl): S182–S187.

29. Burger, H.C. and J.B. Van Milaan, Heart-vector and leads. *Brit. Heart J.*, 1946;**8**: 157–161.
30. Mortara, D.W., Source consistency filtering. Application to resting ECGs. *J. Electrocardiol.*, 1992;**25**(Suppl): 200–206.
31. Kors, J.A. and G. van Herpen, Accurate automatic detection of electrode interchange in the electrocardiogram. *Am. J. Cardiol.*, 2001;**88**: 396–399.
32. Burger, H.C., A. van Brummelen, and G. van Herpen, Compromise in vectorcardiography. II. Alterations of coefficients as a means of adapting one lead system to another. Subjective and mathematical comparison of four systems of VCG. *Am. Heart J.*, 1962;**64**: 666–678.
33. Surawicz, B., H. Uhley, R. Borun, M. Laks, L. Crevasse, K. Rosen, et al., The quest for optimal electrocardiography. Task force I: standardization of terminology and interpretation. *Am. J. Cardiol.*, 1978;**41**: 130–145.
34. The CSE Working Party, Recommendations for measurement standards in quantitative electrocardiography. *Eur. Heart J.*, 1985;**6**: 815–825.
35. Willems, J.L., E.O. Robles de Medina, R. Bernard, P. Coumel, C. Fisch, D. Krikler, et al., Criteria for intraventricular conduction disturbances and pre-excitation. World Health Organizational/International Society and Federation for Cardiology Task Force Ad Hoc. *J. Am. Coll. Cardiol.*, 1985;**5**: 1261–1275.
36. Kadish, A.H., A.E. Buxton, H.L. Kennedy, B.P. Knight, J.W. Mason, C.D. Schuger, et al., ACC/AHA clinical competence statement on electrocardiography and ambulatory electrocardiography: A report of the ACC/AHA/ACP-ASIM task force on clinical competence. *Circulation*, 2001;**104**: 3169–3178.
37. van Bemmel, J.H., J.A. Kors, and G. van Herpen, Methodology of the modular ECG analysis system MEANS. *Methods Inf. Med.*, 1990;**29**: 346–353.
38. Talmon, J.L. and J.H. van Bemmel, The advantage of modular software design in computerized ECG analysis. *Med. Inform.*, 1986;**11**: 117–128.
39. Willems, J.L. and J. Pardaens, Differences in measurement results obtained by four different ECG computer programs, in *Computers in Cardiology 1977*, H.G. Ostrow and K.L. Ripley, Editors. Long Beach: IEEE Comput Soc, 1977, pp. 115–121.
40. Willems, J.L., A plea for common standards in computer aided ECG analysis. *Comput. Biomed. Res.*, 1980;**13**: 120–131.
41. Pipberger, H.V., Comparative evaluation of electrocardiography computer programs, in *Computers in Cardiology 1976*, H.G. Ostrow and K.L. Ripley, Editors. Long Beach: IEEE Computer Society, 1976, pp. 85–88.
42. Crevasse, L. and M.A. Ariet, New scalar electrocardiographic computer program. Clinical evaluation. *JAMA*, 1973;**226**: 1089–1093.
43. Romhilt, D.W. and E.H. Estes, A point-score system for the ECG diagnosis of left ventricular hypertrophy. *Am. Heart J.*, 1968;**75**: 752–758.
44. Bailey, J.J., S.B. Itscoitz, J.W. Hirshfeld, L.E. Grauer, and M.R.A. Horton, Method for evaluating computer programs for electrocardiographic interpretation. I. Application to the experimental IBM program of 1971. *Circulation*, 1974;**50**: 73–79.
45. Bailey, J.J., S.B. Itscoitz, L.E. Grauer, J.W. Hirshfeld, and M.R.A. Horton, Method for evaluating computer programs for electrocardiographic interpretation. II. Application to version D of the PHS program and the Mayo clinic program of 1968. *Circulation*, 1974;**50**: 80–87.
46. Hodges, M., A clinical evaluation of the H-P ECG analysis program: program accuracy and value of adjustable criteria, in *Computers in Cardiology 1979*, K.L. Ripley and H.G. Ostrow, Editors. Long Beach: IEEE Comput Soc, 1979, pp. 167–170.
47. Garcia, R., G.M. Breneman, and S. Goldstein, Electrocardiogram computer analysis. Practical value of the IBM Bonner-2 (V2 MO) program. *J. Electrocardiol.*, 1981;**14**: 283–288.
48. Caceres, C.A. and H.M. Hochberg, Performance of the computer and physician in the analysis of the electrocardiogram. *Am. Heart J.*, 1970;**79**: 439–443.
49. Bourdillon, P.J. and D. Kilpatrick, Clinicians, the Mount Sinai program and the Veterans' Administration program evaluated against clinico-pathological data derived independently of the electrocardiogram. *Eur. J. Cardiol.*, 1978;**8**: 395–412.
50. Willems, J.L., H. Ector, J. Pardaens, J. Piessens, and H. de Geest, Computer and conventional ECG analysis: correlation with cineangiographic data. *Adv. Cardiol.*, 1978;**21**: 177–180.
51. Khadr, N.E., C.L. Bray, D.C. Beton, R.S. Croxson, M. Hughes, C. Jeffery, et al., Diagnosis of left ventricular hypertrophy and myocardial infarction by Bonner/IBM program verified by ECG-independent evidence, in *Computers in Cardiology*, K.L. Ripley and H.G. Ostrow, Editors. Long Beach: IEEE Comput Soc, 1979, pp. 93–97.
52. Macfarlane, P.W., D.I. Melville, M.R. Horton, and J.J. Bailey, Comparative evaluation of the IBM (12-lead) and Royal Infirmary (orthogonal three-lead) ECG computer programs. *Circulation*, 1981;**63**: 354–359.
53. Willems, J.L., E. Lesaffre, and J. Pardaens, Comparison of the classification ability of the electrocardiogram and vectorcardiogram. *Am. J. Cardiol.*, 1987;**59**: 119–124.
54. Zywietz, C. and B. Schneider, Editors. *Computer Application in ECG and VCG Analysis*. Amsterdam: North-Holland, 1973.
55. van Bemmel, J.H. and J.L. Willems, Editors. *Trends in Computer-Processed Electrocardiograms*. Amsterdam: North-Holland, 1977.
56. Wolf, H.K. and P.W. Macfarlane, Editors. *Optimization of Computer ECG Processing*. Amsterdam: North-Holland, 1980.
57. Willems, J.L., P. Arnaud, R. Degani, P.W. Macfarlane, J.H. van Bemmel, and C. Zywietz, *Protocol for the Concerted Action Project "Common Standards for Quantitative Electrocardiography"*. Leuven: ACCO, 1980.
58. The CSE European Working Party, An approach to measurement standards in computer ECG analysis, in *Optimization of Computer ECG Processing*, H.K. Wolf and P.W. Macfarlane, Editors. Amsterdam: North-Holland, 1980, pp. 135–137.
59. Willems, J.L., P. Arnaud, J.H. van Bemmel, R. Degani, P.W. Macfarlane, C. Zywietz, Common standards for quantitative electrocardiography: goals and main results. CSE Working Party. *Methods Inf. Med.*, 1990;**29**: 263–271.
60. Willems, J.L., P. Arnaud, J.H. van Bemmel, P.J. Bourdillon, R. Degani, B. Denis, et al., Establishment of a reference library for evaluating computer ECG measurement programs. *Comput. Biomed. Res.*, 1985;**18**: 439–457.
61. Willems, J.L., P. Arnaud, J.H. van Bemmel, P.J. Bourdillon, R. Degani, B. Denis, et al., A reference data base for multilead electrocardiographic computer measurement programs. *J. Am. Coll. Cardiol.*, 1987;**10**: 1313–1321.
62. Willems, J.L., P. Arnaud, J.H. van Bemmel, P.J. Bourdillon, C. Brohet, S. Dalla Volta, et al., Assessment of the performance of electrocardiographic computer programs with the use of a reference data base. *Circulation*, 1985;**71**: 523–534.

63. Willems, J.L., *Common Standards for Quantitative Electrocardiography. CSE Atlas.* Referee Results First Phase Library – Data Set 1. Leuven: ACCO, 1983.
64. Willems, J.L., *Common Standards for Quantitative Electrocardiography. CSE Multilead Atlas.* Measurement Results – Data Set 3. Leuven: ACCO, 1988.
65. Willems, J.L., *Common Standards for Quantitative Electrocardiography*, 4th Progress Report. Leuven: ACCO, 1984.
66. Willems, J.L., *Common Standards for Quantitative Electrocardiography*, 10th Progress Report. Leuven: ACCO, 1990.
67. Berson, A.S., Analog-to-digital conversion, in: *Computer Application on ECG and VCG Analysis*, C. Zywietz and R. Schneider, Editors. Amsterdam: North-Holland, 1973, pp. 57–72.
68. Berson, A.S., T.A. Ferguson, C.D. Batchlor, R.A. Dunn, and H.V. Pipberger, Filtering and sampling for electrocardiographic data processing. *Comput. Biomed. Res.*, 1977;**10**: 605–616.
69. Bailey, J.J., A.S. Berson, A. Garson, L.G. Horan, P.W. Macfarlane, D.W. Mortara, et al., Recommendations for standardization and specifications in automated electrocardiography: bandwidth and digital signal processing. A report for health professionals by an ad hoc writing group of the Committee on Electrocardiography and Cardiac Electrophysiology of the Council on Clinical Cardiology, American Heart Association. *Circulation*, 1990;**81**: 730–739.
70. Rijnbeek, P.R., J.A. Kors, and M. Witsenburg, Minimum bandwidth requirements for recording of pediatric electrocardiograms. *Circulation*, 2001;**104**: 3087–3090.
71. Hedén, B., M. Ohlsson, L. Edenbrandt, R. Rittner, O. Pahlm, and C. Peterson, Artificial neural networks for recognition of electrocardiographic lead reversal. *Am. J. Cardiol.*, 1995;**75**: 929–933.
72. Schijvenaars, R.J., J.A. Kors, G. van Herpen, and J.H. van Bemmel, A method to reduce the effect of electrode position variations on automated ECG interpretation. *J. Electrocardiol.*, 1995;**28**: 350–351.
73. Brodnick, D., A method to locate electrode placement. *J. Electrocardiol.*, 2000;**33**(Suppl): 211–218.
74. McManus, C.D., K.D. Neubert, and E. Cramer, Characterization and elimination of AC noise in electrocardiograms: a comparison of digital filtering methods. *Comput. Biomed. Res.*, 1993;**26**: 48–67.
75. Cramer, E., C.D. McManus, and D. Neubert, Estimation and removal of power line interference in the electrocardiogram: a comparison of digital approaches. *Comput. Biomed. Res.*, 1987;**20**: 12–28.
76. Lynn, P.A., Online digital filters for biological signals: some fast designs for a small computer. *Med. Biol. Eng. Comput.*, 1977;**15**: 534–540.
77. Weaver, C.S., J. von der Groeben, P.E. Mantey, J.G. Toole, C.A. Cole, J.W. Fitzgerald, et al., Digital filtering with applications to electrocardiogram processing. *IEEE Trans. Audio Electroacoust.*, 1968;**16**: 350–391.
78. Levkov, C., G. Michov, R. Ivanov, and I.K. Daskalov, Subtraction of 50 Hz interference from the electrocardiogram. *Med. Biol. Eng. Comput.*, 1984;**22**: 371–373.
79. Dotsinsky, I. and T. Stoyanov, Power-line interference cancellation in ECG signals. *Biomed. Instrum. Technol.*, 2005;**39**: 155–162.
80. Levkov, C., G. Mihov, R. Ivanov, I. Daskalov, I. Christov, and I. Dotsinsky, Removal of power-line interference from the ECG: a review of the subtraction procedure. *Biomed. Eng. Online*, 2005;**4**: 50.
81. Widrow, B., J.R. Glover, M. McCool, J. Kaunitz, C.S. Williams, R.H. Hearn, et al., Adaptive noise cancelling: principles and applications. *Proc. IEEE*, 1975;**63**: 1692–1716.
82. Glover, J.R., Adaptive noise canceling applied to sinusoidal interferences. *IEEE Trans. Acoust. Speech Signal Process.*, 1977;**25**: 484–491.
83. Thakor, N.V. and Y.S. Zhu, Applications of adaptive filtering to ECG analysis: noise cancellation and arrhythmia detection. *IEEE Trans. Biomed. Eng.*, 1991;**38**: 785–794.
84. Mortara, D.W., Digital filters for ECG signals, in *Computers in Cardiology 1977*, H.G. Ostrow and K.L. Ripley, Editors. New York: IEEE Comput Soc, 1977, pp. 511–514.
85. Talmon, J.L., *Pattern recognition of the ECG. A structured analysis*, dissertation. Amsterdam: Free University, 1983.
86. Ahlstrom, M.L. and W.J. Tompkins, Digital filters for real-time ECG signal processing using microprocessors. *IEEE Trans. Biomed. Eng.*, 1985;**32**: 708–713.
87. Hamilton, P.S., A comparison of adaptive and nonadaptive filters for reduction of power line interference in the ECG. *IEEE Trans. Biomed. Eng.*, 1996;**43**: 105–109.
88. Glover, J.R., Comments on "Digital filters for real-time ECG signal processing using microprocessors". *IEEE Trans. Biomed. Eng.*, 1987;**34**: 962–963.
89. Pipberger, H.V., R.C. Arzbaecher, A.S. Berson, S.A. Briller, D.A. Brody, N.C. Flowers, et al., Recommendations for standardization of leads and of specifications for instruments in electrocardiography and vectorcardiography: report of the Committee on Electrocardiography, American Heart Association. *Circulation*, 1975;**52**: 11–31.
90. Bailey, J.J., The triangular wave test for electrocardiographic devices: a historical perspective. *J. Electrocardiol.*, 2004;**37**(Suppl): 71–73.
91. van Alste, J.A., W. van Eck, and O.E. Herrmann, ECG baseline wander reduction using linear phase filters. *Comput. Biomed. Res.*, 1986;**19**: 417–427.
92. Sörnmo, L., Time-varying digital filtering of ECG baseline wander. *Med. Biol. Eng. Comput.*, 1993;**31**: 503–508.
93. Shusterman, V., S.I. Shah, A. Beigel, and K.P. Anderson, Enhancing the precision of ECG baseline correction: selective filtering and removal of residual error. *Comput Biomed Res* 2000; **33**:144–160.
94. Macfarlane, P.W., J. Peden, G. Lennox, M.P. Watts, and T.D.V. Lawrie, The Glasgow system, in *Trends in Computer-Processed Electrocardiograms*, J.H. van Bemmel and J.L. Willems, Editors. Amsterdam: North-Holland, 1977, pp. 143–150.
95. Boucheham, B., Y. Ferdi, and M.C. Batouche, Recursive versus sequential multiple error measures reduction: a curve simplification approach to ECG data compression. *Comput. Methods Programs Biomed.*, 2005;**78**: 1–10.
96. Douglas, D.H. and T.K. Peucker, Algorithms for the reduction of the number of points required to represent a digitized line or its caricature. *Can. Cartographer*, 1973;**10**: 112–122.
97. Meyer, C.R. and H.N. Keiser, Electrocardiogram baseline noise estimation and removal using cubic splines and state-space computation techniques. *Comput. Biomed. Res.*, 1977;**10**: 459–470.
98. Gradwohl, J.R., E.W. Pottala, M.R. Horton, and J.J. Bailey, Comparison of two methods for removing baseline wander in the ECG, in *Computers in Cardiology 1988*, K.L. Ripley, Editor. Los Angeles: IEEE Comput Soc, 1988, pp. 493–496.

99. Froning, J.N., M.D. Olson, and V.F. Froelicher, Problems and limitations of ECG baseline estimation and removal using a cubic spline technique during exercise ECG testing: recommendations for proper implementation. *J. Electrocardiol.*, 1988;**21**(Suppl): S149–157.
100. Pottala, E.W., J.J. Bailey, M.R. Horton, and J.R. Gradwohl, Suppression of baseline wander in the ECG using a bilinearly transformed, null-phase filter. *J. Electrocardiol.*, 1989;**22**(Suppl): 243–247.
101. Frankel, R.A., E.W. Pottala, R.W. Bowser, and J.J. Bailey, A filter to suppress ECG baseline wander and preserve ST-segment accuracy in a real-time environment. *J. Electrocardiol.*, 1991;**24**: 315–323.
102. Longini, R.L., J.P. Giolma, C. Wall, and R.F. Quick, Filtering without phase shift. *IEEE Trans. Biomed. Eng.*, 1975;**22**: 432–433.
103. dePinto, V., Filters for the reduction of baseline wander and muscle artifact in the ECG. *J. Electrocardiol.*, 1992;**25**(Suppl): 40–48.
104. van Alste, J.A. and T.S. Schilder, Removal of base-line wander and power-line interference from the ECG by an efficient FIR filter with a reduced number of taps. *IEEE Trans. Biomed. Eng.*, 1985;**32**: 1052–1060.
105. Jane, R., P. Laguna, N.V. Thakor, and P. Caminal, Adaptive baseline wander removal in the ECG: comparative analysis with cubic spline technique, in *Computers in Cardiology 1992*, A. Murray and R.C. Arzbaecher, Editors. Los Alamitos: IEEE Comput Soc, 1992, pp. 143–146.
106. Laguna, P., R. Jane, O. Meste, P.W. Poon, P. Caminal, H. Rix, et al., Adaptive filter for event-related bioelectric signals using an impulse correlated reference input: comparison with signal averaging techniques. *IEEE Trans. Biomed. Eng.*, 1992;**39**: 1032–1044.
107. Moody, G.B. and R.G. Mark, The impact of the MIT-BIH arrhythmia database. *IEEE Eng. Med. Biol. Mag.*, 2001;**20**: 45–50.
108. Park, K.L., K.J. Lee, and H.R. Yoon, Application of a wavelet adaptive filter to minimise distortion of the ST-segment. *Med. Biol. Eng. Comput.*, 1998;**36**: 581–586.
109. Taddei, A., G. Distante, M. Emdin, P. Pisani, G.B. Moody, C. Zeelenberg, et al., The European ST-T database: standard for evaluating systems for the analysis of ST-T changes in ambulatory electrocardiography. *Eur. Heart J.*, 1992;**13**: 1164–1172.
110. Chu, C.H. and E.J. Delp, Nonlinear methods in electrocardiogram signal processing. *J. Electrocardiol.*, 1990;**23**(Suppl): 192–197.
111. Sun, Y., K. Chan, and S.M. Krishnan, ECG signal conditioning by morphological filtering. *Comput. Biol. Med.*, 2002;**32**: 465–479.
112. Chu, C.H. and E.J. Delp, Impulsive noise suppression and background normalization of electrocardiogram signals using morphological operators. *IEEE Trans. Biomed. Eng.*, 1989;**36**: 262–273.
113. Haralick, R.M., S.R. Sternberg, and X. Zhuang, Image analysis using mathematical morphology. *IEEE Trans. Pattern Anal. Mach. Intell.*, 1987;**9**: 532–550.
114. Talmon, J.L., J.A. Kors, and J.H. van Bemmel, Adaptive Gaussian filtering in routine ECG/VCG analysis. *IEEE Trans. Acoust. Speech Signal Process.*, 1986;**34**: 527–534.
115. Hodson, E.K., D.R. Thayer, and C. Franklin, Adaptive Gaussian filtering and local frequency estimates using local curvature analysis. *IEEE Trans. Acoust. Speech Signal Process.*, 1981;**29**: 854–859.
116. Wei, D., E. Harasawa, and H. Hosaka, A low-distortion filter method to reject muscle noise in multi-lead electrocardiogram systems. *Front Med. Biol. Eng.*, 1999;**9**: 315–330.
117. Acar, B. and H. Koymen, SVD-based on-line exercise ECG signal orthogonalization. *IEEE Trans. Biomed. Eng.*, 1999;**46**: 311–321.
118. Paul, J.S., M.R. Reddy, and V.J. Kumar, A transform domain SVD filter for suppression of muscle noise artefacts in exercise ECG's. *IEEE Trans. Biomed. Eng.*, 2000;**47**: 654–663.
119. Nikolaev, N., A. Gotchev, K. Egiazarian, and Z. Nikolov, Suppression of electromyogram interference on the electrocardiogram by transform domain denoising. *Med. Biol. Eng. Comput.*, 2001;**39**: 649–655.
120. Raphisak, P., S.C. Schuckers, and A. de Jongh Curry, An algorithm for EMG noise detection in large ECG data, in *Computers in Cardiology 2004*, A. Murray, Editor. Piscataway, NJ: IEEE Comput Soc, 2004, pp. 369–372.
121. Brohet, C.R., C. Derwael, A. Robert, and R. Fesler, Methodology of ECG interpretation in the Louvain program. *Methods Inf. Med.*, 1990;**29**: 403–409.
122. Willems, J.L., *Common Standards for Quantitative Electrocardiography*, 2nd CSE Progress Report. Leuven: ACCO, 1982.
123. Pipberger, H.V., C.D. McManus, and H.A. Pipberger, Methodology of ECG interpretation in the AVA program. *Methods Inf. Med.*, 1990;**29**: 337–340.
124. Helfenbein, E.D., J.M. Lindauer, S.H. Zhou, R.E. Gregg, and E.C. Herleikson, A software-based pacemaker pulse detection and paced rhythm classification algorithm. *J. Electrocardiol.*, 2002;**35**(Suppl): 95–103.
125. Hamilton, P.S. and W.J. Tompkins, Quantitative investigation of QRS detection rules using the MIT/BIH arrhythmia database. *IEEE Trans. Biomed. Eng.*, 1986;**33**: 1157–1165.
126. Friesen, G.M., T.C. Jannett, M.A. Jadallah, S.L. Yates, S.R. Quint, and H.T. Nagle, A comparison of the noise sensitivity of nine QRS detection algorithms. *IEEE Trans. Biomed. Eng.*, 1990;**37**: 85–98.
127. Hu, Y.H., W.J. Tompkins, J.L. Urrusti, and V.X. Afonso, Applications of artificial neural networks for ECG signal detection and classification. *J. Electrocardiol.*, 1993;**26**(Suppl): 66–73.
128. Suppappola, S. and Y. Sun, Nonlinear transforms of ECG signals for digital QRS detection: a quantitative analysis. *IEEE Trans. Biomed. Eng.*, 1994;**41**: 397–400.
129. Afonso, V.X., W.J. Tompkins, T.Q. Nguyen, and S. Luo, ECG beat detection using filter banks. *IEEE Trans. Biomed. Eng.*, 1999;**46**: 192–202.
130. Kohler, B.U., C. Hennig, and R. Orglmeister, The principles of software QRS detection. *IEEE Eng. Med. Biol. Mag.*, 2002;**21**: 42–57.
131. Martinez, J.P., R. Almeida, S. Olmos, A.P. Rocha, and P. Laguna, A wavelet-based ECG delineator: evaluation on standard databases. *IEEE Trans. Biomed. Eng.*, 2004;**51**: 570–581.
132. Pahlm, O. and L. Sörnmo, Software QRS detection in ambulatory monitoring—a review. *Med. Biol. Eng. Comput.*, 1984;**22**: 289–297.
133. Li, C., C. Zheng, and C. Tai, Detection of ECG characteristic points using wavelet transforms. *IEEE Trans. Biomed. Eng.*, 1995;**42**: 21–28.
134. Kadambe, S., R. Murray, and G.F. Boudreaux-Bartels, Wavelet transform-based QRS complex detector. *IEEE Trans. Biomed. Eng.*, 1999;**46**: 838–848.
135. Vijaya, G., V. Kumar, and H.K. Verma, ANN-based QRS-complex analysis of ECG. *J. Med. Eng. Technol.*, 1998;**22**: 160–167.

136. Poli, R., S. Cagnoni, and G. Valli, Genetic design of optimum linear and nonlinear QRS detectors. *IEEE Trans. Biomed. Eng.*, 1995;**42**: 1137–1141.
137. Belforte, G., R. De Mori, and F. Ferraris, A contribution to the automatic processing of electrocardiograms using syntactic methods. *IEEE Trans. Biomed. Eng.*, 1979;26: 125–136.
138. Papakonstantinou, G., E. Skordalakis, and F. Gritzali, An attribute grammar for QRS detection. *Pattern Recog.*, 1986;**19**: 297–303.
139. Skordalakis, E., Syntactic ECG processing: a review. *Pattern Recog.*, 1986;**19**: 305–313.
140. Kors, J.A., J.L. Talmon, and J.H. van Bemmel, Multilead ECG analysis. *Comput. Biomed. Res.*, 1986;**19**: 28–46.
141. Macfarlane, P.W., B. Devine, S. Latif, S. McLaughlin, D.B. Shoat, and M.P. Watts, Methodology of ECG interpretation in the Glasgow program. *Methods Inf. Med.*, 1990;**29**: 354–361.
142. Christov, I., G. Bortolan, and I. Daskalov, Automatic detection of atrial fibrillation and flutter by wave rectification method. *J. Med. Eng. Technol.*, 2001;**25**: 217–221.
143. Balda, R.A., G. Diller, E. Deardorff, J.C. Doue, and P. Hsieh, The HP ECG analysis program, in *Trends in Computer-Processed Electrocardiograms*, J.H. van Bemmel and J.L. Willems, Editors. Amsterdam: North-Holland, 1977, pp. 197–204.
144. Laguna, P., R. Jane, and P. Caminal, Automatic detection of wave boundaries in multilead ECG signals: validation with the CSE database. *Comput. Biomed. Res.*, 1994;**27**: 45–60.
145. Willems, J.L. and H.V. Pipberger, Arrhythmia detection by digital computer. *Comput. Biomed. Res.*, 1972;**5**: 273–278.
146. McManus, C.D., A re-examination of automatic P-wave recognition methods, in *Optimization of Computer ECG Processing*, H.K. Wolf and P.W. Macfarlane, Editors. Amsterdam: North-Holland, 1980, pp. 121–127.
147. Bonner, R.E. and H.D. Schwetman, Computer diagnosis of electrocardiograms. II. A computer program for EKG measurements. *Comput. Biomed. Res.*, 1968;**1**: 366–386.
148. Hengeveld, S.J. and J.H. Bemmel, Computer detection of P-waves. *Comput. Biomed. Res.*, 1976;**9**: 125–132.
149. Schnyders, H.C. and M. Jordan, Energy correlation technique for small P-wave detection in the presence of noise, in *Computers in Cardiology 1980*, K.L. Ripley and H.G. Ostrow, Editors. Los Angeles: IEEE Comput Soc, 1980, pp. 161–164.
150. Gritzali, F., G. Frangakis, and G. Papakonstantinou, Detection of the P and T waves in an ECG. *Comput. Biomed. Res.*, 1989;**22**: 83–91.
151. Talmon, J.L., J.A. Kors, and J.H. van Bemmel, Algorithms for the detection of events in electrocardiograms. *Comput. Methods Programs Biomed.*, 1986;**22**: 149–161.
152. Taha, B., S. Reddy, Q. Xue, and S. Swiryn, Automated discrimination between atrial fibrillation and atrial flutter in the resting 12-lead electrocardiogram. *J. Electrocardiol.*, 2000;**33**(Suppl): 123–125.
153. Giraldo, B.F., P. Laguna, R. Jane, and P. Caminal, Automatic detection of atrial fibrillation and flutter using the differentiated ECG signal, in *Computers in Cardiology 1995*, A. Murray and R.C. Arzbaecher, Editors. Piscataway, NJ: IEEE Comput Soc, 1995, pp. 369–372.
154. Bonner, R.E., L. Crevasse, M.I. Ferrer, and J.C. Greenfield, A new computer program for analysis of scalar electrocardiograms. *Comput. Biomed. Res.*, 1972;**5**: 629–653.
155. van Bemmel, J.H. and S.J. Hengeveld, Clustering algorithm for QRS and ST-T waveform typing. *Comput. Biomed. Res.*, 1973;**6**: 442–456.
156. Simoons, M.L., H.B. Boom, and E. Smallenburg, On-line processing of orthogonal exercise electrocardiograms. *Comput. Biomed. Res.*, 1975;**8**: 105–117.
157. Moraes, J.C.T.B., M.O. Seixas, F.N. Vilani, and E.V. Costa, A real time QRS complex classification method using Mahalanobis distance, in *Computers in Cardiology 2002*, A. Murray, Editor. Piscataway, NJ: IEEE Comput Soc, 2002, pp. 201–204.
158. Lagerholm, M., C. Peterson, G. Braccini, L. Edenbrandt, and L. Sörnmo, Clustering ECG complexes using hermite functions and self-organizing maps. *IEEE Trans. Biomed. Eng.*, 2000;**47**: 838–848.
159. Hu, Y.H., A patient-adaptable ECG beat classifier using a mixture of experts approach. *IEEE Trans. Biomed. Eng.*, 1997;**44**: 891–900.
160. Wieben, O., V.X. Afonso, and W.J. Tompkins, Classification of premature ventricular complexes using filter bank features, induction of decision trees and a fuzzy rule-based system. *Med. Biol. Eng. Comput.*, 1999;**37**: 560–565.
161. de Chazal, P., M. O'Dwyer, and R.B. Reilly, Automatic classification of heartbeats using ECG morphology and heartbeat interval features. *IEEE Trans. Biomed. Eng.*, 2004;**51**: 1196–1206.
162. Christov, I., I. Jekova, and G. Bortolan, Premature ventricular contraction classification by the Kth nearest-neighbours rule. *Physiol. Meas.*, 2005;**26**: 123–130.
163. Zywietz, C., D. Borovsky, G. Gotsch, and G. Joseph, Methodology of ECG interpretation in the Hannover program. *Methods Inf. Med.*, 1990;**29**: 375–385.
164. Rautaharju, P.M., P.J. MacInnis, J.W. Warren, H.K. Wolf, P.M. Rykers, and H.P. Calhoun, Methodology of ECG interpretation in the Dalhousie program; NOVACODE ECG classification procedures for clinical trials and population health surveys. *Methods Inf. Med.*, 1990;**29**: 362–374.
165. Rompelman, O. and H.H. Ros, Coherent averaging technique: a tutorial review. Part 1: Noise reduction and the equivalent filter. *J. Biomed. Eng.*, 1986;**8**: 24–29.
166. Rompelman, O. and H.H. Ros, Coherent averaging technique: a tutorial review. Part 2: Trigger jitter, overlapping responses and non-periodic stimulation. *J. Biomed. Eng.*, 1986;**8**: 30–35.
167. Mertens, J. and D.W. Mortara, A new algorithm for QRS averaging, in *Computers in Cardiology 1984*, K.L. Ripley, Editor. Long Beach: IEEE Comput Soc, 1984, pp. 367–369.
168. Arnaud, P., P. Rubel, D. Morlet, J. Fayn, and M.C. Forlini, Methodology of ECG interpretation in the Lyon program. *Methods Inf. Med.*, 1990;**29**: 393–402.
169. Goetowski, C.R., The Telemed system, in *Trends in Computer-Processed Electrocardiograms*, J.H. van Bemmel and J.L. Willems, Editors. Amsterdam: North-Holland, 1977, pp. 207–210.
170. Degani, R. and G. Bortolan, Methodology of ECG interpretation in the Padova program. *Methods Inf. Med.*, 1990;**29**: 386–392.
171. Willems, J.L., C. Zywietz, P. Arnaud, J.H. van Bemmel, R. Degani, and P.W. Macfarlane, Influence of noise on wave boundary recognition by ECG measurement programs. Recommendations for preprocessing. *Comput. Biomed. Res.*, 1987;**20**: 543–562.
172. Zywietz, C., J.L. Willems, P. Arnaud, J.H. van Bemmel, R. Degani, P.W. Macfarlane, et al., Stability of computer ECG amplitude

measurements in the presence of noise. *Comput. Biomed. Res.*, 1990;**23**: 10–31.
173. Day, C.P., J.M. McComb, and R.W. Campbell, QT dispersion: an indication of arrhythmia risk in patients with long QT intervals. *Br. Heart J.*, 1990;**63**: 342–344.
174. Kors, J.A., G. van Herpen, and J.H. van Bemmel, QT dispersion as an attribute of T-loop morphology. *Circulation*, 1999;**99**: 1458–1463.
175. Rautaharju, P.M., QT and dispersion of ventricular repolarization: the greatest fallacy in electrocardiography in the 1990s. *Circulation*, 1999;**99**: 2477–2478.
176. Malik, M., B. Acar, Y. Gang, Y.G. Yap, K. Hnatkova, and A.J. Camm, QT dispersion does not represent electrocardiographic interlead heterogeneity of ventricular repolarization. *J. Cardiovasc. Electrophysiol.*, 2000;**11**: 835–843.
177. van Herpen, G., H.J. Ritsema van Eck, and J.A. Kors, The evidence against QT dispersion. *Int. J. Bioelectromagn.*, 2003;**5**: 231–233.
178. Ritsema van Eck, H.J., J.A. Kors, and G. van Herpen, The U wave in the electrocardiogram: a solution for a 100-year-old riddle. *Cardiovasc. Res.*, 2005;**67**: 256–262.
179. Nygards, M.E. and L. Sörnmo, Delineation of the QRS complex using the envelope of the e.c.g. *Med. Biol. Eng. Comput.*, 1983;**21**: 538–547.
180. van Bemmel, J.H., C. Zywietz, and J.A. Kors, Signal analysis for ECG interpretation. *Methods Inf. Med.*, 1990;**29**: 317–329.
181. McLaughlin, N.B., R.W. Campbell, and A. Murray, Comparison of automatic QT measurement techniques in the normal 12 lead electrocardiogram. *Br. Heart J.*, 1995;**74**: 84–89.
182. McLaughlin, N.B., R.W. Campbell, and A. Murray, Accuracy of four automatic QT measurement techniques in cardiac patients and healthy subjects. *Heart*, 1996;**76**: 422–426.
183. Vila, J.A., Y. Gang, J.M. Rodriguez Presedo, M. Fernandez-Delgado, S. Barro, and M. Malik, A new approach for TU complex characterization. *IEEE Trans. Biomed. Eng.*, 2000;**47**: 764–772.
184. Rubel, P. and B. Ayad, The true boundary recognition power of multidimensional detection functions. An optimal comparison, in *Computer ECG Analysis: Towards Standardization*, J.L. Willems, J.H. van Bemmel, and C. Zywietz, Editors. Amsterdam: North-Holland, 1986, pp. 97–103.
185. van Bemmel, J.H., J.L. Talmon, J.S. Duisterhout, and S.J. Hengeveld, Template waveform recognition applied to ECG-VCG analysis. *Comput. Biomed. Res.*, 1973;**6**: 430–441.
186. Zhang, Q., A. Illanes Manriquez, C. Medigue, Y. Papelier, and M. Sorine, Robust and efficient location of T-wave ends in electrocardiogram, in *Computers in Cardiology*, A. Murray, Editor. Piscataway, NJ: IEEE Comput Soc, 2005, pp. 711–714.
187. Sörnmo, L., A model-based approach to QRS delineation. *Comput. Biomed. Res.*, 1987;**20**: 526–542.
188. Morlet, D., P. Rubel, P. Arnaud, and J.L. Willems, An improved method to evaluate the precision of computer ECG measurement programs. *Int. J. Biomed. Comput.*, 1988;**22**: 199–216.
189. Kors, J.A., G. van Herpen, A.C. Sittig, and J.H. van Bemmel, Reconstruction of the Frank vectorcardiogram from standard electrocardiographic leads: diagnostic comparison of different methods. *Eur. Heart J.*, 1990;**11**: 1083–1092.
190. Young, T.Y. and W.H. Huggins, Intrinsic component theory of electrocardiograms. *IEEE Trans. Biomed. Eng.*, 1963;**9**: 214–221.
191. Horan, L.G., N.C. Flowers, and D.A. Brody, Principal factor waveforms of the thoracic QRS complex. *Circ. Res.*, 1964;**15**: 131–145.
192. Willems, J.L., Introduction to multivariate and conventional computer ECG analysis: pro's and contra's, in *Trends in Computer-Processed Electrocardiograms*, J.H. van Bemmel and J.L. Willems, Editors. Amsterdam: North-Holland, 1977, pp. 213–220.
193. Smets, P., New quantified approach for diagnostic classification, in *Optimization of Computer ECG Processing*, H.K. Wolf and P.W. Macfarlane, Editors. Amsterdam: North-Holland, 1980, pp. 229–237.
194. Degani, R. and G. Bortolan, Combining measurement precision and fuzzy diagnostic criteria, in *Computer ECG Analysis: Towards Standardization*, J.L. Willems, J.H. van Bemmel, and C. Zywietz, Editors. Amsterdam: North-Holland, 1986, pp. 177–182.
195. Doue, J.C., The role of artificial intelligence in standardizing ECG criteria, in *Computer ECG Analysis: Towards Standardization*, J.L. Willems, J.H. van Bemmel, and C. Zywietz, Editors. Amsterdam: North-Holland, 1986, pp. 53–57.
196. Matthes, T., G. Götsch, and C. Zywietz, Interactive analysis of statistical ECG diagnosis on an intelligent electrocardiograph. An expert system approach, in *Computer ECG Analysis: Towards Standardization*, J.L. Willems, J.H. van Bemmel, and C. Zywietz, Editors. Amsterdam: North-Holland, 1986, pp. 215–220.
197. Edenbrandt, L., B. Devine, and P.W. Macfarlane, Neural networks for classification of ECG ST-T segments. *J. Electrocardiol.*, 1992;**25**: 167–173.
198. Yang, T.F., B. Devine, and P.W. Macfarlane, Use of artificial neural networks within deterministic logic for the computer ECG diagnosis of inferior myocardial infarction. *J. Electrocardiol.*, 1994;**27**(Suppl): 188–193.
199. Kennedy, R.L., A.M. Burton, and R.F. Harrison, Neural networks and early diagnosis of myocardial infarction. *Lancet*, 1996;**347**: 407.
200. Hedén, B., H. Ohlin, R. Rittner, and L. Edenbrandt, Acute myocardial infarction detected in the 12-lead ECG by artificial neural networks. *Circulation*, 1997;**96**: 1798–1802.
201. Olsson, S.E., M. Ohlsson, H. Ohlin, and L. Edenbrandt, Neural networks—a diagnostic tool in acute myocardial infarction with concomitant left bundle branch block. *Clin. Physiol. Funct. Imaging*, 2002;**22**: 295–299.
202. Segall, H.N., The electrocardiogram and its interpretation: a study of reports by 20 physicians on a set of 100 electrocardiograms. *Can. Med. Assoc. J.*, 1960;**82**: 847–850.
203. Simonson, E., N. Tuna, and N. Okamoto, Diagnostic accuracy of the vectorcardiogram and electrocardiogram. A cooperative study. *Am. J. Cardiol.*, 1966;**17**: 829–878.
204. Willems, J.L., C. Abreu-Lima, P. Arnaud, J.H. van Bemmel, C. Brohet, R. Degani, et al., Effect of combining electrocardiographic interpretation results on diagnostic accuracy. *Eur. Heart J.*, 1988;**9**: 1348–1355.
205. Balda, R.A., A.G. Vallance, J.M. Luszcz, F.J. Stahlin, and G. Diller, ECL—a medically oriented ECG criteria language and other clinical research tools, in *Computers in Cardiology 1977*, H.G. Ostrow and K.L. Ripley, Editors. New York: IEEE Comput Soc, 1977, pp. 481–495.
206. Bruce, R.A. and S.R. Yarnall, Reliability and normal variations of computer analysis of Frank electrocardiogram by Smith-Hyde program (1968 version). *Am. J. Cardiol.*, 1972;**29**: 389–396.

207. Willems, J.L., C. Abreu-Lima, P. Arnaud, J.H. van Bemmel, C. Brohet, R. Degani, et al., The diagnostic performance of computer programs for the interpretation of electrocardiograms. *N. Engl. J. Med.*, 1991;**325**: 1767–1773.
208. Breiman, L., J.H. Friedman, R.A. Olshen, and C.J. Stone, *Classification and Regression Trees*. Belmont, CA: Wadsworth, 1984.
209. Kors, J.A. and A.L. Hoffmann, Induction of decision rules that fulfill user-specified performance requirements. *Pattern Recognit. Lett.*, 1997;**18**: 1187–1195.
210. Pipberger, H.V., R.A. Dunn, and J. Cornfield, First and second generation computer programs for diagnostic ECG and VCG classification, in *XIIth International Colloquium Vectorcardiographicum*, P. Rijlant, Editor. Brussels: Presses Académiques Européennes, 1972, pp. 431–439.
211. Willems, J.L. and J. Pardaens, Reproducibility of diagnostic results by a multivariate computer ECG analysis program (AVA 3.5). *Eur. J. Cardiol.*, 1977;**6**: 229–243.
212. Dunn, R.A., R. Babuska, J.M. Wojick, and H.V. Pipberger, Variation in probability levels in electrocardiographic diagnosis. *Comput. Biomed. Res.*, 1978;**11**: 41–49.
213. Cady, L.D., M.A. Woodbury, L.J. Tick, and M.M.A. Gertler, Method for electrocardiogram wave pattern estimation. Example: left ventricular hypertrophy. *Circ. Res.*, 1961;**9**: 1078–1082.
214. Kimura, E., Y. Mibukura, and A. Miura, Statistical diagnosis of electrocardiogram by theorem of Bayes. A preliminary report. *Jpn. Heart J.*, 1963;**4**: 469–488.
215. Young, T.Y. and W.H. Huggins, Computer analysis of electrocardiograms using a linear regression technique. *IEEE Trans. Biomed. Eng.*, 1964;**11**: 60–67.
216. Stark, L., J.F. Dickson, G.H. Whipple, and H. Horibe, Remote real-time diagnosis of clinical electrocardiograms by a digital computer system. *Ann. N.Y. Acad. Sci.*, 1965;**126**: 851–872.
217. Specht, D.F., Vectorcardiographic diagnosis using the polynomial discriminant method of pattern recognition. *IEEE Trans. Biomed. Eng.*, 1967;**14**: 90–95.
218. Yasui, S., M. Yokoi, Y. Watanabe, K. Nishijima, and S. Azuma, Computer diagnosis of electrocardiograms by means of the joint probability. *Jpn. Circ. J.*, 1968;**32**: 517–523.
219. Goldman, M.J. and H.V. Pipberger, Analysis of the orthogonal electrocardiogram and vectorcardiogram in ventricular conduction defects with and without myocardial infarction. *Circulation*, 1969;**39**: 243–250.
220. Kerr, A., A. Adicoff, J.D. Klingeman, and H.V. Pipberger, Computer analysis of the orthogonal electrocardiogram in pulmonary emphysema. *Am. J. Cardiol.*, 1970;**25**: 34–45.
221. Eddleman, E.E. and H.V. Pipberger, Computer analysis of the orthogonal electrocardiogram and vectorcardiogram in 1,002 patients with myocardial infarction. *Am. Heart J.*, 1971;**81**: 608–621.
222. Pipberger, H.V., ECG computer analysis: past, present and future, in *Computer ECG Analysis: Towards Standardization*, J.L. Willems, J.H. van Bemmel, and C. Zywietz, Editors. Amsterdam: North-Holland, 1986, pp. 3–10.
223. Brohet, C.R., A. Robert, C. Derwael, R. Fesler, M. Stijns, A. Vliers, et al., Computer interpretation of pediatric orthogonal electrocardiograms: statistical and deterministic classification methods. *Circulation*, 1984;**70**: 255–262.
224. Willems, J.L., E. Lesaffre, J. Pardaens, and D. de Schreye, Multivariate logistic classification of the standard 12- and 3-lead ECG, in *Computer ECG Analysis: Towards Standardization*, J.L. Willems, J.H. van Bemmel, and C. Zywietz, Editors. Amsterdam: North-Holland, 1986, pp. 203–210.
225. Rios, J., F. Sandquist, D. Ramseth, R. Stratbucker, E. Drazen, and J. Hanmer, The quest for optimal electrocardiography. Tast force V: cost effectiveness of the electrocardiogram. *Am. J. Cardiol.*, 1978;**41**: 175–183.
226. Okajima, M., Current status and future optimization of computerized electrocardiography in Japan, in *Optimization of Computer ECG Processing*, H.K. Wolf and P.W. Macfarlane, Editors. Amsterdam: North-Holland, 1980, pp. 293–307.
227. Moorman, J.R., M.A. Hlatky, D.M. Eddy, and G.S. Wagner, The yield of the routine admission electrocardiogram. A study in a general medical service. *Ann. Intern. Med.*, 1985;**103**: 590–595.
228. Salerno, S.M., P.C. Alguire, and H.S. Waxman, Competency in interpretation of 12-lead electrocardiograms: a summary and appraisal of published evidence. *Ann. Intern. Med.*, 2003;**138**: 751–760.
229. Eisenstein, E.L., Conducting an economic analysis to assess the electrocardiogram's value. *J. Electrocardiol.*, 2006;**39**: 241–247.
230. Willems, J.L., C. Abreu-Lima, P. Arnaud, C.R. Brohet, B. Denis, J. Gehring, et al., Evaluation of ECG interpretation results obtained by computer and cardiologists. *Methods Inf. Med.*, 1990;**29**: 308–316.
231. Farb, A., R.B. Devereux, and P. Kligfield, Day-to-day variability of voltage measurements used in electrocardiographic criteria for left ventricular hypertrophy. *J. Am. Coll. Cardiol.*, 1990;**15**: 618–623.
232. van den Hoogen, J.P., W.H. Mol, A. Kowsoleea, J.W. van Ree, T. Thien, and C. van Weel, Reproducibility of electrocardiographic criteria for left ventricular hypertrophy in hypertensive patients in general practice. *Eur. Heart J.*, 1992;**13**: 1606–1610.
233. de Bruyne, M.C., J.A. Kors, S. Visentin, G. van Herpen, A.W. Hoes, D.E. Grobbee, et al., Reproducibility of computerized ECG measurements and coding in a nonhospitalized elderly population. *J. Electrocardiol.*, 1998;**31**: 189–195.
234. Bailey, J.J., M. Horton, S.B. Itscoitz, A method for evaluating computer programs for electrocardiographic interpretation. 3. Reproducibility testing and the sources of program errors. *Circulation*, 1974;**50**: 88–93.
235. Bailey, J.J., M. Horton, and S.B. Itscoitz, The importance of reproducibility testing of computer programs for electrocardiographic interpretation: application to the automatic vectorcardiographic analysis program (AVA 3.4). *Comput. Biomed. Res.*, 1976;**9**: 307–316.
236. Spodick, D.H. and R.L. Bishop, Computer treason: intraobserver variability of an electrocardiographic computer system. *Am. J. Cardiol.*, 1997;**80**: 102–103.
237. Michaelis, J., S. Wellek, and J.L. Willems, Reference standards for software evaluation. *Methods Inf. Med.*, 1990;**29**: 289–297.
238. Jakobsson, A., P. Ohlin, and O. Pahlm, Does a computer-based ECG-recorder interpret electrocardiograms more efficiently than physicians? *Clin. Physiol.*, 1985;**5**: 417–423.
239. Bernard, P., B.R. Chaitman, J.M. Scholl, P.G. Val, and M. Chabot, Comparative diagnostic performance of the Telemed computer ECG program. *J. Electrocardiol.*, 1983;**16**: 97–103.
240. Willems, J.L., Is human verification of computerized ECGs mandatory? *Adv. Cardiol.*, 1978;**21**: 193–194.
241. Chou, T.C., When is the vectorcardiogram superior to the scalar electrocardiogram? *J. Am. Coll. Cardiol.*, 1986;**8**: 791–799.
242. Kors, J.A., G. van Herpen, J.L. Willems, and J.H. van Bemmel, Improvement of automated electrocardiographic diagnosis by

combination of computer interpretations of the electrocardiogram and vectorcardiogram. *Am. J. Cardiol.*, 1992;**70**: 96–99.
243. Kors, J.A., G. van Herpen, and J.H. van Bemmel, Variability in ECG computer interpretation. Analysis of individual complexes vs analysis of a representative complex. *J. Electrocardiol.*, 1992;**25**: 263–271.
244. Andresen, A., J. Dobkin, C. Maynard, R. Myers, G.S. Wagner, R.A. Warner, et al., Validation of advanced ECG diagnostic software for the detection of prior myocardial infarction by using nuclear cardiac imaging. *J. Electrocardiol.*, 2001;**34**(Suppl): 243–248.
245. Andresen, A., M.D. Gasperina, R. Myers, G.S. Wagner, R.A. Warner, and R.H. Selvester, An improved automated ECG algorithm for detecting acute and prior myocardial infarction. *J. Electrocardiol.*, 2002;**35**(Suppl): 105–110.
246. Wagner, G.S., C. Maynard, A. Andresen, E. Anderson, R. Myers, R.A. Warner, et al., Evaluation of advanced electrocardiographic diagnostic software for detection of prior myocardial infarction. *Am. J. Cardiol.*, 2002;**89**: 75–79.
247. Milliken, J.A., H. Pipberger, H.V. Pipberger, M.A. Araoye, R. Ari, G.W. Burggraf, et al., The impact of an ECG computer analysis program on the cardiologist's interpretation. A cooperative study. *J. Electrocardiol.*, 1983;**16**: 141–149.
248. Hillson, S.D., D.P. Connelly, and Y. Liu, The effects of computer-assisted electrocardiographic interpretation on physicians' diagnostic decisions. *Med. Decis. Making*, 1995;**15**: 107–112.
249. Brailer, D.J., E. Kroch, and M.V. Pauly, The impact of computer-assisted test interpretation on physician decision making: the case of electrocardiograms. *Med. Decis. Making*, 1997;**17**: 80–86.
250. Goodacre, S., A. Webster, and F. Morris, Do computer generated ECG reports improve interpretation by accident and emergency senior house officers? *Postgrad. Med. J.*, 2001;**77**: 455–457.
251. Tsai, T.L., D.B. Fridsma, and G. Gatti, Computer decision support as a source of interpretation error: the case of electrocardiograms. *J. Am. Med. Inform. Assoc.*, 2003;**10**: 478–483.
252. Laks, M.M. and R.H. Selvester, Computerized electrocardiography – an adjunct to the physician. *N. Engl. J. Med.*, 1991;**325**: 1803–1804.
253. Plokker, H.W.M., *Cardiac Rhythm Diagnosis by Digital Computer*, dissertation. Amsterdam: Free University, 1978.
254. Reddy, S., B. Young, Q. Xue, B. Taha, D. Brodnick, and J. Steinberg, Review of methods to predict and detect atrial fibrillation in post-cardiac surgery patients. *J. Electrocardiol.*, 1999;**32**(Suppl): 23–28.
255. Poon, K., P.M. Okin, and P. Kligfield, Diagnostic performance of a computer-based ECG rhythm algorithm. *J. Electrocardiol.*, 2005;**38**: 235–238.
256. Bonner, R.E. and H.D. Schwetman, Computer diagnosis of electrocardiograms. 3. A computer program for arrhythmia diagnosis. *Comput. Biomed. Res.*, 1968;**1**: 387–407.
257. Bonner, R.E., IBM rhythm analysis program, in *Computer Application on ECG and VCG Analysis*, C. Zywietz and B. Schneider, Editors. Amsterdam: North-Holland, 1973, pp. 375–397.
258. Wartak, J., J.A. Milliken, and J. Karchmar, Computer program for diagnostic evaluation of electrocardiograms. *Comput. Biomed. Res.*, 1971;**4**: 225–238.
259. Brohet, C., C. Derwael, R. Fesler, and L.A. Brasseur, Arrhythmia analysis by the Louvain VCG program, in *Computers in Cardiology*, K.L. Ripley, Editor. Los Angeles: IEEE Comput Soc, 1982, pp. 47–51.
260. Shirataka, M., H. Miyahara, N. Ikeda, A. Domae, and T. Sato, Evaluation of five computer programs in the diagnosis of second-degree AV block. *J. Electrocardiol.*, 1992;**25**: 185–195.
261. Thomson, A., S. Mitchell, and P.J. Harris, Computerized electrocardiographic interpretation: an analysis of clinical utility in 5110 electrocardiograms. *Med. J. Aust.*, 1989;**151**: 428–430.
262. Reddy, B.R., B. Taha, S. Swiryn, R. Silberman, and R. Childers, Prospective evaluation of a microprocessor-assisted cardiac rhythm algorithm: results from one clinical center. *J. Electrocardiol.*, 1998;**30**(Suppl): 28–33.
263. Farrell, R.M., J.Q. Xue, and B.J. Young, Enhanced rhythm analysis for resting ECG using spectral and time-domain techniques, in *Computers in Cardiology*, A. Murray, Editor. Piscataway, NJ: IEEE Comput Soc, 2003, pp. 733–736.
264. Guglin, M.E. and D. Thatai, Common errors in computer electrocardiogram interpretation. *Int. J. Cardiol.*, 2006;**106**: 232–237.
265. Pryor, T.A., A.E. Lindsay, and R.W. England, Computer analysis of serial electrocardiograms. *Comput. Biomed. Res.*, 1972;5: 709–714.
266. Macfarlane, P.W., H.T. Cawood, and T.D. Lawrie, A basis for computer interpretation of interpretation of serial electrocardiograms. *Comput. Biomed. Res.*, 1975;**8**: 189–200.
267. Bonner, R.E., L. Crevasse, M.I. Ferrer, and J.C. Greenfield, A new computer program for comparative analysis of serial scalar electrocardiograms: description and performance of the 1976 IBM program. *Comput. Biomed. Res.*, 1978;**11**: 103–118.
268. Schnyders, H.C. and G.A. Kien, Computer-assisted serial comparison of ECGs: the Telemed version, in *Computer Application in Medical Care*, R.A. Dunn, Editor. New York: IEEE Comput Soc, 1979, pp. 652–659.
269. Rubel, P., J. Fayn, J.L. Willems, and C. Zywietz, New trends in serial ECG analysis. *J. Electrocardiol.*, 1993;**26**(Suppl): 122–128.
270. Rubel, P., J. Fayn, G. Nollo, D. Assanelli, B. Li, L. Restier, et al, Toward personal eHealth in cardiology. Results from the EPI-MEDICS telemedicine project. *J. Electrocardiol.*, 2005;**38**(Suppl): 100–106.
271. Willems, J.L., P.F. Poblete, and H.V. Pipberger, Day-to-day variation of the normal orthogonal electrocardiogram and vectorcardiogram. *Circulation*, 1972;**45**: 1057–1064.
272. de Bruyne, M.C., J.A. Kors, S. Visentin, G. van Herpen, A.W. Hoes, D.E. Grobbee, et al., Reproducibility of computerized ECG measurements and coding in a nonhospitalized elderly population. *J. Electrocardiol.*, 1998;**31**: 189–195.
273. Schijvenaars, B.J., *Intra-individual Variability of the Electrocardiogram*, dissertation. Rotterdam: Erasmus University, 2000.
274. Pipberger, H.V., R.A. Dunn, and H.A. Pipberger, Automated comparison of serial electrocardiograms. *Adv. Cardiol.*, 1976;**16**: 157–165.
275. Rubel, P., N. Saccal, J.L. Sourrouille, and M.C. Forlini, Multidimensional techniques for the optimal display of trends in sequential vectorcardiograms, in *Medinfo 1980*, D.A.B. Lindberg and S. Kaihara, Editors. Amsterdam: North-Holland, 1980, p. 274.
276. Zywietz, C., B. Widiger, and R. Fischer, A system for comprehensive comparison of serial ECG beats and serial ECG recordings, in *Computers in Cardiology*, A. Murray, Editor. Piscataway, NJ: IEEE Comput Soc, 2003, pp. 689–692.
277. Hedstrom, K. and P.W. Macfarlane, Development of a new approach to serial analysis. The manufacturer's viewpoint. *J. Electrocardiol.*, 1996;**29**(Suppl): 35–40.

278. McLaughlin, S.C., T.C. Aitchison, and P.W. Macfarlane, Methods for improving the repeatability of automated ECG analysis. *Methods Inf. Med.*, 1995;**34**: 272–282.
279. McLaughlin, S.C., P. Chishti, T.C. Aitchison, and P.W. Macfarlane, Techniques for improving overall consistency of serial ECG analysis. *J. Electrocardiol.*, 1996;**29**(Suppl): 41–45.
280. Fayn, J. and P. Rubel, CAVIAR: a serial ECG processing system for the comparative analysis of VCGs and their interpretation with auto-reference to the patient. *J. Electrocardiol.*, 1988;**21**(Suppl): S173–S176.
281. Sunemark, M., L. Edenbrandt, H. Holst, and L. Sörnmo, Serial VCG/ECG analysis using neural networks. *Comput. Biomed. Res.*, 1998;**31**: 59–69.
282. Bonner, R.E., L. Crevasse, M.I. Ferrer, and J.C. Greenfield, The influence of editing on the performance of a computer program for serial comparison of electrocardiograms. *J. Electrocardiol.*, 1983;**16**: 181–189.
283. van Haelst, A.C., D.K. Donker, F.C. Visser, C.C. de Cock, A. Hasman, and J.L. Talmon, A computer program for the analysis of serial electrocardiograms from patients who suffered a myocardial infarction. *Int. J. Biomed. Comput.*, 1985;**17**: 273–284.
284. Ohlsson, M., H. Ohlin, S.M. Wallerstedt, and L. Edenbrandt, Usefulness of serial electrocardiograms for diagnosis of acute myocardial infarction. *Am. J. Cardiol.*, 2001;**88**: 478–481.
285. Schwartz, P.J., A. Garson, T. Paul, M. Stramba-Badiale, V.L. Vetter, and C. Wren, Guidelines for the interpretation of the neonatal electrocardiogram. A task force of the European Society of Cardiology. *Eur. Heart J.*, 2002;**23**: 1329–1344.
286. Dickinson, D.F., The normal ECG in childhood and adolescence. *Heart*, 2005;**91**: 1626–1630.
287. Davignon, A., P.M. Rautaharju, E. Boisselle, F. Soumis, M. Megelas, and A. Choguette, Normal ECG standards for infants and children. *Pediatr. Cardiol.*, 1979/80;**1**: 123–131.
288. Brohet, C.R., C. Hoeven, A. Robert, C. Derwael, R. Fesler, and L.A. Brasseur, The normal pediatric Frank orthogonal electrocardiogram: variations according to age and sex. *J. Electrocardiol.*, 1986;**19**: 1–13.
289. Perry, L.W., H.V. Pipberger, H.A. Pipberger, C.D. McManus, and L.P. Scott, Scalar, planar, and spatial measurements of the Frank vectorcardiogram in normal infants and children. *Am. Heart J.*, 1986;**111**: 721–730.
290. Macfarlane, P.W., E.N. Coleman, E.O. Pomphrey, S. McLaughlin, A. Houston, and T. Aitchison, Normal limits of the high-fidelity pediatric ECG. Preliminary observations. *J. Electrocardiol.*, 1989;**22**(Suppl): 162–168.
291. Macfarlane, P.W., S.C. McLaughlin, B. Devine, and T.F. Yang, Effects of age, sex, and race on ECG interval measurements. *J. Electrocardiol.*, 1994;**27**(Suppl): 14–19.
292. Rijnbeek, P.R., M. Witsenburg, E. Schrama, J. Hess, and J.A. Kors, New normal limits for the paediatric electrocardiogram. *Eur. Heart J.*, 2001;**22**: 702–711.
293. Horton, L.A., S. Mosee, and J. Brenner, Use of the electrocardiogram in a pediatric emergency department. *Arch. Pediatr. Adolesc. Med.*, 1994;**148**: 184–188.
294. Hamilton, R.M., K. McLeod, A.B. Houston, and P.W. Macfarlane, Inter- and intraobserver variability in LVH and RVH reporting in pediatric ECGs. *Ann. Noninvasive Electrocardiol.*, 2005;**10**: 330–333.
295. Guller, B., P.C. O'Brien, R.E. Smith, W.H. Weidman, and J.W. DuShane, Computer interpretation of Frank vectorcardiograms in normal infants: longitudinal and cross-sectional observations from birth to 2 years of age. *J. Electrocardiol.*, 1975;**8**: 201–208.
296. Guller, B., F.Y. Lau, R.A. Dunn, H.A. Pipberger, and H.V. Pipberger, Computer analysis of changes in Frank vectorcardiograms of 666 normal infants in the first 72 hours of life. *J. Electrocardiol.*, 1977;**10**: 19–26.
297. Francis, D.B., B.L. Miller, and D.W. Benson, A new computer program for the analysis of pediatric scalar electrocardiograms. *Comput. Biomed. Res.*, 1981;**14**: 63–77.
298. Laks, M.M., A computer program for interpretation of infant and children electrocardiograms, in *Computer ECG Analysis: Towards Standardization*, J.L. Willems, J.H. van Bemmel, and C. Zywietz, Editors. Amsterdam: North-Holland, 1986, pp. 59–65.
299. Macfarlane, P.W., E.N. Coleman, B. Devine, A. Houston, S. McLaughlin, T.C. Aitchison, et al., A new 12-lead pediatric ECG interpretation program. *J. Electrocardiol.*, 1990;**23**(Suppl): 76–81.
300. Rijnbeek, P.R., M. Witsenburg, A. Szatmari, J. Hess, and J.A. Kors, PEDMEANS: a computer program for the interpretation of pediatric electrocardiograms. *J. Electrocardiol.*, 2001;**34**(Suppl): 85–91.
301. Zhou, S.H., J. Liebman, A.M. Dubin, P.C. Gillette, R.E. Gregg, E.D. Helfenbein, et al., Using 12-lead ECG and synthesized VCG in detection of right ventricular hypertrophy with terminal right conduction delay versus partial right bundle branch block in the pediatric population. *J. Electrocardiol.*, 2001;**34**(Suppl): 249–257.
302. Guller, B., T. Jones, J. McCloskey, and S.P. Herndon, The Hewlett-Packard pediatric ECG computer program (HP-P3) and independent clinical information. *J. Electrocardiol.*, 1990;**23**(Suppl): 204.
303. Hamilton, R.M., A.B. Houston, K. McLeod, and P.W. Macfarlane, Evaluation of pediatric electrocardiogram diagnosis of ventricular hypertrophy by computer program compared with cardiologists. *Pediatr. Cardiol.*, 2005;**26**: 373–378.
304. Snyder, C.S., A.L. Fenrich, R.A. Friedman, C. Macias, K. O'Reilly, and N.J. Kertesz, The emergency department versus the computer: which is the better electrocardiographer? *Pediatr. Cardiol.*, 2003;**24**: 364–368.

38 Pacemaker Electrocardiography

Thomas Fåhraeus

P. W. Macfarlane, A. van Oosterom, O. Pahlm, P. Kligfield, M. Janse, J. Camm (eds.), *Comprehensive Electrocardiology*, DOI 10.1007/978-1-84882-046-3_38,

38.1 Introduction

When artificial cardiac stimulation was introduced in 1958 as a permanent treatment for bradycardia, the interpretation of pacemaker ECGs posed little difficulty. The indication for pacemaker implantation at that time was total atrioventricular block, and the available pacemaker systems provided fixed-rate ventricular stimulation (➋ Fig. 38.1). Today multiprogrammable pacemaker devices are available with many different pacing modes, sophisticated functions, and diagnostic tools. Pacing during various arrhythmias creates ECG recordings which are sometimes rather difficult to analyze, and nowadays there are many methods for performing an evaluation of the pacemaker function other than looking at a standard surface ECG. To perform a comprehensive examination, an appropriate pacemaker programmer is sometimes necessary where the initial interrogation will allow retrieval of programmed settings and diagnostic data. The ECG (usually both surface ECG and intracardiac electrograms) monitoring capability simultaneously with telemetry event markers on the display facilitates the interpretation.

In this chapter, the basic principles of the interpretation of pacemaker ECGs will be presented, first concerning normal pacemaker function and thereafter concerning malfunction and pseudomalfunction. Most of the ECG recordings have been selected from the daily clinical work at the Skåne University Hospital, Lund, Sweden, and the intention is to provide a survey of the field of pacemaker ECG. Questions concerning the choice of the stimulation mode and clinical application have to be referred to text books dealing with pacemaker treatment.

The different pacing modes used are identified according to the recommendations of the 1987 NASPE and NASPE/BPEG codes. The original three-position letter code (since 1974) was revised to a five-position code in 1981, and has subsequently been expanded further in 2002 (➋ Table 38.1). Usually, the first three letters are used, for example, VVI, DDD, AAI, VDD, etc., but today most pulse generators have a built-in sensor for rate modulation. When this function is activated, the code will be VVIR, DDDR, AAIR, and so on. The use of the letter code facilitates rapid acquisition of information about the programmed basic mode of the pacemaker and how it is pacing and sensing the heart.

38.2 Registration of Pacemaker Impulses

Pacemaker impulses should be clearly visible in the ECG tracings and are commonly called "spikes." An ECG recording of good quality facilitates correct interpretation of pacemaker performance. The typical standard pacemaker ECGs recorded when the lead tip is in the apex area of the right ventricle show broad QRS complexes with an LBBB pattern. New modes with different pacing sites have in some cases changed the QRS morphology. The print-out with data and ECG -markers from a programmer are very useful when difficult and problematic ECG's occur (➋ Figs. 38.2 and ➋ 38.3).

The reproduction of the pacemaker spikes is influenced by many factors. One of the most important is the use of the interference filter of the ECG recorder (➋ Fig. 38.4). While the amplitude of the spikes is huge in unipolar systems, much less stimulus artifact can be seen on the recording when a bipolar lead is used with anode and cathode only about a centimeter apart (➋ Fig. 38.5). Furthermore, the spike amplitude is related to the actual pulse generator output as demonstrated by the example in ➋ Fig. 38.6, where the energy content of the pacemaker impulse is temporarily reduced to 10% for test purposes. The spike amplitude and vector orientation when dealing with unipolar systems are dependent on the choice of ECG lead and the position of the pulse generator (➋ Fig. 38.7).

Digital ECG recorders are responsible for additional variation of the spike amplitude. If the sampling interval is 4 ms, ordinary pacemaker impulses of 0.2–0.5 ms duration may not be registered consistently, resulting in the changes of their reproduction (➋ Fig. 38.8). The variation of the spike amplitude may easily be misinterpreted as a lead or stimulation problem. This can partly be circumvented by sampling the signal at a much higher rate; for example, every 0.25 ms.

38.3 Relationship Between the Stimulation Impulse and Cardiac Activity

In most transvenous ventricular pacing systems with only one electrode, the lead tip is usually positioned in the right ventricle, and during stimulation the ECG will show a left bundle branch block pattern. If the lead tip is located close to the ventricular septum or is positioned on the left ventricle with an epicardial technique, a right bundle branch block pattern will be produced (➋ Fig. 38.9).

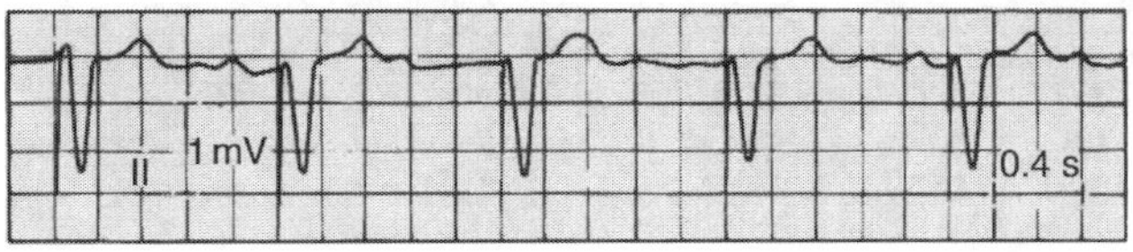

Figure 38.1
Continual asynchronous ventricular pacing at a rate of 60 bpm in the presence of complete AV block. A left bundle block pattern with wide QRS complexes appears due to stimulation first of the right ventricle and then the left.

Table 38.1
The revised NASPBE/BPEG generic code for antibradycardia, adaptive rate, and multisite pacing. Pace, 25(2), 260–264, February 2002

I What is stimulated?	II What is sensed?	III Reaction to sensing	IV Rate modulation	V Multisite pacing
0 = None	0 = None	0 = None	0 = None	0 = None
A = Atrium	A = Atrium	T = Triggered	R = Rate	A = Atrium
V = Ventricle	V = Ventricle	I = Inhibited	Modulation	V = Ventricle
D = Dual (A + V)	D = Dual (A + V)	D = Dual (T + I)		D = Dual (A + V)
S = Single	S = Single			
(A or V)	(A or V)			

The simultaneous occurrence of spontaneous cardiac activity and pacemaker impulses is not uncommon. When the endocardium close to the lead tip is already depolarized spontaneously, the pacemaker impulse will not affect the myocardium. However, the ECG will be distorted by the spike, a phenomenon referred to as a pseudofusion beat. If parts of the myocardium are depolarized by intrinsic activity while others are activated by the pacemaker impulse, fusion beats will occur. When the spontaneous heart rate and the basic stimulation rate are almost the same, fusion and pseudofusion beats are commonly seen (➋ Fig. 38.10).

In atrial pacing, intermittent ventricular extrasystoles may result in a confusing ECG when the atrial spike is followed by a QRS complex, referred to as a pseudopseudofusion beat (➋ Fig. 38.11). Tachycardia-terminating pacemakers provide additional variation to the relationship between spikes and the PQRS complexes; for example, during the termination of reentry tachycardias with rapid overdrive pacing with short sequences (bursts) of stimuli. In ➋ Sect. 38.10, there are more examples of tachycardia treatment (➋ Fig. 38.12).

38.4 The ECG Appearance of Different Pacing Modes

One of the most commonly used modes is VVI pacing, which comprises ventricular stimulation and inhibition. A VVI pacemaker operates with two timing circuitries; the basic rate counter and the technical refractory period, both of which are initiated either by stimulation or sensing events. Possible ECG combinations are demonstrated in the timing diagram shown in ➋ Fig. 38.13. The basic rate interval in a VVI pulse generator is programmable. When the pulse generator detects ventricular depolarization within the basic interval, stimulation will be inhibited and a new basic interval will start (➋ Fig. 38.14).

In the VVT mode, instead of inhibition, the pulse generator delivers an impulse triggered by the QRS complex during the refractory period, thereby leaving the spontaneous depolarization unaffected (➋ Fig. 38.15). This pacing mode is very seldom used but can be useful when sensing problems occur (inhibition is avoided).

A sensing event followed by either inhibition (VVI) or triggering (VVT) occurs when the intracardiac signal reaches a certain amplitude at the site of the electrode. Sometimes the intrinsic deflection of this signal coincides with the beginning of the QRS complex in the surface ECG. Whether or not ectopic ventricular activity causes inhibition of a VVI pulse

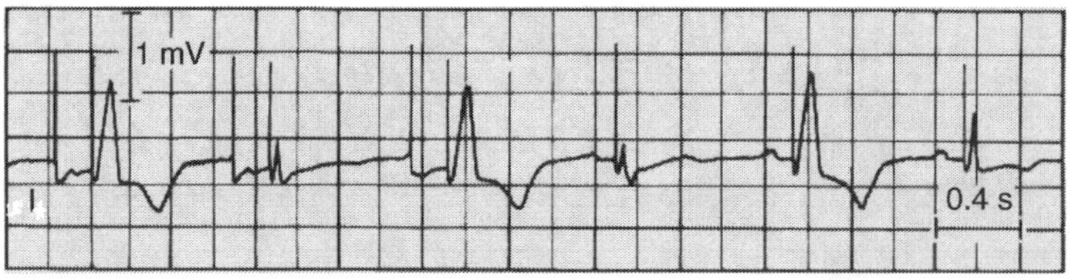

■ **Figure 38.2**
Dual-chamber pacing and sensing (DDD) during second-degree AV block. The basic rate and intrinsic atrial rate are almost the same, which explains the additional variations of the QRS-T morphology.

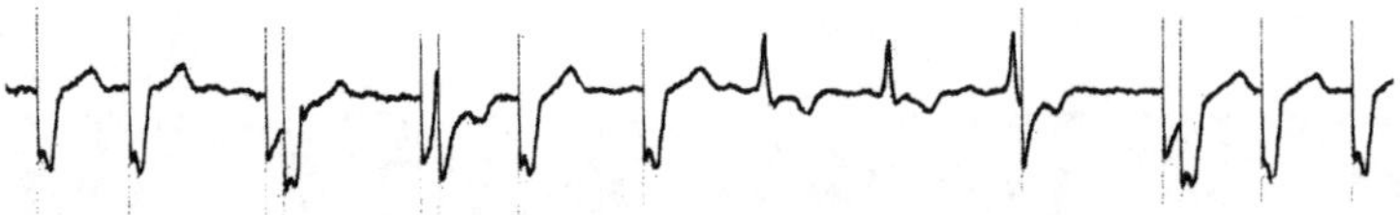

■ **Figure 38.3**
To interpret this ECG from a patient with a DDD pacemaker, a pacemaker programmer is very useful. Parameters such as rate, AV interval, and modes can be changed, diagnostic marker tools are available.

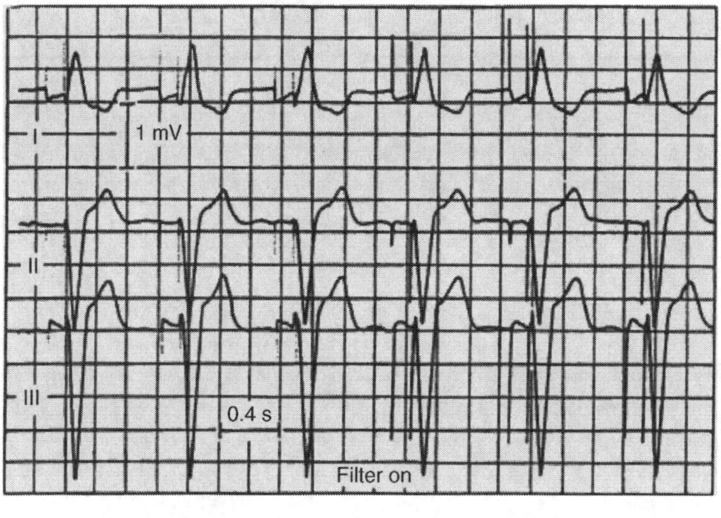

■ **Figure 38.4**
DDD pacing recorded without and with an interference filter. As the spike recording is more clearly seen with the low-pass filter switched on, it is recommended for routine use.

generator depends on the duration of the refractory period (➋ Fig. 38.16). Multiprogrammable pulse generators offer a selection of different refractory periods.

Variation in the rate, because of different timing intervals, is provided in pulse generators where the hysteresis function has been activated. Without hysteresis, the automatic basic interval is equal to the escape interval. With hysteresis, the basic rate interval is constant during stimulation, but after an inhibition the escape interval is prolonged. Thus, a pulse generator with a fixed or programmable hysteresis function operates with an inhibition rate lower than the basic stimulation rate (➋ Fig. 38.17).

Atrial pacing (AAIR) is achieved in a manner quite similar to ventricular pacing. Reliable AV conduction is, however, essential for the use of atrial pacing (➋ Fig. 38.18). As the amplitude of the atrial signal is considerably lower than the corresponding QRS signal in ventricular sensing, the sensitivity of an AAI device must be high enough to allow correct sensing.

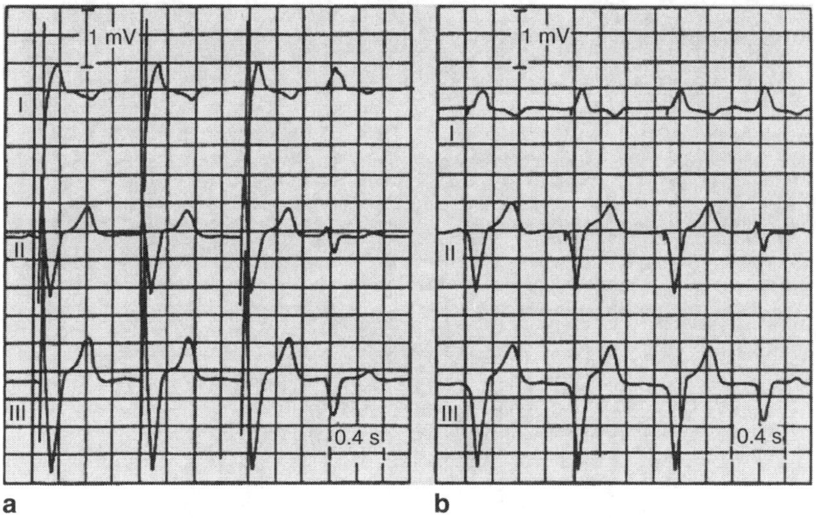

Figure 38.5
Unipolar pacing to the left (a) compared with bipolar (b) ventricular pacing through the same electrode system, which was achieved by means of a pulse generator with programmable polarity.

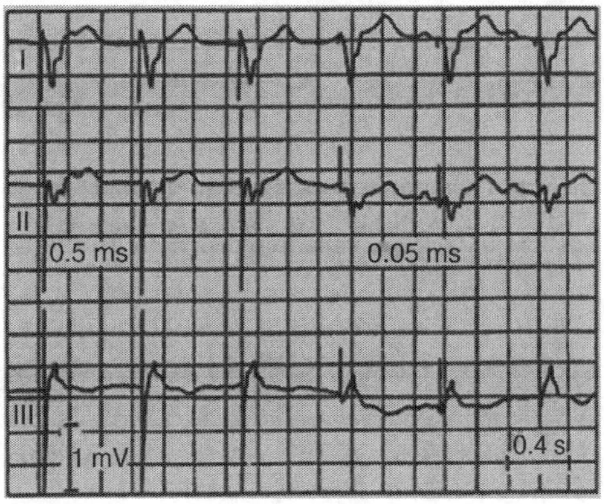

Figure 38.6
Unipolar ventricular pacing, during an automatic stimulation threshold test with reduction of impulse duration from 0.5 to 0.05 ms. Note the change of spike configuration. The output pulse corresponds to the spike on the ECG.

38.5 Dual-Chamber Pacing

The expression *dual-chamber pacing* is used to describe a pacing modality capable of delivering pacing stimuli to two chambers, usually the atrium and the ventricle (DOO, DDD, DDI, and DVI). During the last few years, the term *biventricular pacing* has been used to indicate cardiac resynchronization of the ventricles in heart failure. It is easy to be confused by the terms; resynchronization will be described later in this chapter. In the DVI mode, the stimulation of the atria is followed by ventricular pacing after a technical AV delay. As there is no sensing in the atria, a DVI pulse generator may only be inhibited by intrinsic ventricular activity. Atrial stimulation occurs at the programmed basic rate as long as the pulse generator is not inhibited by spontaneous ventricular beats (➲ Fig. 38.19).

In the VDD mode, atrial sensing is followed by synchronous ventricular stimulation, but without the possibility of stimulating the atria during sinus bradycardia (➲ Fig. 38.20). Ventricular inhibition is accomplished when spontaneous QRS complexes are sensed during the AV delay (➲ Fig. 38.21). Ectopic beats sensed after the end of the technical refractory

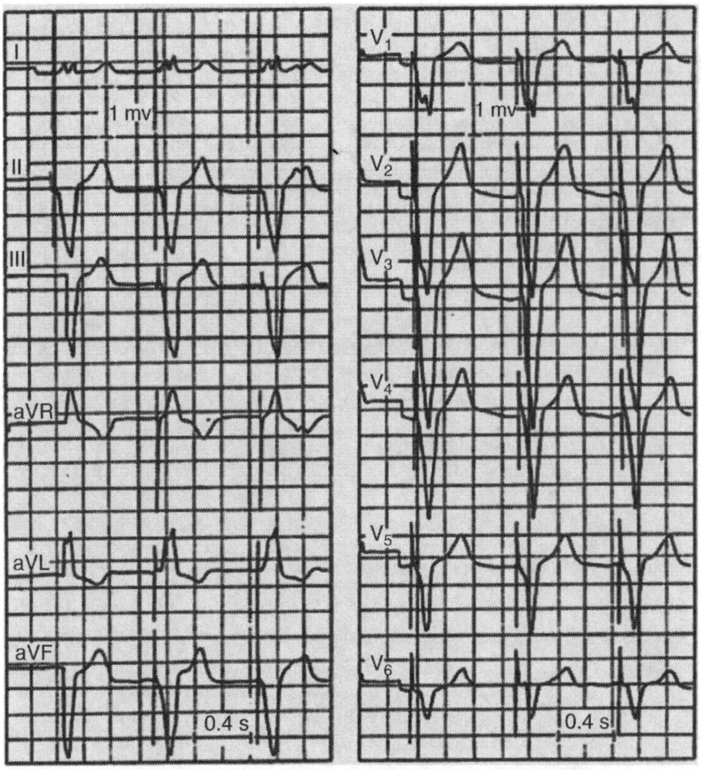

Figure 38.7
The vector orientation and amplitude of the spikes are depended on placement of the pulse generator and the ECG electrodes. The illustration is an example of unipolar right ventricular pacing with the pacemaker pocket located in the left pectoral region.

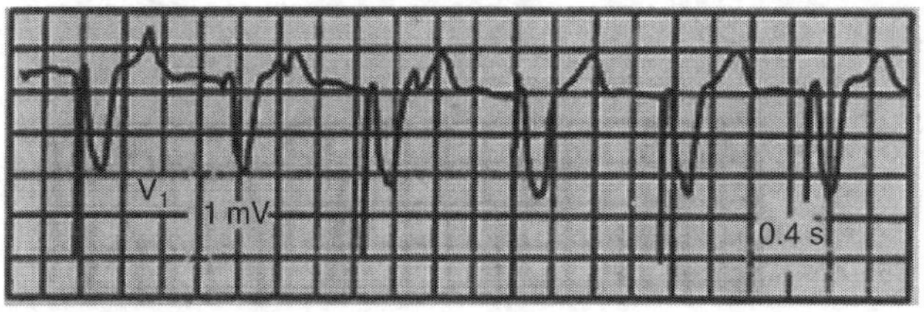

Figure 38.8
A digital ECG recording of unipolar ventricular pacing. The variation in the spike amplitude is caused by the digital sampling technique of this particular ECG recorder.

period may also cause inhibition and resetting of the basic rate interval. The disadvantages of only ventricular stimulation in the absence of spontaneous atrial activity make the VDD mode less attractive from a clinical point of view. The VDD mode is still available as a programmable mode in DDD pulse generators.

In the DDD mode, the pulse generator operates with sensing and stimulation of both the atrial and the ventricular chambers depending on the presence or absence of spontaneous cardiac activity and the programmed time intervals. Thus, it presents a functional combination of AAI, VAT, and VVI principles. The coincidence between dual-chamber

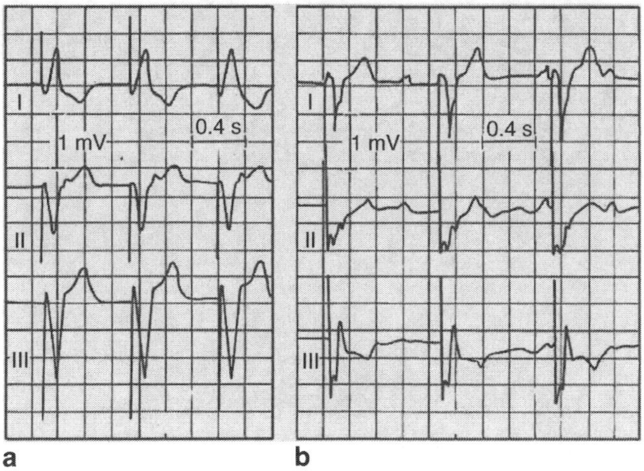

Figure 38.9
(**a**) Unipolar pacing through a transvenous right ventricular lead resembles a left bundle branch block (LBBB) pattern. (**b**) Left ventricular stimulation through a myocardial screw-in electrode produces a right bundle branch block (RBBB) pattern.

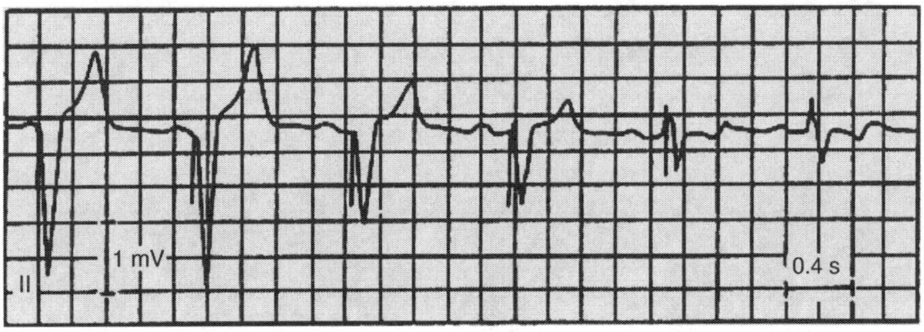

Figure 38.10
Unipolar ventricular-inhibited pacing demonstrating the difference between pacemaker-induced QRS complexes (QRS complexes #1 and #2) and different degrees of fusion beats (QRS complexes #3 and #4) caused by simultaneous depolarization via stimulation and normal AV conduction. The T-wave configuration in complex #5 indicates a spontaneously induced depolarization together with an ineffective pacemaker spike – a pseudofusion beat.

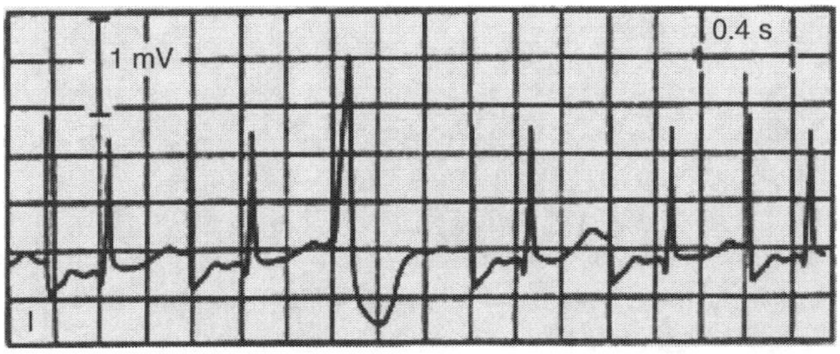

Figure 38.11
Atrial pacing at a rate of 100 min^{-1}. A ventricular ectopic beat (complex #3) coincides with an atrial spike producing a pseudopseudofusion beat.

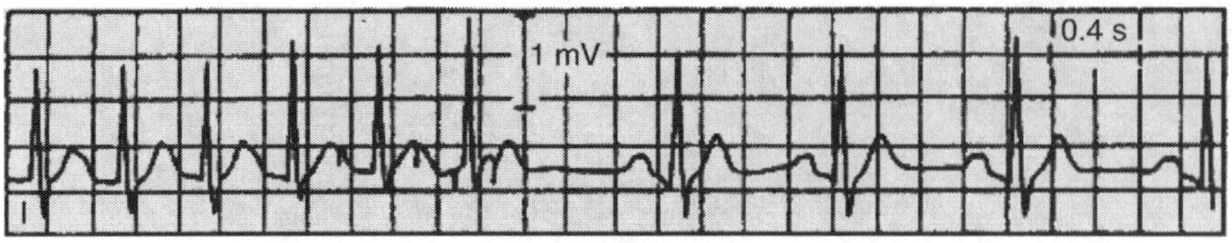

Figure 38.12
Atrial bipolar burst stimulation (300 min^{-1}) terminating a supraventricular tachycardia. The spikes in the burst sequence are clearly visible.

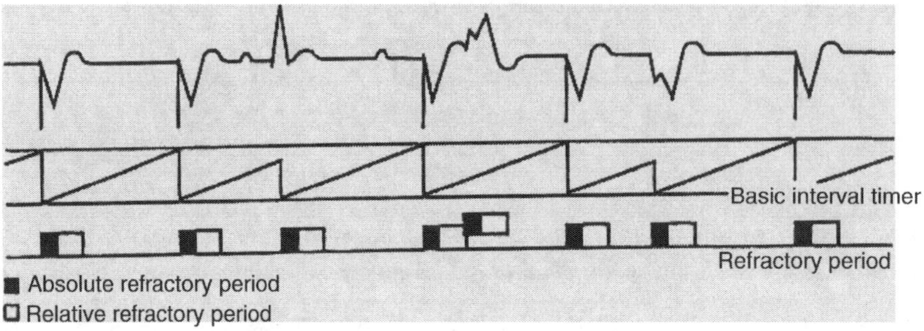

Figure 38.13
Schematic ECG illustration demonstrating the operation of the basic interval timer and the refractory period. In this example the refractory period is programmed to 312 ms. The absolute refractory period is 125 ms and the relative refractory period is thus 187 ms. A conducted ventricular complex which occurs during the relative refractory period (noise-sampling period) does not accomplish inhibition, that is, does not reset the basic interval timer, but restarts a new refractory period, as in the case of complex #5.

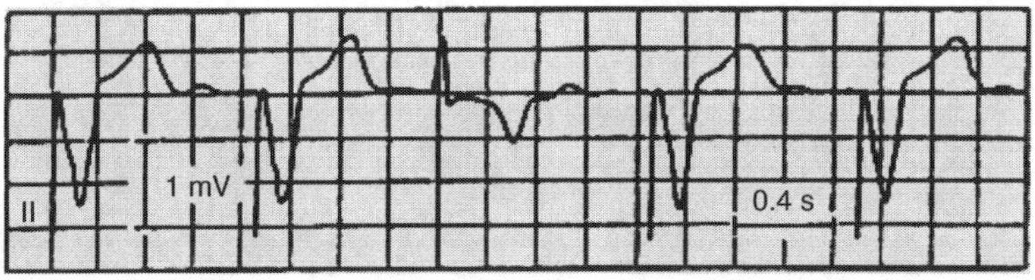

Figure 38.14
Ventricular-inhibited pacing (VVI mode) at a programmed rate of 70 min^{-1}. Inhibition will occur as soon as a spontaneous QRS complex at a higher rate occurs, as in this case.

pacing and the occurrence of spontaneous atrial and/or ventricular activity can cause many different ECG morphologies, including atrial and ventricular fusion or pseudofusion beats.

The resulting ECG appearance is even more confusing when spontaneous AV conduction varies intermittently (➲ Fig. 38.22). Furthermore, the ECG appearance may be altered by different AV delay settings in the presence of AV conduction (➲ Fig. 38.23). It is of clinical interest to have a long AV delay to support possible AV conduction and a normal activation of the ventricles. There are functions in pacemakers that automatically test the AV delay looking for intrinsic conduction and prolong the technical AV delay if the AV conduction is acceptable. During the last few years, the negative effects in some patients have been documented by long-term stimulation through the apex lead with an LBBB pattern. This electrical asynchrony can result in heart failure and in some cases a biventricular (➲ Sect. 38.9) pacemaker system will be implanted.

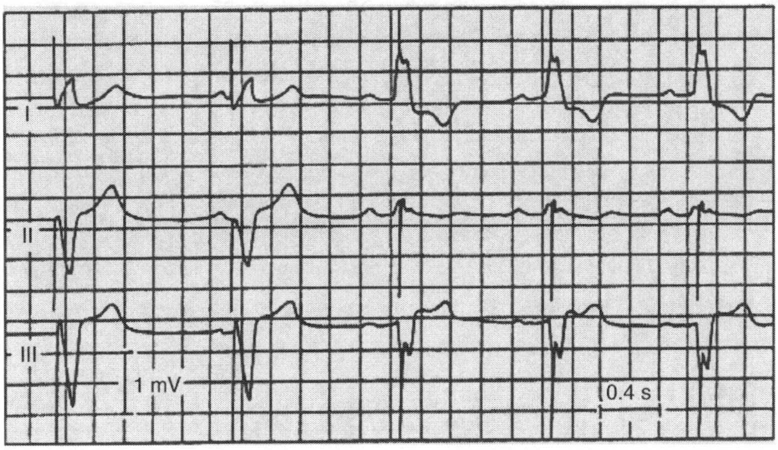

■ Figure 38.15
The ventricular-triggered pacing (VVT mode) at a basic rate of 50 min^{-1} followed by sinus rhythm at a higher rate (60 min^{-1}), and triggered ventricular spikes (pseudofusion beats).

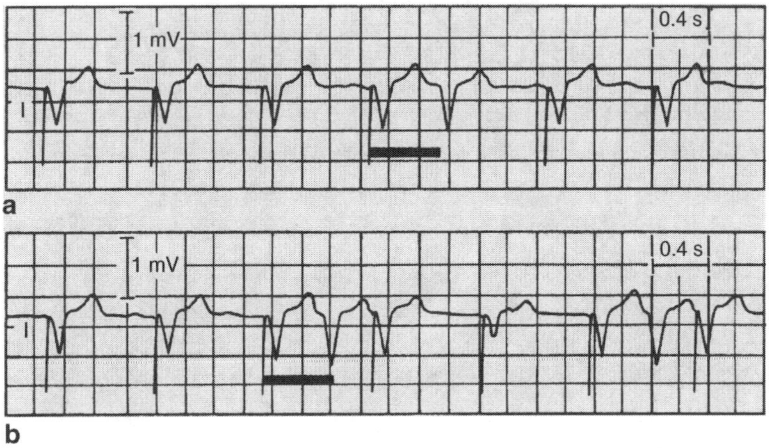

■ Figure 38.16
Ventricular pacing in the VVI mode at a rate of 80 min^{-1}. Ventricular ectopic beats accomplish inhibition as in (**a**) only if they occur after the end of the technical refractory period, denoted here by a horizontal bar. Early ectopic beats in (**b**) are ignored because the technical refractory period is 500 ms in this case.

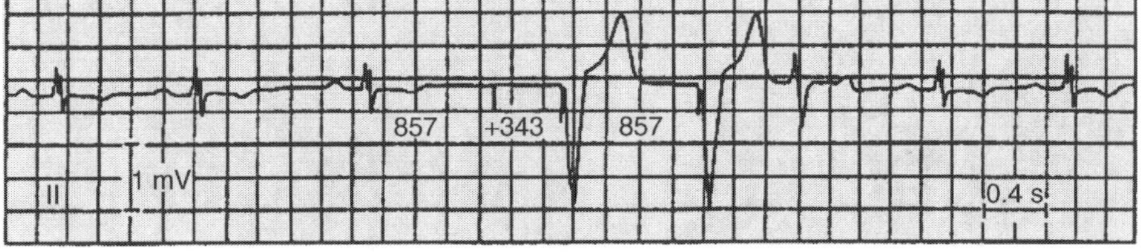

■ Figure 38.17
VVI pacing with a programmed basic rate of 70 min^{-1} and an activated hysteresis function. After an inhibition, a hysteresis delay of 343 ms is added to the basic rate interval of 857 ms thus permitting a slower intrinsic rhythm by consistent inhibition of the pulse generator. Lack of spontaneous rhythm is followed by stimulation at the programmed basic rate until a new inhibition occurs.

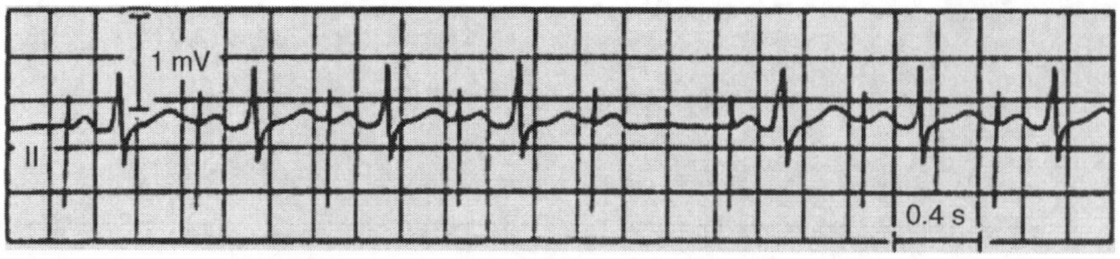

Figure 38.18
Atrial pacing at a programmed rate increased to 100 min^{-1} for test purposes. While consistent atrial capture can be seen, a blocked P wave indicates unreliable AV conduction.

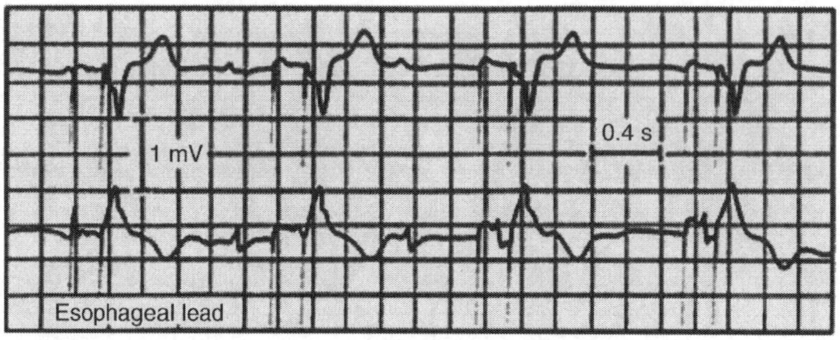

Figure 38.19
Dual-chamber stimulation (DVI mode) at a rate of 50 min^{-1} in the presence of AV block, and an intrinsic atrial rate of approximately 60 bpm. The esophageal ECG clearly indicates the asynchronous atrial stimulation owing to the lack of atrial sensing in a DVI pulse generator. Esophageal recordings are not used so often nowadays, due to availability of atrial electrograms.

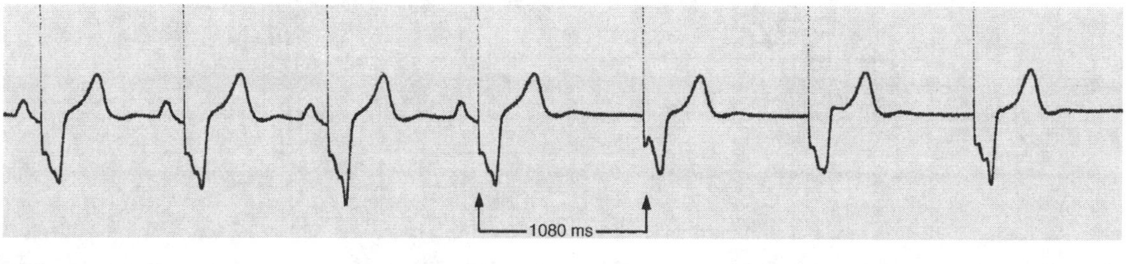

Figure 38.20
The basic function of a VDD pulse generator. In the absence of triggering P waves during sinus arrest or bradycardia, compensation by basic rate stimulation is achieved (The time interval is 1080 ms).

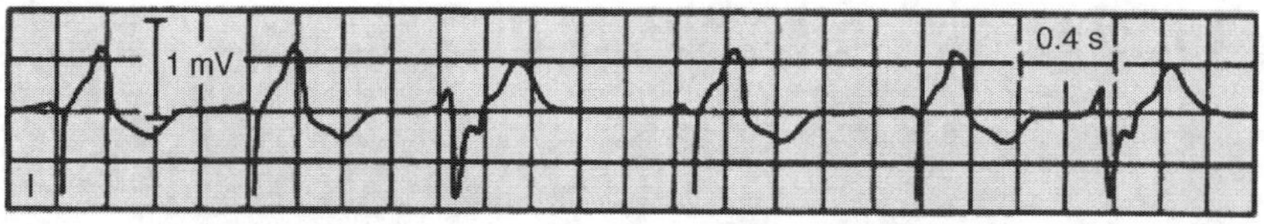

Figure 38.21
Atrial-triggered ventricular stimulation combined with ventricular inhibition (VDD mode). Every P wave is followed by ventricular pacing unless an ectopic beat inhibits the system. Hence, a spontaneous or retrograde P wave within the ectopic ventricular beat cannot initiate a new triggering.

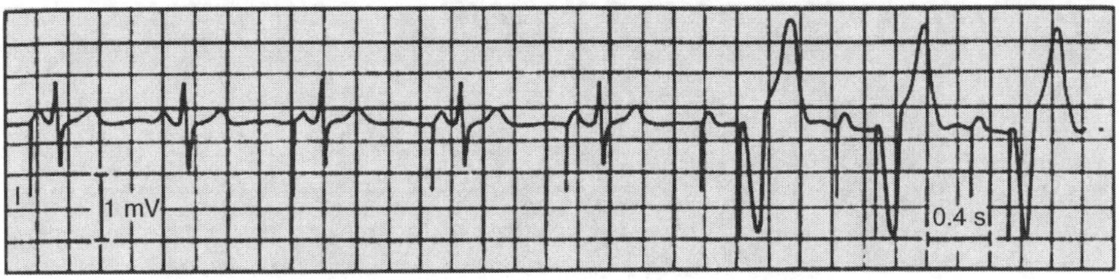

Figure 38.22
Dual-chamber pacing in the DDD mode at a basic rate of 70 min^{-1} alternating with sinus rhythm at a somewhat faster rate. As long as the AV conduction is normal, inhibition of the ventricular output is accomplished (complexes 1–5), but atrial activation is followed by ventricular pacing during impaired AV conduction (complexes 6–8).

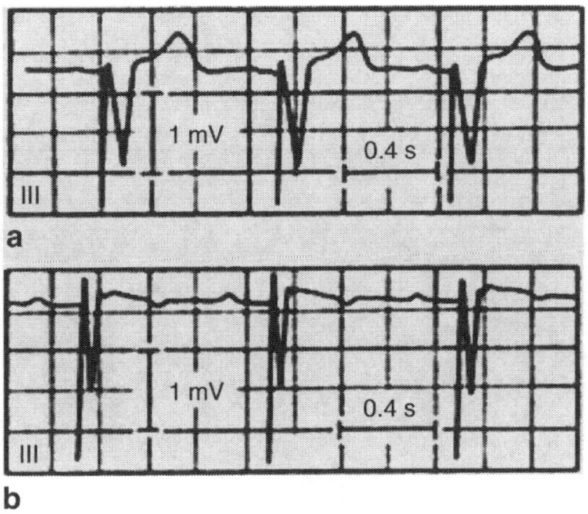

Figure 38.23
Atrial-triggered ventricular stimulation in the same patient as in Fig. 38.22, but with different technical AV delay settings. In the ECG shown in (a), the AV delay is 100 ms followed by typical pacemaker-induced ventricular complexes, while in the ECG in (b), a delay of 200 ms permits AV conduction resulting in pseudofusion beats.

Pathological atrial tachycardia or sensing interference may trigger DDD pulse generators to high rates. In order to avoid unacceptable fast-triggered ventricular pacing, different solutions for rate limitations have been introduced, for example, the technical Wenckebach behavior shown in Fig. 38.24. In atrial fibrillation, the amplitude and the configuration of the intrinsic fibrillation waves may intermittently be too low to be sensed by the atrial amplifier.

If too many high-frequency signals are detected, the pulse generator converts to basic rate stimulation (interference rate stimulation) as a protective countermeasure (Fig. 38.25).

It is therefore impossible to provide a complete survey of all conceivable ECG appearances regarding all different technical solutions, program settings, and changes in the intrinsic cardiac activity (Fig. 38.26).

38.6 Diagnostic Tools

Earlier, standard ECG registration remained the main tool for the evaluation of the pacemaker function. However, complete information about a particular pacemaker system and its function today requires the use of a pacemaker programmer.

Measures have to be taken for assessment of possible stimulation response. Most pulse generators can be converted to fixed-rate stimulation by the application of a magnet. Different manufacturers have chosen different features of the

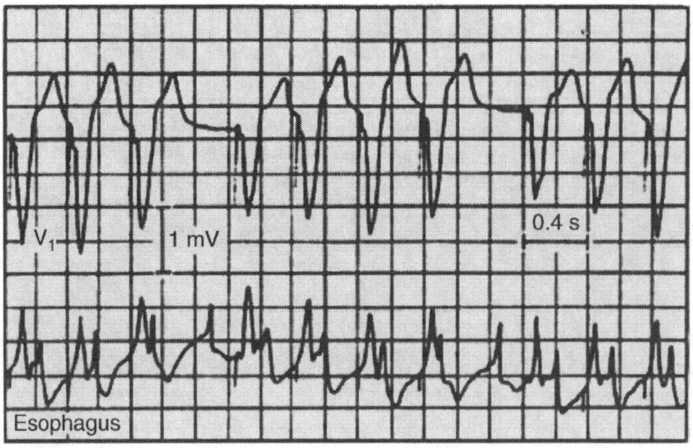

■ Figure 38.24
Atrial-triggered ventricular stimulation (DDD mode) at the highest synchronous rate. Esophageal registration discloses an atrial tachycardia, which runs at a rate of 175 bpm, while ventricular stimulation is running at 160 min^{-1} (the highest synchronous rate) resulting in repetitive Wenckebach block.

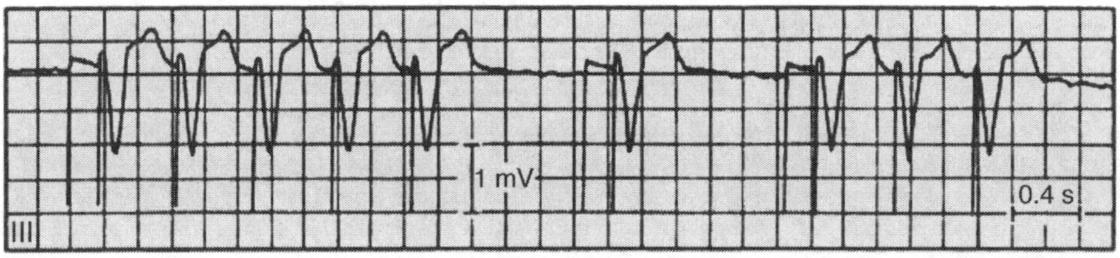

■ Figure 38.25
Atrial-triggered ventricular stimulation (DDD mode) running at different rates alternating with basic rate stimulation. This is a typical example of the impact of atrial fibrillation on a DDD pacemaker system without automatic mode switch when atrial fibrillation occurs.

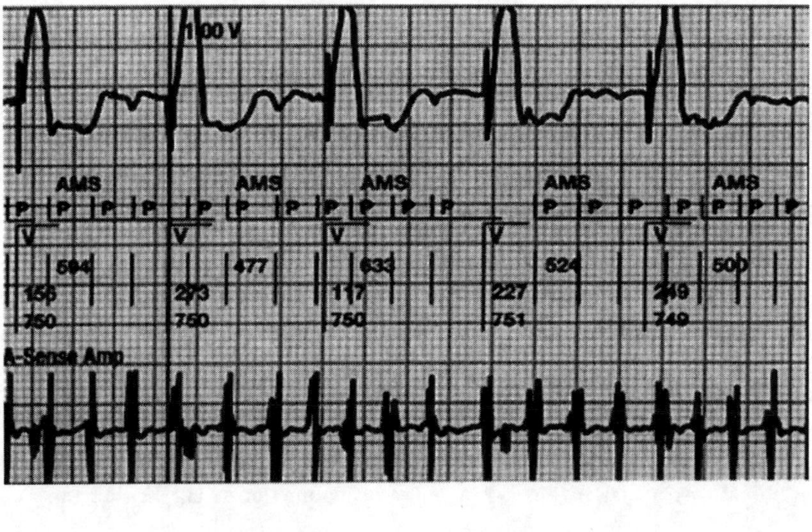

■ Figure 38.26
Print out from a pacemaker programmer with markers and intracardiac recordings. The intracardiac atrial ECG at the bottom reveals atrial fibrillation during DDDR pacing, but the rate of the ventricles is normal. The adequate sensing of the fibrillation waves results in an automatic mode shift (AMS) to DDIR pacing.

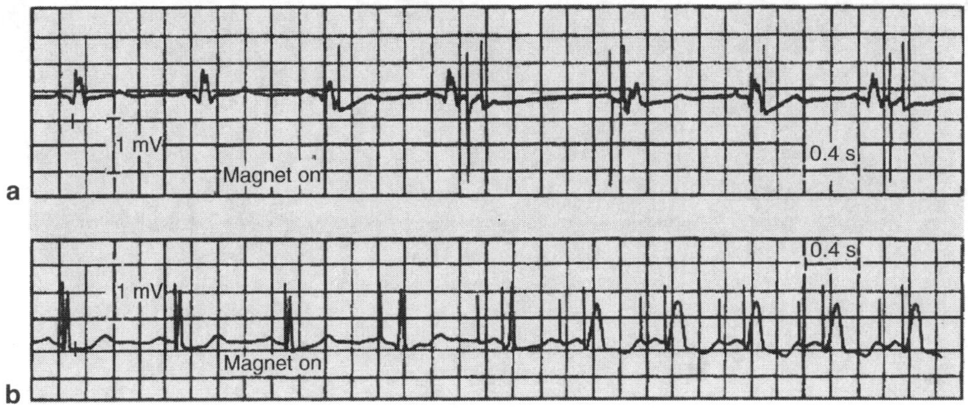

Figure 38.27

Two examples of DDD pacing in the presence of inhibiting atrial rates followed by normal AV conduction, which causes either inhibition of the ventricular output or pseudofusion beats. In (a), magnet application reverts the pulse generator to asynchronous stimulation (DOO) at the programmed basic rate. The interference between intrinsic activity and pacemaker stimulation produces pseudofusion beats without diagnostic information. In (b), the magnet test rate of 100 min^{-1} overdrives both spontaneous atrial activity and conducted ventricular beats thus providing information about atrial and ventricular capture.

magnet test response. The magnet test rate can be the same as the programmed basic rate (➔ Fig. 38.27a). As long as interference between the intrinsic rhythm and pacemaker stimulation is followed by pseudofusion beats, no conclusion can be drawn about capture. Magnet tests running at higher rates facilitate the clinical investigation when overdrive suppression is accomplished (➔ Fig. 38.27b).

Besides the assessment of stimulation response, the magnet test rate also reflects information about the battery condition. Today, telemetric data about the condition of the battery is a more reliable method of controlling the expected lifetime of the battery. Once more, it is necessary to emphasise the individual nature of the technical behavior of different pacemaker models and the need to obtain specific, relevant information before conclusions can be drawn.

The interpretation of a pacemaker ECG is easier when pulse generators with telemetric function are used. By means of corresponding programmers, marker pulses can be recorded indicating which chamber has been involved and whether sensing or stimulation is occurring (➔ Fig. 38.28). In the new models, even the duration of the different refractory periods is displayed, together with the ongoing intracardiac ECG registration.

Besides marker pulses, intracardiac ECG recording can also be obtained via telemetry (➔ Fig. 38.29).

The evaluation of atrial stimulation response in dual-chamber pacing is often difficult because of the distortion of the ECG registration by atrial impulses. It is not uncommon that standard ECG leads fail to permit any conclusion. Earlier, esophageal registration was therefore necessary to analyze atrial activity both concerning stimulation response and spontaneous ectopic beats if intracardiac atrial recording from the programmer is not available.

Nowadays, each patient should be equipped with a pacemaker "passport," including relevant data on the latest program setting. Pacemaker–ECG interpretation can be further facilitated when programmed parameters and information are printed out simultaneously on the ECG strip together with marker pulses.

38.7 Pacemaker Malfunction and Pseudomalfunction

When analyzing a "pathological" pacemaker ECG, it is necessary to differentiate between malfunction owing to lead or pulse generator disorders and pseudomalfunction caused by inappropriate program settings with regard to the underlying arrhythmia.

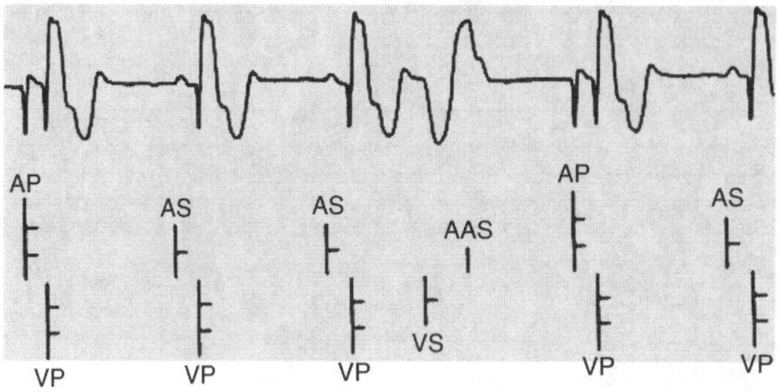

Figure 38.28
An early pacemaker example of the first markers of intracardiac ECG complexes. Dual-chamber pacing and simultaneous display of marker pulses indicating alternation between atrial pacing (AP), atrial sensing (AS), ventricular pacing (VP), ventricular sensing (VS), and abolished atrial sensing (AAS) after ventricular sensing of an ectopic beat.

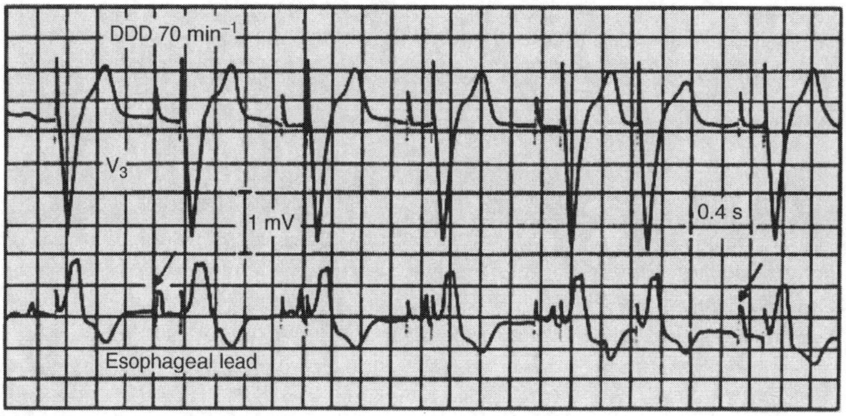

Figure 38.29
Esophageal-lead registration permits atrial activity to be visualized very clearly and discloses alternation between atrial fusion beats (*arrows*), normal atrial capture, and a supraventricular ectopic beat localized within the T wave of the preceding QRS complex. Today, intracardiac recordings and markers from the atrium make esophageal ECG recordings unnecessary.

Technical disorders of pacemaker systems are mainly related to lead problems, but other factors may also disturb the pacemaker operation. A collection of possible causes is given in Fig. 38.30. Some complications are accompanied by typical ECG patterns, and their diagnosis and assessment can be undertaken through the interpretation of standard ECG recordings. However, several malfunctions may cause the same ECG pattern, thereby creating difficulties in making a correct diagnosis and in deciding adequate countermeasures. The disclosure of intermittent and sporadic disorders sometimes requires a closer look at the diagnostic memory of the pacemaker or Holter monitoring.

The most important question regarding pacemaker function is whether there is adequate response (capture) to all pacemaker stimuli. A lack of capture is called *exit block*, a descriptive ECG term, which does not provide any information about the underlying cause. Hence, it is not possible to discriminate between lead dislodgment and pathological threshold rise with an unchanged lead-tip position (Fig. 38.31).

The presence of ventricular exit block is obvious in dual-chamber systems when there is AV block (Fig. 38.32). Sometimes, it is difficult to see if the exit block is from the atrial or the ventricular lead. However, it can be overlooked when spontaneous AV conduction occurs and initiates ventricular depolarization resulting in pseudofusion beats (Fig. 38.33).

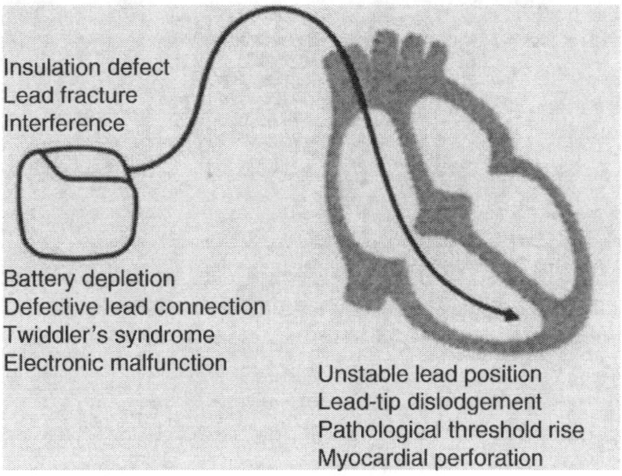

Figure 38.30
Different pacemaker malfunctions detailed concerning cause and location.

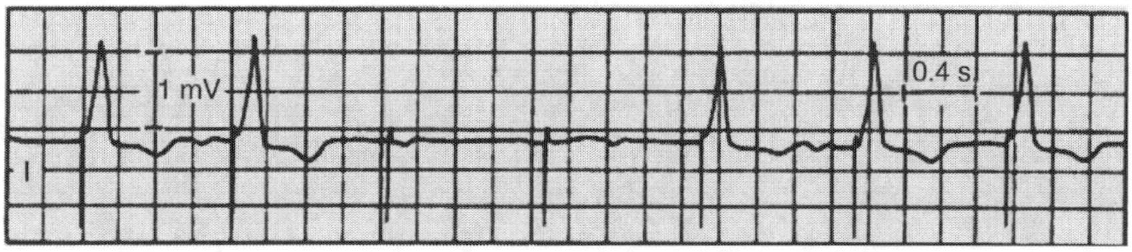

Figure 38.31
VVI pacing at a rate of 70 min^{-1} and intermittent exit block, but without any information about the cause of loss of capture.

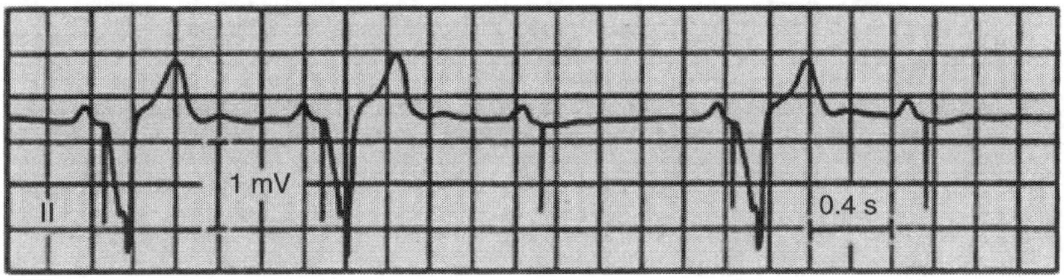

Figure 38.32
DDD mode. Atrial-triggered ventricular stimulation is followed by intermittent ventricular exit block.

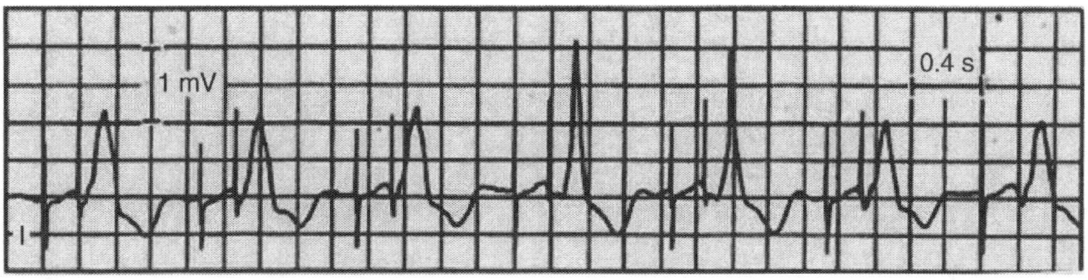

Figure 38.33
Dual-chamber pacing at a rate of 70 min^{-1}. Two of the ventricular complexes differ in configuration and occur approximately 80 ms after the ventricular stimulus is delivered. This indicates the presence of an intermittent ventricular exit block "somewhat concealed" by the spontaneously conducted QRS complexes.

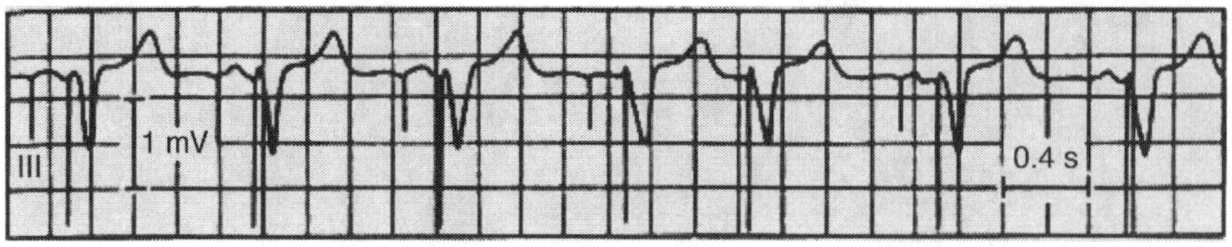

Figure 38.34
Dual-chamber pacing at a rate of 70 min^{-1}. Atrial depolarizations are clearly visible in this recording. The loss of atrial capture in complex #4 is followed by triggered ventricular stimulation, probably because of a spontaneous P wave hidden somewhere in the T wave of the preceding QRS complex.

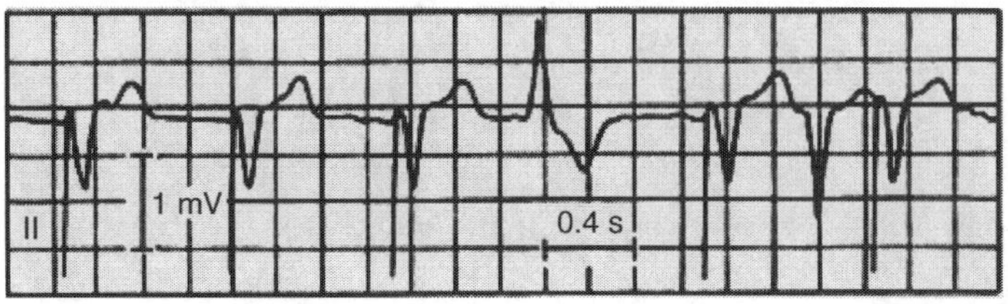

Figure 38.35
VVI pacing at a rate of 80 min^{-1}. The first premature ectopic ventricular beat accomplishes inhibition indicated by the reset of the basic rate interval, while the second one differs in configuration and remains unsensed.

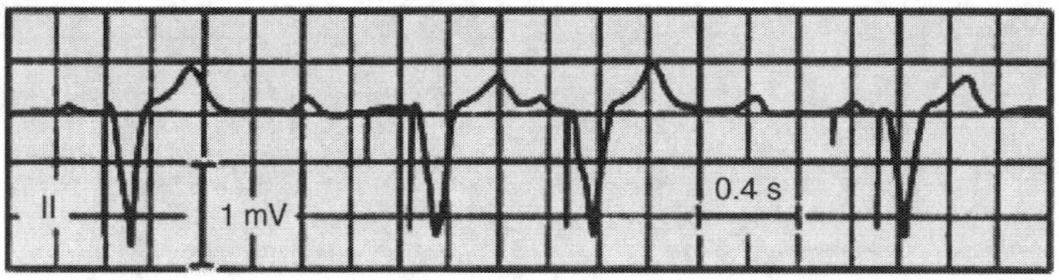

Figure 38.36
Dual-chamber pacing. Intermittent P wave undersensing is followed by atrial stimulation at a basic rate of 50 min^{-1}. Atrial capture can be obtained when the biological refractory period of the unsensed spontaneous P wave has passed.

As mentioned earlier, the assessment of atrial capture demands visualization of atrial activity (➲ Fig. 38.34). If standard ECG leads do not provide sufficient information, intracardiac recording or reprogramming procedures may help to disclose possible atrial exit block.

The sensing function also has to be assessed in an ECG analysis. Ventricular ectopic activity may be sensed with different amplitudes and slew rates. The actual program setting may cover normal ventricular complexes, but not all ectopic beats (➲ Fig. 38.35).

In dual-chamber pacing, one of the major concerns is consistent P-wave sensing. In the presence of AV block, P-wave undersensing is easily detected (➲ Fig. 38.36). However, during normal AV conduction, P-wave undersensing may be overlooked when the pacemaker is correctly inhibited via the ventricular amplifier owing to sensing of spontaneous QRS complexes (➲ Fig. 38.37).

Unfortunately, other signals may affect the pacemaker function; for example, myopotentials. In unipolar pacemaker systems, myopotentials near the pulse generator pocket may be sensed causing inhibition or interference rate pacing, the signals being perceived as cardiac activity. Inhibition by myopotentials is supposed to be a pseudomalfunction, as the pulse generator operates correctly according to the technical specifications (➲ Fig. 38.38). Another example of pseudomalfunction is given in ➲ Fig. 38.39, which demonstrates far-field R wave oversensing in AAI pacing. In unipolar AAI pacing, sometimes rather large intra-atrial R waves can be seen resulting in sensing problems.

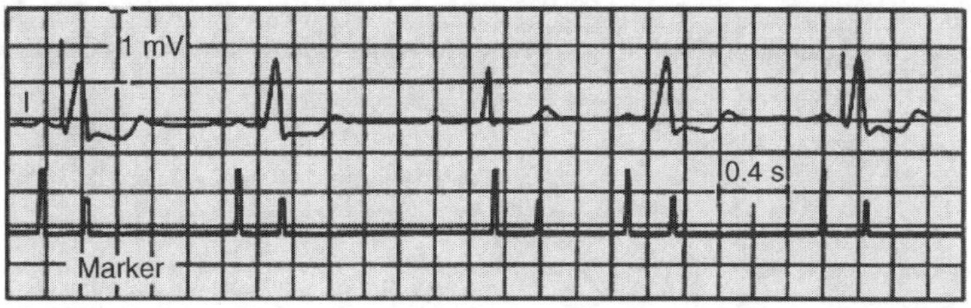

Figure 38.37
Dual-chamber pacing. One P wave (number 3) is not detected by the atrial amplifier. The following spontaneously conducted QRS complex is sensed by the ventricular amplifier as is clearly indicated by the simultaneous marker-pulse recording.

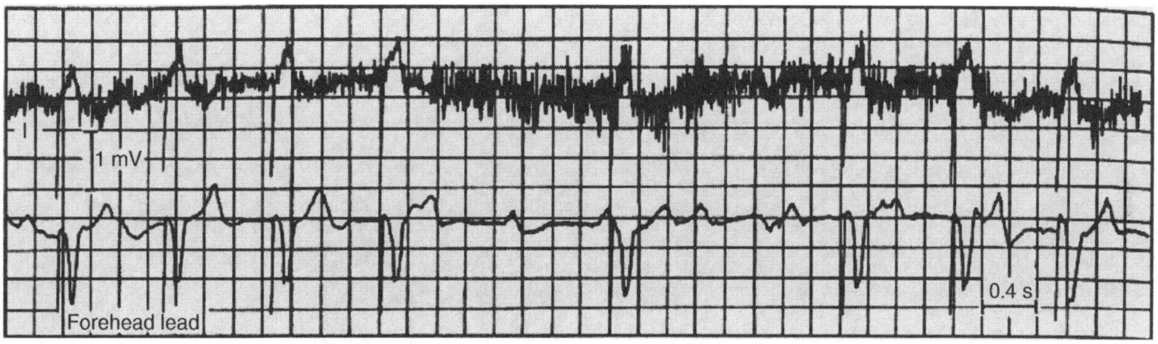

Figure 38.38
VVI pacing at a rate of 80 min^{-1}. The occurrence of myopotentials clearly seen in standard lead I causes intermittent inhibition. The lack of interference in the lower-lead recording is a result of the temporary placement of the ECG lead on the patient's forehead.

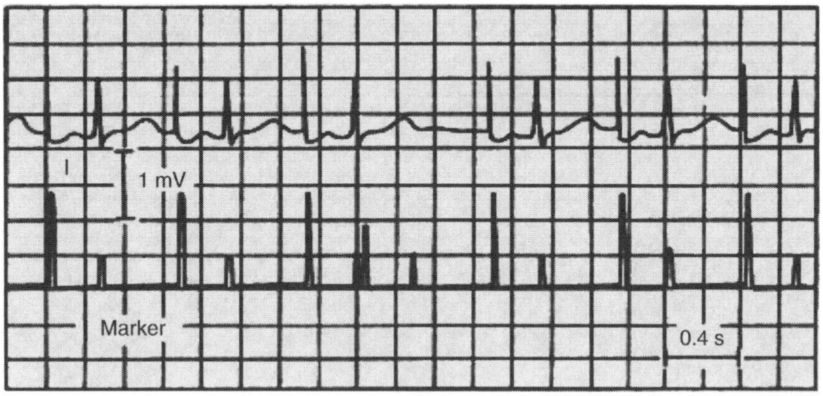

Figure 38.39
AAI pacing at a rate of 90 min^{-1} with an intermittent pause. Marker-pulse registration indicates a reset of the technical refractory period and the basic interval timer. The reset can be related to the QRS complex and the diagnosis is intermittent far-field R-wave oversensing; that is, the actual sensing program detects the "distant" ventricular activity.

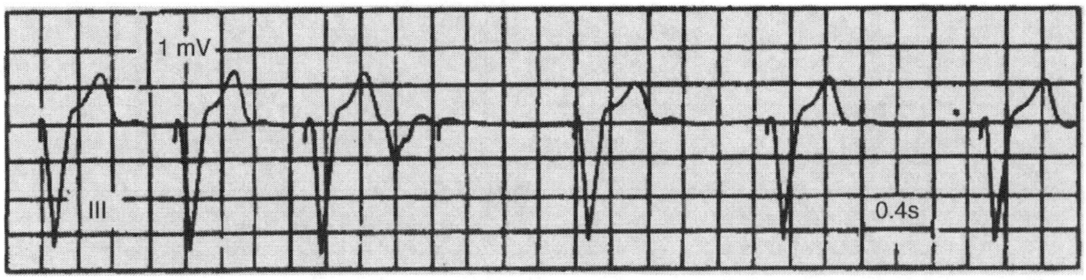

■ **Figure 38.40**

VVI pacing system with a programmed basic rate of 80 min^{-1}. A premature ventricular ectopic beat is not followed by inhibition. Intermittent and irregular variation of pacing intervals indicates inhibition owing to oversensing. Because there are neither signs of external electrical interference nor myopotentials, signals occurring within the lead system can be assumed to cause the intermittent inhibition or conversion to interference rate stimulation thus mimicking undersensing. In this particular patient, an insulation defect of the transvenous lead at the site of the tricuspid valve was found to be the reason for the pathological ECG.

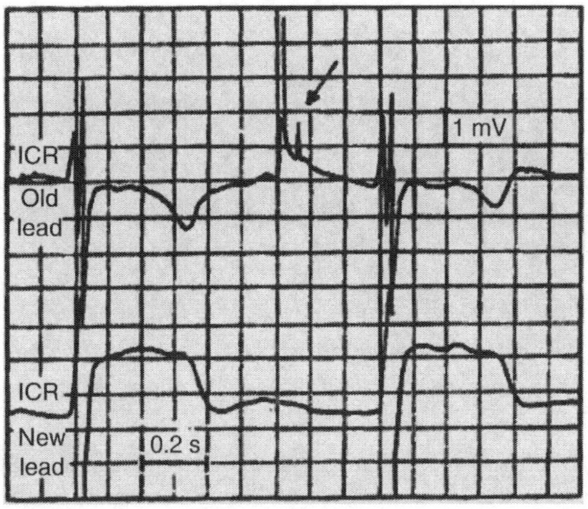

■ **Figure 38.41**

Intracardiac ECG recording (ICR) (paper speed 50 mm s^{-1}) obtained from the same patient as in ➔ Fig. 38.40 during lead replacement. Simultaneous registration discloses pathological signals with maximum amplitude of 5 mV from the old lead, but not from the newly implanted second ventricular lead.

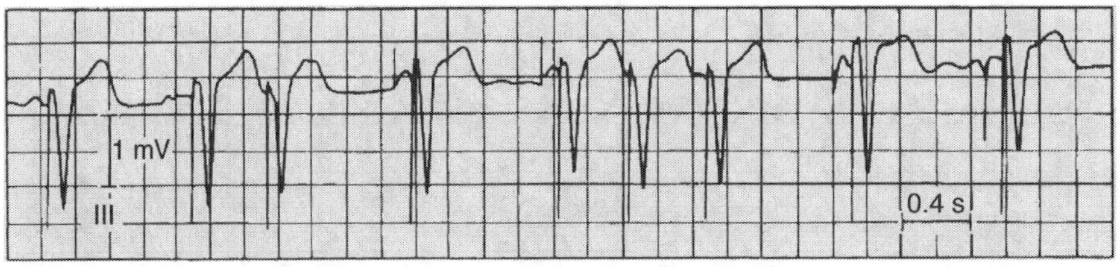

■ **Figure 38.42**

DDD pacing system operating with irregular triggered ventricular stimulation, alternating with basic rate stimulation and occasionally P-wave undersensing. The reason for this confusion ECG turned out to be an insulation defect of the epicardial atrial lead a few centimeters from the lead tip.

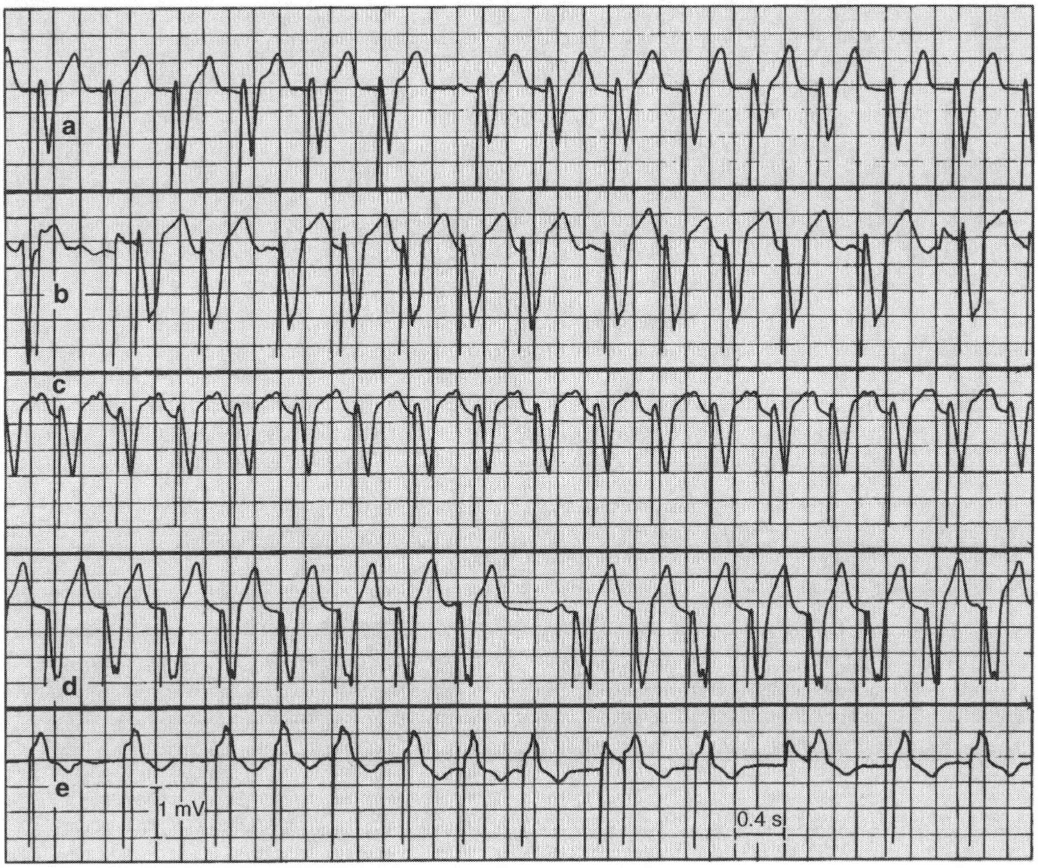

Figure 38.43
DDD-mode tachycardias. Recordings from five different patients, each with a different diagnosis. The recording taken from the patient shown in (**a**) illustrates sinus tachycardia at a rate slightly ahead of the programmed upper rate limit of 110 min^{-1} resulting in prolonged AV delay and a Wenckebach block type of behavior; (**b**) shows a patient with atrial fibrillation with irregular ventricular triggering limited by a programmed upper rate of 128 min^{-1}; (**c**) illustrates pacemaker-mediated reentrant tachycardia owing to sensing of retrograde P waves following ventricular stimulation; (**d**) shows repetitive far-field R-wave oversensing owing to inappropriate program setting with respect to the duration of the atrial refractory period and high atrial sensitivity; and (**e**) illustrates irregular triggered ventricular stimulation owing to "oversensing" of potentials generated by an insulation defect of the atrial lead.

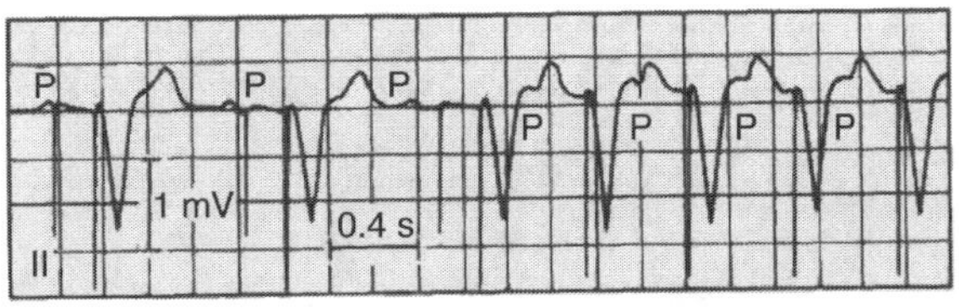

Figure 38.44
DDD pacing with good recording of atrial activity. After three unsensed P waves, ventricular stimulation is followed by a retrograde P wave, which is obviously sensed, subsequently triggering the next ventricular stimulation. Repetitive occurrence and consistent sensing of retrograde P waves maintain the running of an endless-loop reentrant tachycardia.

Rather confusing ECG patterns are obtained when a lead insulation defect appears (➋ Fig. 38.40). Pathological inhibition alternating with interference rate stimulation resulting in pseudoundersensing may occur as intracardiac insulation defects giving rise to pathological potentials (➋ Fig. 38.41). Another puzzling ECG is given in ➋ Fig. 38.42, where an insulation defect in an atrial lead is responsible for pathological triggering of ventricular stimulation.

Tachyarrhythmias in dual-chamber pacing pose challenging diagnostic problems as they may look alike despite different underlying mechanisms (➋ Fig. 38.43). As mentioned earlier, a good ECG recording is necessary for the detection of P waves, which may help to disclose the cause of the arrhythmia (➋ Fig. 38.44).

38.8 Functional Tests

The days when only a magnet was available as a test instrument have more or less disappeared. In multiprogrammable pacemakers, output can be reduced either by shortening the impulse duration or decreasing the amplitude with the programmer. Some manufacturers have chosen to provide an automatic stepwise decrease or a one-step change of the pacing threshold measurement (➋ Fig. 38.45). This is a temporary programming, which necessitates access to the specific programmer and knowledge of how to handle this device. Episodes of asystole may occur during the stimulation tests (➋ Figs. 38.46 and ➋ 38.47).

In dual-chamber pacing, the evaluation of the atrial stimulation threshold by the temporary reduction of voltage or impulse duration is facilitated when the atrial depolarization is clearly visible (➋ Fig. 38.48). In some cases, it is difficult to see the P waves, but if the test is done with the programmer, markers and intracardiac ECG recordings are useful tools.

The clinical information obtained from only reading standard pacemaker ECG registration is limited. Even if the adequate pacemaker function can be assessed, it is advisable to evaluate safety margins for stimulation and sensing, as well as the battery condition during follow-up. Almost all programmers measure sensing automatically or semi-automatically (➋ Fig. 38.48).

The assessment of atrial sensing threshold always requires a reprogramming procedure. P-wave sensing is subjected to changes in the postoperative course without correlation with the variation of stimulation thresholds. Additional provocation tests with ECG recordings in different postures and during maximal inspiration can expose intermittent undersensing (➋ Fig. 38.49).

During the follow up of the AAI/R mode, special attention should be paid to the AV conduction. This can be done by reevaluation of the Wenckebach block by temporary incremental rate increase. Valsalva maneuvers and carotid massage can also provide useful information about the AV conduction properties (➋ Fig. 38.50).

When normal AV conduction is present in the DDD mode, it is possible that – in the presence of long technical AV delay – the ventricular spikes will coincide with spontaneous QRS complexes or be inhibited depending on the critical timing. Hence, no information can be obtained about ventricular stimulation response unless the technical AV delay is reprogrammed to a very short value (➋ Fig. 38.51). A possible exit block may thus be disclosed which is otherwise well-concealed by pseudofusion beats or inhibition.

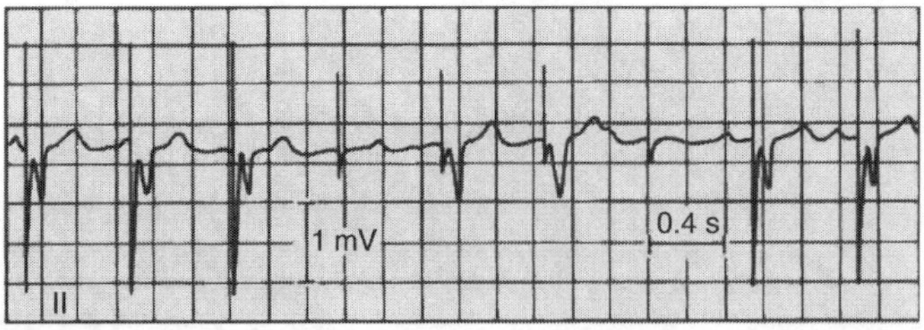

Figure 38.45
VVI pacing during a threshold test where the output of four stimuli is automatically reduced by a decrease in impulse duration from 0.5 to 0.07 ms. Intermittent loss of capture occurs.

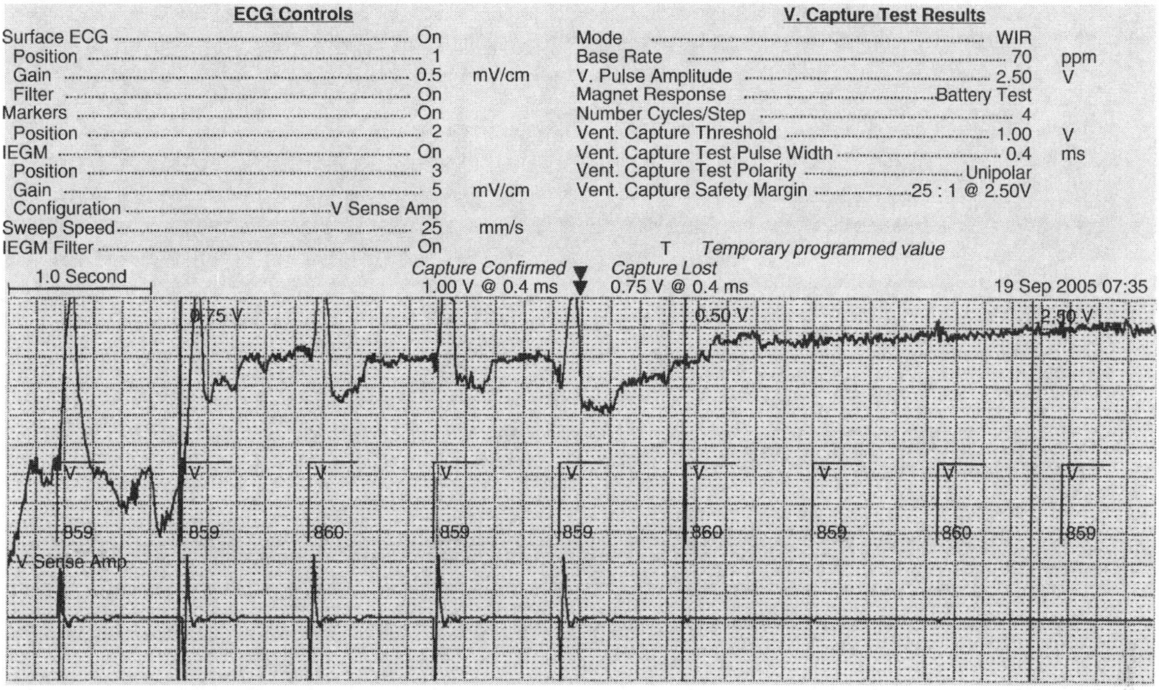

Figure 38.46
An automatic ventricular pacing threshold test with a printout from the programmer is done with reduction of the voltage of the pacing amplitude. The surface ECG has some muscle potential disturbances but loss of capture is clearly seen at 0.75 V. The markers line in the middle has a V sign indicating a ventricular output from the pacemaker. The bottom ECG is an intrinsic recording of the sensed amplitude, which, of course, disappears when capture is lost. No ventricular escape rhythm is detected.

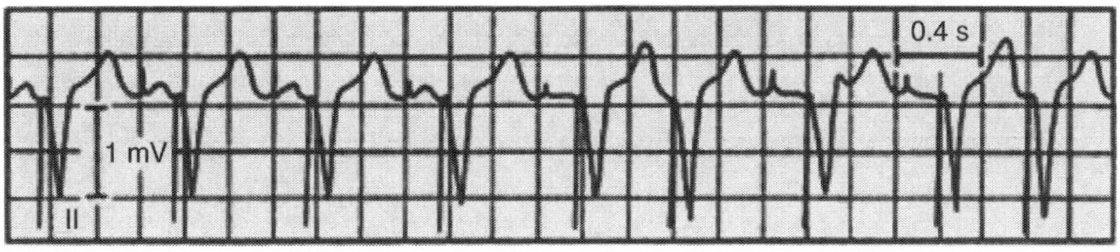

Figure 38.47
Automatic atrial stimulation threshold test of a DDD pacemaker. As the P waves are clearly visible it is simple to estimate when loss of capture occurs.

In unipolar pacing, sensing of myopotentials from muscles close to the pulse generator constitutes a well-known problem. To determine the extent to which an individual patient may experience myopotential inhibition, routine testing is recommended before a final decision on program setting is made. In the DDD pulse generators, both the atrial and the ventricular amplifier may be influenced by myopotentials depending on their amplitude and slew rate in relation to the sensitivity setting (➲ Fig. 38.52).

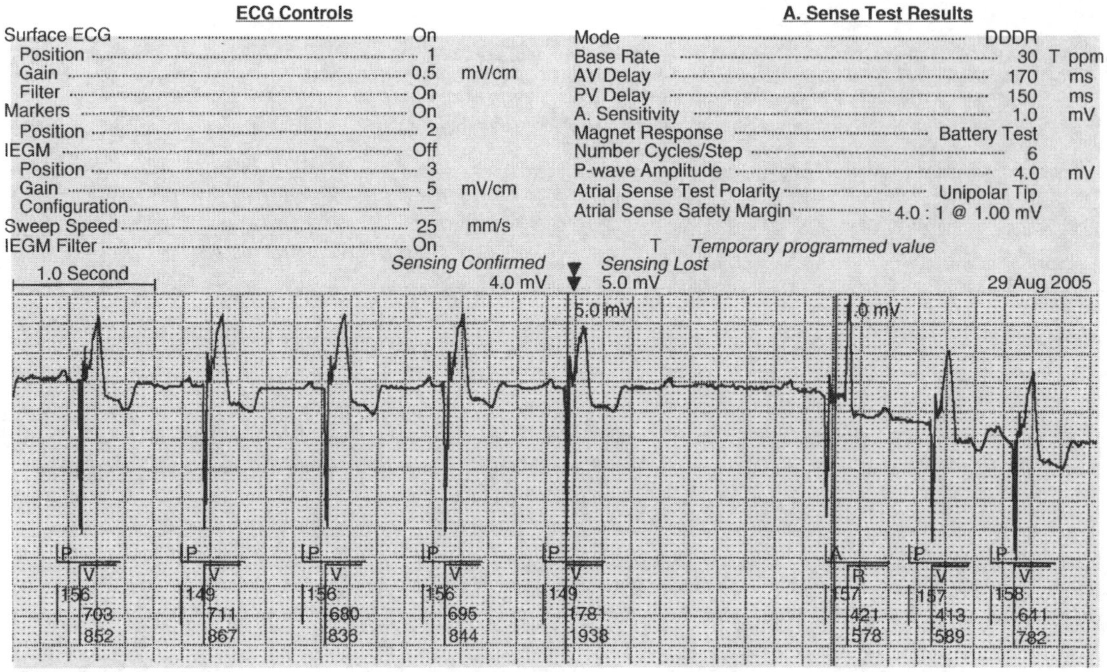

Figure 38.48
Here is an example of an automatic atrial-sensing test in a DDD system. Sensing of the P waves is lost at 5.0 mV, which the two arrows indicate. To simplify the test, the base rate is programmed to 30 bpm temporary.
P spontaneous P wave, *A* paced atrium, *V* paced ventricle, *R* spontaneous ventricular beat.

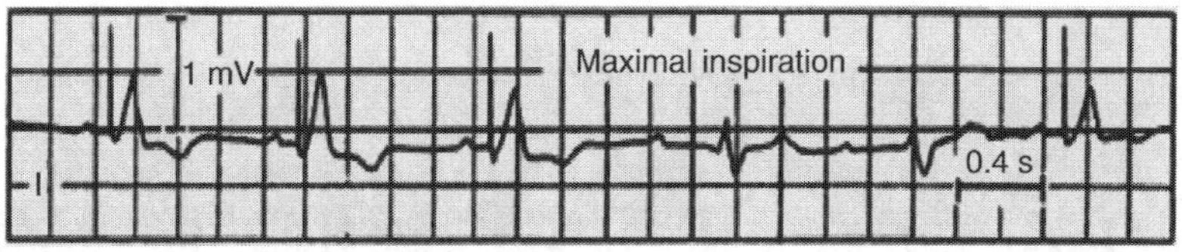

Figure 38.49
DDD pacing with obviously reliable atrial sensing. However, a test during deep inspiration discloses intermittent atrial undersensing.

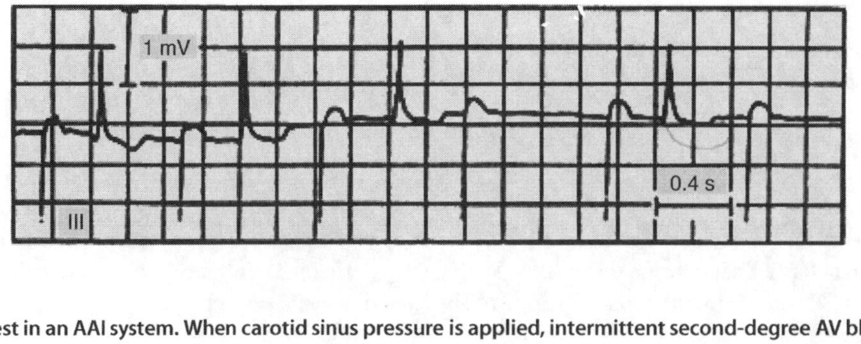

Figure 38.50
AV conduction test in an AAI system. When carotid sinus pressure is applied, intermittent second-degree AV block is exposed.

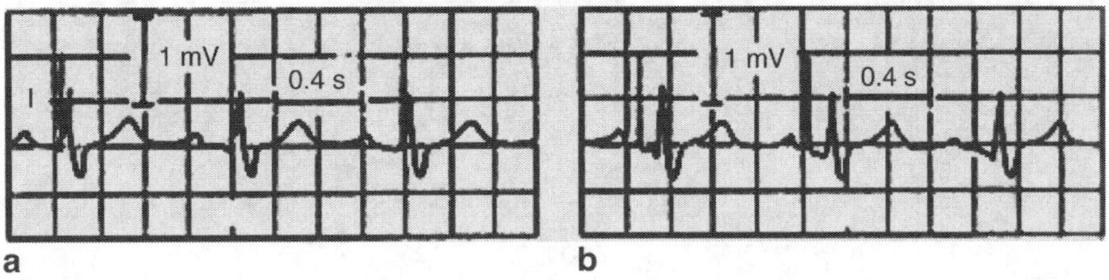

Figure 38.51
A DDD pacemaker with a technical AV delay programmed to 200 ms is shown in (a). Atrial triggered ventricular stimulation coincides with spontaneous conducted beats and the ECG could be interpreted as normal pacemaker function. However, reprogramming to a shorter AV delay of 120 ms discloses an underlying ventricular exit block, which is shown in (b).

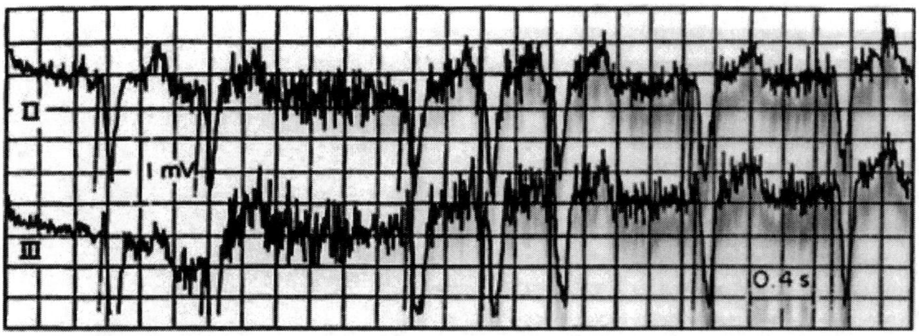

Figure 38.52
Myopotential provocation test. In DDD pacing, myopotentials sensed by the atrial amplifier will cause triggered ventricular stimulation, while sensing by the ventricular amplifier is followed by inhibition or conversion to interference rate stimulation. The type of reaction depends on the program setting and the amplitude of the myopotentials obtained.

38.9 Biventricular Pacing, Cardiac Resynchronization

During the last few years, there has been an enormous interest in treating patients with heart failure with pacing both ventricles at the same time or almost simultaneously by implanting an extra pacing lead on the left ventricle. This is possible by using branches of the coronary sinus vein and specially designed pacemaker leads for this purpose. It is not necessary that the patient has bradycardia; a major intraventricular electromechanical asynchrony with heart failure is the indication for CRT (cardiac resynchronization therapy). Instead of pacing only the right ventricle with bundle branch block as a result, stimulation of both ventricles at the same time is a more optimized treatment for each patient. If chronic atrial fibrillation does not exist, the patient will have three leads implanted: one in the right atrium, another in the right ventricle, and the third one in the coronary sinus (➋ Fig. 38.53). In chronic atrial fibrillation, the atrial lead is skipped.

The ECGs from biventricular pacing can be very similar to the ordinary DDD ECGs, but during different tests it is easier to see how the system works. The pacing ventricular threshold tests can be done with a short AV delay or in the VVI modes for each ventricle (➋ Fig. 38.54).

In the early stages of biventricular pacing, many different systems were available. For example, the ventricular ports were connected together to provide simultaneous pacing of both ventricles. The pacemaker senses RV and LV activity simultaneously. Today, the modern biventricular pulse generator usually has two independent ventricular ports and it is possible to program a short time delay between the LV and RV stimuli and optimize the therapy for each patient.

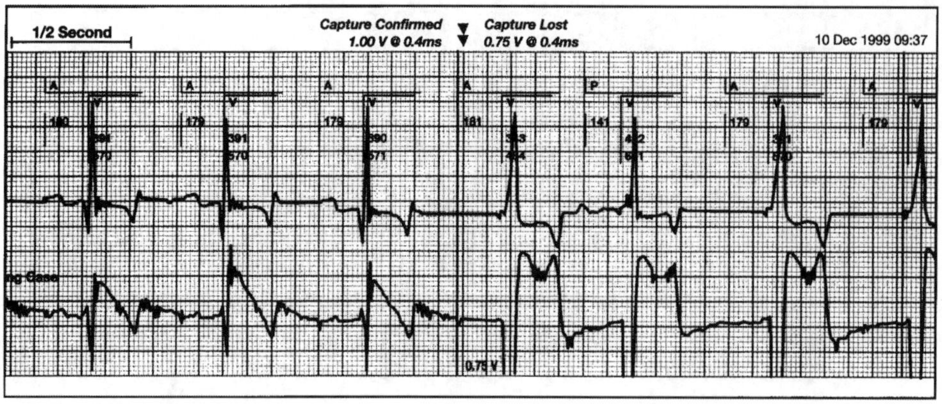

Figure 38.53
In the upper ECG lead, the three first QRS complexes are narrow due to biventricular stimulation. During an automatic threshold test, capture is lost at 0.75 V on one pacemaker channel resulting in a bundle branch block pattern visible both on the surface ECG and the intracardiac electrogram (EGM) in the lower part [7].

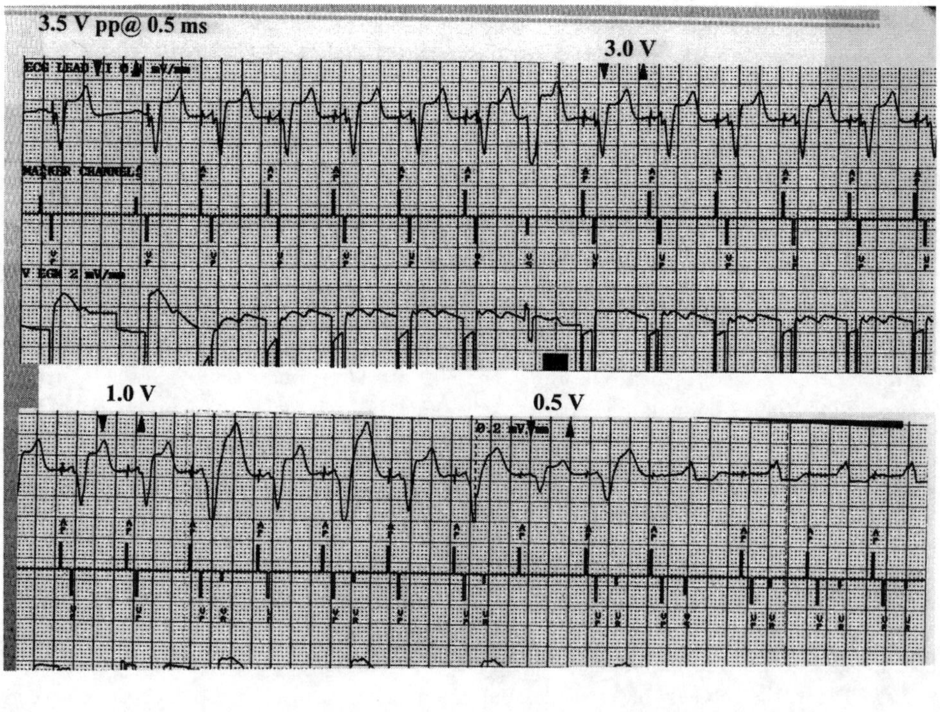

Figure 38.54
Another example of biventricular pacing. The first complexes are similar to ordinary DDD pacing. During the threshold test, bundle branch block occurs on one channel and at 0.5 V the spontaneous intrinsic QRS complex is seen.

38.10 Tachycardia and ICD

Many patients with a life-threatening tachycardia will receive an ICD. The ICD (implantable cardioverter defibrillator) has a built-in pacemaker, and depending on the lead configuration, the pulse generator can pace and sense in single or double chamber mode or even in biventricular configuration. Many patients need pacing just after termination of a tachycardia, when a short asystole can appear. During implantation, ventricular fibrillation is initiated in order to check that the shock therapy terminates the life-threatening tachycardia (➋ Fig. 38.55).

Most implantable defibrillators have the opportunity to terminate a ventricular tachycardia by using a function called ATP (antitachycardia pacing). Instead of a painful shock therapy, a preprogrammed rapid overdrive (bursts) stimuli is automatically delivered with a low-pacing amplitude when the tachycardia starts (➋ Fig. 38.56). If VF is initiated, shocks will be delivered.

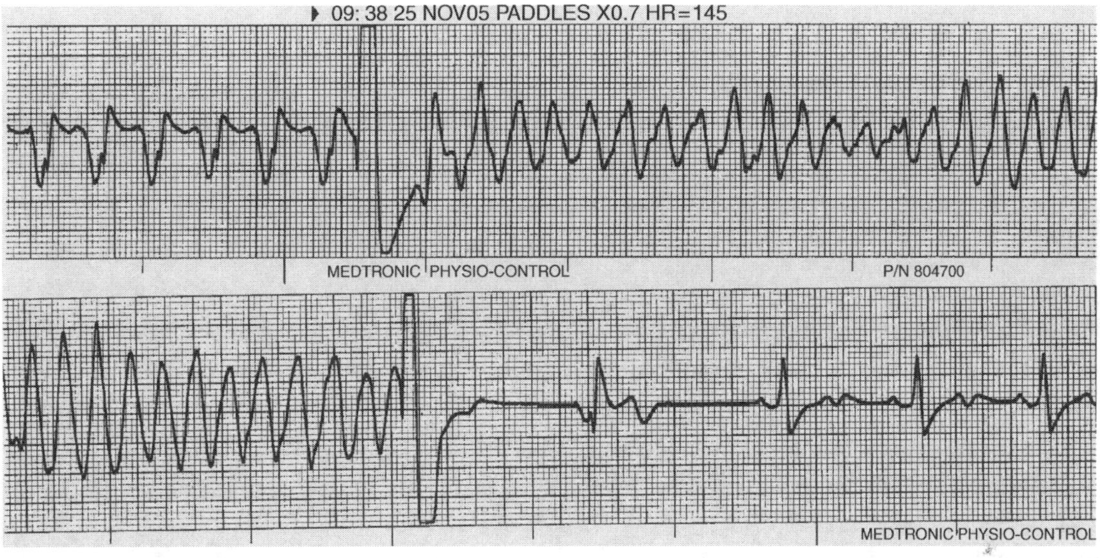

Figure 38.55
Initiation and automatic termination of VF in a test during an implantation of an ICD. The first six beats are overdrive pacing with a shock on the vulnerable phase of the T wave to start the VF. The bottom ECG has a pacemaker beat after the shock has been delivered and then sinus rhythm is established.

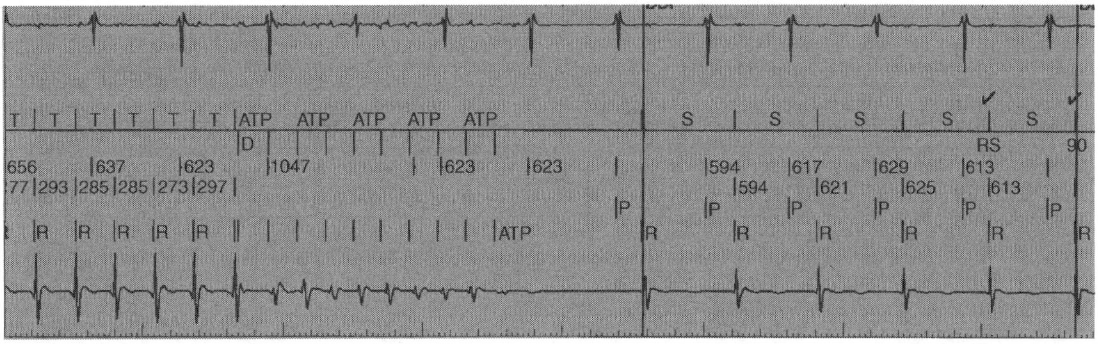

Figure 38.56
Print out from an ICD programmer with a stored ECG, EGM, and markers from an ICD device. The print out demonstrates a ventricular tachycardia and a burst of pacing stimuli is given (ATP) and the VT terminates to sinus rhythm (S). Markers of the P and R waves are also seen.

The pacemaker in the ICD device has a sensor which is very useful when antiarrhythmic drug therapy sometimes results in bradycardia.

The interpretation of stored ECGs in the ICD devices focuses a lot on the correct sensing of VT and VF. Unfortunately, many shocks are delivered due to false sensing (rapid atrial fibrillation, lead insulation defects, etc.). The pacing/sensing ventricular lead is tested in the same way as an ordinary pacing lead regarding thresholds.

References

1. Parsonnet, V., S. Furman, N.P.D. Smyth, and M. Bilitch, Optimal resources for implantable cardiac pacemakers. Pacemaker study group. *Circulation*, 1983;**68**: 227A–244A.
2. Barold, S.S, Editors. *Modern Cardiac Pacing*. Mount Kisco, New York: Futura, 1986.
3. Furman, S., D. Hayes, and D. Holmes, *A Practice of Cardiac Pacing*. Mount Kisco, New York: Futura, 1986.
4. Levine, P.A and R.C. Mace, *Pacing Therapy: A Guide to Cardiac Pacing for Optimum Hemodynamic Benefit*. Mount Kisco, New York: Futura, 1983.
5. Schüller, H. and T. Fåhraeus, *Pacemaker Electrocardiograms – An Introduction to Practical Analysis*. Solna, Sweden: Siemens-Elema, 1983.
6. Serge, B.X.S. Roland, and F.S. Alfons,*Cardiac Pacemakers Step by Step, an Illustrated Guide*. Futura: Blackwell, 2004.
7. Levine, P.A. *Guidelines to the Routine Evaluation, Programming and Follow-Up of the Patient with an Implanted Dual-Chamber Rate-Modulated Pacing System*. St Jude Medical, 2003.
8. Kenneth, E.A., K.G. Neal, and L.W. Bruce, *Clinical Cardiac Pacing*. W.B. Saunders, 1995.

39 The Signal-Averaged Electrocardiogram

Leif Sörnmo · Elin Trägårdh Johansson · Michael B. Simson

P. W. Macfarlane, A. van Oosterom, O. Pahlm, P. Kligfield, M. Janse, J. Camm (eds.), *Comprehensive Electrocardiology*, DOI 10.1007/978-1-84882-046-3_39,

39.1 Introduction

The signal averaging technique is applied to electrocardiographic recordings to reduce extraneous noise, which masks low-amplitude bioelectric signals from the heart. Although modern amplifier design and good recording techniques can minimize certain types of noise, other sources of noise, such as muscle activity, obscure low amplitude potentials. With signal averaging, the noise level can be reduced so that repetitive waveforms at the microvolt level can be reliably detected and analyzed. The noise level after averaging is, in most studies, below 1 μV, the equivalent of 1/100 of a millimeter at a standard ECG display scale.

This chapter discusses the methodology of signal averaging and its use in studying high-frequency components of the QRST complex, manifested either as ventricular late potentials (➋ Sects. 39.3 through ➋ 39.7) or within the QRS complex (➋ Sects. 39.8 and ➋ 39.9). Signal averaging is useful also in other ECG applications such as exercise testing (see ➋ Chap. 36), although the dynamic changes of the ST segment call for recursive signal averaging with a forgetting factor.

39.2 Methods of Signal Averaging

Signal averaging reduces the level of noise that contaminates a repetitive signal such as the ECG [1–3]. The most commonly used form is ensemble averaging in which multiple samples of a repetitive waveform are averaged with equal weights; random noise, which is not synchronized with the waveform of interest, is reduced. The process of averaging begins by measuring the voltage of an ECG fed through a high-gain (×1,000) amplifier with bandwidth ranging from 0.05 Hz to several 100 Hz. The analog signal is converted into digital form at a sampling rate of 1,000 Hz or more and stored. Once all beats of the ECG have been identified and aligned in time, the ensemble average is computed by first summing the samples at a particular instant within the beat and then dividing the sum by the total number of beats. This procedure is repeated for all the samples of the beat.

Several requirements must be met so that the signal averaging technique can reduce noise effectively [1, 3]. First, the waveform of interest must be repetitive so that multiple samples can be obtained to form an average waveform. Second, there must exist a unique feature in the ECG, which can be used as a reference time so that appropriate points of the repetitive signal can be averaged. The time of the maximum amplitude or maximum slope of the QRS complex may be used as reference time; however, such simplistic definitions are useful only when the noise level of the ECG is known to be low. A more robust approach is to cross-correlate each new waveform with a template waveform, determining the reference time as that point in time which corresponds to the highest cross-correlation value [4, 5]. If the algorithm for determining the reference time is inaccurate, that is, causing "trigger jitter," the averaged waveform becomes smoothed and the high-frequency components of the waveform are reduced. For example, it can be shown that a normally distributed jitter with a standard deviation of 1 ms causes the averaging operation to act as a linear, time-invariant low-pass filter with cut-off frequency at about 140 Hz [1]. Similarly, if the QRS complex is used as reference time, then the high-frequency components of an averaged P wave is smoothed because the PR interval varies slightly from beat to beat.

Third, the signal of interest must be uncorrelated with the contaminating noise. If so, the reduction of noise by signal averaging is proportional to the square root of the number of beats contained in the ensemble. For example, averaging of 100 beats will reduce the noise by a factor of 10 [1]. However, the noise must have certain properties to be reduced effectively by signal averaging: it should be random, uncorrelated with the beats, and characterized by an unchanging statistical distribution.

The noise that masks low-level electrical events of the heart has three primary origins:

1. Skeletal muscle noise, typically 5–20 μV even with a relaxed patient
2. Power line interference (50 or 60 Hz and related harmonics) and
3. Electronic and thermal noise from amplifiers and electrodes

Myoelectrical and electrode noise are the most troublesome ones in practice; modern isolation amplifiers have greatly reduced interference originating from power lines. Another noise source is the presence of ectopic beats. These complexes can be eliminated by their premature timing and through matching of a new beat against a template of the desired beat before averaging is performed [5–7].

39.3 Late Potentials and Time Domain Analysis

39.3.1 Introduction

One of the most frequent applications of signal averaging in electrocardiology has been the detection of ventricular late potentials, most often in patients with ventricular tachyarrhythmias. Late potentials are micro-volt level, high-frequency waveforms that are contiguous with the QRS complex and last a variable time into the ST segment. Such potentials are postulated to originate from small areas of delayed and fragmented ventricular depolarization, resulting primarily from myocardial infarction (MI).

39.3.2 Filtering and Time Domain Parameters

Characterization of late potentials is facilitated by linear, time-invariant high-pass filtering of the signal averaged ECG so as to reduce the influence of large-amplitude low-frequency content. Such filtering permits high-frequency components, that is, the late potentials, to pass without loss of amplitude, while low-frequency components are attenuated. High-pass filtering is used because depolarization of cells generates rapid changes in membrane voltage, fast movement of wave fronts of activation, and high frequencies on the body-surface ECG. The plateau or repolarization phases of the action potential generate more slowly changing membrane voltages and lower-frequency signals on the body surface [8]. Microvolt-level waveforms, arising from the late depolarization of small areas of myocardium, would be difficult to perceive if displayed at high gain without filtering. High-pass filters with a cut-off frequency ranging from 25 to 100 Hz have been employed in most studies.

The use of conventional linear, time-invariant high-pass filtering is, however, complicated by the fact that a transient event in the input signal, that is, the QRS complex, produces ringing in the output signal that is subsequent to the event (➋ Fig. 39.1a) [5]. This property is most unfortunate because it impedes the detection of low-amplitude potentials occurring immediately after the QRS complex. While the ringing problem is inherent to linear, time-invariant high-pass filtering, its harm can be largely reduced by employing "bidirectional" filtering–a technique which is implemented in most systems for late potential analysis. The main idea behind this technique is to delay ringing by filtering the signal in different directions so that ringing occurs within the mid-QRS. Forward filtering starts at the P wave and continues until the mid-QRS is reached, whereas backward filtering starts at the T wave and ends at the mid-QRS (➋ Fig. 39.1b) [5]. Evidently, bidirectional filtering suffers from the disadvantage of an undefined output at mid-QRS, and thus no measurements should be done from the bidirectionally filtered signal that involves the entire QRS complex.

➋ Figures 39.2 and ➋ 39.3 show examples of recordings made with signal-averaging technique in patients with and without late potentials. The tracings at the top of each figure are signal-averaged bipolar X, Y, and Z leads (as used in most studies of late potentials) from about 150 beats. Each lead is then high-pass filtered with a bidirectional filter (cut-off frequency at 25 Hz) and combined into a vector magnitude ($\surd(X^2 + Y^2 + Z^2)$), a signal that combines the high-frequency information contained in all three leads. This signal, shown on the bottom of each figure, is here termed the "filtered QRS complex." In patients without late potentials, the filtered QRS complex exhibits an abrupt onset, a peak of high-frequency voltage 40–50 ms after QRS onset, and an abrupt decline to noise level at the end of the QRS complex (➋ Fig. 39.2). There is no high-frequency signal above the noise level in the ST segment. The peak of the T wave is associated with a small amount of high-frequency content. In patients with late potentials (➋ Fig. 39.3), the initial portion of the filtered QRS complex is similar to that in patients without late potentials, except that the voltage tends to be lower [9]. At the end of the filtered QRS complex, however, there is a "tail" of low-amplitude signal which is not present in recordings from patients without late potentials. The low-level signal, a late potential, is contiguous with the QRS complex and corresponds to low-amplitude activity which can be seen at the end of the QRS complex in the unfiltered leads when displayed at high gain. The amplitude of the late potential varies from 1 to 20 μV when 25 Hz filtering is used.

The accuracy of the endpoint of the filtered QRS complex is a crucial parameter in time domain analysis as all other parameters are computed with reference to this point. Poor accuracy of the endpoint definition obviously limits the overall accuracy of the analysis. Clearly, the presence of late potentials causes the transition from signal to noise to be much less clear-cut than when late potentials are absent. The endpoint may be found by a backward search in the vector magnitude for the first samples which exceed a certain threshold; the threshold value is related to the residual noise level

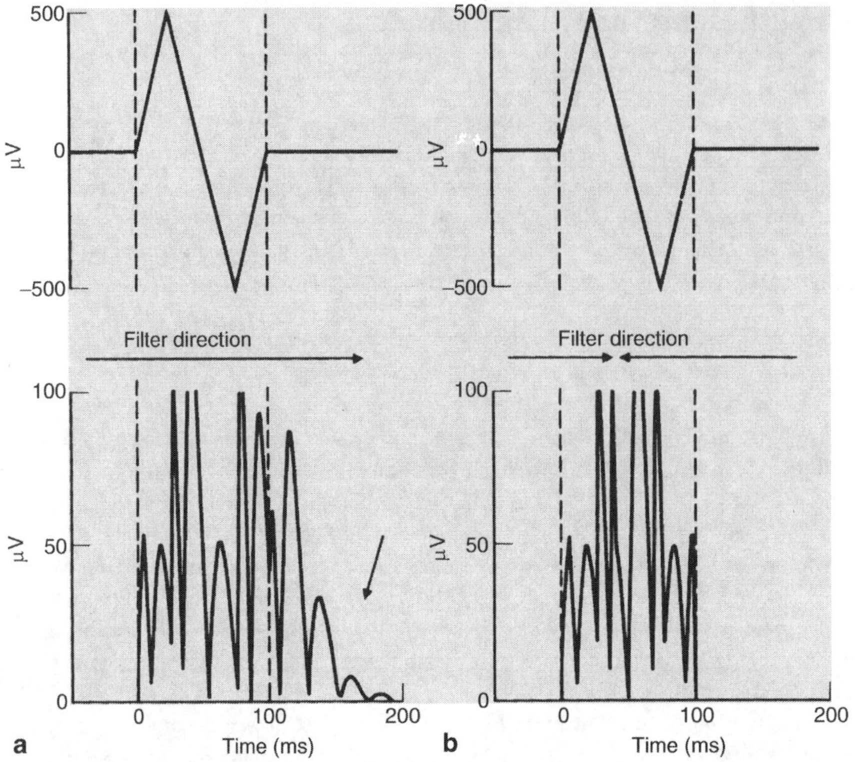

Figure 39.1
In (a), a test ECG signal is shown on top, while below the effect of a unidirectional filter with highpass (>25 Hz) characteristics is shown. Filtering starts at QRS onset and progresses, but it can be clearly seen that undesired signals (*ringing, arrowed*) have been introduced after the end of the QRS complex. In (b), however, the same filter is applied from QRS onset for 40 ms only and then applied from QRS end in the opposite direction, that is, toward QRS onset, again for 40 ms. This time no ringing outside the test ECG is apparent from the use of bidirectional filtering. (After Simson [5]. American Heart Association, Dallas, Texas. Reproduced with permission.)

of the averaged beat [5]; see also [10] for description of a more robust approach in individual leads. Several studies have required that the noise level of the averaged beat is reduced to a root mean square (RMS) level no more than 1 μV. However, it has been shown that the accuracy of the late potential endpoint, as well as the sensitivity and specificity of late potential analysis as such, can be further improved by extending the averaging period so that additional beats are included [11].

Time domain parameters are usually computed from the vector magnitude and include the RMS amplitude of the last 40 ms and the duration of the terminal filtered QRS amplitude <40 μV. These parameters are generally measured in commercially available equipment, as is the filtered QRS duration. The following, commonly used, diagnostic criteria for ventricular late potentials are based on filtering of the QRS complex with pass band 40–250 Hz:

1. RMS amplitude of last 40 ms of filtered QRS complex <20 μV (RMS40)
2. Duration of terminal low amplitude of filtered QRS complex below 40 μV < 38 ms (LAS40)
3. In the absence of a conduction defect, a filtered QRS duration >114 ms (FQRSd)

Using instead filtering with passband 25–250 Hz, the following criteria are commonly used: RMS40 <25μV, LAS40 >30 ms, and fQRSd >100 ms. At least two criteria should be met for late potentials to be considered present. A signal-averaged ECG with the three measurements (RMS40, LAS40, and FQRSd) is displayed in Fig. 39.4.

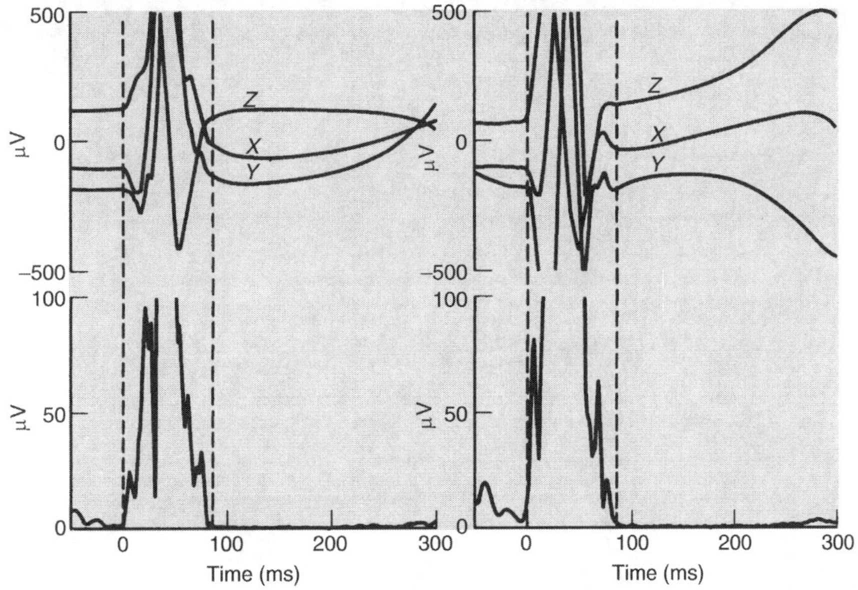

■ **Figure 39.2**
Recordings in patients without late potentials. These patients had anterior (*left*) and inferior (*right*) MI respectively. Bipolar signal-averaged leads are shown in high gain (*top*). On the bottom is the filtered signal-averaged QRS complex which is less than 100 ms in duration in both patients. There is a large RMS amplitude of signal in the last 40 ms of the filtered QRS complex (80 and 129 μV, respectively). The dashed vertical lines denote QRS onset and offset.

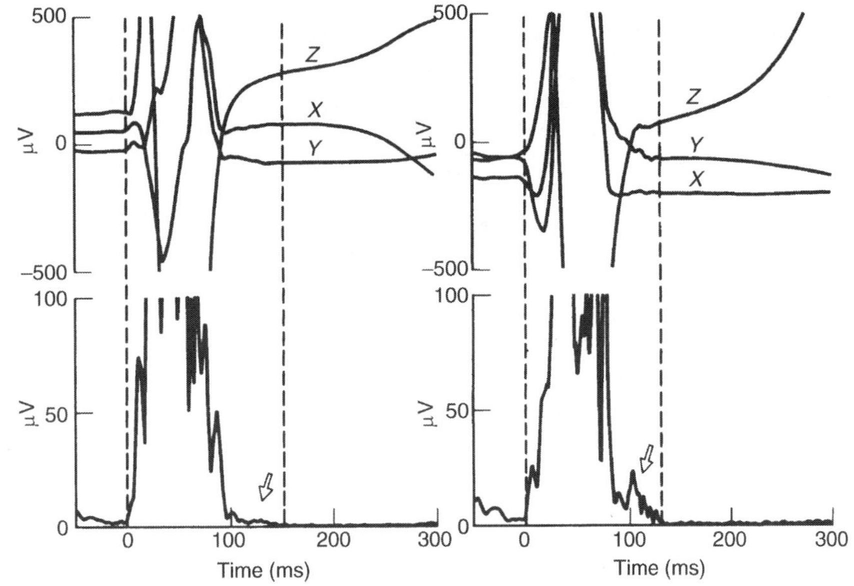

■ **Figure 39.3**
Signal processing in patients with VT and late potentials. The patient on the left had an anterior MI and the patient on the right had an inferior MI. The filtered QRS complex shows a longer duration in both patients and late potentials (*arrows*). The RMS voltages in the last 40 ms of the filtered QRS complex measured 2.0 and 11.2 μV, respectively. The *dashed vertical lines* denote QRS onset and offset.

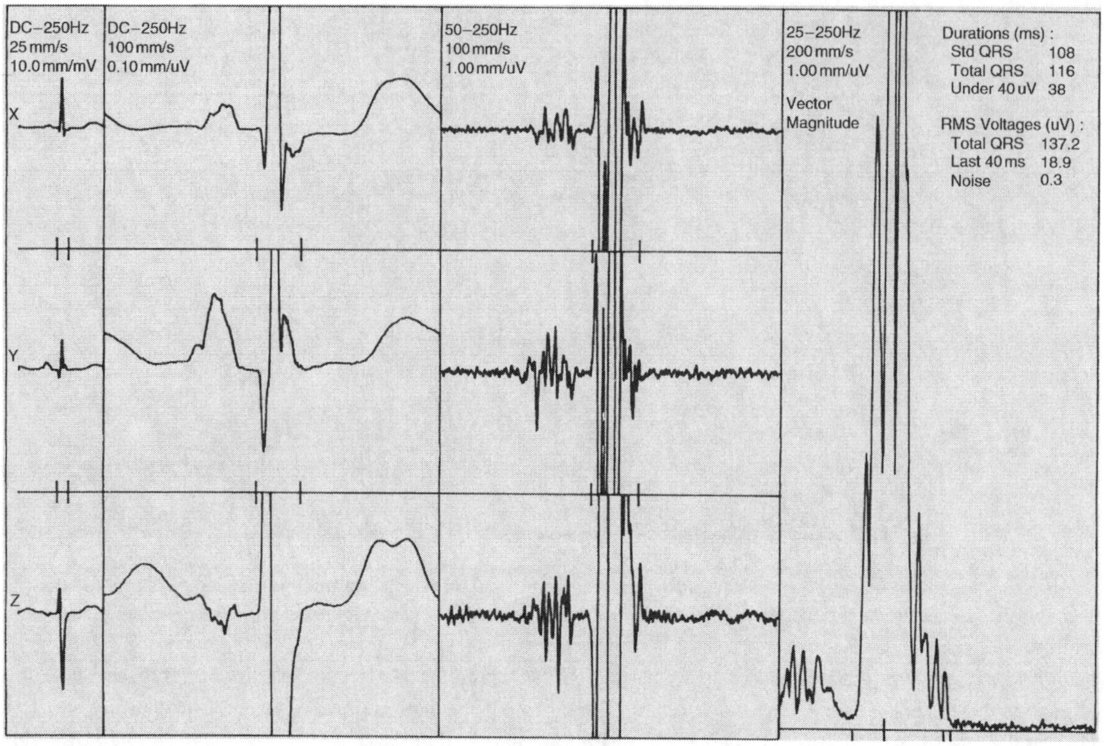

Figure 39.4

An example of the display of the signal-averaged ECG. The ECG is recorded from a 69-year old female with documented VT. The RMS voltage in the last 40 ms (RMS40) is abnormally low at 18.9 μV, the filtered QRS duration (fQRSd) is abnormally long at 116 ms, and the duration of late activity under 40 μV (LAS40) is also abnormally long at 38 ms.

39.3.3 Physiological Background

Late potentials are considered to originate from small areas of delayed ventricular depolarization. Disorganized and late ventricular activation has been recorded directly from infarcted and ischemic myocardium in animals and in humans [12–18]. Studies in animals with experimental infarcts have shown that the delayed ventricular activation can span diastole and that it relates to the occurrence of ventricular arrhythmias [12, 15, 16]. Josephson et al. demonstrated with endocardial recordings in man that fragmented electrical activity can outlast the surface QRS complex, and that the onset and maintenance of ventricular tachycardia (VT) depends on continuous diastolic activity [19]. When VT ceased, the continuous diastolic activity was no longer present. These studies provide evidence that VT after infarction has a reentrant mechanism and involves areas of abnormally slow and protected ventricular activation.

Several groups have reported examples of delayed epicardial ventricular activation in patients with late potentials [20–22]. To establish the time relationship between the body surface late potential and delayed ventricular activation, Simson et al. studied eight patients with signal-averaged ECGs and ventricular mapping [17]. The patients had medically intractable and inducible VT after infarction and underwent surgery for control of the arrhythmia. Twelve to 16 left ventricular endocardial sites were mapped with a catheter and 32–54 epicardial sites were directly recorded in the operating room at normal body temperature. Studies were performed during sinus rhythm. All patients showed evidence of fragmented and low-level electrograms which were prolonged (>60 ms) in duration. These electrograms were recorded from 6.6 ± 3.3 sites per patient; 88% of the fragmented electrograms began during the QRS complex and the latest fragmented electrogram for each patient ended a mean of 161 ms after QRS onset. Six patients with VT had late potentials and the late potentials on the body surface corresponded in time to fragmented and delayed electrogram activity. During the last 40 ms of the filtered QRS complex, when the late potentials were recorded, 68% of the electrograms active showed fragmented activity. In contrast, in earlier segments of the QRS complex only 27% of the active electrograms showed fragmented activity. The mapping studies demonstrated that the late potentials correlated in time with fragmented electrogram activity, which begins during the normal QRS complex but which outlasts normal ventricular activation. The studies are in agreement with those performed in animals with experimental infarcts and late potentials [18, 23].

The mapping studies suggest that fragmented electrogram activity could be detected on the body surface as a low-level waveform only when it outlasted normal ventricular activation. Late potentials were not detected in a patient with left bundle branch block because the late epicardial activation occurred simultaneously with delayed and fragmented endocardial electrograms. When the fragmented electrograms were of brief duration and ended less than 90 ms after QRS onset, then no late potentials could be detected [17]. Because the duration of fragmented ventricular activation is prolonged with premature beats or at rapid heart rates, pacing the heart or inducing a premature beat may enhance the detection of late potentials [15, 16]. The frequency content of late potentials is similar to that of the entire QRS complex and it is unlikely that frequency analysis could detect small areas of fragmented ventricular activation occurring simultaneously with activation of a larger mass of normal myocardium.

39.4 Late Potentials and Time Domain Analysis in Clinical Applications

39.4.1 Recent Myocardial Infarction

Hundreds of articles have been published on clinical applications of analysis of late potentials. Most of these focus on patients after MI. It has been found that the prevalence of late potentials in patients with old MI varies between 71% and 90% in patients with sustained ventricular arrhythmias, compared to 26–34% in patients without ventricular arrhythmias [24–26]. The presence of late potentials after an MI has been associated with increased risk for VT in prospective studies [27, 28]. Breithardt et al. [28] studied 132 patients a mean of 22 days after acute MI. The prevalence of sustained VT on a mean follow-up of 15 months was 11.9% in patients with, and only 2.7% in patients without late potentials. The prevalence of VT increased with the duration of late potentials; VT occurred in 5% of those with late potential duration <40 ms, compared to 25% in whom the late potential lasted >40 ms.

Between 14% and 29% of the patients with late potentials have been shown to have sustained VT within the first year after MI, compared to only 0.8–4.5% of patients without late potentials [29]. The prevalence of late potentials is higher in patients with inferior MI (58%) compared to patients with anterior infarction (31%) [27], which is not surprising when considering the normal activation sequence of the ventricles, but the prognostic value for ventricular tachycardia is higher for anterior infarctions. Analysis of late potentials has higher prognostic value for ventricular tachycardia than offered by long-term ECG and left ventricular ejection fraction [27].

There is a correlation between the prevalence of late potentials and the degree of ventricular dysfunction. Breithardt et al. [25] found that in patients without a history of ventricular arrhythmia, late potentials were detected in only three of 32 patients (9%) with normal ventricular function, but were present in 32 of 69 patients (46%) who had ventricular akinesia or aneurysm. Hombach et al. [2] found a similar relation between the incidence of late potentials and increasing degrees of ventricular dysfunction. Regardless of the degree of ventricular dysfunction, however, late potentials were found more frequently and had a longer duration in patients with ventricular tachycardia or fibrillation [25]. A retrospective study compared the findings on a signal-averaged ECG, prolonged ECG monitoring and cardiac catheterization to determine which combination of findings best characterized patients with VT after MI [26]. A multivariate statistical analysis showed that the abnormalities on the signal-averaged ECG, including the late potentials, provided independent information useful in identifying patients with VT after MI. An abnormal signal-averaged ECG (presence of late potentials or a long-filtered QRS duration), a peak premature ventricular contraction rate >100 h^{-1} and the presence of a left ventricular aneurysm were the only three variables identified which provided significant information to characterize patients with VT. The study suggested that the abnormal signal-averaged ECG may be combined with other clinical information to provide a more specific identification of patients with VT.

The ACC expert consensus document from 1996 [30] concludes that a normal signal-averaged ECG indicates a low risk for developing life-threatening ventricular arrhythmias. The positive predictive accuracy of only 14–29%, however, is not high enough to justify interventions in individual patients with abnormal results.

39.4.2 Myocardial Reperfusion

Successful treatment of acute MI with thrombolytic agents has been shown to reduce the incidence of late potentials (range 5–24% in patients treated with thrombolytic agents compared to 18–43% in patients not treated) [31–40]. Maki et al. [41] investigated the relationship between the time required for reperfusion by percutaneous transluminal coronary angioplasty and the incidence of late potentials in 94 patients with acute MI. They found that the incidence of late potentials in patients undergoing primary angioplasty at ≤4, 4–6, 6–8, 8–10, and >10 h after infarction was 8%, 12%, 14%, 33%, and 43%, respectively. In the control group, consisting of 31 patients who were treated conventionally, the incidence of late potentials was 48%. The presence of late potentials has been shown to have a low positive predictive value in patients who undergo reperfusion after acute MI [42, 43]. The presence of late potentials has also been shown to be a poor predictor of sudden death after surgical revascularization [44]. Thus, signal-averaged ECG has limited value for risk stratification in an unselected postinfarction population, and is currently not recommended as a risk marker for increased mortality [45].

39.4.3 Sudden Death

The incidence of sudden death has been too infrequent in the studies to date to form a firm conclusion on the value of late potentials as a prognostic indicator for that event. Nevertheless, some prospective studies have been undertaken [46, 47] to evaluate whether the presence of late potentials is an independent risk factor for sudden death or serious ventricular arrhythmias after MI, and to establish the role of the signal-averaged ECG along with other noninvasive tests in identifying patients at high risk.

The combination of late potentials with a low left ventricular ejection fraction, as determined by radionuclide techniques, together with complex ventricular ectopy on Holter monitoring have been used to identify those patients at highest risk of sudden death or developing sustained VT from 6 months to 2 years following MI. Kuchar et al. [46] found that an abnormal signal-averaged ECG together with an ejection fraction <40% predicted arrhythmic events with 34% probability. On the other hand, if the patient had abnormal left ventricular function but a normal signal-averaged ECG, there was only 4% risk of the occurrence of an arrhythmia. The signal-averaged ECG, Holter monitoring, and ejection fraction were independently related to outcome, but a left ventricular ejection fraction <40% was the most powerful indicator.

In a similar study, Gomes et al. [47] also found that the presence of late potentials, an abnormal ejection fraction, and high-grade ectopic activity on Holter monitoring were the variables most significantly related to a future arrhythmic event. Their overall conclusion was that the combination of these abnormalities identified the group of patients at highest risk of VT, sudden death or both in the first year after MI.

39.4.4 Unexplained Syncope

The presence of late potentials in patients with unexplained syncope has been associated with high sensitivity and specificity for inducement of VT during electrophysiological studies. Kuchar et al. [48] evaluated 150 patients with syncope. Late potentials were detected in 29 patients, of whom 16 were found to have VT. In the patients with syncope due to other causes than VT, none of the patients had late potentials on signal-averaged ECG. In this study, the sensitivity was 73%, the specificity 55%, the positive predictive value 55%, and the negative predictive value 94%. Another study by Lacroix et al. [49] found that the positive predictive value of late potentials was 39% for predicting the inducibility of sustained VT. The clinical value of analysis of late potentials in patients with unexplained syncope is mainly in its negative predictive accuracy.

39.4.5 Success of Arrhythmia Surgery

Several investigators have reported that late potentials may disappear after a successful operation for VT [20, 22, 50–52]. The operation generally includes an aneurysmectomy and an additional procedure, such as endocardial excision,

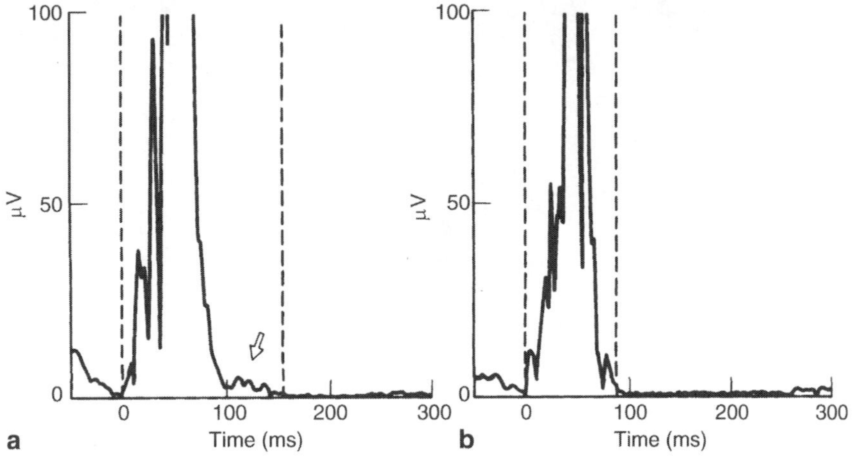

Figure 39.5
The filtered QRS complexes from a patient with VT who underwent aneurysmectomy and endocardial excision for control of the arrhythmia. In (a), before the operation, a 2.8 μV level of LP is present (*arrow*). In (b), after the operation, VT could not be induced and the LPs were no longer present.

to remove or isolate the apparent site of origin of VT. In one study, 24 patients in whom VT could not be induced after operation, the FQRSd decreased (from a mean of 137 to 121 ms) and the incidence of late potentials decreased (71–33%) [52] (➔ Fig. 39.5). Eight patients continued to have late potentials, despite surgical control of the arrhythmias. The incidence of late potentials in the filtered QRS complex was not changed in 13 patients in whom VT could be induced after operation.

Experience indicates that a successful operation for control of VT may cause the late potentials to vanish, but they may persist despite successful control of the arrhythmia. This finding suggests that the operation need not remove all areas of delayed activation in order to control VT. In patients with persistent late potentials, despite surgical control of the arrhythmias, it is hypothesized that either the delayed activation does not outlast the refractoriness of the myocardium or that the interface between the slowly conducting tissue and normal myocardium is sufficiently disrupted so that reentry cannot occur.

39.4.6 Efficacy of Anti-Arrhythmic Drugs

Analysis of late potentials has not yet been established as a method for accurately assessing anti-arrhythmic drug efficacy. Class IA, IC, and II [53–58] anti-arrhythmic drugs have been shown to elicit changes in the late potentials, where especially class IC drugs are associated with a marked increase in FQRSd [58]. Only a few studies have examined the effects of class III drugs [59, 60]. In general, time domain analysis of the signal-averaged ECG has not shown a correlation between the changes induced during drug therapy and anti-arrhythmic drug efficacy [53, 54, 58]. A few studies have shown a correlation between the prolongation of the total FQRSd induced by class I drugs and prolongation of the cycle length of ventricular tachycardia [55–57].

Simson et al. [60] investigated the effect of anti-arrhythmic drug therapy on the signal-averaged ECG in 36 patients with VT after MI. Twenty-nine patients, or 81%, had late potentials on medications. The drugs evaluated, alone and in combination, were procainamide, quinidine, disopyramide, amiodarone, phenytoin, and mexiletine. Electrophysiological stimulation was used to evaluate the control of VT. The duration of the filtered QRS complex was increased by a mean of 8–13 ms by procainamide, quinidine, and amiodarone. Procainamide decreased the voltage in the last 40 ms of the

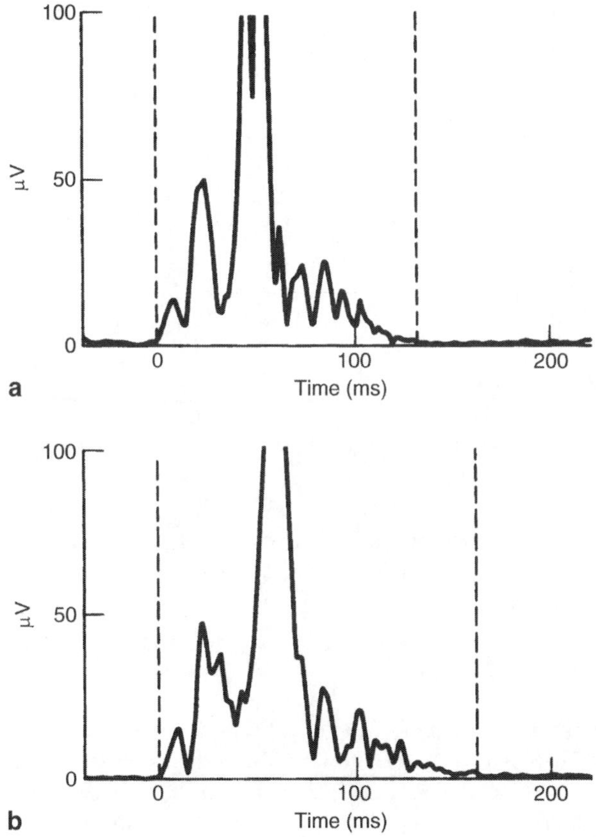

Figure 39.6
The effects of procainamide on the LPs in a patient with VT after MI. In (a), during the period without medication, the patient had a FQRSd of 129 ms and a 4.6 µV level LPs present at the end of the filtered QRS complex. In (b), when receiving the procainamide at a serum level of 16 µg/ml, the patient had a longer FQRSd (162 ms) but the LPs persisted. The VT remained inducible. (After Simson, (c) Saunders, Philadelphia, Pennsylvania. Reproduced with permission.)

filtered QRS complex by 4 µV; however, the incidence of late potentials with anti-arrhythmic drug therapy did not change (Fig. 39.6). Ventricular tachycardia was no longer inducible after anti-arrhythmic drug therapy in ten patients; nor was the late potentials abolished by any agent in these patients. There was no pattern of change in the filtered QRS complex that would indicate a successful response to anti-arrhythmic agents.

39.4.7 Cardiac Transplant

Several studies have investigated possible noninvasive methods for detection of cardiac transplant rejection. A few of them have examined the extent to which late potentials are a measure of heart transplant rejection [61–64]. One report showed a sensitivity of 65%, positive predictive accuracy of 92%, and negative predictive accuracy of 68% for acute rejection [61]. RMS values have also been shown to provide high sensitivity and specificity for rejection [62]. Thus, the potential of signal-averaged ECG analysis to detect acute rejection is promising, but more studies need to be done before the method can be used clinically.

39.4.8 Nonischemic Cardiomyopathy

The occurrence of late potentials in patients with nonischemic congestive cardiomyopathy has been investigated in a group of 41 patients and the findings compared with 55 normal controls [65]. It was found that the FQRSd was longer in patients with sustained ventricular arrhythmia than in those without (130.2 ± 19.5 ms as opposed to 105 ± 13.1 ms); a highly significant difference. The mean control value was 95.9 ms. The RMS voltage in the last 40 ms of the filtered QRS was lower in the group with arrhythmia than in those without (11.3 ± 9.3 as opposed to 53.5 ± 28.3 μV), again a highly significant difference. The corresponding normal group values were 53.7 ± 25.2 μV. Overall, 83% of patients in the group with sustained ventricular arrhythmia had both an abnormally long FQRSd and abnormally low late-potential amplitude; 14% of patients without an arrhythmia and 2% of controls had similar findings. Findings from other [66, 67], but not all [68–70] studies of patients with dilated nonischemic cardiomyopathy have shown similar results.

Arrhythmogenic right ventricular cardiomyopathy was one of the first pathologic entities in which late potentials were identified. Analysis of late potentials can be useful for screening purposes or detection of arrhythmogenic right ventricular cardiomyopathy in family members [71, 72]. Nava et al. [73] evaluated signal-averaged ECGs in 138 patients with arrhythmogenic right ventricular cardiomyopathy and compared the results with those of 146 healthy controls. Late potentials were found in 57% of the patients, and in 4% of the controls. They also found that there is a closer correlation between late potentials and the extent of the disease than with the presence of ventricular arrhythmias. Late potentials have also been found in a significant proportion of patients with the Brugada syndrome. Research suggests that late potentials might be helpful to identify patients at a higher risk of life-threatening arrhythmic events in this population, both in a retrospective [74] and a prospective [75] study, but the results need to be confirmed in larger studies.

39.5 Late Potentials and Frequency Analysis

As described above, late potentials are composed of frequencies which are higher than those of the ST segment, that is, higher than about 25 Hz. However, no specific information is available beforehand on their particular frequency content. By employing frequency analysis based on the discrete-time Fourier transform (and typically implemented by the fast Fourier transform, FFT), it would be possible to pinpoint the range of frequencies which is characteristic of late potentials. The main advantage of the frequency domain approach is that no prior knowledge is required on signal characteristics, whereas the time domain approach requires that the cut-off frequency of the high-pass filter is predetermined. However, this advantage comes at the sacrifice of temporal information as it is not possible to pinpoint the occurrence of late potentials as being late in the QRS complex or early in the ST segment. In practice, frequency analysis is also hampered by certain undesirable properties of the Fourier power spectrum related to frequency leakage and large variance [1, 76]. While these properties can be mitigated to a certain extent by various techniques, care should still be exercised when interpreting the outcome of frequency analysis.

In time domain analysis, a crucial signal processing step is to determine the endpoint of the high-pass filtered QRS complex in order to assure that the diagnostic measurements, that is, RMS amplitude, duration of low-amplitude signals, and filtered QRS duration are accurate. In frequency analysis, the corresponding crucial step is to determine the location and length of the signal segment to be processed. Since the segment begins in the terminal part of the QRS – a part of the heartbeat with drastic changes in amplitude – a displacement of its onset by just a few milliseconds can produce quite different spectra. The segment location should preferably be determined by an algorithmic procedure so as to avoid the subjectivity of manual delineation; one approach is to choose the segment onset as the point when the vector magnitude drops below 40 μV [77].

Different lengths of the segment are associated with different spectral resolution, implying that it is important to keep the length fixed from one patient to another in order to facilitate comparison of results. In the literature, lengths have ranged from 40 ms, thus only including the terminal part of the QRS, to about 200 ms so that a large part of the ST segment is included.

Prior to frequency analysis, it should be standard procedure to subtract the mean value of the ECG samples contained in the segment, that is, the DC component. Windowing is another common time domain operation with which samples at the boundaries of the segment are multiplied with weights smaller than those applied to samples in the middle of the segment, thus reducing the influence of abrupt changes which may occur at the segment boundaries. On the other hand, windowing may have the undesirable effect of reducing the contribution of late potentials when these happen to be located at any of the segment boundaries.

Once the power spectrum has been computed, one or several parameters must be derived which condense its main features. The absolute power can be computed in frequency bands whose limits are determined by some prior knowledge. Alternatively, it may be more appropriate to compute relative power, defined as the ratio of the power in a single frequency band to either the total power or the power contained in certain bands. Relative power measurements are often preferred since the absolute power may be influenced by various nonphysiological factors.

It is sometimes convenient to display and analyze the power spectrum with a logarithmic magnitude scale since it allows detail to be displayed over a wider dynamic range. The scale is defined in units of $20 \cdot \log_{10}$, referred to as decibel (dB). With this scale, 0 dB corresponds to a magnitude equal to 1, −20 dB corresponds to a ten times smaller magnitude, −40 dB corresponds to a 100 times smaller magnitude, and so on. The magnitude function is normalized with respect to its maximum value, thus corresponding to 0 dB. It should be noted that although the power spectrum is commonly employed in the literature, that is, the squared magnitude of the FFT, the (unsquared) magnitude of the FFT is also sometimes employed.

As described earlier, the conventional time domain approach to handle multiple-lead recordings is to simply combine the filtered leads into a vector magnitude from which a set of descriptive parameters is derived. However, the vector magnitude does not lend itself to frequency analysis as this function may obscure the frequency content of the ECG. Instead, the power spectra of individual leads may be averaged into one spectrum from which a set of parameters is derived. Alternatively, parameters can be derived from each spectrum and combined with some suitable technique, for example, averaging of the resulting parameter values.

39.6 Late Potentials and Frequency Analysis in Clinical Applications

The first clinical study exploiting frequency analysis was presented by Cain et al. in 1984 for the purpose of distinguishing patients prone to sustained VT [78]. The analysis method was assessed in three groups of subjects, consisting of 16 patients with previous MI and sustained VT, 35 patients with previous MI but no sustained VT, and ten normal subjects, respectively. The ECG was recorded from the X, Y, and Z Frank leads, digitized, and subjected to signal averaging of 100 beats such that a noise level of 1.5 μV was reached.

In each lead, Fourier-based frequency analysis was performed on a manually delineated 40 ms segment at the terminal part of the QRS, windowed using the Blackman–Harris window. The power spectrum was characterized by two parameters derived from the logarithmic magnitude scale, the spectral value in decibels at 40 Hz and the area enclosed between the 0 and −60 dB level (➲ Fig. 39.7). It should be noted that these two parameters are correlated since a large spectral value, in most cases, implies a large area.

The results for the three groups of subjects suggested that the area was considerably higher in patients with VT than in the other two groups, that is, the amplitude of the high-frequency components of the ECG in patients with VT was much greater than in those without. As one may expect, the spectral value in decibels at 40 Hz was larger in patients with VT than in the other two groups. Besides the terminal part of the QRS, frequency analysis was also performed on segments defined by the QRS complex, the ST segment, and the T wave; the ST segment was associated with results similar to those of the terminal part of the QRS, whereas the other two segments did not offer any discrimination between the VT group and non-VT groups.

In a subsequent study, Cain et al. modified their original way of deriving parameters from the spectrum by instead computing a ratio between the spectral areas defined by the intervals 20–50 Hz and 0–20 Hz [79]. Based on the findings of their previous study, the processed ECG segment was extended from the 40 ms at the terminal part of the QRS to also include the remaining part of the ST segment (though the segment was still manually delineated). The spectral area ratio was tested in 16 patients with a previous MI and sustained VT, 53 patients with a previous MI but without sustained VT, and 11 normal subjects. The results presented in ➲ Fig. 39.8 show that the VT patients have considerably larger ratios than the other two groups. However, the groups could not be discriminated when the 20–50 Hz interval was replaced by 70–120 Hz, a result which is somewhat surprising since the higher frequencies are important to the outcome of time domain analysis.

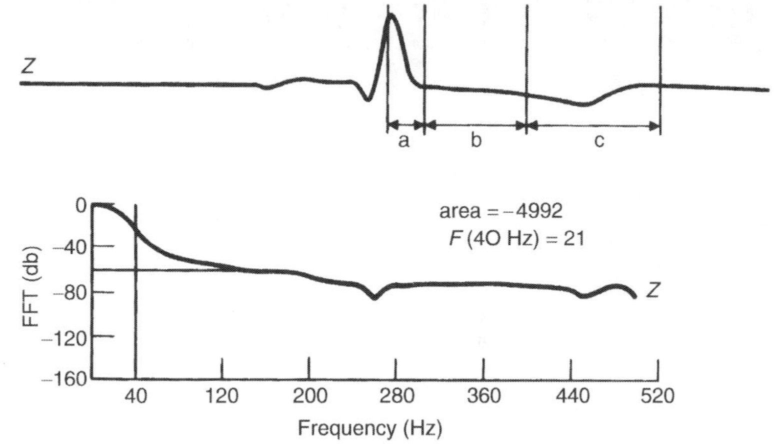

Figure 39.7
The logarithmic power spectrum characterized by the value in dBs at 40 Hz and the area enclosed between the 0 and −60 dB level. (Reprinted from Lindsay et al. [80] with permission.)

It should be noted that the spectral area ratio may account for signal properties unrelated to the presence of late potentials. Since the spectral area ratio involves frequencies in the interval 0–20 Hz, it is obvious that low-frequency components originating from, for example, baseline wander or a slowly changing ST segment will also influence the area ratio. In a later study, however, the lower limit was increased to 10 Hz so as to avoid this problem [80].

The studies by Cain and coworkers were followed by several studies from other groups who also employed frequency analysis to find out if late potentials present during sinus rhythm can be used as a marker for ventricular arrhythmias. The results presented in the literature have been most variable, ranging from "inability of frequency domain parameters to distinguish VT from non-VT patients" to "improved identification with frequency domain parameters." In the remaining part of this section, a number of studies are briefly summarized with variable outcomes. It should be noted that each of these studies has a slightly different approach to the implementation of frequency analysis, for example, the definitions of analyzed signal segment and frequency interval of interest.

In a study by Kelen et al., involving ten patients with spontaneous or inducible VT and ten normal subjects, it was shown that the spectral area ratio was markedly dependent on the length of the analyzed signal segment [81]. In fact, a change of as little as 3 ms in segment length changed the results across proposed boundaries of normalcy in normal subjects and in patients with VT. In contrast, the use of time domain analysis established that the patients had late potentials, whereas the normal subjects did not, using the standard diagnostic criteria.

Similar results were also obtained in a study by Machac et al. where the purpose was to determine whether time domain or frequency domain parameters were better in distinguishing patients with and without sustained VT [82, 83]. The two methods were tested in 26 patients with sustained VT, 18 control patients with organic heart disease but without sustained VT, and 11 normal volunteers. Frequency domain analysis was performed on three different segment lengths (the terminal 40 ms of the QRS complex, either alone or with 216 or 150 ms of the ST segment), and both power and power ratios were calculated in different frequency bands. At a fixed specificity of 78%, the best time domain sensitivity was 85%, whereas the best frequency domain sensitivity was 77%. It was thus concluded that frequency domain analysis did not offer any improvement over time domain analysis in distinguishing patients with VT from those without.

The study performed by Worley et al. was also designed to compare the value of time domain and frequency domain parameters [84]. However, they investigated the spectral area ratio, defined in the same way as in [79], not only for segments related to QRS end but also to QRS onset. Six different 140 ms segments started at 0, 40, 50, and 60 ms after QRS onset, two segments started at 40 and 50 ms before QRS end, and a variable-length segment started 40 ms before QRS end and extended to the T wave. The study included 36 patients with remote MI and sustained VT, 29 asymptomatic patients with remote MI, and 23 normal subjects. The results showed that the spectral area ratios from segments starting at 0, 40, and 60 ms after QRS onset were significantly different between infarct patients with and without VT; however,

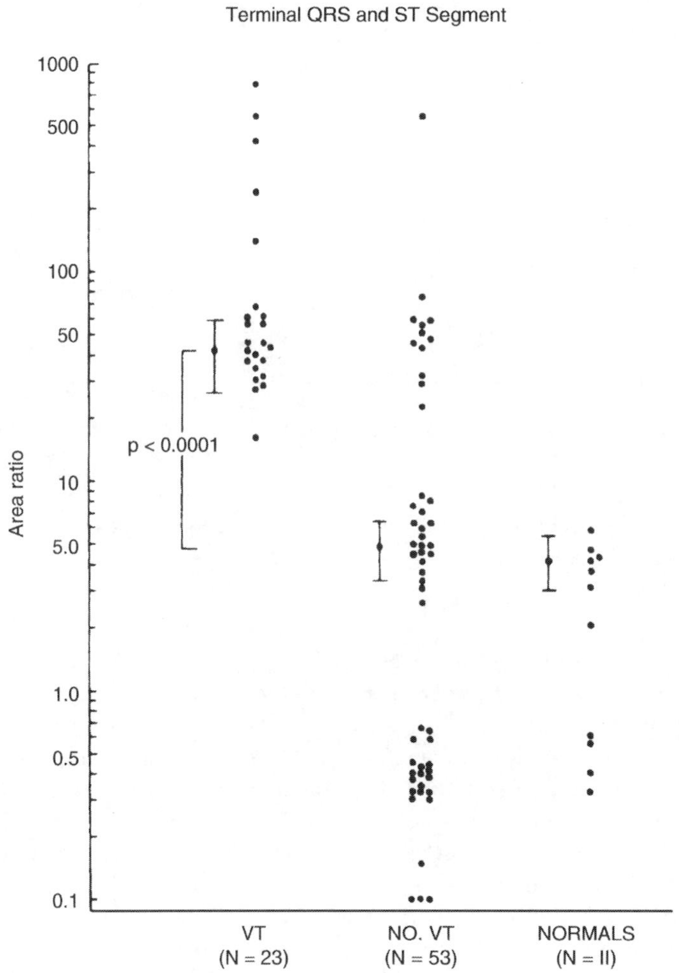

Figure 39.8

Comparison of the spectral area ratio for patients with sustained VT, without sustained VT, and normal subjects. The terminal QRS and the ST segment are analyzed. (Reprinted from Cain et al. [79] with permission.)

this finding did not apply to any of the segments related to QRS end. Thus, the results showed that the spectral content of the QRS complex is a marker to discriminate between the two patient groups. The time domain parameters studied were all found to be significantly different between the groups.

The idea to investigate the influence of the analyzed segment location was also pursued in a study by Buckingham et al. [85]. They used a data set consisting of 84 patients with VT and 150 patients without VT, all patients having had prior MI. All analyzed segments had a length of 140 ms and were located 0, 20, 40, 60, and 80 ms after QRS onset and 20, 40, 60, and 80 ms before QRS end. In each segment the spectral area ratio was computed using the same definition as the one given in [79]. Significant differences between the patient groups were only observed for ratios computed in segments 60 and 80 ms after QRS onset and 80 ms before QRS end.

The importance of higher-frequency components was investigated by Pierce et al. on a data set consisting of 24 patients with coronary artery disease and recurrent VT, 24 control patients with coronary artery disease, and 23 normal subjects [77]. The analyzed segment had its onset at the QRS end and a length of 120 ms. Unlike the studies by Cain et al., the frequency domain parameters involved considerably higher frequencies as contained in the interval 60–120 Hz; normalization was done with the entire interval 0–120 Hz. Quantifying performance in terms of the area under the receiver operating characteristic, the frequency domain parameters were associated with larger areas than the time domain parameters. It was therefore concluded that the higher frequencies in late potentials better identified patients with coronary artery disease being prone to VT than did the time domain parameters.

39.7 Late Potentials and Time–Frequency Analysis

39.7.1 Spectrotemporal Mapping

Frequency analysis for identification of late potentials is, as indicated above, flawed by difficulties to accurately delimit the signal segment to be processed; small changes in segment position may lead to substantial changes in frequency domain parameters. The introduction of a sliding segment, whose initial position is located well before the end of the QRS complex and final position well into the T wave, has been suggested as a means to mitigate such difficulties; "sliding" means that the analysis segment is shifted by one or a few milliseconds at a time. For each segment, the power spectrum is computed so that, when the segment has reached its endpoint, a series of successive power spectra has been produced. Similar to frequency analysis, each segment is windowed and the DC level subtracted. This time-frequency approach is widely known as the short-term Fourier transform (STFT) and represents a standard signal processing tool for characterizing the time-varying spectral properties of a nonstationary signal. In the context of ventricular LPs, the STFT is commonly referred to as spectrotemporal mapping [86–89].

➲ Figure 39.9a illustrates spectrotemporal mapping when implemented using an 80 ms segment whose position slides around the end of the QRS complex so that the presence of late potentials may be identified through spectral changes. In this particular case, the signal-averaged ECG contains low-amplitude activity at the end of the QRS complex, that is, late

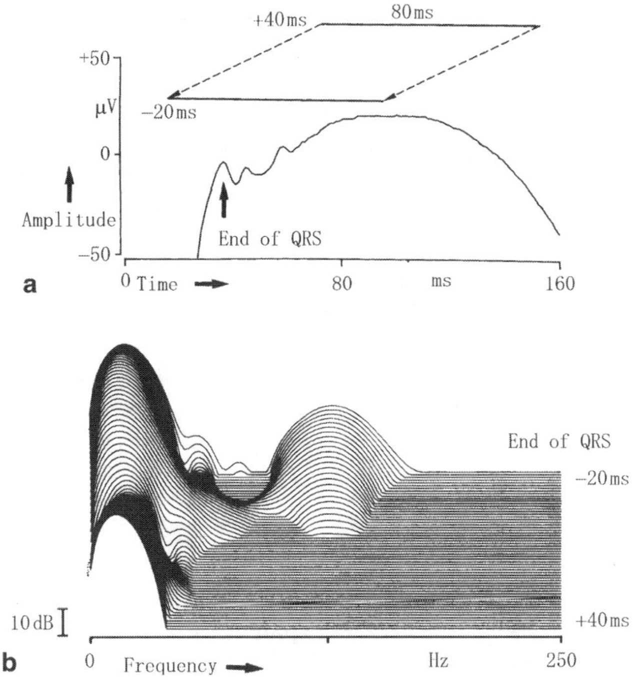

Figure 39.9
(a) Spectrotemporal mapping is created by computing successive Fourier transforms within a sliding window which, in this example, has 80-ms duration. (b) The spectrotemporal map of a patient with late potentials. (Reprinted from Haberl et al. [86] with permission.)

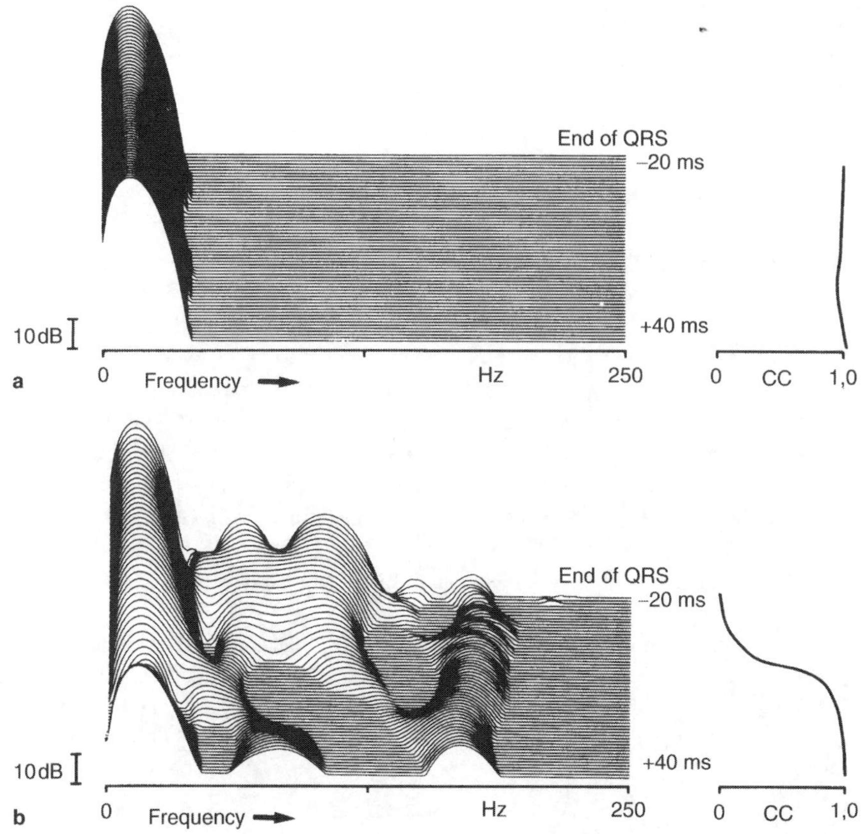

Figure 39.10
Spectrotemporal map for (a) a patient without late potentials, and (b) a patient with late potentials, manifested by a "landscape with ridges." The curve of cross-correlation coefficients is displayed to the right of each map. (Reprinted from Haberl et al. [86] with permission.)

potentials are considered to be present. The corresponding series of successive power spectra is displayed in Fig. 39.9b as a three-dimensional plot where the three axes of the diagram describe frequency, time, and spectral power, respectively. The plot shows that high-frequency components above 50 Hz, contained in segments at the QRS end, gradually vanish as the segment slides toward the T wave, that is, from top to bottom of the plot. The last few spectra are similar in that they do not contain any high-frequency components. In this case, the transition from late potentials to noise is clearly visible using spectrotemporal mapping since the signal-to-noise ratio of the signal-averaged ECG is high.

Although the parameters of frequency analysis can be applied to each individual power spectrum, parameters that exploit the time-dependent properties of spectrotemporal mapping can be expected to add information for identification of late potentials. An early approach to such characterization was the so-called normality factor which determines the similarity of successive spectra through the computation of a series of cross-correlation coefficients [87]; these coefficients have the property of being normalized to the interval [−1,1]. Each individual spectrum is correlated to a template spectrum being determined by averaging the last few spectra positioned at the end of the analysis interval. The curve of cross-correlation coefficients is illustrated for ECGs with and without late potentials in Fig. 39.10a and b, respectively. It is obvious from Fig. 39.10a that the correlation curve increases from almost 0 in segments where high-frequency components are present to a value well outside the QRS complex. On the other hand, Fig. 39.10b displays a correlation curve which remains close to one throughout all spectra, suggesting that late potentials are absent.

In the original work by Haberl et al. [87] an 80 ms segment was analyzed sliding in steps of 3 ms from a start time 23 ms before the QRS end until 52 ms after; the resulting 25 segments were indexed backward in time so that the last segment had index 1. The normality factor was obtained as the ratio between the mean of the cross-correlation coefficients of segments 20–25, where late potentials were expected to occur, and the mean of the cross-correlation coefficients of the segments 1–5. The factor was then multiplied by 100 and interpreted as a percentage. Late potentials were considered to be present when the normality factor dropped below 30%.

The initial results obtained with spectrotemporal mapping surpassed those obtained by conventional time domain analysis when identifying patients prone to sustained VT in the presence of coronary artery disease [87]. Furthermore, most patients with bundle branch block were correctly classified with spectrotemporal mapping, while these patients had to be excluded in time domain analysis. Despite these promising initial results, this technique is not in widespread use, probably due to its poor reproducibility. In a study comparing spectrotemporal mapping to both time domain and frequency domain analysis, the reproducibility of spectrotemporal mapping was substantially lower than that of the other two types of analysis [89].

39.7.2 Spectral Turbulence Analysis

Spectral turbulence analysis defines another approach to time-frequency characterization of the signal-averaged ECG, originally developed to detect abnormal variations in frequency content during depolarization of the diseased myocardium [90]. The word "turbulence" signifies the transient spectral changes that were hypothesized to occur in the ECG as depolarization propagates throughout the ventricles around areas of abnormal conduction, thus resulting in a high degree of spectral turbulence. While spectral turbulence analysis has the short-term Fourier transform in common with spectrotemporal mapping, the analysis embraces the entire QRS complex and the ST segment and is not limited to an interval at the QRS end. Usually, the first segment starts 25 ms before QRS onset and ends 125 ms after QRS end, incremented in 2 ms steps. The resulting power spectra from all available leads are then normalized with respect to the spectrum with the highest magnitude; this magnitude is designated as 100% while the other magnitudes are smaller. A number of parameters are derived from the normalized spectra which describe the similarity between adjacent spectra expressed in terms of the above-mentioned cross-correlation coefficient. It was claimed that spectral turbulence analysis is applicable to patients irrespective of the QRS duration and the presence or absence of bundle branch block.

In the original study by Kelen et al. [90] both spectral turbulence analysis and conventional time domain late potential analysis (40–250 Hz) were employed for identifying patients with inducible sustained monomorphic VT. The signal-averaged ECG was obtained from 144 subjects using a recording configuration of three orthogonal bipolar leads. The data set contained 71 normal control subjects, 33 with both late potentials by time domain analysis and inducible sustained monomorphic VT, 28 with time domain late potentials but no evidence of spontaneous or inducible sustained monomorphic VT, and ten with inducible sustained monomorphic VT but absence of time domain late potentials. The total predictive accuracy for all groups was 94% with spectral turbulence analysis, whereas only 73% could be achieved with time domain analysis.

Spectral turbulence analysis has been employed in several subsequent studies with results that sometimes improve upon those obtained by conventional time domain analysis. In a study by Malik et al. [91], both techniques were considered for risk prediction after acute MI, using a material of 553 survivors of acute MI (bundle branch blocks and other conduction abnormalities were excluded); the patients were followed for at least 1 year. Spectral turbulence analysis provided significantly lower positive predictive accuracy than the time domain analysis for prediction of ventricular tachycardia/fibrillation during 1 year after infarction. On the other hand, it provided significantly higher positive predictive accuracy for the prediction of 1-year all-cause mortality. The study by Copie et al. [92] also made use of both time domain and spectral turbulence analysis for the prediction of cardiac death and arrhythmic events after acute MI in 603 patients (patients with bundle branch block and other conduction abnormalities were excluded). The results showed that spectral turbulence analysis was essentially equivalent to time domain analysis for the prediction of arrhythmic events after MI, but performed significantly better than time domain analysis for the prediction of cardiac death.

Although it was initially claimed that spectral turbulence analysis is equally suitable for patients with bundle branch block, the study by Englund et al. [93] came to the opposite conclusion, that is, spectral turbulence analysis is applicable only to patients without bundle branch block. Their results were based on a material of 169 patients of whom 120 had a QRS duration ≤120 ms; 47 patients had inducible sustained monomorphic VT and were compared to 122 control patients.

The total predictive accuracy for predicting inducible VT was as low as 47% when the whole material was analyzed, however, it increased to 73% when only patients with QRS duration <120 ms were included.

39.7.3 Combined Time Domain and Spectral Turbulence Analysis

The idea to combine time domain and spectral turbulence analysis has been investigated in a number of studies for the purpose of increasing the power of predicting serious arrhythmic events in post-infarction patients [94–96]. The results of these studies indicated that the combined approach may lead to a higher total predictive accuracy than can be achieved using each of the analysis techniques independently. Based on a material of 262 patients with acute MI, Ahuja et al. [89] obtained a total predictive accuracy of 92% when combined analysis was employed for predicting arrhythmic events. The corresponding performance figures for time domain analysis and spectral turbulence analysis were 87% and 78%, respectively.

Improved risk stratification was also found in a study by Mäkijärvi et al. [95] though the levels of predictive accuracy were considerably lower than those reported in [94]. The study comprised a prospective material of 778 males who survived the acute phase of MI. The most powerful prediction of arrhythmic events was achieved by combining time domain and spectral turbulence analysis, reaching a total predictive accuracy of 61% when one of the two techniques showed abnormality and a total predictive accuracy of 87% when both showed abnormality. Spectral turbulence analysis was clearly inferior to time domain analysis in predicting arrhythmic events when used separately.

In yet another study, Vazquez et al. [96] investigated the value of combined analysis on a material of 602 patients after acute MI (of which 38 patients had a major arrhythmic event during the 1-year follow-up period). The total predictive accuracy of combined analysis was 89.9%, significantly higher than achieved separately by time domain analysis (75%) and spectral turbulence analysis (78%).

39.8 High-Frequency QRS Components

39.8.1 Introduction

Frequency analysis of an ECG signal shows that the signal also includes frequencies above the standard frequency range [97]. These high-frequency components are mainly found within the QRS complex, having a low amplitude compared to the components of the standard ECG (μV compared to mV) (➋ Fig. 39.11). Early attempts to visualize high-frequency ECG components employed signal amplification in combination with high paper speed. With this technique, "notches" and "slurs" could be observed within the QRS complex [98, 99]. Studies in the 1970s showed that patients with heart disease had an increased number of notches and slurs compared to healthy individuals. Later analysis of the ECG signal showed that the notches and slurs contained frequencies between 40 and 185 Hz [100].

The development of high-resolution recording techniques combined with digital filtering and signal averaging enabled better analysis of high-frequency components in the ECG signal. Several studies, both on humans and animals, have investigated whether high-frequency QRS components (HF-QRS) contain diagnostic information. Several heart diseases have been studied, such as acute myocardial ischemia [101–109], old MI [109–117] and left ventricular hypertrophy [118, 119]. It is still unknown, however, if the method is useful as a complement to standard ECG for the clinical diagnosis of different heart diseases.

The physiological mechanisms underlying HF-QRS are still not fully understood. One theory is that HF-QRS are related to the conduction velocity and the fragmentation of the depolarization wave in the myocardium. In a three-dimensional model of the ventricles with a fractal conduction system it was shown that high numbers of splitting branches are associated with HF-QRS. In this experiment, it was also shown that the changes seen in HF-QRS in patients with myocardial ischemia might be due to the slowing of the conduction velocity in the region of ischemia [120]. Further electrophysiological studies are needed, however, to better understand the underlying mechanisms of HF-QRS.

There is no standardized method for the recording and quantification of HF-QRS. Both orthogonal leads and all or some of the 12 standard leads have been used in different studies. Several frequency ranges (most commonly 150–250 Hz) and filters have been used as well. Different methods for quantification have also been used. The two most commonly used

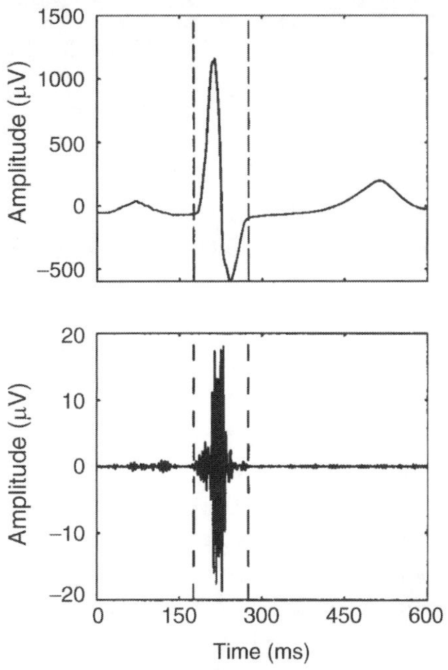

Figure 39.11
A signal-averaged ECG in the standard frequency range (*upper panel*) and the same ECG within the 150–250 Hz frequency range (*lower panel*). *Dashed lines* indicate the QRS duration determined from the standard frequency range. (Reprinted from Pettersson et al. [126] with permission.)

methods for quantification are RMS values during the entire QRS duration and "reduced amplitude zones" (RAZ), a way of describing zones in the filtered QRS complex with reduced amplitudes. Due to the low amplitude of the HF-QRS, noise reduction is necessary before analysis. In order to achieve this in a clinical environment, signal averaging is necessary.

39.8.2 Signal Recording and Analysis

To be able to extract HF-QRS it is important to use recording equipment with high resolution acquisition, both in time and amplitude. The sampling rate should be 1,000 Hz or higher [1], and the amplitude resolution at least 1 μV. Since the amplitudes of HF-QRS are in the microvolt range, a low noise level is required. For example, the influence of noise due to skeletal muscle can be reduced by moving the electrodes on the legs and arms to a more proximal position according to Mason and Likar [121]. More importantly, signal averaging must be performed in order to attain an acceptable noise level, sometimes requiring a recording time of several minutes. If the ECG morphology is subject to dynamic changes, for example, observed during percutaneous transluminal coronary angioplasty, recursive averaging with a forgetting factor must be used instead as conventional ensemble averaging is unsuitable.

The HF-QRS is extracted from the signal averaged beats through band pass filtering, often by employing a Butterworth filter [122]. This filter type has a nonlinear phase response which may distort temporal relationships of the various signal components. Linear phase filtering can be obtained, however, by first filtering the entire signal in a forward direction and then filtered signal once more but backward. Different studies have used different bandwidths when extracting HF-QRS, although the frequency range 150–250 Hz is the most common choice.

Several methods have been used for quantification of HF-QRS of which calculation of RMS values during the entire QRS duration is the most common one. For a correct RMS value it is necessary to correctly identify the onset and end of the QRS complex. It has been shown that the most correct delineation of the QRS complex is obtained when determining the QRS duration in the standard frequency range.

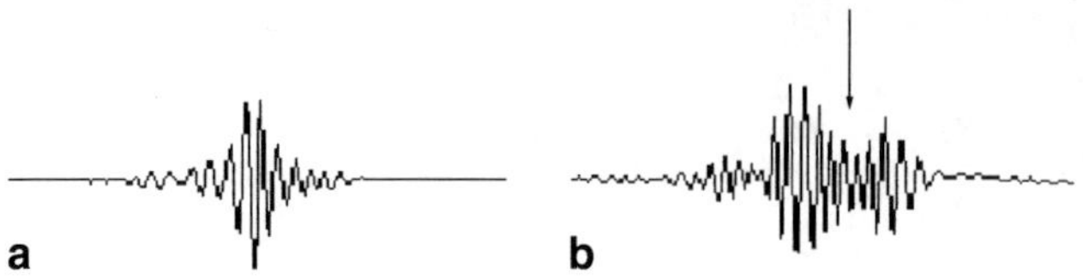

Figure 39.12
HF-QRS (150–250 Hz) from two individuals, with (a) and without (b) RAZ.

Another method for quantification of HF-QRS is calculation of reduced amplitude zones (RAZ). The RAZ measure is a morphological measure and is defined as an interval between two adjacent local maxima or two adjacent local minima in the HF-QRS, where a local maximum or minimum must have an absolute value higher than the three preceding and three following envelope points (Fig. 39.12). The method was first introduced by Abboud in the 1980s [103]. It was discovered that areas with lower amplitude were present in the high-frequency signal in dogs with myocardial ischemia but not in normal dogs. The method has been developed further, and a scoring system has been proposed [123, 124]. Other methods for quantification of HF-QRS include peak-to-peak amplitude and the integral of the signal.

A recent approach to HF-QRS quantification is to compute the upward and downward slopes of the QRS complex [125]. This technique was designed with the aim to reduce the leakage and smearing caused by linear band pass filtering of the QRS complex. It was shown that both slope coefficients are more sensitive to ischemic changes than is the RMS value obtained from the band pass filtered QRS complex.

The amplitude of HF-QRS differs among the 12 standard leads (Fig. 39.13). The largest amplitudes are usually found in the anterior-posterior-oriented leads V2–V4 and in the inferior-superior-oriented leads II, aVF, and III. The lowest amplitudes are found in the left-right-oriented leads aVL, I, -aVR, V1, V5, and V6 [126]. In the transverse plane, the leads V1 and V6 are located furthest away from the left ventricle, which is a possible explanation for the low amplitudes recorded in these leads. In the frontal plane, however, no leads are located close to the heart, but there is still a large difference in amplitude of HF-QRS between these leads.

The correlation between HF-QRS and the QRS amplitudes in standard ECG is generally low. Thus, factors other than the QRS amplitude in standard ECG seem to influence the size of HF-QRS. There is also a large variation in HF-QRS between individuals. During consecutive registrations on the same individual, however, only small variations in HF-QRS are registered [126]. When studying healthy individuals, no significant differences in HF-QRS regarding sex and age have been found [117].

39.9 HF-QRS in Heart Disease

39.9.1 Ischemic Heart Disease

Several studies have compared HF-QRS in patients with old MI to HF-QRS in normal subjects. The vast majority of these studies show reduced amplitudes after old MI. In the frequency range 80–300 Hz HF-QRS are significantly lower in leads V2 and V5 in patients with old anterior MI [112]. In patients with old inferior infarction, HF-QRS are reduced in leads II, aVF, and III [110, 112]. When recording with Frank leads (X, Y, and Z leads), reduced HF-QRS are found in patients with old anterior and/or inferior infarction [114]. There are some studies, however, that show higher HF-QRS in patients with old MI compared to normal individuals, measured with frequencies >90 Hz using Frank leads [115].

Studies on patients with angiographically documented ischemic heart disease, with and without signs of old MI on ECG, do not show any difference in HF-QRS between the two populations [116]. When comparing a healthy population with patients with ischemic heart disease, both with and without old MI, the healthy population shows higher HF-QRS than the patient group [117]. These results indicate that ischemic heart disease, with and without old MI, causes a reduction of HF-QRS. A possible explanation is that chronic ischemic heart disease leads to structural changes in the myocardial tissue. These changes may contribute to abnormal impulse propagation in the ischemic heart. This abnormal propagation might be the reason for the reduced high-frequency content in patients with ischemic heart disease.

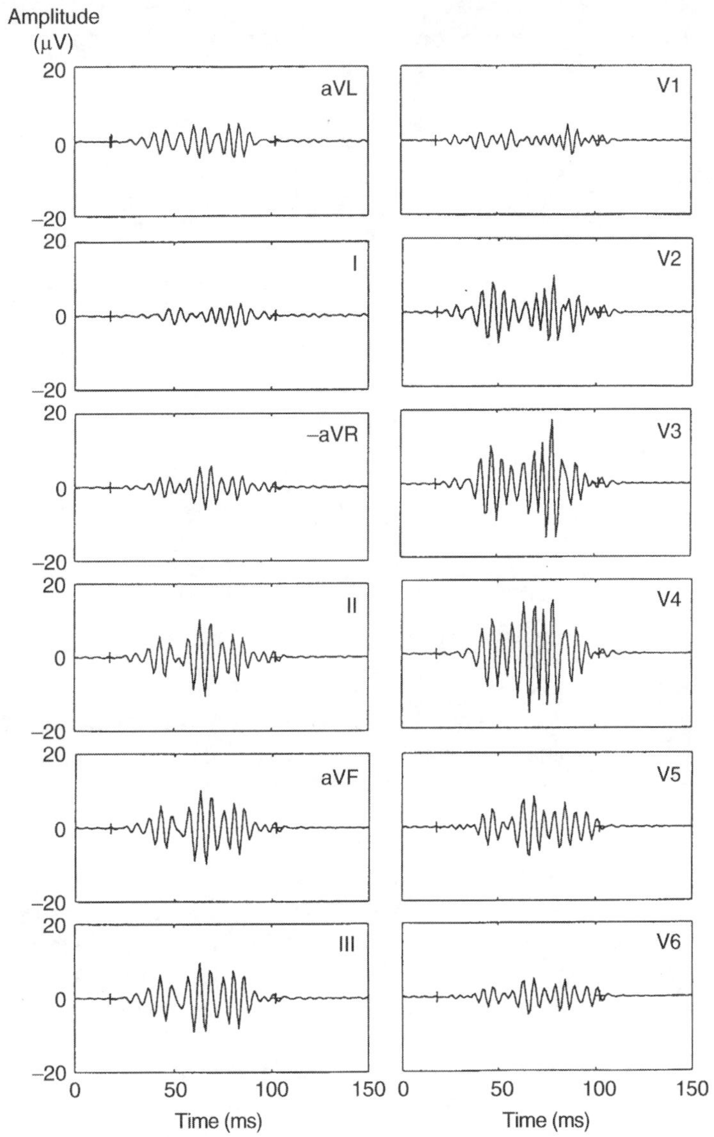

Figure 39.13
Filtered QRS complexes (150–250 Hz) in 12 standard leads from a patient with typical amplitude distribution among the leads. *Tick marks* indicate QRS onset and offset, determined from the standard frequency ECG. (Reprinted from Pettersson et al. [126] with permission.)

39.9.2 Acute Myocardial Ischemia

Several studies have reported lower HF-QRS during acute myocardial ischemia. In a study on dogs, ECG was recorded both from epicardial electrodes and from surface electrodes during occlusion of the left anterior descending coronary artery. HF-QRS recorded from the epicardium of the left ventricle were significantly reduced during the occlusion, while HF-QRS recorded from the nonischemic right ventricle remained unchanged. Reduced HF-QRS from the surface electrodes were also found [102]. Other animal studies have shown similar results [101, 104–106].

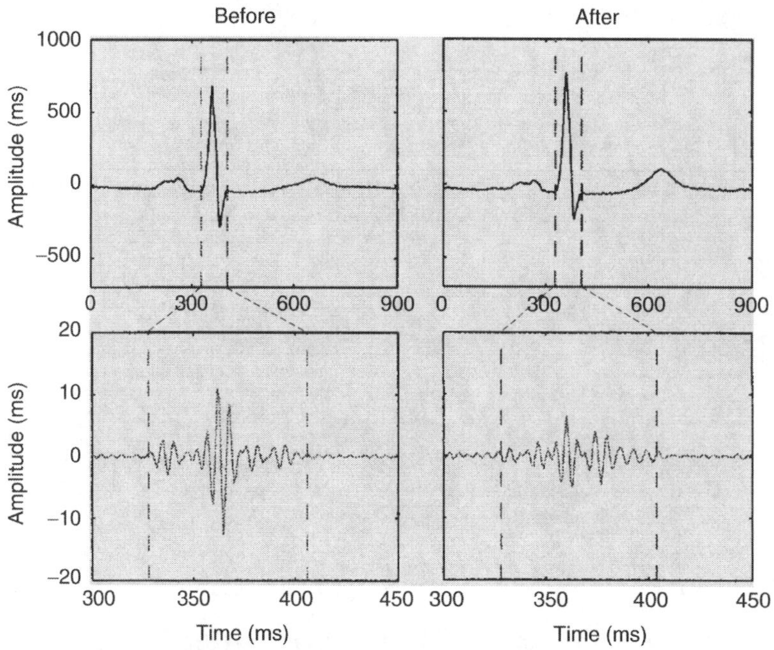

Figure 39.14
Lead V5. *Upper panel*: Pre-inflation and inflation ECGs in the standard frequency range. *Lower panel*: The same ECGs within the HF range (150–250 Hz). The *dashed lines* indicate the QRS duration, determined from the standard frequency ECG. (Reprinted from Pettersson et al. [109] with permission.)

In humans, the ECG has been recorded during percutaneous transluminal coronary angioplasty [108, 109]. The results from these studies show that ischemia leads to changes in HF-QRS in a majority of the patients. These changes can be observed even when no ST changes in standard ECG are seen (➋ Fig. 39.14). The results indicate that acute myocardial ischemia can be detected with higher sensitivity with analysis of HF-QRS compared to conventional analysis with standard ECG. Analysis of HF-QRS might therefore serve as a complement to standard ECG in the detection of myocardial ischemia. The large inter-individual variation in HF-QRS, however, probably makes high-frequency analysis most applicable to monitoring situations when changes from baseline can be identified.

During occlusion of the left anterior descending coronary artery a reduction in HF-QRS is observed in many leads, most commonly in lead V3. During occlusion of other large coronary arteries a reduction in HF-QRS is seen in various leads. Thus, HF-QRS seem to be poorer than standard ECG in detecting the location of ischemia.

39.9.3 Reperfusion

To assess the resolution of ST segment elevation is the most commonly used method for detecting reperfusion during thrombolytic therapy for acute MI. A couple of studies have shown that successful reperfusion also results in a significant increase in HF-QRS (➋ Fig. 39.15) [127, 128]. More and larger studies are needed, however, to determine if analysis of HF-QRS could be a useful method for monitoring patients during treatment of acute MI.

39.9.4 Stress-Induced Ischemia

Analysis of HF-QRS during exercise test has been suggested as a complement to assessment of the ST segment reaction for detection of exercise-induced ischemia. It has been shown that HF-QRS increase during exercise in healthy

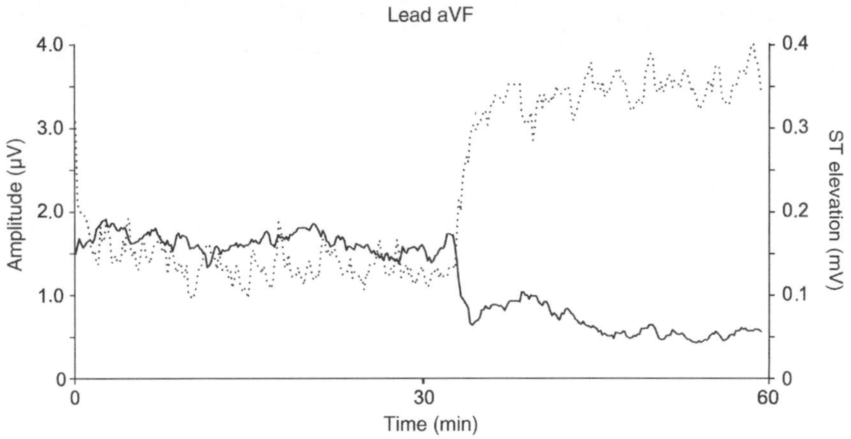

Figure 39.15
Dynamic changes in lead aVF in HF-QRS (*dashed line*) and ST segment (*solid line*) during reperfusion therapy for acute inferior MI. After 30 min of therapy, the ST elevation decreases and the HF-QRS simultaneously increases.

individuals [129]. In a study comparing healthy individuals to patients with ischemic heart disease, it was found that HF-QRS are significantly higher in the healthy population, both during and after exercise [130]. Two other studies have shown that a large relative change in HF-QRS during exercise is more sensitive for detecting ischemia in myocardial perfusion imaging test compared to conventional ST analysis [131, 132]. A recent study investigated changes in HF-QRS during adenosine myocardial perfusion imaging stress tests [133]. It was found that analysis of HF-QRS is highly sensitive and specific for detecting reversible perfusion defects, and significantly more sensitive than conventional ST segment analysis. Another study trying to reproduce these findings found, however, that analysis of HF-QRS was no better than tossing a coin for detecting reversible perfusion defects [134]. A problem with analysis of HF-QRS recorded during exercise is the high noise level generated by skeletal muscle.

39.9.5 Left Ventricular Hypertrophy

Standard ECG is one of the most common methods to detect left ventricular hypertrophy. Several different ECG-based criteria are used clinically. These methods, however, have low sensitivity when a high level of specificity is required. Studies on rabbits, with and without left ventricular hypertrophy, have shown that HF-QRS correlate well with left ventricular mass [119]. In the study, the vector magnitude from orthogonal leads in different frequency ranges was studied. High-pass filtering at 44 Hz showed the best correlation between left ventricular mass and HF-QRS ($r = 0.84$). A high correlation between left ventricular mass and HF-QRS was also found among the healthy rabbits alone.

Studies on humans have shown more diverging results. A study on 15 patients, both with normal and pathologic left ventricular mass, determined by echocardiography, showed that the sum of vectors in orthogonal leads in the 2–250 Hz range had approximately the same correlation with left ventricular mass as do established electrocardiographic criteria for left ventricular hypertrophy [118]. Another study on 60 healthy individuals, using magnetic resonance imaging as gold standard, found no correlation between HF-QRS and left ventricular mass [135]. Thus, it is not certain whether HF-QRS can be of value in the electrocardiographic diagnosis of left ventricular hypertrophy.

39.9.6 Conduction Abnormalities

In dogs, HF-QRS are reduced during slow conduction velocity in the heart [136]. By infusing sodium channel blockers (lidocaine, disopyramide) in the left anterior descending coronary artery at the same time as recording ECGs from the entire ventricular surface, it has been shown that HF-QRS are significantly lower in the areas affected by the sodium channel blockers. These results indicate that HF-QRS is a potent indicator of disturbed local conduction.

39.9.7 Heart Transplantation and Heart Surgery

Allograft rejection is a major cause of morbidity and mortality in patients who have undergone heart transplant. There is no reliable method for detecting rejection except endomyocardial biopsy. Some studies have investigated whether HF-QRS could be used as a noninvasive marker for rejection [137, 138]. The results from the studies have in part showed diverging results, with both an increase and a decrease in HF-QRS at rejection.

In a study on patients after heart surgery, a reduction of HF-QRS correlating with the dysfunction of the heart has been found [139]. It is therefore suggested that HF-QRS could be used as a noninvasive marker of myocardial dysfunction after heart surgery. In children who have undergone heart surgery, the change in HF-QRS during aortic clamping has been investigated [140]. I was found that the recovery time of the HF-QRS significantly correlated with cardioplegic arrest time during surgery.

References

1. Sörnmo, L. and P. Laguna, *Bioelectrical Signal Processing in Cardiac and Neurological Applications*. Amsterdam: Elsevier (Academic Press), 2005.
2. Hombach, V., V. Braun, H.W. Hopp, et al., The applicability of the signal averaging technique in clinical cardiology. *Clin. Cardiol.*, 1982;**5**: 107–124.
3. Ros, H.H., A.S.M. Koeleman, and T.J. Akker, The technique of signal averaging and its practical application in the separation of atrial and His Purkinje activity, in *Signal Averaging Technique in Clinical Cardiology*, V. Hombach and H.H. Hilger, Editors. Stuttgart: Schattauer, 1981, p. 3.
4. Rompelman, O. and H.H. Ros, Coherent averaging technique: a tutorial review. Part 1: noise reduction and the equivalent filter. Part 2: trigger jitter, overlapping responses and nonperiodic stimulation. *J. Biomed. Eng.*, 1986;**8**: 24–35.
5. Laciar, E., R. Jané, and D.H. Brooks, Improved alignment method for noisy high-resolution ECG and Holter records using multiscale cross-correlation. *IEEE Trans. Biomed. Eng.*, 2003;50: 344–353.
6. Simson, M.B., Use of signals in the terminal QRS complex to identify patients with ventricular tachycardia after myocardial infarction. *Circulation*, 1981;**64**: 235–242.
7. Berbari, E.J. and P. Lander, Principles of noise reduction, in *High-Resolution Electrocardiography*, N. El-Sherif and G. Turitto, Editors. Armonk: Futura, 1992, pp. 51–66.
8. Plonsey, R., *Bioelectric Phenomena*. New York: McGraw-Hill, 1969, pp. 281–299.
9. Kienzle, M.G., R.A. Falcone, and M.B. Simson, Alterations in the initial portion of the signal averaged QRS complex in acute myocardial infarction with ventricular tachycardia. *Am. J. Cardiol.*, 1988;**61**: 91–103.
10. Atarius, R. and L. Sörnmo, Maximum likelihood analysis of cardiac late potentials. *IEEE Trans. Biomed. Eng.*, 2006;**43**: 60–68.
11. Steinberg, J.S. and J.T. Bigger, Importance of the endpoint of noise reduction in analysis of the signal-averaged electrocardiogram. *Am. J. Cardiol.*, 1989;**63**: 556–560.
12. Boineau, J.P. and J.L. Cox, Slow ventricular activation in acute myocardial infarction: A source of reentrant premature ventricular contraction. *Circulation*, 1973;**48**(4): 702–713.
13. Waldo, A.L. and G.A. Kaiser, A study of ventricular arrhythmias associated with acute myocardial infarction in the canine heart. *Circulation*, 1973;**47**(6): 1222–1228.
14. El-Sherif, N., B.J. Scherlag, and R. Lazzara, Electrode catheter recordings during malignant ventricular arrhythmias following experimental acute myocardial ischemia. *Circulation*, 1975;**51**(6): 1003–1014.
15. El-Sherif, N., B.J. Scherlag, R. Lazzara, and R.R. Hope, Reentrant ventricular arrhythmias in the late myocardial infarction period. 1. Conduction characteristics in the infarction zone. *Circulation*, 1977;**55**(5): 686–702.
16. El-Sherif, N., R.R. Hope, B.J. Scherlag, and R. Lazzara, Reentrant ventricular arrhythmias in the late myocardial infarction period. 2. Patterns of initiation and termination of re-entry. *Circulation*, 1977;**55**(5): 702–719.
17. Simson, M.B., W.J. Untereker, S.R. Spielman, L.N. Horowitz, N.H. Marcus, R.A. Falcone, et al., The relationship between late potentials on the body surface and directly recorded fragmented electrograms in patients with ventricular tachycardia. *Am. J. Cardiol.*, 1983;**51**(1): 105–112.
18. Berbari, E.J., B.J. Scherlag, R.R. Hope, and R. Lazzara, Recording from the body surface of arrhythmogenic ventricular activity during the S-T segment. *Am. J. Cardiol.*, 1978;**41**(4): 697–702.
19. Josephson, M.E., L.N. Horowitz, and A. Farshidi, Continuous local electrical activity. A mechanism of recurrent ventricular tachycardia. *Circulation*, 1978;**57**(4): 659–665.
20. Breithardt, G., R. Becker, L. Seipel, R.R. Abendroth, and J. Ostermeyer, Non-invasive detection of late potentials in man–a new marker for ventricular tachycardia. *Eur. Heart J.*, 1981;**2**(1): 1–11.
21. Fontaine, G., G. Guiraudon, R. Frank, et al., Stimulation studies and epicardial mapping in ventricular tachycardia. Study of mechanisms and selection for surgery, in *Reentrant Arrhythmias*, H. Kulbertus, Editor. Lancaster: MTP; 1977, p. 334.

22. Rozanski, J.J., D. Mortara, R.J. Myerburg, and A. Castellanos, Body surface detection of delayed depolarizations in patients with recurrent ventricular tachycardia and left ventricular aneurysm. *Circulation*, 1981;**63**(5): 1172–1178.
23. Simson, M.B., D. Euler, and E.L. Michelson, Detection of delayed ventricular activation on the body surface in dogs. *Am. J. Physiol.*, 1981;**241**(3): H363–H369.
24. Denniss, A.R., D.A. Richards, D.V. Cody, P.A. Russell, A.A. Young, D.L. Ross, et al., Correlation between signal-averaged electrocardiogram and programmed stimulation in patients with and without spontaneous ventricular tachyarrhythmias. *Am. J. Cardiol.*, 1987;**59**(6): 586–590.
25. Breithardt, G., M. Borggrefe, U. Karbenn, R.R. Abendroth, H.L. Yeh, and L. Seipel, Prevalence of late potentials in patients with and without ventricular tachycardia: correlation and angiographic findings. *Am. J. Cardiol.*, 1982;**49**(8): 1932.
26. Kanovsky, M.S., R.A. Falcone, C.A. Dresden, M.E. Josephson, and M.B. Simson, Identification of patients with ventricular tachycardia after myocardial infarction: signal-averaged electrocardiogram, Holter monitoring, and cardiac catherization. *Circulation*, 1984;**70**(2): 264–270.
27. Gomes, J.A., S.L. Winters, M. Martinson, J. Machac, D. Stewart, and A. Targonski, The prognostic significance of quantitative signal-averaged variables relative to clinical variables, site of myocardial infarction, ejection fraction and ventricular premature beats: a prospective study. *J. Am. Coll. Cardiol.*, 1989;**13**(2): 377–384.
28. Breithardt, G., M. Borggrefe, and K. Haerten, Role of programmed ventricular stimulation an noninvasive recording of ventricular late potentials for the identification of patients at risk of ventricular tachyarrhythmias after acute myocardial infarction, in *Cardiac Electrophysiology and Arrhythmias*, D.P. Zipes and J. Jalife, Editors. New York: Grune and Stratton, 1985, pp. 553–561.
29. Breithardt, G. and M. Borggrefe, Recent advances in the identification of patients at risk of ventricular tachyarrhythmias: role of ventricular late potentials. *Circulation*, 1987;**75**(6): 1091–1096.
30. Cain, M.E., J.L. Anderson, M.F. Arnsdorf, J.W. Mason, M.M. Scheinman, and A.L. Waldo, Signal-averaged electrocardiography. *J. Am. Coll. Cardiol.*, 1996;**27**(1): 238–249.
31. Gang, E.S., A.S. Lew, M. Hong, F.Z. Wang, C.A. Siebert, T. Peter, Decreased incidence of ventricular late potentials after successful thrombolytic therapy for acute myocardial infarction. *N. Engl. J. Med.*, 1989;**321**(11): 712–716.
32. Eldar, M., J. Leor, H. Hod, Z. Rotstein, S. Truman, E. Kaplinsky, et al., Effect of thrombolysis on the evolution of late potentials within 10 days of infarction. *Br. Heart J.*, 1990;**63**(5): 272–276.
33. Chew, E.W., P. Morton, J.G. Murtagh, M.E. Scott, and D.B. O'Keeffe, Intravenous streptokinase for acute myocardial infarction reduces the occurrence of ventricular late potentials. *Br. Heart J.*, 1990;**64**(1): 5–8.
34. Turitto, G., A.L. Risa, E. Zanchi, and P.L. Prati, The signal-averaged electrocardiogram and ventricular arrhythmias after thrombolysis for acute myocardial infarction. *J. Am. Coll. Cardiol.*, 1990;**15**(6): 1270–1276.
35. Tranchesi, B.J., M. Verstraete, F. van de Werf, C.P. de Albuquerque, B. Caramelli, O.C. Gebara, et al., Usefulness of high-frequency analysis of signal-averaged surface electrocardiograms in acute myocardial infarction before and after coronary thrombolysis for assessing coronary reperfusion. *Am. J. Cardiol.*, 1990;**66**(17): 1196–1198.
36. Aguirre, F.V., M.J. Kern, J. Hsia, H. Serota, D. Janosik, T. Greenwalt, et al., Importance of myocardial infarct artery patency on the prevalence of ventricular arrhythmias and late potentials after thrombolysis in acute myocardial infarction. *Am. J. Cardiol.*, 1991;**68**(15): 1410–1416.
37. Vatterott, P.J., S.C. Hammill, K.R. Bailey, C.M. Wiltgen, and B.J. Gersh, Late potentials on signal-averaged electrocardiograms and patency of the infarct-related artery in survivors of acute myocardial infarction. *J. Am. Coll. Cardiol.*, 1991;**17**(2): 330–337.
38. Santarelli, P., G.A. Lanza, F. Biscione, A. Natale, G. Corsini, C. Riccio, et al., Effects of thrombolysis and atenolol or metoprolol on the signal-averaged electrocardiogram after acute myocardial infarction. Late Potentials Italian Study (LAPIS). *Am. J. Cardiol.*, 1993;**72**(7): 525–531.
39. Moreno, F.L., L. Karagounis, H. Marshall, R.L. Menlove, S. Ipsen, and J.L. Anderson, Thrombolysis-related early patency reduces ECG late potentials after acute myocardial infarction. *Am. Heart J.*, 1992;**124**(3): 557–564.
40. de Chillou, C., N. Sadoul, S. Briancon, and E. Aliot, Factors determining the occurrence of late potentials on the signal-averaged electrocardiogram after a first myocardial infarction: a multivariate analysis. *J. Am. Coll. Cardiol.*, 1991;**18**(7): 1638–1642.
41. Maki, H., Y. Ozawa, N. Tanigawa, I. Watanabe, R. Kojima, S. Yakubo, et al., Effect of reperfusion by direct percutaneous transluminal coronary angioplasty on ventricular late potentials in cases of total coronary occlusion at initial coronary arteriography. *Jpn. Circ. J.* 1993;**57**(3): 183–188.
42. Kawalsky, D.L., K.N. Garratt, S.C. Hammill, K.R. Bailey, and B.J. Gersh, Effects of infarct-related artery patency and late potentials on late mortality after acute myocardial infarction. *Mayo Clin. Proc.*, 1997;**72**(5): 414–421.
43. Savard, P., J.L. Rouleau, J. Ferguson, N. Poitras, P. Morel, R.F. Davies, et al., Risk stratification after myocardial infarction using signal-averaged electrocardiographic criteria adjusted for sex, age, and myocardial infarction location. *Circulation*, 1997;**96**(1): 202–213.
44. Scharf, C., H. Redecker, F. Duru, R. Candinas, H.P. Brunner-La Rocca, A. Gerb, et al., Sudden cardiac death after coronary artery bypass grafting is not predicted by signal-averaged ECG. *Ann. Thorac. Surg.*, 2001;**72**(5): 1546–1551.
45. Bauer, A., P. Guzik, P. Barthel, R. Schneider, K. Ulm, M.A. Watanabe, et al., Reduced prognostic power of ventricular late potentials in post-infarction patients of the reperfusion era. *Eur. Heart J.*, 2005;**26**(8): 755–761.
46. Kuchar, D.L., C.W. Thorburn, and N.L. Sammel, Prediction of serious arrhythmic events after myocardial infarction: Signal-averaged electrocardiogram, Holter monitoring and radionuclide ventriculography. *J. Am. Coll. Cardiol.*, 1987;**9**(3): 531–538.
47. Gomes, J.A., S.L. Winters, D. Stewart, S. Horowitz, M. Milner, and P.A. Barreca, New noninvasive index to predict sustained ventricular tachycardia and sudden death in the first year after myocardial infarction: based on signal averaged electrocardiogram, radionuclide ejection fraction and Holter monitoring. *J. Am. Coll. Cardiol.*, 1987;**10**(2): 349–357.
48. Kuchar, D.L., C.W. Thorburn, and N.L. Sammel, Signal-averaged electrocardiogram for evaluation of recurrent syncope. *Am. J. Cardiol.*, 1986;**58**(10): 949–953.
49. Lacroix, D., M. Dubuc, T. Kus, P. Savard, M. Shenasa, and R. Nadeau, Evaluation of arrhythmic causes of syncope; correlation between Holter monitoring, electrophysiologic testing,

and body surface potential mapping. *Am. Heart J.*, 1991;**122**(5): 1346–1354.

50. Uther, J.B., C.J. Dennett, and A. Tan, The detection of delayed activation signals of low amplitude in the vectorcardiogram of patients with recurrent ventricular tachycardia by signal averaging, in *Management of Ventricular Tachycardia – Role of Mexiletine*, E. Sabndoe, D.J. Julian, and J.W. Bell, Editors. Amsterdam: Excerpta Medica, 1978, p. 80.
51. Breithardt, G., L. Seipel, J. Ostermeyer, U. Karbenn, R.R. Abendorth, M. Borggrefe, et al., Effects of anti-arrhythmic surgery on late ventricular potentials recorded by precordial signal averaging in patients with ventricular tachycardia. *Am. Heart J.*, 1982;**104**(5 Pt 1): 996–1003.
52. Marcus, N.H., R.A. Falcone, A.H. Harken, M.E. Josephson, and M.B. Simson, Body surface late potentials: Effects of endocardial resection in patients with ventricular tachycardia. *Circulation*, 1984;**70**(4): 632–637.
53. Denniss, A.R., A.J. Ross, D.A. Richards, D.V. Cody, P.A. Russell, A.A. Young, et al., Effect of anti-arrhythmic therapy on delayed potentials detected by the signal-averaged electrocardiogram in patients with ventricular tachycardia after acute myocardial infarction. *Am. J. Cardiol.*, 1986;**58**(3): 261–265.
54. Simson, M.B., E. Kindwall, A.E. Buxton, and M.E. Josephson, Signal averaging of the ECG in the management of patients with ventricular tachycardia: prediction of anti-arrhythmic drug efficacy, in *Cardiac Arrhythmias: Where to Go From Here?* P. Brugada and H.J. Wellens, Editors. Armonk, NY: Futura, 1987, p. 299.
55. Hopson, J.R., M.G. Kienzle, A.M. Aschoff, and D.R. Shirkey, Noninvasive prediction of efficacy of type IA anti-arrhythmic drugs by the signal-averaged electrocardiogram in patients with coronary artery disease and sustained ventricular tachycardia. *Am. J. Cardiol.*, 1993;**72**(3): 288–293.
56. Kulakowski, P., Y. Bashir, S. Heald, V. Paul, M.H. Anderson, S. Gibson, et al., Effects of procainamide on the signal-averaged electrocardiogram in relation to the results of programmed ventricular stimulation in patients with sustained monomorphic ventricular tacycardia. *J. Am. Coll. Cardiol.*, 1993;**21**(6): 1428–1439.
57. Freedman, R.A. and J.S. Steinberg, Electrophysiologic Study Versus Electrocardiographic Monitoring Trial (ESVEM) Investigators. Selective prolongation of QRS late potentials by sodium channel blocking anti-arrhythmic drugs: relation to slowing of ventricular tachycardia. *J. Am. Coll. Cardiol.*, 1991;**17**(5): 1017–1025.
58. Greenspon, A.J., G.A. Kidwell, M. DeCaro, and S. Hessen, The effects of type I anti-arrhythmic drugs on the signal-averaged electrocardiogram in patients with malignant ventricular arrhythmias. *Pacing Clin. Electrophysiol.*, 1992;**15**(10 Pt 1): 1445–1453.
59. Goedel-Meinen, L., M. Hofmann, G. Schmidt, W. Maier-Rudolph, P. Barthel, A. Schrag, et al., Amiodarone-efficacy and late potentials during long-term therapy. *Int. J. Clin. Pharm. Ther. Toxicol.*, 1990;**28**(11): 449–454.
60. Simson, M.B., H.L. Waxman, R. Falcone, N.H. Marcus, and M.E. Josephson, Effects of anti-arrhythmic drugs on noninvasively recorded late potentials, in *New Aspects in the Medical Treatment of Tachyarrhythmias*, G. Breithardt and F. Loogen, Editors. Munich: Urban and Schwarzenberg, 1983, pp. 80–86.
61. Keren, A., A.M. Gillis, R.A. Freedman, J.C. Baldwin, M.E. Billingham, E.B. Stinson, et al., Heart transplant rejection monitored by signal-averaged electrocardiography in patients receiving cyclosporine. *Circulation*, 1984;**70**(3 Pt 2): 1124–1129.
62. Lacroix, D., S. Kacet, P. Savard, F. Molin, J. Dagano, A. Pol, et al., Signal-averaged electrocardiography and detection of heart transplant rejection: comparison of time- and frequency-domain analyses. *J. Am. Coll. Cardiol.*, 1992;**19**(3): 553–558.
63. Haberl, R., M. Weber, H. Reichenspurner, B.M. Kemkes, G. Osterholzer, M. Anthuber, et al., Frequency analysis of the surface electrocardiogram for recognition of acute rejection after orthoptic cardiac transplantation in man. *Circulation*, 1987;**76**(1): 101–108.
64. Valentino, V.A., H.O. Ventura, F.M. Abi-Samra, C. van Meter, H.L. Price, The signal-averaged electrocardiogram in cardiac transplantation. *Transplantation*, 1992;**53**(1): 124–127.
65. Poll, D.S., F.E. Marchlinski, R.A. Falcone, M.E. Josephson, and M.B. Simson, Abnormal signal averaged electrocardiograms in patients with nonischemic congestive cardiomyopathy: relationship to sustained ventricular tachyarrhythmias. *Circulation*, 1985;**72**(6): 1308–1313.
66. Mancini, D., K.L. Wong, and M.B. Simson, Prognostic value of an abnormal signal-averaged electrocardiogram in patients with nonischemic congestive cardiomyopathy. *Circulation*, 1993;**87**(4): 1083–1092.
67. Ohnishi, Y., T. Inoue, and H. Fukuzaki, Value of the signal-averaged electrocardiogram as a predictor of sudden death in myocardial infarction and dilated cardiomyopathy. *Jpn. Circ. J.*, 1990;**54**(2): 127–136.
68. Middlekauff, H.R., W.G. Stevenson, M.A. Woo, D.K. Moser, and L.W. Stenvenson, Comparison of frequency of late potentials in idiopathic dilated cardiomyopathy and ischemic cardiomyopathy with advanced congestive heart failure and their usefulness in predicting sudden death. *Am. J. Cardiol.*, 1990;**66**(15): 1113–1117.
69. Denereaz, D., M. Zimmermann, and R. Ademec, Significance of ventricular late potentials in non-ischemic dilated cardiomyopathy. *Eur. Heart J.*, 1992;**13**(7): 895–901.
70. Keeling, P.J., P. Kulakowski, G. Yi, A.K. Slade, S.E. Bent, and W.J. McKenna, Usefulness of signal-averaged electrocardiogram in idiopathic dilated cardiomyopathy for identifying patients with ventricular arrhythmias. *Am. J. Cardiol.*, 1993;**72**(1):78–84.
71. Santangeli, P., F. Infusino, G.A. Sgueglia, A. Sestito, and G.A. Lanza, Ventricular late potentials: a critical overview and current applications. *J. Electrocardiol.*, 2008;**41**: 318–324.
72. Francés, R.J., Arrhythmogenic right ventricular dysplasia/cardiomyopathy. A review and update. *Int. J. Cardiol.*, 2006;**110**: 279–287.
73. Nava, A., A.F. Folino, B. Bauce, P. Turrini, G.F. Buja, L. Daliento, and G. Thiene, Signal-averaged electrocardiogram with arrhythmogenic right ventricular cardiomyopathy and ventricular arrhythmias. *Eur. Heart J.*, 2000;**21**: 58–65.
74. Ikeda, T., H. Sakurada, K. Sakabe, T. Sakata, M. Takami, N. Tezuka, T. Nakae, M. Noro, Y. Enjoji, T. Tejima, K. Sugi, and T. Yamaguchi, Assessment of noninvasive markers in identifying patients at risk in the Brugada syndrome: insight into risk stratification. *J. Am. Coll. Cardiol.*, 2001;**37**: 1628–1634.
75. Ikeda, T., M. Takami, K. Sugi, Y. Mizusawa, H. Sakurada, and H. Yoshino, Noninvasive risk stratification of subjects with a Brugada-type electrocardiogram and no history of cardiac arrest. *Ann. Noninvasive Electrocardiol.*, 2005;**10**: 396–403.
76. Marple, S.J., *Digital Spectral Analysis with Applications*. New Jersey: Prentice-Hall, 1987.

77. Pierce, D.L., A.R. Easley, J.R. Windle, and T.R. Engel, Fast Fourier transformation of the entire low amplitude late QRS potential to predict ventricular tachycardia. *J. Am. Coll. Cardiol.*, 1989;**14**: 1741–1743.
78. Cain, M.E., H.D. Ambos, F. Witkowski, and B.E. Sobel, Fast-Fourier transform analysis of signal-averaged electrocardiograms for identification of patients prone to sustained ventricular tachycardia. *Circulation*, 1984;**69**: 711–720.
79. Cain, M.E., H.D. Ambos, J. Markham, A.E. Fischer, and B.E. Sobel, Quantification of differences in frequency content of signal-averaged electrocardiograms in patients with compared to those without sustained ventricular tachycardia. *Am. J. Cardiol.*, 1985;**55**: 1500–1505.
80. Lindsay, B.D., J. Markham, K.B. Schechtman, H.D. Ambos, and M.E. Cain, Identification of patients with sustained ventricular tachycardia by frequency analysis of signal-averaged electrocardiograms despite the presence of bundle branch block. *Circulation*, 1988;**77**: 122–130.
81. Kelen, G.J., R. Henkin, J. Fontaine, and N. El-Sherif, Effects of analysed signal duration and phase on the results of fast fourier transform analysis of the surface electrocardiogram in subjects with and without late potentials. *Am. J. Cardiol.*, 1987;**60**: 1282–1289.
82. Machac, J., A. Weiss, S.L. Winters, P. Barecca, and J.A. Gomes, A comparative study of frequency domain and time domain analysis of signal averaged electrocardiograms in patients with ventricular tachycardia. *J. Am. Coll. Cardiol.*, 1988;**11**: 284–296.
83. Machac, J. and J.A. Gomes, Frequency domain analysis, in *Signal-Averaged Electrocardiography. Concepts, Methods and Applications*, chapter 6, J.A. Gomes, Editor. Boston, MA: Kluwer, 1993, pp. 81–123.
84. Worley, S.J., D.B. Mark, W.M. Smith, P. Wolf, R.M. Califf, H.C. Strauss, M.G. Manwaring, and R.E. Ideker, Comparison of time domain and frequency domain variables from the signal-averaged electrocardiogram: a multivariable analysis. *J. Am. Coll. Cardiol.*, 1988;**11**: 1041–1051.
85. Buckingham, T.A., C.M. Thessen, D. Hertweck, D.L. Janosik, and H.L. Kennedy, Signal-averaged electrocardiography in the time and frequency domains. *Am. J. Cardiol.*, 1989;**63**: 820–825.
86. Haberl, R., G. Jilge, R. Pulter, and G. Steinbeck, Comparison of frequency and time domain analysis of the signal-averaged electrocardiogram in patients with ventricular tachycardia and coronary artery disease: methodologic validation and clinical relevance. *J. Am. Coll. Cardiol.*, 1988;**12**: 150–158.
87. Haberl, R., G. Jilge, R. Pulter, and G. Steinbeck, Spectral mapping of the electrocardiogram with Fourier transform for identification of patients with sustained ventricular tachycardia and coronary artery disease. *Eur. Heart. J.*, 1989;**10**: 316–322.
88. Lander, P., D.E. Albert, and E.J. Berbari, Spectrotemporal analysis of ventricular late potentials. *J. Electrocardiol.*, 1990;**23**: 95–108.
89. Malik, M., P. Kulakowski, J. Poloniecki, A. Staunton, O. Odemuyiwa, T. Farrell, and J. Camm, Frequency versus time domain analysis of signal-averaged electrocardiograms. I. Reproducibility of the results. *J. Am. Coll. Cardiol.*, 1992;**20**: 127–134.
90. Kelen, G.J., R. Henkin, A.M. Starr, E.B. Caref, D. Bloomfield, and N. El-Sherif, Spectral turbulence analysis of the signal-averaged electrocardiogram and its predictive accuracy for inducible sustained monomorphic ventricular tachycardia. *Am. J. Cardiol.*, 1991;**67**: 965–975.
91. Malik, M., P. Kulakowski, K. Hnatkova, A. Staunton, and A.J. Camm, Spectral turbulence analysis versus time-domain analysis of the signal-averaged ECG in survivors of acute myocardial infarction. *J. Electrocardiol.*, 1994;**27**: S227–S232.
92. Copie, X., K. Hnatkova, A. Staunton, A.J. Camm, and M. Malik, Spectral turbulence versus time-domain analysis of signal-averaged ECG used for the prediction of different arrhythmic events in survivors of acute myocardial infarction. *J. Cardiovasc. Electrophysiol.*, 1996;**7**: 583–593.
93. Englund, A., M. Andersson, and L. Bergfeldt, Spectral turbulence analysis of the signal-averaged electrocardiogram for predicting inducible sustained monomorphic ventricular tachycardia in patients with and without bundle branch block. *Eur. Heart J.*, 1995;**16**: 1936–1942.
94. Ahuja, R.K., G. Turitto, B. Ibrahim, E.B. Caref, and N. El-Sherif, Combined time-domain and spectral turbulence analysis of the signal-averaged ECG improves its predictive accuracy in postinfarction patients. *J. Electrocardiol.*, 1994;**27**: S202–S206.
95. Mäkijärvi, M., T. Fetsch, L. Reinhardt, A. Martinez-Rubio, M. Shenasa, M. Borggrefe, and G. Breithardt, Comparison and combination of late potentials and spectral turbulence analysis to predict arrhythmic events after myocardial infarction in the Post-Infarction Late Potential (PILP) study. *Eur. Heart J.* 1995; **16**: 651–659.
96. Vazquez, R., E.B. Caref, F. Torres, M. Reina, A. Espina, and N. El-Sherif, Improved diagnostic value of combined time and frequency domain analysis of the signal-averaged electrocardiogram after myocardial infarction. *J. Am. Coll. Cardiol.*, 1999;**33**: 385–394.
97. Golden, D.P. Jr., R.A. Wolthuis, and G.W. Hoffler, A spectral analysis of the normal resting electrocardiogram. *IEEE Trans. Biomed. Eng.*, 1973;**20**: 366–372.
98. Reynolds, E.W., B.F. Muller, G.J. Anderson, and B.T. Muller, High frequency components in the electrocardiogram. A comparative study of normals and patients with myocardial disease. *Circulation*, 1967;**35**: 195–206.
99. Flowers, N.C., L.G. Horan, J.R. Thomas, and W.J. Tolleson, The anatomic basis for high-frequency components in the electrocardiogram. *Circulation*, 1969;**39**: 531–539.
100. Mor-Avi, V., S. Abboud, and S. Akselrod, Frequency content of the QRS notching in high-fidelity canine ECG. *Comput. Biomed. Res.*, 1989;**22**: 18–28.
101. Abboud, S., Subtle alterations in the high-frequency QRS potentials during myocardial ischemia in dogs. *Comput. Biomed. Res.*, 1987;**20**: 384–395.
102. Mor-Avi, V., B. Shargorodsky, S. Abboud, S. Laniado, and S. Akselrod, Effects of coronary occlusion on high-frequency components of the epicardial electrogram and body surface electrocardiogram. *Circulation*, 1987;**76**: 237–243.
103. Abboud, S., R.J. Cohen, A. Selwyn, P. Ganz, D. Sadeh, and P.L. Friedman, Detection of transient myocardial ischemia by computer analysis of standard and signal-averaged high-frequency electrocardiograms in patients undergoing percutaneous transluminal coronary angioplasty. *Circulation*, 1987;**76**: 585–596.
104. Abboud, S., J.M. Smith, B. Shargorodsky, S. Laniado, D. Sadeh, and R.J. Cohen, High frequency electrocardiography of three orthogonal leads in dogs during a coronary artery occlusion. *PACE*, 1989;**12**: 574–581.
105. Abboud, S., R.J. Cohen, and D. Sadeh, A spectral analysis of the high frequency QRS potentials observed during acute myocardial ischemia in dogs. *Int. J. Cardiol.*, 1990;**26**: 285–290.

106. Mor-Avi, V. and S. Akselrod, Spectral analysis of canine epicardial electrogram; short term variations in the frequency content induced by myocardial ischemia. *Circ. Res.*, 1990;**66**: 1681–1691.
107. Abboud, S., High-frequency electrocardiogram analysis of the entire QRS in the diagnosis and assessment of coronary artery disease. *Prog. Cardiovasc. Dis.*, 1993;**35**: 311–328.
108. Pettersson, J., P. Lander, O. Pahlm, L. Sörnmo, S.G. Warren, and G.S. Wagner, Electrocardiographic changes during prolonged coronary artery occlusion in man: comparison of standard and high-frequency recordings. *Clin. Physiol.*, 1998;**18**: 179–186.
109. Pettersson, J., O. Pahlm, E. Carro, L. Edenbrandt, M. Ringborn, L. Sörnmo, S.G. Warren, and G.S. Wagner, Changes in high-frequency QRS components are more sensitive than ST segment deviation for detecting acute coronary artery occlusion. *J. Am. Coll. Cardiol.*, 2000;**36**: 1827–1834.
110. Goldberger, A.L., V. Bhargava, V. Froelicher, J. Covell, and D. Mortara, Effect of myocardial infarction on the peak amplitude of high frequency QRS potentials. *J. Electrocardiol.*, 1980;**13**: 367–372.
111. Bhargava, V. and A. Goldberger, Myocardial infarction diminishes both low and high frequency QRS potentials: power spectrum analysis of lead II. *J. Electrocardiol.*, 1981;**14**: 57–60.
112. Goldberger, A.L., V. Bhargava, V. Froelicher, and J. Covell, Effect of myocardial infarction on high-frequency QRS potentials. *Circulation*, 1981;**64**: 34–42.
113. Talwar, K.K., G.S. Rao, U. Nayar, and M.L. Bhatia, Clinical significance of high frequency QRS potentials in myocardial infarction: analysis based on power spectrum of lead III. *Cardiovasc. Res.*, 1989;**23**: 60–63.
114. Berkalp, B., E. Baykal, N. Caglar, C. Erol, G. Akgün, and T. Gürel, Analysis of high frequency QRS potentials observed during acute myocardial infarction. *Int. J. Cardiol.*, 1993;**42**: 147–153.
115. Novak, P., L. Zhixing, V. Novak, and R. Hatala, Time-frequency mapping of the QRS complex in normal subjects and in postmyocardial infarction patients. *J. Electrocardiol.*, 1994;**27**: 49–60.
116. Ringborn, M., O. Pahlm, G.S. Wagner, S.G. Warren, and J. Pettersson, The absence of high-frequency QRS changes in the presence of standard electrocardiographic QRS changes of old myocardial infarction. *Am. Heart J.*, 2001;**36**: 1827–1834.
117. Trägårdh, E., O. Pahlm, G.S. Wagner, and J. Pettersson, Reduced high-frequency QRS components in patients with ischemic heart disease compared to normal subjects. *J. Electrocardiol.*, 2004;**37**: 157–162.
118. Vacek, J.L., D.B. Wilson, G.W. Botteron, and J. Dobbins, Techniques for the determination of left ventricular mass by signal-averaged electrocardiography. *Am. Heart J.*, 1990;**120**: 958–963.
119. Okin, P.M., T.M. Donnelly, T.S. Parker, D.C. Wallerson, N.M. Magid, and P. Kligfield, High-frequency analysis of the signal-averaged ECG. Correlation with left ventricular mass in rabbits. *J. Electrocardiol.*, 1992;**25**: 111–118.
120. Abboud, S., O. Berenfeld, and D. Sadeh, Simulation of high-resolution QRS complex using a ventricular model with a fractal conduction system. Effects of ischemia on high-frequency QRS potentials. *Circ. Res.*, 1991;**68**: 1751–1760.
121. Mason, R.E. and I. Likar, A new system of multiple-lead exercise electrocardiography. *Am. Heart J.*, 1966;**71**: 196–205.
122. Proakis, J.G. and D.G. Manolakis, *Digital Signal Processing – Principles, Algorithms, and Applications*. Upper Saddle River, NJ: Prentice-Hall, 1996.
123. Schlegel, T.T., W.B. Kulecz, J.L. DePalma, A.H. Feiveson, J.S. Wilson, M.A. Rahman, and M.W. Bungo, Real-time 12-lead high-frequency QRS electrocardiography for enhanced detection of myocardial ischemia and coronary artery disease. *Mayo Clin. Proc.* 2004;**79**: 339–350.
124. Schlegel, T.T., B. Arenare, V. Starc, E.C. Greco, G. Poulin, D.R. Moser, and R. Delgado, The best identifiers of cardiomyopathy in short duration ECG recordings: high frequency QRS reduced amplitude zone score, QT interval variability, low frequency RR interval power and heart rate turbulence slope. *Folia Cardiol.*, 2005;**12**(SupplC): 1–4.
125. Pueyo, E., E. Sörnmo, and P. Laguna, QRS slopes for early ischemia detection and characterization. *IEEE Trans. Biomed. Eng.*, 2008;**55**: 468–477.
126. Pettersson, J., E. Carro, L. Edenbrandt, C. Maynard, O. Pahlm, M. Ringborn, L. Sörnmo, S.G. Warren, and G.S. Wagner, Spatial, individual, and temporal variation of the high-frequency QRS amplitude in the 12 standard electrocardiographic leads. *Am. Heart J.*, 2000;**139**: 352–358.
127. Abboud, S., J. Leor, and M. Eldar, High frequency ECG during reperfusion therapy of acute myocardial infarction. *IEEE Comput. Soc. Comput. Cardiol.*, 1990; 351–353.
128. Aversano, T., B. Rudicoff, A. Washington, S. Traill, V. Coombs, and J. Raqueno, High frequency QRS electrocardiography in the detection of reperfusion following thrombolytic therapy. *Clin. Cardiol.*, 1994;**17**: 175–182.
129. Bhargava, V. and A.L. Goldberger, Effect of exercise in healthy men on QRS power spectrum. *Am. J. Physiol.*, 1982;**243**: H964–H969.
130. Beker, A., A. Pinchas, J. Erel, and S. Abboud, Analysis of high frequency QRS potential during exercise testing in patients with coronary artery disease and in healthy subjects. *PACE*, 1996;**19**: 2040–2050.
131. Lipton, J.A., S.G. Warren, M. Broce, S. Abboud, A. Beker, L. Sörnmo, D.R. Lilly, C. Maynard, B.D. Lucas Jr, and G.S. Wagner, High-frequency QRS electrocardiogram during exercise stress testing for detecting ischemia. *Int. J. Cardiol.*, 2008;**124**: 198–203.
132. Toledo, E., J.A. Lipton, S.G. Warren, S. Abboud, M. Broce, D.R. Lilly, C. Maynard, B.D. Lucas Jr, and G.S. Wagner, Detection of stress-induced myocardial ischemia from the depolarization phase of the cardiac cycle – a preliminary study. *J. Electrocardiol.*, 2009;**42**: 240–247.
133. Rahman, M.A., A. Gedevanishvili, Y. Birnbaum, L. Sarmiento, W. Sattam, W.B. Kulecz, and T.T. Schlegel, High-frequency QRS electrocardiogram predicts perfusion defects during myocardial perfusion imaging. *J. Electrocardiol.*, 2006;**39**: 73–81.
134. Trägårdh, E., T.T. Schlegel, M. Carlsson, J. Petterson, K. Nilsson, and O. Pahlm, High-frequency electrocardiogram analysis in the ability to predict reversible reperfusion defects during adenosine myocardial perfusion imaging. *J. Electrocardiol.*, 2007;**40**: 510–514.
135. Trägårdh, E., H. Arheden, J. Pettersson, G.S. Wagner, and O. Pahlm, High-frequency QRS components vs left ventricular mass in humans. *Folia Cardiol.*, 2005;**12**(suppl C): 68.
136. Watanabe, T., M. Yamaki, H. Tachibana, I. Kubota, and H. Tomoike, Decrease in the high-frequency QRS components depending on the local conduction delay. *Jpn. Circ. J.*, 1998;**62**: 844–848.
137. Valentino, V.A., H.O. Ventura, F.M. Abi-Samra, C.H. Van Meter, and H.L. Price, The signal-averaged electrocardiogram in cardiac transplantation. A non-invasive marker of acute allograft rejection. *Transplantation*, 1992;**53**: 124–127.

138. Graceffo, M.A. and R.A. O'Rourke, Cardiac transplant rejection is associated with a decrease in the high-frequency components of the high-resolution, signal-averaged electrocardiogram. *Am. Heart J.*, 1996;**132**: 820–826.
139. Matsushita, S., Y. Sakakibara, T. Imazuru, M. Noma, Y. Hiramatsu, O. Shigeta, T. Jikuya, and T. Mitsui, High-frequency QRS potentials as a marker of myocardial dysfunction after cardiac surgery. *Ann. Thorac. Surg.*, 2004;**77**: 1293–1297.
140. Abe, M., S. Matsushita, and T. Mitsui, Recovery of high-frequency QRS potentials following cardioplegic arrest in pediatric cardiac surgery. *Pediatr. Cardiol.*, 2001;**22**: 315–320.

40 Electrocardiography in Epidemiology

Pentti M. Rautaharju

P. W. Macfarlane, A. van Oosterom, O. Pahlm, P. Kligfield, M. Janse, J. Camm (eds.), *Comprehensive Electrocardiology*, DOI 10.1007/978-1-84882-046-3_40,

40.1 Introduction

Like other areas of electrocardiography, epidemiological electrocardiography has grown and evolved in content since the publication of the first edition of this book. The emphasis in epidemiological electrocardiography has shifted from primarily descriptive population studies to follow-up studies with risk evaluation. Electrocardiographic (ECG) recording technology has evolved, and computer electrocardiography has matured, developments that have greatly enhanced the feasibility of high-volume ECG acquisition and analysis with substantially enhanced precision and reproducibility.

At the time of the preparation of the material for the first volume of this book, computer electrocardiography in epidemiology was still a novelty that required special consideration. This is no longer necessary. The second area of epidemiological electrocardiography that has gone through a similar period of maturation and growth is the use of electrocardiography in clinical trials. This subject area has, in fact, grown so much that it is no longer feasible to cover it any more in the context of this chapter, with a few exceptions. Exercise electrocardiography and ambulatory electrocardiography are covered in separate chapters of this book.

The subject areas covered in this chapter belong in the realm of epidemiological electrocardiography that includes the following broad topics of ECG investigation:

- Estimation of the prevalence and incidence of ECG abnormalities in cross-sectional population studies.
- Determination of the normal limits for ECG intervals, amplitudes, and waveform patterns.
- Assessment of the evolution with age, significance of ECG findings in relation to physiological and anthropometric measurements, and coronary heart disease (CHD) risk factors and natural history of disease processes.
- Assessment of the risk of future adverse events, morbidity, and mortality associated with ECG abnormalities.

Each of these major application areas has certain special requirements which may be different from the needs of the traditional electrocardiographic practice in clinical diagnostic applications. Consideration of normal limits for ECG patterns is covered elsewhere in this book.

There is a scarcity of systematic reviews of epidemiological aspects of electrocardiography. One of the few, published since the first edition of Comprehensive Electrocardiography is the review from 2000 by Ashley et al. [1].

40.1.1 ECG Coding Schemes from Historical Perspective

Modern cardiovascular epidemiological studies were initiated in the USA, several European countries, and Japan during the 1950s when little was still known about the epidemiology of CHD and hypertensive heart disease (HHD). Many of these studies were inspired by the pioneering efforts of Ancel Keys at the Laboratory of Physiological Hygiene, School of Public Health of the University of Minnesota. Dr. Keys initiated a 15-year prospective study in 1947 of business and professional men from Minnesota [2], and in the 1950s, the classic Seven Countries study [3, 4]. These studies led to intensive electrocardiographic research and developmental work in this area of epidemiology. The Framingham study was initiated around 1948–1950 [5]. That study, as well as the studies in Albany, New York [6], Los Angeles [7], and Chicago [8, 9], are other milestones among modern cardiovascular epidemiology. However, the state-of-the-art of ECG coding in the 1950s was appalling. The only common guidelines for ECG wave definitions and measurements were those proposed by the Criteria Committee of the New York Heart Association [10]. Most wave definitions and classification criteria were largely qualitative descriptions and thus were too ambiguous for epidemiological applications. It was evident that the comparability of the reported prevalence of ECG abnormalities from these studies in the 1950s was questionable.

The Minnesota code (MC) was developed between 1956 and 1960. Its primary purpose was to improve the comparability of ECG classification and thus the consistency of the assessment of the coronary heart disease prevalence rates among the cohorts of the Seven Countries study. At that time, it was a formidable task to reach an agreement within any group of cardiologists on any set of ECG criteria, for instance for myocardial infarction (MI) or ventricular hypertrophies. The authors of the Minnesota code were cautious to avoid any reference to diagnostic classification. The code was described strictly as a scheme for objective reporting of morphological ECG wave measurements without prejudice to any interpretation. However, the classification criteria were based on clinical ECG criteria commonly used at that time, many of which were developed during the late 1950s. Dr. Henry Blackburn was in charge of the Minnesota code development. One of the earliest field trials of the new code was carried out in Finland from 1956 to 1959 in epidemiological studies, which assessed the prevalence of arteriosclerotic and hypertensive heart disease among ostensibly healthy working populations [11]. The report concluded that with the aid of the Minnesota code, the ECG can be successfully used as an objective tool in the study of cardiovascular epidemiology. After the publication of the original Minnesota code in Circulation in 1960 [12], it rapidly gained widespread acceptance. The code was systematically used in the Seven Countries study and in a variety of other population studies such as the Tecumseh study of a total natural community in Michigan [13], the Tukisenta study in New Guinea [14], and the National Cooperative Pooling project including Tecumseh and four other longitudinal investigations in the USA [15].

A more recent development is the Novacode [16], which has been used in a variety of epidemiological studies sponsored by the National Institutes of Health. There are several other ECG coding schemes, which have been proposed for morphological description of ECG patterns or for diagnostic ECG classification [17–21]. However, most of these schemes are more suitable for clinical than for epidemiological applications.

40.1.2 Minnesota Code

The Minnesota code belongs to the group of classification systems called linguistic or syntactic. In general, these classifiers extract "morphs" or various feature patterns observed or measured and use a "grammar" to categorize the morphs into a given set of usually, but not necessarily, mutually exclusive diagnostic classes. In most instances, the grammar can be expressed as a branching tree of Boolean decisions. Boolean logic uses the presence or absence of various features in each decision node to determine which branch of the diagnostic decision tree to follow.

40.1.2.1 ECG Measurement Rules

The ECG measurement rules were left fairly ill defined and incomplete in the original version of the Minnesota code. These rules were defined a little more clearly in the 1968 version of the Minnesota code [22]. The Minnesota code ECG measurement rules are best described in the 1982 version of the code [23], which provides detailed illustrations of visual ECG wave measurement devices and procedures. The older hot stylus electrocardiographs, with ECGs recorded on thermosensitive paper produced a tracing that had a thick "baseline." This can considerably bias wave-duration measurements, depending on the writing characteristics of the stylus and the paper speed [24]. Proper measurement procedures can reduce but cannot eliminate these errors, which can have a significant impact on ECG coding.

It was recommended in the original version of the Minnesota code that the majority rule should be applied throughout without exception and a majority of codable complexes was required from each ECG lead coded. The measurement rules for the 1982 version of the Minnesota code specify several exceptions to the majority rule. For instance, an initial R wave exceeding 0.025 mV even in one complex in a given lead (except V_1) rules out codable Q and QS waves in that lead. For some other codes (for instance code 1.2.8), a given feature has to be present in all complexes in order to qualify as a codable item.

Three different reference or baseline points are used for amplitude measurements in the Minnesota code. All QRS-complex and ST-segment amplitude measurements are made with respect to the PR segment immediately preceding the onset of QRS. P-amplitude measurements are referred to the TP segment preceding the P-wave and T-amplitude measurements to the flattest part of the TP segment following the T wave. The amplitudes of positive deflections are measured from the upper margin of the tracing and negative deflections from the lower margin of the tracing at the baseline point.

40.1.2.2 Definition of Codable Waves

An attempt to give unambiguous definitions for codable ECG waves was one of the most important contributions of the Minnesota code. The logic for wave definitions is relatively simple as described in the following sequence:

- The first codable wave is the first deflection within the QRS complex ≥ 0.025 mV or ≤ -0.1 mV. The first codable wave is denoted as R wave if it is positive and a Q or QS wave if it is negative.
- The second codable wave is the first deflection following the first codable wave with an opposite polarity and an absolute amplitude ≥ 0.1 mV. The second wave is an R wave if it is positive and an S wave if it is negative.
- Subsequent codable waves within QRS, with alternating signs, are defined similarly with 0.1 mV amplitude thresholds. The third codable wave is an R′ wave if it is positive and S′ wave if it is negative. The fourth codable wave is an R′ wave if it positive and an S′ wave if it is negative.
- A Q wave is the first codable wave within QRS if it is negative and is followed by a codable R wave.
- A QS wave is the first codable wave if it is negative and is not followed by another codable wave within QRS.

The above definitions are adequate for describing objectively all possible combinations of QRS patterns of type R, RS, RSR′, RSR′S′, QR, QRS, QRSR′, and QS.

40.1.2.3 Coding Criteria for Serial ECG Changes

The basic problem with the initial set of the Minnesota code criteria for serial ECG changes was that trivial changes in ECG amplitudes or wave durations frequently induced artifactual new codable events in myocardial infarction (MI) classification. The coding scheme was vulnerable to considerable coding variation. A new classification scheme for coding of serial ECG changes [25] was developed for determining the incidence rate of reinfarction in the Coronary Drug Project [26], a large secondary prevention trial. This was done when it was noted that up to 27% of ECG changes, coded as significant worsening or reinfarction in the Coronary Drug project, were caused by coding variation alone. Similar problems were encountered in another large intervention study, the Multiple Risk Factor Intervention Trial (MRFIT) [27].

The serial change criteria for the Minnesota code are not based on changes in ECG measurements or simultaneous comparison of successively recorded ECGs acquired from periodic examinations. Instead, the coding scheme is based on changes in the severity level of the coding category of independently coded records. For major Q–QS changes, a jump over one severity level in code 1 is required (e.g., from 1.3 to 1.1). For minor Q–QS changes, a worsening by one severity level in code 1, plus a worsening by one severity level in code 5 (T waves) is needed (excluding code 5.4).

A significant worsening for ST changes requires a transition over one severity level (e.g., from 5.3 to 5.2). There are no serial change criteria for items related to ventricular hypertrophies.

The coding rules suitable for use in algorithms for classification of serial changes in Q–QS patterns are summarized in Table 40.1. There are additional rules for certain other items. For instance, a codable new left bundle branch block (LBBB) (7.1) requires a QRS-duration increase of 0.020 s or more from the baseline ECG.

Table 40.1
Serial change comparison rules for significant worsening of Minnesota Code (MC) Q and QS waves

Follow-up ECG Minnesota Code	Qualifying conditions for a significant change
1.1.6, 1.1.7, 1.2.3, 1.2.7, 1.3.2, 1.3.6, 1.1.1, 1.1.4, 1.1.5, 1.2.1, 1.2.2, 1.2.4	Initial R amplitude decrease ≥0.1 mV or Q:R amplitude ratio increase ≥50%
1.1.2	Initial R-amplitude decrease ≥0.15 mV and Q: R ratio increase ≥50%
1.1.3	Initial R-amplitude decrease ≥0.10 mV or Q: R ratio increase ≥75%
1.2.6	(Initial R-amplitude decrease >0.10 mV or Q: R ratio increase ≥75%) and appearance of a new codable Q wave in a VF
1.2.8	R-amplitude decrease in the "lead to the left" causing new code ≥0.10 mV

40.2 Prevalence of ECG Abnormalities

Most cardiovascular epidemiological studies have primarily focused on coronary heart disease (CHD) and hypertensive heart disease. ECG coding of Q waves and associated ST-T abnormalities has been considered as an index of old myocardial infarction (MI) in comparison with contrasting populations. Left ventricular hypertrophy (LVH) by ECG criteria (ECG-LVH), in turn, has been considered as an index of true anatomical LVH. Limited diagnostic accuracy of more moderate ECG-MI criteria and low prevalence of more strict ECG-MI criteria limits the utility of ECG-MI prevalence estimation in free-living populations. As will be shown later, similar or even more severe limitations are encountered with the application of ECG-LVH criteria. Prognostic evaluation in early epidemiological studies was done for broad categories of ECG abnormalities, in part because of limited sample size in categories considered disease specific.

40.2.1 Contrasting Prevalence and Age Trends of ECG Abnormalities in Middle-Aged Men and Women

The coronary heart disease study of the Social Insurance Institution of Finland is one of the best-documented recent prevalence surveys conducted in free-living male and female populations [28]. The study population of 5,738 men and 5,224 women consisted of whole or random samples of rural or semi-urban populations, with an overall participation rate of 90%.

➲ Table 40.2 summarizes the prevalence of selected Minnesota code (MC) abnormalities for various age-groups of men and women ranging from 30 to 59 years. In Finnish men, the prevalence of major and minor Q waves (MC 1.1 to 1.3) was nearly four times higher than in Finnish women. These prevalence rates are nine times higher than in a 1965 report from Tecumseh population in the corresponding age range (30–59 years and less) [13]. The linear gradient in increase of the prevalence of Q waves with age was not significant in Finnish men and only of borderline significance in women. The prevalence of Q waves was higher in men than in women. However, the gender difference was no longer significant for the age-group 55–59 years.

In the Finnish study group, the prevalence of left axis deviation (MC 2.1) was 4.1% in men and 2.3% in women. The prevalence increased significantly with age in a fashion similar to that in the Tecumseh population and several others. Interestingly, such left axis trend with age was not present in men in a Polynesian community of Pukapuka [29]. Blood pressure levels in Pukapuka men have low mean values and show no rise with age and the prevalence of obesity is reportedly low.

Among the most intriguing observations in the study of Finnish communities was the contrasting age trend between men and women in the prevalence of high-amplitude R waves in leads V_5, V_6, or the frontal plane limb leads (MC 3.1). The prevalence was very high (25.9%) among Finnish men and showed no significant age trend. Evidently, the decrease of R-wave amplitudes with age observed among normotensive men is offset by R-wave amplitude increase with hypertension. The prevalence of high blood pressure ($\geq$ 160 mm Hg systolic, or $\geq$95 mm Hg diastolic) among men increased from 12.9% in the age-group 30–39 years to 33.1% in the age-group 50–59 years.

The prevalence of high-amplitude QRS waves in Finnish women was substantially lower than in men, and particularly in women younger than 50 years. The gradient with age in women is significant. The striking sex differences in the prevalence indicate the inappropriateness of using identical criteria for men and women. These differences also reflect the influence on these voltage criteria of factors other than LVH associated with hypertension, such as physical activity level, age, sex, and other anthropometric extracardiac factors.

The inclusion of high-voltage criteria of Sokolow and Lyon (MC 3.1 and 3.3) causes a substantial increase in the LVH prevalence estimates. With these criteria, the prevalence was 41.9% in men and 19.5% in women. The prevalence of LVH according to the high-voltage criteria combined with ST-T changes is very low in most populations studied and these criteria are likely to be too insensitive to be used as an index for LVH in hypertension.

Abnormal ST depression in the resting ECG (MC 4.1, 4.2 or 4.3) was present significantly more often in women (4.3%) than in men (2.2%), and the prevalence increased markedly with age in both sex groups. The inclusion of the borderline abnormal ST category (J depression in excess of 0.1 mV with upsloping ST where the J point denotes the end of the QRS complex) added merely 0.8% to the prevalence in men and 1.4% in women. This reflects a deficiency in the hierarchical structure of abnormal ST codes regarding the severity of the ST abnormality, since prevalence would be expected to

Table 40.2

Prevalence (%) of Minnesota Codes among a Finnish cohort representing total or random samples of middle-aged rural or semi-urban community dwellers or factory employees by age and gender*

Age and Gender	MC Category N	Q, QS Waves 1.1–1.3	Left Axis 2.1	High R Waves 3.1	Abnormal ST 4.1–4.3	Abnormal T 5.1–5.3	PR Prolonged 6.3	LBBB 7.1	RBBB 7.2	*R′* Wave 7.3	Atrial Fibrillation 8.3	ST Elevation 9.2
30–34												
M	1,073	6.1	2.4	26.6	0.7	2.5	1.4	0.0	0.4	0.6	0.0	22.3
F	865	3.0	0.9	3.9	1.5	7.4	0.6	0.0	0.1	0.2	0.2	0.4
35–39												
M	1,076	5.7	3.0	25.1	0.9	3.2	1.3	0.1	0.2	0.5	0.1	18.8
F	889	4.6	1.2	5.0	1.1	8.2	0.0	0.1	0.1	0.3	0.0	0.3
40–44												
M	1,099	6.6	2.8	25.8	0.9	3.1	1.8	0.5	0.0	1.0	0.1	15.2
F	997	4.0	1.8	5.3	2.7	10.7	0.2	0.1	0.1	0.3	0.0	0.5
45–49												
M	906	5.7	5.0	23.1	1.6	6.2	1.9	0.3	0.6	0.9	0.2	14.6
F	927	3.8	3.0	7.7	3.9	13.1	0.5	0.1	0.2	0.5	0.0	0.4
50–54												
M	791	8.2	6.6	25.2	3.2	9.2	1.4	0.3	0.9	1.1	0.4	9.2
F	793	5.2	3.3	11.1	7.8	18.5	0.8	0.5	0.4	0.9	0.3	0.9
55–59												
M	793	8.1	5.8	29.5	6.3	13.1	1.3	1.0	1.3	0.8	1.0	10.6
F	753	6.9	3.9	16.1	9.7	24.0	1.3	0.9	0.4	0.5	0.4	0.9
All												
M	5,738	6.6	4.1	25.9	2.2	5.8	1.5	0.4	0.5	0.8	0.3	15.6
F	5,224	4.5	2.3	7.9	4.3	13.3	0.6	0.3	0.2	0.5	0.1	0.6

M = males; F = females; MC = Minnesota code. Data from the Finnish Social Insurance Institution's Coronary Heart Disease Study, Reunanen et al. [28], Acta Med Scand 1983;673 (suppl.):1–120.

increase with decreasing severity of the code. For instance, for men, the prevalences for codes 4.1 to 4.4 were 0.9%, 0.7%, 0.6%, and 0.8%, respectively.

Abnormal T waves (MC 5.1 to 5.3) were coded in 5.8% of men and 13.3% of women. The prevalence increased with age and was significantly higher in women. In women aged 55–59 years, the prevalence of abnormal T waves at rest was as high as 24%. Over 60% of abnormal T waves were flat or biphasic (MC 5.3).

The high prevalence of T-wave abnormalities in women observed in the Finnish populations, Tecumseh, and other studies, evidently indicates that repolarization abnormalities in women are often associated with factors not related to CHD. Similar relatively high prevalence rates have been reported in nonindustrialized native female populations relatively free from CHD. For instance, about one quarter of the women aged 40–59 years had T-wave abnormalities (MC 5.1 to 5.3) in a non-urbanized population of Tukisenta, New Guinea [14].

The prevalence of other coded ECG abnormalities in men and women was low, with the exception of ST elevation in men. Prolonged PR (MC 6.3) was present in 1.5% of men and in 0.6% of women, left bundle branch block in 0.4% of men and 0.3% of women, right bundle branch block (RBBB) in 0.5% of men and 0.2% of women, incomplete right bundle branch block in 0.8% of men and 0.5% of women, and atrial fibrillation in 0.3% of men and 0.1% of women.

One striking contrast between the prevalence rates of men and women was the high prevalence of ST elevation in Finnish men (15.6%). The corresponding prevalence in women was only 0.6%. The prevalence of ST elevation in men was seen to decrease significantly with age. An opposite, less-pronounced trend was observed in women. The high prevalence of ST elevation in men is apparently associated with tall T waves in the chest leads, which is a common finding especially in younger men. For other abnormalities of interest, the prevalence of Wolff–Parkinson–White syndrome (MC 6.4) was 0.2% in men and 0.1% in women. Frequent ventricular ectopic complexes (over 10% of QRS complexes) were present in 1.0% of men and 1.4% of women. Occasional ventricular ectopic complexes were observed in 1.2% of men and 1.4% of women. The prevalence of frequent supraventricular ectopic complexes (over 10%) was 0.5% in both sex groups, and the prevalence of occasional supraventricular complexes was 0.8% in men and 1.0% in women.

The reported prevalence of abnormal ECG findings in three major coding categories of the Minnesota code in five male and female population studies [13, 28, 30–32] is summarized in ➋ Table 40.3. The data confirm the uniform trend toward higher prevalence of ST depression and lower prevalence of significant Q waves in women. The prevalence of significant Q waves varies little between male populations, as a contrast to quite dramatic variations in the prevalence of high-amplitude QRS waves.

The prevalence of high-amplitude QRS waves in Finnish men is unusually high. Comparable prevalence rates have been reported only for black Jamaican males [33], with a reported prevalence of 29.9% for MC 3.1. Prevalence rates for code 3.1 as low as 0.6% have been reported for British men aged 50–59 years employed as civil servants [34]. However, this Whitehall study of Rose et al. was based on limb-lead ECGs only. As pointed out by Reunanen et al. [28], differences in occupational distributions may partly explain these differences in prevalence that, however, remain largely unresolved discrepancy.

Table 40.3

Contrasting prevalence (%) of major Q waves, high-amplitude R waves, and ST depression in five studies on male and female populations aged 50–59 years

	No. Subjects		Major Q Waves MC 1.1, 1.2		High-Amplitude R Waves MC 3.1		ST Abnormalities MC 4.1–4.3	
Study/Ref.	Male	Female	Male	Female	Male	Female	Male	Female
Tecumseh [13]	331	327	4.2	0.6	3.3	3.4	3.3	12.2
Framingham [30]	650	808	3.5	1.9	9.5	5.0	5.2	6.7
Busselton [31]	310	375	2.3	0	6.8	1.1	3.2	2.7
Copenhagen [32]	2,014	2,791	2.2	0.7	12.0	4.1	4.0	4.7
Finland's Soc. Ins. Inst. [28]	1,584	1,546	3.9	2.5	27.3	13.5	4.7	8.7

MC = Minnesota code.

As a contrast to the variability of the high R-wave amplitude prevalence, the ST-depression prevalence among male populations was relatively uniform.

40.2.2 Prevalence of ECG Abnormalities in Adult Male Populations

Table 40.4 summarizes prevalence data for ECG findings from the classic Seven Countries study of Keys et al. [3]. Twelve of these 17 cohorts of men aged 40–59 years represent total populations of males in each geographical area. Four occupational groups included in the study consisted of rail employees from the USA and Italy. One of the European populations (Zutphen) was drawn as a random (five out of nine) subsample.

The prevalence of any codable Q, QS waves, and related items ranged from 0.6% in the Japanese fishing village of Ushibuka to 6% among US railroad executives. An even larger range of variation was found in the prevalence of high-amplitude R waves (code 3.1): 1.2% among US railroad executives and 17.9% in Karelia, Finland. However, the prevalence of MI and LVH according to more stringent criteria, which gives a higher specificity, was very low. The total prevalence of diagnostic Q waves (code 1.1), lesser Q waves plus negative T waves (code 1.2 or 1.2 with code 5.1 or 5.2) or ventricular conduction defects (7.1, 7.2, or 7.4) ranged from 0.9% in Crete to 7.1% among non-sedentary US railroad clerks. The range for the prevalence of high-amplitude R waves combined with ST changes (code 3.1 and any of the codes 4.1 to 4.4) was from 0.13% for the Italian railwaymen to 1.98% in the farming village of Tanushimaru in Japan.

Rose et al. [35] compared prevalence rates for Minnesota code ECG abnormalities in six cohorts of middle-aged male clerical workers from five European countries (Belgium, Denmark, Italy, The Netherlands, and the USSR). The prevalence of Q, QS waves (MC 1.1 to 1.3) ranged from 3.4% to 5.5%, ST abnormalities (MC 4.1 to 4.3) from 2.6% to 3.6%,

Table 40.4
Prevalence (%) of Minnesota Code ECG abnormalities among 17 cohorts of adult male populations in the Seven Countries Study of Keys et al.[a]

MC Category/Cohort	N	Major Q 1.1	Any Q 1.1–1.3	Minor Q + Abn. T 1.3 + 5.1, 5.2	Major Abn. T 5.1	High R 3.1	High R + Abn. ST 3.1 + 4.1–4.4	Abn. ST, Isolated 4.1–4.4	Abn. T, Isolated 5.1–5.3	Ventricular Conduction 7.1, 7.2, 7.4
Crete, Greece	683	1 (0.2)	9 (1.3)	0 (0.0)	0 (0.0)	35 (5.1)	2 (0.3)	4 (0.6)	5 (0.7)	5 (0.7)
Corfu, Greece	529	3 (0.6)	17 (3.2)	0 (0.0)	1 (0.2)	40 (7.6)	5 (1.0)	6 (1.1)	3 (0.6)	11 (2.1)
Velikakrsna, Yugoslavia	510	5 (1.0)	14 (2.7)	0 (0.0)	0 (0.0)	62 (12.2)	6 (1.2)	1 (0.2)	0 (0.0)	6 (1.2)
Dalmatia, Yugoslavia	669	0 (0.0)	14 (2.1)	1 (0.2)	0 (0.0)	19 (2.8)	1 (0.2)	5 (0.8)	5 (0.8)	7 (1.1)
Slavonia, Yugoslavia	694	3 (0.4)	13 (1.9)	0 (0.0)	0 (0.0)	79 (11.4)	5 (0.7)	2 (0.3)	9 (1.3)	6 (0.9)
Finland, West	857	6 (0.7)	15 (1.8)	0 (0.0)	0 (0.0)	139 (16.2)	10 (1.2)	1 (0.1)	21 (2.5)	12 (1.4)
Finland, Karelia	814	7 (0.9)	18 (202)	4 (0.5)	0 (0.0)	146 (17.9)	12 (1.5)	2 (0.3)	25 (3.1)	5 (0.6)
Italy, Crevalcore	993	5 (0.5)	35 (3.5)	3 (0.3)	0 (0.0)	55 (5.5)	13 (1.3)	1 (0.1)	18 (1.8)	14 (1.4)
Italy, Montegiorgio	717	4 (0.6)	15 (2.1)	1 (0.1)	0 (0.0)	24 (3.6)	3 (0.4)	4 (0.6)	3 (0.4)	5 (0.7)
Rome, Railwaymen	766	4 (0.5)	22 (2.9)	0 (0.0)	0 (0.0)	26 (3.4)	1 (0.1)	4 (0.5)	13 (1.7)	9 (1.2)
Netherlands, Zuthen	877	8 (0.9)	33 (3.8)	0 (0.0)	0 (0.0)	38 (4.3)	8 (0.9)	2 (0.2)	5 (0.6)	21 (2.4)
Japan, Tanushimaru	504	3 (0.6)	11 (2.2)	0 (0.0)	0 (0.0)	38 (7.5)	10 (2.0)	38 (7.5)	4 (0.8)	4 (0.8)
Japan, Ushibuka,	484	2 (0.4)	3 (0.6)	0 (0.0)	0 (0.0)	71 (14.7)	7 (1.5)	32 (6.6)	2 (0.4)	6 (1.2)
USA, Switchmen	835	8 (1.0)	31 (3.7)	0 (0.0)	0 (0.0)	13 (1.6)	2 (0.2)	4 (0.5)	16 (1.9)	11 (1.3)
USA Clarks Sedentary	847	10 (1.2)	29 (3.5)	1 (0.1)	0 (0.0)	29 (3.4)	4 (0.5)	6 (0.7)	15 (1.8)	12 (1.4)
USA Clarks, Nonsedentary	155	2 (1.3)	8 (5.2)	2 (1.3)	0 (0.0)	10 (6.5)	1 (0.7)	2 (1.3)	1 (0.7)	7 (4.5)
USA, Executives	250	4 (1.6)	15 (6.0)	1 (0.4)	0 (0.0)	3 (1.2)	1 (0.4)	1 (0.4)	2 (0.8)	7 (2.8)
Totals	11184	75 (0.7)	302 (2.7)	13 (0.1)	1 (0.0)	827 (7.4)	91 (0.8)	115 (1.0)	147 (1.3)	148 (1.3)

[a] From Keys et al., Acta Medica Scand 1966; Suppl 460, pp 1–392, Tampereen Kirjapaino, Tampere, Finland.

and T-wave abnormalities from 3.4% to 5.9%. No significant heterogeneity was evident between the cohorts except for T-wave abnormalities in the younger age-group (40–49 years). Rose et al. also concluded that the codable ECG abnormalities did not directly reflect national CHD mortality except the prevalence of major Q, QS waves (MC 1.1, 1.2), and even this association may have been owing to the small number of men with these changes (1.6%).

The World Health Organization (WHO) European Collaborative Group [36] reported contrasting prevalence rates for major Q, QS waves, and ST-T segment abnormalities in industrial populations (mainly factories) in five countries (Belgium, Italy, Poland, Spain, and the UK). The prevalence estimates for the intervention group of this large multi-factorial prevention trial (63,732 men aged 40–59 years, employed in 88 factories) varied significantly between centers in various countries. The prevalence of major Q, QS waves (MC 1.1 and 1.2) ranged from 0.74% in Italy to 1.34% in Poland.

There was no apparent parallelism between the prevalence of major Q, QS waves, and the mean CHD risk for each center estimated using multiple logistic coefficients derived from the Seven Countries study based on age, number of cigarettes smoked per day, systolic blood pressure, plasma or serum cholesterol, and body mass index (BMI). The prevalence of other ECG abnormalities considered to be related to suspect myocardial ischemia (ST depression, MC 4.1 to 4.3, T-wave abnormalities, MC 5.1 to 5.3 or complete bundle branch block, MC 7.1) also varied significantly between centers, with prevalence rates for the UK and Belgian centers more than double those of the Polish and Spanish centers. The authors of the report suggest that these independent variations in the prevalence of Q, QS, and ST-T changes may indicate that populations may differ in the type of CHD as well as the extent, possibly in relation to the severity, chronicity, or age at the onset of ischemia.

The determination of prevalence estimates for ECG abnormalities for clinically normal populations as well as the establishment of normal ECG standards based on unselected samples of the general population can be problematic because of difficulties in setting up criteria for "clinically normal." This is evident, for instance, by considering the problem of clinically unrecognized "silent" MI. On the other hand, results from studies in highly selected populations cannot easily be extrapolated to general populations as pointed out by Barrett et al. [37]. However, some of these studies in selected populations (e.g., aviators) have provided information regarding the significance of ECG abnormalities. This holds true particularly for the natural history studies on conduction defects and other evolutionary investigations. One key question is whether given ECG abnormalities are indicators of latent CHD in asymptomatic individuals such as healthy aviators.

The reported prevalence rates in various occupational groups vary widely, evidently dependent on the selection criteria and differences in ECG classification criteria. For instance, the prevalence of ST-segment abnormality in the resting ECG was 0.08% among 3,983 aviators with a mean age of 27 years [38]. In contrast, the prevalence of ST abnormalities (myocardial ischemia) was 11.7% in a large random sample of Israeli male permanent civil service employees aged 40 years or over [39]. It appears that in the latter study, the ischemic category included negative T waves (−0.1 mV or more negative) and incomplete left bundle branch block [40]. On the other hand, nearly 3% of the aviators in the former study had T-wave abnormalities (primary T-wave changes). Differences in classification criteria make it difficult to compare reported prevalence rates in different occupational groups.

There is increasing evidence that repolarization abnormalities in the resting ECG have significant association with latent CHD in asymptomatic men. For instance, Froelicher et al. have reported coronary angiographic findings in 58 asymptomatic aviators with repolarization abnormalities in their resting ECG [41]. Twenty six (45%) of these men had 50% or greater obstruction in one or more coronary arteries and 55% had some evidence of coronary lesions.

40.2.3 Prevalence of ECG Abnormalities in Ostensibly CHD-Free US Male Populations

The 1978 report of the Pooling Project Research group [15] provides detailed information on the prevalence of codable ECG abnormalities from three studies on samples of working male populations drawn from employment groups and from two studies based on community populations samples (➒ Table 40.5). The report is limited to white males aged 40–59 years. Excluded were all men with definite, probable, or suspect MI, definite angina pectoris by history, or Minnesota code 1.1 or 1.2 Q or QS waves.

Table 40.5
Prevalence (percent) of major and minor ECG abnormalities in five adult populations of US males considered free from coronary heart disease at study baseline

Population	N	No Codable Abnormality	Minor Abnormality[a]	Major Abnormality[b]	Minor of Major Abnormality
Albany	1,765	90.4	7.1	2.5	9.6
Chicago Peoples Gas Co.	1,264	92.2	4.7	3.1	7.8
Chicago Western Electric Co.	1,981	86.4	11.1	2.5	13.6
Framingham	1,375	90.0	5.1	4.9	10.0
Tecumseh	691	73.0	20.7	6.4	27.1
Total	7,076	88.0	8.7	3.5	12.2

[a] Minnesota code 1.3; 4.3; 5.3; 6.3; 3.1; 9.1; 2.1, 2.2.
[b] Minnesota code 5.5, 5.2; 6.1, 6.2; 7.1, 7.2, 7.4; 8.1; 8.3.

The following Minnesota code items were coded as major abnormalities:

MC 4.1	Significant ST depression
MC 5.1 and 5.2	Negative T waves
MC 6.1 and 6.2	Complete and second-degree atrioventricular (AV) block
MC 7.1, 7.2 and 7.4	Complete left bundle branch block, complete right bundle branch block. and intraventricular (IV) block
MC 8.3	Atrial flutter or fibrillation and
MC 8.1	Frequent ventricular ectopic complexes

The prevalence of major ECG abnormalities ranged from 2.5% to 3.1% in the employee groups and 5.0% and 6.4% in the two community samples in ➋ Table 40.5. ST depression, T-wave inversion, and frequent premature ventricular complexes accounted for the majority of major abnormalities, each category having an individual prevalence of approximately 1%.

The wide prevalence difference of minor ECG abnormalities among these five male populations is of interest. The Framingham and Tecumseh study groups represent male populations of total communities. The 5.1% prevalence rate of minor ECG abnormalities in Framingham contrasts with 20.7% prevalence in Tecumseh. A closer examination of the data reported reveals that this highly significant difference in the prevalence is caused by left axis deviations and flat or biphasic T waves in Tecumseh in comparison with Framingham and other populations. The prevalence of code 5.3 (flat T waves) was 11.6% in Tecumseh and 4.2% in Framingham. The prevalence of left axis deviation (code 2.1) was 7.2% in Tecumseh and 0.9% in Framingham. The fact that the ECGs in Tecumseh were recorded during a glucose tolerance test may explain the higher prevalence of minor T-wave changes in that population. There is no obvious reason for the high prevalence of left axis deviation in Tecumseh. Perhaps systematic coding differences may at least in part be responsible.

40.3 Broad Abnormal ECG Categories and Mortality Risk

Early epidemiological studies commonly reported mortality risk for broad categories of ECG abnormalities, mainly because sample size and endpoint events were too low for a meaningful risk evaluation of disease-specific abnormalities.

40.3.1 ECG Abnormalities and Mortality Risk in General Male Populations

Blackburn et al. reported detailed analyses of the 5-year risk of incident CHD for various ECG findings by Minnesota code among men aged 40–59 years [42, 43] The groups compared for risk were matched for age, skinfold thickness, systolic blood pressure, serum cholesterol, smoking, and physical activity. Diagnostic Q waves (code 1.1), negative T waves (code 5.2), and atrial fibrillation (code 8.3) were all highly significant predictors of CHD death. The ratio of observed to expected number of deaths ranged from 10 to 15 for these three categories. Moderately large Q waves associated with T-wave inversion (1.2 and (5.1 or 5.2)) and bundle branch blocks (7.1, 7.2, 7.4) were not significant predictors of CHD death.

In prediction of all CHD events, both fatal and nonfatal, minor T-wave abnormalities (code 5.3), first-degree AV block (code 6.3), and frequent ventricular ectopic complexes (code 8.1) were significant predictors of subsequent clinical CHD in cohorts outside the USA but not in the US cohorts. Any level of ST depression (codes 4.1 to 4.3) carried a significant association with future CHD. Smaller Q waves (code 1.3) and high-amplitude R waves (code 3.1) were not significant independent predictors of CHD; nor was sinus tachycardia (heart rate >100).

Blackburn et al. also reported a more detailed breakdown of the association of ST depression with 5-year mortality from all causes in their pooled European populations (without any exclusion and irrespective of other ECG findings). The results are best summarized by expressing the risk ratio (RR), or the ratio of death rate in the subgroup with a certain ST code to the rate in the subgroup with no codable ST items. The ratios were 10.1, 4.3, 4.0 and 3.3 for codes 4.1, 4.2, 4.3 and 4.4, respectively.

Keys has subsequently reported 10-year follow-up data for a subgroup of the Seven Countries study populations [44]. This subgroup was considered CHD-free at the initial examination. However, nonspecific resting ECG abnormalities in the absence of clinical judgment of definite or possible heart disease did not qualify for exclusion, thus men with any of the following codes were included in the CHD-free subgroup: 1.2, 1.3, 4.1, 4.2, 5.1, 5.2, 6.1, 6.2, 7.1, 7.2, 7.4. and 8.3. The observed-to-expected ratio of the death rate from all causes among men with any of these nonspecific ECG abnormalities ranged from 1.0 to 2.7 in different populations, with the average value of 1.7. Adjusting for the age difference between men with and without ECG abnormalities reduced to the observed-to-expected ratio to 1.48. Thus, the all-causes 10-year death rate among men with nonspecific ECG abnormalities was 48% greater than that for the other men. High-amplitude criteria for LVH showed no predictive significance among the CHD-free cohorts of the Seven Countries study.

Risk of incident CHD for newly acquired ECG-LVH was reported by Kannel et al. in the 14-year follow-up of 5,127 men and women in the Framingham study [45]. Life-table analysis from that study showed that persons who acquired definite LVH by their ECG criteria at any time during the first seven biennial examinations, had a threefold increase in risk of clinical CHD after adjustment for coexisting hypertension. A twofold increase in risk of incident CHD was found among persons with possible LVH by ECG. However, the latter association was no longer significant after adjustment for hypertension.

Criteria for definite LVH in the above-mentioned Framingham study included primarily high voltage associated with ST-segment and T-wave changes or left axis deviation or increased left ventricular activation time. Criteria for possible LVH were based solely on QRS-amplitude criteria without ST-T changes. Thus, in both the Framingham study and the Seven Countries study, high-voltage criteria alone did not carry predictive information independently of other CHD risk factors, in contrast to criteria that included ST-T changes.

40.3.2 Major and Minor ECG Abnormalities and CHD Risk in Industrial Populations

Findings from three longitudinal studies on white, middle-aged, male populations screened from industrial companies and organizations in Chicago provide new information on the importance of major and minor ECG abnormalities for subsequent risk of death from CHD and other causes [46]. Excluded were men who had major Q waves (codes 1.1, 1.2) at the entry examination. Two of these studies, the Chicago Peoples Gas Company study and the Chicago Western Electric Company study are follow-up projects included in the Pooling project described earlier. The third major study reported is the Chicago Heart Association detection project in industry. The combined population of the three projects is 11,204 men aged 40–59 years at baseline, with 5–20 years of follow-up.

40.3.2.1 Major Abnormalities

The following Minnesota codes were included among the major abnormalities: ST depression (codes 4.1, 4.2), T-wave inversion (codes 5.1, 5.2), complete atrioventricular (AV) block (code 6.1), second-degree AV block (code 6.2), complete left bundle branch block (code 7.1), complete right bundle branch block (code 7.2), intraventricular (IV) block (code 7.4), frequent ectopic ventricular beats (code 8.1), and atrial flutter/fibrillation (code 8.3).

Major abnormalities were found in 11.2% among subgroups, which had a full 12-lead ECG recorded. The yield was significantly lower when only 5-lead or 6-lead ECGs were used. The risk ratio for CHD death for men with major ECG abnormalities relative to men with normal ECG was 1.57 for the Chicago Peoples Gas Company, 2.85 for the Western Electric Company, and 6.89 for the men in the Chicago Heart Association detection project in industry. The corresponding risk ratios for death from all causes were 1.64, 1.79, and 3.85. Thus, the death rates for men with major ECG abnormalities were considerably higher than those among men with a normal ECG. ECG abnormalities retained a significant association with death rates when baseline age, diastolic blood pressure, serum cholesterol, relative weight, and number of cigarettes smoked per day were taken into consideration in multivariate analysis.

40.3.2.2 Minor ECG Abnormalities

The following items were included among minor ECG abnormalities: Borderline Q waves (code 1.3), borderline ST depression (code 4.3), flat or biphasic T waves (code 5.3), first-degree AV block (code 6.3), low-voltage QRS (code 9.1), high-amplitude R waves (code 3.1), left axis deviation (code 2.1), and right axis deviation (code 2.2). The risk ratio for CHD death for men with minor ECG abnormalities in relation to men with normal ECG was 1.04 for the Chicago Peoples Gas Company, 1.63 for the Chicago Western Electric Company, and 3.67 for the Chicago Heart Association detection project in industry. The corresponding risk ratios for death from all causes were 1.29, 1.35, and 2.42. Minor ECG abnormalities were shown to carry significant predictive information independently from CHD risk factors in the Chicago Western Electric Company study and the Chicago Heart Association study groups. The association with excess mortality and ECG abnormalities was demonstrated for deaths, which occurred within the first 10 years of follow-up as well as for deaths that occurred more than 10 years after entry.

The overall CHD death risk ratio for men with major or minor ECG abnormalities in relation to men with a normal ECG in the Chicago studies was 1.79 and the risk ratio for death from all causes was 1.57. These risk ratios are similar to those reported by Keys from the Seven Countries study (and the follow-up data from the pooling project [44]). The Pooling Project Research group reported an average risk ratio of 1.7 for a first major or minor ECG abnormality defined as in the Chicago study report summarized above.

40.3.3 Comparative Value of the ECG in Prediction of CHD Risk in Men and Women

The coronary heart disease study of the Finnish Social Insurance Institution [28] provides well-documented information on the relative risk of CHD death and total mortality within 5 years in men and women for various severity levels of Minnesota code abnormalities. The results of the study concerning prognostic value of ECG-MI are summarized later in the section 40.4.1. Gender comparison of more nonspecific categories is summarized here.

The Finnish study consisted of whole or random samples of rural or semi-urban dwellers. The reported relative risk for CHD death and total mortality within 5 years for a subgroup of 3,589 men and 3,470 women aged 40–59 years is of particular interest here. The relative risk was expressed as the age-adjusted risk ratio for subgroups with defined CHD-related abnormalities compared to those without such abnormalities. In judging the practical utility of risk ratios, it is important to consider what fraction of the total population has such abnormalities. It is possible to achieve, by strict classification criteria, very high-risk ratios in subgroups so small that they would be of limited value in screening and prevention because of an insufficient fraction of the total risk in a given population for a preventive effort with an adequate efficacy.

The combined category of other ischemic abnormalities (smaller Q, QS waves, ST changes, significant conduction defects, or atrial fibrillation) had a substantial increase in relative risk in men for CHD death (risk ratio 7.2) and death from all causes (risk ratio 3.1). Nonspecific T-wave changes (MC 5.3, flat or biphasic T) had over threefold excess risk for CHD death and nearly twofold risk ratio for death from all causes.

The results from the Finnish study illustrate the difficulties encountered in the evaluation of the prognostic significance of ECG abnormalities in female populations. Only 16 of the 3,470 women (0.46%) aged 40–59 years had ECG findings compatible with a probable old MI, and none of them died in 5 years of follow-up. The prevalence of ischemic findings in the resting ECG was quite high (12.8%) in women aged 40–59 years. Ten of these 445 women died within 5 years (risk ratio of 2.2 for death from all causes compared to women without CHD-related ECG abnormalities). There were only two CHD deaths in this group. Similarly, there was only one CHD death in the group of

509 women in this age range who had nonspecific T-wave changes. Thus, although there was some indication of a trend toward an increased risk ratio for death from all causes among women with increasing severity level of ECG abnormalities, the predictive power of such ECG changes for CHD deaths appears weak, and the low overall incidence of CHD deaths does not permit meaningful analysis of a possible association between ECG abnormalities and coronary mortality.

Similar problems were encountered in other studies in attempts to compose risk ratios for ECG abnormalities in men and women owing to the relatively low incidence of CHD death in women during the follow-up period involved. The report from the Framingham study [30] compared the risk of CHD (8-year incidence of CHD including angina pectoris, MI, and sudden death) in 2,336 men and 2,873 women with various combinations of Minnesota code ECG abnormalities. This cohort, aged 32–62 years at entry to the study, included only those persons who were considered CHD-free at the initial examination. The relative risk of CHD for a given category of ECG abnormalities was determined by calculating the ratio of the observed and expected number of cases with CHD. For both men and women without codable ECG abnormalities, the risk was 0.9. For 241 men with any Q waves (MC 1.1 to 1.3), ST or T abnormalities (MC 4.1 to 4.3, 5.1 to 5.3), the risk was 1.8, and for 375 women with any of these abnormalities, the risk was 1.1. Thus, there was a twofold risk for men with these abnormalities compared to men without codable abnormalities, and the risk ratio for women was 1.2.

The Copenhagen City heart study [32] evaluated the risk of death from all causes in a random sample of men and women for eight major categories of the Minnesota code. This prognostic evaluation was done for age-groups 50 years or older, and the association with mortality was assessed by comparing the ratio of the observed and expected number of deaths in a given coding category with the ratio in the group with a completely normal ECG. There was a total of 325 deaths among males and 135 among females during the variable follow-up period between February 1, 1976 and March 31, 1980. The log-rank test in the age-group 50–59 years indicated significant association with mortality in men for high-amplitude R waves, bundle branch blocks (MC 7.1, 7.4), and arrhythmias (MC 8.1 to 8.3). In women, aged 50–59 years the expected number of deaths in all these categories was too low to permit a meaningful evaluation of risk. In the older age-groups of women, there was a significant association with mortality for abnormal T waves in the age-group 70–79 years.

40.3.4 Contrasting Prognostic Significance of ECG Abnormalities in Symptomatic and Asymptomatic Men

Rose et al. reported on the 5-year incidence of CHD death according to the initial ECG findings in standard limb-lead ECGs among 18,403 male civil servants aged 60–64 years (➋ Table 40.6) [34]. The men were divided into subgroups according to the presence or absence of symptomatic heart disease, defined at the initial examination as either a positive response to the chest-pain questionnaire (angina or history of possible MI) or being under medical care for heart disease or high blood pressure. Age-adjusted CHD mortality incidence rates were given in the report so that the outcome in men with and without given ECG abnormalities can be compared in spite of differences in age distribution.

Of interest is the overlap between predictive information from different sources. As expected, men who were under medical care, had a history of angina or possible MI and also had an ischemic ECG had the highest mortality (20%) but the prevalence was also low (0.6%), and this category contributed only 7% of all deaths. An ischemic ECG included the following abnormalities: any Q waves, any ST abnormality, negative or flat T waves, or left bundle branch block. The prevalence of an isolated ischemic ECG was 6.5%, and this category contributed 27 of the total of 274 deaths (9.9%), a relatively small fraction. The 5-year age-adjusted CHD mortality for men with ischemic ECG only was 3.3%, slightly higher than the mortality rate for men with chest pain only (2.5%).

For asymptomatic men, only three categories of Minnesota code abnormalities were associated with excess CHD deaths at ≤5% level of statistical significance (in a two-tailed test): Q, QS waves (MC 1.1 to 1.3), T-wave abnormalities (MC 1.1 to 1.3), and left bundle branch block (MC 7.1). The 5-year CHD mortality rate was in general low for asymptomatic men. It exceeded 5% only for men with prominent Q waves (MC 1.1 to 1.2) and atrial fibrillation (MC 8.3). For the total group of asymptomatic men, the 5-year CHD mortality was 1%. It was 2% for asymptomatic men with left axis deviation (MC 2.1) and 2% for men with ventricular conduction defects (MC 7.1, 7.3, 7.4). There was no excess CHD mortality among asymptomatic men with prolonged PR interval (MC 6.3).

Table 40.6
Five year coronary heart disease mortality among 18,228 men according to mutually exclusive combinations of baseline findings

Status Combinations	N (%)	Number of CHD Deaths	Age-adjusted Percentage
Chest pain[a] only	1,453 (8.0)	38	2.5
Ischemic ECG[b] only	791 (6.5)	27	3.3
Under medical care, no chest pain or ischemic ECG	313 (1.7)	16	4.5
Chest pain and under medical care, no ischemic ECG	201 (1.1)	17	6.0
Ischemic ECG and under medical care, no chest pain	98 (0.5)	11	7.6
Ischemic ECG and chest pain, not under medical care	147 (0.8)	22	10.2
Ischemic ECG and chest pain and under medical care	107 (0.6)	20	20.2
Remainder of men	15,118 (82.9)	123	0.84

[a]Positive response to questionnaire (angina or history of possible myocardial infarction.
[b]Minnesota codes 1.1–1.3; 4.1–4.4; 5.1–5.3; 7.1.
Recomposed from Fig. 40.2, Rose et al., Ref. [34], British Heart J 1978;40:636–643, © 1978 BMJ Publishing Group, amended with permission.

The Israel ischemic heart disease project [39, 40] introduced new information on the value of nonspecific ECG abnormalities in predicting future symptomatic and asymptomatic ischemic heart disease. The results were based on a 5-year follow-up of a random sample of 10,000 Israeli male permanent civil service employees aged 40 years and over in 1963.

Two categories of resting ECG abnormalities were associated with a significant increase in the 5-year incidence of angina pectoris: nonspecific T-wave abnormalities, and probable infarct by computer criteria when two electrocardiographers disagreed with the computer interpretation! The prevalence of these nonspecific T waves by computer was close to 6% of the Israeli study population chosen for the follow-up, and the prevalence of probable MI was 3.3%.

40.3.5 Contrasting Racial Differences in Prognostic Significance of ECG Abnormalities

Substantial differences have been reported in many studies in the prevalence of ECG abnormalities between black and white populations. The increased prevalence of ST-T abnormalities in black males compared to white males has been known a long time [33, 47, 48].

The Evans County study gave a comprehensive description of the rather drastic differences in the prevalence of Minnesota code ECG abnormalities in black and white males and females, and also differences in the prognostic significance of various ECG abnormalities [49]. The Evans County study population screened between 1960 and 1962 consisted of 3,009 persons 15 years or older who were considered free of evidence of past or present CHD at the time of the initial examination. Also excluded from the CHD incidence comparisons were individuals with diagnosed hypertensive heart disease in the initial examination. The reported age-adjusted prevalence of abnormal ECGs (any specified abnormality) was considerably higher in blacks than in whites. The excess prevalence of ECG abnormalities in black men and women was largely due to high prevalence of high-amplitude R waves and T-wave abnormalities (both major and minor).

In sharp contrast to the highly significant increased prevalence of ECG abnormalities in black men, the 1971 follow-up report showed a significantly lower CHD incidence associated with an abnormal ECG in black males. Small sizes of the subgroups and small number of CHD events does not permit a meaningful comparison in specific categories of abnormalities in men and particularly in women. The comparability of the prognostic significance of ECG abnormalities from this study is also limited by the age structure of the study population. Approximately 39% of the black makes, 22.2% of the white males, and about 43% of the black and white females were from the age-group 15–45 years.

A more recent 20-year mortality follow-up report from the Evans County heart study clarified the role of the racial differences regarding the prognostic value of ECG abnormalities in this cohort. That newer report was restricted to black and white men who were aged 40–64 years of age at entry. Men with major or moderate Q waves (MC 1.1, 1.2) and men with a history of angina pectoris or myocardial infarction at the initial examination were excluded.

Baseline ECG abnormalities were categorized into major and minor abnormalities using a similar grouping of the Minnesota codes as was done in the pooling project cited above [15]. The association between baseline ECG abnormalities and time to death was examined using the Cox proportional hazards model to adjust for standard CHD risk factors. Relative risks were estimated for CHD, cardiovascular disease (CVD), and all cause mortality. They were not significantly increased for black or for white men with respect to minor ECG abnormalities, which included minor Q waves (MC 1.3) and high-amplitude R waves (MC 3.1).

For major ECG abnormalities, the unadjusted relative risk of CHD, CVD, and all-cause mortality was significantly increased in both race groups. The relative risks of CVD and all-cause mortality remained significant in both races after adjustment for age at entry, systolic blood pressure, cholesterol, current and past smoking, and body mass index (BMI). The adjusted relative risk was over twofold for CVD mortality (2.3 for black men and 2.2 for white men, with 95% CI 1.1–4.5 and 1.2–4.2, respectively) and the relative risk of CHD mortality was nearly as high although no longer significant after adjustment for standard CHD risk factors. It thus appears that the adjusted relative risks of CVD mortality for major ECG abnormalities were of the same order of magnitude in black and white men.

40.4 ECG-MI and Mortality Risk

CHD prevalence is known to differ drastically in populations from different geographic locations, as was initially demonstrated by the Seven Countries study. The massive WHO MONICA project used carefully standardized procedures for data entered into population registers by MONICA centers selected for the study for documenting death rates in men and women aged 35–64 years [50]. The survey was conducted during the 1985–1987 period in 38 selected communities from 21 countries. The data revealed a 12-fold range in age-standardized annual event rates (coronary deaths and definite MI) in men and an 8.5-fold range in women. Event rates for men were highest in North Karelia, Finland (907/100,000) and lowest in Beijing, China (73/100,000). In women, event rates were highest in Glasgow, UK (241/100,000) and the rates were lowest in Toulouse, France (24/100,000) and Catalonia, Spain (25/100,000).

40.4.1 Mortality Risk in Q Wave Myocardial Infarction

Data from three well-documented studies reporting mortality risk for Q wave MI are listed in ➋ Table 40.7. In a CHD-free cohort of the Seven Countries study, there was a drastic, over 15-fold increase in short-term mortality risk (up to 5 years) for diagnostic Q waves (large Q waves or major Q waves with ST-T abnormalities) and an over fourfold increased risk for the combined category of any Q waves with or without ST-T abnormalities [51, 52]. The long-term (6–25 years) mortality risk was increased over threefold for diagnostic Q waves but was not significant for the combined category of any Q waves.

In Finland's Social Insurance Institute's study [28], the mortality risk was increased in men for diagnostic Q wave MI (defined similarly as in the Seven Countries study) for all mortality endpoints: RR was 19.5 for CHD mortality, 13.3 for CVD mortality, and 5.6 for all-cause mortality. There were too few events in women for risk evaluation in this ECG category. The risk for Minnesota code 1.1 large Q waves in men was 13.4, and for smaller Q waves in codes 1.2, 1.3 approximately threefold compared to men without significant Q waves. The prevalence of diagnostic Q wave myocardial infarction was quite low. The prevalence of the category labeled as "other ischemic abnormalities" was considerably higher, and also the risk of CHD, CVD, and all-cause mortality was markedly increased in all men, in men with an apparent silent MI, and also in women for CVD and all-cause mortality.

The Italian RIFLE Pooling project [53] reported mortality risk for large group of CHD-free men ($n = 12,180$) and women ($n = 10,373$) aged 30–69 years. CHD mortality risk for combined category of any Minnesota code Q waves was not significant in men or in women. There was a profound, over 17-fold increased risk of CHD mortality in men with major Q waves combined with major T waves. Again, there were too few events in women for risk evaluation.

As seen from ➋ Table 40.7, the low prevalence of diagnostic Q wave codes in all three of the populations is too low for population screening for preventive efforts, in spite of the extremely high mortality risk associated with them. However, when identified in any other connection, they warrant consideration for intensified secondary prevention. The Italian study suggests that the prevalence and the mortality risk for the combined category of ischemic abnormalities is at least potentially high enough for identification of subgroups for intervention.

Table 40.7
Three population studies with evaluation of the risk for ECG manifestations of an old myocardial infarction (MI)

Study/Ref.	Seven Countries Study [15, 16]	Finland's Social Insurance Institution's CHD study [17]	Italian Rifle Pooling Project [18]
Geographic Location	Finland, Greece, Italy, the Netherlands, former Yugoslavia, Japan, the USA	South-western, western, central, and eastern districts of the country	Eight regions in Italy.
Source population	12,763 men (11 cohorts from rural areas of Finland, Greece, Japan, and Yugoslavia; three other more diverse cohorts from Yugoslavia; the town of Zutphen, the Netherlands; and cohorts of railroad employees from the USA and Italy.	Men and women of the districts. Participation rate 90% (5,738 men and 5,224 women).	Reported four out of nine studies, 23 cohorts total. Participation rate 65–70%.
Baseline	1958–1964	1966–1972	1978–1987
Exclusions	Sampling goal: total male population from selected geographic areas. Participation rate ≥60%. For CHD-free sample, excluded 854 with prevalent CHD at baseline; no ECG exclusions.	All participants included; risk evaluated separately for subjects with ischemic ECG free from symptoms typical of MI.	History of angina pectoris by Rose questionnaire, hospitalization with discharge diagnosis of MI, history of other heart disease.
CHD-free sample	11,860		12,180 men, 10,373 women.
Age range	40–59 years	30–59 years	30–69 years.
Follow-up	Initial 5-years and subsequent 6–25 years	5 years	6 years
Q wave MI-related ECG findings evaluated for risk	1. Any Q wave codes, with or without other ECG abnormalities (MC 1.1–1.3) 2. Diagnostic Q wave MI (MC 1.1 OR (MC 1.2 and MC 5.1–5.2)).	1. Diagnostic Q wave MI (MC 1.1 OR (MC 1.2 and MC 5.1–5.2). 2. Other "ischemic" categories (MC 1.2–1.3, 4.1–4.3, 5.1–5.3, 6.1–6.2, 7.1, 7.2, 7.4, 8.3))	Any Q wave codes (MC 1.1–1.3) Any Q wave codes with any T wave codes (5.1–5.3) 3. Q codes 1.1–1.2 with T codes 5.1, 5.2

Table 40.7 (Continued)

Study/Ref.	Seven Countries Study [15, 16]	Finland's Social Insurance Institution's CHD study [17]	Italian Rifle Pooling Project [18]
Baseline prevalence	MC 1.1–1.3 (all men combined): 2.7% CHD-free men: 2.1%	1. Diagnostic Q wave MI Men 40–59 years: 1.3% Women 40–59 years: 0.5% Other "ischemic" ECG Men 40–59 years: 10.4% Women 40–59 years: 12.8%	Men Any Q codes 30–49 years: 0.6% 50–69 years: 1.1% Women Any Q codes 30–49 years: 0.5%; 50–59 years: 1.2%
Follow-up (Years)	25	5	6
Associated risk (hazard ratio HR)	Short-term (Years 0–5), Risk for CHD death (CHD-free men) Any Q: HR 4.1 (2.20, 7.61) Large Q or major Q with ST-T: HR 15.6 (7.2, 34.1) (not significant if large Q category excluded) Long-term (Years 6–25), Risk for CHD Death 1. Any Q: HR 1.2 (0.84, 1.58) 2. Large Q or major Q with ST-T: HR 3.4 (1.9, 6.2)	1. Diagnostic Q wave MI (5-year, age-standardized risk for age-group 40–49) Men: CHD mortality RR = 19.5 CVD mortality RR = 13.3 All-cause mortality RR = 5.6 Women: too few events for risk evaluation 2. "Other ischemic" ECG Men: CHD mortality RR = 7.2 CVD mortality RR = 6.4 All-cause mortality RR = 3.1 Women: CVD mortality RR = 3.0; All-cause mortality 2.2 Symptom-free men with "other ischemic" ECG CHD mortality, RR = 4.4 CVD mortality, RR = 3.9	CHD mortality Any Q waves: not significant in men or in women Any Q waves with any T waves: not significant for men or for women Major Q waves with major T waves: Men: RR 17.2 (1.54, 1.92) Women: Too few events for RR calculations

40.4.2 Other Studies with Mortality Risk Assessed for Major and Minor Q Waves

Mortality risk for various categories of Q waves has been evaluated in many other diverse populations. The Whitehall study on 18,403 male civil servants in the age-group 40–60 years conducted in early 1970s evaluated the 5-year CHD mortality risk for Q waves coded from limb-lead ECGs [34]. The study population contained a subgroup of 15,974 non-symptomatic men with no history of angina pectoris or MI and who were not under medical care for heart disease or hypertension. The prevalence of codes 1.1–1.3 was 2.0% in the whole group and 1.6% in the non-symptomatic group. Approximately two thirds of the Q waves were minor, Minnesota code 1.3. The age-adjusted CHD mortality ratio for men with any Q waves was 6.1 with all men in the study as the reference. In the non-symptomatic group, the mortality ratio was 4.0.

The Busselton study on 2,119 unselected subjects reported that the 13-year standardized CVD mortality rate was significantly higher in the pooled group of men and women with Minnesota code 1.1–1.3 Q waves than among those with a normal ECG [54].

The cross-sectional data from the Copenhagen City Heart Study involved a random sample of 9,384 men and 10,314 women aged 20 years and older [32]. In the age-group 60–69 years, 92 men (5.9%) and 38 women (2.4%) had Minnesota code 1.1–1.3 Q waves. Among 25 men who died, 21 had major Q waves (Minnesota code 1.1, 1.2), a highly significant difference between the observed and expected number of deaths compared to men without Q waves. There was only one death among women with Q waves.

40.4.3 Unrecognized Compared to Recognized Myocardial Infarction

Three studies listed in ➋ Table 40.8 evaluated mortality risk for recognized versus unrecognized MI: the Reykjavik study, the Framingham study, and the Honolulu study. Unrecognized MI was defined in these studies as either totally asymptomatic or associated with symptoms atypical for acute MI.

40.4.3.1 The Reykjavik Study

This study in a large male cohort (N = 9,141) was performed in five stages, 3–5 years apart, with the last stage conducted in 1983–1987 [55]. The overall prevalence of unrecognized MI at the first (0.5%) stage of the study increased sharply with age, from 0.5% at age 50 years to over 5% at age 75 years. The prevalence also increased at the later phases of the study, to 2.8% in 1987. About 30% of all MIs in the pooled data from all phases of the study were unrecognized. In logistic regression analysis, angina, age, smoking, serum cholesterol level, cardiomegaly, and diuretic therapy were associated with both types of MI. Impaired glucose tolerance and digoxin therapy were significantly associated only with recognized MI. Factors with predictive power for future unrecognized and recognized MI from Poisson regression included age, diastolic blood pressure, and hypertension medication. Digoxin therapy was significantly predictive for future unrecognized MI. Current smoking had a more consistent association with future recognized MI than with unrecognized MI.

Relative risk of CHD mortality for MI without angina pectoris was 4.6 (2.4, 8.6) for unrecognized MI and 6.3 (3.7, 10.6) for recognized MI. For all-cause mortality, the corresponding relative risks were 2.7 (1.5, 4.8) for unrecognized MI and 2.9 (1.8, 4.6) for recognized MI. The survival probabilities for both types of MI from life table analysis were relatively similar for subjects with and without MI in the first four stages of the study. Ten-year survival probabilities for unrecognized and recognized MI were 49% and 45%, respectively, and 15-year survival probabilities 62% and 48%, respectively.

40.4.3.2 Framingham Study

In a 1984 report from the Framingham study population, subjects with unrecognized MI were as likely as those with recognized MI to be at increased risk of death, heart failure, or strokes [56]. In subjects initially free from CHD, three successive 10-year periods since the start of the Framingham study in 1948 were used as the baseline for comparing the risk for the two types of MI. Of all baseline MIs in 5,127 subjects at the initial examination, 130 of 469 (27.7%) among

Table 40.8
Mortality risk for recognized versus unrecognized myocardial infarction

Study/Ref.	The Reykjavik Study [21]	Framingham [22]	Hawaii Heart Program [24]
Geographic Location	Reykjavik area, Iceland	Framingham, Massachusetts	Oahu Island, Hawaii
Source population	Male residents born between 1907 and 1934. Response rate 64–75%.	Town's representative adult population, 2,336 men, 2,873 women	Men of Japanese ancestry born between 1900 and 1919. Participation rate 73%.
Baseline	1967–1968	Onset of each of the three successive 10-year periods starting 1948	1965–1968
Selection criteria	Whole cohort included	Free from overt CHD at baseline, excluding from next follow-up subjects with MI during the preceding period	CHD-free at baseline, with at least one follow-up examination
Sample size	9,141	5,127 men and women	7,331
Age range	33–80 years	35–62 years	45–68 years
Follow-up period	4–24 years (in four stages, at 3–5 year intervals)	30 years	10 years from the examination where incident MI was detected
Definition of MI	1. Recognized MI: MONICA criteria (Patients with MC 1.1,1.2 Q waves, or with MI by enzymes, chest pain, borderline Q waves) 2. Unrecognized MI: MC 1.1, 1.2 with no history of MI	Appearance of pathologic (≥40 ms) Q waves or loss of initial R waves Unrecognized MI: Incident MI when neither the patient nor the attending physician had considered heart attack.	1. Serial ECG changes at follow-up considered diagnostic for old or age-undetermined MI; 2. Hospitalization with acute chest pain with evolving ECG changes and/or elevated enzymes 3. Serial ECG changes at hospital surveillance considered as old or age-undetermined MI
Prevalence/incidence of MI	Prevalence of unrecognized MI 0.5% at first stage, increased at later stages to 2.8% in 1980. Prevalence increased sharply with age, from 0.5% at age of 50 years to over 5% at age of 75 years. About 30% of all MI unrecognized (in pooled data from all phases).	Incidence of all MIs (during three 10-year periods): Increased in men from 7.1%, in women from 1.3% at age 45–54 years to 12.8% at age 75–84 years. Over one quarter of all MIs in men and over one third in women were unrecognized.	Average annual rate/1,000 of recognized MI (in category 1 above) 1.46, unrecognized MI 0.71, with significant age trend in both. Unrecognized MIs 32.6%.

Associated Risk	CHD death, MI without angina: RR = 4.6 (2.4, 8.6) for unrecognized MI, RR = 6.3 (3.7, 10.6) for recognized MI CHD death, MI with angina: RR = 16.9 (9.4, 30.3) for unrecognized MI, RR = 8.5 (5.8, 12.6) for recognized MI	Ten-year age-adjusted mortality risk ratio (unrecognized MI versus recognized MI) from proportional hazards model Men CHD death 1.0 CVD death 1.2 Sudden coronary death 0.6 All deaths 1.2 Women CHD death 0.6 CVD death 0.5 Sudden coronary death 0.8 All deaths 0.7	Ten-year CHD mortality 35% versus 25% ($p = 0.107$), CVD mortality 39% versus28% ($p = 0.045$), total mortality 45% versus 35% ($p = 0.045$), for unrecognized versus recognized MI, respectively.
Multivariate adjustment	Not clearly stated. Age, cholesterol level, antihypertensive medication, diastolic pressure, and smoking were associated with risk of unrecognized MI	Age-adjusted, controlling also for the effect of the loss of subjects to follow-up because of death from unrelated causes.	Unadjusted (age distribution similar in both groups).
Comments	Ten-year survival probabilities for CHD death: 49% for unrecognized MI, 62% for recognized MI; 15 years: 45% for unrecognized MI, 48% for symptomatic MI.	$p < 0.05$; not significant difference for other recognized versus non-recognized MI	

men and 83 of 239 (34.7%) among women were unrecognized, almost half without any symptoms. A similar proportion as mentioned above was unrecognized among incident MI during the three 10-year follow-up periods. ➔ Table 40.8 shows that the cause-specific mortality and all-cause mortality in men was similar for unrecognized and recognized MI. In women, the mortality was lower for unrecognized compared to recognized MI, and the difference was significant ($p < 0.05$) for CVD mortality.

A 1986 report from the Framingham study compared the relative 10-year risk of clinical CHD for unrecognized, asymptomatic ECG-MI and ECG-LVH [57]. The evaluation group was CHD-free at the baseline and new asymptomatic ECG-MI and ECG-LVH was detected as a 2-year incidence before the start of the follow-up. There was a profound two- to fourfold increase in CHD mortality, particularly sudden death, for both ECG abnormalities, with similar rates for both. Of interest was also the finding that ECG-LVH carried a significantly greater risk than ECG-MI for CHD death in women.

40.4.3.3 Honolulu Heart Program

The prognosis of recognized and unrecognized MI was also evaluated in the Honolulu Heart Program in a 10-year follow-up among 7,331 men who were CHD-free at baseline examination and who had serial ECG changes classified as incident Q-wave MI at the second or third examinations (2 and 6 years following the baseline) [58]. There was a total of 89 Q-wave MIs classified from serial ECG changes, 33% of them asymptomatic. Among men who were classified as MI from hospital surveillance of the study, the proportion of silent MIs among all nonfatal MIs was 22%. (The average annual incidence rate (per 1,000) for all nonfatal MIs was 3.29). The unrecognized MI group had a consistently higher (60–70%) total CVD and CHD mortality than the group with recognized MI. The number of events in groups compared in the Honolulu study was small and the difference in risk was not significant. However, consistent with the other studies mentioned above, the results suggest that the mortality risk for silent MI is at least as high as for recognized, symptomatic MI.

40.4.3.4 Other Studies Comparing Recognized Versus Unrecognized MI

A Finnish cohort of 697 men aged 65–84 years of the Seven Countries study, who had survived at the time of the 23-year follow-up examination, was followed up for the next 5-year period [59]. At the time of the beginning of the follow-up, 98 of the men (14.1%) had Minnesota code 1.1–1.3 Q waves. Q waves combined with ST depression (codes 4.1–4.3) or negative T waves (codes 5.1 or 5.2) were significantly associated with excess risk of fatal and nonfatal MI and total mortality. Isolated Q waves were not associated with independent risk with any of the endpoints.

The Bronx Aging Study assessed the prognosis of recognized and unrecognized MI in 390 elderly (75–85 years) community-based men and women in an 8-year prospective evaluation [60]. In this older group, baseline prevalence of MI was 18.5%, over one third (34.7%) of them unrecognized (completely silent MI and MI with atypical symptoms were included in this category). During the follow-up, the proportion of unrecognized MIs among all incident or recurrent MIs was 44%. The total mortality rate in subjects with recognized or unrecognized MI was 5.9/100 person–years compared with 3.9/100 person–years in the group without MI ($p = 0.059$). The sample size is small and the event rates in study subgroups are low, but the mortality rates were similar among those with recognized and unrecognized MI. The report reviewed other studies on unrecognized MI and noted that in newer reports (often from prospective studies) the average proportion of unrecognized MI was 30%.

In the older Israel ischemic heart disease project [61], the mortality follow-up was conducted 5 years from the initial examination in 1963. The second clinical examination took place in 1965 and the third in 1968. The occurrence of the MI was assumed to have taken place halfway between the two closest examinations before and after the abnormality developed. The average annual mortality rate in the group of men without any signs of MI during the follow-up period was 4.6/1,000.

There was a total of 170 men with clinically unrecognized MI during the follow-up and the annual mortality rate in this group of men was 17.3/1,000, or 3.8 times the mortality rate in men without any sign of MI. One half of all unrecognized MIs were silent and classified strictly on the basis of ECG findings in the 1965 or 1968 reexamination. The other half of clinically unrecognized MIs was not asymptomatic but their complaints were considered atypical for MI. The annual mortality rate of 17.3/1,000 for men with unrecognized MI can be compared with the rate of 36.3/1,000 for the group of 120 men with clinically recognized MI.

The incident (or recurrent) MI rate, particularly that of unrecognized MI, was high among men with possible or probable MI at baseline according to computer criteria but who were considered as non-MI by electrocardiographers reviewing the computer interpretation. A sizable fraction of these men may have had latent CHD already at the onset of the study.

No serial ECG change classification criteria were evidently used in the study and it is possible that relatively minor changes from the baseline ECG were more likely to cause a classification as a new unrecognized MI in this subgroup. In any case, ECG signs of unrecognized MI seemed to be associated with excess mortality, with a prognosis perhaps half as serious as that for clinically recognized MI. It is noted that in the Israel study, 62% of the unrecognized MIs were categorized as possible old anterior MI, defined by the presence of a Q wave in V_2 or V_3 or R wave ≤ 100 mV in any two of the leads V_2–V_5. Multivariate analyses also indicated that the increased incidence of unrecognized MIs was associated with age, cigarette smoking, blood pressure (systolic and diastolic), left axis deviation, and LVH on the ECG. Poor progression of R waves in anterior chest leads is often seen in hypertensives with LVH, and it is possible that a significant proportion of these unrecognized MIs were false positives because of the limited specificity of the criteria used.

40.4.4 ECG Risk Predictors in Heart Attack Survivors

The Coronary Drug project was the first large-scale clinical trial that reported on systematic analyses of the prognostic value of ECG abnormalities among survivors from a first heart attack during a 3-year period of a longer follow-up [62]. Baseline ECGs, including the placebo group, were recorded at least 3 months after the acute phase.

ST depression had the strongest association of any ECG items with excess mortality. The risk ratio for major ST depression (>0.1 mV J depression with horizontal or downsloping ST segment) relative to normal ST was 4.0, and any degree of ST depression, except J-point depression with upsloping ST had a significant association with excess mortality. Actually, any degree of ST depression, except J-point depression with upsloping ST, was associated with a significant excess mortality. T waves more negative than −0.1 mV were also associated with over twofold excess mortality, and there was also a significant excess among men with minor T-wave abnormalities (flat or biphasic T waves).

Of particular importance was the observation that resting ECG ST depression was related to excess mortality also among men with no history of heart failure and among men not taking digitalis.

Extensive multivariate analyses were performed using Coronary Drug project data to elucidate possible independent prognostic value of ECG findings after simultaneous adjustment for all major known CHD risk factors and clinical risk indicators. Minnesota code items that contain independent prognostic information included the following categories: Q, QS waves (any code 1), ST depression (codes 4.1 to 4.4), ventricular conduction defects (e.g. codes 7.1 and 7.4) except complete right bundle branch block (code 7.2), atrial flutter/fibrillation (code 8.3), and any ventricular premature beats (code 8.2).

ST depression in the resting ECG turned out to be the strongest single ECG predictor of mortality and as strong an independent risk predictor as cardiac enlargement and functional class. This is clinically important because codable ST changes are common among survivors of a first heart attack (25% of men in the Coronary Drug project had codable ST changes) and a considerable proportion of all deaths occur among this subgroup.

40.4.5 Heart Attack Prevention Programs and Prognostic Value of Rest and Exercise ECG Abnormalities

Numerous studies cited in previous sections have demonstrated the adverse prognostic value of several resting ECG abnormalities. Relatively little is known, however, on how the presence of these abnormalities may influence heart attack prevention efforts, for instance, whether persons with defined ECG abnormalities might respond particularly favorably to intervention. The MRFIT project was designed to test the effect of a multifactor intervention program on mortality from CHD in 12,866 high-risk men aged 35–57 years. The two a priori ECG-based subgroup hypotheses were formulated with the expectation that intervention would be especially beneficial in men with a normal resting ECG and a normal exercise

ECG, respectively. No significant difference in CHD mortality was found between the special intervention and usual care (UC) groups of the trial [63]. The findings from the MRFIT study even brought up the possibility that hypertensive high-risk men with resting ECG abnormalities may have experienced increased CHD mortality in the special intervention group of the trial, an observation that created a great deal of interest because of its potential implications on diuretic therapy [64]. Similar adverse trends in coronary events were reported from the Oslo study trial on mild hypertension although the number of events involved was too low to reach nominal statistical significance [65].

The Hypertension Detection and Follow-up Program (HDFP) [66] compared CHD and total mortality in a subgroup of mild hypertensives (diastolic pressure 90–104 mm Hg) with resting ECG abnormalities similar to the MRFIT study population. CHD mortality for persons receiving systematic antihypertensive therapy in special program centers (stepped care group) was compared to that for persons referred to existing community medical care (referred care group.) The CHD mortality rate was slightly but not significantly higher in the stepped care group than in the referred care group in white men and black women, but not in black men. However, the rates for all-cause mortality were consistently lower in the stepped care group than in the referred care group. This 24% difference in favor of the stepped care group did not reach a nominal level of statistical significance because the subgroups with ECG abnormalities of HDFP cohort were small, again demonstrating problems in reaching adequate statistical power even in very large clinical trials.

In primary prevention trials, the emphasis is in demonstrating the effectiveness of risk-factor intervention in subgroups without ECG abnormalities. This is based on the assumption that intervention should be most beneficial in an early, asymptomatic phase of the disease process before end-organ damage develops.

As a contrast to the unexpected negative overall outcome of the intervention among men with a normal ECG response to exercise in MRFIT, men with an abnormal exercise ECG seemed to benefit substantially from risk-factor reduction [67]. There was a 57% lower rate of CHD deaths among men in the special intervention group with an abnormal ECG response to exercise compared with men in the usual care group (22.2 versus 51.8/1,000). Abnormal ECG response to exercise was defined as an ST-depression integral measured by a computer as 16 μVs or more in peak exercise or immediate recovery records in any of leads CS5, aVL, aVF, or V_5, with the ST-depression integral being less than 6 μVs in the above leads in the pre-exercise sitting record. These abnormal responses to a submaximal heart-rate-limited treadmill exercise test represented mainly early repolarization abnormalities with upsloping ST in lead CS5 during peak exercise. At the baseline of the study, 12.2% of the usual care men of the study had an abnormal exercise ECG. There was a nearly fourfold increase in 7-year coronary mortality among men with an abnormal response to exercise compared with men with a normal response [68].

Changing trends in CHD mortality can cause formidable problems for primary prevention trials. Unexpectedly, low CHD and total mortality particularly among subgroups or high-risk men selected on the basis of normal resting and exercise ECG can become problematic. In MRFIT, the CHD mortality rate was only about 2 per 1,000 per year in the large subgroup (61%) of these high-risk men with a normal resting and exercise ECG. It is difficult to demonstrate a significant reduction of CHD deaths below this level because a very large sample size is required for attaining adequate statistical power.

MRFIT mortality follow-up data were later extended to cover a 10.5-year follow-up period [69]. Ischemic response to exercise remained the only abnormality in the UC men with a significant association with CHD mortality, confirming the results from the initial 7-year mortality data. In the SI group, ST-T abnormalities at rest, absent or low-amplitude U waves and an abnormal cardiac infarction injury score (CIIS) [20] were all significantly associated with 10.5-year CHD mortality. However, absent or low-amplitude U waves were the only ECG abnormalities with a significant difference between the special intervention and usual care men in the relative risk estimates ($p = 0.004$).

In the Belgian heart disease prevention project [70], there was no significant difference in CHD incidence between intervention and control groups among men with a normal resting ECG. However, reduction of CHD risk factors among men with ischemic changes in their resting ECG by the Minnesota code criteria was associated with a significant reduction in 6-year CHD incidence and total mortality.

40.4.6 Time Trends: Are Risk Evaluation Data from Older Studies Still Valid?

To what extent are risk evaluation results from the older studies still valid in view of the reported decline in-hospital MI mortality rate? For instance, a survey of patients hospitalized for acute Q-wave MI in Worcester, Massachusetts, metropolitan area hospitals, compared two periods 1 decade apart (1995–1997 versus 1986–1988) [71]. The in-hospital case fatality rate had declined from 19% to 14%. Controlled clinical trials have demonstrated reduced mortality with more common use of primary angioplasty and improved treatment of acute MI with more widespread use of coronary reperfusion and antiplatelet therapy.

It is apparent that the short-term risk of acute MI patients has improved at least in industrialized countries that have benefited from improved acute care. The question remains to what extent the long-term prognosis has improved. Already before the introduction of the major improvements in the care of CHD patients, factors other than ECG evidence of old MI seemed to determine the long-term outcome. Unrecognized MI may be less likely to benefit from improved care until a later phase of the evolution of the disease. The proportion of unrecognized MI has been approximately one third of all MIs in the studies cited above, including the Western Collaborative Group Study [72].

It takes a prolonged period of time and a large sample size to produce results from long-term studies, and by the time the results come in, the question of obsolescence often arises. The popularity of the traditional observational population studies has declined, in part because of funding problems and because clinical drug trials have taken a higher priority for funding as well as for the acceptance of manuscripts for publication in high-impact medical journals.

40.5 ECG-LVH: A Spectrum of Connotations

40.5.1 Age Trends and Ethnic Differences in ECG-LVH

The Copenhagen City Heart Study [32] conducted from 1976 to 1978, reported in 1981, combined Minnesota code 3.1, 3.3 prevalence data of 6,505 men and 7,713 women. The data showed the well-known drop in LVH prevalence in young adult men until age 40–49 years. There was little subsequent variation with age in men but in women, there is a steady increase in ECG-LVH prevalence by these criteria after age 40–49 years.

In US populations, there are rather striking differences in age trends of ECG-LVH by Cornell voltage and Sokolow-Lyon voltage criteria [73]. The Cornell voltage increases by age in men and in women (➋ Figs. 40.1 and ➋ 40.2). Cornell voltage patterns are relatively similar in Hispanic and white men and women, and the amplitudes are drastically higher in African-American men and women. In contrast to Cornell voltage, Sokolow-Lyon voltage decreases with age in all three ethnic groups except African-American women (➋ Figs. 40.3 and ➋ 40.4).

Many studies have reported a higher ECG-LVH prevalence in blacks compared to whites. In the Evans County study, LVH prevalence by Sokolow-Lyon criteria was threefold in blacks compared to whites, and also the ECG estimate of LV mass was significantly higher [49, 74, 75]. By Sokolow-Lyon criteria, LVH prevalence was over fourfold in the Charleston study [76]. In the Chicago Heart Study, LVH was defined by Minnesota code 3.1 criteria combined with repolarization abnormalities (MC 4.1–4.3 or MC 5.1–5.3) [77]. Although the overall LVH prevalence by this combination was substantially lower, the black and white differences were still pronounced in all age-groups from 20 to 64 years. ECG-LVH prevalence in Nigerian civil servants by Minnesota code 3.1–3.3 criteria was reported as 36.3% in men and 16.9% in women [78].

40.5.2 Time Trends in ECG-LVH Prevalence

A notable decline was found in ECG-LVH prevalence from 1950 to 1989 in the predominantly white combined original and offspring cohorts of the Framingham study [79]. ECG-LVH by combined high R and abnormal ST criteria had decreased from 4.5% to 2.5% in men and from 3.6 to 1.1% in women. The mean age-adjusted Cornell voltage amplitude had declined 80 μV per decade in men ($p = 0.03$) and 60 μV per decade in women ($p = 0.06$). Similar, profound decline has been observed in other US populations, including data from the National Health and Nutrition Surveys [80]. Although increasing use and improved effectiveness of antihypertensive medications parallel this decrease in ECG-LVH prevalence, questions remain about the reliability of ECG-LVH criteria in general and possible confounding factors.

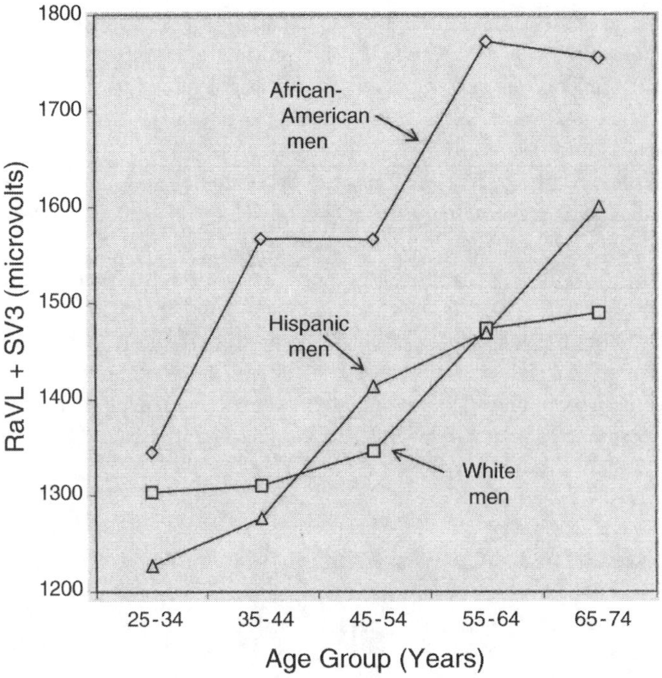

Figure 40.1
Cornell voltage (RaVL + SV3) by age in white (*squares*), Hispanic (*triangles*), and African–American men (*diamonds*). Note consistent increasing trend with age in all three ethnic groups and the substantially higher mean values in African–American compared to white and Hispanic men. Data modified from ref. [73], Rautaharju et al., J Electrocardiol 1994; 27(suppl): [20–30]. © 1994 Churchill Livingstone, reproduced with permission.

40.5.3 Echo-LVH Versus ECG-LVH: Gender and Racial Differences

In the Treatment of Mild Hypertension Study (TOMHS) [81], Echo-LVH (LV mass index ≥134 g/m^2 for men and ≥110 g/m^2 for women) was present in 13% of men and in 20% of women. ECG-LVH was reported to be "virtually absent" by Minnesota code 3.1 criteria combined with abnormal repolarization (Minnesota code 4.1–4.3 or 5.1–5.3).

Okin et al. concluded that gender differences in body size and LV mass do not completely account for gender differences in voltage measurements and QRS duration [82].

Although available echocardiographic data are limited, it has become evident that standard electrocardiographic criteria overestimate racial differences in LVH prevalence [83]. There was no notable difference in the echocardiographic LV mass between white and African–American men or women in CHD-free subgroups of the CHS population of men and women 65 years old and older [84]. In that report, relatively strict selection criteria were used to establish upper normal limits for LV mass (116 g/m^2 for men and 104 g/m^2 for women). LVH prevalence was 18.1% in white men, 15.5% in African–American men, 14.7% in white women, and 12.8% in African–American women. Thus, racial differences in Echo-LVH were relatively small. These relative differences are not overly dependent on the LV mass cut points chosen.

40.5.4 LVH and Overweight

The role of overweight and obesity in relation to ECG-LVH and Echo-LVH is a relatively complex issue, and space limitations do not permit presentation of any data here. Classification accuracy of ECG-LVH by Sokolow–Lyon criteria is limited in general, and in particular in the presence of obesity. Revaluation of the CHS data indicated that there is a

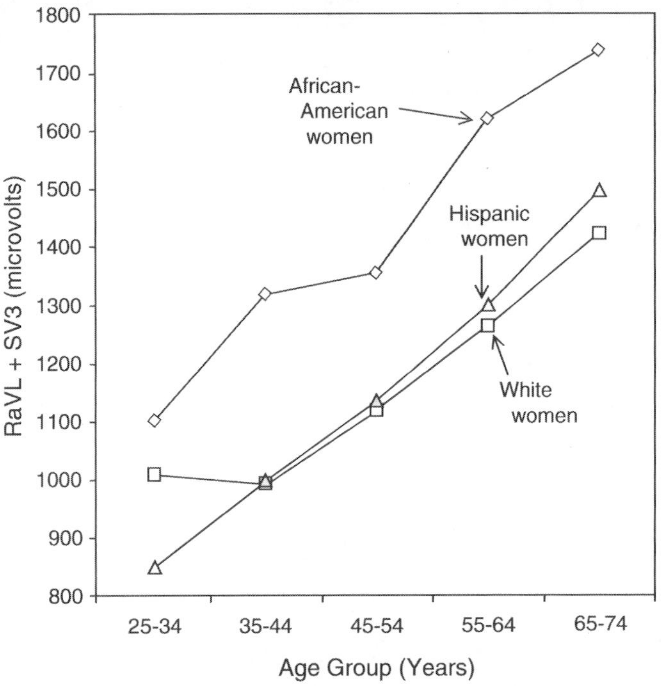

Figure 40.2
Cornell voltage (RaVL + SV3) by age in white (*squares*), Hispanic (*triangles*), and African-American women (*diamonds*). Note consistent increasing trend with age in all three ethnic groups and the substantially higher mean values in African-American compared to white and Hispanic men. Data modified from ref. [73], Rautaharju et al., J Electrocardiol 1994;27(suppl):20–30. © 1994 Churchill Livingstone, reproduced with permission.

substantial underestimate in ECG-LVH by the Cornell voltage criteria particularly in white men. In white and in African-American women, being overweight is associated with a notably higher Echo-LVH [85]. The overall LVH prevalence estimates by both methods may not differ substantially but the fraction of cases where both methods agree with the classification is relatively small.

Various studies have produced differing results about the role of overweight and LVH. The results differ depending on the method of indexing of LV mass to body size [82–86]. The availability of lean body weight data may be necessary to resolve the role of overweight in LVH.

40.5.5 ECG-LVH Prevalence in Hypertensive Cohorts

Higher ECG-LVH prevalence in blacks than in whites has also been reported in hypertensive cohorts. The Hypertension Detection and Follow-up Program (HDFP) evaluated ECGs of 10,940 hypertensive men and women with diastolic blood pressure (fifth phase) at the second screening visit of 90 mm Hg or above [54]. By Minnesota code 3.1, 3.3 plus 4.1–4.3, 5.1–5.3 criteria, the prevalences were 2.7% and 8.6% for white and black men, and 1.7% and 7.7% for white and black women, respectively. With high QRS amplitudes combined with ST-T abnormalities, the specificity of the criteria is very high, but the sensitivity is very low. The question of the need for improved LVH criteria arises again.

The Italian PIUMA study [87] reported Cornell voltage sensitivity as 16% and specificity as 97%. The authors reported that for the Perugia score for LVH [88], the sensitivity was 34% and specificity 93%. The operating points for various criteria can be expected to be quite different in hypertensive hospital populations compared to community-dwelling populations.

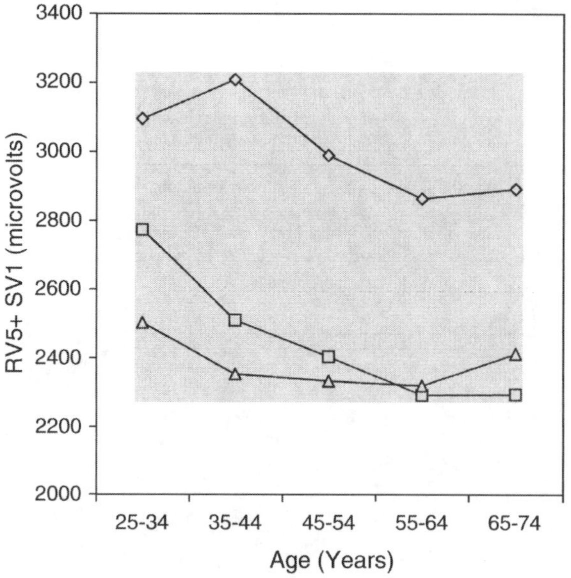

Figure 40.3
Sokolow–Lyon voltage (RV5 + SV1) by age in white (*squares*), Hispanic (*triangles*), and African–American men (*diamonds*). Note the opposite age trend in Sokolow–Lyon voltage in all three ethnic groups in comparison to the systematic increase in Cornell voltage with age in Fig. 40.1. The mean values of the Sokolow–Lyon voltage in African–American men are substantially higher compared to white and Hispanic men. Rautaharju, PM, unpublished data.

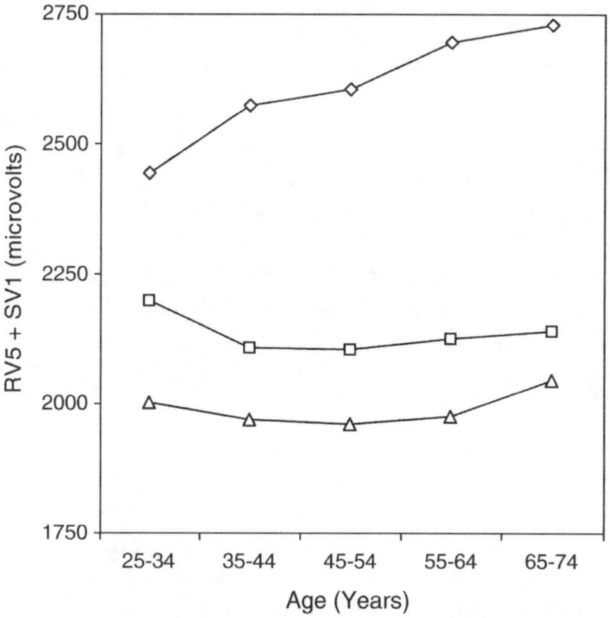

Figure 40.4
Sokolow–Lyon voltage (RV5 + SV1) by age in white (*squares*), Hispanic (*triangles*), and African–American women (*diamonds*). Increasing age trend in Sokolow–Lyon voltage is present only in African–American women and their mean values are substantially higher compared to white and Hispanic women. From NHANES 3 and HHANES ECG data, ref. [73], Rautaharju, PM, unpublished.

In summary, there are profound problems in using ECG-LVH criteria for estimating LVH prevalence in contrasting populations. LVH prevalence estimates by echocardiographic criteria will face similar, although not quite as severe, problems when some more accurate evaluation method will become available as an independent standard.

40.5.6 Visual Coding Errors as Source for Limited Sensitivity

The primary reason for visual coding errors is the complexity of the Minnesota code coding rules, particularly for serial comparison. All significant findings, particularly in code 1 category, are usually verified by an experienced supervisor in case there are any coding disagreements after duplicate reading by two coders (the procedure usually followed). Thus, false coding of significant abnormalities is actually rare. From experience in large clinical trials such as the Multiple Risk Factor Intervention Trial, the major problem in visual coding is the relatively frequent miss rate of truly codable items in spite of duplicate or even triplicate coding of each record. This miss rate can be 20% or even 30% when high volumes of records have to be coded. These error rates were found when visually coded items were verified with a computer program and all disagreements were again arbitrated.

Rautaharju et al. used a statistical model in an attempt to elucidate the reasons for the high miss rate with visual coding [89]. The authors concluded that a high miss rate in certain difficult categories of Code 1 will explain the high overall miss rate, and they also suggested that an initial screening by a computer and possible visual verification of selected items may substantially improve the accuracy and efficiency of ECG coding.

40.5.7 Prognostic Value of ECG-LVH in General Populations

Population characteristics and ECG-LVH criteria used differ from study to study, and the reported CVD mortality risk data differ considerably as seen from data derived from five diverse populations in ➋ Table 40.9. In the Framingham cohort, ECG-LVH by high QRS amplitude with LV strain was associated with a substantial excess of CVD, CHD, and all-cause mortality [51], and the mortality rates for ECG-LVH were similar as for MI by ECG, and in women they were always higher for ECG-LVH. Two-year age-adjusted incidence of ECG-LVH increased sharply both in men and in women with hypertensive status classified as mild and definite, compared with non-hypertensive groups.

LVH by Cornell voltage criteria was associated with an over threefold increased risk in black men and with an over twofold increased risk in white women in the NHANES one survey from the early 1970s. The risk was not significantly increased in any of the subgroups for Sokolow–Lyon criteria (PM Rautaharju, unpublished observations).

The Belgian Inter-University Research on Nutrition and Health (BIRNH) found a significantly increased age-adjusted and multivariately adjusted relative risk for ECG-LVH by high QRS amplitude criteria of the Minnesota code in men for CVD mortality but not for CHD mortality or total mortality [70]. The risk model included an adjustment for other major ECG abnormalities. The multivariately adjusted risk ratio for ECG-LVH was 3.14 (1.36–7.26). The risk was not significantly increased in women for any of the three major endpoints.

Data in ➋ Table 40.9 includes data from two elderly cohorts, namely an older Finnish cohort of men aged 65–84 years of the Seven Countries study [59] and the Bronx Longitudinal Aging Study that included men and women aged 75–85 years [90]. In the Finnish study with a 5-year follow-up of 697 survivors aged 65–85 years subsequent to the 25-year examination of the initial Finnish cohort of the Seven Countries study, the mortality risk for high-amplitude QRS waves and other Minnesota code items was evaluated, first separately for each abnormality and then according to a clearly defined hierarchic scheme. In the latter scenario, high QRS codes (MC 3.1, 3.3) without significant Q, ST, and T codes were entered into logistic regression models also adjusted for major CHD risk factors. High QRS amplitude codes entered without considering other coexisting codes were associated with a significant excess risk of all-cause mortality, and also with risk of fatal and nonfatal MI. The risk for isolated high R waves alone in the absence of ST-T abnormalities was not significant for any of the study endpoints. This finding again suggests that the inclusion of repolarization abnormalities with high-amplitude QRS variables is essential, not only for diagnostic applications but in particular for improved risk identification.

Table 40.9
ECG-LVH and CVD Mortality risk in general adult populations, including two old cohorts

Study/Baseline years	Age (years)	Follow-Up	Criteria	Endpoint	Gender/Race	RR (95% CI)	Comments
Framingham 1948	> 28	18 biennials	High QRS, ST Strain	CVD mortality	Men Women	3.4 ($p < 0.05$) 3.4 ($p < 0.05$)	RR = age-adjusted mortality rate versus general Framingham sample.
BIRNH 1981–1984	25–74	≥ 10 years	High R MC	CVD mortality	Men Women	3.14 (1.36–7.26) 2.20 (0.53–9.16)	CHD-free at baseline;
NHANES 1 1971–1975	35–74	7–13 years	High QRS + ST-T MC		White Men Black Men White Women Black Women	3.26 (1.91, 5.56) 4.29 (2.00, 9.18) 2.59 (1.29, 5.19) 1.90 (0.64, 5.65)	RR multivariately adjusted RR age-adjusted
RIFLE Pooling Project 1978–1987	30–69	6 years	High R MC High QRS + ST-T MC	CVD mortality CVD mortality	Men Women Men Women	1.86 (1.13, 3.07) 3.66 (0.96, 14.0) 6.33 (3.02, 13.3) 5.91 (0.70, 49.9)	Reference group free from LVH; multivariately adjusted, also for other ECG abnormalities
Bronx Longitudinal Aging Study 1980	75–85	10 years	High QRS, ST-T MC	CVD mortality All-cause mortality	Men and women Men and women	2.72 (0.53, 2.09) 2.18 (1.06, 4.45)	
Finnish cohort of the Seven Countries Study 1984	65–84	5 years	High QRS MC With or without Q, ST-T codes High QRS alone MC	All-cause mortality All-cause mortality	Men Men	1.27 ($p = 0.034$) 0.68 (0.22, 1.45)	p value from chi-square test adjusted for age and geographic area Reference group men with no major ECG abnormalities

In the Bronx Longitudinal Aging Study, ECG-LVH (MC 3.1, 3.3 with 4.1–4.3 or 5.1–5.3) had a significantly higher risk of CVD mortality than no ECG-LVH, with risk ratio 2.65 (1.58–4.41). In a multivariate model adjusting for common risk factors, CVD mortality risk was increased but not statistically significant.

One of the most informative reports from evaluation of the risk of LVH with various combinations of the relevant Minnesota codes comes from the Copenhagen City Heart Study [91]. In that report, ECG abnormalities for risk evaluation were classified as normal (reference group) together with five hierarchic, mutually exclusive abnormal categories: high-voltage QRS alone, negative T wave, ST depression with negative T wave, high-voltage QRS with negative T, and high-voltage QRS with ST depression and negative T. Each abnormality was an isolated finding, with no other abnormalities. Short-term (7-year) risk and long-term (21-year) risk were estimated for three endpoints: fatal and nonfatal MI, ischemic heart disease (ICD 8:410–414), and CVD mortality. The short-term risks for ischemic heart disease and CVD mortality are listed in ➲ Table 40.10, reproduced from ➲ Tables 4 and ➲ 5 of the Copenhagen study report. The study found no evidence of significant interaction between gender and ECG abnormalities in the Cox risk models when evaluated as individual categories or as a combined group.

The results confirm the findings from other studies that high QRS voltage alone is of little importance for risk identification, particularly when adjusted for blood pressure. Negative T waves and ST depression as isolated findings and not associated with high-voltage QRS were important predictors. The highest relative risk was found for high QRS voltage combined with negative T waves and high QRS voltage with ST depression was the strongest risk predictor of all five abnormal combinations. The report did not try to identify dominant predictors by entering all ECG abnormal categories simultaneously into the multivariately adjusted risk model. It is most likely, however, that such a model would have identified high-voltage QRS with ST depression and possibly also with negative T wave as dominant predictors among the abnormal categories.

Using ECG predictors as continuous variables will, in principle, improve the risk prediction power. An older ECG model for estimation of LV mass indexed to body surface area was used in an older study to evaluate mortality risk using NHANESI data [92]. Comparing relative risk for an increment from 20th to 80th percentile, age-adjusted risks for CVD mortality were 1.39 (1.21, 1.60) for white men, 1.67 (1.21, 2.29) for black men, 1.62 (1.17, 2.24) for white women, and 2.08 (1.27, 3.42) for black women. With an additional adjustment for systolic blood pressure and history of heart attack, CVD mortality risk remained significant in white men (RR = 1.21 (1.03, 1.43)), white women (RR = 1.36 (1.08, 1.70)), and in black women (RR = 1.95 (1.44, 2.66)) but not for black men (RR = 1.26 (0.81, 1.96)). These data suggest that the risk is graded across a wide range of estimated LV mass values.

40.5.8 Incident ECG-LVH

The Framingham study is among the very few with adequate documentation of CHD risk in persons developing new ECG evidence of LVH during a long follow-up period [93]. It is of practical importance that the incidence of ECG-LVH is higher than anticipated from the cross-sectional prevalence data. In the Framingham population, one in ten people developed some evidence of ECG-LVH in the first 12 years of follow-up. In about 3% of the cohort, ECG findings were categorized as definitive new LVH (mostly high-amplitude R waves combined with repolarization abnormalities).

There was a pronounced increase in the risk of every manifestation of CHD in men and women with definite LVH, including stroke and heart failure. The incidence of angina, MI, and sudden death in this group was about as high as in persons surviving a first MI. Data from 20-year follow-up of the Framingham study indicate a risk ratio for the age-adjusted overall mortality of about five for both males and females with definitive LVH compared to those without ECG evidence of LVH. For men, this risk ratio was nearly six for CHD death and sudden death. For women, the risk ratio for total cardiovascular mortality was nearly ten. The excess mortality associated with definite LVH on ECG carried a greater risk of cardiovascular events than cardiac enlargement. The risk was three times the risk associated with hypertension alone. As a contrast, electrocardiographic LVH based on high-amplitude R-wave criteria alone carried a risk for CHD mortality, which was about half of the risk for definite LVH. Furthermore, this excess risk was no longer manifest when adjustment was made for the coexisting hypertension.

Table 40.10

Age-standardized incidence and relative risk of ischemic heart disease events and of cardiovascular disease mortality during 7 years of follow-up in relation to ECG findings among those without ischemic heart disease at baseline, for the age range 35–74 years in Copenhagen City Heart Study

	Ischemic Heart Disease Events			Cardiovascular Disease Mortality		
	N (Inc.)	Age-adjusted RR	Multivariately adjusted RR[a] (95% CI)	N (Inc.)	Age-adjusted RR	RR[a] (95% CI)
Normal ECG (n = 8,460)	345 (6.7)	1	1	223 (4.5)	1	1
Voltage-only LVH (n = 1,197)	66 (10.2)	1.16 (0.89–1.52)	1.15 (0.88–1.51)	45 (6.6)	1.23 (0.89–1.69)	1.28 (0.92–1.77)
T, n = 799	73 (11.6)	1.69 (1.31–2.18)*	1.56 (1.21–2.03)†	55 (8.5)	1.82 (1.35–2.45)*	1.61 (1.19–2.18)†
ST/T, n = 257	27 (14.1)	2.25 (1.52–3.34)*	2.07 (1.39–3.08)*	17 (9.1)	1.98(1.21–3.26)†	1.68 (1.02–2.77)†
LVH with neg. T (n = 137)	17 (16.2)	1.95 (1.20–3.19)†	1.89 (1.15–3.09)††	14 (13.4)	2.28 (1.32–3.92)†	2.16 (1.25–3.74)†
LVH with ST/T (n = 132)	32 (30.5)	4.27 (2.95–6.16)*	3.62 (2.47–5.30)*	22 (17.8)	3.75 (2.41–5.85)*	2.96 (1.87–4.68)*

CI = confidence interval; RR = relative risk; ECG = electrocardiographic; LVH = left ventricular hypertrophy; ST/T = LVH with ST depression and negative T wave; Inc. = age-standardized number of endpoint events per 1,000 years of follow-up.

[a]Adjusted for age, systolic and diastolic blood pressure, heart rate, body mass index, cholesterol, smoking, diabetes, alcohol, physical exercise, and family history of ischemic heart disease.

* $P < 0.001$.

† $P < 0.01$.

†† $P < 0.05$.

From Larsen et al., ref [90], Eur Heart J 2002;23:315–324. ©2001 the European Society of Cardiology, reproduced with permission.

40.6 Incident Bundle Branch Blocks: Prognostic Value

The Framingham study cited above has also reported on the prognostic significance of newly acquired left and right bundle branch block [93]. There were 55 individuals in the Framingham population who acquired a left bundle branch block and 70 who acquired a right bundle branch block during 18 years of follow-up. The CVD mortality in these two subgroups of men and women was compared to the mortality in a group of age-matched members of the Framingham population who were presumed free from bundle branch block. Within 10 years after the onset of the block, the cumulative cardiovascular disease mortality was more than four times greater in those with left bundle branch block and more than three times greater in those with right bundle branch block than in the age-matched group of the study population. The proportion of sudden deaths was similar in those with left and right bundle branch block. In women, there was no indication of a different trend in cardiovascular disease mortality rates for left and right bundle branch block.

The acquired bundle branch blocks were all associated with prospective cardiovascular abnormalities during the 18-year follow-up. In men with right bundle branch block and in women with either kind of block, the presence of a block did not contribute to the increased risk of death from cardiovascular disease independently from associated cardiovascular abnormalities, whereas the appearance of a new left bundle branch block in men contributed important independent predictive information.

The University of Manitoba follow-up study report on a 29-year follow-up of 3,983 young pilots included observations on 28 men who developed complete left bundle branch block [94]. Excluded were blocks associated with ischemic or valvular heart disease at the time of the occurrence of the conduction defect. The 5-year incidence of sudden death as the first manifestation of heart disease was at least ten times higher among men with acquired left bundle branch block than among the remainder of the men without left bundle branch block (and apparently also free from ischemic heart disease). However, this marked excess risk of sudden death was not manifest for men who were less than 45 years old at the occurrence of the left bundle branch block.

It is uncertain to what extent observations from these highly selected special occupational groups can be extrapolated to general populations. The low incidence of these incident abnormalities makes risk assessment for them difficult.

40.7 ADDENDUM

40.7.1 New Reports on Repolarization Abnormalities as Mortality Predictors from Large Population-Based Cohorts

Several new reports have been published since the preparation of the manuscript for this chapter five years ago. Some of these reports on large population-based cohorts have brought new information about the risk associated with repolarization abnormalities as mortality predictors. Summary tables describing most salient results from these studies can be found in a 2007 monograph Investigative Electrocardiography in Epidemiological Studies and Clinical Trials by Rautaharju and Rautaharju [80].

Of particular interest are the results from the Women's Health Initiative (WHI) involving nearly 40,000 women aged 50 years and older. One of the WHI reports evaluated the risk of CHD and all-cause mortality for ECG abnormalities and a second report the risk of incident CVD and congestive heart failure (CHF) [95, 96]. With all significant individual risk predictors entered simultaneously into a multivariably-adjusted CHD mortality risk model, QRS-T angle was associated with an over two-fold increase in risk, and the rate-adjusted QT interval also remained a significant predictor [95]. QRS nondipolar voltage, possibly reflecting fragmented excitation, was a dominant predictor in these women, together with an old ECG-MI. Wide QRS-T angle, ST V5 depression, high T V1 amplitude and prolonged QT were dominant predictors of incident CHF [96]. The investigators concluded that ventricular repolarization abnormalities are as important as an old ECG-MI as predictors of incident CHD, CHF and mortality.

CHF is one of the leading causes of mortality and morbidity in the USA, and the prevalence of diastolic dysfunction has been reported to be higher in women than in men [97, 98]. In the Cardiovascular Health Study (CHS), the prevalence of CHF was 8.8% and was associated with increased age, particularly for women [99]. In women with CHF, systolic ventricular function was normal significantly more often than in men (67% vs. 42%). Diagnosis of diastolic dysfunction is presently done by clinical exclusion of other cardiac conditions. Potential importance of repolarization abnormalities as markers of CHS and for monitoring its evolution is obvious..

A report from CHS compared the relative risk of CHD and all-cause mortality during a 9-year follow-up in 4,912 men and in women aged 65 years old and older [100]. In men and in women, the relative risk of CHD mortality was increased 60% for wide QRS-T angle and there was a two-fold increase in risk of CHD mortality for ST depression. These risk levels were as high as for an old ECG-MI. Relative risk for left ventricular mass (LVM) from an ECG model with Cornell voltage and body weight as model covariates was significant in women only, as was QRS nondipolar voltage. These investigators concluded that the association of ECG abnormalities with mortality risk in women was consistently as strong as in men.

Acknowledgement

Dr. Farida Rautaharju has contributed to the contents and the preparation of this chapter.

References

1. Ashley, E.A., V.K. Raxwal, and V.F. Froelicher, The prevalence and prognostic significance of electrocardiographic abnormalities. *Curr. Probl. Cardiol.*, 2000;**25**: 1–72.
2. Keys, A., H.L. Taylor, H. Blackurn, J. Brozek, J.T. Anderson, and E. Simonson, Coronary heart disease among the Minnesota business and professional men followed fifteen years. *Circulation*, 1965;**28**: 381–395.
3. Keys, A., C. Aravanis, H.W. Blackburn, et al., Epidemiological studies related to coronary heart disease: characteristics of men aged 40–59 in seven countries. *Acta Med. Scand.*, 1967;**460**(Suppl): 1–392.
4. Keys, A., Editor. *Coronary Heart Disease in Seven Countries*. Heart Association Monograph Number 29. New York: American Heart Association, Inc., 1970.
5. Dawber, T.R., F.E. Moore, and G.V. Mann, Coronary heart disease in the Framingham study. *Am. J. Public Health*, 1957;**47**(Suppl. 1): 4–24.
6. Doyle, J.T., A.S. Heslin, H.E. Hilleboe, P.F. Formel, and R.F.A. Korns, prospective study of degenerative cardiovascular disease in Albany. Report of three years' experience –I. Ischemic heart disease. *Am. J. Public Health*, 1957;**47**(Suppl. 1): 25–32.
7. Chapman, J.M., L.S. Goerke, W. Dixon, D.B. Loveland, and E. Phillips, The clinical status of a population group in Los Angeles under observation for two to three years. *Am. J. Public Health*, 1957;**47**(Suppl. 1): 33–42.
8. Stamler, J., H.A. Lindberg, D.M. Berkson, A. Shaffer, W. Miller, and A. Poindexter, Prevalence and incidence of coronary heart disease in strata of the labor force of a Chicago industrial corporation. *J. Chronic Dis.*, 1960;**11**: 405–420.
9. Paul, O., M.H. Lepper, W.H. Phelan, et al., A longitudinal study of coronary heart disease. *Circulation*, 1963;**28**: 20–31.
10. New York Heart Association, Subcommittee on Electrocardiographic Criteria. *Nomenclature and Criteria for Diagnosis of Diseases of the Heart and Blood Vessels*, 5th edn. New York: New York Heart Association, 1953.
11. Rautaharju, P.M., M.J. Karvonen, and A. Keys, The frequency of arteriosclerotic and hypertensive heart disease among ostensibly health working populations in Finland. *J. Chronic Dis.*, 1961;**13**: 426–438.
12. Blackburn, H., A. Keys, E. Simonson, P. Rautaharju, and S. Punsar, The electrocardiogram in population studies. A classification system. *Circulation*, 1960;**21**: 1160–1175.
13. Ostrander, L.D. Jr, R.L. Brandt, M.O. Kjelsberg, and F.H. Epstein, Electrocardiographic findings among the adult population of a total natural community, Tecumseh, Michigan. *Circulation*, 1965;**31**: 888–898.
14. Sinnett, P.F. and H.M. Whyte, Epidemiological studies in a total highland population. Tukisenta, New Guinea, Cardiovascular disease and relevant clinical, electrocardiographic, radiological and biochemical findings. *J. Chronic Dis.*, 1973;**26**: 265–290.
15. The Pooling Project research Group, Relationship of blood pressure, serum cholesterol, smoking habit, relative weight and ECG abnormalities to incidence of major coronary events: Final report of the Pooling Project. *J. Chronic Dis.*, 1978;**31**: 201–306.
16. Rautaharju, P.M., H.P. Calhoun, and B.R. Chaitman, Novacode serial ECG classification system for clinical trials and epidemiological studies. *J. Electrocardiol.*, 1992;**24**: 179–187.
17. Burch, G.E. and T. Winsor, *A Primer of Electrocardiography*, 5th edn. Philadelphia, PA: Lea and Febiger, 1966.
18. Robles de Medina, E.O., *A New Coding System for Electrocardiography*. Assen: Royal van Gorcum, 1966.
19. Schamroth, L. and H.D. Friedberg, A coding system for cardiac arrhythmias. *J. Electrocardiol.*, 1970;**3**: 169–172.
20. Rautaharju, P.M., J.W. Warren, U. Jain, H.K. Wolf, and C.L. Nielsen, Cardiac infarction injury score: an electrocardiographic coding scheme for ischemic heart disease. *Circulation*, 1981;**64**: 249–256.
21. Pipberger, H.V., E. Simonson, E.A. Lopez Jr, A. Araoye, and H.A. Pipberger, The electrocardiogram in epidemiologic investigations. A new classification system. *Circulation*, 1982;**65**: 1456–1464.
22. Rose, G.A. and H. Blackburn, *Cardiovascular Survey Methods*. Geneva: World Health Organization, 1968. Monograph Series, no. 56
23. Prineas, R.J., R.S. Crow, and H. Blackburn, *The Minnesota Code Manual of Electrocardiographic Findings. Standards and Procedures for Measurement and Classification*. Boston, MA/Bristol/London: John Wright PSG Inc, 1982.

24. Rautaharju, P.M., D. Seale, R. Prineas, H. Wolf, R. Crow, and J. Warren, Changing electrocardiographic recording technology and diagnostic accuracy of myocardial infarction criteria. Improved standards for evaluation of ECG measurement precision. *J. Electrocardiol.*, 1978;**11**: 321–230.
25. Crow, R., R.J. Prineas, D.R. Jacobs, and H. Blackburn, A new epidemiological classification system for interim myocardial infarction from serial electrocardiographic changes. *Am. J. Cardiol.*, 1989;**64**: 454–461.
26. The Coronary Drug Project Research group, The coronary drug project: design, methods, and baseline results. *Circulation*, 1973;**47**(Suppl. 1): 11–50.
27. Rautaharju, P.M., S.K. Broste, R.J. Prineas, W.J. Eifler, R.S. Crow, and C.D. Furberg, Quality control procedures for the resting electrocardiogram in the multiple risk factor intervention trial. *Controlled Clin. Trials*, 1986;7(Suppl. 3): 46S–65S.
28. Reunanen, A., A. Aromaa, K. Pyörälä, S. Punsar, J. Maatela, and P. Knekt, The Social Insurance Institution's Coronary Heart Disease Study. Baseline data and 5 year mortality experience. *Acta Med. Scand.*, 1983;**673**(Suppl): 1–120.
29. Evans, J.G., I.A.M. Prior, and W.M.G. Turnbridge, Age-associated change in QRS axis: intrinsic or extrinsic aging? *Gerontology*, 1982;**28**: 132–137.
30. Higgins, I.T.T., W.B. Kannel, and T.R. Dawber, The electrocardiogram in epidemiological studies: reproducibility, validity and international comparison. *Br. J. Prev. Soc. Med.*, 1965;**19**: 53–68.
31. Cullen, K.J., B.P. Murphy, and G.N. Cumpston, Electrocardiograms in the Busselton population. *Aust. N.Z. J. Med.*, 1974;**4**: 325–330.
32. Ostor, E., P. Schnohr, G. Jensen, J. Nybe, and A.T. Hansen, Electrocardiographic findings and their association with mortality in the Copenhagen city heart study. *Eur. Heart J.*, 1981;**2**: 317–328.
33. Miall, W.E., E. Campo, J. Fodor, et al., Longitudinal study of heart disease in a Jamaican rural population. 1. Prevalence, with special reference to ECG findings. *Bull. W.H.O.*, 1972;**46**: 429–441.
34. Rose, G., P.J. Baxter, D.D. Reid, and P. McCartney, Prevalence and prognosis of electrocardiographic findings in middle aged men. *Br. Heart J.*, 1989;**1**: 73–80.
35. Rose, G.A., M. Ahmeteli, L. Checcacci, et al., Ischemic heart disease in middle-aged men: Prevalence comparisons in Europe. *Bull. W.H.O.*, 1968;**38**: 885–895.
36. World Health Organization European Collaborative Group, Multifunctional trial in the prevention of coronary heart disease: 1. Recruitment and initial findings. *Eur. Heart J.*, 1980;**1**: 73–80.
37. Barrett, P.A., C.T. Peter, H.J.C. Swan, B.N. Singh, and W.J. Mandel, The frequency and prognostic significance of electrocardiographic abnormalities in clinically normal individuals. *Prog. Cardivasc. Dis.*, 1981;**23**: 299–319.
38. Mathewson, F.A.L. and G.S. Varnam, Abnormal electrocardiograms in apparently healthy people. I. Long term follow-up study. *Circulation*, 1960;**21**: 196–203.
39. Medalie, J.H., M. Snyder, J.J. Croen, H.N. Neufeld, U. Goldbourt, and E. Riss, Angina pectoris among 10,000 men: 5 year incidence and univariate analysis. *Am. J. Med.*, 1973;**55**: 583–594.
40. Medalie, J.H., H.A. Khan, H.N. Neufeld, et al., Myocardial infarction over a five-year period. I. Prevalence, incidence and mortality experience. *J. Chronic Dis.*, 1973;**26**: 63–84.
41. Froelicher, V.F. Jr, F.G. Yanowitz, A.J. Thomson, and M.C. Lancaster, The correlation of coronary arteriography and the electrocardiographic response to maximal treadmill testing in 76 asymptomatic men. *Circulation*, 1973;**48**: 597–604.
42. Blackburn, H., H.I. Taylor, and A. Keys, The electrocardiogram in prediction of five-year coronary heart disease incidence among men aged forty through fifty-nine. *Circulation*, 1979;41(Suppl. 1): 154–161.
43. Blackburn, H., The importance of electrocardiograms in populations outside the hospital. *Can. Med. Assoc. J.*, 1973;**108**: 1262–1265.
44. Keys, A., *Seven Countries. A Multivariate Analysis of Death and Coronary Heart Disease.* Cambridge, MA: Harvard University Press, 1960.
45. Kannel, W.B., T. Gordon, W.P. Castelli, and J.R. Margolis, Electrocardiographic left ventricular hypertrophy and risk of coronary heart disease. The Framingham study. *Ann. Int. Med.*, 1970;**72**: 813–822.
46. Cedres, B.L., K. Liu, J. Stamler, et al., Independent contribution of electrocardiographic abnormalities to risk of death from coronary heart disease, cardiovascular diseases and all causes. Findings of three Chicago epidemiological studies. *Circulation*, 1982;**65**: 146–153.
47. Walker, A.R.P. and B.F. Walker, The bearing of race, sex, age, and nutritional state on the precordial electrocardiograms of young South African Bantu and Caucasian subjects. *Am. Heart J.*, 1969;**77**: 441–459.
48. Gottschalk, C.W. and E. Craige, A comparison of the precordial S-T and T waves in the electrocardiograms of 600 healthy young negro and white adults. *South. Med. J.*, 1956;**49**: 453–457.
49. Beaglehole, R., H.A. Tyroler, J.C. Cassel, D.C. Deubner, A.G. Bartel, and C.G. Hames, An epidemiological study of left ventricular hypertrophy in the biracial population of Evans County, Georgia. *J. Chron. Dis.*, 1975;**28**: 549–559.
50. WHI MONICA Project, Myocardial infarction and coronary deaths in the World Health Organization MONICA Project. Registration procedures, event rates, and case-fatality rates in 38 populations from 21 countries in four continents. *Circulation*, 1994;**90**: 583–612.
51. Menotti, A. and H. Blackburn, Electrocardiographic predictors of coronary heart disease in the seven countries study, in *Prevention of Coronary Heart Disease. Diet, Lifestyle and Risk Factors in the Seven Countries Study*, D. Kromhout, A. Menotti, and H. Blackburn, Editors. Norwell, MA: Kluwer, 2002, pp. 199–211.
52. Menotti, A., H. Blackburn, D.R. Jacobs, et al., *The predictive value of resting electrocardiographic findings in cardiovascular disease-free men. Twenty-five-year follow-up in the Seven Countries Study.* Internal document, Division of Epidemiology, School of Public Health, University of Minnesota, 2001.
53. Menotti, A., F. Seccaraccia, and the RIFLE Research Group, Electrocardiographic Minnesota Code findings predicting short-term mortality in asymptomatic subjects. The Italian RIFLE Pooling Project (Risk Factors and Life Expectancy). *G. Ital. Cardiol.*, 1997;**27**: 40–49.
54. Cullen, K., N.S. Stenhouse, K.L. Wearne, and G.N. Cumpston, Electrocardiograms and 13 year cardiovascular mortality in Busselton study. *Br. Heart J.*, 1982;**47**: 209–212.
55. Sigurdson, E., M. Sigfusson, H. Sigvaldason, and G. Thorgeirsson, Silent ST-T changes in an epidemiologic cohort study – A marker of hypertension or coronary artery disease, or both: The Reykjavik study. *J. Am. Coll. Cardiol.*, 1996;**27**: 1140–1147.

56. Kannel, B.W. and R. Abbott, Incidence and prognosis of unrecognized myocardial infarction. An update on the Framingham study. *N. Engl. J. Med.*, 1984;**311**: 1144–1147.
57. Kannel, W.B. and R.A. Abbott, Prognostic comparison of asymptomatic left ventricular hypertrophy and unrecognized myocardial infarction: The Framingham Study. *Am. Heart J.*, 1986;**111**: 391–397.
58. Yano, K. and C.J. MacLean, The incidence and prognosis of unrecognized myocardial infarction in the Honolulu, Hawaii, Heart Program. *Arch. Intern. Med.*, 1989;**149**: 1526–1532.
59. Tervahauta, M., J. Pekkanen, S. Punsar, and A. Nissinen, Resting electrocardiographic abnormalities as predictors of coronary events and total mortality among elderly men. *Am. J. Med.*, 1996;**100**: 641–645.
60. Nadelmann, J., W.H. Frishman, W.L. Ooi, et al., Prevalence, incidence and prognosis of recognized and unrecognized myocardial infarction in persons aged 75 years or older: The Bronx Aging Study. *Am. J. Cardiol.*, 1990;**6**: 533–537.
61. Medalie, J.H. and U. Goldbourt, Unrecognized myocardial infarction: five-year incidence, mortality, and risk factors. *Ann. Intern. Med.*, 1976;**84**: 526–531.
62. The Coronary Drug Project Research Group, The prognostic importance of the electrocardiogram after myocardial infarction. Experience from the Coronary Drug Project. *Ann. Intern. Med.*, 1972;**77**: 677–679.
63. The MRFIT Research Group, Relationship between baseline risk factors and coronary heart disease and total mortality in the Multiple Risk Factor Intervention Trial. *Prev. Med.*, 1986;**15**: 254–273.
64. The Multiple Risk Factor Intervention Trial Research Group, Baseline rest electrocardiographic abnormalities, antihypertensive treatment and mortality in the Multiple Risk Factor Intervention Trial. *Am. J. Cardiol.*, 1985;**55**: 1–15.
65. Holme, I., A. Helgeland, I. Hjermann, P. Leren, and P.G. Lund-Larsen, Treatment of mild hypertensives with diuretics: the importance of ECG abnormalities in the Oslo Study and in MRFIT. *J. Am. Med. Assoc.*, 1984;**251**: 1298–1299.
66. The Hypertension Detection and Follow-up Program Cooperative Research Group, The effect of antihypertensive drug treatment on mortality in the presence of resting electrocardiographic abnormalities at baseline: the HDFP experience. *Circulation*, 1984;**70**: 996–1003.
67. The Multiple Risk Factor Intervention Trial Research Group, Exercise electrocardiogram and coronary heart disease mortality in the Multiple Risk Factor Intervention Trial. *Am. J. Cardiol.*, 1985;**55**: 16–24.
68. Rautaharju, P.M., R.J. Prineas, W.J. Eifler, C.D. Furberg, J.D. Neaton, R.S. Crow, J. Stamler, and J.A. Cutler for the Multiple Risk Factor Intervention Trial Research Group, Prognostic value of exercise ECG in men at high risk of future coronary heart disease. *J. Am. Coll. Cardiol.*, 1986;**8**(1): 1–10.
69. Rautaharju, P.M. and J.D. Neaton for, the MRFIT Research Group, Electrocardiographic abnormalities and coronary heart disease mortality among hypertensive men in the Multiple Risk Factor Intervention Trial. *Clin. Invest. Med.*, 1987;**10**: 606–615.
70. De Bacquer, D., G. De Backer, M. Kornitzer, and H. Blackburn, Prognostic value of ECG findings for total, cardiovascular disease, and coronary heart disease death in men and women. *Heart*, 1998;**80**: 570–577.
71. Dauerman, H.L., D. Lessard, J. Yarzebski, M.I. Furman, J.M. Gore, and R.J. Goldberg, Ten-year trends in the incidence, treatment, and outcome of Q-wave myocardial infarction. *Am. J. Cardiol.*, 2000;**86**: 730–735.
72. Rosenman, R.H., M. Friedman, C.D. Jenkins, R. Straus, M. Wurm, and R. Kosichek, Clinically unrecognized myocardial infarction in the Western Collaborative Group Study. *Am. J. Cardiol.*, 1967;**19**: 776–782.
73. Rautaharju, P.M., S.H. Zhou, and H.P. Calhoun, Ethnic differences in electrocardiographic amplitudes in North American white, black and hispanic men and women: effect of obesity and age. *J. Electrocardiol.*, 1994;**27**(Suppl): 20–30.
74. Strogatz, D.S., H.A. Tyroler, L.O. Watkins, and C.G. Hames, Electrocardiographic abnormalities and mortality among middle-aged black men and white men of Evans County, Georgia. *J. Chron. Dis.*, 1987;**40**: 149–155.
75. Arnett, D.K., D.S. Strogatz, S.A. Ephross, C.G. Hames, and H.A. Tyroler, Greater incidence of electrocardiographic left ventricular hypertrophy in black men than in white men in Evans County, Georgia. *Ethn. Dis.*, 1992;**2**: 10–17.
76. Arnett, D.K., P. Rautaharju, S. Sutherland, B. Usher, and J. Keil, Validity of electrocardiographic estimates of left ventricular hypertrophy and mass in African Americans (The Charlston Heart Study). *Am. J. Cardiol.*, 1997;**79**: 1289–1292.
77. Xie, X., K. Liu, J. Stamler, and R. Stamler, Ethnic differences in electrocardiographic left ventricular hypertrophy in young and middle-aged employed American men. *Am. J. Cardiol.*, 1994;**73**: 564–567.
78. Huston, S.L., C.H. Bunker, F.A.M. Ukoli, P.M. Rautaharju, and H.K. Lewis, Electrocardiographic left ventricular hypertrophy by five criteria among civil servants in Benin City, Nigeria: prevalence and correlates. *Int. J. Cardiol.*, 1999; **70**: 1–14.
79. Mosterd, A., R.B. D'Agostino, H. Silbershatz, P.A. Sytkowski, W.B. Kannel, D.E. Grobbee, and D. Levy, Trends in the prevalence of hypertension, antihypertensive therapy, and left ventricular hypertrophy from 1950 to 1989. *N. Engl. J. Med.*, 1999;**340**: 1221–1227.
80. Rautaharju P, Rautaharju F. *Investigative Electrocardiography in Epidemiological Studies and Clinical Trials.* Springer-Verlag London Limited, London, 2007, pp 1:289.
81. Liebson, P.R., G. Grandits, R. Prineas, S. Dianzumba, J.M. Flack, J.A. Cutler, R. Grimm, and J. Stamler, Echocardiographic correlates of left ventricular structure among 844 mildly hypertensive men and women in the Treatment of Mild Hypertension Study (TOMHS). *Circulation*, 1993;**87**: 476–486.
82. Okin, P.M., J. Sverker, R.B. Devereux, S.E. Kjeldsen, and B. Dahlof, Effect of obesity on electrocardiographic left ventricular hypertrophy in hypertensive patients: the Losartan Intervention for Endpoint (LIFE) Reduction in Hypertension Study. *Hypertension*, 2000;**35**: 13–18.
83. Lee, D.K., P.R. Marantz, R.B. Devereux, P. Kligfield, and M.H. Alderman, Left ventricular hypertrophy in black and white hypertensives. Standard electrocardiographic criteria overestimate racial differences in prevalence. *J.A.M.A.*, 1992;**267**: 3294–3299.
84. Rautaharju, P.M., L.P. Park, J.S. Gottdiener, D. Siscovick, R. Boineau, V. Smith, and N.R. Powe, Race-and sex-specific ECG models for left ventricular mass in older populations. Factors influencing overestimation of left ventricular hypertrophy prevalence by ECG criteria in African-Americans. *J. Electrocardiol.*, 2000;**33**: 205–218.

85. Rautaharju, P.M., T.A. Manolio, D. Siscovick, S.H. Zhou, J.M. Gardin, R. Kronmal, C.D. Furberg, N.O. Borhani, and A. Newman, for the Cardiovascular Health Study Collaborative Research Group. Utility of new electrocardiographic models for left ventricular mass in older adults. *Hypertension*, 1996;**28**: 8–15.
86. Levy, D., K.M. Anderson, D.D. Savage, W.B. Kannel, J.C. Christiansen, and W.P. Castelli, Echocardiographically detected left ventricular hypertrophy: prevalence and risk factors. The Framingham study. *Ann. Intern. Med.*, 1988;**108**: 7–13.
87. Verdecchia, P., G. Schillaci, C. Borgioni, A. Ciucci, R. Gattobigio, I. Zampi, G. Reboldi, and C. Porcellati, Prognostic significance of serial changes in left ventricular mass in essential hypertension. *Circulation*, 1998;**97**: 48–54.
88. Schillaci, G., P. Verdecchia, Borgioni, A. Ciucci, M. Guerrieri, I. Zampi, M. Battistelli, C. Bartoccini, and C. Porcellati, Improved electrocardiographic diagnosis of left ventricular hypertrophy. *Am. J. Cardiol.*, 1994;**74**: 714–719.
89. Rautaharju, P.M., J. Warren, R.J. Prineas, and P.h. Smets, Optimal coding of electrocardiograms for epidemiological studies. The performance of human coders – a statistical model. *J. Electrocardiol.*, 1979;**13**: 55–59.
90. Kahn, S., W.H. Frishman, S. Weissman, W.L. Ooi, and M. Aronson, Left ventricular hypertrophy on electrocardiogram: prognostic implications from a 10-year cohort study of older subjects: a report from the Bronx longitudinal aging study. *J. Am. Geriatr. Soc.*, 1996;**44**: 524–529.
91. Larsen, C.T., J. Dahlin, H. Blackburn, H. Scharling, M. Appleyard, B. Sigurd, and P. Schnohr, Prevalence and prognosis of electrocardiographic left ventricular hypertrophy, ST segment depression and negative T-wave. *Eur. Heart J.*, 2002;**23**: 315–324.
92. Rautaharju, P.M., A.Z. LaCroix, D.D. Savage, S. Haynes, J.H. Madans, H.K. Wolf, W. Hadden, J. Keller, and J. Cornoni-Huntly, Electrocardiographic estimate of left ventricular mass vs. Radiographic cardiac size and the risk of cardiovascular disease mortality in the epidemiologic follow-up study of the First National Health and Nutrition Examination Survey. *Am. J. Cardiol.*, 1988;**62**: 59–66.
93. Levy, D., M. Salomon, R.B. D'Agostino, A.J. Belanger, and W.B. Kannel, Prognostic implications of baseline electrocardiographic features and their serial changes in subjects with left ventricular hypertrophy. *Circulation*, 1994;**90**: 1786–1793.
94. Mathewson, F.A.L., J. Manfreda, R.B. Tate, and T. Cuddy, The University of Manitoba Follow-up Study-an investigation of cardiovascular disease with 35 years of follow-up (1948–1983). *Can. J. Cardiol.*, 1987;**3**: 378–382.
95. Rautaharju, P.M., C. Kooperberg, J.C. Larson, and A. LaCroix, Electrocardiographic abnormalities that predict coronary heart disease events and Mortality in Postmenopausal Women. The Women's Health Initiative. *Circulation*, 2006;**113**: 473–480.
96. Rautaharju, P.M., C. Kooperberg, J.C. Larson, and A. LaCroix, Electrocardiographic Predictors of Incident Congestive Heart Failure and All-cause Mortality in Postmenopausal Women. The Women's Health Initiative. *Circulation*, 2006;**113**: 481–489.
97. Zile, M.R. and D.L. Brutsaert, New concepts in diastolic dysfunction and diastolic heart failure: Part I: diagnosis, prognosis, and measurements of diastolic function. *Circulation*, 2002;**105**: 1387–1393.
98. Zile, M.R. and D.L. Brutsaert, New concepts in diastolic dysfunction and diastolic heart failure: Part II: causal mechanisms and treatment. *Circulation*, 2002;**105**: 1503–1508.
99. Kitzman, D.W., J.M. Gardin, J.S. Gottdiener, A. Arnold, R. Boineau, G. Aurigemma, E.K. Marino, M. Lyles, M. Cushman, and P.L. Enright, Importance of heart failure with preserved systolic function in patients ≥65 years of age. CHS Research Group. Cardiovascular Health Study. *Am. J. Cardiol.*, 2001;**87**: 413–419.
100. Rautaharju, P.M., S.G. Ge, J. Clark Nelson, E.K. Marino Larsen, B.M. Psaty, C.D. Furberg, Z.M. Zhang, J.A. Robbins, MD, MHS, J.S. Gottdiener, MD, and P. Chaves, Comparison of Mortality risk for Electrocardiographic Abnormalities in Men and Women With and Without Coronary Heart Disease (From the Cardiovascular Health Study). *Am. J. Cardiol.*, 2006;**97**: 309–315.

41 The Dog Electrocardiogram: A Critical Review

David K. Detweiler[†]

[†]For this 2nd Edition of "Comprehensive Electrocardiology," Dr. Sydney Moise has updated this 1st Edition chapter, which was originally written by the late Dr. Detweiler.

P. W. Macfarlane, A. van Oosterom, O. Pahlm, P. Kligfield, M. Janse, J. Camm (eds.), *Comprehensive Electrocardiology*, DOI 10.1007/978-1-84882-046-3_41,

41.1 History and Literature

Electrocardiographic studies in dogs date back to the pioneering investigations of Augustus Waller [1] with the capillary electrometer and Willem Einthoven's development of the string galvanometer electrocardiograph [2–4]. As late as 1914 [5], Waller considered electrocardiography (ECG) as an experimental method, useful to physiologists rather than as a clinical tool for physicians. However, clinical application in man had already started and advanced rapidly (Lewis, 1909–1925 [6]; Rothberger, 1912–1930 [7]; Winterberg, 1912–1930 [9]; Scherf, 1921–present [8]; Wenckebach, 1899–1930 [9]; and Wilson, 1919–1945 [10]). Clinical use in canine medicine was modest in those early days (Nörr, 1913–1931 [11]; Roos, 1925 [12]; Haupt, 1929 [13]; Ludwig, 1924 [14]; and Gyarmati, 1939 [15]) and subsequently, until Nils Lannek's systematic study and statistical analysis of clinical records from healthy and diseased dogs [16]. Lannek also introduced a precordial-lead system that is still in use. Scherf and Schott's encyclopedic monograph, *Extrasystoles and Allied Arrhythmias* [8], reviews much of the electrocardiographic literature on experimental cardiac arrhythmias and drug effects in dogs. Burch and DePasquale's *A History of Electrocardiography* [17], Sir Thomas Lewis's classical *The Mechanism and Graphic Registration of the Heart Beat* [6], and Wilson's collected works (edited by Johnston and Lepeschkin [10]) are rich sources of information on earlier canine studies.

Modern imaging of the heart with echocardiography, angiography, magnetic resonance imaging (MRI), endocardial mapping, computer-assisted tomography (CT) and other modalities potentially provides more valuable information of the structure, function, and electrical competency than the routine surface electrocardiogram. However, the ECG is the mainstay for the diagnosis of arrhythmias in all animals. In the complete diagnosis of disease, the ECG must be supplemented with other technologies. The ECG remains a cornerstone for the initial screening and recognition of disease. Moreover, in pharmacological studies, the dog remains a key animal of study for which the review of ECG changes for the treatment effect or toxicity involves routine examinations.

Besides the routine analysis of the ECG in the dog, more thorough approaches to the clues of the disease, drug effect, or toxicity that the ECG offers are in use today. These include 24-h ambulatory ECG monitoring (Holter monitoring), telemetry recordings via implantable recording devices, loop-recording devices to capture arrhythmias, and heart rate variability.

41.1.1 Canine Electrocardiography

An enormous amount of literature has been published on experimental electrocardiographic studies in dogs. Historical and useful reviews are found in textbooks and monographs on arrhythmias and conduction disorders such as Bellet [18, 19], Scherf and Schott [8], and Schamroth [20]. Normal values for dog ECGs have been summarized in several textbooks (Ettinger and Suter [21], Detweiler et al. [22], Bolton [23], and Tilley [24]). In early papers and textbooks, generalized statements were frequently made with regard to the interpretation of ECG measurements to heart size based on the amplitude and duration of a specific waveform. Today, we recognize that some of these conclusions were too specific because of breed, age, and body conformation. Some examples will be addressed in the following discussions on the ECG waveforms.

41.1.2 Beagle Electrocardiogram

The ECG of the beagle is of special interest because of its widespread use as a research animal [25]. The published normal values for the ECG waveforms with regard to time and amplitude serve as a guideline to the evaluation of the ECG for the beagle. ➔ Table 41.1 lists findings that are commonly found when evaluating the electrocardiogram of the research beagle.

Table 41.1

Electrocardiographic findings of the research beagle that are not within the usual normal range for dogs and that are not likely pathologic[a]

Number	Electrocardiographic finding	Comments	Level of concern for use of dog in cardiovascular studies
1	Deep (not wide) S waves in lead III (1.0 mV > S wave > 0.7 mV)	Common singular finding without known association to structural or electrical abnormality	Low
2	Deep (not wide) S waves in lead III (S wave ⩾ 1 mV) only	Less common than #1. May be an insignificant finding	Low-medium
3	Deep (not wide) S waves in leads II, III, and aVF (0.5 mV > S wave > 0.3 mV in lead II, 1 mV > S wave > 0.5 mV in lead III, and 1 mV ⩾ S wave > 0.5 mV in lead III)	Common cluster finding without known association to structural or electrical abnormality or may be associated with right ventricular enlargement, incomplete right bundle branch block, or left anterior fascicular block	Medium
4	Very deep (not wide) S waves in II, III, and aVF (S wave > 0.5 mV in lead II, S wave > 1 mV in lead III, and S wave > 1 mV in lead III)	Much less common cluster finding without known association to structural or electrical abnormality but more likely than #3 to be associated with right ventricular enlargement, incomplete right bundle branch block, or left anterior fascicular block	High
5	T_a wave (Tsub "a" wave)	Occasionally seen. A negative deflection immediately following the P wave. Indicates the T wave of the P wave (repolarization of the atria)	None unless associated with large P wave
6	Tall (not wide) R wave (3.5 mV > R wave > 3.0 mV)	Common. Mild elevation in the amplitude of the R wave is usually not associated with structural or functional abnormality, but could indicate left ventricular hypertrophy	Low
7	Very tall (not wide) R waves (R wave > 3.5 mV)	Less common than #6. Moderate elevation in the amplitude may still not be associated with structural or functional abnormality, but more likely to be associated with left ventricular hypertrophy than #6	Medium
8	Splintered or notched QRS complex	Occasionally seen. The appearance of this finding can be affected by the filter settings of the ECG recording device. Splintered R waves are associated with tricuspid dysplasia in the dog, but this congenital anomaly has not been reported in the beagle	Low, but if marked could affect ease of interpretation
9	Low amplitude QRS complex (QRS complex < 0.7 mV)	Occasionally seen. Although low R waves are reported for a variety of conditions (e.g., pericardial effusion, ascites, pleural effusion, obesity, hypothyroidism, pulmonary embolism) in the beagle this is seen with a normal heart	Low
10	Deep Q waves (Q wave > 1.2 mV)	Common. May be a singular finding or with a tall or very tall R wave. Can be a normal variation or indicative of septal hypertrophy or right ventricular hypertrophy	Low as a singular finding
11	Large T wave (>25% of the R wave)	Occasionally seen as a singular finding. May be present with tall and very tall R wave. In the latter situations the T wave is large as a negative deflection	Low as a singular finding

Table 41.1 (Continued)

Number	Electrocardiographic finding	Comments	Level of concern for use of dog in cardiovascular studies
12	Second-degree heart block (low grade with only single P waves not associated with QRS complex)	Occasionally seen in dogs more than 4 months of age. Common in dogs less than 2 months of age. In these situations usually not associated with disease, but with high vagal tone	Low as a singular finding, but most investigators do not want dogs with this finding[b]
13	Sinus bradycardia or long sinus pauses (heart rate < 60 bpm, PP interval > 1.5 s)	Occasionally seen. Dogs bred for calm personality tend to have slower heart rates due to higher vagal tone and less sympathetic tone. Many of these dogs have heart rates that approach the lower limit of 60 bpm	Low to medium depending on the degree of the bradycardia and the length of the sinus pause. Extreme bradycardia or pauses would have a high concern
14	Presence of a J wave. The J wave is a positive deflection at the very terminal point of the downstroke of the R wave. It may be a complete secondary positive deflection with amplitudes of 0.2 mV or just a widening of the R wave usually beginning at the amplitude of 0.1–0.2 mV	Common finding and is due to the current density of I_{to}. Seen in other breeds too	None, but in some cases can make the determination of the QRS duration problematic because the end of the QRS is difficult to determine Some include this wave in the duration of the QRS

[a]This table is not a listing of electrocardiographic abnormalities per se, but is a listing of findings often found in the beagles that usually are not associated with pathology. However, as described, the findings may be associated with an abnormality although frequently they are not

[b]In 2,232 beagles on which two 1-min recordings were taken was about 1% [26], while that in 11 resting dogs monitored for about 6 h by radiotelemetry was about 64%. This figure increased to 100% in 12 puppies, 8–11 weeks old [47]

41.2 Recording Techniques

41.2.1 Lead Systems

A variety of lead systems has been used for decades in the dog. Most commonly used is the six-lead limb system which includes leads I, II, III, aVR, aVL, and aVF. However, the addition of other leads may be beneficial for some studies.

Historically, Waller [1] initiated the limb-lead system of recording from dogs when he taught his pet bulldog "Jimmie" to stand in beakers filled with a conducting solution into each of which an electrode was fixed (➲ Fig. 41.1). In Waller's earliest experiments, wires connected these electrodes to the capillary manometer (➲ Fig. 41.2). This method of recording was also used with other species including humans and anticipated the technique that Einthoven adopted for his string galvanometer.

Because of the offset potentials generated at the metal–skin surface interface, nonpolarizable electrode systems were required with the capillary electrometer and the early string galvanometers of the Einthoven type. This was accomplished by immersing the limb in a bath containing a salt of the metal used in the electrode; for example, silver chloride solution with silver electrodes or zinc chloride with zinc or nickel silver (German silver, a silver white alloy of copper, zinc, and nickel) electrodes. Soon, the bath was replaced by cloth strips that were saturated with the solution and wrapped about the limb to form contact between the electrode and the skin. Later, the material was replaced by conducting electrolyte pastes and gels placed between the skin and the electrodes. In animals, a variety of needle and clip electrodes was used to fix the leads firmly in place despite inadvertent movement of the subjects.

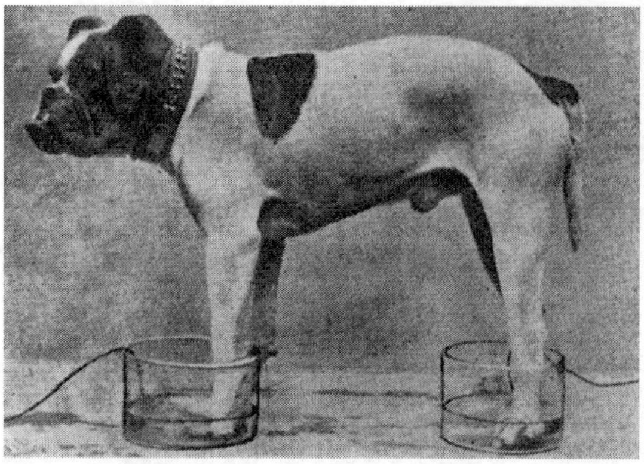

Figure 41.1
Waller's pet dog "Jimmie" patiently standing with his left foreleg and connected by wires to an electrometer. (A. Waller. *Physiology, the Servant*. London Press/Hodder & Stoughton, London, 1888. Reproduced with permission.)

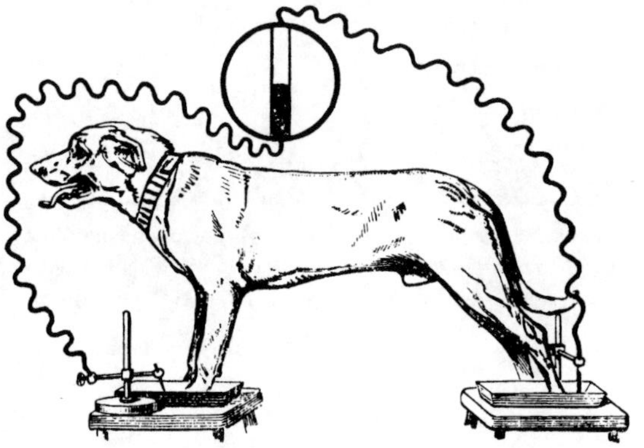

Figure 41.2
Probably the first picture of an ECG being recorded from a dog. The instrument is a capillary electrometer. The line drawing depicts a dog standing with the left foreleg and hindleg in pans of saline solution and wires leading from electrodes immersed in the solution to a schema of the electrometer. (Waller [1]. © British Medical Association, London. Reproduced with permission.)

The idea prevailed for some years [13, 26] that a bipolar lead along the imaginary anatomical longitudinal axis of the heart would be best for animals. A single lead with one electrode attached to the left precordium over the cardiac apex and the other electrode attached over the base of the heart at the junction of the neck and thorax, the scapular spine, or anterodorsal edge of the scapula on the right side, became popular. It was not realized that this was essentially a precordial lead with a neck or scapular electrode acting more or less as the indifferent electrode. A later modification of this was a three-lead triangular system similar to that of Nehb [27] with the three limb electrodes placed at the cardiac apex (left leg [LF] electrode), the base of the neck (anterodorsal edge of the scapula) on the right side (right arm [RA] electrode), and the sacral region (left arm [LA] electrode). The leads thus obtained are termed dorsal (RA to LA), axial (RA to LF), and inferior (LA to LF) [28]. This system never gained popularity in the dog, but has been used more often in large farm animals and the laboratory rat [29].

Lannek [16] developed a precordial-lead system that has been used in the dog. This system utilized anatomical criteria to position one electrode over the right ventricle and two electrodes over the left ventricle. In taking these precordial leads, he paired the exploring electrode with the right foreleg electrode and used the symbol CR for these chest leads.

Wilson's central terminal soon replaced the right leg as the indifferent electrode and by early 1960, Hamlin [30] initiated the use of a lead corresponding to V_{10} in man to the chest leads for the dog. V_{10} is uncommonly used today.

The original lead symbols introduced by Lannek were CR_5RL, CR_6LL, and CR_6LU. With the introduction of Wilson's central terminal, these symbols were changed to CV_5RL, CV_6LL, and CV_6LU. In 1977, the Committee of the American Academy of Veterinary Cardiology introduced new lead symbols [31] similar to those used for man. Although these electrode positions in the dog roughly approximate, they do not correspond, accurately, to the identically named lead positions in man. Both nomenclatures are given here. The electrode positions for each lead are as follows:

(a) V_2 (CV_6LL): sixth left intercostal space near the edge of the sternum at the most curved part of the costal cartilage
(b) V_4 (CV_6LU): sixth left intercostal space at the costochondral junction
(c) V_{10}: over the dorsal spinous process of the seventh thoracic vertebra (on the dorsal midline vertically above the V_4 position)
(d) rV_2 (CV_5RL): fifth right intercostal space near the edge of the sternum at the most rounded part of the costal cartilage

The abbreviations in the parentheses are the terms originally used by Lannek [16] to designate these electrode positions. The terminology and corresponding equivalent in humans has been questioned [32]. The conformation of the thorax of the dog is dissimilar to humans. Thus, the positioning of the leads at the same points on the thorax does not correspond exactly to that of the humans. Also, the location and effect of the diaphragm on cardiac position is different between the two species. Importantly, the morphology of the thorax varies greatly amongst different somatotypic breeds of dogs, such that even within the canine species, variability must be expected when trying to make anatomical comparisons to the location and direction of the electrical depolarization. Moreover, consistent positioning of these leads and additional precordial leads (V_1, V_2,, V_3, V_4, V_5, and V_6) is critical for an acceptable amount of variability between recordings.

Three-lead (X, Y, and Z)-corrected orthogonal systems such as those of McFee and Parungao [33] (➋ Fig. 41.3) and Frank [34, 35] (➋ Fig. 41.4) have been used historically for vectorcardiography. Vectorcardiography has been replaced by more sophisticated means of electrical mapping of the heart. Such systems are beyond the scope of this review. The X-, Y-, and Z-lead system has been used extensively in the Holter Laboratory of Cornell University, College of Veterinary Medicine. Such a system typically provides excellent recordings for analysis in the dog.

A more complex lead system is used in Japan [37]. Takahashi [38], on the basis of experimental studies in the dog, introduced an elaborate 12-lead precordial system with 6 leads on each side of the thorax as follows (➋ Fig. 41.5): C_1, C_2, and C_3 at the left costochondral junction anterior to rib one and at the second and fifth intercostal space, respectively; C_4, C_5, and C_6 at the right costochondral junctions in the seventh, fifth, and third intercostal spaces, respectively; M_1, M_2 at the widest portion of the thorax in the third and sixth left intercostal spaces, respectively; M3 at the left of the xiphoid process; M_4 at the right of the xiphoid process; and M_5, M_6 at the widest portion of the thorax in the right seventh and third intercostal spaces, respectively.

In 1966, the Japanese Association of Animal Electrocardiography recommended a bipolar base-apex (termed A-B) lead similar to that recommended in 1929 by Haupt [13], one of Nörr's [11] pupils, for use in dogs. The positive electrode (A) is placed at the costochondral junction of the left sixth rib and the negative electrode at the right scapular spine (Japanese Association of Animal Electrocardiography, 1975) [36]. In addition, they recommend the use of the Takahashi precordial leads $C_1 - C_6$, M_1, and M_6.

In summary, although multiple lead systems have been proposed, the most common one used for the evaluation of the electrical activity of the dog is the six-lead limb system. Importantly, in modern times, a critical evaluation demanded in pharmacological studies is the evaluation of the QT interval. Most of time and amplitude measurements are done in lead II; however, in approximately 10–20% of the canine recordings, the clarity of the T wave and particularly the offset point is not clear in lead II or other limb leads. In such situations, other leads may be better suited to more definitively make the QT interval measurement (see below).

41.2.2 Position and Restraint

In quadrupeds, the magnitude and direction of electrocardiographic vectors determined from limb leads can be vastly altered by changes in the position of the muscular attachments of the shoulder girdle to the thorax (➋ Fig. 41.6).

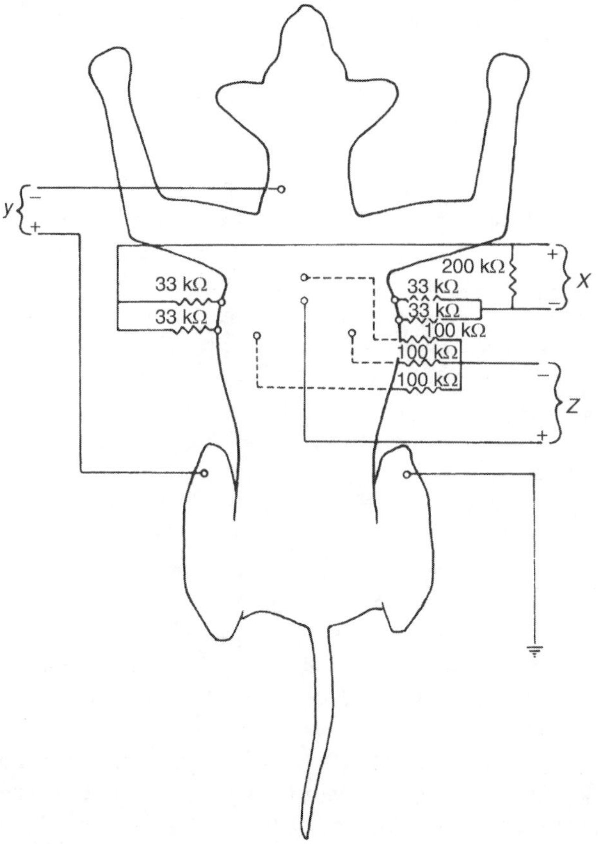

Figure 41.3

The corrected orthogonal-lead system designed for the canine thorax by McFee and Parungao [33]. Electrode placement for transverse (*X*), longitudinal (*Y*), and sagittal (*Z*) axis leads. Viewed from the dorsal aspect of the dog. Note that the Z+ electrode is on the back of the dog. (After Chastain et al. [83]. ©American Veterinary Medical Association, Schaumburg, Illinois. Reproduced with permission.)

Thus consistent positioning of the forelimbs, and especially the scapulae, is crucial to obtaining reproducible vectors in serial ECGs. This was not known until the early 1950s, while Lannek [16] reported that in dogs, the mean manifest QRS vector in the frontal plane could be highly variable. This finding was later confirmed independently by Cagan et al. [39, 40], Hulin and Rippa [41], and was studied systematically by Hill [42, 43]. The following technique is recommended: The dog is placed in right lateral recumbency. The head and neck are held flat on the table in line with the long axis of the trunk. The forelegs are positioned parallel to one another and perpendicular to the long axis of the body so that the point of the left shoulder (anterior aspect of the scapulohumeral joint) is vertically above the point of the right shoulder. The complexes in lead aVL are often those most sensitive to changes in foreleg position. Therefore, in serial records, complexes in lead aVL can be compared to verify the consistency in foreleg positioning.

To restrain the dogs on a table, the handler should face the right side of the standing animal with its head to his right side, reach over the animal's back, grasp the forelegs in his right hand and the hind legs in his left hand. The dog is then laid on its right side, the head and neck pressed flat against the table with the right forearm, and its back restrained against the handler's body. The technician operating the electrocardiograph then attaches the electrodes and arranges the forelimbs and head and neck as described. The handlers must avoid touching moist surfaces or electrodes in case AC interference is introduced.

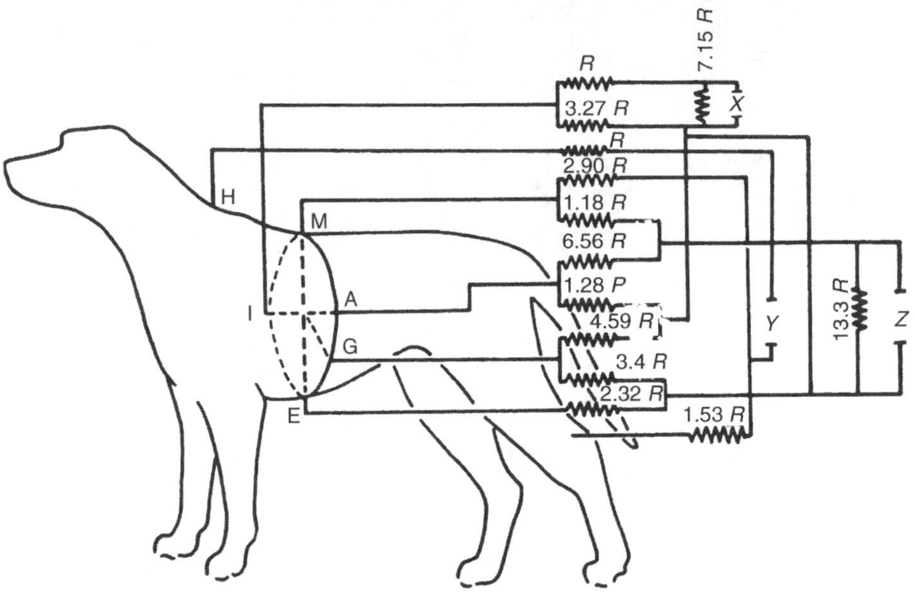

Figure 41.4
Frank's orthogonal-lead system applied to the dog. *R* = 100, 000Q. (After Bojrab et al. [35]. © American Veterinary Medical Association, Schaumburg, Illinois. Reproduced with permission.)

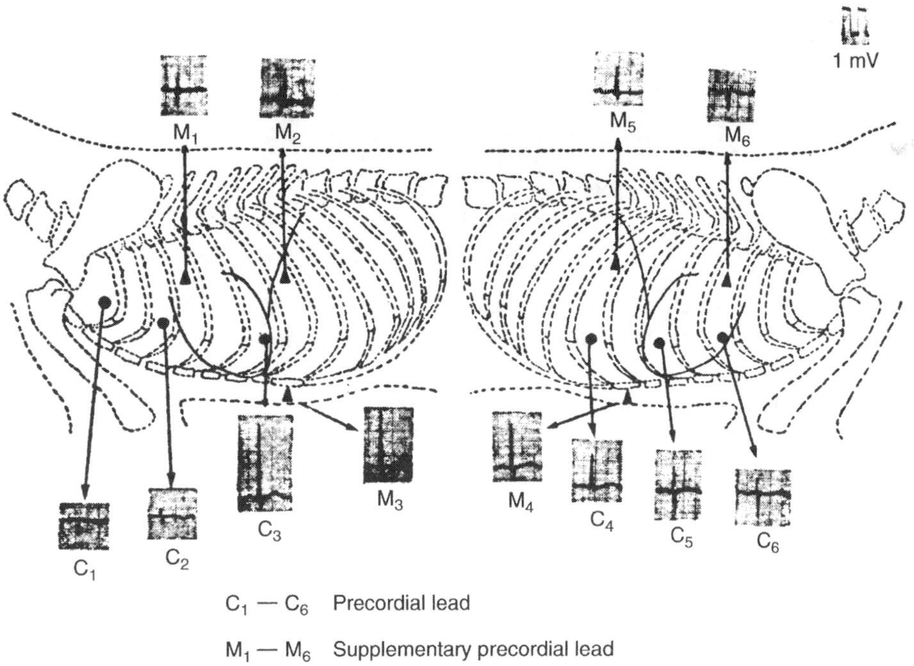

Figure 41.5
The precordial-lead system proposed by Takahashi for clinical use. (After Takahashi [38]. Society of Veterinary Science, Tokyo. Reproduced with permission.)

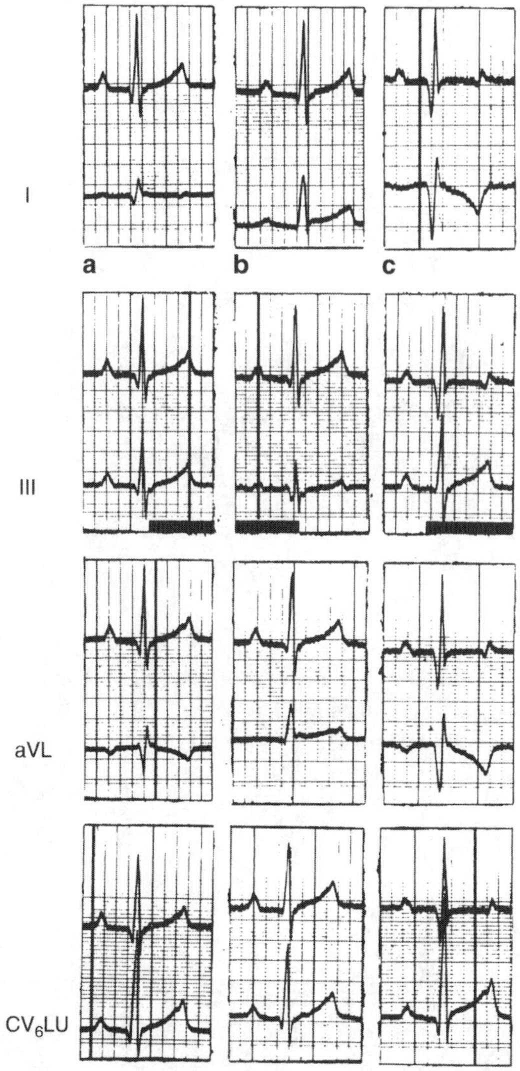

Figure 41.6

The effect of foreleg position on the ECG. The upper tracing in each recording is lead II. The dog was positioned in right lateral recumbency. In column (a) the forelegs are parallel and at right angles to the long axis of the body. In column (b) the right foreleg is pulled forward and the left foreleg pulled backward. In column (c) the right foreleg is pulled backward and the left forward. Note the dramatic alterations in the QRS complexes and T waves in leads I, III, and aVL that falsely suggest changes in ventricular excitation and recovery pathways. See also Table 41.5.

41.2.3 Electrodes

From the standpoint of their electrical characteristics, plate electrodes are superior to electronic alligator clips, while needle electrodes are unacceptable for use in conscious dogs [44]. Electronic alligator clips are recommended, however, because their somewhat inferior electrical characteristics compared to those of plate electrodes are offset by their convenience in application and their tendency to remain in place when dogs struggle. They must all be of the same metal (copper is preferred) to avoid offset voltages that can cause baseline instability. The serrations on the jaws should be flattened to reduce discomfort and maximize the area of contact with the skin. The contact area (including both jaws of the clip) should be approximately 1.0 cm^2. The skin and hair under the electrodes are saturated with a conductive gel or solution. The site of foreleg attachment must be well below the ventral thoracic surface (e.g., halfway between the olecranon and carpus) to avoid the influence of precordial potentials.

41.2.4 Duration of Recording

It is generally possible to restrain untrained and unsedated dogs in the recommended recording position (right lateral recumbency) for about 1 min before struggling becomes excessive. Electronic ECG systems are available now that have enhanced the ability to keep recording until the dog is relaxed. Such systems also allow the electronic storage of the ECG. Because arrhythmias are sporadic events, if it is required to compare arrhythmia prevalence rates in groups such as in the chronic toxicity testing, the duration of recording should be the same in each individual dog and a standard recording time of 1 min has been proposed for such studies [26]. However, it must be emphasized that such a duration actually evaluates less than 0.1% of the QRS complexes that a dog has in 24 h. Most normal beagles have approximately 150,000–175,000 beats in 24 h, and realizing that a 1-min recording only documents approximately 75–100 of these beats gives a perspective for conclusions.

As is well known, for the comparison of the incidence of cardiac arrhythmias, 24-h monitoring of the ECG (Holter monitoring) is optimal [45–47]. In veterinary practice, Holter monitoring is commonly performed. Despite its common use clinically, an appreciation of the need for validation of the accuracy is lacking. When arrhythmias are infrequent, the accuracy of modern analyzing systems is acceptable. However, when complex, frequent, and rapid arrhythmias are present, the accuracy of the analysis suffers. Importantly, quality control and validation of the accuracy of the reports is required, yet few labs have such oversight for canine recordings. Consequently, if Holter monitoring is required for studies, confirmation of high standards in the evaluation of the recordings should be sought.

41.2.5 Artifacts

Since the occurrence of artifacts (see reference [26] for a description of common artifacts) is nearly universal in dog ECGs, they must be identified as such in screening for abnormalities. The most common artifacts that cause problems in diagnosis are:

(a) Skeletal muscle movements
(b) Baseline drift (including oscillations associated with normal breathing and with panting)
(c) Fifty- or 60-Hz electrical inference (➋ Fig. 41.7)

Somatic muscle artifacts appear in most ECGs taken from unanesthetized dogs. Three general types may be recognized:

(a) *Somatic muscle tremor*. The frequency of skeletal muscle tremor artifact is irregular, ranging from about 15 to 35 Hz. Often a constant baseline "jiggle" is caused in long stretches of the record. Typically, such trembling intensifies on inspiration and diminishes on expiration so that it may wax and wane, or appear and disappear periodically.
(b) *Intermediate frequency somatic muscle artifacts*. These are caused by more sporadic muscle twitching that are discontinuous and occur at a frequency of about 1–15 Hz in tense subjects. Their amplitude is generally from 0.1 to 0.5 mV. These artifacts are likely to distort electrocardiographic complexes and may mimic the morphology of P waves.
(c) *Gross muscle movement artifact*. High-amplitude deflections with rapid voltage change (high Vmax) may resemble bizarre QRS complexes and mimic ventricular ectopic beats, or when rapid, paroxysmal ventricular tachycardia. Their amplitudes are often from 1.0 to 5.0 mV or greater and may exceed the excursion limit of the electrocardiographic stylus. Rhythmic tail wagging sometimes transmits movement to electrode–skin-surface interfaces, mimicking a run of ventricular tachycardia (VT), while panting can cause baseline oscillations with the frequency and amplitude of those seen in atrial flutter or atrial fibrillation.

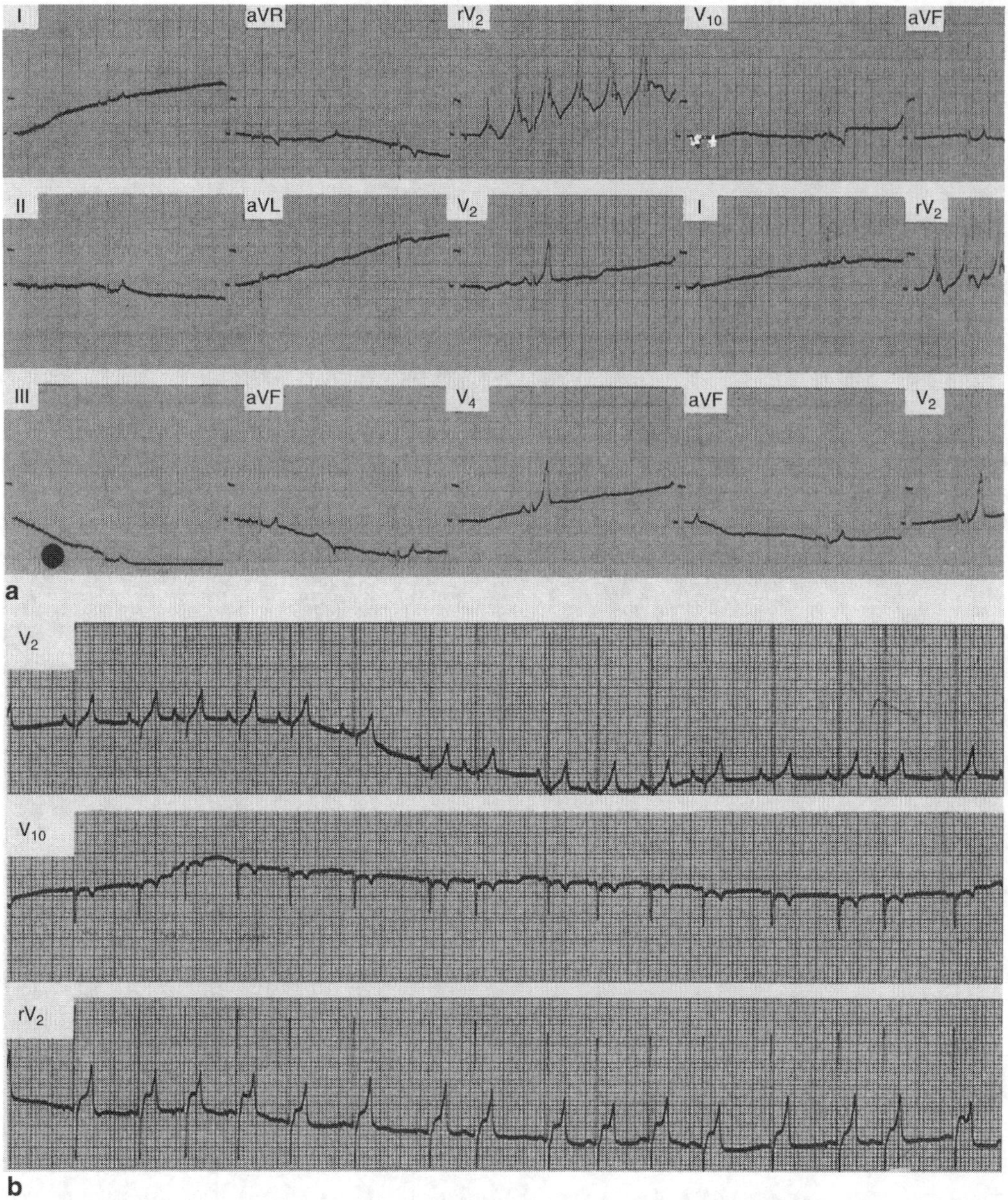

Figure 41.7
The ECG in (**a**) illustrates several artifacts caused by skeletal muscle movement in dogs. Five sets of three leads, each are recorded automatically in each of four 2.6-s panels. Baseline drift is present in all panels. Muscle tremor artifact at 30–35 Hz and intermediate frequency muscle movement artifact at a frequency of about 15 Hz are present in the limb leads. "Tail-wag" artifact simulates paroxysmal ventricular rhythm in lead rV_2; this is caused by the movement of the rV_2 electrode (which is situated between the right thorax wall and the table surface) as the chest is moved synchronously with the wagging tail. The R wave in lead V_2 in the third panel from the left is clipped. In (**b**), the rV_2 lead illustrates simulated ST-segment elevation caused by the apex beat artifact in which the thrust of the heart against the chest wall at the fifth right intercostal space moves the electrode. Note the variability in the ST-segment elevation and contour.

Artifacts present should be identified and the records examined more closely if they:

(a) Resemble electrocardiographic abnormalities
(b) Interfere with interval measurement
(c) Distort many complexes

41.3 The Normal Electrocardiogram

41.3.1 Values

The chief sources summarizing normal values are Lannek [16], Grauwiler [48], Hill [42, 43], Ettinger and Suter [21], Bolton [23], Hahn et al. [31], Tilley [24], and Detweiler [25]. A representative normal ECG is shown in Fig. 41.8. It should be emphasized that there is breed variability in the normal values for the dog. Most dogs "fit" within the "established normals"; however, some breeds have their own particular standard. For example, giant large-boned dogs with a somatotype similar to a Saint Bernard will have R waves of lower amplitude than dogs of similar weight such as a Great Dane. Also, dogs such as the Saint Bernard can have P waves with a duration that exceeds the "usual" standard in most breeds. Most research dogs are beagles; however, in recent years, pharmaceutical and device companies have requested larger mongrels. Obviously, this stresses the importance of pretrial electrocardiographic recordings for comparisons to be made.

Importantly, it should be stated that although most of the differences in the ECG discussed are the result of the conformation of the dog, there are likely differences in the current density of certain ion channels amongst breeds, particularly those of repolarization.

41.3.1.1 Amplitude

Because foreleg position affects limb-lead potentials, the limb-lead data from Hill [42], who standardized foreleg position, are given in Tables 41.2 and 41.3.

41.3.1.2 Intervals

Representative lead II time intervals are 0.03–0.06 s for P, 0.06–0.14 s for PR, 0.03–0.07 s for QRS, and 0.15–0.23 s for QT, all at heart rate 60–180 bpm [23, 26, 33]. In Table 41.4, the available regression formulae are given, and Table 41.5 relates PR and QT intervals and heart rate in beagles. The values represented by these three sources do not agree, probably because the authors used different criteria to determine where the time intervals begin and end.

Today the ECG measurement that garnishes the greatest attention is the QT interval. This duration measured from the first depolarization deflection to the end of the T wave must be considered in the context of heart rate. Thus, the QT interval is often corrected using formulas. The Bazett formula [49] has historically been the one most commonly used across species, although a variety of others have also been proposed [50–57]. However, the limitations and inaccuracies of this formula should be stressed. The Bazett formula was specifically developed by Bazett to correct for the influence of heart rate under specific conditions. Unfortunately, for decades, his original work was applied widely across species and conditions that were not in the original intent for heart rate correction of the QT interval. The majority of formulas either over or under correct the QT interval depending on the rate. This problem has been addressed with a proposal to specifically adjust the QT to the HR [50–57]. The latter is the only one of many proposals for the evaluation of the QT interval. Because the accurate interpretation of the QT interval is imperative to the evaluation of drugs, international conferences have been held to address the proper approach in the pharmacological evaluations of drugs.

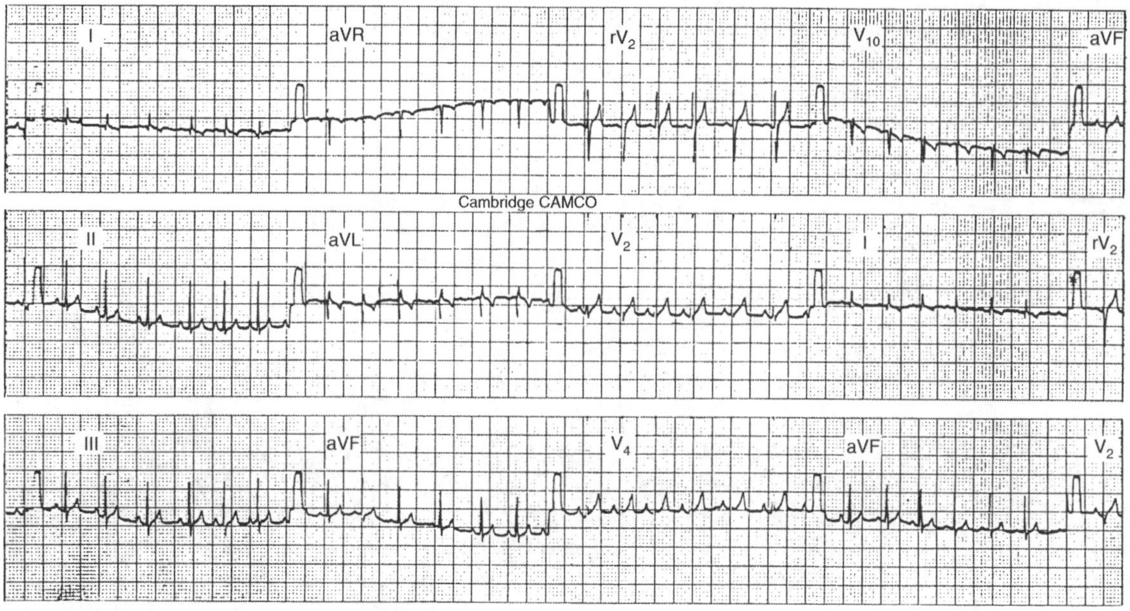

Figure 41.8
A ten-lead ECG from an 8-month-old male beagle recorded with the dog in right lateral recumbency with the forelegs held parallel and at a right angle to the long axis of the body. The recording format for each 2.8-s panel separated by 1.0 mV = 10 divisions sensitivity standardization signal is shown. Panel 5 is the beginning of a rhythm strip of 48-s duration to complete a standard (approximately) 1-min recording interval, which is the duration recommended for routine ECGs taken on groups of dogs in toxicological studies. Note that the R waves in leads V2 and V4 are clipped (by 0.2 mV for lead V2), because the true amplitude determined when the baseline was adjusted exceeds the maximum positive excursion of the stylus. Because such clipping often goes unnoticed, the true amplitude of R waves in these leads frequently present in young dogs (sometimes exceeding 6.0 mV) has been underestimated in the literature except by Lannek [16], who recorded at half sensitivity (10 divisions = 2.0 mV). The T-wave polarity in the dog changes with body position and age in addition to physiologic, pharmacologic, and pathaologic alterations. See also Fig. 35.7.

41.3.2 P-Wave Amplitude and Configuration

The form and amplitude of P waves are more variable in canines than in most other animals. The variability is generally far more pronounced in limb leads than in the conventional thoracic leads (rV_2, V_2, V_4, and V_{10}), where it may be minimal or absent. When the P wave varies in association with respiration (RR interval), it is referred to as a "wandering pacemaker." The P wave also may vary within a dog during serial recordings. The P-wave amplitude can change quickly with changes in heart rate. The faster the heart rate, the taller the P wave. The T wave also changes with heart rate. Both are related to the changes in parasympathetic/sympathetic tone.

41.3.2.1 Wandering Pacemaker

As the heart rate speeds and slows with respiratory sinus arrhythmia, there are cyclic P-wave changes, both in form and amplitude. These are most pronounced in the limb leads in which the P waves are usually of greatest amplitude (e.g., II, III, and aVF). Ordinarily, there is no change in the PR interval, except when the initial part of the P wave becomes isoelectric, or having been isoelectric, becomes positive or negative.

Table 41.2

The amplitude of positive and negative P and T waves (in millivolts) from 70 normal dogs with standardized body and limb positions [43]

	I	aVF	V_{10}	V_4	rV_2
Positive P waves					
Relative frequency	97.1	100.0	47.1	100.0	72.9
Range	0.05–0.15	0.05–0.25	0.05–0.10	0.05–0.25	0.05–0.15
Median	0.05	0.15	0.05	0.10	0.05
Mean	0.063	0.133	0.055	0.126	0.065
Variance	0.00084	0.0037	0.00023	0.0029	0.00073
Negative P waves					
Relative frequency			38.6		10.0
Range			0.05–0.10		0.05–0.10
Median			0.05		0.05
Mean			0.059		0.066
Variance			0.0004		
Positive T waves					
Relative frequency	44.3	45.7	8.6	65.7	98.5
Range	0.05–0.25	0.05–0.45	0.05–0.20	0.05–0.90	0.05–0.80
Median	0.05	0.20	0.05	0.25	0.30
Mean	0.071	0.186	0.075	0.320	0.346
Variance	0.002	0.012		0.048	0.037
Negative T waves					
Relative frequency	51.4	45.7	87.1	15.7	1.5
Range	0.05–0.25	0.05–0.60	0.05–0.40	0.05–0.35	0.15
Median	0.05	0.25	0.20	0.25	
Mean	0.095	0.236	0.211	0.236	
Variance	0.003	0.019	0.006		

The cause of these changes is currently attributed to shifting of the pacemaker within the sinoatrial node, induced by variations of vagal tone (wandering pacemaker within the sinoatrial node). As is well known, experimental vagal stimulation can move the apparent pacemaker site from the upper to the lower end of the sinoatrial node [58–60]. The effect of such vagally induced shifts in pacemaker location in experimental dogs has been confirmed and related to P-wave configurational changes in leads I, aVF, and V_{10} [61]. The pacemaker location was established by plotting wave-front vectors from two sets of bipolar electrodes located close to the sinoatrial node. The inspiration was accompanied by an increase in the heart rate and a shift in the pacemaker location toward the head of the sinoatrial node and the expiration (or electrical stimulation of the right vagus or carotid sinus pressure) by slowing of the heart rate and movement of the pacemaker location toward the tail of the sinoatrial node. The shift of the plotted pacemaker location amounted to about 1.4 cm. With slowing of the heart rate, P waves increase slightly in amplitude in leads I and V_{10} and decrease (often markedly) in lead aVF, frequently becoming notched or bifid in lead aVF. Rarely, P-wave polarity may reverse in leads III (becoming negative) and aVL (becoming positive) during expiration. When pacemaker locations were determined [61], the P-wave in lead aVF was:

(a) Peaked when the pacemaker was located in the head of the sinoatrial node
(b) Of lower amplitude and bifid when the pacemaker was located in the middle area of the sinoatrial node and a low amplitude
(c) Flat when the pacemaker was low in the tail of the sinoatrial node

Table 41.3

Amplitudes of the QRS complex (in millivolts) from 70 normal dogs with standardized body and limb positions

	I	II	III	aVR	aVL	aVF	V2	V4	V10	rV2
Q wave										
Relative frequency	90.0	98.6	87.1	1.5	37.1	97.1	40.0	81.5	100.0	None
Range	0.05–0.60	0.05–1.20	0.05–1.00	0.30	0.05–0.65	0.05–1.05	0.05–0.35	0.05–0.55	0.15–1.20	
Median	0.15	0.30	0.25		0.30	0.25	0.05	0.15	0.65	
Mean	0.22	0.38	0.27		0.30	0.31	0.09	0.17	0.65	
Variance[a]	0.022	0.075	0.046		0.0324	0.052	0.004	0.013	0.057	
R wave										
Relative frequency	100.0	100.0	100.0	100.0	100.0	100.0	100.0	100.0	98.6	100.0
Range	0.10–1.50	0.45–3.00	0.25–2.35	0.05–0.70	0.05–0.90	0.20–2.40	0.50–4.00	0.25–5.40	0.05–0.80	0.15–3.60
Median	0.60	1.60	1.10	0.25	0.15	1.35	1.90	1.90	0.30	1.00
Mean	0.64	1.61	1.12	0.27	0.23	1.37	1.96	1.97	0.31	1.11
Variance[a]	0.117	0.334	0.247	0.032	0.042	0.366	0.593	0.828	0.0286	0.449
S wave										
Relative frequency	8.6	32.9	34.3	100.0	64.3	35.7	80.0	54.3	None	100.0
Range	0.05	0.05–0.35	0.05–0.70	0.15–2.30	0.10–1.10	0.05–0.55	0.05–1.60	0.05–1.30		0.05–1.60
Median		0.15	0.20	1.10	0.40	0.10	0.30	0.20		0.60
Mean		0.16	0.22	1.09	0.42	0.16	0.39	0.28		0.67
Variance[a]		0.009	0.029	0.223	0.054	0.014	0.093	0.067		0.156

[a]Standard deviations are not listed because of the number of asymmetrical distributions found in this series

Table 41.4

Linear regression formulae for time intervals (R-R in milliseconds)

	Grauwiler [48]	Lannek [16]
PQ (ms)	82 + 0.023 (R-R) ± 30 (3SD)	
QT (ms)	136 + 0.084 (R-R) ± 10 (5SD)	160 + 0.03 (R-R) ± 38 (2SE)

It appeared that in each dog there were several specific locations in the sinoatrial node for pacemaker activity, and that a change in vagal tone caused a sudden shift in the location of pacemaker activity from one of these preferred sites to another. This is in accordance with the view that major changes in sinoatrial node discharge rate result from, say, suppression of a faster pacemaking cell and shift of pacemaking dominance to a slower discharging cell, rather than from a slowing in rate of the original pacemaker cell. Stellate ganglion stimulation or norepinephrine injection can also cause shifts in pacemaker site and alterations in P-wave morphology [62]. Mapping studies [63] have demonstrated three regions of origin of atrial depolarization located at anterior, middle, and posterior regions of the anterior vena cava-right atrial junction. The middle position was near the sinoatrial node. The anterior and posterior positions were 1–2 cm away from the sinoatrial node area. Changes in P-wave morphology resulted from abrupt shifts in the region of earliest activation among these three points, two of which were outside the sinoatrial node. The multiple points of origin of atrial activation might represent a trifocal, distributed pacemaker or the epicardial exits of three specialized pathways that rapidly conduct the impulse from a single focus. Higher heart rates and taller P waves in lead II were associated with the higher centers of early activation, while lower rates and lower-amplitude notched P waves were linked with the lower centers of early activation. Others have demonstrated that in the dog, the sinus node is long (4 cm in length), providing multiple areas for discharge that are influenced by autonomic tone and the affected spontaneous discharge rate. In summary, when ECGs

Table 41.5
Variation of PR and QT intervals with heart rate in the beagle ECG [165]

Group number	Heart rate (bpm)	Sex	Number of ECGs	Interval (s)	
				PR	QT
1	61–80	Male	16	0.105 ± 0.029	0.214 ± 0.037
		Female	16	0.100 ± 0.027	0.213 ± 0.041
2	80–100	Male	96	0.102 ± 0.028	0.206 ± 0.031
		Female	86	0.104 ± 0.032	0.203 ± 0.028
3	101–120	Male	210	0.096 ± 0.025	0.195 ± 0.030
		Female	219	0.101 ± 0.028	0.192 ± 0.027
4	121–140	Male	353	0.095 ± 0.020	0.183 ± 0.031
		Female	324	0.097 ± 0.028	0.182 ± 0.029
5	141–160	Male	295	0.092 ± 0.026	0.173 ± 0.028
		Female	326	0.095 ± 0.026	0.174 ± 0.028
6	161–180	Male	172	0.088 ± 0.020	0.167 ± 0.026
		Female	187	0.078 ± 0.020	0.165 ± 0.028
7	181–200	Male	51	0.086 ± 0.020	0.160 ± 0.020
		Female	47	0.083 ± 0.021	0.160 ± 0.023
Student's t for groups 1 vs 7		Male		3.366*	5.036**
		Female		3.198*	4.942*

Values for PR and QT intervals are the means and 95% range for the numbers of ECGs shown. The asterisks indicate the degree of significance associated with the differences between the values for the lowest and highest heart-rate groups
$^{*}p < 0.001$; $^{**}p < 0.01$

Table 41.6
QRS patterns in various leads. For each lead the patterns are given in the order of frequency of occurrence. Patterns that are encountered less often or rarely are placed in parentheses. Refer to Fig. 41.6 and note how these patterns are altered by changing forelimb position

Lead	Pattern	Lead	Pattern	Lead	Pattern
I	QR, qR, (qRS)	aVR	rS, rSr′, (qrS)	rV_2	RS, (Rs)
II	qR, zRs, Rs, (QrS)	aVL	Qr, QR, (qRs)	V_2	Rs, qRs, (qR)
III	qR, qR, Rs, (rS)	aVF	qR, Rs, (QrS)	V_4	Rs, qRs, (qR)
				V_{10}	Qr, rS, (qR, QR)

are monitored in the dog, the autonomic tone will alter the P wave, and thus, this parameter is not one which is likely to be reliable in detecting differences between recordings that have impact on a study.

Given the above discussion, if the P wave is measured, it has not to our knowledge been stated which P wave to measure on an ECG recording (e.g., tallest, shortest, average). This in itself is problematic. We have arbitrarily decided to always measure the tallest P wave. It should also be emphasized that in an evaluation, if the heart rate changes, the P-wave values are likely to change as a consequence of rate, and not because of pathology of the atria.

41.3.3 QRS Complex

Typical QRS patterns in the various leads are given symbolically in Table 41.6. The QR pattern in lead I corresponds to the high incidence of counterclockwise rotation of the QRS vector loops in dogs (see Sect. 41.5).

41.3.4 ST-T Wave

In dogs, the ST segment is seldom horizontal and often the onset of T is imperceptible. Except for leads rV_2 and V_{10}, the polarity of the T waves may be positive or negative. In ➋ Fig. 41.9, a series of ST-T patterns observed in lead II are depicted. ST segment or ST junction (STj) deviations, especially in leads II, aVF, V_2, and V_4 are common. These ST deviations seldom remain consistent in serial records, but vary or disappear entirely from record to record. Also, the degree of ST deviation often varies cyclically with the changing R ~ R intervals of respiratory sinus arrhythmia. In this case, the degree of deviation increases with shorter preceding R-R intervals. Recent work in serial recordings of beagles and mongrels studied at Cornell University has demonstrated further that the polarity of the T wave is dependent on the age of the dog. The younger the dog, the more likely the T wave is to be positive. The T wave does not "stabilize" in the dog until approximately 12 months of age. This is an important notation for long-term studies in the dog in which the evaluation of repolarization as assessed by the ECG is important. The T wave is changing during the first year of life, so if a study is begun when the dog is 3–6 months of age, changes at 12 months of age must be assessed in conjunction with the expected changes seen developmentally with a more negative T wave at the older age. Again, the importance of control dogs is obvious for all studies.

41.3.5 U Wave

When present, U waves are generally most prominent in leads rV2, V_2, and V_4. They may vary in amplitude or disappear in consecutive beats and are usually less than 0.05 mV in amplitude and concordant with positive T waves. The TU interval approximates 0.04 s and U-wave duration approximates 0.07–0.09 s. The electrophysiological genesis of U waves is controversial [64]. They have been attributed to:

(a) Purkinje cell repolarization potentials
(b) An afterpotential gradient from endocardium to epicardium created during ventricular relaxation [65]

Tall U waves appear in hypokalemia, and in humans, have been described in hypomagnesemia, and in patients receiving digoxin, phenothiazines or tricyclic antidepressants [64]. U-wave inversion in humans has been described in hypertension, valvular, pulmonary, congenital, and ischemic heart disease [64]. In dogs, U-wave amplitude exceeding 0.05 mV has been associated with hypertension and anemic heart disease, but this finding is not consistent and is therefore diagnostically unreliable. U-wave values of 0.08 ~ 0.15 mV may occur in otherwise normal dogs.

41.3.6 Evolution During the First 3 Months of Life

Consistent with the functional and morphological cardiovascular changes during the first few weeks after birth [66–68], the mean (sometimes termed modal) QRS vector moves from a rightward, cranial and ventral orientation to the leftward, caudal and ventral direction [69–71]. The principal cause of this change is that the two ventricles are of about equal mass at birth, and this 1:1 ratio of right to left ventricular mass changes to a ratio of 1:2 or 1:3 [67] depending on how the ventricles are separated.

The mean QRS axis (derived from leads I and aVF) in the frontal plane is directed toward the right and either cranially or caudally at birth. By the 12th week, the mean QRS axis is directed to the left and usually resides in the left caudal quadrant of the frontal plane. In the transverse plane (leads I and V_{10}), the mean QRS axis changes from the right ventral quadrant at birth to the left ventral quadrant at week 12. The shift in the sagittal plane is from a more cranial ventral direction to a more caudal ventral orientation.

The R/S ratio in the left precordial leads increases primarily between the first and second week. It is less than 1.0 at birth and exceeds 1.0 at the sixth week. Later, this ratio becomes larger and may reach infinity because the S wave disappears. There is little change in the R/S ratio in lead rV2. In the frontal plane, the mean QRS vector shift may be from the right cranial through the left cranial to the left caudal quadrant (i.e., clockwise), or from the right cranial through the right caudal to the left caudal quadrant (i.e., counterclockwise).

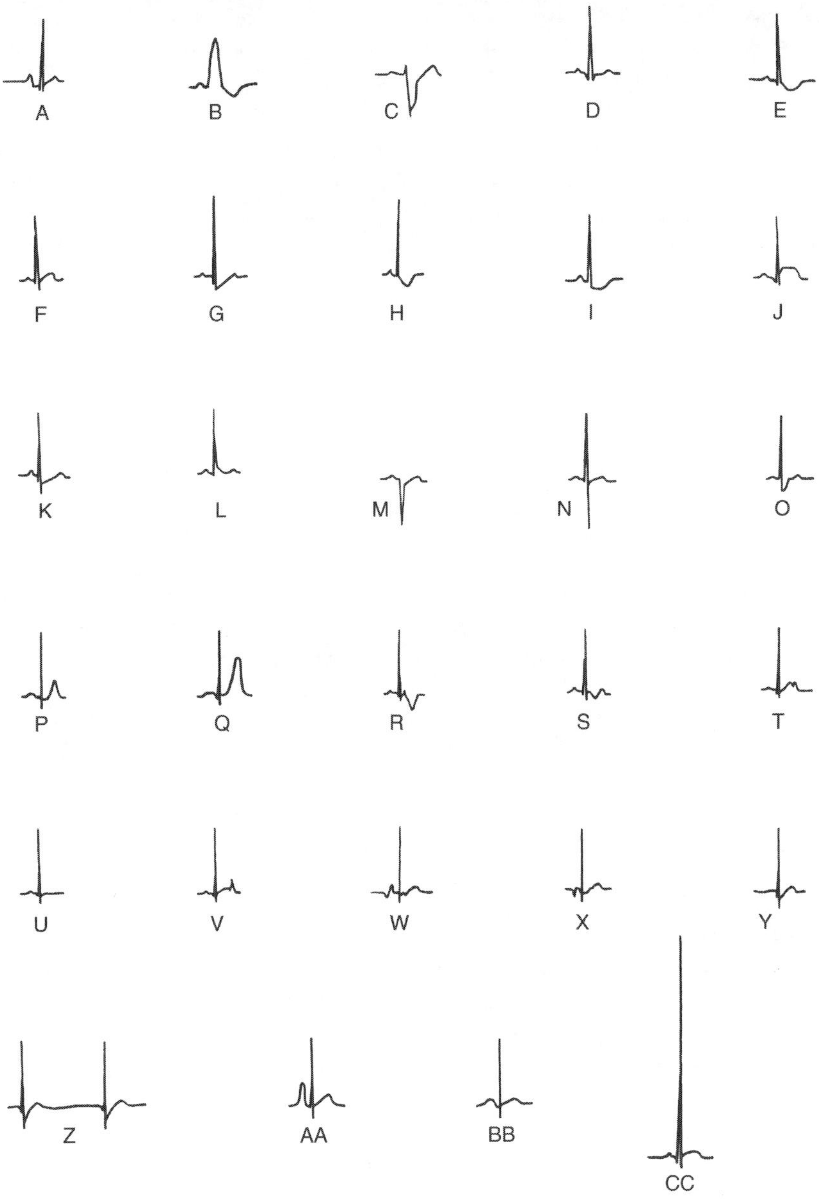

Figure 41.9

Pattern code: A, increased Ta amplitude; B, left bundle branch block like pattern (LBBB-like); C, right bundle branch block like pattern (RBBB-like); D, flat ST segment; E, coved ST segment; F, arched ST segment; G, ascending ST segment; H, descending ST segment; I, depressed ST segment; J, elevated ST segment; K, STj depressed; L, STj elevated; M, QS complex; N, deep S wave; 0, broad S wave; P, symmetrical T wave; Q, tall T wave; R, deep negative T wave; S, diphasic T wave: −/+ or +/−; T, notched T wave: – or −t; U, flat T wave; V, dome-dart ST-T complex; W, diphasic P wave: −/+ or +1/−; X, negative P wave; Y, flat P wave; Z, absent P waves; AA, tall P waves: >0.4 mV; BB, broad P waves: >60 ms; and CC, tall RV_2, V_4: >5.0 mV.

41.3.7 Classification

All canine ECGs fall into one of the following three general categories: normal, normal variant, or abnormal.

(a) *Normal ECGs.* These are ECGs in which all variables (heart rate, time intervals, amplitude and polarity of waves in all leads, and rhythm) fall within the specified ranges given in Tables 41.2–41.6.
(b) *Normal variant ECGs.* These ECGs have characteristics as described in Table 41.1 without pathology of the heart.
(c) *Abnormal ECGs.* These include any electrocardiographic feature that is outside normal limits and is definitely more prevalent in the presence of heart disease or cardiotoxicity.

41.3.8 Normal/Abnormal ECG Screening

Screening becomes necessary when records of large numbers of experimental dog ECGs are examined to eliminate any dogs with abnormal or questionable ECGs. Since the beagle is the breed normally used in such studies (e.g., chronic toxicity trials in preclinical drug testing), the criteria listed apply primarily to this breed, although they would apply equally well to most mongrels and other breeds of similar size and conformation.

Since over 95% of experimental beagle ECGs will ordinarily be within normal limits, the initial goal in examining pretest records is to identify the few abnormal records rapidly. This is done by quickly scanning the records for a series of features that establish normality. The criteria should be conservative so that borderline records will be scrutinized more closely.

41.3.8.1 Normal Criteria

The following criteria are customized from our experiences in the interpretation of the ECG recorded from the beagle. These criteria are not exactly the same as the normal published values for dogs in general.

(a) Intervals and rhythm
  (i) Heart rate: 60–190 bpm (R-R interval 316–1,000 ms)
  (ii) P interval: less than 50 ms
  (iii) PR interval: greater than 70 ms, less than 130 ms
  (iv) QRS interval: less than 60 ms
  (v) QT interval: less than 220 ms
  (vi) Sinus arrhythmia present
  (vii) No abnormal arrhythmias
(b) Wave amplitudes and patterns
  (i) T wave amplitude: 0.05–1.0 mV in any lead
  (Flat T waves are not necessarily abnormal especially in leads I, aVR, aVL, and V_{10})
  (ii) ST deviation: less than 0.20 mV in limb leads and less than 0.25 mV in chest leads
  (iii) ST-T complex: no dome-dart complexes
  (iv) R wave:
    - Less than 3.0 mV in limb leads rV_2 and V_{10}
    - Less than 5.0 mV in leads V_2 and V_4
    - More than 0.5 mV in II, III, and aVF
  (v) QRS axis in frontal plane: −20° to +110°
  (vi) P wave: less than 0.4 mV
  (vii) R/S amplitude ratio:
    - Greater than 0.5 in rV_2; and
    - Greater than 0.9 in V_2 and V_4

(viii) QRS: no rs, rS, Rs or RS patterns
(ix) S/QRS duration ratio: less than 50%

Records that do not meet these normal criteria require closer examination to determine if they are abnormal or normal variants.

41.3.9 Normal Variants

A large number of electrocardiographic characteristics are sufficiently rare or puzzling to attract attention in records from dogs not known to have cardiac disease. These are grouped together here as normal variants. They are based primarily on findings in young dogs (5–18 months old) of a single breed, the beagle [26].

41.3.9.1 QRS Complex

(a) *Variant frontal plane QRS vector loop*. This is also known as the "butterfly" QRS vector loop and requires a special mention. In a small number of beagles and in some other breeds as well, the mean QRS frontal plane axis is negative, ranging from −30° to −110° (➋ Figs. 41.11 and ➋ 41.12). The QRS complexes in thoracic leads are within normal limits. In leads II, III, and aVF, the QRS pattern is characterized by having a small r wave (e.g., qrS or QrS pattern) or a W-shaped pattern in which the "r" wave fails to reach the isoelectric line. In lead I, there is usually a deep Q and pronounced R wave, the descending limb of which is slurred or notched as it approaches the isoelectric level. The QRS in lead aVL is usually markedly positive. If a frontal plane vector diagram is constructed from leads I and aVF, the early forces are directed counterclockwise to the right and craniad, then return close to their starting point before passing counterclockwise to the left and craniad. Thus, two connected vector loops are formed vaguely resembling butterfly wings, hence the name "butterfly" QRS vector loop (➋ Figs. 41.10 and ➋ 41.11).

This variant has neither been studied systematically nor investigated electrophysiologically. Since the thoracic lead complexes are essentially normal with tall R waves in leads V_2 and V_4 and typical configuration in leads rV2 and V_{10}, there does not appear to be an electrophysiological abnormality. It is obvious, however, that the vector forces producing tall R waves in leads V_2 and V_4 do not do so in the limb leads. This pattern, when seen in 5–7-month-old dogs, may persist into adulthood, or within a few months or a year, tall R waves may appear in leads II, III, and aVF, while the chest lead complexes remain the same. The hearts are normal at necropsy.

The conclusion that can be drawn from the information at hand is that these low-amplitude R waves in leads II, III, and aVF and the "butterfly" QRS frontal plane vector loop result from the position of the heart in the thorax relative to the limbs, such that the major vector forces during the inscription of QRS are directed largely craniad and ventrad. As shown in ➋ Figs. 41.10 and ➋ 41.11, the two components of the QRS frontal plane-constructed vector loops may be more or less symmetrically disposed, with the initial part of the loop being directed toward the right arm and the final part of the loop directed toward the left arm.

(b) Other QRS complex variants (see ➋ Table 41.1 for most common variants):
(i) Circular frontal plane QRS vector loop
(ii) QRS slurring and notching
(iii) Delayed terminal forces causing broad SII, III, and aVF and broader RaVL; patterns variously (and often erroneously) described as incomplete RBBB, left anterior fascicular block
(iv) Rr' pattern in II, III, aVF, and rV_2 unaccompanied by broad S waves (see also (v))
(v) "Foot" or slur terminating QRS
(vi) Variable SrV_2 and V_2 amplitude

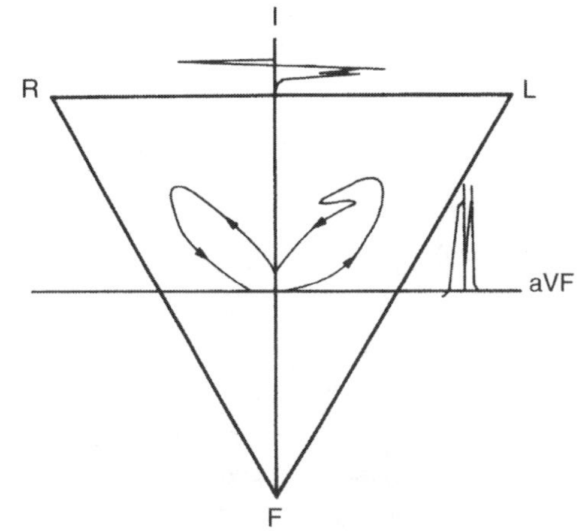

Figure 41.10

Butterfly QRS complex. A vector diagram constructed from QRS complexes of leads I and aVF. The QRS complexes are each traced on their respective zero potential lines, which intersect at zero potential in the center. The winged appearance of the vector diagrams of the variant QRS complexes gave rise to the term "butterfly" QRS complex. Note that the QRS complexes from leads I and aVF are drawn on their respective zero or isopotential lines which intersect at the middle of the triangle. To view lead aVF in its conventional orientation as recorded, the diagram must be rotated 90° counterclockwise. To view lead aVF in its conventional orientation, the diagram must be rotated 180° (i.e., lead aVF appears upside down here). The positive pole for lead I is toward L from its zero potential line and that for lead aVF is toward F, in the diagram.

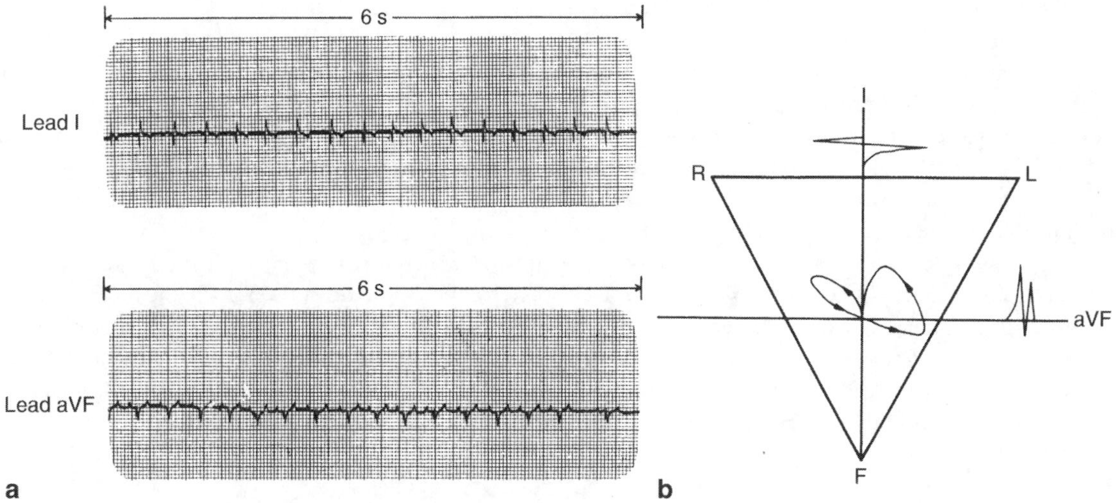

Figure 41.11

Butterfly QRS complex: (a) leads I and aVF and (b) vector diagram constructed from the QRS complex as in leads I and aVF. Note that the QRS complexes from leads I and aVF are drawn on their respective zero or isoelectric lines which intersect at the middle of the triangle. To view lead I in its conventional orientation as recorded, the diagram must be rotated 90° counterclockwise. To view lead aVF in its conventional orientation, the diagram must be rotated 180° (i.e., lead aVF appears upside down here). The positive pole for lead I is toward L from its zero potential line and that for lead aVF is toward F.

41.3.9.2 T wave and ST-T Complex: U Wave

(a) T wave in V_{10} flat, diphasic −/+ or positive
(b) T wave in rV_2 negative terminally or diphasic +/−
(c) Biphasic T waves in other leads
(d) T wave variable in amplitude and form (including polarity reversal) related to varying R-R intervals
(e) Tall-peaked T waves
(f) Dome-dart ST-T complex without QT prolongation
(g) Prominent U wave (>0.5 mV)
(h) TU fusion

41.3.9.3 P Wave and T_a Wave

(a) P-wave reversal in limb leads except lead I (wandering pacemaker within the sinoatrial node)
(b) Notched P waves
(c) High-amplitude (negative deflection) T_a waves (>0.05 mV; up to 0.15 mV in lead II)

41.3.9.4 Amplitude

(a) ST deviation (with accompanying ST-segment deviation or followed by rapid return of ST segment to baseline level) up to 0.20 mV in limb leads and 0.25 mV in chest leads. Variable deviation related to R-R-interval duration
(b) Amplitude variation of QRS and T waves with breathing cycle and R-R-interval duration
(c) Tall P wave in leads II and aVF (greater than 0.4 mV). Tall R wave in leads V_2 and V_4 (> 5.5 mV)

41.3.9.5 Rhythm and Rate

(a) Bradycardia (40–55 bpm)
(b) Tachycardia (rate > 200 bpm)
(c) First-degree AV block (PR > 0.14 s)
(d) Second-degree AV block

41.4 Normal Rhythm

41.4.1 Sinus Rhythm and Rate

In the denervated heart of the dog, the sinoatrial node discharges regularly at an intrinsic rate of about 90–120 bpm (see reference [72] for literature). Parasympathectomy alone results in a persistently rapid rate of 140–160 bpm. In intact, resting, and adult dogs, heart rates of 70–120 bpm are representative; with exercise, rates of 220–325 bpm are recorded; and in newborn and young puppies, the range is 140–275 bpm. In ECGs, the heart rates tend to be more rapid than in the resting dog owing to the effect of restraint. The ranges given in ➋ Sect. 41.3.1 and ➋ Table 41.5 are characteristic of those obtained from ECGs and show that the usual range in ECGs from beagles is 100–180 bpm. At the more rapid frequencies, respiratory sinus arrhythmia is decreased or abolished.

41.4.2 Respiratory Sinus Arrhythmia

In resting dogs, the heart rate decreases and increases with the breathing cycle, owing to a waxing and waning of vagal activity with respiration, although the primary reason is a central effect on the heart rate with variation in autonomic tone. The usual pattern is acceleration with inspiration and deceleration with expiration, but this relationship is not absolute [73].

The control systems that interact to cause sinus arrhythmia include the radiation of respiratory center activity to medullary cardiovascular centers, the cardiac component of the Hering–Breuer reflex, the Bainbridge reflex (reflex cardioacceleration caused by increased stretch of the right atrium and great veins during inspiration), and the baroreceptor reflex (reflex cardiac slowing when the blood pressure rises during the accelerating phase of sinus arrhythmia). The degree of respiratory sinus arrhythmia decreases as vagal tone is decreased with exercise or excitement. The degree of sinus arrhythmia can be expressed as an index I [74]:

$$I = (\text{standard deviation of R-R} \times 100)/\text{mean R-R}.$$

This index is variable in serial records. For accuracy, 100 R-R intervals should be averaged for the calculation. Certain drugs can decrease this index or abolish sinus arrhythmia altogether at heart rates where it is ordinarily present. Other methods to evaluate heart rate variability have been studied extensively. These include time and frequency domains to assess the balance of autonomic tone [75].

In resting dogs, the rapid beats in respiratory sinus arrhythmias often occur in sets of two (bigeminy), three (trigeminy), or sets of three beats with a fourth beat during expiration [76]. Sets of four or five beats are rare. These cyclic changes in the R-R-interval duration are often accompanied by cyclic alterations in the amplitude of electrocardiographic complexes and, sometimes, in the form (and occasionally, polarity) of T waves. In puppies, Lange [77] found that respiratory sinus arrhythmia does not appear before 4 weeks of age.

Usually, the respiratory sinus arrhythmia tends to become more pronounced (i.e., the disparity between R-R intervals increases) as the heart rate slows. Rarely, this fails to occur and a slow heart rate with a relatively regular rhythm may be present. Certain drugs appear to depress the brain stem and ameliorate or abolish respiratory arrhythmia in dogs (see ➲ Sect. 41.10.1.3).

Over the years, the breeding of beagles for research purposes has led to some dogs being particularly calm with slow rates and more pronounced sinus arrhythmias.

41.5 Normal Vectorcardiogram

Starting in 1957 with the study of Horan et al. [78], vectorcardiography became a useful investigation tool in canine cardiology. Various lead systems have been employed: for example, the Wilson equilateral tetrahedral reference system [31, 78–82], the Frank corrected lead system [35, 37] (➲ Fig. 41.4), and the orthogonal-lead system of McFee and Parungao [23, 34, 43, 82–84] (➲ Fig. 41.3). Of these, the only lead system specifically corrected for the dog is that of McFee and Parungao [34]. Rarely is vectorcardiography performed today in clinical veterinary patients, in screening research dogs, or as a part of the evaluation of treatment. We continue a discussion of this method here because, although not performed, it is a valuable concept to grasp in the often two-dimensional representation of the frontal limb leads to which we are limited. In electrophysiology laboratories, extensive mapping has replaced the vectorcardiogram.

41.5.1 Normal Values

Vectorcardiographic data from 50 mongrel dogs using the axial-lead system of McFee and Parungao are given in ➲ Tables 41.7–41.9 [84]. In these tables, the term prevalent direction or θ expressed in degrees is used (see ➲ Chap. 13) rather than the mean of observed angular directions. This variable is an estimate of the average angle of the prevalent direction, and the range of distribution of the individual points is given in the tables for 5–95% of the grouping.

Table 41.7
Angles and magnitudes of P, QRS, and T maximal vectors [84]

	Maximal P vector	Maximal QRS vector	Maximal T vector
Frontal plane			
Amplitude (mV)[a]	0.42 ± 0.12	2.17 ± .0.44	0.28 ± 0.12
Prevalent direction $\hat{\theta}$ (°)	71	69	84
5–95% directional range (°)	44–85	38–95	–55 to –124
Horizontal plane			
Amplitude (mV)[a]	0.21 ± 0.11	1.72 ± 0.36	0.36 ± 0.19
Prevalent direction $\hat{\theta}$ (°)	9	71	87
5–95% directional range (°)	–55 to 72	–66 to 137	30 to –163
Left sagittal plane			
Amplitude (mV)[a]	0.41 ± 0.11	2.27 ± 0.43	0.42 ± 0.22
Prevalent direction $\hat{\theta}$ (°)	95	117	161
5–95% directional range (°)	70–125	47–153	119 to –104

[a]Mean ± standard deviation

Table 41.8
Maximal projections of QRS vectors on W, Y, and Z [84]

	Mean ± SD (mV)	5–95% range (mV)
X axis		
Initial right	–0.21 ± 0.30	0.0 to –0.60
Left	0.91 ± 0.43	0.20–0.160
Terminal right	–0.18 ± 0.28	0.00 to –0.80
Y axis		
Initial cranial	–0.04 ± 0.08	0.00 to –0.20
Caudad	1.9 ± 0.63	1.00–2.75
Terminal cranial	–0.21 ± 0.16	0.00 to –0.50
Z axis		
Sternal	–1.50 ± 0.54	–0.50 to –2.25
Vertebral	0.98 ± 0.44	0.20–1.80

Table 41.9
Spatial angles and magnitudes of instantaneous QRS vectors [84]

			Magnitude[c] (mV)	
Time after onset of QRS (ms)	$\hat{\alpha}^a$ (°)	$\hat{\beta}^b$ (°)	Mean ± SD	5–95% range
5	97	86	0.29 ± 0.17	0.12–0.46
15	81	128	1.81 ± .0.45	0.63–2.59
25	29	154	2.11 ± 0.53	1.03–2.89
35	292	128	1.28 ± 0.63	0.39–2.31
45	262	84	0.60 ± 0.48	0.08–1.62
55	257	97	0.27 ± 0.22	0.08–0.98

[a] $\hat{\alpha}$, longitude or angular deviation from the left in the horizontal plane (0°–360°) for the spatial prevalent direction
[b] $\hat{\beta}$, colatitude or angular deviation from the cranial (0°–180°) for the spatial prevalent direction
[c]Magnitude = $(X^2 + Y^2 + Z^2)^{1/2}$

41.5.2 P and T Vector Loops

With the lead system of McFee and Parungao, the P loop is a thin ellipse pointing sternally and to the left. The mean P-wave axis is within ±90° of the maximal QRS vector in 100% of the cases in the frontal plane, 82% in the horizontal plane, and 98% in the left sagittal plane [84]. With the McFee and Parungao lead system, the T loop is a moderately open ellipse with the mean axis within ±90° of the maximal QRS axis in 80% of the cases in the frontal plane, 77% in the horizontal plane, and 82% in the left sagittal plane. With the Frank system, the T vector loop is usually concordant with the QRS vector loop in the frontal (86%), transverse (90%), and sagittal (97%) planes [34]. Thus, the major T vector axis points either cranially or caudally in the frontal plane, ventrally in the horizontal plane, and sternocranially or sternocaudally in the left sagittal plane.

41.5.3 QRS Vector Loops

The incidence of counterclockwise rotation of QRS vector loops in frontal, left sagittal, and transverse ("horizontal") planes is very high with the three (Wilson tetrahedral, Frank, and McFee) lead systems. For example, with the McFee system, counterclockwise rotation of QRS in the frontal, transverse, and sagittal planes, respectively, was 52–60%, 80–98%, and 98–100% [43, 83, 84]. These percentages are similar to those reported for the frontal and left sagittal planes (85% and 100%) with the Wilson tetrahedron reference system and in all three planes with the Frank system (66%, 97%, and 100%, respectively) [35]. This predominance of counterclockwise QRS rotation in the dog frontal plane is in sharp contrast to the clockwise rotation more commonly found (65%) in the human VCG, which has a similar mean frontal axis. Since the spread of excitation through the heart in both species is similar, the difference must relate to the differing orientation of the heart within the thoracic cavity of man and dog.

Chest conformation has a distinct effect on the electrical axis of the heart such that narrow-chested breeds (e.g., Doberman pinschers, German shepherds) have more ventrally oriented QRS axes in the transverse plane and more caudally oriented QRS axes in the frontal plane than do broad-chested breeds (e.g., cocker spaniels and boxers) [83].

Vectorcardiographic appearances have been reported in dogs with a variety of spontaneous and experimental abnormal cardiac conditions such as right ventricular hypertrophy [23, 82, 85, 86], imperforate ventricular septal defect [87], congenital peritoneopericardial diaphragmatic hernia [88], patent ductus arteriosus [23, 89], idiopathic cardiomyopathy [23], coronary occlusion and other localized myocardial destruction [72, 73], ventricular ectopic beats [90], and bundle branch block patterns [23, 91–94].

41.5.4 The Vector Diagram

Despite the knowledge gained from these various investigations, vectorcardiography has been little used in the routine diagnostic cardiology in dogs. The vector diagram, however, derived from the limb-lead scalar ECG has a special usefulness in evaluating the QRS vector changes in the canine ECG [85] (➋ Fig. 41.12). This is because the normal range of mean QRS axis values is broad and the normal limits for the various breeds have not been established statistically. For example, in right ventricular hypertrophy, the usual counterclockwise rotation of the constructed QRS vector loops in the frontal, transverse, and sagittal planes is reversed. In the frontal plane, early forces are usually directed cephalically or caudally toward the left and late forces toward the right and cranially. In left ventricular hypertrophy, in contrast, the QRS axis and loop are not usually altered in form or in direction of inscription.

41.5.5 Evolution During the First 3 Months of Life

With the Wilson tetrahedron system, the QRS loops shortly after birth are directed chiefly toward the right (frontal and transverse planes) and cranially (frontal and sagittal planes); at week 12, the major orientation is toward the left and caudally. The inscription of the QRS vector loops in all three planes changes from clockwise just after birth to a counterclockwise rotation at week 12 [69, 70].

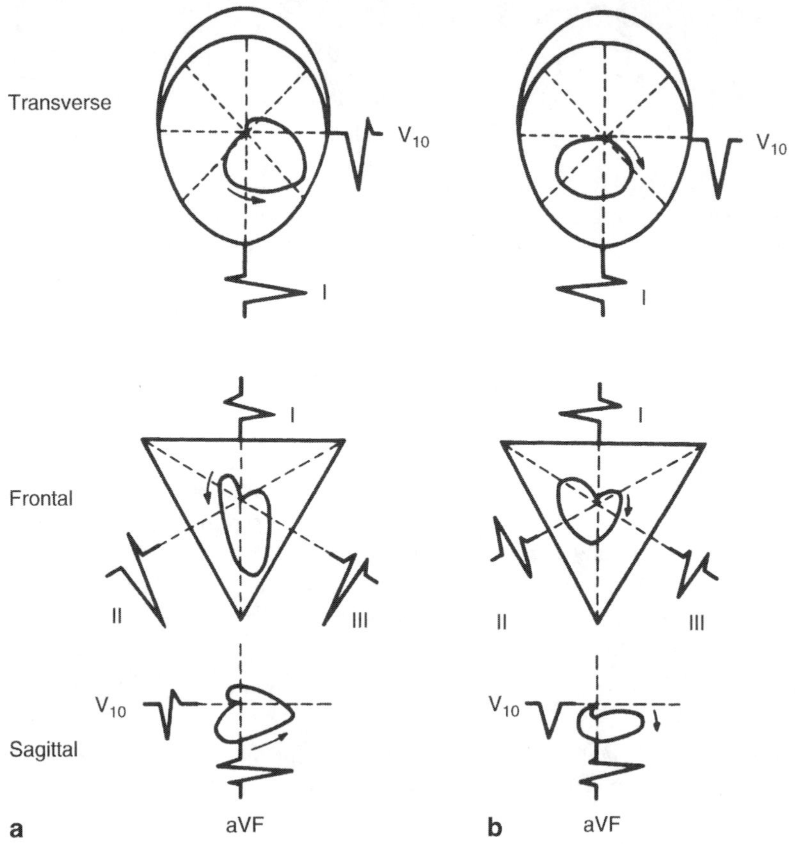

Figure 41.12

Vector diagrams in three planes constructed from limb-lead scalar ECGs (leads I, II, III, aVF, and V_{10}): (a) normal pattern and (b) right ventricular hypertrophy pattern. Note that the direction of rotation is counterclockwise in the normal record (a) and clockwise in right ventricular hypertrophy (b) in all three planes. (After Detweiler et al. [78]. © New York Academy of Sciences, New York. Reproduced with permission.)

The direction of the T-wave vector loop at birth in some dogs produces negative T waves in rV_2 and positive T waves in V_{10}. This changes with aging, so that by the fourth to the eighth week, the adult pattern (positive T wave in rV_2 and negative T wave in V_{10}) is present in most dogs. Positive T waves in V_{10} may persist in some dogs beyond the third month and become negative only between months 5 and 7. This T-wave polarity change from negative to positive in the right precordial lead (rV_2) is opposite to that observed in human infants, where the right precordial T waves are generally positive at birth and become negative after the first week after birth.

41.6 ECG Descriptors and Pattern Code

A number of electrocardiographic features that change spontaneously or are altered by drugs and disease are often described rather than quantified. Thus, wave contour, polarity, and amplitude characteristics require descriptive terms that are sufficiently specific to convey the intended meaning. Many terms in use are part of the jargon of electrocardiography, and few have found their way into medical dictionaries. Consequently, changes in meaning, elimination from use, or modification of terminology are continuing processes. The terms offered here have been found useful as descriptors for various recognizable characteristics of the dog ECG. Some are self explanatory while others require definition or examples. They are compiled here as an approach toward standardizing descriptive terms.

41.6.1 Descriptors Applicable to All Leads

(a) *Amplitude variable.* Amplitude variation of any of the electrocardiographic waves (P, Q, R, S, and T) from beat to beat. Example of use: P-wave amplitude variable from beat to beat in all limb leads.

(i) Related to previous R-R interval. The form and amplitude of P and T waves and the degree of deviation of the STj or ST segment may vary predictably with changes in the R-R interval during respiratory sinus arrhythmia. The P-wave changes have been discussed in previous sections. The changes in T waves and STj or segment deviations occur because the degree of recovery of conductivity in the heart is time dependent and less complete with shorter R-R intervals. When the R-R intervals are short enough, the conduction and recovery processes are somewhat aberrant and the form of the ST-T complex changes.

(ii) Electrical alternation. While this may occur in diseased hearts, minor variations (e.g., of the S wave in one of the chest leads) occurring in alternate beats are found in occasional records from the same dog and are not associated with any other abnormality.

(b) *Low amplitude.* Although all wave amplitudes vary in serial records, this term is used when the recorded potential is far below the average for a particular wave in a given lead. When cardiomyopathy is extensive, the entire QRST complex may be of reduced amplitude. The term is relative and the amplitudes may not be below the normal range when this term is applied. It is used primarily to describe P waves and R, R+S, or T waves. In the case of P and T waves, their polarity must be indicated as positive (+) or negative (−).

(c) *High amplitude.* The reverse of (b)

(d) *Polarity reversal.* The term is used when P and T waves change their polarity from beat to beat or from record to record in serial ECGs.

(e) *STj deviation.* The J (or junction) point is the sharp inflection marking the end of the QRS complex and the beginning of the ST segment. It deviates to some extent from the baseline in normal records. Increased STj deviation may accompany cardiac hypoxia, necrosis and epicardial inflammation. Depending on the polarity of the deviation, it is described as elevation or depression of the STj.

(f) *ST-segment deviation.* This term is used when the entire ST segment is elevated or depressed above or below the isoelectric line. The ST segments are seldom horizontal in dogs. They either ascend or descend from the STj to the T wave and a clear inflection point separating the ST segment from the beginning of T is often not present. Accordingly, it is necessary to measure the degree of ST deviation at some defined point in the ventricular complex. It is recommended that this point be located 0.04 ms after the J point.

(i) Variable ST-segment deviation (see (a) (i) above)

(g) *Coved ST segment.* Concave upward

(h) *Arched ST segment.* Convex upward

(i) *Ascending ST segment*

(j) *Descending ST segment*

(k) *Flat ST segment*

The terms (g)–(k) are used alone or together with items (e) and (f).

(l) *Dome-dart ST-T complex.* A relatively stereotyped drug effect (rarely also seen in control records) consists of ST-T complexes in which the ST segment and first limb of the T wave form a rounded arch. This terminates in a spike that peaks and is followed by the terminal limb of the T wave. These dome-dart T waves are positive in leads rV_2 and V_2, and negative in lead V_{10} (see Fig. 41.13).

(m) *Symmetrical T waves.* Ordinarily, the first limb of the T wave has a more gradual slope than the final limb. T waves that increase in amplitude may also become symmetrical (e.g., with excitement or during hyperkalemia).

(n) *Concordant T waves.* These have the same polarity as the predominant direction of the QRS complex and discordant T waves have the opposite polarity.

(o) *Prolonged terminal QRS forces.* Delays in intraventricular conduction or heritable patterns of ventricular excitation produce broad S waves.

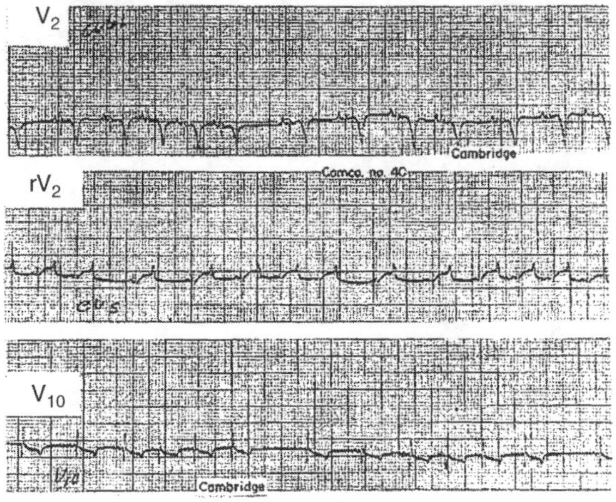

Figure 41.13
Dome-dart T waves in leads rV_2 and V_{10} accompanied by deep negative T waves in lead V_2 and QT prolongation (0.28 s). This beagle dog was receiving toxic doses of a tricyclic antidepressant neuroleptic drug with antihistaminic and local anesthetic properties. (After Detweiler [25]. ©CRC Press. With permission.)

(p) *Prominent U wave.* U waves are ordinarily absent or less than 0.05 mV in amplitude. They are regarded as prominent when their amplitude exceeds 0.05 mV.
(q) *R/S ratio.* This amplitude ratio is usually greater than 0.5 in rV2 and greater than 0.9 in V2 and V4. In disease (e.g., cardiomyopathy, focal cardiac hypertrophy), the ratio changes ordinarily by a reciprocal decrease in R and increase in S amplitudes.
(r) *High-amplitude Ta wave.* The Ta waves are usually absent or of low amplitude. A distinct increase may be apparent after certain drugs.
(s) *Afterpotential.* This is a distinct positive or negative potential change following the T wave that is of far greater amplitude and duration than the U wave.

41.6.2 Pattern Code for Various Types of PQRST Complexes

Various influences including autonomic and electrolyte changes, drugs, and disease affect the contour of the ECG. For descriptive purposes, frequently occurring patterns may be coded by capital letters as shown in Fig. 41.3.

41.7 Electrocardiographic Abnormalities: Diagnostic Criteria

The diagnostic criteria for specific cardiac lesions in dogs are far less reliable than those for humans for several reasons:

(a) The database for normal ranges is small.
(b) The great variation in body size and chest conformation among the various breeds of dogs increases the morphological extremes of electrocardiographic complexes in chest leads.
(c) The T waves are labile, change morphology spontaneously, and may reverse polarity in all conventional leads except rV_2 and V_{10}.
(d) ST-segment and STj-point deviation from isopotential is variable and in some normal dogs is large.
(e) Changes in the forelimb position produce marked swings in frontal plane vectors.
(f) Limb-lead P-wave morphology is normally extremely variable.

41.7.1 Hypertrophy

The electrocardiographic criteria for chamber enlargement are rather reliable for right ventricular hypertrophy (RVH), but unreliable for left ventricular hypertrophy (LVH) and for atrial enlargement.

41.7.1.1 Right Ventricular Hypertrophy

The 12 electrocardiographic criteria for the diagnosis of RVH are given in ➋ Table 41.10 [86]. Three or more of these criteria were present in 93% of dogs with RVH, the corresponding false-positive rate being about 1% [86]. An example is given in ➋ Fig. 41.14.

Table 41.10

Electrocardiographic criteria in order of relative frequency in which they occurred in 70 dogs with right ventricular hypertrophy and in 70 normal dogs matched by breed, age, and sex [85]. The first column lists the rank order of 12 diagnostic electrocardiographic criteria for right ventricular hypertrophy. Two of the 12 criteria have the same rank order of 1, since their relative frequencies (given in the fourth column) are the same. The numbers listed in the sixth column are totals representing the number of abnormal dogs out of 70 which had three or more of the electrocardiographic criteria listed by rank order *above* and row in which the number appears. Thus, for example, 57 dogs out of 70 had three or more of the six criteria listed under rank orders 1–5

Rank	Criteria	Lead	Relative frequency in dogs with right ventricular enlargement	Relative frequency in normal dogs [a]	Number of abnormal more dogs with three or ECG abnormalities	Relative frequency
1	S > 0.80 mV	V2	79.4	7.1		
1	Frontal plane QRS mean electric axis from +103° clockwise to 0°	I and aVF	79.4	0		
2	S > 0.70 mV	V4	78.8	2.9		
3	S > 0.05 mV	I	77.1	0		
4	R/S ratio < 0.87	V4	62.1	0		
5	Transverse plane QRS mean electric axis from +105° clockwise to −31°	I and V10	60.7	1.4		
6	QRS algebraic sum > −0.20 mV	I	55.7	0	57	81.4
7	S > 0.35 mV	II	50.0	0	60	85.7
8	Sagittal plane QRS mean electric axis from +91° clockwise to −12°	aVF and V10	32.8	0	61	87.1
9	A > 0.30 mV	aVR	29.4	0	62	88.6
10	Positive T > 0.25 mV	I	12.9	0	63	90.0
11	R′	II	11.4	0	64	91.5
					65	92.9

[a] None of the normal dogs had ECG abnormalities

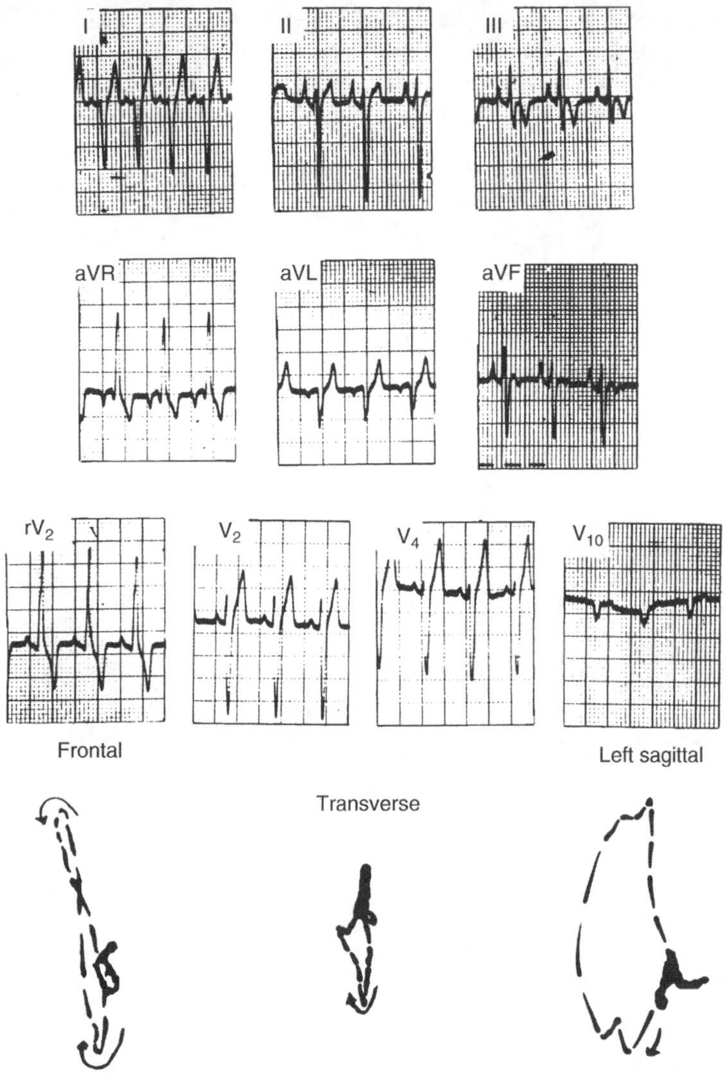

Figure 41.14

Right ventricular hypertrophy electrocardiographic patterns and VCGs from a 6-month-old Dalmatian with congenital pulmonic stenosis. The mean QRS axis is shifted far to the right (about −145 degrees), TrV_2 is negative, TV_{10} is positive and the vector loops rotate in a clockwise direction except for the final part of the figure-of-eight loop in the frontal plane (Wilson equilateral tetrahedron lead system). (Courtesy of D. F. Patterson.)

41.7.1.2 Left Ventricular Hypertrophy

Reliable criteria for the electrocardiographic diagnosis of LVH in dogs are not available. This is unfortunate since LVH secondary to mitral insufficiency is the most common form of chamber enlargement seen clinically in dogs [24].

The electrocardiographic findings compatible with LVH are tall R waves, wide QRS complexes, deep negative T waves in leads II, III, aVF, and V_{10} and a mean QRS axis in the frontal plane directed toward the left. The reason these findings are not diagnostic is that they may be present in normal dogs and absent in dogs with marked LVH [24]. The QRS axis and loop are usually not altered in dogs with confirmed LVH. Young dogs between 6 and 24 months of age are especially likely to have tall R waves, exceeding 3 mV in leads II, III, and aVF, and exceeding 4 mV in leads V_2 and V4. On the other hand, such tall R waves in older dogs in the presence of QRS intervals greater than 0.07 s, a frontal mean QRS axis of less than 45°, deep negative T waves in leads I, II, III, and aVF, and the notching or slurring of the QRS complex are findings that indicate the possibility of LVH. In some cases of LVH, enormous R waves are present (e.g., 7.7 mV in lead III) [95]. On the other hand, in a series of 41 Newfoundlands with congenital subaortic stenosis, 31 had normal ECGs and the electrocardiographic features found in 10 cases with the most severe stenoses (atrial fibrillation in 4, ventricular extrasystoles in 3, RII greater than 2.5 mV in 3, and notching of QRS in 1) were not diagnostic of LVH [96].

41.7.1.3 Atrial Enlargement

Commonly in textbooks, a tall P wave is described as indicating right atrial enlargement and a wide P wave as indicating left atrial enlargement. However, these criteria are too often incorrect. It is likely better to just assume that if the P wave is either too tall or too wide that one of the atria is enlarged. Also, with atrial enlargement, the large P wave is sometimes accompanied by exaggerated T_a waves [23, 97]. Tall P waves are common in certain breeds (e.g., greyhounds) and P waves exceeding 0.4 mV are seen occasionally in young beagles as well as other breeds. Also, remember that a high sympathetic tone with fast rates results in tall P waves. This is also true if adrenergic drugs or parasympatholytic drugs are given.

41.7.2 Bundle Branch Block

The QRS interval is 0.08 s or longer in both right and left bundle branch block. In the right bundle branch block (RBBB) (➋ Fig. 41.15), the widest part of the QRS complex is directed downward in leads I, II, III, aVF, the left precordial leads V_2 and V_4, and V_{10}, while it is upright in lead aVL, the right precordial lead rV_2, and in aVR. In the left bundle branch block (LBBB) this pattern is reversed. The widest part of the QRS complex is upright in the standard limb leads and aVF, as well as the left precordial leads V_2, V_4, and V_{10}. It is negative in the leads aVR, aVL, and rV_2.

The electrocardiographic pattern of RBBB can occur in the absence of organic heart disease in dogs [78, 95]. Complete RBBB (QRS complex > 0.08 s) or incomplete RBBB (same distribution of S waves but the QRS complex < 0.08 s) can exist (➋ Fig. 41.15). When an S wave is identified in the anterior leads and left precordial leads, the following are potential reasons:

(a) Right bundle branch block (clinical and experimental)
(b) Left anterior hemiblock (experimental section of the left anterior fascicle of the left bundle branch [96])
(c) Localized hypertrophy of the free right ventricular wall at the outflow tract [94, 95]

Although the ECG criteria for bundle branch blocks are generally accepted amongst veterinary cardiologists, there is some descent as to their validity. Rosenbaum et al. [98] found that in experimental dogs, a section of the left anterior or the left posterior fascicles of the His bundle failed to produce characteristic changes in the limb-lead ECG when the heart was in the normal position, although this could be achieved by placing the heart in a more "horizontal" position by constructing a sling from the pericardium. Similarly, in a study by Okuma [99], the changes in the limb-lead and leads V_1 and V_5 of dogs caused by sectioning these fascicles were minor.

41.7.3 Bypass Conduction

(a) *Preexcitation.* Preexcitation occurs when the more distal heart is prematurely activated because supraventricular impulses travel via an accessory pathway to the distal AV node or the ventricles. This additional path of conduction is in addition to the normal conduction pathway. The accessory pathways are indeed anatomical structures [100]. In the dog, preexcitation is usually associated with a short PR interval. If this preexcitation has episodes of supraventricular tachycardia, the Wolff–Parkinson–White (WPW) syndrome is the most common diagnosis (➋ Fig. 41.16). This conduction disorder is rare in dogs, occurring in approximately 1 in 2,000 experimental beagles [25] and in 1 in 3,000 clinic patients [101]. Bypass tachycardias can be concealed such that during normal heart rates, the premature activation is not identified by a short PR interval.

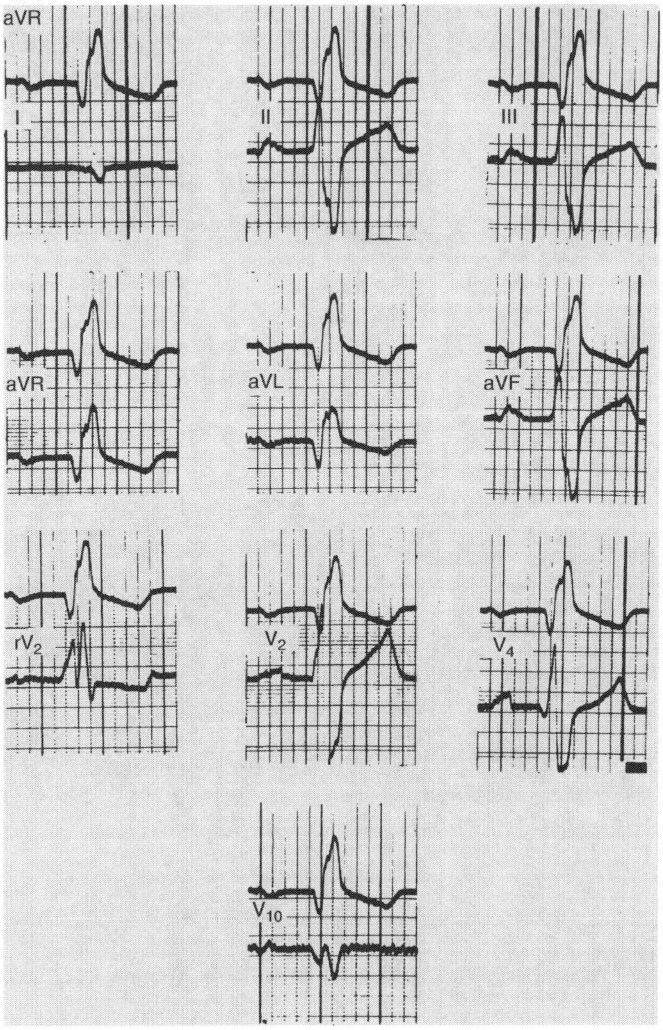

Figure 41.15

Right bundle branch block pattern from a 12-year-old male poodle with chronic mitral insufficiency, myocardial disease, and congestive heart failure. The upper trace is lead aVR in each record. Note the RR′ complex in lead rV_2; the wide part of the QRS complex is negative in leads I, II, III, aVR, V_2, V_4, and V_{10}, while it is positive in leads aVR and aVL. Paper speed is 75 mm s^{-1}.

41.8 Rhythm Abnormalities

41.8.1 Supraventricular Arrhythmias

41.8.1.1 Sinus Rhythms

The normal heart rate for dogs under clinical conditions is about 60–120 bpm. The heart rates given in data from dog ECGs [25] vary from 34 to 238 bpm. Such extremes would be abnormal if sustained, but when observed in a short ECG strip, are frequently not representative of the unperturbed heart rate. Heart rates determined from the ECG represent values obtained under some degree of duress. They must be interpreted with care.

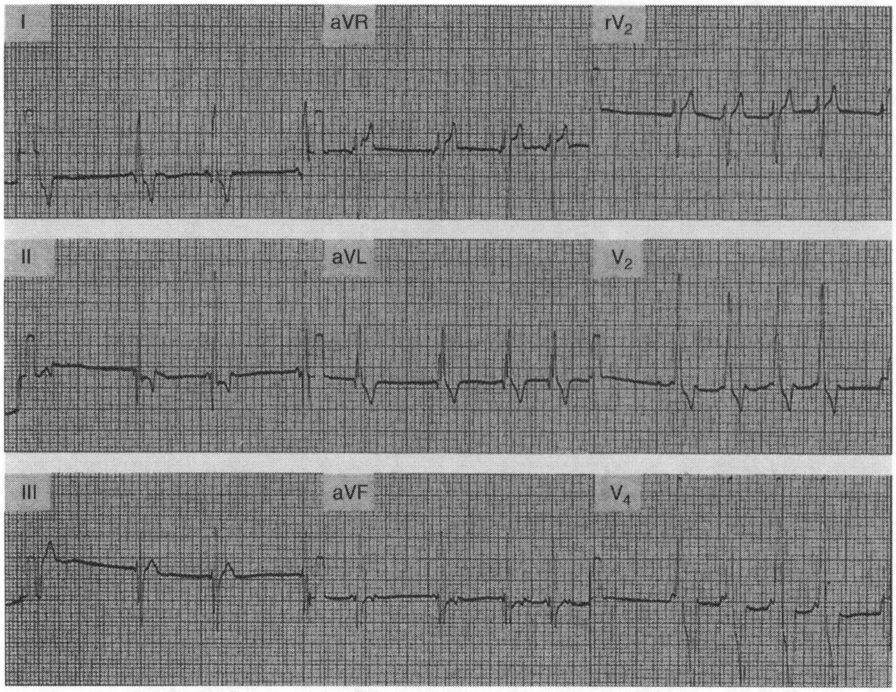

Figure 41.16
Wolff-Parkinson-White syndrome in a research beagle control record. The time intervals are: PR, 0.04; QRS, 0.08; and PJ, 0.12 s. The delta wave is positive in all three precordial leads.

Sinus bradycardia and tachycardia. Heart rates in individual ECGs below 60 and above 180 bpm may be arbitrarily designated bradycardia and tachycardia, respectively.

(a) *Sinus bradycardia. Sinus bradycardia* can be found occasionally in otherwise normal dogs. Long and associates [102, 103] demonstrated a substantial reduction in the average heart rate of groups of dogs (e.g., from a mean of 86 bpm upon first examination to 66 bpm for a group of 60 dogs) with restraint for 2 h in a quiet, isolated environment. Under these circumstances heart rates as low as 48 bpm were observed. Sick sinus syndrome is a disorder that is characterized by sinus bradycardia that often has long sinus pauses that can extend for more than 8 s and result in syncope. Moreover, some dogs are afflicted with episodes of supraventricular tachycardias in addition to the bradycardia and pauses. Sick sinus syndrome is most common in miniature schnauzers, cocker spaniels, dachshunds, and West Highland white terriers.
(b) *Sinus tachycardia.* It can be physiologic, in response to a pathologic condition or actually pathologic in and of itself. During the recording of an ECG in some dogs, the heart rate can be as high as 180 bpm, but rates exceeding 190–200 bpm are suspect. Based on Holter recordings, normal healthy dogs can obtain heart rates as high as 300 bpm for very brief times (<10 s) associated with excitement, fear, or acute pain. Kennel activities such as feeding time and the presence of strangers are likely to induce rapid heart rates in groups of dogs, even though the ECGs are taken at a distant location. Sinus tachycardia is expected in certain disease states; for example, anemia, fever, hemorrhage, shock, and congestive heart failure. As mentioned before, it has been found in toxicological studies that sustained (e.g., several hours daily over several days) sinus tachycardia exceeding 190–200 bpm results in myocardial damage with lesions (hemorrhage and necrosis, followed by fibrosis) in left ventricular subendocardium and papillary muscles.

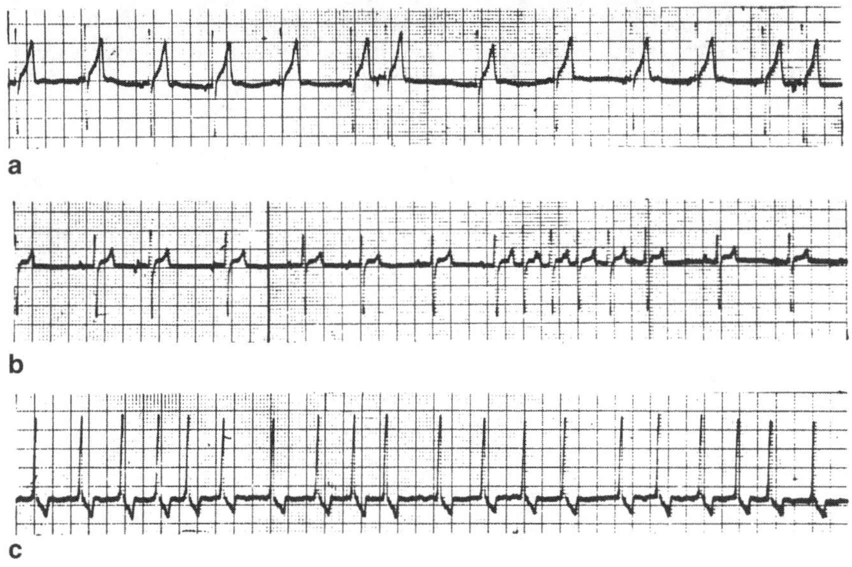

Figure 41.17

Atrial arrhythmias: (a) atrial premature beats. Two premature atrial beats with deeply inverted P waves are present, one in the middle and one at the end of the record (lead rV_2 of an eleven-year-old male cocker spaniel with mitral insufficiency). Part (b), a run of atrial tachycardia occurs after the seventh complex The P waves vary in shape from beat to beat, the PR interval is prolonged (0.16 s, first-degree AV block) and P waves are partially buried in the preceding T waves during the tachycardia (lead rV_2 of a nine-year-old male boxer with aortic body tumor infiltrating the right atrium). Part (c), atrial fibrillation. The ventricular rate has been slowed to 140 bpm with digoxin therapy (lead V_2 of a thirteen-year-o1d male Doberman pinscher with mitral insufficiency and congestive heart failure).

41.8.1.2 Atrial Rhythms

(a) *Atrial extrasystole* (see Fig. 41.17a). Premature atrial complexes are also known as atrial extrasystoles. They can originate from any location above the AV node. They may be singles, couplets, or triplets. If excessively premature, atrial extrasystoles can arrive at the AV node during the refractory period and not result in ventricular activation. In such cases, only a nonconducted P wave is identified.(b) *Atrial tachycardia* (see Fig. 41.17b). As with other atrial arrhythmias, the normally variable P waves complicate this diagnosis. Paroxysmal atrial tachycardia (PAT) with (AV) block occurs in digitalis intoxication in dogs especially when complicated by hypokalemia [98]. Such arrhythmias secondary to digitalis intoxication are far less common today because dosing (including the total dose used) is more cautious than in the past.

(c) *Atrial flutter*. It occurs at an atrial rate of usually 340–440 bpm in dogs. Although mechanistically the atrial flutter may vary from atrial tachycardia, difficulty can exist in clearly separating the two rhythms. Atrial flutter classically is characterized by undulating waves (F wave) that result in no baseline. There may be 2:1 AV block which results in a constant RR interval, although more commonly the conduction is varied and an irregular RR pattern prevails.

(d) *Atrial fibrillation*. It occurs at an atrial rate of 500–750 bpm in the dog. See Fig. 41.17c. The atrial rhythm results in fine undulations of the baseline identified as "f waves." Classically atrial fibrillation is a tachyarrhythmia in which the RR interval is very irregular.

41.8.1.3 Atrioventricular Junctional (Nodal) Rhythms

While studies in the rabbit [103] indicate that there are no pacemaker cells in the AV node, pacemaker cells have been found in the AV node of dogs [104]. Because of earlier findings in the rabbit, the term atrioventricular nodal rhythms was changed to AV junctional rhythms or His-bundle rhythms since pacemaker cells are found in the region of the AV node, especially in the His bundle [105]. Because the precise origin of these ectopic beats cannot be determined in the ECG, the more general term, AV junction, is preferable. The AV junctional escape beats occur as a subsidiary pacemaker when the sinus rate falls below 60 bpm. The inherent escape rhythm rate of the AV junctional region is 40–60 bpm. Single escape complexes can occur after long sinus pauses or after AV nodal block. The AV junctional escaperhythms occur as a sustained rhythm to "rescue" the heart when the sinus node fails to fire sufficiently or the AV node is blocked.

41.8.2 Ventricular Rhythms

41.8.2.1 Ventricular Escape Rhythm

When the sinus node and AV junctional subsidiary pacemakers fail to fire, the heart is rescued by a ventricular escape beat or, if sustained for more than three complexes in a row, a ventricular escape rhythm. For most dogs, the rate of the secondary pacemakers from the Purkinje system in the ventricle is between 20 and 30 bpm. These can be seen on occasion during Holter recordings of normal dogs that are sleeping.

41.8.2.2 Ventricular Extrasystoles

Ectopic premature complexes originating from the ventricle are termed as: (1) ventricular extrasystoles, (2) ventricular premature complexes, or (3) premature ventricular complexes. Their prevalence in control records from beagles varies from 0.6% to 1.0% [25], which approximates to that reported in human ECGs [21]. Accordingly, the occurrence of ventricular extrasystoles in an occasional animal may not be related to drugs or disease. When ventricular extrasystoles alternate with normal complexes, the term to describe the pattern is ventricular bigeminy. See ➋ Fig. 41.18.

41.8.2.3 Ventricular Parasystole

For this and more complex arrhythmias, long recording times of the cardiac rhythm are required. Of the electrocardiographic criteria for parasystole, the most controversial is the mathematical relation of the interectopic intervals. As this type of ventricular rhythm has been studied, the initial "rules" to make this diagnosis have been altered as the complexity has been more fully understood. In the simplified understanding of ventricular parasystole, the ventricular ectopic pacemaker discharges with complete regularity, and the interectopic intervals (the intervals between two consecutive ectopic beats that are separated by intervening sinus beats) are nearly an exact multiple of the ectopic cycle length (the time interval between two consecutive ectopic beats without any intervening sinus beats). This is the case in humans with this arrhythmia [8], but is more difficult to identify in dogs. The ectopic cycle length can vary for a variety of reasons including:

(a) Changes in the fundamental discharge rate
(b) Delay in conduction from the ectopic focus to responsive myocardium

In dogs, the most likely cause of alterations in the ectopic cycle length is the presence of sinus arrhythmia associated with the parasystole. It has been demonstrated experimentally [106–108] that a parasystolic focus is protected, but not insulated by a surrounding area of depressed excitability, and can be modulated by electrical events in surrounding tissues; the ectopic cycle length can be lengthened or shortened by electrotonic influences [109]. In man, the parasystolic focus is seldom completely regular and can be altered by vagal reflexes (e.g., carotid sinus pressure), while the timing of normotopic beats during interectopic intervals [110] and the ectopic cycle length can change abruptly without intervention [111]. Also, in the intact heart, neurogenic, endocrine, and hemodynamic factors can change the discharge rate of parasystolic foci

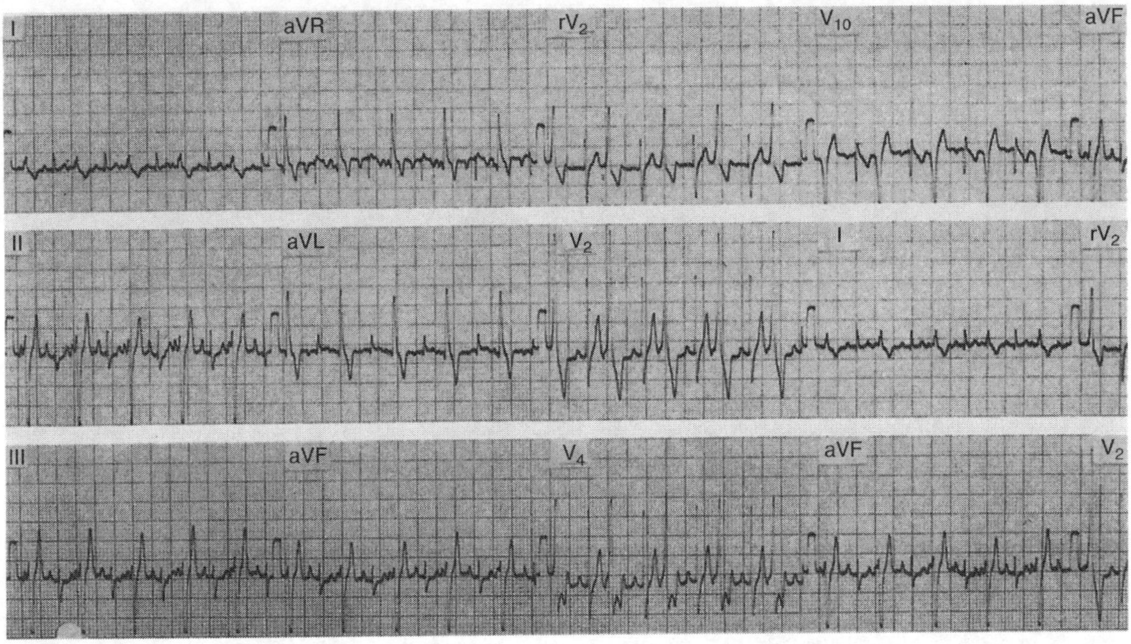

Figure 41.18
Ventricular bigeminy. This arrhythmia appeared in a control group beagle bitch in a routine ECG taken on day 1813 of a chronic drug trial. A record taken 16 days later was normal and no cardiac lesions were observed at necropsy 14 days later.

[110]. Further, the human heart can give rise to coupled ventricular extrasystoles and ventricular parasystole at different times from the same focus [111].

41.8.2.4 Ventricular Tachycardia

Ventricular tachycardia (VT) is a rapid succession of impulses originating from below the AV node. The rate of VT is one which is greater than the inherent rate of the Purkinje fibers in the ventricle which is approximately 30 bpm. In the dog, the usual rate of VT can be one that approximates a normal sinus rhythm (named idioventricular tachycardia) or be as rapid as 500 bpm. The latter rate usually can be maintained for less than a second.

The causes of VT in the dog are varied and include noncardiac and cardiac diseases. Noncardiac diseases include trauma both directly to the heart and indirectly such as to the cranium, gastric torsion, neoplasia, and varied causes of hemodynamic compromise. Several breeds have inherited ventricular arrhythmias that can cause sudden death [112–138]. Boxers are afflicted with arrhythmogenic right ventricular cardiomyopathy [112–118]. This disease results in a distinctive VT that is characterized by a positive QRS complex VT in the anterior (leads II, III, and aVF) leads since the arrhythmia originates from the right ventricle. Rate of the VT in the boxer is very high (frequently 250–300 bpm) and often results in syncope when it is sustained. Some boxers die suddenly while others do not, either due to treatment or chance, but do develop congestive heart failure as the disease progresses. German shepherds are afflicted with inherited ventricular arrhythmias that primarily are manifested in young dogs between the ages of 13 and 70 weeks. Sudden death can occur without prodromes that give evidence of the disease [119? –138]. Most commonly, the VT in the German shepherd is usually characterized as a nonsustained polymorphic rapid ventricular tachycardia. Ventricular tachycardia also is documented in Dobermans with dilated cardiomyopathy and other dogs with myocardial failurc.

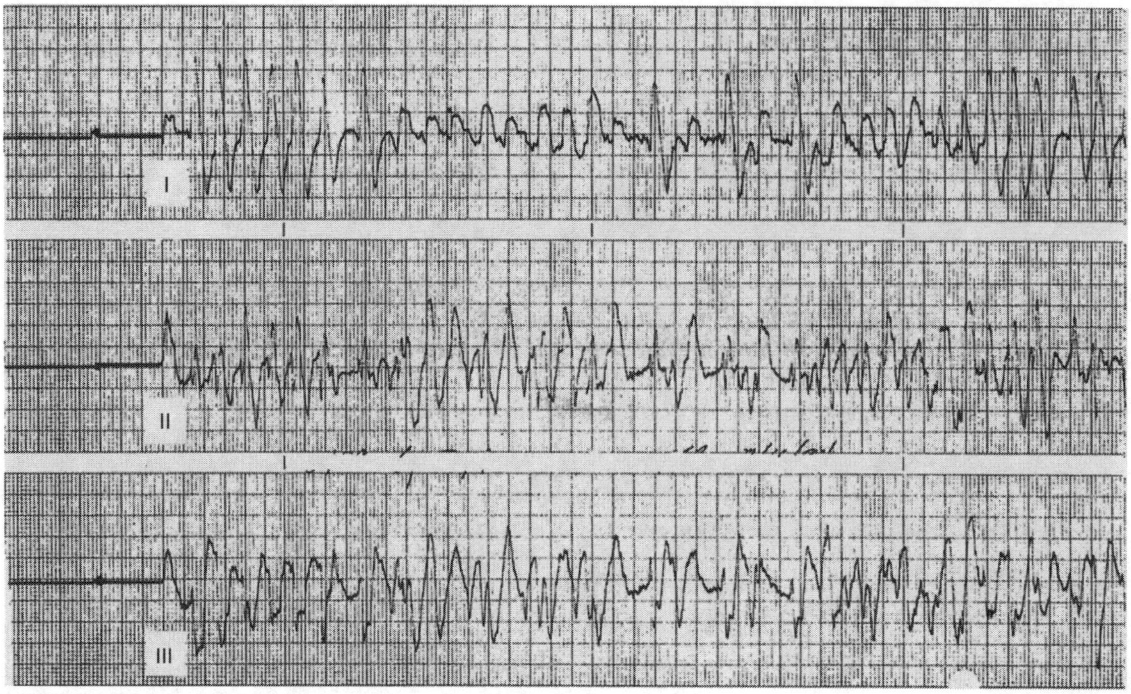

Figure 41.19
Multiform ventricular flutter (torsades de pointes). This arrhythmia was induced during a subchronic toxicity study with an experimental class I antiarrhythmic agent in an experimental beagle. The ventricular rate varies from 240 to 260 bpm.

Drugs can induce VT in dogs with structurally normal hearts, but more commonly when ischemia or other disorders are present. When drugs are screened, the development of VT is considered a marked adverse response to a drug.

(a) *Torsades des pointes.* Torsades des pointes is a specific type of VT that is characterized by a rapid polymorphic VT that shows the complexes changing polarity and morphology as though around a line similar to the twisting of a fence (thus the name). This particular arrhythmia has been the reason for extensive studies of many drugs, both cardiac and noncardiac to assure that the investigated drug does not induce torsades des pointes. This rhythm is induced most likely because of the drug's effects on repolarization currents that result in prolongation of the QT interval. Many drugs have been identified to induce prolongation of the QT and torsades des pointes, and an international registry exists for the logging of cases of prolonged QT (see www.qtdrugs.org). Examples of drugs known to prolong the QT interval include quinidine, disopyramide, lidocaine, procainamide, prenylamine, phenothiazines, and tricyclic antidepressants [139]. It can be induced experimentally by burst pacing hearts with coronary occlusion in dogs, given 30 mg kg^{-1} of quinidine [140] (Fig. 41.19).

41.8.2.5 Atrioventricular Dissociation

The term AV dissociation often should not be considered an "arrhythmia," but only the result of an arrhythmia [19]. AV dissociation occurs when the atria and ventricles are depolarizing independently. This general category includes arrhythmias such as the third-degree heart block and ventricular tachycardia. Isorhythmic dissociation (same rhythm rate but not associated) occurs when the rates of the sinus and ventricular discharge are almost exact (Fig. 41.20).

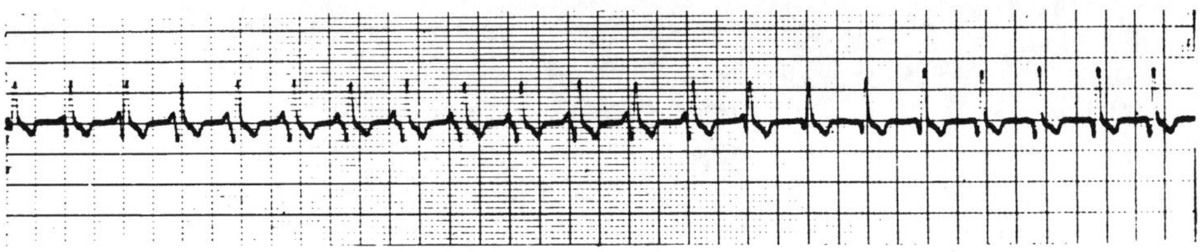

Figure 41.20
Isorhythmic atrioventricular dissociation: starting with the second complex, the p waves are seen to separate sequentially from the Q wave then move forward to fuse with the Q wave again, decreasing its amplitude for three beats, after which it moves further forward to fuse with the R wave, increasing its amplitude in the final five complexes (accrochage). Lead V_2 of a six-year-old male Labrador retriever with cor pulmonale caused by heartworm infestation and congestive heart failure is shown. The arrhythmia was caused by digoxin intoxication.

41.8.3 Atrial and Atrioventricular Conduction Disorders

41.8.3.1 Intra-atrial Conduction Disorders

(a) *Sinoatrial block.* Although this arrhythmia occurs spontaneously in dogs [23], the diagnosis of SA block usually cannot be made with certainty at resting heart rates because of the ubiquitous presence of sinus arrhythmia.
(b) *Sinus arrest, sinus standstill.* This electrocardiographic diagnosis indicates that the sinus P waves are absent for an interval that should have included several heartbeats. Such periods of pacemaker arrest are occasionally caused by reflex vagal effects (carotid sinus reflex, vasovagal reflex from the esophagus as induced by passing a stomach tube, etc.) or drug effects that increase baroreceptor sensitivity, and cause central vagal nucleus stimulation (e.g., digitalis, morphine, tranquilizers). In the clinic, it is observed sometimes in the presence of intracranial tumors, atrial disease, and as a hereditary abnormality in certain breeds [25].
(c) *Intra-atrial block.* This term is applied when there is evidence of a conduction delay in the atria causing widening and deformation of the P waves beyond normal limits of variation. As discussed before, the almost universal respiratory sinus arrhythmia is accompanied by marked P-wave variation in amplitude in certain limb leads. The duration of P waves, however, varies little. The notched P waves, in the absence of increased P-wave duration beyond 0.06 s, are not abnormal per se. Conduction defects are indicated when the P-wave duration increases beyond 0.06 s in limb leads. Drugs that slow myocardial conduction, such as type I antiarrhythmic agents, are likely to delay atrial depolarization. Hyperkalemia above 7.5 mEq l^{-1} will widen P waves and, at higher levels, abolish the P waves entirely. Widening of P waves sometimes accompanies atrial enlargement.

41.8.3.2 Atrioventricular Block

(a) *First-degree AV block.* In this case, the PR interval is prolonged beyond 0.13 s and rarely exceeds 0.20 s. The normal PR interval shortens with increasing heart rate (➋ Table 41.5). When sinus tachycardia is present (i.e., at rates exceeding 150 bpm), the PR interval ordinarily is shortened to less than 0.12 s, so that PR intervals greater than this may be considered first-degree AV block at more rapid heart rates. Not infrequently, the PR interval varies in first-degree AV block, so that only some of the intervals exceed these upper limits.
(b) *Second-degree AV block.* The block may be sporadic or frequent, regular or irregular. It may be preceded by progressively increasing PR intervals, fixed PR intervals, or PR intervals that lengthen and then shorten. The ratio of blocked to conducted beats is only rarely constant in dogs, probably because of the superimposed respiratory sinus arrhythmia that is usually also present: Type I second-degree AV block (Wenckebach phenomenon or Mobitz type I) is more common in dogs than type II. Type II second-degree AV block (Mobitz type II) occurs in otherwise normal dogs only sporadically and often at fairly rapid heart rates. Although this conduction disorder (see ➋ Chap. 28) is reputed to occur in man only in the presence of organic heart disease [21], this is not true in dogs. Most likely, it occurs in

dogs as a more-or-less physiological phenomenon when, for some reason, there is a sudden, transient increase in vagal tone that blocks AV conduction for just a single cardiac cycle. When encountered, however, the presence of organic heart disease should be ruled out by further cardiac examination. In some dogs, type I second-degree AV block develops at slower sinus rates, when type II has occurred as an isolated event at a more rapid rate.

Except for type II (Mobitz) second-degree AV block in dogs, first-degree and second-degree AV blocks are generally simply normal physiological variants.

(c) *Third-degree AV block.* Third-degree AV block is a common bradyarrhythmia in the dog that demands the implantation of a permanent pacemaker. Although underlying myocardial disease that will eventually affect hemodynamic function exists, most dogs respond well to this treatment. Usually, the atrial rate is rapid (e.g., 105–145 bpm) and the ventricular rate is slow (e.g., 25–30 bpm). Syncope or weakness is a common clinical sign. If not treated, heart failure or sudden death will result in spontaneous cases [23].

In experimental AV block, following the crushing of the His bundle, ventricular rates were 49 ± 2 bpm after surgery and reached a plateau by day 31 at 44 ± 2 bpm [141]. This rate continued to the end of the 3-year observation period. One to 4 weeks after creating the AV block, other authors have reported similar rates, for example, Hurwitz [142] found an atrial rate of 120–200 bpm and a ventricular rate of 26–54 bpm; Reynolds and DiSalvo [143] found a ventricular rate of 55 ± 13 bpm; and Robinson et al. [144] gave an atrial rate of 125 ± 10 bpm and a ventricular rate of 37 ± 3 bpm.

41.8.4 Frequency of Arrhythmias in the Human and the Dog

The ideal way to evaluate the frequency of arrhythmias in any species is with a 24-h Holter. However, with a large number of subjects, an appreciation of the number of arrhythmias in a population can be gleaned. It must be understood that if the arrhythmia number is restricted to ECG recordings for a brief period, also in the situation whereby an animal is stressed by the procedure, the amount of arrhythmias will be different than in circumstances that do not provoke the autonomic nervous system as much. These latter conditions would include the 24-h Holter and telemetry recordings. Consequently, comparisons must be made with the same techniques. Also, the age of the animal has a bearing on the arrhythmia number.

Comparable electrocardiographic series in the dog and the human are not available. The most relevant data from the human that might be compared with data from young experimental beagles of both sexes are the findings among 67,375 asymptomatic young adult male air force officers from whom 12-lead records were taken [145]. The dog ECGs are from pretest records on 5,513 young (usually 5–7 months old) experimental beagles from which pretest records were taken before being used in toxicity trials. In about 90% of the beagles, two ten-lead ECGs taken several days apart were available. The overall arrhythmia prevalence rates in the two groups were similar (27.9 per thousand for the human against 25.6 per thousand for the dog). The prevalence of ventricular ectopic beats for the human (6.2 per thousand) was about the same as for the dog (8.0 per thousand). Note that the first-degree AV block is more common in the human than in the dog, while the reverse is true for the second-degree AV block. Ventricular parasystole may be more prevalent in beagles than in humans, as may be ventricular escape rhythms, and electrical alternans. Wolff–Parkinson–White syndrome and the bundle branch block may be more common in humans. However, because of the disparities between the data and their collection and analysis, comparison is, although interesting, of questionable validity.

41.9 Comparing Serial Electrocardiogram Records

Either in clinical heart disease or in experimental cardiotoxicity, serial ECGs may permit the recognition of electrophysiological changes that presage definite electrocardiographic abnormalities. This requires that the latest record from each dog should be compared with the previous records to detect subtle changes that are not necessarily outside normal limits. Ideally, adequate numbers of dogs are included in control and treated groups such that a statistical evaluation can be made; however, often studies are undertaken with very low numbers of animals (particularly pilot investigations), and this necessitates comparisons among dogs for treatment effects. Because of the variability in recording the ECG, however, enough of a difference must be documented in order to say that the change is less likely to be due to chance. The following is a list of changes that, if exceeded, require further examination. Note that the following are suggested from experience and not based on the variability analysis. In toxicity trials, their significance is increased if several individuals in a dose group are similarly affected. It should be emphasized that these differences are based on testimonial experience of the author. Ideally, studies to document repeatability and variability between recordings for each study group are needed for precision. However, these can serve as a beginning for comparisons.

(a) Increases in time intervals
P wave + 20 ms
PR interval + 30 ms (when not rate related, see Table 41.5)
Second-degree AV block
QRS interval + 15 ms
QT + 40 ms (when not rate related, see Table 41.5)
(b) QRS frontal plane axis: change of 30°
(c) Increases in amplitude
R-wave amplitude increased by 0.7 mV
T-wave amplitude increased by 50%
ST-segment deviation greater than 0.15 mV
(d) Sinus arrhythmia index (I = 100 [R-R standard deviation/mean R-R]):
substantial decrease in the absence of a commensurate increase in heart rate
(e) Contour and polarity
P reversal in any limb lead
ST-segment slurring

Reversal of T-wave polarity (caution is advised when the evaluation involves a very young dog and also when a dog is close to 1 year of age)

41.10 Cardiotoxic and Drug Effects on the Electrocardiogram

An important sphere of canine electrocardiography is the detection of myocardial injury and of altered regulation of cardiac activity in drug trials, nutritional studies, immunopathological responses, and in the veterinary clinic. Although the injurious agents are legion, the types of electrophysiological changes possible are limited, so that the diagnosis of specific chemical or biological agent effects is not possible through electrocardiography. Certain classes of drugs and chemicals, however, produce characteristic constellations of electrophysiological changes, so that when the test agent is known or suspected, its presence may often be predicted from the electrocardiographic findings.

Reviews of cardiotoxicity and myocardial injury include those of Bohle [146], Wenzel [147], Selye [148], Chung and Dean [149], Davies and Gold [150], Bristow [151], Balazs [152], Van Stee [153], and Spitzer [154]. Reviews dealing principally with electrocardiological drug effects include those of Scherf and Schott [8], Bellet [18, 19], Surawicz and Lasseter [155], Surawicz [156, 157], Vaughan Williams [158], Singh et al. [159], Schamroth [20], Detweiler [26], Harrison [160], and Lazdunski and Renaud [161].

41.10.1 Drug Effects on Transmembrane Action Potentials and ECG Changes

The relationship between transmembrane action potential changes and electrocardiographic changes is known for a number of drugs and electrolytes. From simultaneous records of ventricular transmembrane potentials and ECGs, the influence of transmembrane potential (TMP) effects on the ECG can be shown [161].

(a) When the slope of phase 0 of the TMP is decreased, conduction velocity is slowed, and in the ECG the QRS increases in duration (e.g., quinidine and other type I antiarrhythmics).

(b) The duration of the TMP approaches that of the QT interval, so that shortening or lengthening of TMP duration has a directly corresponding effect on the QT interval (e.g., digitalis glycosides shorten and hypocalcemia lengthens the TMP).
(c) Lengthening of phase 2 of the TMP lengthens the ST segment of the ECG and shortening phase 2 has the opposite effect (e.g., hypocalcemia prolongs and digitalis shortens the ST segment).
(d) Increased velocity of repolarization during phase 2 of the TMP reduces or abolishes this plateau and the ST segment of the ECG is abbreviated (e.g., digitalis glycosides).
(e) A more acute transition from the slope of phase 2 to that of phase 3 of the TMP, together with increased slope of phase 3, produces symmetrical peaking of the T waves in the ECG (e.g., hyperkalemia).
(f) A less acute transition from the slope of phase 2 of the TMP to that of phase 3 such that the repolarization configuration approaches a straight line and causes reduction of the T-wave amplitude in the ECG (e.g., barbiturates).
(g) Prolongation and decreased slope of phase 3 of the TMP exaggerates the U wave of the ECG (e.g., hypokalemia).

41.10.1.1 Drug Effects on the ECG

There are some electrocardiographic alterations that are somewhat less well known [26, 162]. This consists of a characteristic configurational change in the form of the ST-T segments in lead rV_2 especially and sometimes also in leads V_2, V_4, and V_{10}. In the precordial leads (rV_2, V_2, and V_4), the ST segment and first portion of the T wave form a convex upward curve and the terminal portion of the T wave forms a positive spike, hence the descriptive term "dome-dart T waves." In lead V_{10}, the ST curve is concave upward and the terminal spike is negative; that is, more or less the reciprocal of the configuration in lead rV_2. This morphological change is usually accompanied by QT-interval prolongation. The duration of QT may vary, depending on the length of the preceding R-R interval in sinus arrhythmia (e.g., from 0.30 to 0.34 s, with the longer QT intervals following the shorter R-R intervals).

This combination of electrocardiographic effects has been observed in toxicological studies with tricyclic antidepressants, phenothiazine derivative and butyrophenone derivative antipsychotic agents, and certain antihistaminic compounds. When the various pharmacological properties of these test agents were known, they combined the central nervous system, antihistaminic, and local anesthetic actions. Often, however, only their central nervous system or antihistaminic effects had been identified at the time of the toxicity trial.

One of these experimental drugs, a butyrophenone derivative neuroleptic, in addition to dome-dart T waves and QT interval prolongation, produced cardiac slowing, first-degree and second-degree AV block, left and right bundle branch block (LBBB and RBBB), and ventricular ectopic beats. Marked sinus arrhythmia was present and the LBBB and RBBB appeared sporadically following short R-R intervals. Frequently, the PR intervals preceding these bizarre complexes were not prolonged beyond normal limits. Ventricular reentrant beats with RBBB configuration followed occasional RBBB beats, and LBBB reentrant beats followed occasional LBBB complexes. Since BBB occurred in the absence of prolonged PR intervals, sometimes it appears that the depression of conduction was relatively greater on the bundle branch Purkinje fibers than on the atrioventricular junctional tissues.

41.10.1.2 The QT Interval

The important thing about QT prolongation is not its possible relation to negative inotropy, but rather its relation to arrhythmia vulnerability. It is one manifestation of increased inhomogeneity of ventricular refractory periods. The long QT syndrome has been an extensive study and the measurement of the QT interval, the subject of thousands of papers. Moreover, the correct evaluation of the QT interval in the dog for toxicological studies is one for continued discussion [51–54], and drugs that prolong the QT interval are associated with sudden death [163]. Evaluation of the QT interval is a mainstay for drug studies; however, adjusting for the QT interval relative to heart rate remains problematic for studies that involve small numbers of animals with brief recordings. The formulas derived for use in humans are not suitable for the dog [50, 51]. Full conferences have been held with this as the focus of attention. As this is a changing field of standards, it is advised to search via the Internet for the latest information at the time of design for studies in the dog for which the QT interval is an integral part of the examination.

41.10.1.3 ST-T and T-Wave Changes

The T waves in dog ECGs are markedly labile. They may be either positive or negative in most leads, and their polarity may change from record to record taken sequentially. The exceptions are lead rV_2 in which the T waves are almost always positive and lead V_{10} in which they are almost always negative. This characteristic of lead V_{10} applies only if the records are taken with the dogs in right lateral recumbency since, for example, V_{10} is frequently positive in beagles placed in the supine position.

The contour, slope, and deviation from the isopotential line of the ST segment in dog ECGs are also quite variable and change spontaneously from record to record taken at different times. Serum electrolyte alterations and cardioactive drugs often reveal their presence by changing the relative durations and contours of the ST segment and T wave. Depending on the magnitude of the change induced, such drug effects may merely alter the form of ST-T complexes in a characteristic way, but not produce a distinctly abnormal record. Thus, the complexes may be judged to be within normal limits of form and duration for the dog ECG, but because similar changes occur in several animals in a given dose group they can be attributed to drug action. For example, some drugs cause the ST-T complexes in several leads to assume a similar contour and polarity. Reversal of T waves in rV_2 and V_{10} is a reliable sign of left ventricular subendocardial and papillary muscle damage (e.g., ischemia, hemorrhage, necrosis) in the dog and, when present in toxicity trials, is accompanied by demonstrable histological lesions in myocardial tissue or small intramural coronary arteries in these regions in about 80% of the cases [164].

41.11 Interpretative Statements

When evaluating ECGs for drug studies, the interpretation is important not only to point out any relevant changes, but also to stress when the changes are unclear because of the vulnerability of the ECG. This is important when a small number of dogs are being studied. The interpretation should include an electrocardiographic diagnosis and relate this to possible electrophysiological mechanisms. It should not go beyond the scope of electrocardiography into possible cardiodynamic or hemodynamic consequences or effects on other organs and tissues. This correlative step can be made only after all the clinical or toxicological information has been assembled.

The interpretative statements should cover the following three areas:

(a) Electrophysiological characteristics detected by the ECG. This will include changes in conductivity and rhythmicity and speculations about the anatomical sites and possible effects on the transmembrane action potentials of the cardiac cell.
(b) Description of electrocardiographic features that may or may not be related to heart disease, cardiotoxicity, or physiological state. These may have a low order of significance in themselves, but may relate to other clinical or toxicological findings. Examples are nonspecific contour changes in ST segments and T waves in various leads.
(c) The likelihood that heart disease or a cardiotoxic effect is present because the physiological state of the heart has been sufficiently altered or because pathological myocardial lesions may be present.

Acknowledgement

Dr. Sydney Moise at Cornell University contributed updates to this chapter.

References

1. Waller, A., Introductory address on the electromotive properties of the human heart. *Br. Med. J.*, 1888;**2**: 751–754.
2. Einthoven, W., Enregistrement galvanométrique de l'électrocardiogramme humain et contrôle des résultats obtenus par l'emploi de l'électromètre capillaire en physiologie. *Arch. Néerland. Sci. Not.*, 1904;**11**(9): 202–209.
3. Einthoven, W., Weiteres über das Elektrokardiogramm. *Pfluegers Arch.*, 1908;**122**: 517–584.
4. Einthoven, W, G. Fahr, and A. de Waart, Über die Richtung und die manifeste Grösse **der** Potentialschwankungen immenschlichen Herzen und über den Einfluss der Herzlage auf die Form des Elektrokardiogramms. *Pfluegers Arch.*, 1913;**150**: 275–315.
5. Waller, A.D., *A Short Account of the Origin and Scope of Electrocardiography. The Harvey Lectures, 1913/14.* Philadelphia, PA: Lippincott, 1915, pp. 17–33.
6. Lewis, T., *The Mechanism and Graphic Registration of the Heart Beat*, 3rd edn. London: Shaw, 1925.
7. Rothberger, C.J., Normale und pathologische Physiologie der Rhythmik unct Koordination des Herzens. *Erg. Physiol.*, 1931;**32**: 472–820.
8. Scherf, D. and A. Schott, *Extrasysto/es and Allied Arrhythmias*, 2nd edn. Chicago, IL: Year Book Medical, 1973.
9. Wenckebach, K.F. and H. Winterberg, *Die unregelmässige-Herztätigket.* Leipzig: Engelman, 1927.
10. Wilson, F.N., F.D. Johnston, and E. Lepeschkin, Editors. *Selected Papers.* Ann Arbor, MI: Heart Station, University Hospital, 1954.
11. Nörr, J., Über Herzstromkurvenaufnahmen an Haustieren. Zur Einführung der Elektrokardiographie in die Veterinärmedizin. *Arch. Wiss. Prakt. Tierheilkd.*, 1922;**48**: 85–111.
12. Roos, J., Vorhofflimmern bei den Haustieren. *Arch. Wiss. Prakt. Tierheilkd.*, 1924;**51**: 280–293.
13. Haupt, K., *Die Aufnahmetechnik des Hundeelektrokardiogramms in der Veterinärklinik und ihre Ergebnisse*, dissertation. Giessen: University of Giessen, 1929.
14. Ludwig, K-H., *Die Elektrokardiographie beim gesunden Hund unter besonderer Berücksichtigung ihrer Anwendung in der Klinik*, dissertation. Leipzig: University of Leipzig, 1924.
15. Gyarmati, E., *Klinische elektrokardiographische Untersuchungen beim Hunde*, dissertation. Budapest: University of Budapest, 1939.
16. Lannek, N.A., *Clinical and Experimental Study on the Electrocardiogram in Dogs*, dissertation. Stockholm: Royal Veterinary College, 1949.
17. Burch, G.E. and N.P. DePasquale, *A History of Electrocardiography.* Chicago, IL: Year Book Medical, 1964.
18. Bellet, S., *Clinical Disorders of the Heart Beat*, 2nd edn. Philadelphia, PA: Lea & Febiger, 1963.
19. Bellet, S., *Essentials of Cardiac Arrhythmias: Diagnosis and Management.* Philadelphia, PA: Saunders, 1972.
20. Schamroth, L., *The Disorders of Cardiac Rhythm*, vols. 1, 2, 2nd edn. London: Blackwell Scientific, 1980.
21. Ettinger, S.J. and P.F. Suter, *Canine Cardiology.* Philadelphia, PA: Saunders, 1970.
22. Detweiler, D.K., D.F. Patterson, J.W. Buchanan, and D.N. Knight, The cardiovascular system, in *Canine Medicine*, vol. 2, 4th edn., E.J. Catcott, Editor. Santa Barbara, CA: American Veterinary, 1979, pp. 813–949.
23. Bolton, G.R., *Handbook of Canine Electrocardiography.* Philadelphia, PA: Saunders, 1975.
24. Tilley, L.P., *Essentials of Canine and Feline Electrocardiography*, 2nd edn. Philadelphia, PA: Lea & Febiger, 1985.
25. Detweiler, D.K., The use of electrocardiography in toxicological studies with Beagle dogs, in *Cardiac Toxicology*, vol. 3, T. Balazs, Editor. Boca Raton, FL: CRC Press, 1981, pp. 33–82.
26. Lautenschlager, O., *Grundlagen der Aufnahmetechnik des Elektrokardiogrammes von Pferd und Rind und ihre Ergebnisse*, dissertation. Giessen: University of Giessen, 1928.
27. Nehb, W., Zur Standardisierung der Brustwandableitungen des Elektrokardiogramms. *Klin. Wchnschr.*, 1938;**17**: 1807–1811. Cited by Lepeschkin, E., *Modern Electrocardiography*, vol. 1. Baltimore, MD: Williams & Wilkins, 1951.
28. Spörri, H., Der Einfluss der Tuberkulose auf das Elektrokardiogramm. (Untersuchungen an Meerschweinchen und Rindern.) *Arch. Wiss. Prakt. Tierheilkd.*, 1944;**79**: 1–57.
29. Detweiler, D.K., The use of electrocardiography in toxicological studies with rats, in *The Rat Electrocardiogram in Pharmacology and Toxicology*, R. Budden, D.K. Detweiler, and G. Zbinden, Editors. Oxford: Pergamon, 1981, pp. 83–115.
30. Hellerstein, H.K. and R. Hamlin, QRS component of the spatial vectorcardiogram and of the spatial magnitude and velocity electrocardiograms of the normal dog. *Am. J. Cardiol.*, 1960;**6**: 1049–1061.
31. Hahn, A.W., R.L. Hamlin, and D.F. Patterson, Standards for canine electrocardiography. *Academy of Veterinary Cardiology Committee Report*, 1977.
32. Kraus, M.S., N.S. Moise, M. Rishniw, et al. Morphology of ventricular arrhythmias in the boxer as measured by 12-lead electrocardiography with pace-mapping comparison. *J. Vet. Intern. Med.*, 2002;**16**(2): 153–158.
33. McFee, R. and A. Parungao, An orthogonal lead system for clinical electrocardiography. *Am. Heart J.*, 1961;**62**: 93–100.
34. Bloch, W.N. Jr., K.A. Busch, and T.R. Lewis, The Frank vectorcardiogram of the Beagle dog. *J. Electrocardiol.*, 1972;**5**: 119–125.
35. Bojrab, M.J., J.E. Breazile, and R.D. Morrison, Vectorcardiography in normal dogs using; the Frank lead system. *Am. J. Vet. Res.*, 1971;**32**: 925–934.
36. Morita, H., Electrocardiograms of conscious Beagle dogs by apex-base bipolar lead. *Adv. Anim. Cardiol.*, 1984;**17**: 19–23.
37. Sugano, S., Electrocardiographic studies in the beagle as an experimental dog. *Adv. Anim. Electrocardiography*, 1977;**10**: 45–50.
38. Takahashi, M., Experimental studies on the electrocardiogram of the dog. *Jpn. J. Vet. Sci.*, 1964;**24**: 191–210.
39. Cagan, S., Prespevok k elektrokardiogramu, psa. *Bratisl. Lek. Listy.*, 1959;**39**: 540–545.
40. Cagan, S. and E. Barta, Die Bedingungen des konstanten Elektrokardiogrammes beim Hund. *Z. Kreislaufforsch.*, 1959;**48**: 1101–1105.
41. Hulin, I. and S. Rippa, Why does the electrocardiogram of the dog change with a change in the foreleg position? *Am. Heart J.*, 1970;**79**: 143.
42. Hill, J.D., The significance of foreleg positions in the interpretation of electrocardiograms and vectorcardiograms from research animals. *Am. Heart J.*, 1968;**75**(4): 518–527.

43. Hill, J.D., The electrocardiogram in dogs with standardized body and limb positions. *J. Electrocardiol.*, 1968;**1**: 175–182.
44. Almasi, J.J., O.H. Schmitt, and E.F. Jankus,. Electrical characteristics of commonly used canine ECG electrodes. *Proc. Annu. Conf. Eng. Med. Biol.*, 1970;**12**: 190.
45. Rydén, L., A. Waldenström, and S. Holmberg, The reliability of intermittent ECG sampling in arrhythmia detection. *Circulation*, 1975;**52**: 540–545.
46. Morganroth, J., Ambulatory monitoring: the impact of spontaneous variability of simple and complex ventricular ectopy, in *Cardiac Arrhythmias*, D.G. Harrison, Editor. Boston, MA: Hall, 1981, pp. 479–492.
47. Moïse, N.S., Diagnosis and management of canine arrhythmias, in *Canine and Feline Cardiology*, 2nd edn., P.R. Fox, D.D. Sisson, N.S. Moïse, Editors. Philadelphia, PA: W. B. Saunders, 1999, pp. 331–385.
48. Grauwiler, J., *Herz und Kreislauf der Säugetiere: Vergleichend-Funktionelle Daten*. Batiel: Birkhäuser,1965.
49. Bazett, H.C., An analysis of the time relations of electrocardiograms. *Heart*, 1920;**7**: 353–370.
50. Chiang, A.Y., D.L. Holdsworth, and D.J. Leishman, A one-step approach to the analysis of the QT interval in conscious telemetrized dogs. *J. Pharmacol. Toxicol. Methods*, 2006 Mar 6; E-print.
51. Miyazaki, H., H. Watanabe, T. Kitayama, M. Nishida, Y. Nishi, K. Sekiya, H. Suganami, and K. Yamamoto, QT PRODACT: sensitivity and specificity of the canine telemetry assay for detecting drug-induced QT interval prolongation. *J. Pharmacol. Sci.*, 2005;**99**(5): 523–529.
52. Gauvin, D.V., L.P. Tilley, F.W. Smith Jr,, and T.J. Baird, Electrocardiogram, hemodynamics, and core body temperatures of the normal freely moving laboratory beagle dog by remote radiotelemetry. *J. Pharmacol. Toxicol. Methods*, 2006;**53**(2): 128–139.
53. Watanabe, H. and H. Miyazaki, A new approach to correct the QT interval for changes in heart rate using a nonparametric regression model in beagle dogs. *J. Pharmacol. Toxicol. Methods*, 2006;**53**(3): 234–241.
54. Camm, A.J., Clinical trial design to evaluate the effects of drugs on cardiac repolarization: current state of the art. *Heart Rhythm*, 2005;**2**(2 Suppl): S23–29. Review.
55. Harada, T., J. Abe, M. Shiotani, Y. Hamada, and I. Horii, Effect of autonomic nervous function on QT interval in dogs. *J. Toxicol. Sci.*, 2005;**30**(3): 229–237.
56. Batchivarov, V.N. and M. Makik, There is little sense in "common" QT correction methods. *J. Cardiovasc. Electrophysiol.*, 2005;**16**(7): 809.
57. Tattersall, M.L., M. Dymond, T. Hammond, and J.P. Valentin, Correction of QT values to allow for increases in heart rate in conscious Beagle dogs in toxicology assessment. *J. Pharmacol. Toxicol. Methods.*, 2006;**53**: 11–19.
58. Lewis, T., J. Meakins, P.D. White, The excitatory process in the dog's heart. Part I. The auricles. *Philos. Trans. R. Soc. Lond. Ser. B*, 1914;**205**: 375–420.
59. Meek, W.J. and J.A.E. Eyster, Experiments on the origin and propagation of the impulse in the heart. IV. The effect of vagal stimulation and of cooling on the location of the pacemaker within the sino-auricular node. *Am. J. Physiol.*, 1914;**34**: 368–383.
60. Hinds, M.H., D.R. Clark, J.D. McCrady, and L.A. Geddes, The relationship among pacemaker location, heart rate, and P-wave configuration in the dog. *J. Electrocardiol.*, 1972;**5**: 56–64.
61. Goldberg, J.M. and M.H. Lynn-Johnson, Changes in canine P wave morphology observed with shifts in intra-SA nodal pacemaker localization. *J. Electrocardiol.*, 1980;**13**: 209–217.
62. Boineau, J.P., R.B. Schuessler, and C.R. Mooney, et al., Multicentric origin of the atrial depolarization wave: the pacemaker complex. *Circulation*, 1978;**58**: 1036–1048.
63. Anonymous (editorial). U waves: unimportant undulations? *Lancet*, 1983;**2**: 776–777.
64. Watanabe, Y. and H. Toda, The U wave and aberrant intraventricular conduction. Further evidence for the Purkinje repolarization theory on genesis of the U wave. *Am. J. Cardiol.*, 1978;**41**: 23–31.
65. Kishida, H., J.S. Cole, and B. Surawicz, Negative U wave: A highly specific but poorly understood sign of heart disease. *Am. J. Cardiol.*, 1982;**49**: 2030–2036.
66. Rudolph, A.M., P.A.M. Auld, R.J. Golinko, and M.H. Paul, Pulmonary vascular adjustments in the neonatal period. *Pediatrics*, 1961;**28**: 28–34.
67. Averill, K.H., W.W. Wagner Jr., and J.H.K. Vogel, Correlation of right ventricular pressure with right ventricular weight. *Am. Heart J.*, 1963;**66**: 632–635.
68. Kirk, G.R., D.M. Smith, D.P. Hutcheson, and R. Kirby, Postnatal growth of the dog heart. *J. Anat.*, 1975;**119**: 461–469.
69. Trautvetter, E., *Untersuchungen zur EKG-Entwicklung an gesunden Welpen und Welpen mitangeborenen Pulmonalstenosen, Habilitationsschrift*. Berlin: Freie Universität Berlin, 1972.
70. Trautvetter, E., D.K. Detweiler, and D.F. Patterson, Evolution of the electrocardiogram in young dogs during the first 12 weeks of life. *J. Electrocardiol.*, 1981;**14**: 267–273.
71. Trautvetter, E., D.K. Detweiler, F.K. Bohn, and D.F. Patterson, Evolution of the electrocardiogram in young dogs with congenital heart disease leading to right ventricular hypertrophy. *J. Electrocardiol.*, 1981;**14**: 275–282.
72. Detweiler, D.K., The cardiovascular system, in *Duke's Physiology of Domestic Animals*, chaps. 5–12, 10th edn., M.J. Swenson, Editor Ithaca, NY: Cornell University Press, 1984, pp. 68–225.
73. Amend, J.F. and H.E. Hoff, Analysis of patterns and parameters of the respiratory heart rate response in the unanesthetized dog. *Southwest. Vet.*, 1970;**22**: 301–311.
74. Fuller, J.L., Genetic variability in some physiological constants of dogs. *Am. J. Physiol.*, 1951;**166**: 20–24.
75. Kleiger, R.E., P.K. Stein, and J.T. Bigger Jr., Heart rate variability: measurement and clinical utility. *Ann. Noninvasive Electrocardiol.*, 2005;**10**(1): 88–101. Review.
76. Werner, J., A. von Recum, E. Trautvetter, and H. Sklaschus, Über den Ruherhythmus des Herzens beim Hund. *Z. Kreislaufforsch.*, 1969;**58**: 593–600.
77. Lange, H., *Ober den Eintritt der Atmungsarrhythmie in der ersten Lebenszeit des Hundes*, dissertation. Munich: University of Munich, 1937.
78. Horan, L., G.E. Burch, and J.A. Cronvich, Spatial vectorcardiograms in normal dogs. *Circ. Res.*, 1957;**5**: 133–136.
79. Horan, L.G., G.E. Burch, and J.A. Cronvich, A study of the influence upon the spatial vectorcardiogram of localized destruction of the myocardium of dog. *Am. Heart J.*, 1957;**53**: 74–90.
80. Horan, L.G., G.E. Burch, and J.A. Cronvich, Spatial vectorcardiogram in dogs with chronic localized myocardial lesions. *J. Appl. Physiol.*, 1960;**15**: 624–628.

81. Hamlin, R.L., F.S. Pipers, and C.R. Smith, Computer methods for analysis of dipolar characteristics of the electrocardiogram. *Am. J. Vet. Res.*, 1968;**29**: 1867–1881.
82. Boineau, J.P., J.D. Hill, M.S. Spach, and E.N. Moore, Basis of the electrocardiogram in right ventricular hypertrophy: relationship between ventricular depolarization and body surface potentials in dogs with spontaneous RVH-contrasted with normal dogs. *Am. Heart J.*, 1968;**76**: 605–627.
83. Chastain, C.B., D.H. Riedesel, and P.T. Pearson, McFee and Parungao. Orthogonal lead vectorcardiography in normal dogs. *Am. J. Vet. Res.*, 1974;**35**: 275–280.
84. Bruninx, P. and H.E. Kulbertus, The McFee-Parungao vectorcardiogram in normal dogs. *J. Electrocardiol.*, 1974;**7**: 227–236.
85. Detweiler, D.K. and D.F. Patterson, The prevalence and types of cardiovascular disease in dogs. *Ann. N. Y. Acad. Sci.*, 1965;**127**: 481–516.
86. Hill, J.D., Electrocardiographic diagnosis of right ventricular enlargement in dogs. *J. Electrocardiol.*, 1971;**4**: 347–357.
87. Clark, D.R., J.G. Anderson, and C. Paterson, Imperforate cardiac septal defect in a dog. *J. Am. Vet. Med. Assoc.*, 1970;**156**: 1020–1025.
88. Bolton, G.R., S. Ettinger, and J.C. Roush II, Congenital peritoneopericardial diaphragmatic hernia in a dog. *J. Am. Vet. Med. Assoc.*, 1969;**155**: 723–730.
89. Patterson, D.F., Animal models of congenital heart disease (with special reference to patent ductus arteriosus in the dog), in *Animal Models for Biomedical Research.* Washington, DC: National Academy of Sciences Publication 1594, 1968, pp. 131–156.
90. Boineau, J.P., M.S. Spach, and J.S. Harris, Study of premature systoles of the canine heart by means of the spatial vectorcardiogram. *Am. Heart J.*, 1960;**60**: 924–935.
91. Bolton, G.R. and S.J. Ettinger, Right bundle branch block in the dog. *J. Am. Vet. Med. Assoc.*, 1972;**160**: 1104–1119.
92. Blake, D.F. and P. Kezdi, Vectorcardiography in uncomplicated canine bundle branch block. *Circulation*, 1961;**24**: 888–889.
93. De Micheli, A., G.A. Medrano, and D. Sodi-Pallares, Etude électro-vectocardiographique des blocs de branche chez le chien àla lumière du processus d'activation ventriculaire. *Acta Cardiol.*, 1963;**18**: 483–514.
94. Moore, E.N., J.P. Boineau, and D.F. Patterson, Incomplete right bundle-branch block. An electrocardiographic enigma and possible misnomer. *Circulation*, 1971;**44**: 678–687.
95. Littlewort, M.C.G., Canine electrocardiography; some potentialities and limitations of thetechnique. *J. Small Anim. Pract.*, 1967;**8**: 437–458.
96. Pyle, R.L., *A Study of Certain Clinical, Genetical, and Pathological Aspects of Congenital Fibrous Subaortic Stenosis in the Dog*, thesis. Philadelphia, PA: University of Pennsylvania, 1971.
97. Tilley, L.P., *Essentials of Canine and Feline Electrocardiography: Interpretation and Treatment*, 2nd edn. Philadelphia, PA: Lea & Febiger, 1985.
98. Rosenbaum, M.B., M.V. Elizari, and J.O. Lazzari, *The Hemiblocks*. Oldsmar, FL: Tampa Tracings, 1970.
99. Okuma, K., ECG and VCG changes in experimental hemiblock and bifascicular block. *Am. Heart J.*, 1976;**92**: 473–480.
100. Glomset, D.J. and A.T.A. Glomset, A morphologic study of the cardiac conduction system in ungulates, dog, and man. I and II. *Am. Heart J.*, 1940;**20**: 389–98, 677–701.
101. Patterson, D.F., D.K. Detweiler, K. Hubben, and R.P. Botts, Spontaneous abnormal cardiac arrhythmias and conduction disturbances in the dog. A clinical and pathologic study of 3,000 dogs. *Am. J. Vet Res.*, 1961;**22**: 355–369.
102. Long, D.M., R.C. Truex, K.R. Friedmann, A.K. Olsen, and S.J. Phillips, Heart rate of the dog following autonomic denervation. *Anat. Rec.*, 1958;**130**: 73–89.
103. Hoffman, B.F. and P.F. Cranefield, The physiological basis of cardiac arrhythmias. *Am. J. Med.*, 1964;**37**: 670–684.
104. Tse, W.W., Evidence of presence of automatic fibers in the canine atrioventricular node. *Am. J. Physiol.*, 1973;**225**: 716–723.
105. Damato, A.N. S.H. Lau, His bundle rhythm. *Circulation*, 1969;**40**: 527–534.
106. Jalife, J. and G.K. Moe, Effect of electrotonic potentials on pacemaker activity of canine Purkinje fibers in relation to parasystole. *Circ. Res.*, 1976;**39**: 801–818.
107. Jalife, J. and G.K. Moe, A biologic model of parasystole. *Am. J. Cardiol.*, 1979;**43**: 761–712.
108. Moe, G.K., J. Jalife, W.J. Mueller, and B. Moe, A mathematical model of parasystole and its application to clinical arrhythmias. *Circulation*, 1977;**56**: 968–979.
109. Furuse, A., G. Shindo, H. Makuuchi, et al. Apparent suppression of ventricular parasystole by cardiac pacing. *Jpn. Heart J.*, 1979;**20**: 843–851.
110. Castellanos, A., E. Melgarejo, R. Dubois, and R.M. Luceri, Modulation of ventricular parasystole by extraneous depolarizations. *J. Electrocardiol.*, 1984;**17**: 195–198.
111. Soloff, L.A., Parasystole, in *Cardiac Arrhythmias*, L.S. Dreifus and W. Likoff, Editors. New York: Grune & Stratton, 1973, pp. 409–415.
112. Meurs, K.M., Boxer dog cardiomyopathy: an update. *Vet. Clin. North Am. Small Anim. Pract.*, 2004;**34**(5): 1235–1244.
113. Baumwart, R.D., K.M. Meurs, C.E. Atkins, et al., Clinical, echocardiographic, and electrocardiographic abnormalities in Boxers with cardiomyopathy and left ventricular systolic dysfunction: 48 cases (1985–2003). *J. Am. Vet. Med. Assoc.*, 2005;**226**(7): 1102–1104.
114. Basso, C., P.R. Fox, K.M. Meurs, et al., Arrhythmogenic right ventricular cardiomyopathy causing sudden cardiac death in boxer dogs: a new animal model of human disease. *Circulation*, 2004;**109**(9): 1180–1185.
115. Kraus, M.S., N.S. Moise, M. Rishniw, et al., Morphology of ventricular arrhythmias in the boxer as measured by 12-lead electrocardiography with pace-mapping comparison. *J. Vet. Intern. Med.*, 2002;**16**(2): 153–158.
116. Meurs, K.M., A.W. Spier, N.A. Wright, et al., Comparison of the effects of four antiarrhythmic treatments for familial ventricular arrhythmias in Boxers. *J. Am. Vet. Med. Assoc.*, 2002;**221**(4): 522–527.
117. Moise, N.S., From cell to cageside: autonomic influences on cardiac rhythms in the dog. *J. Small Anim. Pract.*, 1998;**39**(10): 460–468.
118. Harpster, N.K., Boxer cardiomyopathy. A review of the long-term benefits of antiarrhythmic therapy. *Vet. Clin. North Am. Small Anim. Pract.*, 1991;**21**(5): 989–1004.
119. Moïse, N.S., V. Meyers-Wallen, W.J. Flahive, et al., Inherited ventricular arrhythmias and sudden death in German shepherd dogs. *J. Am. Coll. Cardiol.*, 1994;**24**: 233–243.
120. Moïse, N.S., P.F. Moon, W.J. Flahive, et al., Phenylephrine induced ventricular arrhythmias in dogs with inherited sudden death. *J. Cardiovasc. Electrophysiol.*, 1996;**7**: 217–230.

121. Gilmour, R.F. Jr. and N.S. Moïse, Triggered activity as a mechanism for inherited ventricular arrhythmias in German shepherd dogs. *J. Am. Coll. Cardiol.*, 1996;**27**: 1526–1533.
122. Moïse, N.S. and R.F. Gilmour Jr., and M.L. Riccio, An animal model of sudden arrhythmic death. *J. Cardiovasc. Electrophysiol.*, 1997;**8**: 98–103.
123. Moïse, N.S., D.A. Dugger, D. Brittain, et al., Relationship of ventricular tachycardia to sleep/wakefulness in a model of sudden cardiac death. *Ped. Res.*, 1996;**40**: 344–350.
124. Moïse, N.S., R.F. Gilmour Jr., M.L. Riccio, et al., Diagnosis of inherited ventricular tachycardia in German shepherd dogs. *Am. J. Vet. Med. Assoc.*, 1997;**210**: 403–410.
125. Freeman, L.C., L.M. Pacioretty, and N.S. Moïse, et al., Decreased density of I_{to} in left ventricular myocytes from German shepherd dogs with inherited arrhythmias. *J. Cardiovasc. Electrophysiol.*, 1997;**8**: 872–883.
126. Dae, M., P. Ursell, R. Lee, C. Stilson, M. Chin, and N.S. Moïse, Heterogeneous sympathetic innervation in German shepherd dogs with inherited ventricular arrhythmias and sudden death. *Circulation*, 1997;**96**: 1337–1342.
127. Moïse, N.S., M.J. Riccio, W.J. Flahive, et al., Age dependent development of ventricular arrhythmias in a spontaneous animal model of sudden cardiac death. *Cardiovasc. Res.*, 1997;**34**: 483–492.
128. Riccio, M.L., N.S. Moïse, N.F. Otani, et al., Vector quantization of T wave abnormalities associated with a predisposition to ventricular arrhythmias and sudden death. *Ann. Noninvasive Electrocardiol.*, Jan 1998;**3**(1): 46–53.
129. Moïse, N.S., From cell to cageside cardiac rhythms in the dog: autonomic influence. *J. Small Anim. Pract.*, 1998;**39**: 460–468.
130. Sosunov, E.A., E.P. Anyukhovsky, A. Shvilkin, M. Hara, S.F. Steinberg, P. Danilo Jr., M.R. Rosen, N.S. Moïse, et al., Abnormal cardiac repolarization and impulse initialization in German shepherd dogs with inherited ventricular analysis and sudden death. *Cardiovasc. Res.*, 1999;**42**: 65–79.
131. Moïse, N.S., Inherited arrhythmias in the dog: potential experimental models of cardiac disease. *Cardiovasc. Res.*, 1999;**44**: 37–46.
132. Merot, J., V. Probst, M. Debailleul, U. Gerlacin, N.S. Moïse, et al., Electropharmacological characterization of cardiac repolarization in German shepherd dogs with an inherited syndrome of sudden death. *J. Am. Coll. Cardiol.*, 2000;**36**: 939–947.
133. Sosunov, E.A., R.Z. Gainullin, N.S. Moïse, et al., β_1 and β_2-Adrenergic receptor subtype effects in German shepherd dogs with inherited lethal ventricular arrhythmias. *Cardiovasc. Res.*, 2000;**48**: 211–219.
134. Steinberg, S.F., S.A. Alcott, E. Pak, D.H. Hu, L. Protas, N.S. Moïse, et al., Beta-receptors increase in cAMP and include abnormal CAI cycling in the German shepherd sudden death model. *Am. J. Physiol. Heart Circ. Physiol.*, 2002;**282**: H1181–H1188.
135. Obreztchikova, M.N., E.A. Sosunov, E.P. Anyukhovsky, N.S. Moïse, et al., Heterogeneous ventricular repolarization provides a substrate for arrhythmias in German shepherd model of spontaneous arrhythmic death. *Circulation*, 2003;**108**: 1389–1394.
136. Sosunov, E.A., M.N. Obreztchikova, E.P. Anyukhovsky, N.S. Moïse, et al., Mechanisms of alph adrenergic potentiation of ventricular arrhythmias in German shepherd dogs with inherited arrhythmic sudden death. *Cardiovasc. Res.*, 2004;**61**: 715–723.
137. Protas, L., E.A. Sosunov, E.P. Anyukhovsky, N.S. Moïse, et al., Regional dispersion of L-type calcium current in ventricular myocytes of German shepherd dogs with lethal cardiac arrhythmias. *Heart Rhythm*, 2005;**2**: 172–176.
138. Gelzer, A.R.M., N.S. Mo¿se, and M.L. Koller, Defibrillation of German shepherds with inherited ventricular arrhythmias and sudden death. *J. Vet. Cardiol.*, 2005;**7**(2): 97–107.
139. Wald, R.W., M.B. Waxman, and J.M. Colman, Torsade de pointes ventricular tachycardia: a complication of disopyramide shared with quinidine. *J. Electrocardiol.*, 1981;**14**: 301–307.
140. Bardy, G.H., R.M. Ungerleider, W.M. Smith, and R.E. Ideker, A mechanism of Torsades de Pointes as observed in a dog model. *Circulation*, 1981;**64**(Suppl. 4): 218.
141. Boucher, M., C. Dubray, and P. Duchene-Marullaz, Long-term observation of atri, fl and ventricular rates in the unanesthetized dog with complete atrioventricular block. *Pfluegers Arch.*, 1982;**395**: 341–343.
142. Hurwitz, R.A., Effect of glucagon on dogs with acute and chronic heart block. *Am. Heart J.*, 1971;**81**: 644–649.
143. Reynolds, R.D. and J. Di Salvo, Effects of dl-propranolol on atrial and ventricular rates in unaesthetized atrioventricular blocked dogs. *J. Pharmacol. Exp. Ther.*, 1978;**205**: 374–381.
144. Robinson, J.L., W.C. Farr, and G. Grupp, Atrial rate response to ventricular pacing in the unanesthetized A-V blocked dog. *Am. J. Physiol.*, 1973;**224**: 40–45.
145. Averill, K.H. and L.E. Lamb, Electrocardiographic findings in 67,375 asymptomatic subjects. I. Incidence of abnormalities. *Am. J. Cardiol.*, 1960;**6**: 76–83.
146. Böhle, E., Blutgefässe, in *Erkrankungen durch Arzneimittel*, R. Heinz, Editor. Stuttgart: Thieme, 1966, pp. 170–187.
147. Wenzel, D.G., Drug induced cardiomyopathies. *J. Pharm. Sci.*, 1967;**56**: 1209–1224.
148. Selye, H., *Experimental Cardiovascular Diseases*, vols. 1, 2. Berlin: Springer, 1970.
149. Chung, E.K. and H.M. Dean, Diseases of the heart and vascular system due to drugs, in *Drug-Induced Diseases*, vol. 4, L. Meyler and H.M. Peck, Editors. Amsterdam: Excerpta Medica, 1972, pp. 345–381.
150. Davies, D.M. and R.G. Gold, Cardiac disorders, in *Textbook of Adverse Drug Reactions*, D.M. Davies, Editor. Oxford: Oxford University Press, 1977, pp. 81–102.
151. Bristow, M.R., Editor, *Drug-Induced Heart Disease*. Amsterdam: Elsevier, 1980.
152. Balazs, T., Editor, *Cardiac Toxicology*, vols. 1, 2, 3. Boca Raton, FL: CRC Press, 1981.
153. Van Stee, E.W., Editor, *Cardiovascular Toxicology*. New York: Raven, 1982.
154. Spitzer, J. J., Editor, Myocardial injury. *Adv. Exp. Med. Biol; Ser.*, New York: Plenum, 1983;**161**: 421–443.
155. Surawicz, B. and K.C. Lasseter, Effect of drugs on the electrocardiogram. *Prog. Cardiovasc. Dis.*, 1970;**13**: 26–55.
156. Surawicz, B., Relationship between electrocardiogram and electrolytes. *Am. Heart J.*, 1967;**73**: 814–834.
157. Surawicz, B., The pathogenesis and clinical significance of primary T-wave abnormalities, in *Advances in Electrocardiography*, R.C. Schlant and J. Hurst, Editors. New York: Grune & Stratton, 1972, pp. 377–421.
158. Vaughan Williams, E.M., Classification of anti-arrhythmic drugs, in *Symposium of Cardiac Arrhythmias*, E. Sandøe, E. Flensted-Jensen, and K.H. Olesen, Editors. Södertälje, Sweden: Astra, 1970, pp. 449–468.

159. Singh, B.N., J.T. Collett, and C.Y. Chew, New perspectives in the pharmacologic therapy of cardiac arrhythmias. *Prog. Cardiovasc. Dis.*, 1980;**22**: 243–301.

160. Harrison, D.G., Editor, *Cardiac Arrhythmias: A Decade of Progress*. Boston, MA: Hall, 1981.

161. Lazdunski, M. and J.F. Renaud, The action of cardiotoxins on cardiac plasma membranes. *Annu. Rev. Physiol.*, 1982;**44**: 463–473.

162. Detweiler, D.K., Electrocardiographic monitoring in toxicological studies: principles and interpretations, in *Myocardial Injury*, J.J. Spitzer, Editor, *Adv. Exp. Med. Biol.*, 1983;**161**: 579–607.

163. Reynolds, E.W. and C.R. Vander Ark, Quinidine syncope and the delayed repolarization syndromes. *Mod. Concepts Cardiovasc. Dis.*, 1976;**45**: 117–122.

164. Detweiler, D.K., *Reversal of T Waves in Leads* rV_2 *and* V_{10} *in Toxicity Trials Indicates Left Ventricular Subendocardial and Papillary Muscle Ischemia or Damage*, Personal Observation. Philadelphia, PA: University of Pennsylvania, 1985.

165. Osborne, B.E. and B.D.H. Leach, The Beagle Electrocardiogram. *Food Cosmet. Toxicol.*, 1971;**9**: 857–864.

42 The Mammalian Electrocardiogram: Comparative Features

D.K. Detweiler†

†For this 2nd Edition of "Comprehensive Electrocardiology," Dr. Sydney Moise has updated this 1st Edition chapter, which was originally written by the late Dr. Detweiler.

P. W. Macfarlane, A. van Oosterom, O. Pahlm, P. Kligfield, M. Janse, J. Camm (eds.), *Comprehensive Electrocardiology*, DOI 10.1007/978-1-84882-046-3_42,

42.1 Introduction

The purpose of this chapter is to summarize the characteristics that distinguish electrocardiograms recorded from different mammalian species rather than reviewing the contributions of experimental animal research to cardiac electrophysiology.

In 1888, using the capillary electrometer, Waller [1] was the first to obtain electrocardiograms from mammals (human, horse, dog, cat, and rabbit). Since that time, aside from the enormous literature on animal experimentation, a modest literature has accumulated on applied mammalian electrocardiography, written by investigators interested in the comparative aspects of electrocardiography, the use of electrocardiography to record cardiac activity in various types of laboratory animal experiments, and the application of electrocardiography in veterinary medicine. While these investigations were generally not designed to add to the mainstream of basic electrocardiographic thought and concept, they have been useful in providing a database for the interpretation of animal electrocardiograms in veterinary medicine and in a large variety of animal research applications in which electrocardiograms are monitored to detect the effects of various experimental interventions on the heart. In more recent years, the mouse has emerged as a vital animal for investigation. Although the surface electrocardiogram is frequently assessed, more detailed electrophysiologic studies have been at the forefront.

Among prominent physician cardiologists, whose interests and work have given impetus to study in this field, are Tawara (anatomy of the conduction system [2]), Paul Dudley White (elephant [3] and whale [4] electrocardiograms), Rudolph Zuckermann (atlas of animal electrocardiograms [5]), Bruno Kisch (electrocardiographic and electrographic studies in animals [6]), Eugene Lepeschkin (literature survey and analysis [7, 8]), Pierre Rijlant (pacemaker function [9]), Thomas Lewis (atrial fibrillation in horses [10, 11]), Luisada (electrocardiograms and phonocardiograms of domestic animals [12]), and Jane Sands Robb (comparative anatomy, histology, embryology, and electrophysiology [13]).

Veterinary electrocardiography had its beginnings with studies in the horse. As a prelude to this, the first normal equine electrocardiogram published (1910) was a record, which von Tschermak obtained from Einthoven [14]. Shortly thereafter, Lewis (1911) published three abnormal records from a horse with atrial fibrillation [10]. In 1911, Kahn [15] published 26 records taken from six horses and in the same year, Waller [16, 17], in two short notes, briefly discussed the cardiac electrical axis of the horse and the relationship between the duration of mechanical systole and the electrocardiogram. Also in 1913, Norr published his inaugural dissertation on the electrocardiogram of the horse and a companion article in the *Zeitschrift fur Biologie* [18]. Norr was to become the leading figure in veterinary electrocardiography for the next 25 years [19–21], contributing by himself or through his students to the literature on animal electrocardiography and comparative pathophysiology of the circulation.

Following these beginnings, until the early 1940s, in addition to Norr and Kahn, only about 20 authors contributed articles on the electrocardiogram of the horse, recommending some five different lead systems [22]. During this period, only a handful of veterinary authors published articles on the canine electrocardiogram [23] and that of other domestic species [24].

During the 1940s, owing to World War II and its aftermath, scientific publication of all kinds diminished. Notable among the few publications that appeared were those of Sporri (ox, guinea pig [25, 26]) in Switzerland; Alfredson and Sykes (cattle [27]) in the United States; Charton, Minot, and Bressou (horse [28–30]) in France; Kelso (lamb [31]) in Australia; Krzywanek and Ruud (swine [32, 33]) in Germany; and Voskanyan and Filatov (horses, cattle [34, 35]) in Russia. Roshchevsky [36] credited Voskanyan with having developed the clinical use of electrocardiography in cattle in the Soviet Union in 1938–1939. This decade was followed by a 30-year period of ever increasing publications on applied electrocardiography in veterinary medicine, toxicological studies, and records from nondomesticated mammals. Modern day electrocardiography is becoming more "paper-free" with the advent of electronic electrocardiograph systems. These systems can store continuous recordings and post-process the leads shown, with adjustable filtering, speed, and gain on the display. Storage of recordings can be facilitated with electronic retrieval, and data bases are searchable by identification and diagnosis.

42.2 Literature Reviews

Today, internet searching for electrocardiographic data on numerous species is the standard. Wide searches can result in hits not only from published papers but also from other sources of information indexed on the web. Although such means are the standard, some references may be missed from the past. From a historical perspective, the leading sources of references, listed chronologically, in comparative mammalian electrocardiography are the reviews written by Lepeschkin [7, 8], Grauwiler [37], and Roshchevsky [24, 38]. The following summarizes some of these early and important works.

Lepeschkin, in his book published in 1951 [8], covered publications on invertebrates, fish, amphibia, birds, and mammals, citing over 200 references published from 1934 to 1950 as well as publications prior to 1934 cited in his previous book [7]. Grauwiler [37] cites some 125 publications on mammalian electrocardiograms covering the period 1913–1962. Fourteen classes of mammals are represented from monotremes to primates with many illustrations of electrocardiograms and tables of electrocardiographic data for each of the 70 species.

Comparative electrocardiography was reviewed in the publications of Roshchevsky [24, 38]. His earliest effort [38] covers the entire animal kingdom and cites over 970 references. Electrocardiographic, anatomical, heart rate/body weight, and related data are reported. Roshchevsky's monograph [24] emphasizes the spread of ventricular excitation through analysis of intracardiac (endocardial), intramural, and epicardial electrograms, body-surface mapping, vectorcardiography, and the study of various electrocardiographic lead systems. This publication extended an earlier monograph published in 1958 [36]. Monographs on comparative electrocardiology edited by Roshchevsky include the *Physiological Basis of Animal Electrocardiography* (1965) [39] and proceedings of the first (1979) [40] and second (1985) [41] International Symposia on Comparative Electrocardiology held in Syktyvkar, Komi Republic, USSR.

In 1952, Kisch and his coworkers [6] summarized their studies of electrograms recorded from vertebrate hearts, including those of the rabbit, cat, dog, calf, and human. Data are given on electrocardiographic time intervals, heart rate, and pattern of ventricular excitation from unipolar epicardial and endocardial ventricular electrocardiograms recorded from five mammalian species.

In 1959, Zuckermann, as part of the third edition of his textbook on electrocardiography [5], published an extensive atlas of animal electrocardiograms from arthropods, fish, amphibians, reptiles, birds, and mammals.

Other more recent review publications that contain extensive bibliographies and illustrations, which are not listed in the bibliographies of the works already cited, are on the following animals: horse [42], dog [43–45], monkey [46], and rat [47]. ➲ Table 42.1 lists some influential publications on mammalian electrocardiography that appeared between 1888 and 1970. Today, the internet serves as a means to identify electrocardiographic data for many species. Still, compared to the data for other commercial studies on animals, not much data are available to the public on animals used in nonacademic research.

42.3 Classification of Mammalian Electrocardiograms

42.3.1 Bases for Classification

Mammalian ECGs from different species can be classified in accordance with the following general characteristics:

1. Relative duration of QT interval and ST segment
2. QRS-vector direction and sense
3. Constancy of T-wave polarity (T-wave lability) [49, 50]

These differences, in turn, are determined by three electrophysiological properties that are genetically governed: the relative duration of the ventricular cell action potential as determined by the presence or absence of phase 2 (plateau); the pattern of spread of excitation throughout the ventricles; and constancy of the pattern of repolarization, that is, the constancy of the ventricular gradient.

42.3.2 QT Duration and ST Segment

Many species (rodents, insectivores, bats, and kangaroos) have short QT intervals relative to the duration of mechanical systole [51]. The ST segment is essentially absent. The QRST complex consists of rapid QRS deflections that merge with the slower T wave, and its duration is about half that of mechanical systole. The transmembrane action potentials of these

Table 42.1
Contributions important to the development of comparative mammalian electrocardiography

Year	Subject	Author	Reference
1988/9	First horse ECG (capillary electrometer).	Waller	[66]
1910	Einthoven provided horse ECG for veterinary textbook.	von Tschermak	[14]
1911	Visual proof that fibrillating atria produce f waves in ECG (from horses with atrial fibrillation)	Lewis	[10, 11]
1913	Veterinary clinical electrocardiography initiated.	Nörr	[18]
1921	Fetal electrocardiograms obtained from mares.	Nörr	[95]
1921	First elephant ECG	Forbes et al.	[216]
1924	Cardiac arrythmias in horses.	Nörr	[218]
1924	Atrial fibrillation in dogs and horses.	Roos	[219, 220]
1930	Electrocardiographic monitoring in chronic rat (nutritional) studies.	Agduhr, Drury et al.	[221, 222]
1935	Respiratory sinus arrhythmia in *canidae* (dog, fox).	Nörr	[223]
1942	Electrocardiography in dairy cattle.	Alfredson, Sykes	[27]
1949	Clinical chest leads for dogs (rV_2, V_2, V_4).	Lannek	[23]
1951	Survey of animal electrocardiography.	Lepeschkin	[8]
1952	Comparative electrocardiography in mammals.	Kisch et al.	[6]
1953	First whale ECG	King et al.	[4]
1953	First WPW in lower animals (cow).	Spörri	[161]
1953	His-bundle potentials recorded (dog).	Scher	[224]
1953	No ST segment and QT shorter than mechanical systole in adult mouse.	Richards et al.	[185]
1955	Ventricular activation pattern in the dog.	Scher	[225]
1956	QT shorter than mechanical systole and ST segment absent in adult Bennett Kangaroo but not in young in pouch.	Spörri	[186]
1957	Clinical electrocardiography in horse and cattle.	Brooijmans	[67]
1958	Right ventricular hypertrophy pattern in dog.	Detweiler	[226]
1958	Electrocardiography in cattle.	Roschchevsky	[36]
1959	Atlas of animal ECGs.	Zuckerman	[5]
1960	ECG finback whale.	Senft, Kanwischer	[181]
1960	Lead V_{10} introduced in dog.	Hamlin, Hellerstein	[84]
1965	Large atrial mass favors atrial fibrillation.	Moore et al.	[160]
1965	Survey of mammalian ECGs.	Grauwiler	[37]
1966	Monkey ECG.	Malinow	[80]
1967	Definitive criteria for right ventricular hypertrophy in the dog.	Hill	[228, 229]
1968	First journal of animal electrocardiography published (Japan).		[48]
1970	Accessory AV (Kent) bundle conduction identified in WPW syndrome (monkey, dog).	Boineau, Moore	[233]
1970	First monograph on canine cardiology.	Ettinger, Suter	[227]

species do not have a distinct plateau, which explains the absence of the ST segment (Fig. 42.1). This is in contrast to the QRST complex of the remainder of mammals that have an ST segment, a QT interval equivalent to mechanical systole, and an action potential with a distinct plateau. The reason for the differences in the QT interval, ST segment, and the T wave rests in the differences in the specific ion channels that are responsible for repolarization in these varied species. The differences in the ion channels are important not only to understand from the point of recognizing the different ECG

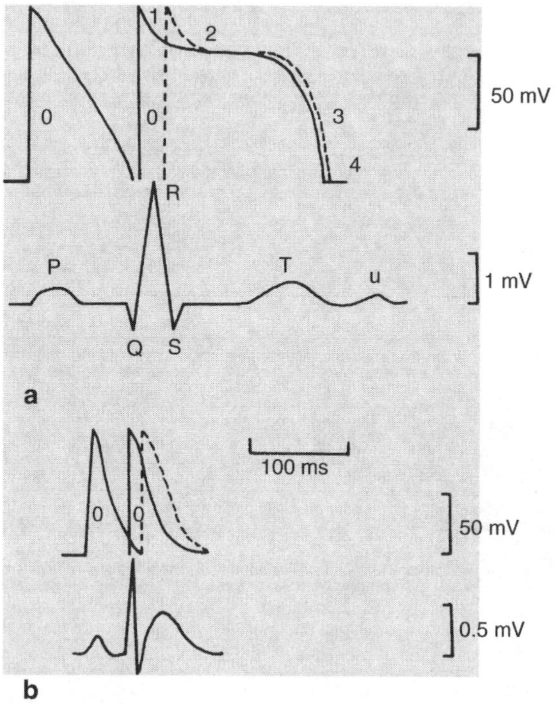

Figure 42.1
Schema of atrial and ventricular transmembrane action potentials (TMAP) and the ECG, drawn on the same time scale. In (a), the five phases of the dog TMAP are shown; 0, initial rapid depolarization or spike; 1, initial rapid repolarization; 2, slow repolarization or plateau; 3, final rapid repolarization; 4, resting or diastolic transmembrane potential. The rat TMAP is shown in (b). Note that the rat and dog atrial and the rat ventricular TMAPs have no plateaux. The TMAPs drawn with the *solid line* represent excitation of ventricular cells early during the QRS interval; those drawn with a *dashed line* represent excitation of ventricular cells later during the QRS. Note that in the dog the plateaux of the TMAPs overlap so that there is little difference in charge between groups of cells. This period of overlap coincides with the isopotential ST segment of the ECG. The T wave of the dog ECG is generated during the final rapid repolarization, phase 3, when at any given instant in time the charges of different masses of cells are not the same. In the rat, the ventricular TMAPs do not have a plateau and there is no period during repolarization when most cells are isopotential. Thus, no ST segment appears in the ECG.

patterns, but also for comparative medicine and the effects of drugs. The latter is vital in understanding the impact that ion channel type has on the response to a pharmacologic intervention that may have little or no effect in one species, yet have a profound consequence in another. This point is of particular importance with regard to the screening for QT prolongation [52].

42.3.3 Ventricular Activation Patterns

The ventricular activation patterns of various species fall into two general classes of QRS-vector direction and sense [24, 53] as follows:

1. Class A includes those animals with QRS vectors which, generally, are directed along the long axis of the body, caudally and ventrally, and produce a largely negative deflection in lead V_{10} and a positive deflection in lead aVF; for example, dogs, humans, monkeys, cats, and rats.

2. Class B includes those animals with QRS vectors which, generally, are directed from sternum toward the spine and which produce largely positive deflections in lead V_{10} and negative deflections in lead aVF; hoofed mammals and dolphins, for example.

These differences are associated with the distributive characteristics of the Purkinje network. In animals belonging to class A, it is primarily a subendocardial network. In class B animals, the Purkinje network is more elaborate and penetrates deeply into the ventricular myocardium.

42.3.4 T-Wave Lability

In humans, primates, and many hoofed mammals, T wave amplitude and polarity tend to be fairly constant in serial records. In dogs and especially in horses, T wave vectors are quite labile. The T waves vary in polarity and amplitude in limb leads and some thoracic leads in serial records, and sometimes change during the course of recording of a given lead. In the dog, there are two conventional thoracic leads in which the T wave polarity is remarkably consistent: the T wave is normally positive in lead rV_2 (this lead approximates that of V_1 in humans and negative in V_{10} in about 90% of individuals [54, 55]).

42.3.5 Effect of ECG Characteristics on Choice of Lead Systems and Terminology

For animals having QRS vector and sense of class A in ➲ Sect. 42.3.3, the conventional limb leads are oriented favorably for recording QRS potentials in the frontal plane projection. In animals with class B QRS vectors, on the other hand, the major QRS-vector forces are directed more-or-less perpendicular to the frontal plane and limb leads are not favorably disposed for recording these potentials. Hamlin and Smith's [53] use of lead V_{10} to compare species made this amply clear, but not until there had been a great deal of experimentation to find favorable lead systems, especially in large domestic animals (reviewed by Roshchevsky [24, 36]). Today the base-apex lead is the most common lead used in horses and cattle.

The discovery that another group of mammals has virtually no ST segment in the electrocardiogram caused another terminological dilemma, since ST-segment deviation is an important diagnostic term in classical electrocardiography. Actually, a flat or isopotential ST segment is abbreviated in limb leads in many mammals with the ST segment of class A (e.g., the dog) and is virtually absent in precordial leads of all species. The presence of a stable isoelectric ST segment depends on two factors: first, the occurrence of a distinct plateau and consequent long duration of the transmembrane action potential and second, in large hearts, the presence of a rapid conduction system that spreads the excitation quickly so that the plateaus of many ventricular cells overlap in time. Ventricular ectopic beats, for example, result in slowing of the depolarization throughout the ventricle such that synchrony of the plateau period in various parts of the heart is reduced and the ST segment is consequently abbreviated.

The terminological dilemma arises because the same perturbations that produce ST segment deviations in species with a distinct ST segment (for example, regional ischemia or hypoxia, "injury current") produce a similar shift of the slow-wave portion of the QRST complex in species with no ST segment (e.g., the rat). The phrase ST deviation is so ingrained in diagnostic electrocardiographic parlance as a term with definite diagnostic significance that it represents a concept. Therefore, it has been suggested that the term should be placed in quotation marks (e.g., "ST" segment elevation or depression) when used to indicate the effect of an injury potential on the immediate post-QRS portion of the ventricular complex in species lacking an ST segment [50].

Finally, for species with labile T waves in normal electrocardiograms (e.g., dog and horse), the significance of T-wave reversal in most leads in disease or cardiotoxicity cannot be evaluated. Fortunately, in the dog the T waves in leads rV_2 (V_1) and V_{10} are consistent in polarity and these leads can be used to detect T wave reversal associated with myocardial damage. No such leads have been established for the horse. Additionally, in dogs, rV_2 (V_1) may be a more reliable lead than lead II for the measurement of the QT interval because of the more consistent T wave morphology.

42.4 Technique

42.4.1 Electrocardiographs, Lead Lines, and Electrodes

Since the time of Einthoven, electrocardiographs have been designed primarily to record the cardiac potentials of humans. As the electrocardiograms of smaller mammals have frequency components that are higher than those of humans, electrocardiographs of an earlier era were found to have an inadequate frequency response for these small species [57, 58]. In the rat electrocardiogram, frequency components up to 400 Hz have been reported [59] but little distortion occurs when recording with equipment having a frequency response of 200 Hz [60]. Most modern electrocardiographs are reputed to approach an upper frequency response limit of 200 Hz. Electronic electrocardiographs now permit a longer recording time, as well as post-processing of rate, amplitude, filtering, and speed. The advantage for data storage and review is far superior performance compared to the use of paper recordings. However, these machines are often equipped with optional electronic filters that reduce response to well below that needed to record electrocardiograms of small mammals with fidelity [50]. Thus, to avoid distortion, an understanding of the filtering methods must be appreciated by the operator.

Lead lines connecting the animals to the electrocardiograph must be flexible, made of shielded cable and long enough to accommodate the size of the subject. Fine-wire extensions that are not shielded are satisfactory when dealing with very small species. The major problem to avoid is any tugging on the electrodes by swinging lead lines since this will shift the electrodes at the skin interface and cause baseline artifacts in the electrocardiogram at frequencies or harmonics of the lead-line motion that can vary from slow drift to rapid waves that mimic atrial flutter or fibrillation.

There has been considerable experimentation with electrodes in animal electrocardiography over the years but few studies of the electrical conductive properties of the various types recommended. One exception is the study of Almasi et al. [61]. Clip electrodes, such as electronic crocodile clips, are satisfactory if the total area of contact is ~1 cm^2. Clipping the fur and applying patches similar to the method used in humans is better for intermediate length recordings. The conductive properties of copper are superior to those of stainless steel. Importantly, the medal used for the different leads should not be mixed but uniform.

42.4.2 Restraint

42.4.2.1 Physical Restraint

For farm animals, the usual methods of tethering or use of stanchions are satisfactory. The main requirement is adequate control of foreleg position. For monkeys, flat restraint boards, V-boards or cradles (supine position), and monkey "chairs" (sitting position) are commonly used. Dogs and cats can generally be handled. The use of slings or support in a "begging position" [62] is not recommended. For small laboratory animals (rat, guinea pig, etc.) a variety of restraint devices have been used including the usual commercially available laboratory animal holders. Some investigators have constructed special holders for rats and guinea pigs, which confine the animal so that each limb dangles in separate beakers of saline [63] or each foot rests on electrodes covered with conducting paste [64]. A simple method is to attach the electrodes, cover the animal with a cloth to exclude light and visual stimuli, and hold by hand. For example, satisfactory records were obtained from moles by restraining by hand in a cloth, or by allowing freedom in a box containing earth under which they burrowed and became quiet [65]. For extended studies, implantable recorders are now possible for long-term recording of quality data in very small mammals.

42.4.2.2 Chemical Restraint

Whenever possible, anesthetics, narcotics, or tranquilizers should be avoided because all such agents affect heart rate, rhythm, and electrocardiographic time intervals while some are arrhythmogenic. On the other hand sedation may be required in order to obtain a quality recording. The final decision depends on the question to be answered.

42.4.3 Positioning

In all quadrupeds, it is necessary to control foreleg position to obtain consistent direction and magnitude of cardiac vectors from limb-lead scalar electrocardiograms. It must be emphasized that this precaution applies to large domestic animals as well as small laboratory species [50].

42.4.4 Leads and Lead Systems

42.4.4.1 Limb Leads

Since the time of Waller [1, 66] limb leads have been recorded in animals for scalar electrocardiograms and calculation of vector forces. They were used in Einthoven's laboratory for horses and dogs [67]. For years, some investigators rejected limb leads, considering them unsuitable for various quadrupeds. This was on the spurious ground that the heart is not in the center of an equilateral triangle as in man, not realizing that Einthoven's triangle concept relating the potentials of the three bipolar limb leads depends on Kirchhoff's second law of circuits rather than the geometry of the thorax or the location of limb attachment to the thorax.

Nörr [18] and Kahn [15], on the other hand, favored chest leads over limb leads for sounder reasons. Both investigators considered limb leads unsuitable for horses because the electrocardiographic waves were small and variable compared to those of man. For species belonging to the QRS vector and sense class A (primates, carnivores, certain rodents, etc.), the limb leads have always been considered satisfactory.

42.4.4.2 Chest Leads

In accordance with the aforementioned, Nörr [18] and Kahn [15] devised two similar bipolar electrode positions for the horse. Both placed one electrode on the ventral thoracic surface near the region of the ventricular apex. Nörr located the other electrode on the right anterior breast at the level of the scapulohumeral joint (point of the shoulder) while Kahn placed the other electrode on the right side of the base of the neck at the level of the middle of the anterior border of the scapula. The direction of the deflection was positive when the electrical vector was directed toward the anterior breast or neck electrode and vice versa. Nörr [19] later chose for cattle a location for the neck electrode similar to that used by Kahn for horses because of the more vertical anatomical axis of the ox heart.

Later, in 1928, Lautenschlager [69] investigated thoracic bipolar lead combinations systematically in horses and cattle by mapping a grid over the entire trunk, dividing the surface (right and left sides) into 106 squares measuring 15 × 15 cm. His essential conclusion was that electrode positions like those used by Kahn for horses and by Nörr for cattle were most suitable; that is, one electrode at the cardiac apex region on the left side of the chest at the level of the olecranon, and the other electrode on the right side at the base of the neck in the region of the right scapula. This bipolar thoracic lead, often referred to as a base-apex (or apex-base) lead because the electrodes are placed on the body surface where an extension of the anatomic longitudinal axis of the heart would meet the skin, has been used as a monitoring lead in hoofed mammals and is recommended by the Japanese Association of Animal Electrocardiography for use in dogs [70].

Since 1928, a large number of lead systems have been devised for domestic hoofed mammals in which the three limb-lead electrodes have been placed in some triangular configuration on the torso. These have all been detailed in Roshchevsky's book on hoofed mammal electrocardiography [24]. Sporri's [26] modification of these systems has had wide use in many species, including small animals such as the rat [50], since his 1944 publication. In this system, the Lautenschlager base-apex lead, that is, right arm electrode at the base of the neck (cervical) on the right side paired with the left leg electrode placed in the region of the cardiac apex (apical) on the left ventral chest wall at the level of the olecranon, is recorded when conventional electrocardiographs are switched to the lead II recording position. The left arm electrode is placed on the back over the last thoracic vertebra. When lead I is recorded a longitudinal lead is recorded between the neck and the last thoracic vertebral electrodes and when lead III is recorded, the recording is between the cardiac apical and last thoracic vertebral electrodes. Sporri's designation for these leads is as follows:

1. Lead D for dorsal (neck to thoracic vertebral electrode, lead I electrodes)
2. Lead A for axial (neck to cardiac apical electrode, lead II electrodes) and
3. Lead J (or I) for inferior (sacral to cardiac apical electrode, lead III electrodes)

This D, A, J lead system, therefore, is equivalent to lead II (dorsal (0) lead) and two precordial leads with the different or "exploring" electrode at the cardiac apex region and the distant electrode at the base of the neck (axial (A) lead) or posterior thoracic region (inferior (1) lead). A variant of this system is to place the left arm electrode at the cardiac apex and the left leg electrode at the last thoracic vertebra (Sander [71]).

In 1954, Brooijmans [67, 72] proposed a thoracic unipolar lead system (employing Wilson's central terminal) in which nine (horse) or seven (cattle) equidistant chest leads encircle the chest vertically at the sixth (horse) or fifth (cattle) intercostal space and five (horse) or four (cattle) electrodes encircle the anterior chest in the horizontal plane. A more complicated thoracic lead system with 28 lead positions was studied by Sellers et al. [73] for dairy cattle. None of the more elaborate chest-lead systems has come into use.

Other triangular thoracic lead systems for large hoofed mammals have been proposed to represent the electrical forces of the cardiac cycle as projected on the frontal (F), transverse (T), and sagittal (S) planes. Roshchevsky [24, 36, 39] devised such a system for hoofed animals (cattle, reindeer, etc.) in which the frontal plane is defined by placing the right arm electrode at the right scapulohumeral (shoulder) joint, the left arm electrode at the left shoulder joint (lead IF), and the left leg electrode on the ventral abdominal midline at the level of the 13th (last thoracic) vertebra (leads I IF and IIIF). The sagittal plane is defined by placing the right arm electrode at the cranial end of the manubrium sterni, the left arm electrode on the withers at the fourth vertebra (lead IS), and the left electrode on the ventral abdomen at the same site as for the frontal plane (leads IIS and IIIS). The electrocardiograph is then switched to obtain six leads in each plane (that is, IF, IIF, IIIF, aVR_F, aVL_F, aVF_F, for the frontal plane and IS; IIS; IIIS, aVR_S, aVL_S, aVF_S, for the sagittal plane). Roshchevsky, from his detailed studies of the cardiac potential over the torso surface of ungulates, considered the sagittal lead system most suitable and it was adopted by [74], for clinical studies with cattle. Sugeno et al. [75] also adopted a three-lead triangular recording system with one lead at the cardiac apex for their earlier studies. Too et al. [76] investigated a three-bipolar-lead sagittal system in cattle (placing the right arm electrode at the cardiac apex, the left arm electrode at the scapulohumeral joint, and the left leg electrode at the withers) and a bipolar transverse three-lead system (placing the right arm electrode on the right olecranon, the left arm electrode on the left olecranon, and the left leg electrode at the withers). Kusachi and Sato [77] experimented with four bipolar systems in horses, finally recommending one in which three electrodes were placed on the right foreleg, left foreleg, and withers, respectively.

In horses, Lannek and Rutqvist [78] introduced a unipolar thoracic read system (employing Wilson's central terminal) patterned after the system Lannek had used in the dog [23]. This system, with the addition of leads V_{10} and CV_6RU, has been used in horses and cattle[42, 79]. These chest leads are as follows:

1. Lead CT_1 in which the exploring electrode was placed on the right side of the thorax 3–5 cm above a horizontal line through the highest point of the olecranon and behind the posterior edge of the triceps brachii muscle
2. Lead CT_2, where the exploring electrode was placed on the left side of the thorax at the same height as a horizontal line through the highest point of the olecranon and behind the triceps brachii muscle
3. Lead CT_3, where the exploring electrode was placed on the left side of the thorax about 8 cm above the CV_6LL (or CT_2) electrode

From this system the following thoracic unipolar leads evolved [79]:

1. Lead CV_6RU, with the electrode on the right side of the thorax at the sixth rib at the level of a horizontal line drawn through the point (scapulohumeral joint) of the shoulder
2. Lead CV_6RL, with the electrode on the right side of the thorax, just above the highest point of the olecranon, behind the triceps brachii muscle at the sixth rib

3. Lead CV_6LL, where the electrode is on the left side of the thorax at the same height as a horizontal line drawn through the highest point of the olecranon, behind the triceps brachii at the sixth rib
4. Lead CV_6LU, with the electrode on the left side of the thorax directly above CV_6LL at the level of a horizontal line drawn through the point of the shoulder
5. Lead V_{10}, where the electrode is placed over the vertebral column vertically above CV_6LL (at about the dorsal spinous process of the seventh thoracic vertebra)

In primates, two systems of thoracic leads have emerged. The most common practice is to use counterparts of the six precordial leads employed in humans [46, 80, 81]. Atta and Vanace [82] introduced a precordial three-lead system for the rhesus monkey (*Macaca mulatta*) with electrodes placed as follows: MV_1 fourth right intercostal space at the midclavicular line; MV_2, fourth left intercostal space at the midclavicular line; and MV_3, fifth left intercostal space at the left, midaxillary line. These leads correspond roughly to leads V_1, V_3, and V_5 in humans [80]. A thoracic lead system found satisfactory for squirrel monkeys (*Saimiri sciurcus*) was devised by Wolf et al. [83]. The precordial electrodes were placed as follows: V_6 over the third rib 1 cm to the right of the midline; V_4, over the ninth rib 1 cm to the left of the midline; and V_6 over the eighth rib at the midaxillary line.

In drug studies, lead V_{10} is often added to these precordial leads [49]. For cynomolgus monkeys (*Macaca fascicularis*) used in drug studies, the Atta and Vanace [82] system is often used [49]. Often however, for primates the system used for humans is used.

42.4.4.3 Orthogonal Leads and Cardiac Vectors

Hamlin and associates [53, 84] used leads I, aVF, and V_{10} as X, Y, and Z leads, respectively, to define three recording planes that could be considered somewhat orthogonal to one another: frontal, leads I and aVF (or Z and Y); sagittal, leads V_{10} and aVF (or Y and Z); and transverse, leads I and V_{10} (or X and Z). While arguably a nonorthogonal system in the true sense, this approach was useful and led to the current classification of mammalian electrocardiograms on the basis of QRS direction and sense [53].

Hamlin et al. later modified the orthogonal lead system of McFee and Parungao for dogs for use in studies with the horse [85] and miniature pig [86]. Roshchevsky [24] similarly employed an uncorrected lead system to obtain approximate orthogonality in cattle. The lead pairs for each axis were: X, right scapulohumeral joint to left scapulohumeral joint; Y, cranial end of the manubrium sterni to the ventral abdominal midline a handsbreadth in front of the navel; and Z, between the withers and the left foreleg.

Holmes et al. [87, 88] developed an orthogonal system as did Grauerholz [89] for vectorial evaluation of the horse electrocardiogram.

Grauerholz adopted the lead system of Baron [89]. The lead pairs for each axis were: X, right scapulohumeral joint to left scapulohumeral joint; Z, middle of the back, halfway between the forelimbs and hindlimbs to a point vertically under this on the abdominal midline; Y, determined trigonometrically from the Z-axis lead and a lead between the dorsal back electrode of lead Z and the right shoulder electrode of lead X on the right scapulohumeral joint.

42.4.4.4 Cardiac Electric Fields and Generation of Cardiac Potentials in Hoofed Mammals

In hoofed mammals (horses, cattle, sheep, swine) that are categorized by the QRS direction and sense into class A, reliable electrocardiographic criteria for the diagnosis of left and right ventricular hypertrophy and bundle branch block have not been established. It seems that the most likely reason for this is that the elaborate Purkinje conduction system is capable of maintaining an orderly spread of excitation despite considerable disruption of the ventricular myocardium by disease in these species.

42.4.4.5 Vectorcardiography

Vectorcardiography is not commonly used today. Mapping systems that evaluate the endocardial, epicardial, and even the transmural patterns of initiation and propagation of depolarization and repolarization are used in electrophysiological studies. Historically, the vectorcardiographic approach in mammalian electrocardiology was used primarily to determine the direction of electrical forces responsible for the form of the electrocardiogram in various species and for the study of appropriate lead systems [24, 85–93]. In the horse, for example, such vectorcardiographic investigations have used a Duchosal or similar cube system [79, 96], a form of tetrahedron [53], a triangulation of leads in the horizontal (frontal) and transverse planes [90], or some similar arrangement. Holmes et al. [87, 88] first experimented with an *XYZ* system in which the *X* lead consisted of five electrodes on the left and five on the right side of the chest, distributed over the shoulders and forelimbs so as to cover the heart area anatomically [87]. In a second study, these multiple electrodes were replaced with two malleable tin plates (0.6 mm thick) that could be molded to the body surface, covering an equivalent area [88]. The *Y* lead consisted of an electrode at the xyphoid and one at a point on the anterior chest on the midline between and on the level of the points of the shoulders (scapulohumeral joints). The Z lead consisted of two electrodes, one just to the left of the withers and the other on the upper left foreleg [90]. The resultant vector loops were displayed using an *XY* plotter [94].

42.4.4.6 Fetal Electrocardiography

Fetal electrocardiography has been practised in mares since 1921 [95]. In this species, the fetal QRS complexes may appear in limb leads II and III in advanced pregnancy (for instance, about 1 month before birth) and P and QRS complexes identified in abdominal leads [96]. In middle gestational stages (for example, 150 days or so), in both cows and mares, the fetal QRS complexes may be detected with appropriate abdominal and rectal leads [97].

In the mare the most favorable bipolar electrode positions are on the back at the midline in the midlumbar position and 6 in. (~15 cm) anterior to the udder on the midline of the ventral abdominal wall [96]. In the cow, favorable bipolar electrode positions are the right side of the lower abdomen or flank paired with an electrode located on the right side of the anterior abdomen or right paralumbar fossa region, or paired with a rectal or vaginal electrode [97]. The superiority of right-sided abdominal leads in cows has also been confirmed by other investigators (for example, Golikov and Vershinina [99]).

In pregnant ewes, fetal electrocardiograms from abdominal lead positions as used in cattle were obtained in only one of five animals tested [99].

As in other species, the heart rate of fetuses in cows and mares tends to decrease as pregnancy advances (for instance, in cows, fetal heart rate varies from about 140 beats per minute [bpm] at 160–191 days of pregnancy to about 120 bpm at days 251–281 of pregnancy and in mares, fetal heart rate varies from about 120 bpm in midstages of pregnancy to about 80 or 90 bpm toward the end of term) [96, 97, 99]. The fetal heart rate, however, is not stable and may vary from moment to moment independent of the maternal heart rate, possibly on account of fetal movement.

There has not been a great deal of clinical application of fetal electrocardiography in veterinary medicine although it is useful in cows and mares for monitoring fetal well-being, for making the diagnosis of fetal death after the midstage of gestation and, by identifying three spike rhythms (maternal and two fetal QRS complexes), for the diagnosis of twin pregnancies [96, 97]. In the case of twin pregnancy in cows, different bipolar lead positions were found best at different stages of pregnancy: at 5 months, vertical right-sided or longitudinal right-sided bipolar abdominal leads; at seven months bipolar leads between rectum or flank and midabdominal line on each side of lower abdomen; and at 9 months, bipolar leads from the right side to the left side of the lower abdomen [100].

Fetal electrocardiography has found another interesting application in teratological studies of toxins with rats [50, 101]. In this case, pregnant rats from the group used in a teratological study are selected and anesthetized. The fetuses are delivered surgically, and three-bipolar-lead electrocardiograms are recorded with intramuscular electrodes attached to both shoulders and one thigh of the fetus (leads I, II, III) while placental attachment is maintained. Since fetal and maternal rat electrocardiograms have different electrophysiological properties and sensitivity to such drugs as cardiac glycosides, simultaneous recording of maternal as well as fetal electrocardiograms should have other applications in toxicology and pharmacology [50].

42.4.5 Types and Duration of Recording

42.4.5.1 Clinical Records

Electrocardiography for the diagnosis of disease in clinical medicine is commonly used in veterinary practice. Here, the selection of leads and duration of records will depend on the purpose being served. In the veterinary clinic the leads selected will depend on their usefulness in diagnosis for the species being examined; in dogs, for instance, the six- or 12-lead system can be used. If the detection or diagnosis of an arrhythmia is needed, the records may be quite long, depending on the tractability of the subject, the patience of the investigator, and the pertinence of the findings, such as for clinical diagnosis in a veterinary clinic. In an epidemiological study of 4,831 dogs from a clinic population, cardiac rate and rhythm were monitored over 5 min by auscultation, palpation of the precordium, palpation of the femoral pulse, and a short casual single-lead electrocardiogram [102]. This standardized examination procedure permitted an analysis of the incidence of abnormal arrhythmias in a clinical population of dogs under these circumstances [103]. It should be stressed that today 24-h ambulatory electrocardiographic monitoring is vital in the detection of arrhythmias and the management of treatment in these animals.

42.4.5.2 Serial Records

Repeated electrocardiograms taken at predetermined intervals are useful in following clinical cases and in monitoring experimental investigations such as chronic or subchronic toxicity studies [45, 49]. In the case of toxicological studies, one important question is whether the test animal has arrhythmogenic properties. Since arrhythmias are typically episodic or sporadic events, it is important that the duration of heart rhythm monitoring be standardized for groups and individuals within groups in such studies. Ideally, this would require continuous monitoring over 24 h.

42.4.6 Telemetry and Holter Monitoring

Electrocardiographic telemetry is almost as old as the electrocardiograph itself. Einthoven, in 1903, reported using the wires of the Leiden telephone system to transmit electrocardiographic signals from a hospital patient to his laboratory, a distance of about 1.6 km [104, 105]. From these early beginnings, remote recording of the electrocardiogram developed along two general lines: telemetry by telephone and radiotelemetry [106] and then, the early Holter monitors (24 h electrocardiographic monitoring) [107].

Extensive use of radiotelemetry for measuring biological variables began with the availability of transistors in 1954 [108] and a strong stimulus for work in this area was the telemetering of biological data from a dog in Sputnik II by the USSR in 1957 [109].

Radiotelemetry was soon applied to unrestrained domestic [110–117], wild [118] and laboratory [108, 119, 120] animals and veterinary patients [121]. With miniaturization and other technological advances, totally implantable biotelemetry systems are now available that record seven channels of physiological data and are small enough (for example, $2.5 \times 2.5 \times 1.0$ cm) to insert into animals the size of infant bonnet monkeys [122–124].

Holter monitoring is commonly used in the clinical care of veterinary patients in order to diagnose arrhythmias, determine their frequency, and verify a successful treatment. In some studies, continuous 24 h ECG monitoring may be of value in determining whether arrhythmias develop as a result of a drug. It is not realistic to think that the brief rhythm strips of a routine ECG, whether it is a casual or serial ECG, could identify accurately the frequency of arrhythmias in response to a drug. Conversely, in the evaluation of the response to therapy for arrhythmias, it is essential that a 24-h Holter be used to most fully determine the benefits or proarrhythmic effects.

42.5 Interspecies Correlations

Electrocardiography is a routine examination undertaken as part of many investigations that seek to evaluate drugs for safety. Moreover, selected species are used in research of all types, including those specific to the cardiovascular system. The most common biomedical research animals today include the mouse [125], rabbit [126], dog [127], minipig [128], and nonhuman primates [129, 130].

42.5.1 Body Size, Heart Rate, and Time Intervals

Many biological variables are related to the size of animals and such allometric relations have been described for various measures of cardiac function including the electrocardiogram [131]. The resting heart rate is related to body size, which is in turn related to metabolic rate, while both heart rate and metabolic rate are regulated by autonomic balance and humoral agents. Broadly considered, the heart weight of mammals is nearly proportional to body mass (approximately 0.6% of body mass) and heart rate per minute is inversely related to both [130–133]. The relationship between heart rate and body weight can be expressed by equations, in which HR is heart rate per minute and W (in kg) is body weight. Two such equations proposed (see [133] and [131], respectively) are in essential agreement:

$$HR = 241 \times W^{-0.25}$$

$$HR = 360 \times W^{-0.26}$$

Individual species or strains may, however, differ substantially from such generalized relationships because of special physiological adaptations. For example, the resting heart rate of the horse ranges from 28 to 48 min^{-1} while that of dairy cattle is 48–84 min^{-1}; the heart rate of the domestic rabbit is 180–350 min^{-1} while that of the hare is 70–80 min^{-1} [132]. Likewise heart weight/body weight ratios may differ in species or strains of the same general size; the wild hare heart weight/body weight ratio is four times that of the Texas jackrabbit and 1.5 times that of the domestic rabbit [135, 136]; that of wild rats is twice the value of laboratory rats [135]; the adult greyhound heart weight/body weight ratio is about 1.3 times that of the mongrel [137]; and thoroughbred race horses have larger hearts than other breeds [135]. With these exceptions to allometric generalizations in mind, it is useful to examine such electrocardiographic relationships [130].

The nearly universal positive correlation between the R-R interval (heart rate) and PR, QRS, and QT durations observed with changes in heart rate in an individual is seen to be preserved in the intraspecies correlation, that is, as heart rates increase and heart sizes decrease, PQ, QRS, and QT intervals decrease linearly, although there are some species that do not fit the regression lines, for example, dolphin and draft horse for QT and draft horse for QRS. The ratio of QRS/QT duration is seen to be about the same among the different species at about 0.20–0.25 while that of PQ/QT is approximately 0.6.

42.5.2 Heart Rate Variability and Acceleration

Different species vary enormously in their respective degree of sinus arrhythmia and some species have a remarkable ability to increase heart rate over resting levels [130, 131].

There has been considerable recent interest in using changes in the degree of heart rate variability (HRV, sometimes termed HPV for heart period variability) as a measure of psychological stress in man and animals [137–139]. In man, HRV is suppressed during sustained attention and in certain psychiatric conditions. In drug testing, agents depressing the brain stem or the sinoatrial node (for instance, calcium-channel blockers) can decrease the degree of respiratory sinus arrhythmia [45] and this variable has been incorporated into a computer-assisted electrocardiographic analysis program for drug studies [140]. Among domestic and laboratory mammals, respiratory sinus arrhythmia is most pronounced in the dog. In this species the R-R intervals may vary as much as eightfold in a given record (that is, from 250 to 2,000 ms).

While sinus arrhythmia is known to be extreme in certain exotic animals, for example, the European mole [65] and the California ground squirrel [118], there is insufficient information on occurrence of this characteristic among the various mammalian species to make valid comparisons between species.

Table 42.2
Prevalence rates per thousand of common spontaneous arrhythmias and conduction disorders in pretest electrocardiograms for unanesthetized beagles (sample population *N* = 8,977), cynomolgus monkeys (*N* = 1,165) and albino rats (*N* = 442)

Arrhythmias or conduction disorders	Beagle	Cynomolgus monkey	Albino rat
Ventricular extrasystoles.	8.0	14.6	67.9
Second-degree AV block.	10.0	0.0	88.2
First-degree AV block.	1.9	0.0	2.3
Right bundle branch block.	0.0	42.1	0.1
Left bundle branch block.	0.0	1.7	2.3
Wolff–Parkinson–White Syndrome.	0.2	0.0	0.0
Any of the above.	20.1	38.6	160.6

With respect to cardiac acceleration, of the animals for which data are available, the horse has the greatest capacity to increase heart rate with exercise [137], for example, from 30 min^{-1} at rest to 240 min^{-1} with strenuous exercise, that is, about an eightfold increase. In man and dog the usual increase in heart rate with strenuous exercise is only about threefold although exceptional racing greyhounds may rival the horses' six to eightfold increase. There is both theoretical and experimental evidence (dog) that even at these high heart rates, continued increase in cardiac output with increase in rate prevails and that this positive relationship between increased heart rate and cardiac output is favored in thoroughbred horses and greyhounds because of their relatively larger hearts and stroke volumes [141].

42.5.3 Arrhythmias and Conduction Disorders

Table 42.2 compares the prevalence of cardiac arrhythmias in pretest (control) electrocardiograms from experimental beagles, cynomolgus monkeys, and albino rats. Respiratory sinus arrhythmia is not included in this table because it is ubiquitous in normal dogs at rest, although it is less frequent in restrained monkeys and rats. Except for Wolff–Parkinson–White (WPW) syndrome, the conduction disturbances are generally considered normal (see below).

In monkeys and rats (Table 42.2), and in pigs as well, ventricular extrasystoles appear to be induced by the excitement and struggling caused by restraint, because their prevalence decreases with training and because their appearance will occur in a few additional subjects during a study in which serial records are taken over a sustained period.

In minipigs, premature sinus beats with aberrant ventricular conduction may be fairly common in untrained groups of animals. In cats under halothane anesthesia, atrioventricular dissociation occurs frequently and may appear occasionally merely as the result of excitement [49].

42.5.4 Heart-Rate Dependence of Electrocardiographic Time Intervals PR, QRS, and QT

The change in PR, QRS, and QT intervals with change in R-R (that is, cycle length or heart rate) makes it necessary to correct these values when comparisons of time intervals are made at different heart rates. If this is not done, the effects of disease, drugs, or toxins, for example, on the QT interval (or on PR or QRS) may be masked by changes in the interval that are dependent on heart rate.

Various linear, cubic or square root, and logarithmic or exponential equations have been proposed to describe these relations [8, 144, 145]. Most of the adjustment formulae are flawed.

Probably the most useful approach at present is for each clinic or laboratory to develop its own database, as has been done for experimental beagles [146, 147]. Either a new formula that adequately describes the relations between the heart rates and time intervals over the entire data range, or a frequency distribution table of these values may be used (see Table 41.4). The practice of correcting the observed QT interval for heart rate by dividing it by the square root of the length of R-R interval serves very little purpose. With this formula, $QT_c = QT/(RR)^{1/2}$, the observed value of QT at a given heart rate over 60 is increased and, for a rate under 60 it is decreased, but the resultant QT_c does not necessarily remain constant over the heart-rate range for the same animal. Hence, it does not increase the comparability of interval durations at different heart rates.

Since QRS is the shortest of these intervals, its absolute variation with heart-rate change is least and ordinarily is not considered, although the relationship is definitely present [8].

Determination of PR interval and QT interval change with R-R duration poses special problems with rapid heart rates in smaller species such as the rat [50]. There appear to be three reasons for this:

1. At high heart rates (475–600 min^{-1}) and paper speed of 50 mm s^{-1} or less, any changes in PR-interval and QT-interval duration with rate are small (that is, a few milliseconds) and therefore, difficult to measure accurately. This problem can be resolved with high-frequency sampling and electronic digital recordings that permit measurements at high speeds.
2. Also, at rapid rates (e.g., >450 min^{-1}) the succeeding P waves are superimposed on the descending limb of the previous T wave and the true QT interval cannot be determined accurately.
3. At lower heart rates (~250–350 min^{-1}) which occur frequently with anesthesia, these interval changes with rate are greater and easily measurable, but it is uncertain whether this is a rate effect or the result of anesthetic action on the myocardium.

The QT versus R-R relationship exhibits hysteresis; that is, with sudden changes in heart rate, the QT interval changes its duration gradually, requiring several heart beats at the same heart rate to attain a new steady state. Thus, with moderate variations in R-R intervals, as in respiratory sinus arrhythmia of the dog, the associated QT intervals do not undulate. This is also true with second-degree AV block in the dog in which the post-block PR interval is ordinarily shortened but the QT interval is little changed. Horses behave differently, however, because in this species both the PR and QT intervals are shortened in the post-block complex [148].

In premature ventricular extrasystoles, although the preceding R-R interval is shortened, the extrasystolic QT interval is usually longer because of the increase in QRS duration. Exceptionally, in premature ventricular extrasystoles, the QT interval may be shortened, perhaps because of the effect of the shortened R-R interval on conduction velocity and action-potential duration. This latter circumstance appears to be more common in standard swine and minipigs than in other species studied.

42.6 Normal Values

For each species, the lead II time intervals and amplitudes are presented. In these tables, only representative values (mean [above] and range [below]) for time intervals and amplitudes will be given for lead II. Unless otherwise stated, heart-rate and time-interval data are from unanesthetized and unsedated animals. Vectors (frontal plane unless otherwise designated), configuration, normal variants, and special features will be outlined in text.

Normal variants are electrocardiographic rhythm or conduction disturbances and uncharacteristic wave configurations, occurring with a sufficiently high incidence in otherwise healthy individuals of a given species that they cannot be considered abnormal [143]. Some examples are (➲ Table 42.2): first-degree and second-degree atrioventricular block in dogs, horses, and rats; RBBB and (more rarely) LBBB in monkeys; rate-dependent LBBB in rats; premature sinus beats with aberrant ventricular conduction in minipigs; QRS complexes in limb leads only, with broad S waves and low-amplitude R waves, dome-dart ST-T complexes, exaggerated U waves, tall P waves (>0.4 mV), tall R waves (RV4 up to 6 mV), and P reversal in leads II, III, and a VF in dogs.

Special features mentioned include maximal heart rates with exercise, degree of sinus arrhythmia, and fetal or newborn versus adult electrocardiographic patterns when these facts are known.

42.6.1 Primates

The ECG classification and lead II time intervals and amplitudes for the primates discussed here, namely, the rhesus (➲ Fig. 42.2), cynomolgus and squirrel monkeys (➲ Fig. 42.3), the baboon, and the chimpanzee, are listed in ➲ Table 42.3.

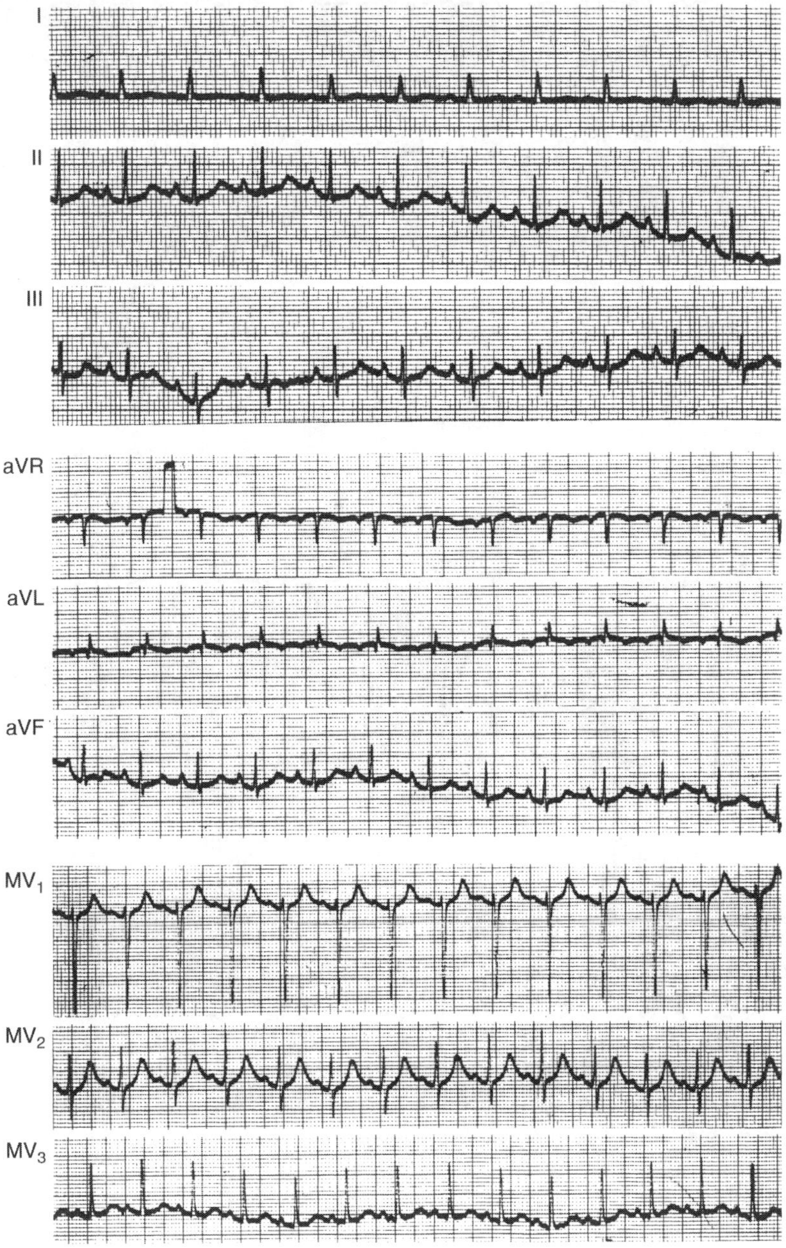

Figure 42.2

Electrocardiogram from an adult rhesus monkey (*Macaca mulatta*) restrained in a monkey chair: paper speed 50 div s^{-1}. The leads labeled MV_1, MV_2, and MV_3 are the thoracic leads recommended by Atta and Vanace [82]. MV_1 is located in the fourth right intercostals space, 4 cm from the midsternal line; MV_2 is on the left side symmetrical with MV_1; and MV_3 is registered at the left midaxillary line in the fifth intercostals space, approximately 1 cm below the level of MV_2. MV_1 is located over the right ventricle, MV_2 often records "transitional" type QRS potentials like those registered over the intraventricular septum and MV_3 is located over the left ventricle.

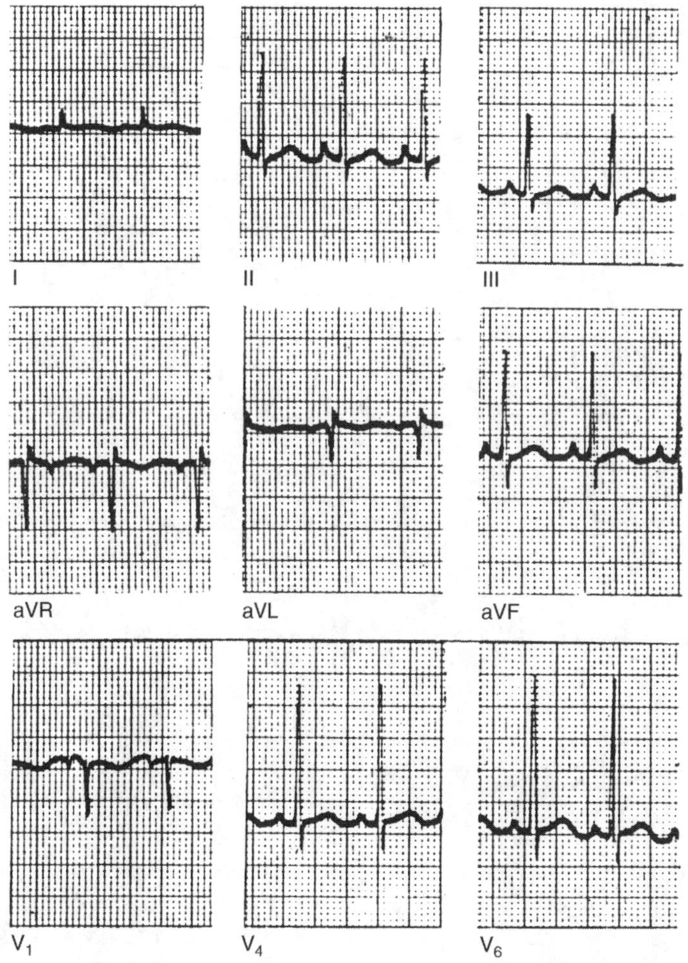

Figure 42.3

Electrocardiogram from a squirrel monkey (*Saimiri sciureus*) taken under sodium thiopental anesthesia in the supine position. Paper speed 50 div s^{-1}, standardization 0.1 mV div^{-1}. The lead positions are: V_1, third right rib 1 cm to the right of the midline; V_4, ninth left rib 1 cm to the left of the midline; V_6, eighth left rib at the midaxillary line. (After Wolf et al. [83]. @ American Physiological Society, Bethesda, Maryland. Reproduced with permission.)

The ECG configuration of the rhesus monkey commonly has peaked P waves in II, III, and aVF, and notchings in V_1 and V_3. The QRS complex is usually positive in standard limb leads and the ST segment is generally isoelectric in limb leads. The T wave is usually positive in standard and precordial leads.

Normal variants in the rhesus monkey include RBBB, which is present in about 1.4% and LBBB, which is present in about 0.8% [80]. QRS vectors suggesting incomplete RBBB occur in about 5% [80]. Negative T waves occur in bipolar limb leads in about 6% [80]. Negative T waves in thoracic leads are less common and, in serial electrocardiograms, reversal of T wave polarity may occur occasionally in limb or thoracic leads. Coupled ventricular extrasystoles may be expected in about 1.4%.

A special feature of the rhesus monkey is that the heart rates found in electrocardiograms are far more rapid than basal rates obtained by telemetry. In telemetered electrocardiograms from rhesus monkeys resting and isolated from man, the heart rates are 80–100 min^{-1} and sinus arrhythmia is pronounced. Electrocardiograms of the cynomolgus monkey (*Macaca fascicularis*) have an ECG configuration that is the same as that of the rhesus monkey. Normal variants

Table 42.3

ECG classification and normal parameters in primates

Common English name (*species name*)	ECG classification			Heart rate	Lead II time interval (s)				Lead II amplitude (mV)				
	QRS[a]	QT[b]	T[c]	(bpm)	P	PR	QRS	QT	P	Q	R	S	T
Rhesus monkey (*Macaca Mulatta*) [37, 38, 46, 78, 80]	A	A	A	257 160–350	0.05 0.03–0.06	0.07 0.05–0.10	0.03 0.02–0.05	0.14 0.11–0.19	0.25 0.12–0.37	0.10 0.00–0.50	0.90 0.00–2.04	0.15 0.00–0.40	0.10 −0.10–0.20
Cynomolgus monkey (*Macaca fascicularis*) [82, 130, 149]	A	A	A	180 120–220	0.03 0.02–0.05	0.08 0.06–0.11	0.03 0.03–0.05	0.19 0.16–0.25	0.07 0.04–0.26	0.06 0.00–0.30	0.30 0.10–0.95	0.09 0.00–0.46	0.07 0.00–0.50
Squirrel monkey (*Saimiri Sciureus*) [83]	A	A	A	248 160–340	0.03 0.025–0.035	0.05 0.042–0.058	0.03 0.023–0.037	0.15 0.11–0.20	0.18 0.05–0.30	0.05 0.0–0.1	0.90 0.10–1.70	0.20 0.00–0.50	0.05 −0.10–0.20
Baboon (*Papio spp*) [37, 81]	A	A	A	127 80–190	0.05 0.04–0.07	0.12 0.05–0.15	0.06 0.04–0.07	0.27 0.20–0.31	0.19 0.09–0.30	0.03 0.00–0.15	1.10 0.80–1.30	0.11 0.00–0.25	0.08 −0.10-0.25
Chimpanzee (*Pan Troglodytes*) [37]	A	A	A	159 100–249	0.071 0.04–0.10	0.11 0.07–0.15	0.045 0.03–0.07	0.233 0.18–0.32	0.17 0.05–0.40	0.06 0.20–1.70	0.82 0.20–1.70	0.25 0.05–0.90	0.04 −0.10–0.20

[a] QRS-vector and sense: A, along the long axis of the body, caudally and ventrally; B, from sternum toward spine.
[b] Relative QT-interval, ST-segment and action-potential duration: A, long; B, short.
[c] T-wave liability: A, constant; B, labile.

again include RBBB, which is present in about 4% of cases, and LBBB, which is present in about 0.2%. Coupled ventricular extrasystoles also occur with an incidence of 1.5%. Negative T waves may be present in bipolar limb leads and thoracic leads and reversal of previously positive T waves may occur spontaneously in serial records. An electrocardiogram (taken under sodium thiopental anesthesia in the supine position) of the squirrel monkey (Saimiri sciureus) as shown in ➲ Fig. 42.3 has a mean A QRS direction of 62° and A QRS range −60°–115°. The P wave of this species is usually positive in bipolar limb leads, occasionally tall and narrow (0.35 mV) in lead II, and sometimes its amplitude is variable in limb leads. The QRS is occasionally of low amplitude in limb leads. The ST segment is usually isoelectric but deviations of 0.1 mV are sometimes present. T is usually concordant and positive or diphasic in bipolar limb leads.

As in other monkeys, RBBB is not uncommon in its occurrence and is therefore considered a normal variant. A special feature concerning the squirrel monkey is that coupled ventricular extrasystoles occur occasionally in otherwise healthy subjects. Also, since two individuals with WPW syndrome have been found among a relatively small number of squirrel monkeys, a higher prevalence of Wolff–Parkinson–White syndrome may occur in this species than in others.

Electrocardiographic data have been obtained from several monkey species sedated with ketamine [149] including the three discussed above.

Two special features of the baboon should be mentioned. First, WPW syndrome was described among one group of 170 baboons and second, with telemetry the heart rate is 80–100 min^{-1} at rest.

ECGs from chimpanzees (Pan troglodytes) have a P wave that is usually positive in standard limb leads, but occasionally may be isoelectric in either lead I or III. The QRS complex is both of lower amplitude and lower duration than in humans and the T wave is usually concordant and positive in standard limb leads but may be isoelectric in about 30% and occasionally is discordant and negative.

Characteristics particular to the chimpanzee include sinus arrhythmia, which is frequent at heart rates below 125 min^{-1}. Also, wandering pacemaker from the SA node to the AV junction and premature atrial beats occur rarely.

42.6.2 Perissodactyla

The domestic horse (*Equus caballus*) (➲ Fig. 42.4) is the only member of the Perissodactyla discussed here. ➲ Table 42.4 lists the ECG classifications and lead II time intervals and amplitudes. In the domestic horse, the area vector *A* QRS in the frontal plane lies between −64° and 102°. The range of its direction in the transverse plane is −32°–100° and in the left sagittal plane is −34°–173°. The P wave in bipolar leads is usually bifid, is most often positive, is sometimes diphasic and may change its form spontaneously (for instance, from bifid to single peak) for a series of beats. Such changes in form occur in about 30% of horses. Group B ventricular activation pattern results in major QRS vector forces being directed from the sternum toward the spine. Thus, maximal QRS deflections are rarely recorded in the frontal plane limb leads but rather in the thoracic leads, especially CV_6LL, V_{10}, Z, or the base apex bipolar lead. The ST segment is usually isoelectric in limb leads and is often arched or coved in chest leads. With regard to the T lability, amplitude may change and polarity reverses with changes in heart rate. In the post-block beat in second-degree AV block, the QT interval shortens, and there is reduction in T wave amplitude or change in form, for instance, from diphasic to positive.

Normal variants in the domestic horse include periodic changes in P-wave configuration (considered as wandering pacemaker in the sinoatrial node), sinoatrial block, first-degree and second-degree atrioventricular block, and nonrespiratory sinus arrhythmia, which all occur commonly in otherwise healthy horses. All these changes are considered to be largely the result of increases in vagal tone in this "vagotonic" species.

Atrial fibrillation is more common in horses and occurs with less-severe underlying heart disease (or with no evidence of heart disease) than in other domestic species. This arrhythmia is treated with quinidine sulfate. Electrical cardioversion is another successful means of treatment. This apparent increased susceptibility to atrial fibrillation in horses is considered to be related to two predisposing factors: large atrial mass and high vagal tone.

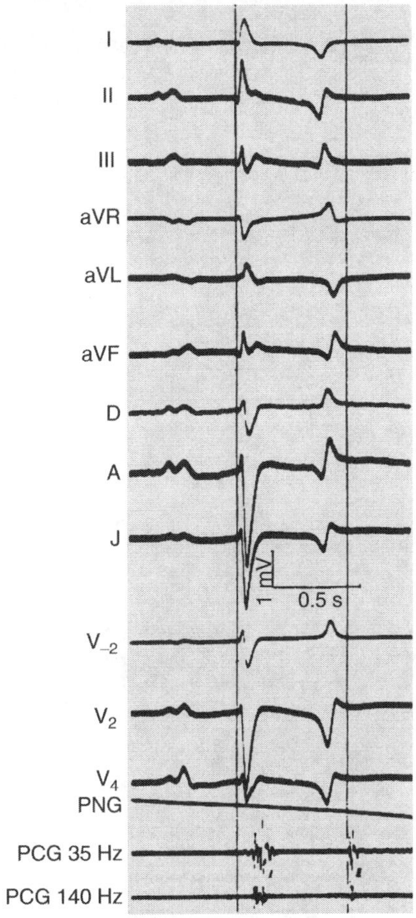

Figure 42.4
Electrocardiogram of a horse. Conventional bipolar and unipolar limb leads. D, A, and J leads after Sporri [26] as described in Sect. 42.4.4.2. The V leads are after Brooijmans [67, 72]: V_{-2} right sixth intercostal space at the junction of the lower and middle thirds of the vertical distance between the sternum and the level of the scapulohumeral joint; V_2, symmetrical with V_{-2} in the left sixth intercostal space; V_4 left sixth intercostal space at the level of the scapulohumeral joint. A pneumogram (PNG) and phonocardiograms (PCG) with cutoff filters set at two frequencies, 35 and 140 Hz, are shown at the bottom. The vertical lines are placed at the beginning of QRS and at the beginning of the second heart sound. (From Grauwiler [37]. © Birkhäuser, Basel. Reproduced with permission.)

42.6.3 Artiodactyla

The members of the Artiodactyla family discussed below are domestic cattle, sheep, goats, and pigs, giraffes, and camels. The ECG classifications, lead II time intervals and amplitudes, and references for further discussion are given in Table 42.4.

In domestic cattle (*Bos Taurus*), the mean *A* QRS is 70° with a range 30–90°. The P wave is usually positive in standard limb leads and may be bifid in left chest leads. QRS complexes are generally of low amplitude in the standard limb leads. T waves may be concordant or discordant in standard limb leads and are sometimes diphasic negative/positive.

P waves in the domestic pig (*Sus domesticus*) are usually upright in standard limb leads and may be bifid in V_{10} and the chest leads. The ST segment is isoelectric. T waves are often discordant in lead I and usually positive and sometimes diphasic in leads II and III.

Table 42.4
ECG classification and normal parameters in Perissodactyla and Artiodactyla

Common English name (*species name*)	ECG classification QRS[a]	QT[b]	T[c]	Heart rate (bpm)	Lead II time interval (s) P	PR	QRS	QT	Lead II amplitude (mV) P	Q	R	S	T
Perissodactyla Domestic horse (*Equus Caballus*) [37, 42, 65, 72, 75, 76, 83, 86–89, 159, 160]	B	A	B	35 26–50	0.14 0.08–0.20	0.33 0.22–0.56	0.13 0.08–0.17	0.51 0.32–0.64	0.28 0.1–0.5	0.12 0.025–0.35	1.13 0.2–2.5	0.15 0.025–0.45	0.20 −0.20-0.90
Artiodactyla Domestic cattle (*Bos Taurus*) [19, 24, 26, 27, 35–37, 53, 67–69, 71–76, 92, 93, 100, 161]	B	A	A	70 48–98	0.06 0.03–0.08	0.19 0.1–0.3	0.095 0.065–0.120	0.40 0.29–0.47	0.10 0.03–0.18	0.16 0.03–1.00	0.37 0.03–2.60	0.07 0.03–0.10	0.31 0.03–1.10
Domestic sheep (*Ovis aries*) [37, 162–168]	B	A	A	107 60–197	0.05 0.04–0.07	0.10 0.06–0.14	0.046 0.025–0.080	0.26 0.17–0.34	0.20 0.1–0.5	0.17 0.1–0.3	0.17 0.08–0.40	0.10 0.10	0.44 0.10–1.60
Domestic goat (*Capra hircus*) [37, 169–171]	B	A	A	96 70–120	0.04 0.02–0.06	0.12 0.08–0.16	0.045 0.03–0.06	0.30 0.24–0.36	0.08 0.02–0.15	0.48 0.07–0.95	0.19 0.02–0.50	0.08 0.02–0.25	0.20 0.05–0.50
Domestic pig (*Sus domesticus*) [37, 172–176]	B	A	A	135 100–180	0.04 0.03–0.05	0.09 0.08–0.12	0.04 0.03–0.06	0.24 0.21–0.26	0.13 0.05–0.30	0.09 0.00–0.14	0.61 0.40–0.78	0.39 0.14–0.50	0.38 −0.20–0.70
Giraffe (*Giraffa camelopardalis* reticulate) [177, 178]	B	A	A	70, 83	0.10, 0.08	0.18, 0.15	0.08, 0.10	0.45, 0.33	0.10, 0.08	0.00, 0.11	0.15, 0.70	0.00, 0.00	−0.30, −0.25
Camel (*Camelus dromedarius*) [179, 180]	B	A	A	30 24–49	0.10 0.08–0.10	0.25 0.24–0.26	0.09 0.04–0.09	0.50 0.48–0.52	0.10 0.06–0.20	0.00	0.40 0.20–1.20	0.45 0.40–0.50	−0.20 −0.40–0.10

[a] QRS-vector direction and sense; A, along the long axis of the body, caudally and ventrally; B, from sternum toward spine.
[b] Relative QT-interval, ST-segment and action-potential duration: A, long; B. short.
[c] T-wave lability: A, constant; B, labile.
[d] Two animals under etorphine (M99).

Ventricular extrasystoles are fairly frequent in excited pigs. In swine, the influence of body weight on electrocardiographic time intervals is separable from the influence of heart rate, as shown by holding the latter constant. When this is done, all electrocardiographic intervals can be shown to increase relatively with increasing body weight [150] (the effect of aging is not separable from that of increased body weight). When the influence of heart rate (R-R interval) alone on the electrocardiographic intervals was examined by holding the body weight constant, the R-R interval was found to have no effect on P-wave duration and a weak effect on PR and QRS intervals, but a strong effect on the QT interval [150]. Thus, both RR interval, that is, heart rate, as well as body weight *W* (in kg) serve to determine the expected QT interval. The following equations show these relations as calculated from data on 71 swine weighing 11–300 kg [150]:

$$\mathrm{P} = 61 + 0.11\mathrm{W}$$

$$\mathrm{PR} = 107 + 0.17\mathrm{W}$$

$$7\mathrm{QRS} = 58.8 + 0.14\mathrm{W}$$

and

$$\mathrm{QT} = 159 + 0.15\,(\mathrm{R} - \mathrm{R}) + 0.26\mathrm{W}$$

where P, PR, QRS, and QT are given in milliseconds.

In the ECGs of two giraffes, taken under etorphine, the QRS vectors were −25° and 95° (an example is shown in ➋ Fig. 42.5). The R wave was either positive or negative in leads I, II, and III. The QRS was positive in leads I and II in both animals and negative in lead III in one subject. The ST segment was isoelectric and the T waves were usually discordant.

➋ Figure 42.6 shows the ECG of a dromedary camel (*Camelus droinedarius*) taken in the standing position and under no drugs. The mean *A* QRS is +250° (with a range between 90° and 280°). In standard limb leads, the P waves are positive, QRS complexes may be chiefly positive or negative and the ST segment is isoelectric. T waves are usually discordant. A special attribute of the camel is that sinus arrhythmia is present.

42.6.4 Cetacea

➋ Table 42.5 lists the ECG classifications, lead II time intervals and amplitudes, and references of the finback, beluga and killer whales, and the dolphin. ECGs from bottle-nosed dolphins are shown in ➋ Fig. 42.7

The ECG of a finback whale (*Balaenoptera physalus*), beached 23 h before the recordings were made, has a broad P wave with low amplitude, a QRS complex which is chiefly negative in leads II and III and an ST segment which is isoelectric. T waves are discordant in most leads. In the beluga whale (*Delphinapterus leucas*), the bipolar lead between an electrode on the back at the pectoral girdle and one about the midportion of the back did not record the P wave distinctly. The QRS pattern was qR and the T wave was negative (discordant). In the killer whale (*Orcinus orca*), no P wave was recorded, the QRS had a qR configuration, and T was discordant.

42.6.5 Marsupialia

➋ Table 42.5 lists the ECG classifications, lead II time intervals and amplitudes, and references of the Bennett's kangaroo and the opossum.

Electrocardiograms from the Bennett's kangaroo (*Macropus bennetti*) have P waves that are positive and sometimes bifid in standard limb and chest leads. In leads II, III, and in chest leads over the left ventricle, QRS complexes are somewhat similar with an RS configuration. The ST segment is absent in adults. T waves are positive, concordant in limb and chest leads, beginning immediately after QRS with no intervening ST segment.

Bennett's kangaroo was the first larger species in which it was shown that when the ST segment is absent and the QT, therefore, short, there is marked dissociation between the end of the T wave and the end of mechanical systole.

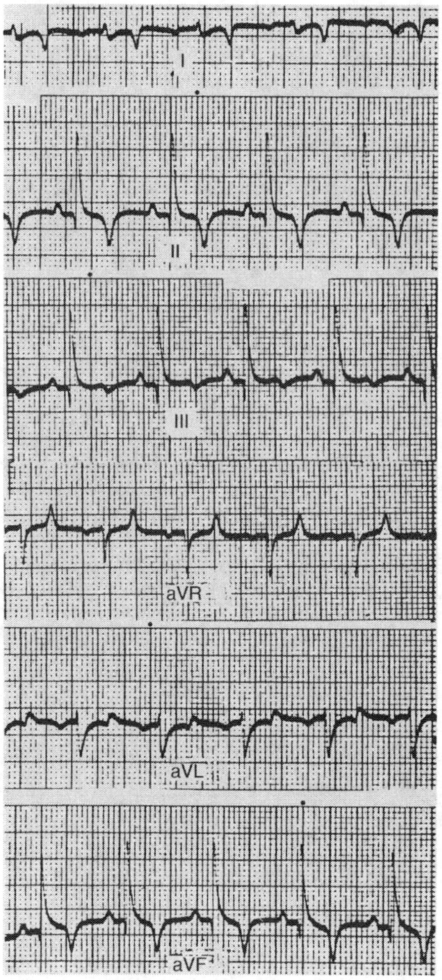

Figure 42.5

Electrocardiogram of a reticulated giraffe taken under etorphine HCl (M99) sedation in the right lateral recumbent position: paper speed 25 div s^{-1}; standardization 0.05 mV div^{-1}. (After Jefferson [154]. © American Veterinary Medical Association, Schaumburg, Illinois. Reproduced with permission.)

Correct interpretation of such fusion of QRS and T and mechanical and electrical dissociation in the mouse had been made a few years earlier in 1953 [151], The single record available from a wallaby (*Macropus walabatus*) had a short ST segment in limb leads and QT intervals of 0.16–0.18 s at a heart rate of 160 bpm. In another single record labeled "kangaroo" with no species designation, a short ST segment in limb leads was present and the QT interval was 0.23 s [5]. In 14 adult Australian rock kangaroos (*Macropus robustus*) under ketamine and pentobarbital anesthesia, the mean QT intervals of 0.214 s were far shorter than the mean mechanical systole of 0.393 s [152].

In the young Bennett's kangaroos still in the maternal pouch, the ST segment is present, as is true in the fetal or newborn rat [50] and mouse [151], and the duration of QT and mechanical systole are also similar. The adult conformation appears at the time the young kangaroos leave the pouch.

In the opossum (*Didelphis marsupialis*), the P waves are small and positive in standard limb leads. QRS complexes are chiefly positive in standard limb leads. Unlike the case of the adult Bennett's kangaroo, the ST segment is present. T waves are positive and concordant in standard limb leads.

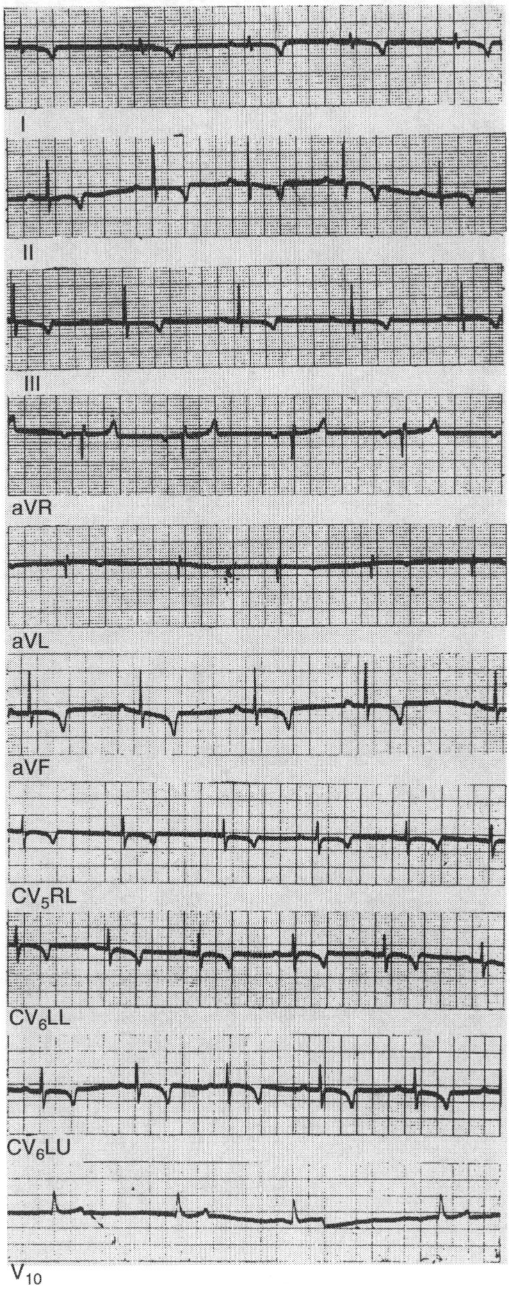

Figure 42.6
Electrocardiogram of a dromedary camel taken in the standing position. (Ind with no drugs: paper speed 25 div s^{-1}; standardization 0,1 mV div^{-1}) (See Sect. 42.4.4.2 for description of thoracic lead system) (After Jayasinghe et al. [158], © American Veterinary Medical Association, Schaumburg, Illinois, Reproduced with permission.)

■ Table 42.5

ECG classification and normal parameters in the Cetacea and Marsupialia

Common English name (*species name*)	ECG classification QRS[a]	QT[b]	T[c]	Heart rate (bpm)	Lead II time interval (s) P	PR	QRS	QT	Lead II amplitude (mV) P	Q	R	S	T
Cetacea Finback whale[d] (*Balaenoptera physalus*) [181]	B	A	A	27–32	0.40	0.68–0.73	0.32–0.34	0.96–1.08	0.10	0.30	0.00	0.00	0.50
Beluga whale (*Delphinapterus leucas*) [4]	B	A	A	16 12–24	not measurable	0.32	0.09–0.12	0.36–0.40	0.03	0.10	0.3–0.4	0.00	0.10–0.15
Killer whale (*Orcinus orca*) [182]	B	A	A	20–60	not visible P waves	not visible P waves	0.12	0.36	0.00	0.05	0.60	0.00	−0.30 0.10–0.25
Dolphin [183, 184]	B	A	A	100 60–137	0.82 0.06–0.10	0.178 0.120–0.210	0.065 0.05–0.08	0.257 0.05–0.320	0.03 0.03–0.20	0.03 0.00–0.15	0.30 0.20–0.40	0.20 0.03–0.40	0.15 0.10–0.25
Marsupialia Bennett's kangaroo (*Macropus bennetti*) [37, 186–231]	A	B	A	119 108–152	0.04 0.03–0.05	0.107 0.090–0.118	0.045 0.040–0.050	0.142 0.120–0.165	0.15	0.00	2.30	1.40	1.20
Opossum[e] (*Didelphis marsupialis*) [189]	A	A	A	200	not given	0.08	0.02–0.03	0.14	0.01	0.02	1.5	0.00	0.10

[a] QRS-vector direction and sense; A, along the long axis of the body, caudally and ventrally; B, from sternum toward spine.
[b] Relative QT-interval, ST-segment and action-potential duration: A, long; B, short.
[c] T-wave lability: A, constant; B, labile.
[d] Beached 23 h before recordings were made.
[e] Under pentobarbital sodium anesthesia.

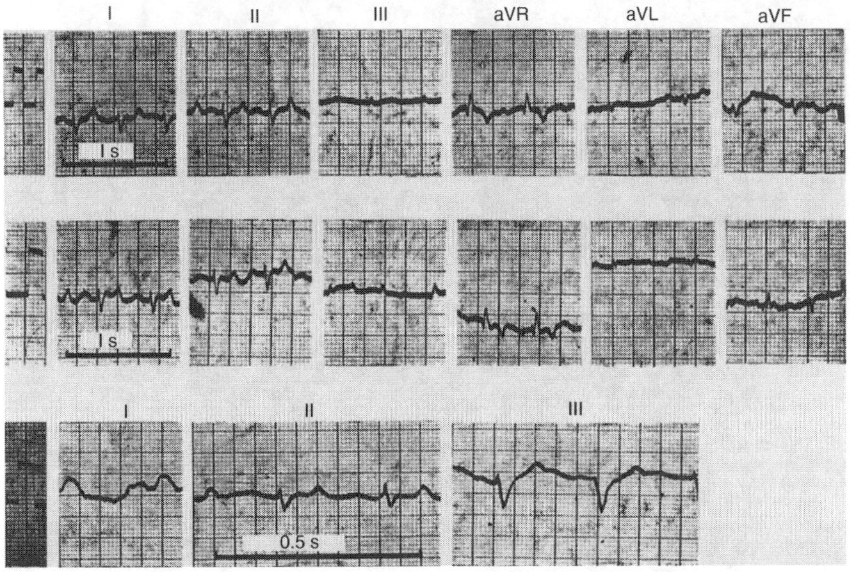

Figure 42.7
Electrocardiograms from a bottle-nosed dolphin: (a) ECG recorded immediately after loading on a truck at the port of Arari Bay: heart rate (HR) = 136.3 min^{-1}, interval times are R-R = 0.44 s, PQ = 0.16 s, QRS = 0.14 s, OT = 0.26 s, and T = 0.1 s. (b) ECG recorded after 4 h on the truck at Mishima: HR = 120 min^{-1}, interval times are R-R = 0.5 s, PQ = 0.2 s, QRS = 0.08 s, QT = 0.28 s, T = 0.14 s. (c) ECG recorded after 8 h on the truck at Enoshima: HR = 125 min^{-1}, R-R interval is 0.48 s. Needle electrodes or plate electrodes were placed at sites corresponding to the attachment of limbs in man. The pulses shown on the left of each panel indicate the scale of 1 mV. (After Tokita et al. [156]. © Geirui Kenkyosho, Tokyo, Japan. Reproduces with permission.)

42.6.6 Lagomorpha

The ECG classifications, lead II time intervals and amplitudes, and references of the rabbit (*Orytolagus cuniculus*) are given in Table 42.6.

The mean *A* QRS in the rabbit is 64° with a range of 0–180°. The P waves in rabbits are positive in standard limb leads and may be rather pointed in some strains. QRS complexes are generally positive, ST segments are usually isoelectric, and T waves are usually positive in standard limb leads. While a number of investigators have noted changes in T-wave polarity, especially in lead III and in chest leads, the amount of variability is only about the same as seen in nonhuman primates and is insufficient to classify this species as T labile or class B.

42.6.7 Rodentia

In recent years mice have been used extensively in genetic research. The expansion of this area of research, in addition to advances in the field of electrophysiology and the use of the mouse to match cardiac phenotype to genotype, has expanded the role of the mouse as a model. Despite this increase, criticism continues about the conclusions from the murine cardiac responses to that of other mammals and, importantly, humans [52]. The mouse has been evaluated by cardiac electrophysiological mapping studies, induction of particular arrhythmias, conduction abnormalities, infarction changes, and infectious responses by methods beyond the ECG.

The ECG classifications, lead II amplitudes and time intervals of the white laboratory rat and mouse and the guinea pig are listed in Table 42.6.

In the electrocardiograms of a white laboratory rat (*Rattus norvegicus*) for various foreleg positions the mean *A* QRS is 50° (with a range from −22° to 120°). There is some evidence that the *A* QRS shifts to the left with aging, but this has not

Table 42.6
ECG classification and normal parameters in Lagomorpha and Rodentia

Common English name (*species name*)	ECG classification			Heart rate (bpm)	Lead II time interval (s)				Lead II amplitude (mV)				
	QRS[a]	QT[b]	T[c]		P	PR	QRS	QT	P	Q	R	S	T
Lagomorpha Rabbit (*Orytolagus cuniculus*) [37, 190–195]	A	A	A	240 190–300	0.30 0.25–0.40	0.70 0.5–0.8	0.35 0.10–0.15		0.10 0.05–0.20	0.00	0.40 0.30–0.80	0.15 0.05–0.30	0.15 0.05–0.30
Rodentia White laboratory rat (*Ratus norvegicus*) [37, 47, 50, 196–200]	A	B	A	460 228–600	0.013 0.010–0.016	0.040 0.033–0.050	0.017 0.012–0.026	0.066 0.038–0.080	0.11 0.02–0.20	0.00	1.06 0.22–1.50	0.20 0.00–0.05	0.15 0.05–0.30
White laboratory mouse (*Mus musculus*) [37, 185, 201–204]	A	B	A	632 500–750	0.010	0.038 0.028–0.50	0.010 0.006–0.020	0.035 –0.030-0.050	0.05 0.04–0.06	0.00	0.30 0.10–0.40	0.15 0.00–0.30	0.20 0.10–0.30
Guinea pig (*Cavia porcellus*) [26, 37]	A	A	A	226 200–300	0.036 0.022–0.048	0.061 0.046–0.077	0.023 0.020–0.030	0.146 0.110–0.185	0.17 0.10–0.22	0.02 0.00–0.11	1.65 0.82–2.63	0.19 0.00–0.58	0.21 0.08–0.38

[a] QRS-vector direction and sense; A, along the long axis of the body, caudally and ventrally; B, from sternum toward spine.
[b] Relative QT-interval, ST-segment and action-potential duration: A, along; B, short.
[c] T-wave lability: A, constant, B, labile.

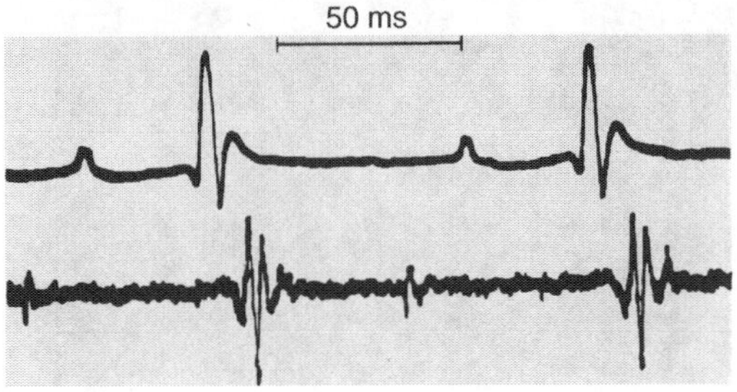

Figure 42.8
Electrocardiogram (*top*) and phonocardiogram (*bottom*) from a white laboratory mouse (*Mus musculus*), lead A (right arm electrode right side at base of neck, left leg electrode at cardiac apex). The second heart sound occurs 40 ms after the end of the T wave and no ST segment is present. (After Grauwiler [37]. © Birkhauser, Basel. Reproduced with permission.)

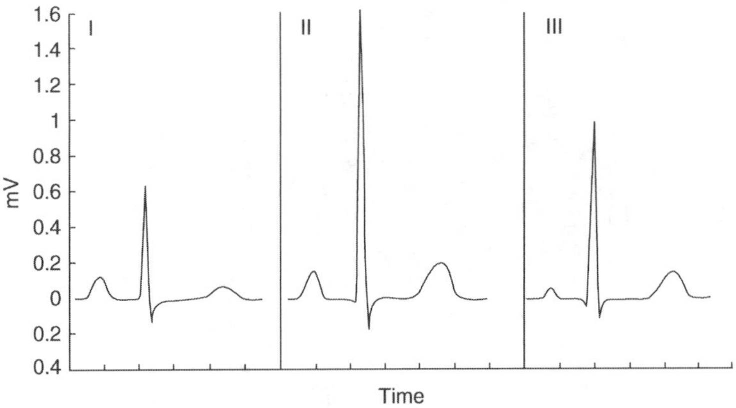

Figure 42.9
Limb leads I. II, III from a guinea pig (*Cavia parcellus*) (After Sporri H. *Habilitatiansschrift*. Zurich: University of Zurich, 1944 [26]. Reproduced with permission.)

been studied. In the white laboratory rat, the P wave is normally positive in leads I, II, III, and aVF, negative in aVR and flat or negative in aVL. Negative P waves are common in lead III. The Ta wave is usually discordant and appears to terminate or become isoelectric prior to the onset of the QRS complex. The Q of the QRS complex is usually absent in standard bipolar limb leads while R is prominent and S may be either prominent or absent. The ST segment is absent and the S-wave termination in bipolar leads is often difficult to separate from the onset of T. There is often no distinct isoelectric line during the electrocardiographic complex; that is, the points at which P, Ta, QRS, and T waves originate are often on different levels. This is especially true at more rapid heart rates when the P wave originates on the descending limb of the preceding T wave. The T wave is usually positive and concordant with QRS in leads II and III, but may be negative and discordant in lead I. Its ascending limb is typically steeper than its descending limb. The latter usually approaches the isoelectric line gradually, such that its termination is difficult to identify. At more rapid heart rates (450 bpm), the P wave interrupts the descent of the T wave. QT intervals are difficult to measure accurately on account of the gradual descent of T and because the succeeding P wave may interrupt the descent of T before it reaches the baseline. This has led to measuring the interval from the beginning of QRS to the apex of T (the QTa interval) instead of QT. Since the T apex can shift within the QT interval if the shape of T changes, this measurement is not a satisfactory determinant of the true QT interval.

Table 42.7
ECG classification and normal parameters in Carnivora and Proboscidea

Common English name (*species name*)	ECG classification			Heart rate rate (bpm)	Lead II time interval (s)				Lead II amplitude (mV)				
	QRS[a]	QT[b]	T[c]		P	PR	QRS	QT	P	Q	R	S	T
Carnivora Domestic cat (*Felis catus*) [37, 205–212, 232]	A	A	A	190 110–300	0.03 0.02–0.05	0.07 0.05–0.10	0.03 0.015–0.04	0.15 0.1–0.2	0.16 0.00–0.20	0.15 0.00–0.30	0.50 0.00–1.05	0.27 0.00–0.70	0.18 −0.05–0.40
Proboscidea Indian elephant (*Elephas maximus*) [3, 37, 213–216]	B	A	A	38 30–46	0.16 0.12–0.30	0.44 0.36–0.48	0.16 0.12–0.18	0.64 0.60–0.70	0.05 0.03–0.10	0.05 0.03–0.10	0.70 0.60–0.80	0.15 0.10–0.20	0.10 0.05–0.20
African elephant (*Loxodonto Africana*) [37, 216, 217]	B	A	A	46 42–53	0.11 0.10–0.12	0.28 0.20–0.32	0.13 0.12–0.19	0.55 0.48–0.60	0.10	0.05	0.30	0.35	0.20

[a] QRS-vector direction and sense; A, along the long axis of the body, caudally and ventrally; B, from sternum toward spine.
[b] Relative QT-interval, ST-segment and action-potential duration: A, along; B, short.
[c] T-wave lability: A, constant, B, labile.

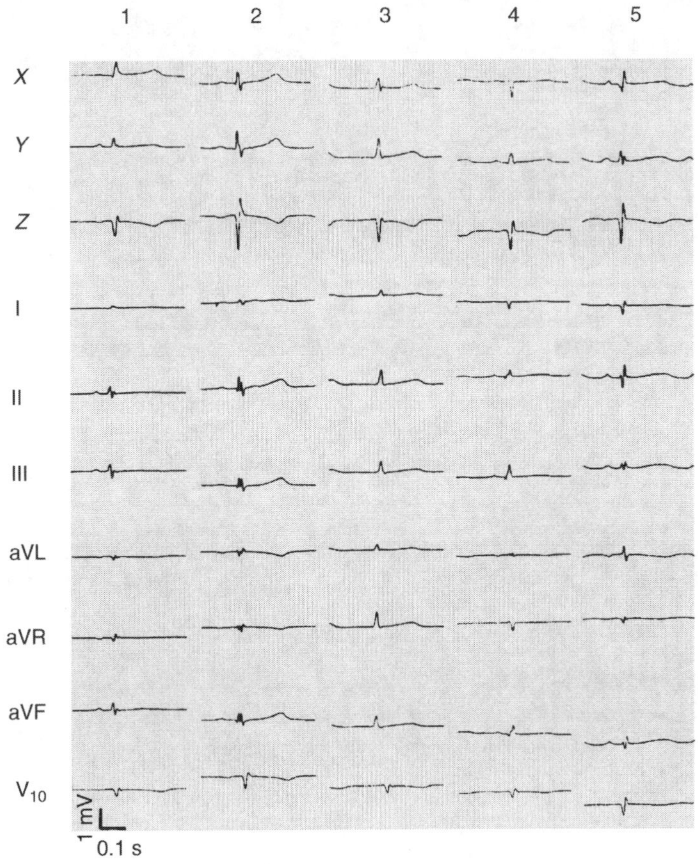

Figure 42.10
Electrocardiogram from five cats (1–5). *X, Y, Z* are the McFee leads described in Chap. 41 (for the dog). Note that the P waves are often small and difficult to identify, for example, as in ECG 3. (After Rogers and Bishop [157]. © Churchill Livingstone, New York. Reproduced with permission.)

As shown in Table 42.2, ventricular extrasystoles (about 7%) and second-degree atrioventricular block (about 8%) are sufficiently frequent in control electrocardiograms from rats to be considered normal variants. This is also true for sinus arrhythmia since in some series this has been present in over half the subjects.

A shift of *A* QRS in the frontal plane has been noted by some observers during the first 4 or 5 months of life. Others have found that the percentage of rats with a leftward QRS axis increased in older and postpartum groups of rats.

The maximum heart rate found in rat electrocardiograms is usually no higher than 600 bpm. In some individuals when rates exceed 600 bpm by a few beats (e.g., 620–635 bpm) an LBBB pattern appears in the electrocardiogram. Also, in a group of 75 rats entered in a drug study, LBBB appeared spontaneously in 32% after 1 year of age (week 65) unrelated to the test agent (that is, the distribution of the bundle branch block was the same in control and treated animals). This has not been seen in other studies and probably represents an inherited characteristic of this rat strain. As mentioned previously, the ST segment is present in newborn rats but disappears during the first three to four weeks of life.

An ECG and phonocardiogram of a white laboratory mouse (*Mus musculus*) is shown in Fig. 42.8. This species of rodent has a mean *A* QRS of 31° (with a range from −10° to 70°). P waves are usually positive in the standard limb leads. QRS complexes are chiefly positive in the standard limb leads with no Q wave and the S wave is absent or, if present, is usually small. The ST segment is absent and the T waves are concordant.

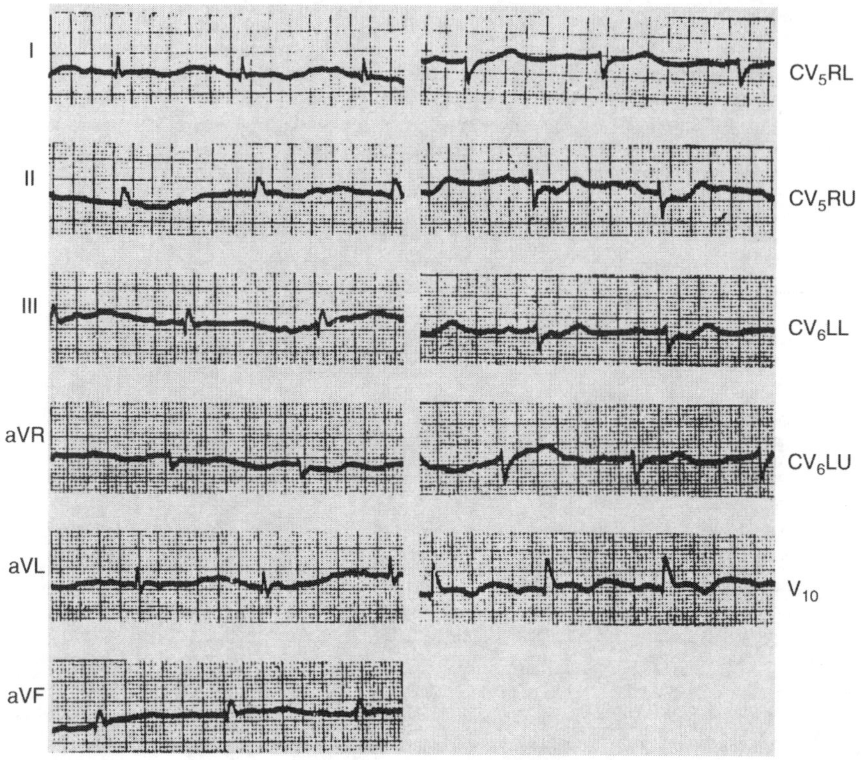

Figure 42.11
Electrocardiogram from the Indian elephant (*Elephas maximus*). The lead system is as described in Sect. 42.4.4.2. (After Jayasinghe et al. [158]. Bailliere Tindall, London. Reproduced with permission.)

In the ECG of the guinea pig (*Cavia porcellus*) (Fig. 42.9), the mean *A* QRS is 6° (range −20° to 60°). The ECG configuration has, in the standard limb leads, P waves which are positive and QRS complexes which generally have small q, large R and small S waves. The ST segment is present and usually isoelectric in limb leads. T waves are chiefly positive and concordant in the standard limb leads.

42.6.8 Carnivora

The domestic cat is the only member of the Carnivora discussed here. Table 42.7 lists the ECG classification lead II time intervals and amplitudes of the cat.

The ECG of the domestic cat (*Felis catus*) (Fig. 42.10) has a mean *A* QRS of 70° (the range is −20°–170°). P waves are usually positive in standard limb leads. When P waves are absent in limb leads, the recording should be inspected closely for signs of AV dissociation when P waves periodically become buried in the QRS. QRS complexes are dominated by R waves in standard limb leads. The ST segment is usually isoelectric and T waves are usually concordant and stable in limb leads.

Cardiomyopathy is a common disease seen in the cat. Such afflicted cats can have a normal ECG or evidence of left anterior fascicular block or left ventricular enlargement. The ECG in the kitten has also been studied from birth to 30 days [153].

42.6.9 Proboscidea

An example of the ECG of the Indian elephant (*Elephas maximus*) is shown in Fig. 42.11. Table 42.7 gives the ECG classification and lead II time intervals and amplitudes for the Indian elephant and the African elephant (*Loxodonto africana*).

Acknowledgements

This work was supported in part by the US National Institutes of Health Grant Number LM 01660 and by Smith, Kline and French Laboratories, Philadelphia, Pennsylvania. Dr. Sydney Moise at Cornell University contributed updates to this chapter.

References

1. Waller, A.D., Introductory address on the electromotive properties of the human heart. *Br. Med. J.*, 1888;2: 751–754.
2. Tawara, S., *Das Reizleitungssystem des Säugetierherzens*. Jena: Fischer, 1906.
3. White, P.D., J.L. Jenks Jr, and F.G. Benedict, The electrocardiogram of the elephant. *Am. Heart J.*, 1938;**16**: 744–750.
4. King, R.L., J.L. Jenks Jr, and P.D. White, The electrocardiogram of a Beluga whale. *Circulation*, 1953;**8**: 387–393.
5. Zuckermann, R., *Grundriss und Atlas der Elektrokardiographie*, 3rd edn. Leipzig: Thieme, 1959.
6. Kisch, B., F.M. Groedel, and P.R. Borchardt, *Comparative Direct Electrography of the Heart of Vertebrates*. New York: Fordham University Press, 1952.
7. Lepeschkin, E., *Das Elektrokardiogramm. Ein Handbuchfiir Theorie und Praxis*, 2nd edn. Dresden: Steinkopff, 1947.
8. Lepeschkin, E., *Modern Electrocardiography*. Baltimore, MD: Williams and Wilkins, 1951.
9. Rijlant, P.B., Origin of the heart beat in the mammalian heart (in Russian), in *Comparative Electrocardiology International Symposium* (in Russian), M.P. Roshchevsky, Editor. Leningrad: Akademiia Nauk SSSR, 1981, pp. 20–23.
10. Lewis, T., Irregularity of the heart's action in horses and its relationship to fibrillation of the auricles in experiment and to complete irregularity of the human heart. *Heart*, 1912;3: 161–171.
11. Lewis, T., *The Mechanism and Graphic Registration of the Heart Beat*, 3rd edn. London: Shaw, 1925.
12. Luisada, A. L. Weisz, and H.W. Hartman, A comparative study of electrocardiogram and heart sounds in common and domestic animals. *Cardiologia*, 1944;**8**: 63–84.
13. Robb, J.S., *Comparative Basic Cardiology*. New York: Grune and Stratton, 1965.
14. von Tschermak, A., Lehre von den bioelektrischen Strömen (Elektrophysiologie), in *Lehrbuch der Vergleichenden Physiologie der Haussäugetiere*, W. Ellenberger and A. Scheunert, Editors. Berlin: Parey, 1910, p. 520.
15. Kahn, R.H., Das Pferde-Ekg. *Pfluegers Arch. Gesamle Physiol. Menschen Tiere*, 1913;**154**: 1–15.
16. Waller, A.D., Electrocardiogram of horse. *J. Physiol.*, 1913;**47**(Proc. Physiol. Soc. London): xxxii–xxxiv.
17. Waller, A.D., A short account of the origin and scope of electrocardiography. *Harvey Lect.*, 1913–1914;**9**: 17–33. (*N. Y. Med. J.* 1914;97: 719).
18. Nörr, J., Das Elektrocardiogramm des Pferdes; seine Aufnahme und Form. *Z. Biol.*, 1913;**61**: 197–229.
19. Nörr, J., Elektrokardiogrammstudien am Rind. *Z. Biol.*, 1921;**73**: 129–140.
20. Nörr, J., Über Herzstromkurvenaufnahmen an Haustieren. Zur Einfiihrung der Elektrokardiographie in die Veteriniirmedizin. *Arch. Wiss. Prakl. Tierheilkd.*, 1922;**48**: 85–111.
21. Gylstorffl. Prof. Dr. Nörr 85 Jahre. Berlin u. München. *Tieraerztl. Wochenschr.*, 1971;**4**: 240.
22. Lannek, N. and L. Rutqvist, Electrocardiography in horses. A historical review. *Nord. Veterinaermed*, 1951;**3**: 435–447.
23. Lannek, N., *A Clinical and Experimental Study on the Electrocardiogram in dogs*, Thesis. Stockholm, Sweden: Royal Veterinary College, 1949.
24. Roshchevsky, M.P., *Electrocardiology of Hoofed Animals* (in Russian). Leningrad: Nauka, 1978.
25. Spörri, H., Veränderungen der Systolendauer im Elektrokardiogramm von Rind und Meerschweinchen. *Arch. Wiss. Prakl. Tierheilkd.*, 1941;**76**: 236–247.
26. Spörri, H., Der Einfluss der Tuberkulose auf das Elektrokardiogramm. (Untersuchungen an Meerschweinchen und Rindern.) *Arch. Wiss. Prakl. Tierheilkd.*, 1944;**79**: 1–57.
27. Alfredson, B.V. and J.F. Sykes, Electrocardiographic studies in normal dairy cattle. *J. Agric.Res. (Washington D.C.)*, 1942;**65**: 61–87.
28. Charton, A. and G. Minot, Electrocardiogramme normal du cheval. *C. R. Seances Soc. Biol. Ses Fil.*, 1943;**137**: 150–152.
29. Charton, A., G. Minot, and M. Bressou, Electrocardiogramme normal du cheval. *Bull. Acad Vet. Fr.*, 1943;**16**: 141–148.
30. Bressou, M., *L 'Electrocardiogramme et le Phonocardiogramme du Cheval Normal*. Paris: Foulon, 1944.
31. Kelso, W.T., Electrocardiograms of young lambs. *Queensl. J. Agric. Sci.*, 1947;**4**: 60–71.
32. Krzywanek, F.W. and G. Ruud, Die Veränderungen des Elektrokardiogramms beim sogenannten Herztod des Schweines. *Tieraerztl. Umsch.*, 1944;**2**: 23–25.

33. Ruud, G., *Das Elektrokardiogramm des Schweines und die Veränderungen desselben beim sogenannten Herztod des Schweines*, Veterinary dissertation. Berlin, 1945.
34. Voskanyan, P.M., Electrocardiography of horses (in Russian). *Trudi XV Plenum Vet. Sektsii Vses. Akad. Moskva, Nauka* 1941; 250–256.
35. Filatov, P.B., Electrocardiography of cattle, in *Non-infectious Internal Diseases of Animals* (in Russian). *Sb. Rab. Voenno-Vel. Fakulteta Mosk. Vet. Akad. Moskva. Nauka* 1949;**6**: 207–222.
36. Roshchevsky, M.P., *Electrical Activity of the Heart and Methods of Recording Electrocardiograms from Large Livestock(Cattle)* (in Russian). Sverdlovsk: Uralskii Gosudarstvennyi Universitet, 1958.
37. Grauwiler, J., *Herz und Kreislauf der Säugetiere; Vergleichend-Funktionelle Daten*. Basel: Birkhäuser, 1965.
38. Roshchevsky, M.P., *Evolutional Electrocardiology* (in Russian). Leningrad: Nauka, 1972.
39. Roshchevsky, M.P., Editor. *Physiological Basis of Animal Electrocardiography*. Moscow: Nauka, 1965.
40. Roshchevsky, M.P., Editor. *Comparative Electrocardiology. Proceedings of the 1st International Symposium* (in Russian). Leningrad: Nauka, 1981.
41. Roshchevsky, M.P., Editor. *Comparative Electrocardiology. Proceedings of the 2nd International Symposium* (in Russian).
42. G.F. Fregin, The cardiovascular system, in *Equine Medicine and Surgery*, vol. 1, 3rd edn, RA Mansmann, ES McAllister, and PW Pratt, Editors. Santa Barbara: American Veterinary Publishing, 1982, pp. 645–704.
43. Edwards, N.J., *Bolton's Handbook of Canine and Feline Electrocardiography*, 2nd edn. Philadelphia, PA: Saunders, 1987.
44. Tilley, L.P., *Essentials of Canine and Feline Electrocardiography*, 2nd edn. Philadelphia, PA: Lea and Febiger, 1985.
45. Detweiler, D.K., The use of electrocardiography in toxicological studies with Beagle dogs, in *Cardiac Toxicology*, vol. 3, T. Balazs, Editor. Boca Raton, FL: CRC Press, 1981, pp. 33–82.
46. Malinow, M.R., The electrocardiogram (ECG) and vectorcardiogram (VCG) of the rhesus monkey, in *Anatomy and Physiology*, G.H. Bourne, Editor. New York: Academic Press, 1975, pp. 77–105. (*The Rhesus Monkey*; vol. 1.)
47. Budden, R., D.K. Detweiler, and G. Zbinden, Editors. *The Rat Electrocardiogram in Pharmacology and Toxicology*. Oxford: Pergamon, 1981.
48. Kachiku No Shindenzu *(Advances in Animal Electrocardiography)*. Tokyo: Japanese Association of Veterinary Cardiology. (Japanese language journals in existence since 1968. English translations of titles, legends, and summaries often included.)
49. Detweiler, D.K., Electrocardiographic monitoring in toxicological studies: Principles and interpretations, in *Myocardial Injury, Advances in Experimental Medicine and Biology*, vol. 161, J.J Spitzer, Editor. New York: Plenum, 1983, pp. 579–607.
50. Detweiler, D.K., The use of electrocardiography in toxicological studies with rats, in *The Rat Electrocardiogram in Pharmacology and Toxicology*, R. Budden, D.K. Detweiler, and G. Zbinden, Editors. Oxford: Pergamon, 1981, pp. 83–115.
51. Grauwiler, J. and H. Spörri, Fehlen der ST-Strecke im Elektrokardiogramm von verschiedenen Saugetierarten. *Helv. Physiol. Pharmacol. Acta.*, 1960;**C18**: 77–78.
52. Nerbonne, J.M., Studying cardiac arrhythmias in the mouse-a reasonable model for probing mechanisms? *Trends Cardiovasc. Med.*, 2004;**14**: 83–93.
53. Hamlin, R.L. and C.R. Smith, Categorization of common domestic mammals based upon their ventricular activation process. *Ann. N. Y. Acad. Sci.*, 1965;**127**: 195–203. (See also Hamlin, R.L., D.L. Smetzer, and C.R. Smith, Analysis of QRS complex recorded through a semiorthogonal lead system in the horse. *Am. J. Physiol.*, 1964;**207**: 325–333.)
54. Hill, J.D., The electrocardiogram in dogs with standardized body and limb positions. *J. Electrocardiol.*, 1968;**1**: 175–182.
55. Hill, J.D., The significance of foreleg positions in the interpretation of electrocardiograms and vectorcardiograms from research animals. *Am. Heart J.*, 1968;**75**: 518–527.
56. Detweiler, D.K., Electrophysiology of the heart, in *Dukes' Physiology of Domestic Animals*, 10th edn, M.J. Swenson, Editor. Ithaca, NY: Cornell University Press, 1984, pp. 103–130.
57. Rappaport, M.B. and I. Rappaport, Electrocardiographic considerations in small animal investigations. *Am. Heart J.*, 1943;**26**: 662–680.
58. Werth, G. and S. Wink, Das Elektrokardiogramm der normalen Ratte. *Arch. Kreislaufforsch.*, 1967;**54**: 272–308.
59. Angelakos, E.T. and P. Bernardini, Frequency components and positional changes in electrocardiogram of the adult rat. *J. Appl. Physiol.*, 1963;**18**: 261–263.
60. Godwin, K.O. and F.J. Fraser, Simultaneous recording of ECGs from disease-free rats, using a cathode ray oscilloscope and a direct writing instrument. *Q. J. Exp. Physiol.*, 1965;**50**: 277–281.
61. Almasi, J.J., O.H. Schmitt, and E.F. Jankus, Electrical characteristics of commonly used canine ECG electrodes. *Proc. Annu. Conf. Eng. Med. Biol.*, 1970;**12**: 190.
62. Osborne, B.E., A restraining device facilitating electrocardiogram recording in dogs. *Lab. Anim. Care*, 1970;**20**: 1142–1143.
63. Beinfield, W.H. and D. Lehr, Advantages of ventral position in recording electrocardiogram of rat. *J. Appl. Physiol.*, 1956;**9**: 153–156.
64. Richtarik, A., T.A. Woolsey, and E. Valdivia, Method for recording ECG's in unanesthetized guinea pigs. *J. Appl. Physiol.*, 1965;**20**: 1091–1093.
65. Detweiler, D.K. and H. Spörri, A note on the absence of auricular fibrillation in the European mole (Talpa europaea). *Cardiologia*, 1957;**30**: 372–375.
66. Waller, A.D., Ueber die den Pulsbegleitende elektrische Schwankung des Herzens (Ausserordentliche) Sitzung am 27. Dec. 1889. *Arch. Physiol. Physiologische Abtheilung des Arch. Anat. Physiol.*, 1890; 186–190.
67. Brooijmans, A.W.M., *Electrocardiography in Horses and Cattle: Theoretical and Clinical Aspects, Vet. Proefschrift Rijksuniversiteit*. Utrecht: Cantecleer, 1957.
68. Balbo, T. and U. Dotta, Gli effetti dell'eta sull'elettrocardiogramma del bovino. 1 and 2. *La Nuova Vet.*, 1966;**62**: 307–355 and Suppl. 6.
69. Lautenschlager, O., *Grundlagen der Aufnahmetechnik des Elektrokardiogrammes von Pferd und Rind und ihre Ergebnisse*, Veterinary dissertation. Giessen, 1928.
70. Morita, H., Electrocardiograms of conscious Beagle dogs by apex-base bipolar lead. *Adv. Anim. Cardiol.*, 1984;**17**: 19–23.
71. Sander, W., Das Elektrokardiogramm des Rindes. *Zentralbl. Veterinaermed.*, 1968;**15**: 587–634.
72. Brooijmans, A.W.M., Standardization of leads in veterinary clinical electrocardiography. *Tijdschr. Diergeneeskd.*, 1954;**79**: 801–811.

73. Sellers, A.F., A. Hemingway, E. Simonson, and W.E. Petersen, Unipolar and bipolar electrocardiographic studies in dairy cattle. *Am. J. Vet. Res.*, 1958;**19**: 620–624.
74. Junge, G., Über die Elektrokardiographie in der Veterinärmedizin unter besonderer Berücksichtigung der allgemeinen Elektrophysiologie und der Ableitung des Rinderelektrokardiogramms. *Arch. Exper. Veterinaermedizin*, 1967;**21**: 835–866.
75. Sugeno, H., Y. Yasuda, H. Nishikawa, and T. Takeya, Studies of the electrocardiogram of normal healthy cows. *J. Fac. Agr. Iwate Univ.*, 1956;**3**: 114–125.
76. Too, K., R. Nakamura, and K. Hirao, Studies on the applications of electrocardiogram in cattle. *Jpn. J. Vet. Res.*, 1958;**6**: 230–244.
77. Kusachi, R. and H. Sato, Fundamental studies on electrocardiograms of the horse. II. Bipolar lead. *Jpn. J. Vet. Res.*, 1955;**3**: 195–208.
78. Lannek, N. and L. Rutqvist, Normal area variation for the electrocardiogram of horses. *Nord. Veterinaermed.*, 1951;**3**: 1094–1117.
79. Detweiler, D.K. and D.F. Patterson, The cardiovascular system, in *Equine Medicine and Surgery*, 2nd edn, E.J. Catcott and J.F. Smithcors, Editors. Wheaton, IL: American Veterinary Publishing, 1972, pp. 277–347.
80. Malinow, M.R., An electrocardiographic study of *Macaca mulatta*. *Folia Primarol*, 1966;**4**: 51–65.
81. Herrmann, G.R. and A.H.W. Herrmann, The electrocardiographic patterns in 170 baboons in the domestic and African colonies at the primate center of the Southwest Foundation for Research and Education, in *The Baboon in Medical Research*, vol. I, H. Vagtborg, Editor. Austin, TX: University of Texas Press, 1965, pp. 251–264.
82. Atta, A.G. and P.W. Vanace, Electrocardiographic studies in the *Macaca mulatta* monkey. *Ann. N.Y. Acad. Sci.*, 1960;**85**: 811–818.
83. Wolf, R.H., N.D.M. Lehner, E.C. Miller, and T.B. Clarkson, Electrocardiogram of the squirrel monkey *(Saimiri sciureus)*. *J. Appl. Physiol.*, 1969;**26**: 346–351.
84. Hellerstein, H.K. and R. Hamlin, QRS component of the spatial vectorcardiogram and of the spatial magnitude and velocity electrocardiograms of the normal dog. *Am. J. Cardiol.*, 1960;**6**: 1049–1061.
85. Hamlin, R.L.,J.A. Himes, H. Guttridge, and W. Kirkham, P wave in the electrocardiogram of the horse. *Am. J. Vet. Res.*, 1970;**31**: 1027–1031.
86. Harolin, R.L., R.R. Burton, S.D. Leverett, andJ.W. Burns, The electrocardiogram from miniature swine recorded with the McFee-axial reference program. *J. Electrocardiol.*, 1974;**7**: 155–162.
87. Holmes, J.R. and P.G.G. Darke, Studies on the development of a new lead system for equine electrocardiography. *Equine Vel. I.*, 1970;**2**: 12–21.
88. Holmes, J.R. and R.W. Else, Further studies on a new lead system for equine electrocardiography. *Equine Vel J*, 1972;**4**: 81–87.
89. Grauerholz, G., Eine Methode zur vektoriellen Auswertung des Elektrokardiogramms beim Pferd. *Zentralbl. Velerinaermed. A*, 1974;**21**: 188–197. (This article cites: Baron M. Contribution a l'etude du vectorcardiogramme du cheval de sport. Applications dans l'examen preoperative, veterinary dissertation. Paris, 1970.)
90. Holmes, J.R. and B.J. Alps, Studies into equine electrocardiography and vectorcardiography. I. Cardiac electric forces and the dipole vector theory: II. Cardiac vector distribution in apparently healthy horses. III. Vector distribution in some cardiovascular disorders. IV. Vector distributions in some arrhythmias. *Can. I. Comp. Med. Vel. Sci.*, 1967;**31**: 92–102; 150–155; 207–212; 219–225.
91. Darke, P.G.G. and J.R. Holmes, Studies on the equine cardiac electric field. I. Body surface potentials. II. The integration of body surface potentials to derive resultant cardiac dipole moments. *I. Electrocardiol*, 1969;**2**: 229–234; 235–244.
92. Van Arsdel, W.C., III, *Lead Selection, Cardiac Axes, and the Interpretation of Electrocardiograms in Beef Cattle.* Corvallis, OR: Oregon State College, 1959.
93. Sugeno, H., Analytical study on bovine electrocardiogram. *Ipn. Circ. I.*, 1959;**23**: 1193–1203.
94. Holmes, J.R.A., Method of vectorcardiogram: loop portrayal. *Equine Vel. J.*, 1970;**2**: 27–34.
95. Nörr, J., Fötale Elektrokardiogramme von Pferd. *Z. Biol.*, 1921;**73**: 123–128.
96. Holmes, J.R. and P.G.G. Darke, Fetal electrocardiography in the mare. *Vel. Rec.*, 1968;**82**: 651–655.
97. Too, K., H. Kanagawa, and K. Kawata, Fetal electrocardiogram in dairy cattle. I. Fundamental Studies. III. Variations in fetal QRS pattern. *Jpn. J. Vel. Res.*, 1965;**13**: 71–83; 1966;**14**: 103–113.
98. Too, K., H. Kanagawa, K. Kawata, and H. Ono, Fetal electrocardiogram in dairy cattle. II. Diagnosis for twin pregnancies. *Jpn. J. Vet. Res.*, 1965;**13**: 111–119. Too, K., H. Kanagawa, K. Kawata, T. Inoue, T.F. Odajima, et al., Electrocardiogram in cattle. V. Findings at parturition. *Jpn. J. Vel. Res.* 1967;**15**: 21–30.
99. Golikov, A.N. and R.S. Vershinina, Electrocardiographic control of the stage of pregnancy in the cow (in Russian). *Velerinariya*, 1973;**2**: 87–88.
100. Larks, S.D., L.W. Holm, and H.R. Parker, A new technic for the demonstration of the fetal electrocardiogram in the large domestic animal (cattle, sheep, horse). *Cornell Vet*, 1960;**50**: 459–468.
101. Grabowski, C.T. and D.B. Payne, An electrocardiographic study of cardiovascular problems in Mirex-fed rat fetuses. *Teratology*, 1980;**22**: 167–177.
102. Detweiler, D.K. and D.F. Patterson, The prevalence and types of cardiovascular disease in dogs. *Ann. N.Y. Acad. Sci.*, 1965;**127**: 481–516.
103. Patterson, D.F., D.K. Detweiler, K. Hubben, and R.P. Botts, Spontaneous abnormal cardiac arrhythmias and conduction disturbances in the dog. A clinical and pathologic study of 3,000 dogs. *Am. I. Vet. Res.*, 1961;**22**: 355–369.
104. Einthoven, W., Die galvanometrische Registrierung des menschlichen Elektrokardiogramms. *Pfluegers Arch. Gesamte Physiol. Menschen Tiere.*, 1903;**99**: 472–480.
105. Einthoven, W., Le Télécardiogramme. *Arch. Int. Physiol.*, 1906;**4**: 132–164.
106. Caceres, C.A., Editor. *Biomedical Telemetry.* New York: Academic Press, 1965.
107. Holter, N.J., New method for heart studies. *Science*, 1961;**134**: 1214–1220.
108. Sandler, H., H.L. Stone, T.B. Fryer, and R.M. Westbrook, Use of implantable telemetry systems for study of cardiovascular phenomena. *Circ. Res.*, 1972;**31**(Suppl. II): 85–100.
109. Denton, D., Recording of physiological functions of a suitable experimental animal during protracted space flight. *Ausl. J. Sci.*, 1958;**20**: 202–207.
110. Benazet, P., R. Bordet, A. Brion, M. Fontaine, and J. Sevestre, Étude télémétrique de l'éléctrocardiogramme du cheval de sport. *Rect. Med. Vel.*, 1964;**140**: 449–459.

111. Nomura, S., Adaptation of radiotelemetry to equestrian games and horse racing. *Jpn. J. Vel. Sci.*, 1966;**28**: 191–203.
112. Fregin, G.F., Radioelectrocardiography in horses. *Pennsylvania Vel*, 1967;**9**: 6–10.
113. Banister, E.W. and A.D. Purvis, Exercise electrocardiography in the horse by radiotelemetry. *J. Am. Vet. Med. Assoc.*, 1968;**152**: 1004–1008.
114. Senta, T., D.L. Smetzer, and C.R. Smith, Effects of exercise on certain electrocardiographic parameters and cardiac arrhythmias in the horse. A radiotelemetric study. *Cornell Vet*, 1970;**60**: 552–569.
115. Börnert, D., H. Seidel, R. Maiwald, and G. Börnert, Drahtlos übertragene EKG-Ableitungen vom freibeweglichen Rind. *Arch. Exp. Velerinaermed.*, 1964;**18**: 701–712.
116. Dracy, A.E. and J.R. Jahn, Use of electrocardiographic radiotelemetry to determine heart rate in ruminants. *J. Dairy Sci.*, 1964;**47**: 561–563.
117. Fregin, G.F. and D.P. Thomas, Cardiovascular response to exercise in the horse: a review, in *Equine Exercise Physiology*, D.H. Snow, S.G.B. Persson, and R.J. Rose, Editors. Cambridge: Granta, 1983, pp. 76–90.
118. Adams, L., R.E. Wetmore, R.L. Limes, and H.J. Hauer, Automated analysis of heart rate in ground squirrels using radiotelemetry and computers. *BioScience*, 1971;**21**: 1040–1042.
119. Essler, W.O. and G.E. Folk Jr, Determination of physiological rhythms of unrestrained animals by radiotelemetry. *Nature*, 1961;**190**: 90–91.
120. Essler, W.O., *Radiolelemetry of Electrocardiograms and Body Temperatures from Surgically Implanled Transmitters*, State University of Iowa Studies in Natural History; vol. 20, no. 4. Iowa City, IA: State University of Iowa, 1961.
121. Branch, C.E., S.D. Beckett, and B.T. Robertson, Spontaneous syncopal attacks in dogs. A method of documentation. *I. Am. Anim. Hosp. Assoc.*, 1972;**13**: 673–679.
122. Reite, M., Implantable biotelemetry and social separation in monkeys, in *Animal Stress*, G.P. Moberg, Editor. Bethesda, MD: American Physiological Society, 1985, pp. 141–160.
123. Patiley, J.D. and M. Reite, A microminiature hybrid multichannel implantable biotelemetry system. *Biolelem. Patient Monit.*, 1981;**8**: 163–172.
124. Kimmich, H.P. and J.W. Knutti, Editors. Implantable telemetry systems based on integrated circuits. *Bioelem. Patient Monit.*, 1979;**6**: 91–170.
125. Mohler, P.J., I. Splawski, C. Napolitano, G. Bottelli, L. Sharpe, K. Timothy, S.G. Priori, M.T. Keating, and V. Bennett, A cardiac arrhythmia syndrome caused by loss of ankyrin-B function. *Proc. Natl. Acad. Sci. USA*, Jun 15; 2004;**101**(24): 9137–9142.
126. Farkas, A., A.J. Batey, and S.J. Coker, How to measure electrocardiographic QT interval in the anaesthetized rabbit. *J. Pharmacol. Toxicol. Methods*, 2004 Nov–Dec;**50**(3): 175–185.
127. Ryu, K., S.C. Shroff, J. Sahadevan, N.L. Martovitz, C.M. Khrestian, and B.S. Stambler, Mapping of atrial activation during sustained atrial fibrillation in dogs with rapid ventricular pacing induced heart failure: evidence for a role of driver regions. *J. Cardiovasc. Electrophysiol.*, Dec 2005;**16**(12): 1348–1361.
128. Kano, M., T. Toyoshi, S. Iwasaki, M. Kato, M. Shimizu, and T. Ota, QT PRODACT: usability of miniature pigs in safety pharmacology studies: assessment for drug-induced QT interval prolongation. *J. Pharmacol. Sci.*, 2005;**99**(5): 501–511.
129. Spach, M.S., R.C. Barr, C.F. Lanning, and P.C. Tucek, Origin of body surface QRS and T wave potentials form epicardial potential distributions in the intact chimpanzee. *Circulation*, Feb 1977;**55**(2): 268–278.
130. Gauvin, D.V., L.P. Tilley, F.W. Smith Jr, and T.J. Baird, Electrocardiogram, hemodynamics, and core body temperatures of the normal freely moving cynomolgus monkey by remote radiotelemetry. *J. Pharmacol. Toxicol. Methods*, Mar–Apr 2006;**53**(2): 140–151.
131. Sawazaki, H. and H. Hirose, Comparative electrocardiographical studies on the conduction time of heart in vertebrates. *Jpn. J. Vel. Sci.*, 1974;**36**: 421–426.
132. Detweiler, D.K., Regulation of the heart, in *Dukes' Physiology of Domestic Animals*, 10th edn, M.J. Swenson, Editor. Ithaca, NY: Cornell University Press, 1984, pp. 150–162.
133. Schmidt-Nielsen, K., *Animal Physiology: Adaptation and Environment*. London: Cambridge University Press, 1975.
134. Stahl, W.R., Scaling of respiratory variables in mammals. *J. Appl. Physiol.*, 1967;**22**: 453–460.
135. Schaible, T.F. and J. Scheuer, Response of the heart to exercise training, in *Growth of the Heart in Health and Disease*, R. Zak, Editor. New York: Raven, 1984, pp. 381–419.
136. Hermann, G.R., T.E. Vice, A.R. Rodriguez, and A.W. Herrmann, Heart weight to body weight, left to right ventricular weights, and ratios of baboon hearts, in *The Baboon in Medical Research*, vol. I, H. Vagtborg, Editor. Austin, TX: University of Texas Press, 1965, pp. 269–274.
137. Detweiler, D.K., Normal and pathological circulatory stresses, in *Dukes' Physiology of Domestic Animals*, 10th edn, M.J. Swenson, Editor. Ithaca, NY: Cornell University Press, 1984, pp. 207–225.
138. Porges, S.W., P.M. McCabe, and B.G. Yongue, Respiratory-heart-rate interactions: psychophysiological implications for pathophysiology and behavior, in *Perspectives in Cardiovascular Psychophysiology*, J. Cacioppo and R. Petty, Editors. New York: Guilford, 1982, pp. 223–264.
139. Porges, S.W., Spontaneous oscillations in heart rate; potential index of stress, in *Animal Stress*, G.P. Moberg, Editor. Bethesda, MD: American Physiological Society, 1985, pp. 97–111.
140. Kitney, R.I. and E. Rompelman, Editors. *The Study of Heart-Rate Variability*. Oxford: Clarendon Press, 1980.
141. Kwatny, E., D. Peltzman, and D.K. Detweiler, Automated analysis of Beagle electrocardiograms, in *Proceedingsof the 13th North England Bioengineering Conference*, K.R. Foster, Editor. New York: IEEE, 1987.
142. Melbin, J., D.K. Detweiler, R.A. Riffle, and A. Noordergraaf, Coherence of cardiac output with rate changes. *Am. J. Physiol.*, 1982;**243**: H499–H504.
143. Detweiler, D.K., Electrocardiographic detection of cardiotoxicity in preclinical studies, in *Safely Evaluation and Regulation of Chemicals*, vol. 3, F. Homburger, Editor. Basel: Karger, 1986, pp. 76–86.
144. Bauer, K. and O. Nehring, Betrachtungen zu den mathematischen Formuliertingen der Abhängigkeit der QT-Zeit von der Frequenz und die Deutung derselben. *Z. Kreislaufforsch*, 1968;**57**: 430–436.
145. Schoenwald, R.D. and V.E.Q.T. Isaacs, Corrected for heart rate: a new approach and its application. *Arch. Int. Pharmacodyn. Ther.*, 1974;**211**: 34–48.

146. Grauwiler, J., Das normale Elektrokardiogramm des Beagle-Hundes. *Naunyn-Schmiedebergs Arch. Pharmakol.*, 1970;**266**: 337.
147. Osborne, B.E. and G.D.H. Leach, The Beagle electrocardiogram. *Food Cosmet. Toxicol.*, 1971;**9**: 857–864.
148. Patterson, D.F., D.K. Detweiler, and S.A. Glendenning, Heart sounds and murmurs of the normal horse. *Ann. N.Y. Acad. Sci.*, 1965;**127**: 242–305.
149. Gonder, J.C., E.A. Gard, and N.E.I.I.I. Lott, Electrocardiograms of nine species of nonhuman primates. *Am. J. Vel. Res.*, 1980;**41**: 972–975.
150. von Mickwitz, G., *Herz- und Kreislaufuntersuchungen beim Schwein mit Berücksichtigung des Elektrokardiogramms und des Phonokardiogramms.* Hannover: Habilitationsschrift, 1967, pp. 1–246.
151. Richards, A.G., E. Simonson, and M.B. Visscher, Electrocardiogram and phonogram of adult and newborn mice in normal conditions and under the effect of cooling, hypoxia and potassium. *Am. J. Physiol.*, 1953;**174**: 293–298.
152. O'Rourke, M.F., A.P. Avolio, and W.W. Nichols, The kangaroo as a model for the study of hypertrophic cardiomyopathy in man. *Cardiovasc. Res.*, 1986;**20**: 398–402.
153. Lourenco, M.L. and H. Ferreira, Electrocardiographic evolution in cats from birth to 30 days of age. *Can. Vet. J.*, 2003 Nov;**44**(11): 914–917.
154. Jefferson, J.W., Electrocardiographic and phonocardiographic findings in a reticulated giraffe. *J. Am. Vet. Med. Assoc.*, 1971;**159**: 602–604.
155. Rossof, A.H., An electrocardiographic study of the giraffe. *Am. Heart J.*, 1972;**83**: 142–143.
156. Tokita, K., Electrocardiographical studies on bottle-nosed dolphin (Tursiops truncatus). *Sci. Rep. Whales Res. Inst.*, 1960;**15**: 159–165.
157. Rogers, W.A. and S.P. Bishop, Electrocardiographic parameters of the normal domestic cat: A comparison of standard limb leads and an orthogonal system. *I. Electrocardiol.*, 1971;**4**: 315–321.
158. Jayasinghe, J.B., S.D.A. Fernando, and L.A.P. Brito-Babapulle, The electrocardiographic patterns of *Elephas maxim us* – the elephant of Ceylon. *Br. Vel. I.*, 1963;**119**: 559–564.
159. Stewart, J.H., R.J. Rose, P.E. Davis, and K. Hoffman, A comparison of electrocardiographic findings in racehorses presented either for routine examination or poor racing performance, in *Equine Exercise Physiology* D.H. Snow, S.G.B. Persson, and R.J. Rose, Editors. Cambridge: Granta, 1983, pp. 135–143.
160. Moore, E.N., G. Fisher, D.K. Detweiler, and G.K. Moe, The importance of atrial mass in the maintenance of atrial fibrillation, in *International Symposium on Comparative Medicine*, R.J. Tashjian, Editor. Norwich, CT: Eaton Laboratory, 1965, pp. 229–238.
161. Spörri, H., Die ersten Fälle von sog. Wolff-Parkinson-White Syndrom, einer eigenartigen Herzanomalie, bei Tieren. *Schweiz Arch. Tierheilkd.*, 1953;**95**: 13–22.
162. Giuliano, G., G. Angrisani, P.P. Campa, M. Condorelli, and V. Pennetti, L'elettrocardiogram: na nella pecora. *Boll. Soc. Ital. Biol. Sper.*, 1958;**34**: 1785–1787.
163. Mullick, D.N., B.V. Alfredson, and E.P. Reineke, Influence of thyroid status on the electrocardiogram and certain blood constituents of the sheep. *Am. J. Physiol.*, 1948;**152**: 100–105.
164. Walper, F., *Elektrokardiographische und andere kardiographische Studien am Schaf*, Dissertation. Munich, 1932.
165. Unshelm, J., H.H. Thielscher, F. Haring, H. Hohns, U.E. Pfleiderer, and W. von Schutzbar, Elektrokardiographische Untersuchungen bei Schafen unter Berücksichtigung der Rasse, des Lebensalters und anderer Einflussfaktoren. *Zentralbl. Veterinaermed. A.*, 1974;**21**: 479–491.
166. Schultz, R.A., P.J. Pretorius, and M. Terblanche, An electrocardiographic study of normal sheep using a modified technique. *Onderstepoort J. Vet. Res.*, 1972;**39**: 97–106.
167. Rozanova, T.V., Electrocardiographic indicators of sheep of various constitutional types (in Russian). *Dokl. Skh. Akad. Sofia*, 1958;**32**: 369–377.
168. Roshchevsky, M.P., Vector- i elektrokardiograficheskie iccledovaniya cerdechnoi deyatel'nosti ovetz. *Skh. Biol.*, 1969;**4**: 594–600.
169. Szabuniewicz, M. and D.R. Clark, Analysis of the electrocardiograms of 100 normal goats. *Am. J. Vet. Res.*, 1967;**28**: 511–516.
170. Upadhyay, R.C. and S.C. Sud, Electrocardiogram of the goat. *Indian J. Exp. Biol.*, 1977;**15**: 359–362.
171. Senta, T., Experimental investigation of electrocardiogram in the goat. *Exp. Rep. Equme Health Lab.*, 1967;**4**: 37–72.
172. Dukes, T.W. and M. Szabuniewicz, The electrocardiogram of conventional and miniature swine (Sus scrofa). *Can. J. Comp. Med.*, 1969;**33**: 118–127.
173. Thielscher, H.H., Elektrokardiographische Untersuchungen an Deutschen veredelten Landschweinen der Landeszucht und der Herdbuchzucht. *Zentralbl. Veterinaermed. A.*, 1969;**16**: 370–383.
174. von Mickwitz, G., *Herz- und Kreislaufuntersuchungen beim Schwein mit Berücksichtigung des Elektrokardiogramms und des Phonokardiogramms.* Hannover: Habilitationsschrift, 1967, pp. 1–246.
175. Hausmann, W.O., *Das Elektrokardiogramm des Hausschweines*, Dissertation. München, 1934, pp. 1–28.
176. Cox, J.L., D.E. Becker, and A.H. Jensen, Electrocardiographic evaluation of potassium deficiency in young swine. *J. Anim. Sci.*, 1966;**25**: 203–206.
177. Jefferson, J.W., Electrocardiographic and phonocardiographic findings in a reticulated giraffe. *J. Am. Vet. Med. Assoc.*, 1971;**159**: 602–604.
178. Rossof, A.H., An electrocardiographic study of the giraffe. *Am. Heart J.*, 1972;**83**: 142–143.
179. Jayasinghe, J.B., D.A. Fernando, and L.A.P. Brito-Babapulle, The electrocardiogram of the camel. *Am. J. Vet. Res.*, 1963;**24**: 883–885.
180. Geddes, L.A., W.A. Tacker, J. Rosborough, A.G. Moore, and P. Cabler, The electrocardiogram of a dromedary camel. *J. Electrocardiol.*, 1973;**6**: 211–214.
181. Senft, A.W. and J.K. Kanwisher, Cardiographic observations on a fin-back whale. *Circ. Res.*, 1960;**8**: 961–964.
182. Spencer, M.P., T.A. Gornall III, and T.C. Poulter, Respiratory and cardiac activity of killer whales. *J. Appl. Physiol.*, 1967;**22**: 974–981.
183. Tokita, K., Electrocardiographical studies on bottle-nosed dolphin (Tursiops truncatus). *Sci. Rep. Whales Res. Inst.*, 1960;**15**: 159–165.
184. Hamlin, R.L., R.F. Jackson, J.A. Himes, F.S. Pipers, and A.C. Townsend, Electrocardiogram of bottle-nosed dolphin *(Tursiops truncatus)*. *Am. J. Vet. Res.*, 1970;**31**: 501–505.

185. Richards, A.G., E. Simonson, and M.B. Visscher, Electrocardiogram and phonogram of adult and newborn mice in normal conditions and under the effect of cooling, hypoxia and potassium. *Am. J. Physiol.*, 1953;**174**: 293–298.
186. Spörri, H., Starke Dissoziation zwischen dem Ende der elektrischen und mechanischen Systolendauer bei Känguruhs. *Cardiologia*, 1956;**28**: 278–284.
187. Siegfried, J.P., *Elektrokardiographische Untersuchungen an ZOO-Tieren*, Dissertation. Zürich: University of Zurich, 1956, pp. 1–57.
188. Jayasinghe, J.B. and S.D.A. Fernando, Electrocardiograms of zoo animals, II. Leopard and Wallaby. *Ceylon Vet. J.*, 1964;**12**: 21–22.
189. Wilbur, C.G., Electrocardiographic studies on the opossum. *J. Mammal., 1955*;**36**: 284–286.
190. Szabuniewicz, M., D. Hightower, and J.R. Kyzar, The electrocardiogram, vectorcardiogram, and spatiocardiogram in the rabbit. *Can. J. Comp. Med.*, 1971;**35**: 107–114.
191. Levine, H.D., Spontaneous changes in the normal rabbit electrocardiogram. *Am. Heart J.*, 1942;**24**: 209–214.
192. Massmann, W. and H. Opitz, Das normale Kaninchen-EKG. Z. *Gesamte Exp. Med.*, 1954;**124**: 35–43.
193. Slapak, L. and P. Hermanek, Beobachtungen über das Elektrokardiogramm des Kaninchens. I: Das normale Extremitäten-Elektrokardiogramm des Kaninchens. *Z. Kreislaufforsch.*, 1957;**46**: 136–142.
194. Slapak, L. and P. Hermanek, Beobachtungen uber des Elektrokardiogramm des Kaninchens. II: Das normale Brustwand-Elektrokardiogramm des Kaninchens. *Z. Kreislaufforsch.*, 1957;**46**: 143–166.
195. Jacotot, B., L'electrocardiogramme du lapin. Analyse des tracés de 75 animaux sains. *Recl. Med. Vet.*, 1965;**141**: 1095–1107.
196. Angelakos, E.T. and P. Bernardini, Frequency components and positional changes in electrocardiogram of the adult rat. *J. Appl. Physiol.*, 1963;**18**: 261–263.
197. Beinfeld, W.H. and D. Lehr, QRS-T variations in the rat electrocardiogram. *Am. J. Physiol.*, 1968;**214**: 197–204.
198. Beinfeld, W.H. and D. Lehr, P-R interval of the rat electrocardiogram. *Am. J. Physiol.*, 1968;**214**: 205–211.
199. Werth, G. and S. Wink, Das Elektrokardiogramm der normalen Ratte. *Arch. Kreislaufforsch.*, 1967;**54**: 272–308.
200. Langer, G.A., Interspecies variation in myocardial physiology: the anomalous rat. *Environ. Health Perspect.*, 1978;**26**: 175–179.
201. O'Bryant, J.W., A. Packchanian, G.W. Reimer, and R.H. Vadheim, An apparatus for studying electrocardiographic changes in small animals. *Tex. Rep. Bioi. Med.*, 1949;**7**: 661–670.
202. Lombard, E.A., Electrocardiograms of small mammals. *Am. J. Physiol.*, 1952;**171**: 189–193.
203. Giordano, G. and G. Nigro, Caratteristiche dell' elettrocardiogramma normale del Mus musculus albus. *Sperimentale*, 1957;**107**: 63–68.
204. Goldbarg, A.N., H.K. Hellerstein, J.H. Bruell, and A.F. Daroczy, Electrocardiogram of the normal mouse, *Mus musculus*: general considerations and genetic aspects. *Cardiovasc. Res.*, 1968;**2**: 93–99.
205. Callsen, A.D., *Elektrokardiographische Untersuchungen an wachen und anästhesierten Katzen mit einem System von zehn Standardableitungen*, Dissertation. Berlin: Freie University, 1983, Journal No.1130, pp. 1–105.
206. Tilley, L.P. and R.E. Gompf, Feline electrocardiography. *Vet. Clin. North Am.*, 1977;**7**: 257–272.
207. Rogers, W.A. and S.P. Bishop, Electrocardiographic parameters of the normal domestic cat: a comparison of standard limb leads and an orthogonal system. *I. Electrocardiol.*, 1971;**4**: 315–321.
208. Blok, J. and J.Th.F. Boeles, The electrocardiogram of the normal cat. *Acta Physiol. Pharmacol. Neerl.*, 1957;**6**: 95–102, 209.
209. Massmann, W. and H. Opitz, Das Katzen-Ekg. *Cardiologia*, 1954;**24**: 54–64.
210. Rothlin, E. and E. Suter, Glykosidwirkung auf Elektrokardiogramm und Myokard. I Vergleichende elektrokardiographische Untersuchungen verschiedener herzwirksamer Glykoside an der Katze bei intravenöser Infusion. *Helv. Physiol. Pharmacol. Acta*, 1947;**5**: 298–321.
211. Purchase, I.F.H., The effect of halothane on the isolated cat heart. *Br. J. Anaesth.*, 1966;**38**: 80–91.
212. Purchase, I.F.H., Cardiac arrhythmias occurring during halothane anaesthesia in cats. *Br. I. Anaeslh.*, 1966;**38**: 13–22, 213.
213. Jayasinghe, J.B., S.D.A. Fernando, and L.A.P. Brito-Babapulle, The electrocardiograpic patterns of *Elephas maxim us* - the elephant of Ceylon. *Br. Vel. I.*, 1963;**119**: 559–564.
214. Jayasinghe, J.B. and L.A.P. Brito-Babapulle, A report on an electrocardiogram of the Ceylon elephant. *Ceylon Vel. J.*, 1961;**9**: 69–70.
215. Grauweiler, J., Beobachtungen am Elektrokardiogramm von nicht-domestizierten Säugetieren. *Schweiz. Arch. Tierheilkd.*, 1961;**103**: 397–417.
216. Forbes, A. and S. Cobb, Cattell McK. An electrocardiogram and an electromyogram in an elephant. *Am. I. Physiol.*, 1921;**55**: 385–389.
217. Siegfried, J.P., *Elektrokardiographische Untersuchungen an Zoo-Tieren*, Dissertation. Zürich: University of Zürich, 1956.
218. Nörr, J., Hundert klinische Fälle von Herz und Pulsarrhythmien beim Pferde. *Mh. Prakt. Tierheilkd. Slullgarl*, 1924;**34**: 177–232.
219. Roos, J., Vorhoffiimmern bei den Haustieren. *Arch. Wiss. Prakt. Tierheilkd.*, 1924;**51**: 280–293.
220. Roos, J., Auricular fibrillation in the domestic animals. *Heart*, 1924;**11**: 1–7.
221. Agduhr, E. and N. Stenström, The appearance of the electrocardiogram in the heart lesions produced by cod liver oil treatment: electrocardiogram in rats treated with cod liver oil. *Acla Paediatr. (Uppsala)*, 1930;**9**: 280–306.
222. Drury, A.N., L.J. Harris, C. Maudsley, and B. Vitamin, deficiency in the rat. Bradycardia as a distinctive feature. *Biochem. J.*, 1930;**24**: 1632–1649.
223. Nörr, J., Über Atmungsarrhythmie bei Caniden. *Verh. Dtsch. Ges. Kreislaufforsch.*, 1935;**7**: 144–148.
224. Scher, A.M., Direct recording from the A-V conducting system in the dog and monkey. *Science*, 1955;**121**: 398–399.
225. Scher, A.M. and A.C. Young, The pathway of ventricular depolarization in the dog. *Circ. Res.*, 1956;**4**: 461–469.
226. Detweiler, D.K., Perspectives in canine cardiology. *Univ. Pennsylvania Bull. Vet. Ext. Q.*, 1958;**149**: 27–48.
227. Ettinger, S.J. and P.F. Suter, *Canine Cardiology*. Philadelphia, PA: Saunders, 1970.
228. Hill, J.D.A., *Correlative Study of Right Ventricular Conduction Disturbances in the Dog*, Veterinary thesis. Philadelphia, PA: University of Pennsylvania, 1967.

229. Hill, J.D., Electrocardiographic diagnosis of right ventricular enlargement in dogs. *J. Electrocardiol.*, 1971;**4**: 347–357.
230. Pruitt, R.D., Electrocardiogram of bundle-branch block in the bovine heart. *Circ. Res.*, 1962;**10**: 593–597.
231. O'Rourke, M.F., A.P. Avolio, and W.W. Nichols, The kangaroo as a model for the study of hypertrophic cardiomyopathy in man. *Cardiovasc. Res.*, 1986;**20**: 398–402.
232. Gompf, R.E. and L.P. Tilley, Comparison of lateral and sternal recumbent positions for electrocardiography of the cat. *Am. J. Vet. Res.*, 1979;**40**: 1483–1486.
233. Boineau, J.P., E.N. Moore, J.F. Spear, and W.C. Sealy, Basis of static and dynamic electrocardiographic variations in Wolff-Parkinson-White syndrome. *Am. J. Cardiol*, 1973;**32**: 32–45.

43 12 Lead Vectorcardiography

Peter W. Macfarlane · Olle Pahlm

P. W. Macfarlane, A. van Oosterom, O. Pahlm, P. Kligfield, M. Janse, J. Camm (eds.), *Comprehensive Electrocardiology*, DOI 10.1007/978-1-84882-046-3_43,

43.1 Vectorcardiography

43.1.1 What Is a Vector?

The term vector can have different meanings, but for the purposes of the study of vectorcardiography, the relevant definition states that a vector is an entity possessing a magnitude and a direction. For example, if a wind blows in an easterly direction at 10 km/h, it could be represented by the vector in ➋ Fig. 43.1a. On the other hand, if a light breeze blows at 5 km/h in a northeasterly direction, it would be represented using the same scheme by the vector of ➋ Fig. 43.1b. It can be seen that the length of the vector is proportional to the strength of the wind and the direction of the vector is that of the wind.

Of course, there can be many different forces that are represented by a vector. Within the context of electrocardiography (ECG), it is the cardiac electromotive force that is desired to be represented by a vector. It was Einthoven et al. [1] in their classic paper of 1913 who suggested that the electrical forces of heart could be summed and represented by a single vector.

43.1.2 Concept of Resultant Force

While a vector can be used to represent an individual force as shown in ➋ Fig. 43.1, a series of vectors can be used simultaneously to represent a variety of forces acting together or in opposition. It is possible to use some simple mathematical techniques to calculate the resultant effect of the different forces, and it is instructive to consider an example.

Imagine that a rower sets out to cross a river. He is able to row constantly at 4 km/h directly across the water but has to contend with a current that is flowing at a rate of 3 km/h. This is depicted in ➋ Fig. 43.2. It should be clear that if the rower consistently pulls directly across the river he will not reach the bank at a point directly opposite his starting point but will be carried some way downstream by the current. In fact, the distance can be calculated by what is known as the "Triangle of Forces," which shows that he would travel at a net speed of 5 km/h. The exact point at which the rower reaches the opposite bank, of course, depends on the width of the river but this can be calculated from the triangle. For example, if the river is 200 m wide, the boat should reach the opposite side 150 m downstream on the opposite side.

The combined velocity of the boat and the current produces a resultant velocity of 5 km/h depicted by the hypotenuse on the triangle. Conversely, it can be said that if there is a resultant velocity of 5 km/h, it has components of 3 and 4 km/h at right angles to each other in keeping with ➋ Fig. 43.2. Thus, there exists the concept that a resultant velocity has a component in a particular direction. The size of each component can be obtained by drawing a perpendicular from the tip of the resultant vector to a line indicating the direction in which it is desired to measure the component.

A similar concept applies in electrocardiography. Consider that in the frontal plane of the body, there is a resultant cardiac electromotive force of 2 mV acting at 45° to the horizontal, that is, approximately similar to the path of the rowing boat in ➋ Fig. 43.2. ➋ Figure 43.3 shows that there is a component of 1.41 mV in the direction of lead I. Similarly, it can be shown that there is a component of approximately 1.93 mV in the direction of lead II. This estimate assumes that the equilateral Einthoven triangle is a valid model, which in reality is not the case. However, the potential measured by lead I can be considered as the component of the resultant cardiac electromotive force acting in that direction in the frontal plane. It follows from Einthoven's Law (see ➋ Chap. 11) that the potential in lead III would be 0.52 mV at the same instant in the cardiac cycle.

43.1.3 Spatial Vector

➋ Sections 43.1.1 and ➋ 43.1.2 have dealt with the two-dimensional situation of a vector or vectors acting effectively in a plane. A more realistic situation is a force or a vector having the ability to be directed at any point in space. ➋ Figure 43.4 illustrates the concept of a spatial vector drawn within a three-dimensional coordinate reference system with axes denoted

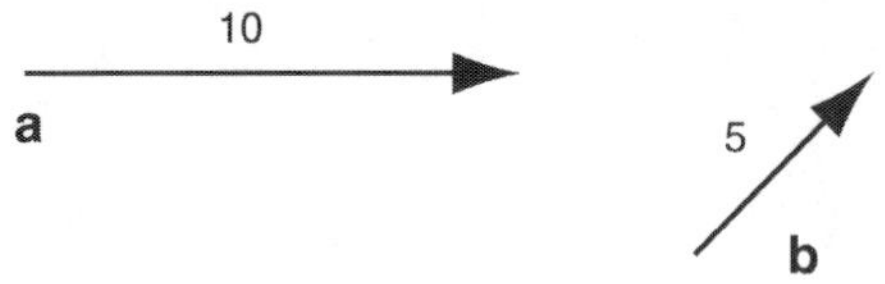

Figure 43.1
The direction and speed of the wind represented as a vector when the (a) wind blows in an easterly direction at 10 km/h (b) wind blows in a northeasterly direction at 5 km/h.

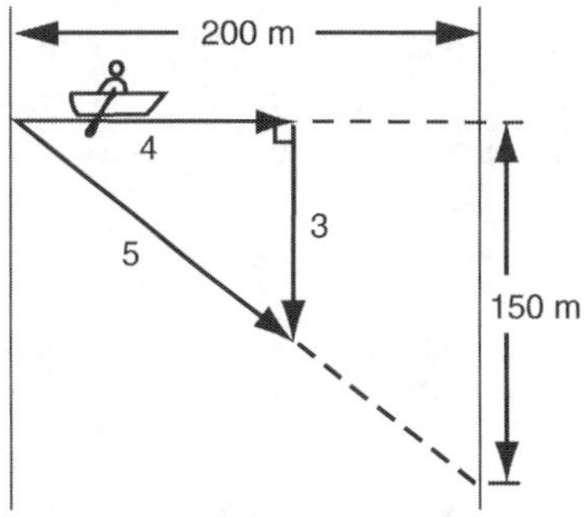

Figure 43.2
A rower crossing a river 200 m wide. For explanation, see text.

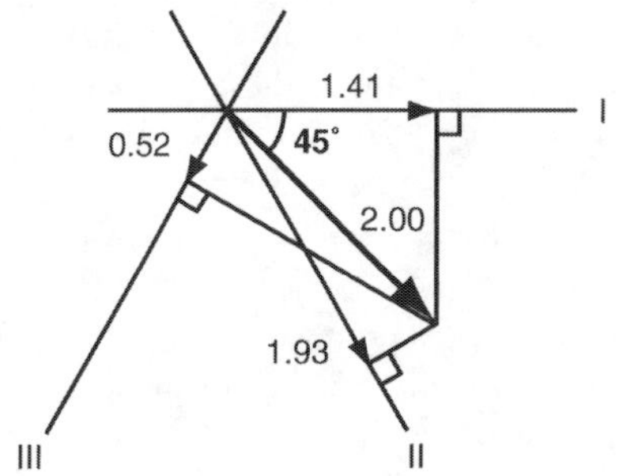

Figure 43.3
The cardiac electromotive force and its components in the direction of leads I, II, and III.

X, *Y*, and *Z*. It can be seen from the illustration that the magnitude of the vector can be calculated by using triangle OMA if the length of sides OA and AM can be determined.

However, from the Figure, it follows that:

$$OA^2 = x^2 + z^2$$

From right angled triangle OMA,

$$OM^2 = OA^2 + AM^2 = x^2 + y^2 + z^2$$

Thus, for a point M with coordinates *x*, *y*, *z* in three-dimensional space, the length of the vector OM is given by the above expression. By analogy with the two-dimensional situation, where it was shown that a vector lying in a plane could have

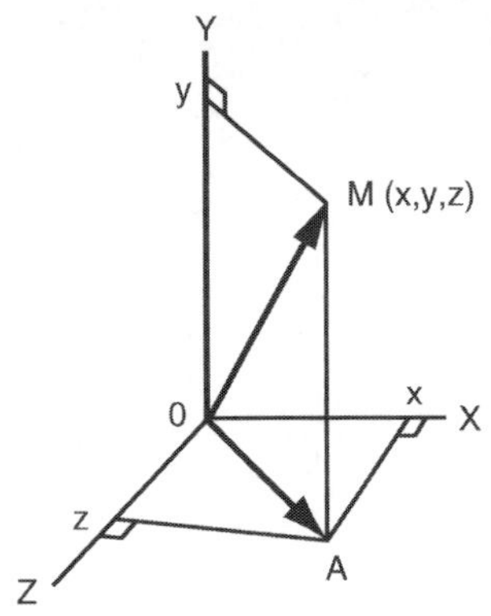

Figure 43.4
A spatial vector OM and its components *x, y, z* in a three-dimensional coordinate system.

components calculated in any particular direction by drawing a perpendicular to that line, it can also be shown that in the spatial situation, a vector, for example, OM, can have components in the three mutually perpendicular directions X, Y, Z. In the forward situation, if the components *x*, *y*, *z* can be measured at a particular instant in the cardiac cycle, then a resultant OM can be calculated. This is the basis of an orthogonal lead system.

43.1.4 Orthogonal Lead Systems

43.1.4.1 Theoretical Considerations

Section 43.1.3 suggests that if three leads can be designed to record components of a resultant cardiac electromotive force in three mutually perpendicular directions, then the problem of deriving the resultant cardiac electromotive force is solved. A considerable amount of research went into designing such lead systems over the past 50 years. The theory is detailed but a few simple concepts are worthy of discussion at this point. Further aspects are considered in Chap. 11, Sect. 11.5.

Assume that the potential measured by any electrocardiographic lead is represented by V. Then, assume that the resultant cardiac electromotive force is denoted by **H** or, as it is sometimes known, "The heart vector." Then from mathematical considerations, it can be shown that $V = \mathbf{H} \cdot \mathbf{L}$ where **L** is the vector representing the strength of the lead being used to measure the potential. In fact, it is one of the basic rules of vector mathematics that the dot product of two vectors is a scalar, that is, the potential or voltage does not have an associated direction but only a magnitude whereas the heart vector **H** and the lead vector **L** each has its own direction. This basic rule of vector mathematics can also be expanded to the following:

$$V = H_X L_X + H_Y L_Y + H_Z L_Z$$

where H_X, H_Y, H_Z are the three components of the heart vector and L_x, L_y, L_z are the three components of the lead vector. From this observation, it follows that if it is desired to measure the component of the heart vector in the X direction, then a lead should be designed that has components $(L_X, 0, 0)$. In that case,

$$V_X = H_X L_X$$

If the strength L_X of the lead is known, then when the potential V_X is measured, H_X can be calculated.

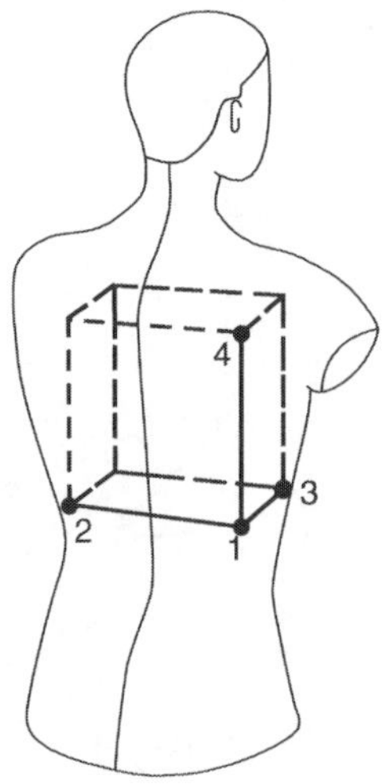

Figure 43.5
The cube lead system introduced by Grishman [2]. © American Heart Association, Dallas,Texas. Reproduced with permission.

43.1.4.2 Uncorrected Lead Systems

For historical reasons, it is worth noting that the earliest attempts at designing orthogonal lead systems were made on the basis of constructing leads such that lines joining the electrodes were essentially mutually perpendicular. This is most easily understood by considering the cube system introduced by Grishman [2] (Fig. 43.5). However, as experience was gained and mathematical modeling advanced, it was found that these lead systems did not accurately measure the required components.

43.1.4.3 Corrected Orthogonal Lead Systems

As a result of considerable modeling, both mathematical and physical, such as using model torsos filled with conducting solution, corrected orthogonal lead systems were gradually introduced.

The most notable and the one that is generally used wherever vectorcardiography is currently studied using an orthogonal lead system, is that of Frank [3]. This lead system is shown in Fig. 43.6. As can be seen, the lead positions are different from those of the 12-lead system, although the C and A electrodes are indeed close to the V_4 and V_6 positions, respectively.

Mathematical modeling showed that the following equations represent the derivation of the three potentials V_X, V_Y, V_Z:

$$V_X = 0.610V_A + 0.171V_C - 0.781V_I$$
$$V_Y = 0.655V_F + 0.345V_M - 1.0V_H$$
$$V_Z = 0.133V_A + 0.736V_M - 0.264V_I - 0.374V_E - 0.231V_C$$

These contributions to the individual leads from the different electrodes correspond to the resistor network also seen in Fig. 43.6. The major disadvantage of using this type of orthogonal lead system is the need to apply a completely different set of electrodes to the patient compared to that required for recording the conventional 12-lead ECG. There have been attempts to minimize the differences by doubling the C and A electrodes as V_4 and V_6, for example, and using a common left leg electrode but this still leaves four additional electrodes to be positioned on the thorax and neck.

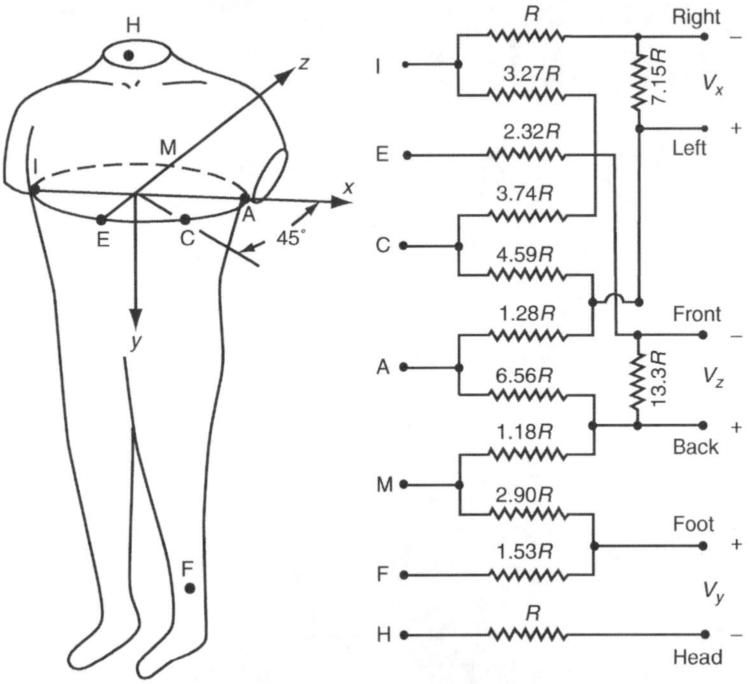

Figure 43.6
The Frank lead system (After Frank [3]). © American Heart Association, Dallas, Texas. Reproduced with permission.

43.1.5 Cardiac Activation

43.1.5.1 Vectorial Spread of Cardiac Activation

The concept of resultant cardiac electromotive force by now should be gaining hold. It is possible to consider the various resultant forces acting throughout ventricular depolarization, for example. Figure 43.7a shows a series of individual vectors each of which represents the resultant cardiac electromotive force at one particular instant during the process of ventricular depolarization. For example, the first small vector shows the initial septal activation from left to right. Note that all of these vectors are depicted in the two-dimensional situation, which in this case is the frontal plane of the body. Figure 43.7b indicates how each of these resultant vectors can be translated to a common origin and a loop drawn to connect the tips of the vectors. This loop is a form of vectorcardiogram. Indeed, the first form of planar vectorcardiogram was introduced by Mann in 1920 [4].

43.1.5.2 Spatial Vector Loop

The previous Sect. 43.1.5.1 showed a planar loop. However, the concept of spatial vector has been introduced in Sect. 43.1.3 and it follows that if a resultant vector varies in magnitude and direction throughout the cardiac cycle, then it can be imagined that the tip of this vector will trace out a three-dimensional path. This observation is illustrated in Fig. 43.8 for the case of ventricular depolarization, that is, a spatial QRS loop is shown. In addition, the projection of this loop onto three mutually perpendicular planes is also illustrated. The collection of three planar loop projections is known as the vectorcardiogram. While only the QRS loop is depicted here for clarity, it follows that P and T loops can also be derived.

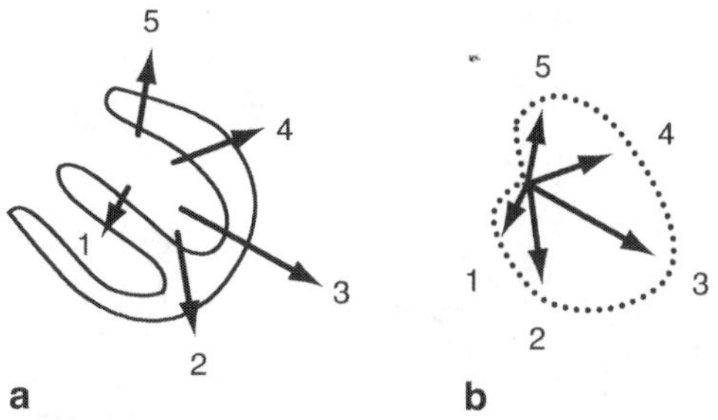

Figure 43.7
(a) Ventricular depolarization illustrated as a sequence of vectors, 1–5. (b) After translation of vectors to a common origin, a vectorcardiographic loop can be constructed.

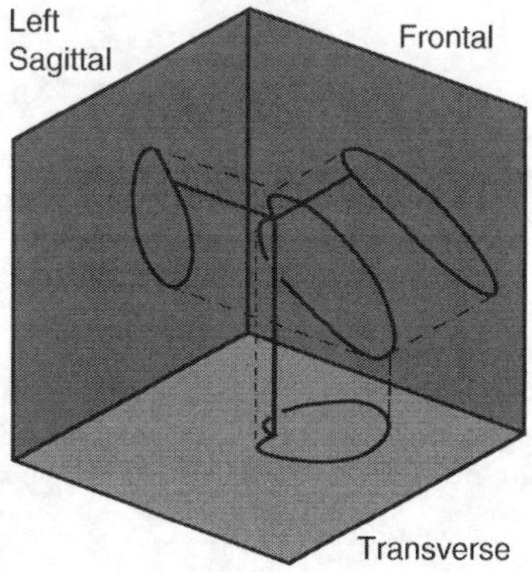

Figure 43.8
A vectorcardiographic loop in space projected onto the three orthogonal planes.

43.1.6 Vector Loop Presentation

43.1.6.1 Nomenclature

The American Heart Association committee on electrocardiography (1975) [5] published a set of recommendations for vectorcardiographic terminology. The committee recommended that the lead Z be directed positively to the posterior thorax, although this does mean that the scalar presentation of lead Z is essentially opposite to that of lead V_2. This recommendation certainly causes much confusion when describing scalar lead appearances and for this reason, in this chapter, lead Z is directed positively to the anterior to be similar to V_2. Thus, R_Z can be thought of in the same way as R_{V2}. Of more contention is the choice of whether to view the left or right sagittal planes, that is, the sagittal plane as viewed from the left or right. Figure 43.9 shows the left sagittal projection. The Committee did not make a particular recommendation, but for the purposes of illustrations in this book, the left sagittal view has been chosen in keeping with Fig. 43.9 so as to have a uniform collection of reference axes.

43.1.6.2 Display Techniques

About 20 years ago, the most common method of displaying the vectorcardiogram was via an oscilloscope. Pairs of leads such as X and Y were used to deflect the electron beam horizontally and vertically, respectively, and in this case, the

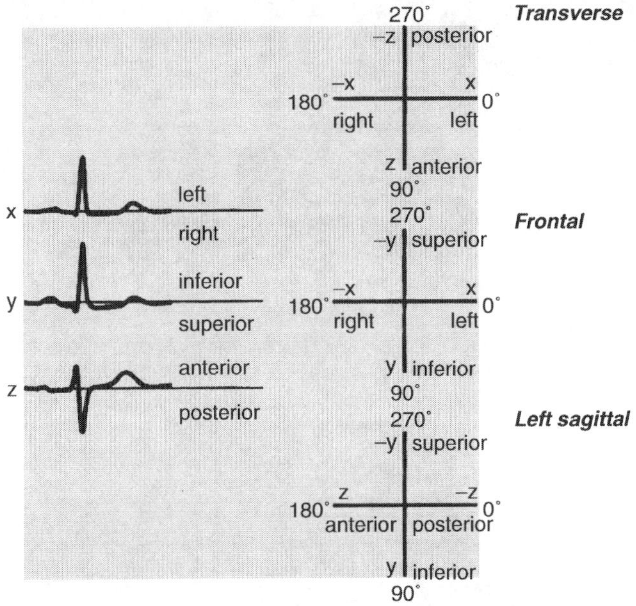

Figure 43.9
Polarity of leads X, Y, and Z and angular reference frame of the frontal, transverse, and left sagittal planes.

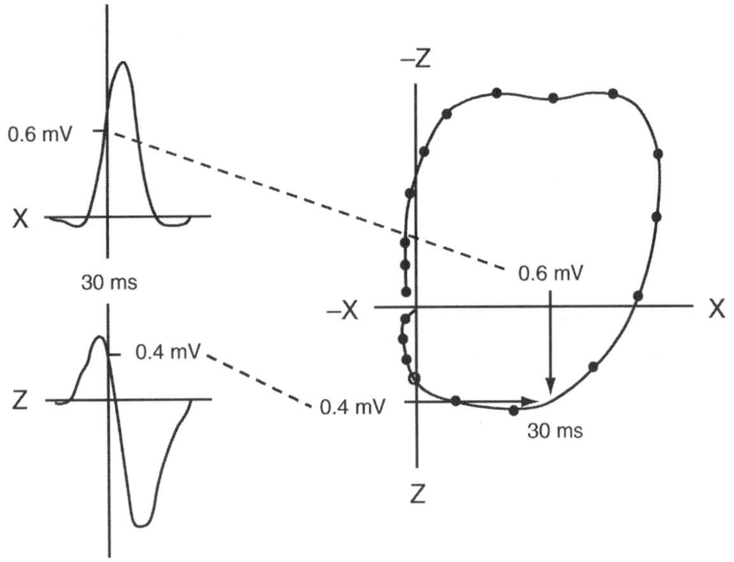

Figure 43.10
Derivation of transverse plane loop from leads X and Z. Dots are at 4 ms intervals and the 20 ms vector is indicated with an open circle.

frontal plane loop would be generated. Consider that leads X and Y are recorded simultaneously. At any instant in time, an amplitude for each lead is known, that is, an (x, y) coordinate pair of values is available. These values could be plotted simply on XY axes. If this is repeated throughout the cardiac cycle, then a complete frontal plane set of P, QRS, and T loops can be generated. Figure 43.10 shows the derivation of a QRS loop in the transverse plane.

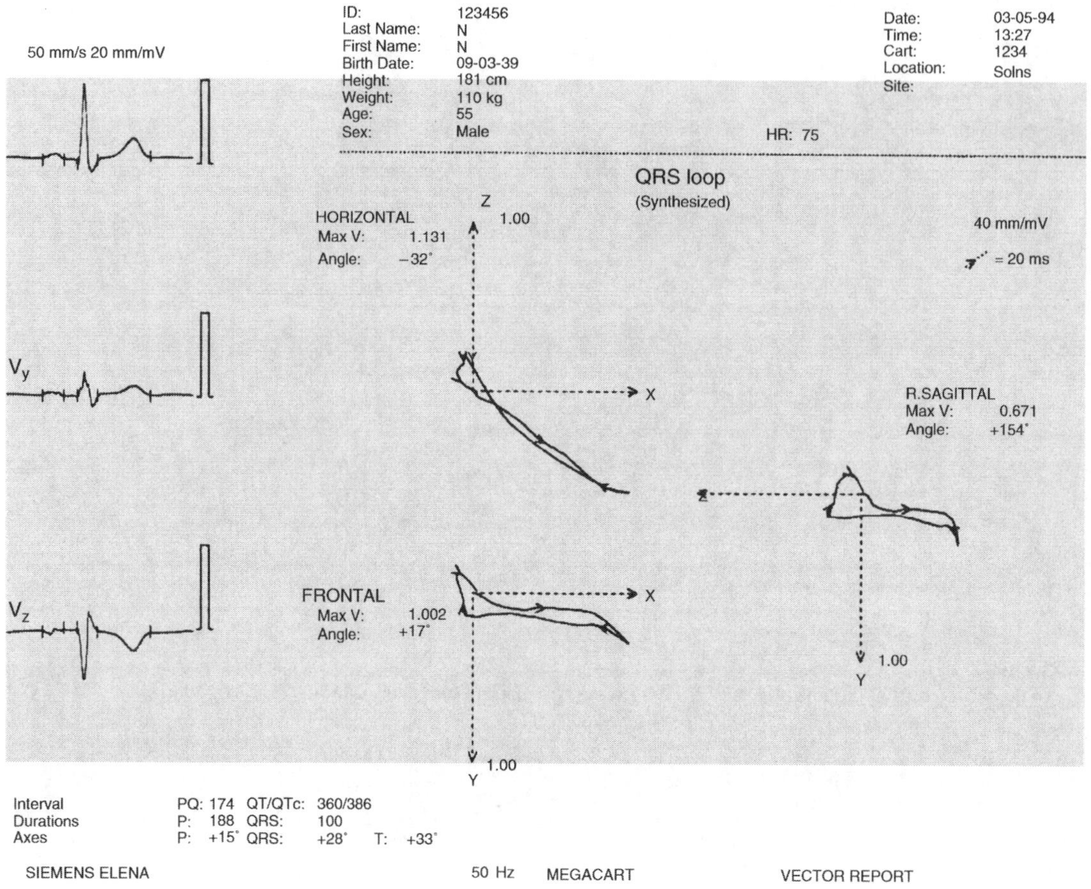

Figure 43.11

Vectorcardiogram printout from a commercially available electrocardiograph. Note that lead Z is inverted in comparison with other illustrations in this book.

If the leads X, Y, Z are considered in pairs, then by plotting leads XY, the frontal plane vectorcardiographic loop can be obtained. Similarly, XZ plots produce the transverse plane loop and ZY plots produce the sagittal plane vectorcardiographic loop. Nowadays, with the widespread availability of computer technology, these plots can be produced in a straightforward fashion. It is somewhat more complex to program a thermal writer to produce an XY display but nevertheless this is readily attainable. Thus, vectorcardiographic loops can now be produced even on small 4″ paper displays or on the larger A4 writers such as are common at the bedside. An illustration of a typical vectorcardiographic display from a computer-based electrocardiograph is shown in Fig. 43.11.

It is important that vectorcardiographic loops have some indication of the speed of inscription as this can contain diagnostic information. A number of methods have been used. Conventionally, the vectorcardiographic loop has been interrupted so that time intervals can actually be measured by counting the number of dots between two points. Generally, 2 or 4 ms intervals have been used. An alternative is to produce a continuous loop and mark a number such as 1, 2 indicating 10, 20 ms from the onset of the QRS complex. This is helpful but creates difficulties around the onset and termination of the QRS loop, which is often the most interesting part in terms of looking for conduction problems. Either way, the direction of inscription of the loop is also of vital clinical significance. Thus, if a numbering system is not used, some indication must be given to make it quite clear to the viewer in which direction the different planar loops are inscribed. With a knowledge of the theory of vectorcardiography, the experienced cardiologist can always determine the direction of inscription but it is certainly easier if this is made obvious in a good display.

In this book, loops are presented with an arrow indicating the direction of inscription. Further, black dots indicate 4 ms intervals and the 20 ms vector is indicated with an open circle.

43.2 Normal Ranges

43.2.1 Introduction

The aim of good diagnostic criteria is to separate normal from abnormal with the highest possible sensitivity and specificity. It may be superfluous to repeat well-known definitions but for the avoidance of doubt, the following apply:

$$\text{Specificity} = A/B$$
$$\text{Sensitivity} = C/D$$

where

A = number of normals correctly reported as normal
B = total number of normals
C = number of abnormals correctly reported as abnormal
D = total number of abnormals.

Figure 43.12 shows the distribution of Q wave duration in lead Y for normals and for a group of patients with inferior myocardial infarction. If a value of 20 ms is chosen as the upper limit of normal, it can be seen that the specificity of the

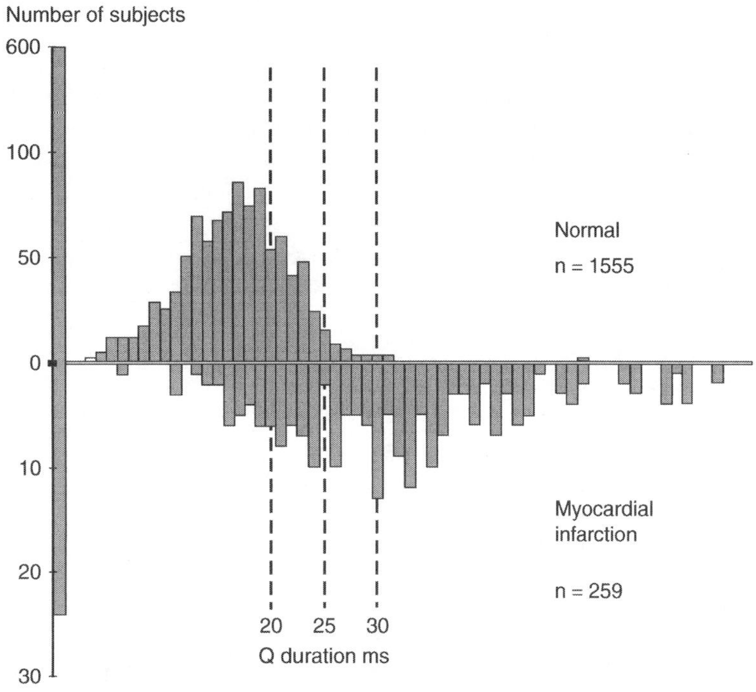

Figure 43.12
Shows distribution of Q duration in lead Y in a group of 1,555 normal subjects and in a group of 259 patients with inferior myocardial infarction. For explanation, see text.

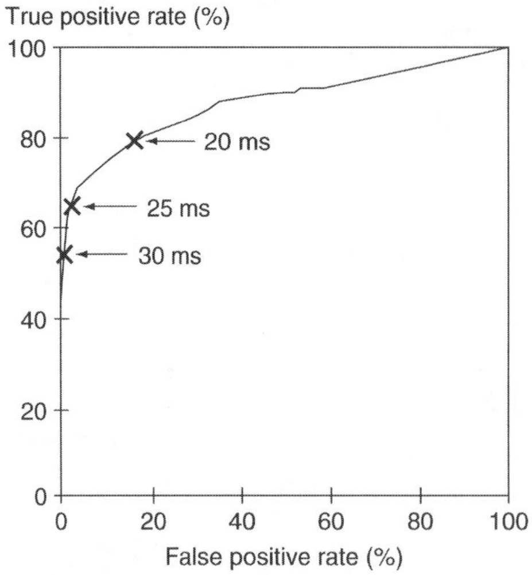

Figure 43.13
Receiver operating characteristic (ROC) curve showing the relationship between true positive rate (sensitivity) and false positive rate (100 – specificity) for different limits of Q duration in lead 1′: The curve is based on the data in Fig. 43.12.

criteria would be approximately 83%. The sensitivity for inferior infarction would be the order of 79%. However, some would argue that 83% specificity, that is, close to one in five normals reported as abnormal, is not high enough and might adjust the borderline value to 25 ms. In this case, the specificity would increase to 97% and the sensitivity for infarction would decrease to 65%. This process can be continued. For example, with a borderline of 30 ms, specificity is over 99% but sensitivity decreases to 54%.

It is possible to plot the relationship between sensitivity and specificity on what is known as a receiver operating characteristic (ROC) curve. This is shown in Fig. 43.13. The point on the curve, which approaches closest to 100% sensitivity and specificity, that is, the top left-hand corner, is often regarded as being optimum. In the case of electrocardiography, it is often preferable to choose a point on the curve with higher specificity (e.g., 25 ms point in Fig. 43.13).

While it is clear that ROC curves are dependent on the knowledge of measurements from a normal population and patients with a particular abnormality, it should also be noted that if a user decides that 95% specificity is the desired level, then the knowledge of the abnormal data are not required. This is perhaps an extreme view but it emphasizes the value of having well-defined normal data.

In Glasgow, every effort has been made over the past 25 years to gather a population of controls from birth upward, and of different ethnic origin, in order to meet the objectives outlined above. The remainder of this chapter will deal with the techniques involved and present the results obtained from the Caucasian cohort.

43.2.2 Data Acquisition

43.2.2.1 Techniques

Data have been gathered using two separate types of electrocardiographs each with a common factor of sampling electrocardiographic waveforms at 500 samples/s. All of the recordings made outside Glasgow Royal Infirmary, for example, on infants and children, were gathered using a Mingorec 4 from Siemens-Elema AB, Solna, Sweden. This acquires eight leads simultaneously, converts from analog to digital form at a rate of 500 samples/s for further analysis. More recent work has involved the use of the Burdick Atria 6100 electrocardiograph.

For ECGs recorded within Glasgow Royal Infirmary, an electrocardiograph designed and constructed within the Department of Medical Cardiology was used ([6], 1987). This device was connected by a broadband network from wards and clinics to the central computing facility within the Department.

The methods for analyzing the ECGs have been described in detail elsewhere [7] but are summarized very briefly here.

Up to 8 s of ECG with all leads sampled simultaneously, are processed initially to remove baseline wander, if present, and also any AC interference. Thereafter, QRS detection is undertaken. The same methods apply whether the ECG is recorded from a neonate or an adult. The QRS complexes so detected are then typed into different morphologies and logic selects one particular morphology for analysis. All PQRST cycles of that morphology are then averaged to form a single synthesized beat (with all 12 leads effectively recorded simultaneously). The derived leads X, Y, Z are then obtained using the methods outlined in detail in ➋ Chap. 11. In particular, the adult X, Y, Z leads were calculated using equation and the pediatric x, y, z leads were derived using equation. The wave measurement program then locates the onsets and terminations of the various P, QRS, and T components in order to measure amplitudes and durations.

Rhythm analysis is then undertaken. This uses some measurements from the average beat matrix but also three complete leads from the initial recording. Generally, these would be II, III, and V_1. When rhythm has been determined, diagnostic logic is then entered to interpret the measurements from the average beats. The program can output several hundred diagnostic statements. The same program can now be used for adults as well as for children [8]. In other words, the ability to interpret ECGs from children is not an optional add-on to the logic but is integral to the diagnostic criteria. Some details are presented elsewhere [9].

43.2.2.2 Sampling Methods

Different approaches to the selection of an apparently healthy population can be adopted. For example, age and sex registers can be used [10], but the technique adopted for adults in the Glasgow data was essentially to seek volunteers from the different departments of local government, for example, teaching, administration, and building. All individuals admitted to the normal group were seen by a physician who obtained a complete history and undertook a physical examination. The usual blood tests were performed and in the initial part of the study, chest x-rays were obtained. It was found that a positive yield from chest x-rays was essentially nil and latterly they were discontinued as part of the screening procedure.

For sampling from children, the procedure was different. Recordings were obtained from a maternity hospital with the full consent of parents. In preschool children, it was necessary to visit postnatal clinics, health centers, and play groups in order to obtain recordings from children, again with the permission of parents. Recordings were obtained from schoolchildren by installing the Siemens electrocardiograph for periods in different schools with the permission of the local health and education authority. Volunteers were sought subject to parental permission.

43.2.2.3 Population Data

A total number of 1,555 adults were entered into the normal database. The age and sex distribution is shown in ➋ Fig. 43.14. There tends to be a preponderance of males over 30 years of age, probably because of the predominance of male workers of that age group.

➋ Figure 43.15 shows similar data for 1,782 healthy neonates, infants, and children whose ECGs were recorded. It should be pointed out that in this age group, lead V_{4R} was used in preference to lead V_3 as is the custom in many countries.

Actual numbers of males and females in the total group of 3,337 are given in ➋ Table 43.1.

In Taiwan, in collaboration with the Veterans' General Hospital in Taipei, it was possible to obtain 12-lead ECGs from 503 apparently healthy Chinese individuals [11]. From these, the 12-lead vectorcardiogram was obtained [12]. Essentially, methods used were similar to those for the collection of adult data in Caucasians although a higher percentage of individuals were in hospital with noncardiac problems.

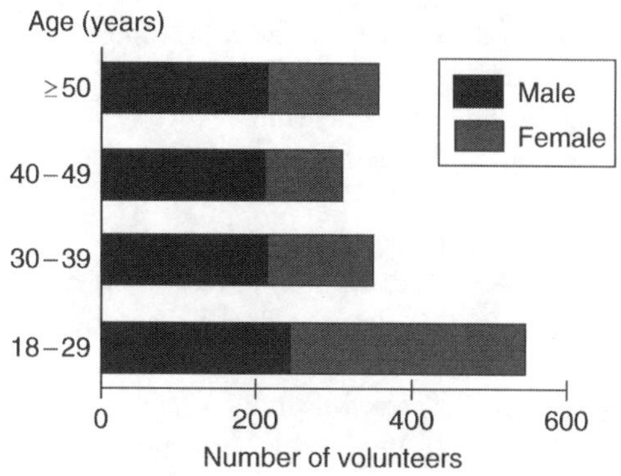

Figure 43.14
Age/sex distribution of the normal adult database. Further details are provided in Table 43.1.

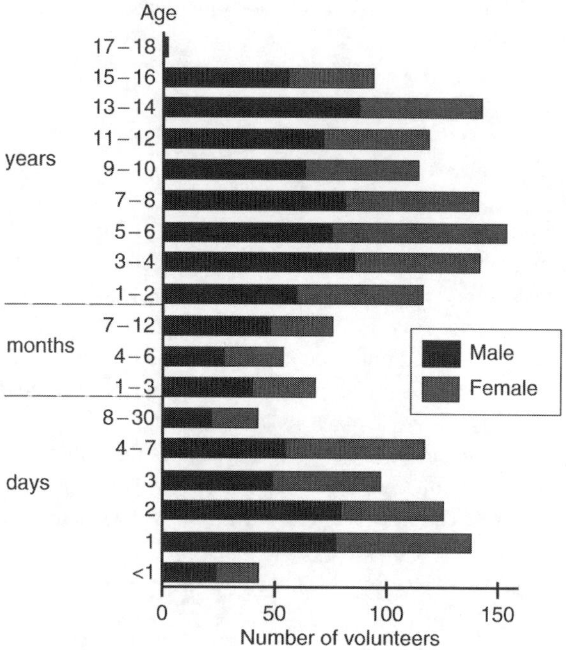

Figure 43.15
Age/sex distribution of the normal children's database. Further details are provided in Table 43.1.

43.2.2.4 Methods of Analysis

All measurements from each recording were stored on a computer file and added to a database. The adult and pediatric data were kept separate.

The BMDP suite of statistical programs was available for obtaining the basic data such as mean and standard deviation. The programs mainly used were P2D and P6D.

Table 43.1
Numbers of males and females in the total population group

Age	Male	Female	Total
⩽24 h	23	19	
1 day	77	61	
2 days	79	46	
3 days	48	49	
4–7 days	54	63	
⩽1 month	21	21	
⩽3 months	39	29	
⩽6 months	27	26	
⩽1 year	47	29	
1–2 years	59	57	
3–4 years	85	57	
5–6 years	75	79	
7–8 years	81	60	
9–10 years	63	51	
11–12 years	71	48	
13–14 years	87	56	
15–16 years	55	39	
17–18 years	0	1	
Children ∑	*991*	*791*	*1782*
18–29 years	242	304	
30–39 years	217	131	
40–49 years	210	97	
⩽50 years	215	139	
Adult ∑	*884*	*671*	*1555*
Total ∑	1875	1462	*3337*

43.2.2.5 Statistical Considerations

It has been known for many years that in general terms, ECG data are not normally distributed but tend to be skewed. An illustration is shown in ➋ Fig. 43.16. For this reason, normal ranges are best described not by using the mean ± twice the standard deviation but by 96 percentile ranges, that is, by excluding 2% of measurements at the top and bottom end of a particular set of measurements. Wherever possible this has been done in analyzing the data. Only in the case of small numbers such as, for example, healthy males with a Q wave in V_1, was it necessary to include the complete range because total numbers were too small. One other point that should be noted is that the calculation of the mean is based only on measurements that were present. In other words, if there were 100 patients in a particular group but only 40 had an S wave in a selected lead, then the mean amplitude was derived using only the 40 measurements and the 60 values of 0 mV for the remaining patients were excluded from the calculation of the mean.

With respect to angular data, care was taken to ensure that all measurements were in a meaningful range. In other words, the recommendations for measuring angles would suggest that the direction in the frontal plane of the X-axis would be labeled as 0°. If a vector measurement inferior to that was perhaps 20° and one superior to the X-axis was 340°, the mean value would certainly not be 180° but 0°. In other words, 340° would be converted to -20° before calculating the mean value. Alternative methods for dealing with angular data were elaborated many years ago by Downs et al. [13].

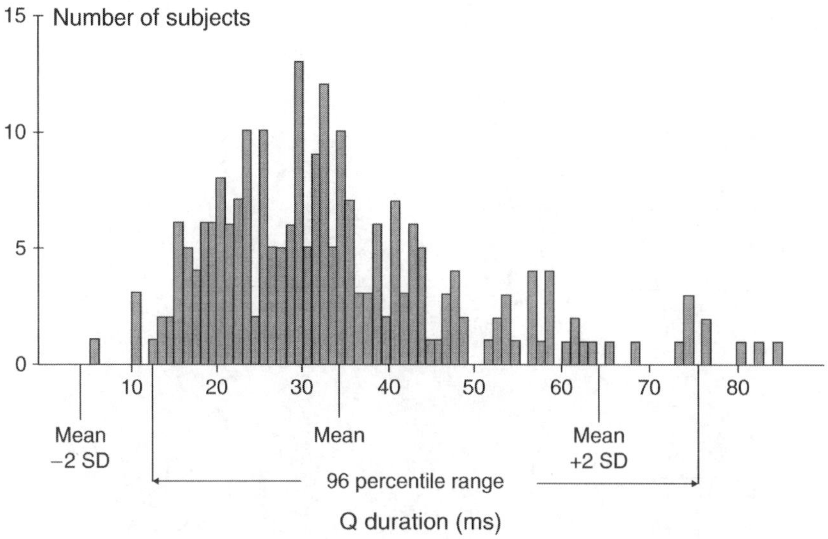

Figure 43.16
An example of a skewed distribution where there is a long tail of measurements at the upper end of the distribution. The figure shows a histogram of Q wave duration in lead Y in a group of 259 patients with proven inferior myocardial infarction. A total number of 24 patients with no Q waves in lead Y are excluded from the calculation of the mean.

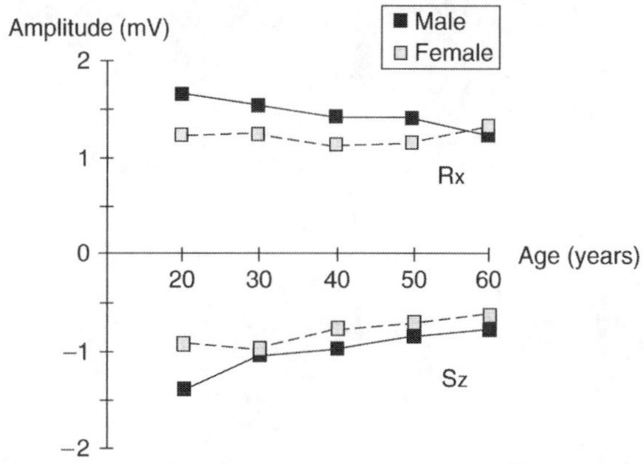

Figure 43.17
Effect of age and sex on mean R wave amplitude in lead X and mean S wave amplitude in lead Z.

43.2.3 Results: Scalar Data

43.2.3.1 Wave Amplitudes and Durations

The relevant tables of normal limits of PQRST amplitudes and durations in the derived leads X, Y, and Z are presented in Appendix 5A. Again, it can be confirmed that the effects of age and sex on the amplitudes of waveforms are significant. This can be seen in Fig. 43.17 where the mean S wave amplitude in lead Z is presented.

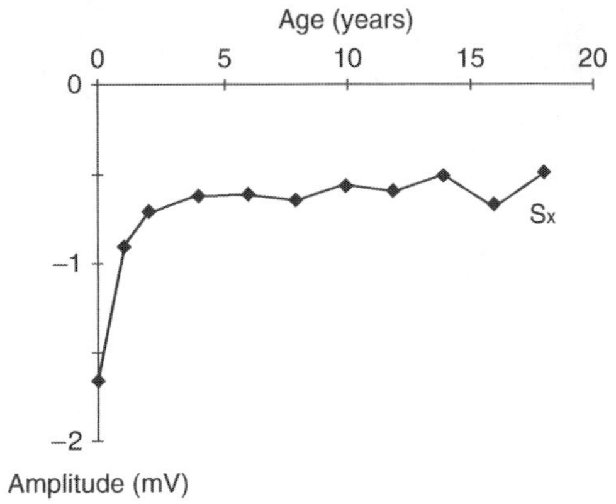

Figure 43.18
The upper limit of normal S_X amplitude in children from birth to adolescence.

Here the amplitude in young males is significantly higher than in young females, although the difference diminishes as age increases. These findings are similar to measurements of the S wave in V_2. The effect is not so marked for the mean R wave in lead X, which is also shown in the same figure.

In contrast, durations tended to show little difference between different age and sex groups with the exception of the QRS duration, which, as is well known, is approximately 7 ms longer in males than in females although strangely almost no cognizance is taken of this in any diagnostic criteria. This point is discussed in Chap. 13 (see Sect. 13.7.8).

43.2.3.2 Pediatric Data

It goes without saying that dramatic changes in pediatric measurements can be seen from birth onward. Again, appropriate tables are presented in Appendix 5B. As an example, the upper limit of normal S wave amplitude in lead X is shown in Fig. 43.18. There is a rapid decrease in S_X amplitude in the first year of life corresponding to a counterclockwise shift of the maximum frontal plane QRS vector.

43.2.3.3 Comparative Vectorcardiography

Yang and Macfarlane [14] reported on a comparison of the 12-lead vectorcardiogram in apparently healthy Caucasians and Chinese. To the database of 1,555 Caucasians, 503 Chinese were added giving a total of 2,058 individuals whose vectorcardiograms were derived from the conventional 12-lead ECG.

The trend of the influence of age and sex on the magnitude and direction of the derived QRS and T vectors was found to be similar in both cases. In the younger age groups, the magnitude of the maximal spatial vector was essentially greater in Caucasians than in Chinese while in the older age groups over 40, the reverse was the case. This was a somewhat surprising finding for which there is no clear explanation. Figures 43.19 and 43.20 show a comparison between the QRS and T vector magnitude in both races. It can be seen that it is essential to include the effect of age, sex, and race when interpreting 12-lead vectorcardiographic appearances.

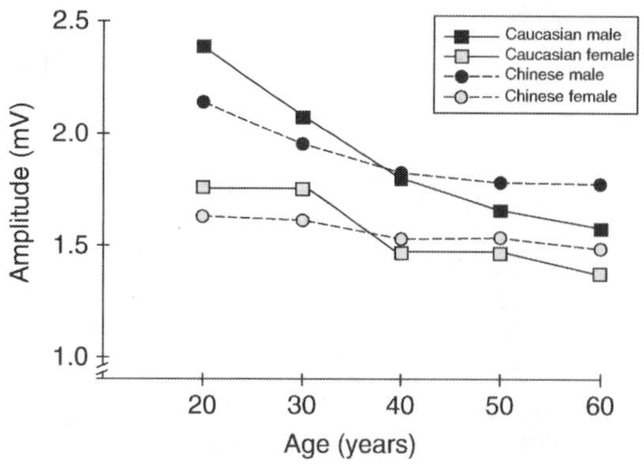

Figure 43.19
Mean magnitude of the maximal QRS vector amplitude in Caucasians and Chinese.

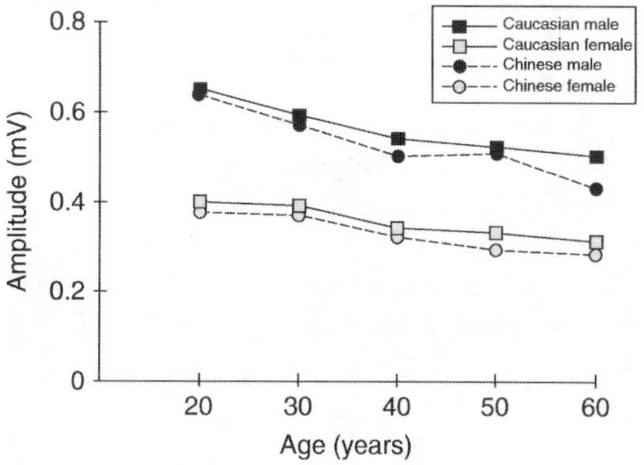

Figure 43.20
Mean magnitude of the maximal T vector amplitude in Caucasians and Chinese.

43.2.4 Results: Vector Data

43.2.4.1 P Loops

In the infant, the P loop tends to be directed vertically at birth but it soon rotates superiorly in the frontal plane and remains around 55°. In the transverse plane, the P loop in children at birth is approximately 20–25° and subsequently shifts a little toward the adult value of around 0°. Sex differences between the mean P wave vector in the transverse plane are significant with the mean direction for males being 349° and for females 14° (Draper et al. [15], Nemati et al. [16]). It should be noted that the results obtained by these authors were derived using the Frank system.

43.2.4.2 QRS Loops

There is, of course, a considerable change from birth to adulthood in the QRS loop in the 12-lead vectorcardiogram. Figure 43.21 shows a 12-lead QRS loop from a neonate where the maximum QRS vector is oriented around 150° in the frontal plane. On the other hand, in the normal adult QRS in the frontal plane, Fig. 43.22, the QRS loop is oriented at around 50°. In general terms, in the frontal plane, the QRS loop is in the vast majority of cases inscribed in a

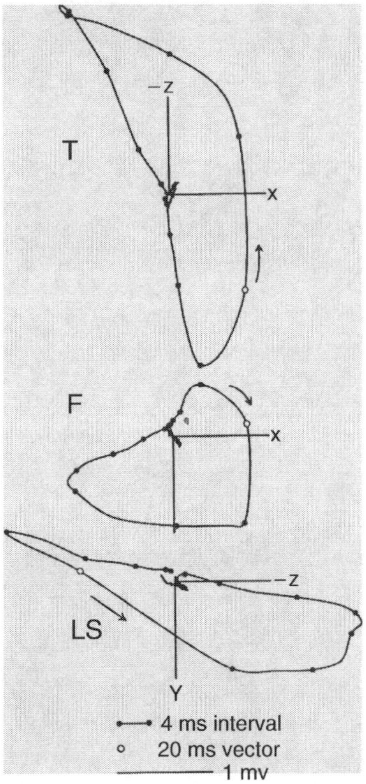

Figure 43.21
12-lead QRS and T loops from a healthy neonate.

clockwise direction in the infant. In the adult, the frontal plane loop can be inscribed either in a clockwise or counter-clockwise direction, although the clockwise loop tends to predominate again. In the transverse plane, in the neonate, the inscription around birth is 43% clockwise and 45% counterclockwise but soon changes to being almost totally counterclockwise as it is in the normal adult. A figure-of-eight loop can be found in 20–40% of children up to 8 months of age using the Frank system [17] although our own 12-lead data suggest a lower incidence. Table 43.2 shows the results derived from the Glasgow data in respect of direction of inscription of the QRS loops in the frontal and transverse planes.

43.2.4.3 Left Axis Deviation

It may seem superfluous to consider a discussion on left axis deviation but the normal range of the maximum QRS vector in the frontal plane is, indeed, quite different in the 12-lead vectorcardiogram from that in the standard 12-lead ECG. A full list of ranges is given in Appendix 5.

It can be seen that in healthy females, for example, the maximum QRS vector is never superior to 0°. In males, this seems to occur in a few individuals in the 30–59 years age group but, by and large, the vast majority of individuals have a maximum QRS vector inferior to 0°. This observation suggests the following criteria:

Borderline/left axis deviation: $0^\circ \rightarrow -15^\circ$ $(360^\circ \rightarrow 345^\circ)$
Left axis deviation: $-15^\circ \rightarrow -90^\circ$ $(345^\circ \rightarrow 270^\circ)$

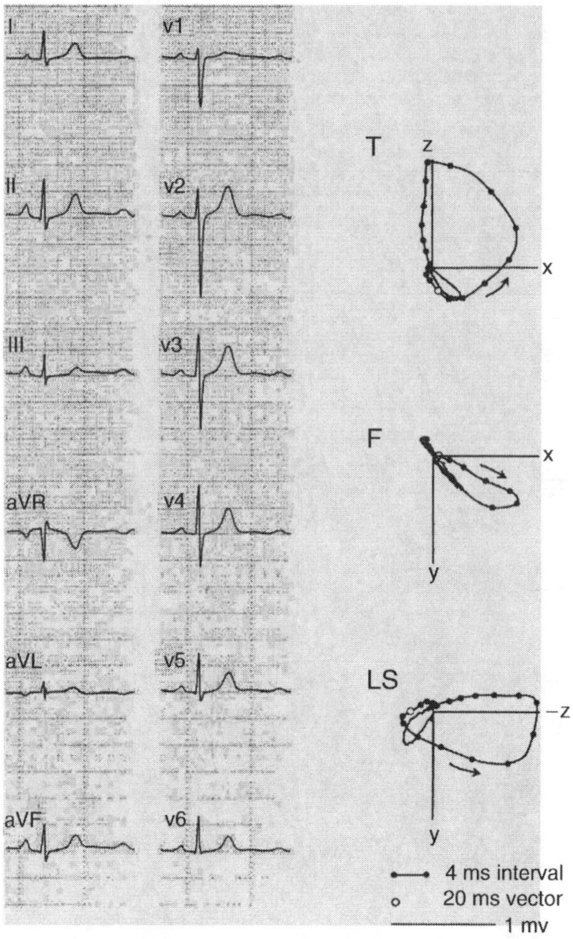

Figure 43.22
12-lead QRS and T loops from a healthy adult.

Table 43.2
Direction of inscription of the QRS vector loop in transverse and frontal planes expressed as a percentage from 1,555 adult Caucasian normals (CCW; Counterclockwise Rotation, CW; Clockwise Rotation)

Planes		CCW	Figure of 8	CW
Transverse	Male	98.1	0.9	1.0
	Female	97.9	1.2	0.9
Frontal	Male	22.9	19.1	58.0
	Female	21.8	25.3	52.9

43.2.4.4 Right Axis Deviation

The tables in Appendix 5 indicate that with the exception of young males <30 years of age, the normal maximum QRS vector in the frontal plane is always <65°. This indication is considerably different from the normal frontal plane vector calculated from the 12-lead ECG. It does suggest the use of the following criteria:

Borderline right axis deviation:
Females and (males > 30 years) 65° → 75°
Males < 30 years 90° → 100°

Right axis deviation:
Females and (males > 30 years) > 75°
Males < 30 years > 100°

43.2.4.5 T Loops

From Figs. 43.21 and 43.22 discussed above, the direction of the maximum T vector in the frontal and transverse planes can be seen for the average newborn and the average adult. It follows that there is a gradual change of T vector orientation from one position to another with increasing age. Details of some T vector measurements can be found in the Appendices.

43.3 Hypertrophy

43.3.1 Introduction

43.3.1.1 What is Hypertrophy?

Electrocardiographers tend to report the pattern of increased voltage in the lateral leads together with accompanying ST-T changes as left ventricular hypertrophy (LVH). The advent of echocardiography as well as cardiac catheterization has meant that the variety of pathologies, which constitute the generic term "hypertrophy" is now better known. In general terms, hypertrophy can be taken to mean an increase in mass. All four chambers of the heart can demonstrate hypertrophy or enlargement either in isolation or in combination. Enlargement is a term that is perhaps more associated with an increase in volume whereas hypertrophy strictly may relate to an increase in muscle mass.

The subdivision of different types of hypertrophy is essentially based around left ventricular geometry. Where there is an overall increase in mass with a dominant increase in muscle thickness without an increase in cavity volume, the term concentric hypertrophy is used. Where there is an increase in mass predominantly due to an increase in volume, the term eccentric hypertrophy is used.

Huwez, Pringle, and Macfarlane [18] have introduced a new classification for hypertrophy on the basis of mass and volume. The different types and the criteria are given in Table 43.3. A salutary lesson from that study was that a patient with apparently normal mass and volume could demonstrate the ECG changes of LVH described above. In addition, both concentric and eccentric hypertrophy could produce similar ECG changes or none at all.

Notwithstanding the above, the remainder of this chapter is generally concerned with the classical description of vectorcardiographic changes accompanying hypertrophy.

43.3.1.2 Effects of Age, Sex and Race

It will be apparent from the previous Sect. 43.2 on normal ranges, that QRS voltage in some leads such as X and Z increases with age until early adulthood and then decreases again. Likewise, it has also been shown that, in many cases,

Table 43.3
Left ventricular geometry classification

Left ventricular		
Mass	Volume	Type
Normal	Normal	Normal
Normal	Increased	Isolated left ventricular volume overload
Increased	Normal	Concentric LVH
Increased	Increased	Eccentric LVH

sex differences particularly with respect to voltage can be demonstrated in vectorcardiographic parameters. Similar effects can be seen within different races.

All these observations suggest that the criteria for ventricular hypertrophy have to be based on a knowledge of all of these three variables.

43.3.1.3 Value of Vectorcardiography

It goes without saying that the echocardiogram gives a detailed picture of left ventricular geometry. On the other hand, it is the ECG that may demonstrate secondary ST-T changes, which are well known to be associated with a poor prognosis [19, 20]. If a reasonable specificity of 95% is desired, then the best ECG criteria should have a sensitivity around 50%. Data such as these vary from one study to another depending on the gold standard, which may be postmortem weights, on the one hand, or echocardiographic measurements, on the other. The reader should therefore be aware, from the outset, that the vectorcardiographic diagnosis of LVH is somewhat insensitive.

43.3.2 Atrial Enlargement

In the normal 12-lead vectorcardiogram, the P wave may exhibit two distinct components, which probably result from asynchronous depolarization of the left and right atrium. In this event, the bifid nature of the P wave is best seen in the inferior lead Y. In view of the fact that normal atrial depolarization commences in the right atrium before spreading to the left atrium, it follows that the first component is due to right atrial excitation and the second due to left atrial excitation.

43.3.2.1 Right Atrial Enlargement

One of the manifestations of right atrial enlargement is a P wave in the inferior lead Y with an amplitude > 0.3 mV. As this abnormality not infrequently occurs in respiratory disorders, it is sometimes called P pulmonale (Fig. 43.23). Occasionally, P_Y may be of normal amplitude but there may be a prominent $P_Z > 0.15$ mV on account of right atrial enlargement.

Chou and Helm [21] have pointed out that the P-wave changes in the inferior lead Y occasionally resemble P pulmonale when there are no clinical findings to support such a diagnosis. The cause of the abnormal P wave may be left atrial hypertrophy and the term "pseudo P pulmonale" is used to describe such a phenomenon. Clearly, this diagnosis should be made on the basis of the clinical findings taken in conjunction with the ECG appearances. The normal ranges

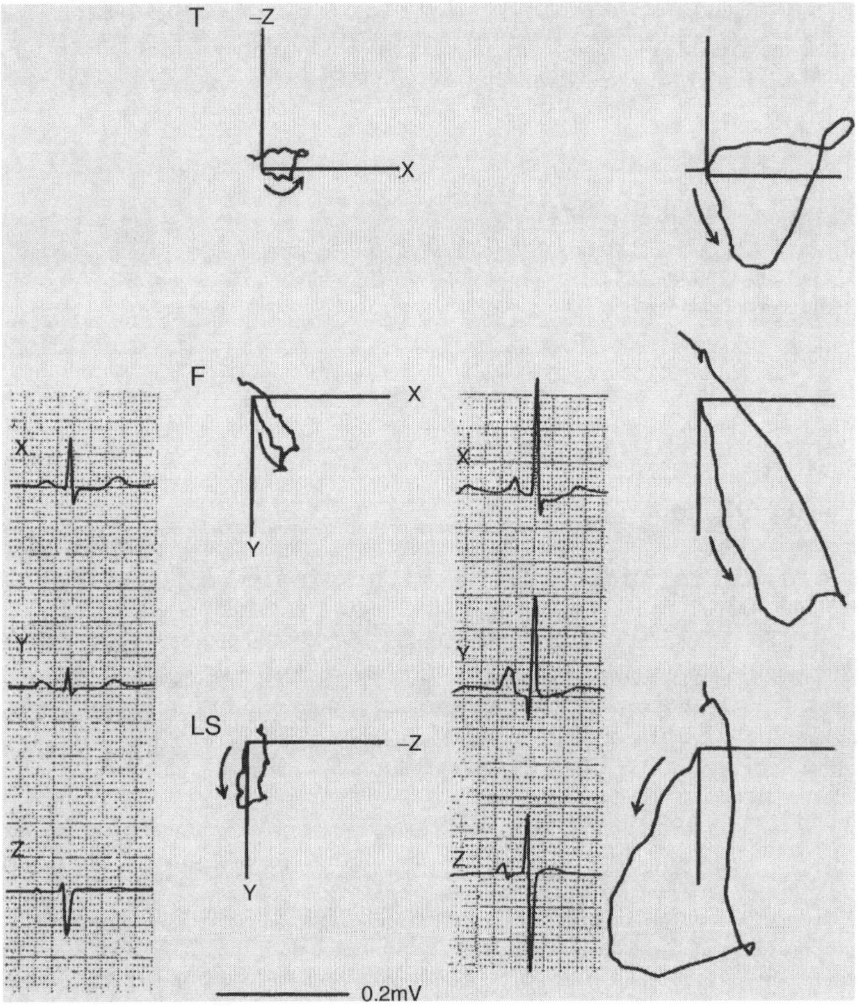

Figure 43.23
Normal P loops (*left*) and an example of combined atrial enlargement (*right*).

of the projections of the maximum P vector on to the three planes are so wide that little diagnostic advantage can be gained from a study of these parameters in respect of right atrial enlargement.

43.3.2.2 Left Atrial Enlargement

When left atrial enlargement occurs, several abnormalities may result. First, the P-wave duration may be increased beyond 120 ms. Second, bifid or M-shaped P waves of greater than normal duration may be found in the inferior lead Y and this pattern is often referred to as P mitrale because of its common association with mitral stenosis. From the vectorcardiographic point of view, enlargement of the left atrium may also cause an increase in the left atrial vectors, which in turn produce a rotation of the terminal portion of the P-wave vector posteriorly and to the left (Fig. 43.23). Thus, the terminal component of the scalar P wave in lead Z can be increased in negativity and in duration.

The maximum P vector amplitude may be increased but the normal range of P-vector orientation is so wide as to be of little value. On occasions, the maximum P-vector magnitude may exceed 0.3 mV when the individual scalar components are within normal. In this case, left atrial enlargement is the most likely cause.

43.3.2.3 Combined Atrial Enlargement

In general terms, a combination of the individual criteria for left and right atrial enlargement when present would be suggestive of combined atrial enlargement (➋ Fig. 43.23).

43.3.3 Left Ventricular Hypertrophy

43.3.3.1 Diagnostic Criteria

When LVH or left ventricular enlargement produces alterations in QRS morphology, these relate generally to an increase in QRS vector amplitude. On occasions, the maximum QRS vector will be rotated posteriorly and this is seen best in the transverse plane where the resultant effect is to produce an increase in the amplitude of S_Z. Consideration of the basic principles of vectorcardiography indicates that if a vectorcardiographic loop rotates posteriorly increasing S_Z, then R_X would decrease simultaneously. Thus, one of the commonly used vectorcardiographic criteria is based on $R_X + S_Z$. The data from the 1,555 patients in the Glasgow database suggest that the upper limit of $R_X + S_Z = 3.95$ mV for males over 40 years of age. It is, however, possible to express the upper limit of normal for males as a continuous age dependent equation as follows:

$$R_X + S_Z = [72 : 81 - 0.02074\,\text{age}\,(\text{months})]^2\ \mu\text{V}$$

Similar equations apply for women and for other races.

The typical vectorcardiographic appearances in LVH are shown in ➋ Fig. 43.24. In this case, there is increased magnitude of the QRS vector, which is more posteriorly oriented than the mean maximum QRS vector in normals in the transverse plane while the T loop is oppositely directed to the QRS loop. This is equivalent to the secondary ST-T pattern in the lateral leads.

It should be noted that the individual component amplitudes of the scalar leads may be normal while the resultant vector amplitude can be abnormal. For example, if the amplitude of R_X is 2.4 mV and, at the same instant, R_Y is 1.5 mV (although corresponding peak amplitudes might be a little higher), it follows that the amplitude of the maximum QRS vector would be

$$3.2\ \text{mV} = \sqrt{(2.4^2 + 1.5^2 + 1.5^2)}\,\text{mV}$$

Each of the scalar amplitudes is within the normal range for a 45-year-old male, but the vector magnitude is outside normal (see Appendix 5A).

Various types of QRS loop may be seen in the vectorcardiogram. These are best differentiated by appearances in the transverse plane. Type I may simply resemble a normal QRS loop but be of increased magnitude. In Type II, there is a rotation of the projection of the maximum QRS vector posteriorly beyond 310° in the transverse plane (➋ Fig. 43.25), possibly with a normal QRS voltage.

It is not common, however, to find an abnormal orientation of the QRS vector loop in LVH without voltage evidence in addition. In Type III, where LVH is very marked, there may be a figure-of-eight loop in the transverse plane with the distal part of the loop inscribed in a clockwise direction (➋ Fig. 43.26). In Type IV, there may be slightly increased QRS

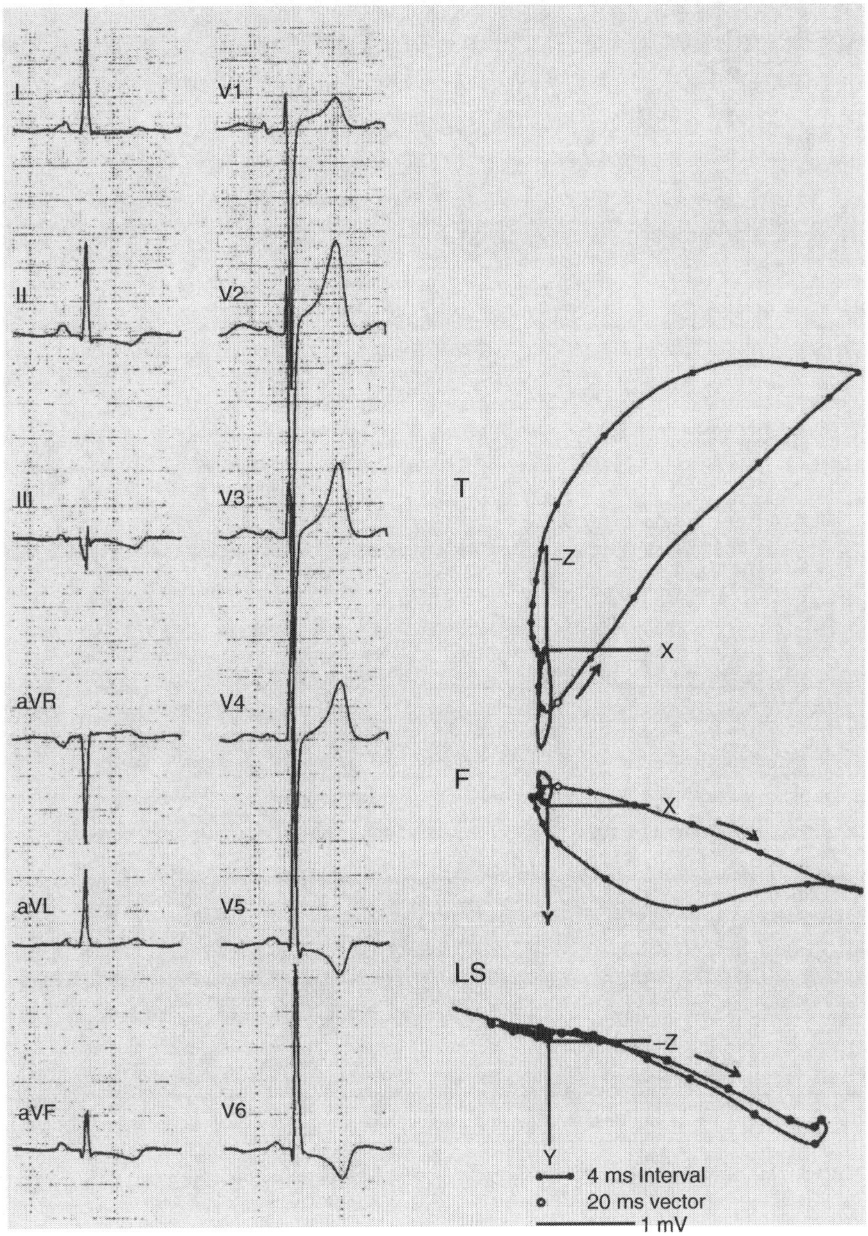

Figure 43.24
An example of left ventricular hypertrophy (LVH) Type I. Note the increased magnitude of the QRS loops best seen in the transverse plane where the T loop is essentially oppositely directed to the QRS loop.

duration, and an abnormally large QRS maximum vector, but the rate of inscription of the loop is slower than normal as manifested by the closeness of the dots. This pattern is sometimes called "incomplete left bundle branch block (LBBB)" that often accompanies LVH.

Reference has already been made to secondary STT changes (Fig. 43.24) sometimes called left ventricular strain or overload pattern. In a series of patients [22], it was found that this pattern of ST depression with asymmetric T wave

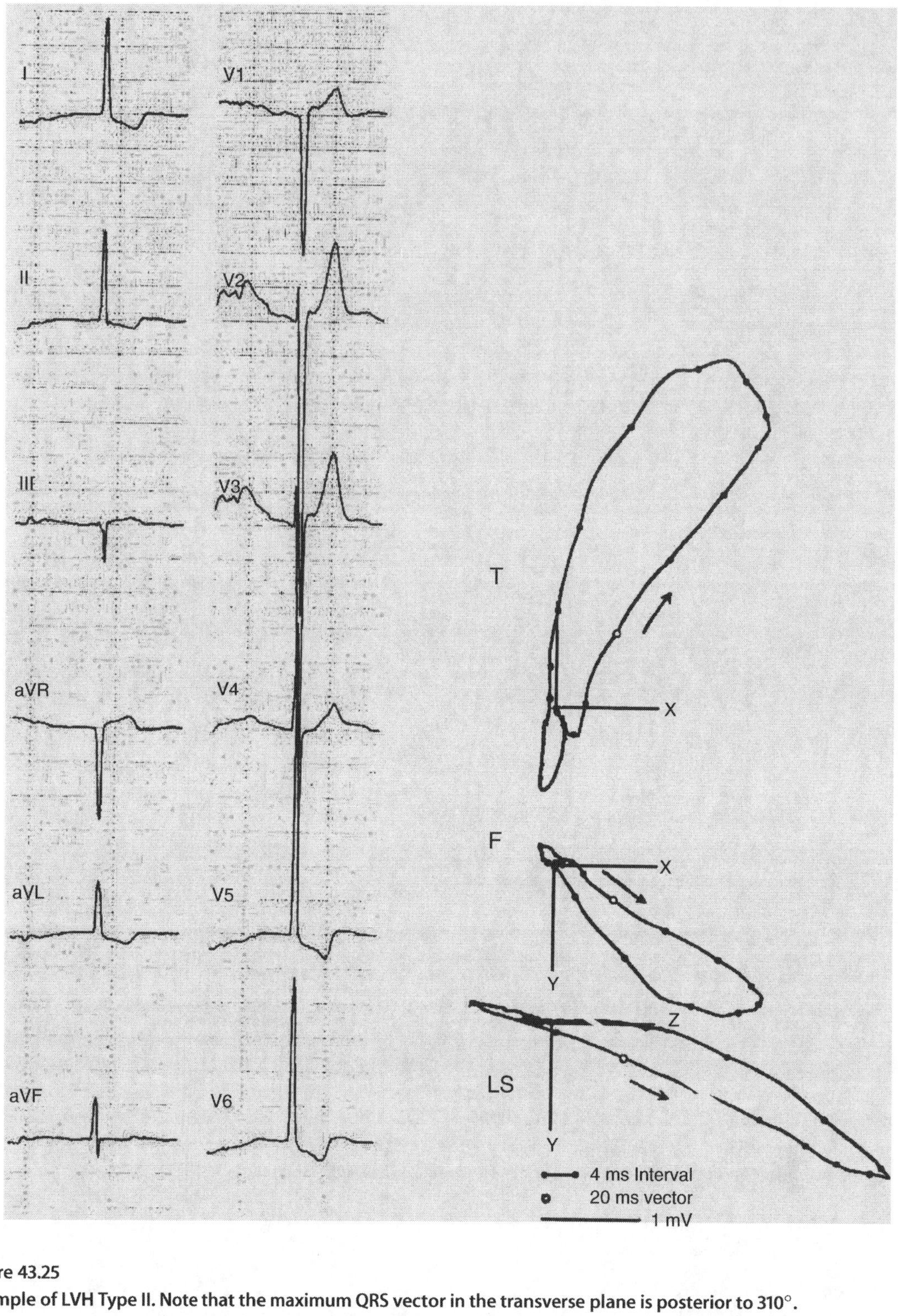

Figure 43.25
An example of LVH Type II. Note that the maximum QRS vector in the transverse plane is posterior to 310°.

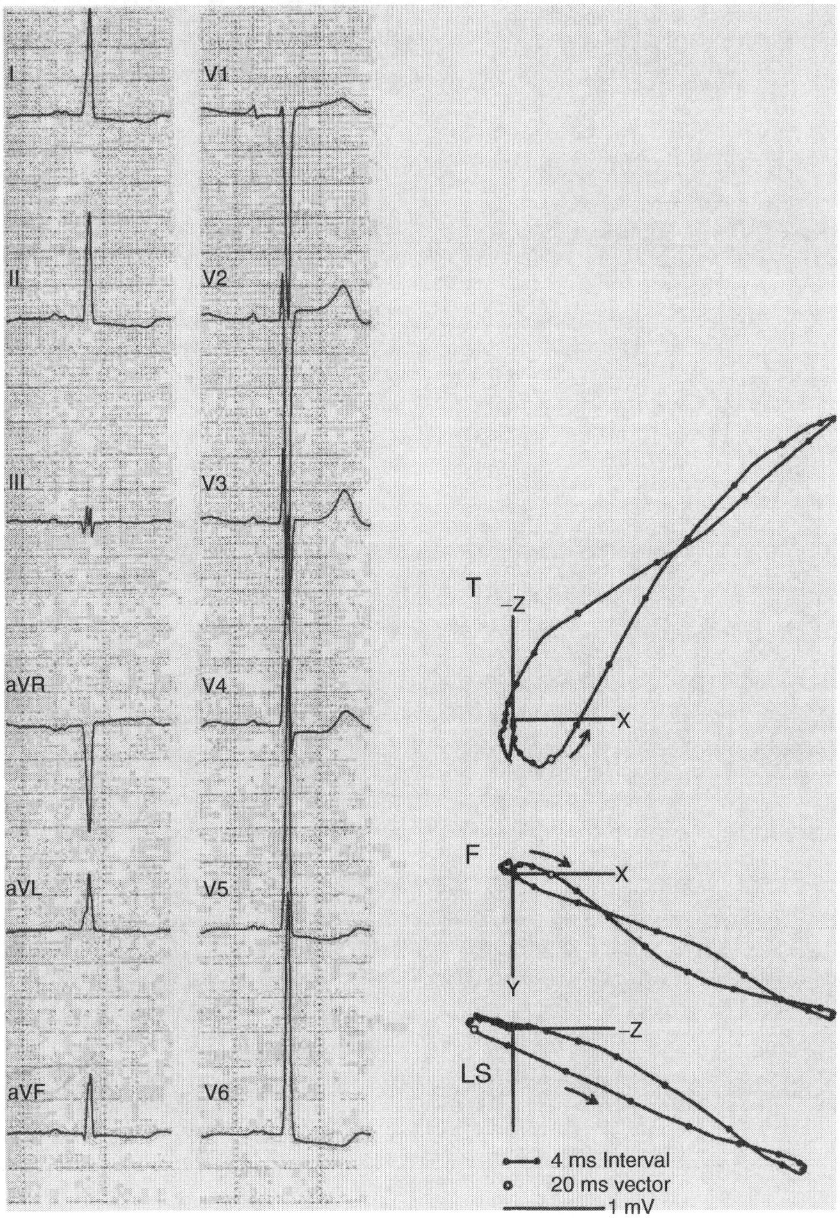

Figure 43.26
An example of LVH Type III. In this case, a figure-of-eight loop can be seen clearly in the transverse plane while the maximum QRS vector exceeds 4 mV.

inversion was 94% sensitive for LVH. Thus, even in the absence of high voltage, this ECG finding in the scalar lead X should be regarded as a pointer toward the diagnosis of LVH.

One further point can be made concerning the vectorcardiographic appearances in severe LVH. It can happen that the initial QRS vectors are directed posteriorly, that is, there is a Q wave in the anteroseptal lead Z or V_2. This makes the differential diagnosis of anteroseptal infarction from LVH difficult unless the clinical picture is relatively clear-cut. For example, Fig. 43.27 shows such a pattern in a 73-year-old male with hypertension and aortic stenosis and insufficiency.

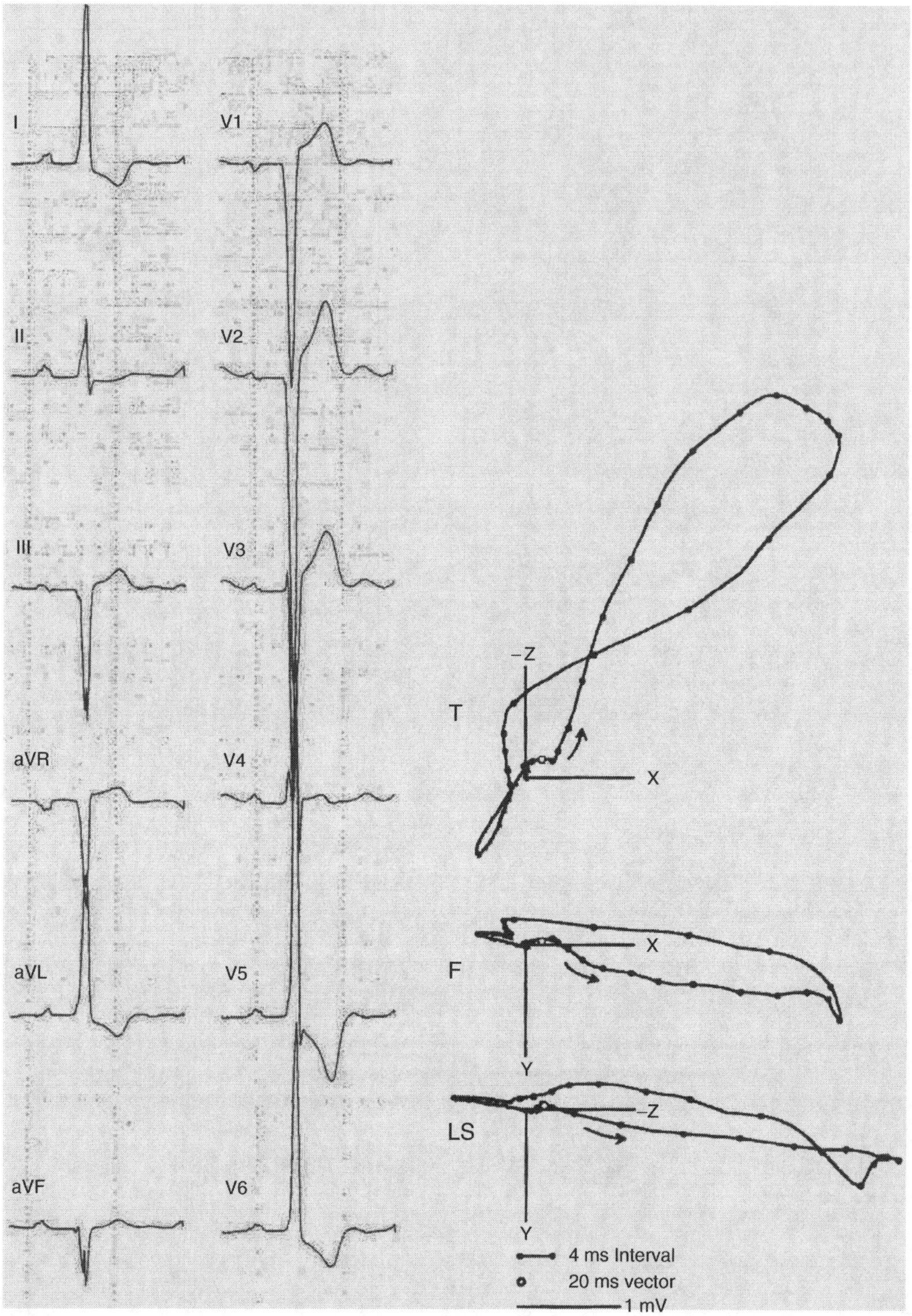

Figure 43.27

A 12-lead ECG and vectorcardiogram recorded from a 73-year-old male with hypertension and aortic stenosis and insufficiency. Note that the initial QRS vectors are directed posteriorly, that is, there is a Q wave in lead Z. Other typical features of LVH can also be seen.

The reasons for the presence of the Q wave are not fully understood although many hypotheses have been put forward. Possibly on account of there being three areas of initial activation in the left ventricle, the concept of the genesis of the Q wave may be revised. It could be postulated that the electrical activation of the area of the left ventricle adjacent to the posterior wall, which is among the first to be depolarized, predominates on account of left ventricular wall thickness, thereby leading to initial QRS forces oriented posteriorly producing a Q wave in the anteroseptal lead.

The other criteria that point to LVH include left axis deviation, which, in the vectorcardiogram, is a QRS axis superior to 0° in the frontal plane. Indeed, apart from the specific criteria of increased vector magnitude and posteriorly rotated maximum QRS vector orientation, other criteria mirror those for the conventional 12-lead ECG, for example, a delayed intrinsicoid deflection in lead X and increased P terminal force in lead Z.

43.3.3.2 Bundle Branch Block and LVH

The diagnosis of LVH from an ECG that shows bundle branch block is controversial. Some authors have claimed it is possible to make such a diagnosis while others have noted in various series that in the presence of LBBB, LVH is always found. Thus, no specific criteria for the diagnosis of LVH and LBBB are presented here. However, one or two points can be made.

As will be seen in ➋ Sect. 43.5, the vectorcardiographic appearances in LBBB are characteristic in having a narrow QRS loop in the transverse plane. If the scalar lead appears to have the LBBB pattern with a tall R_X but does not have the narrow bundle branch block loop, then LVH could be considered as a possible cause (incomplete LBBB/LVH pattern).

43.3.4 Right Ventricular Hypertrophy

Increased right ventricular excitation forces or increased right ventricular volume can produce varying ECG patterns depending on the time of the occurrence of the abnormal electrical activity compared to that of the left ventricle. Hypertrophy of the free wall of the right ventricle will produce abnormal anterior forces in early ventricular depolarization whereas basal hypertrophy will produce abnormal posterior forces late in ventricular depolarization. In the vectorcardiogram, there are essentially two presentations of right ventricular hypertrophy (RVH), namely, a prominent R wave in lead Z and a deep S wave in lead X. The presence of these abnormalities either singly or together produces four patterns of RVH or enlargement. These are as follows.

43.3.4.1 RVH: Type A

Type A RVH is manifested as an increase in the ratio of anterior/posterior forces in the transverse plane with counterclockwise inscription of the QRS loop (➋ Fig. 43.28). Often the abnormal QRS loop may have a T loop directed posteriorly corresponding to the secondary ST–T abnormalities sometimes seen in V_1 and V_2.

One criterion of value in Type A is the projection of the maximum QRS vector onto the transverse plane >30°. Occasionally, this may be present when voltage and ratio measurements in the anteroseptal leads are normal.

43.3.4.2 RVH: Type B

One of the major advantages of the vectorcardiogram is preservation of the timing relationships between the different scalar leads. Thus, while two different scalar patterns may have similar appearances, the vectorcardiogram can be normal in one and abnormal in another. This is often apparent in Type B RVH. In this case, the QRS loop in the transverse plane

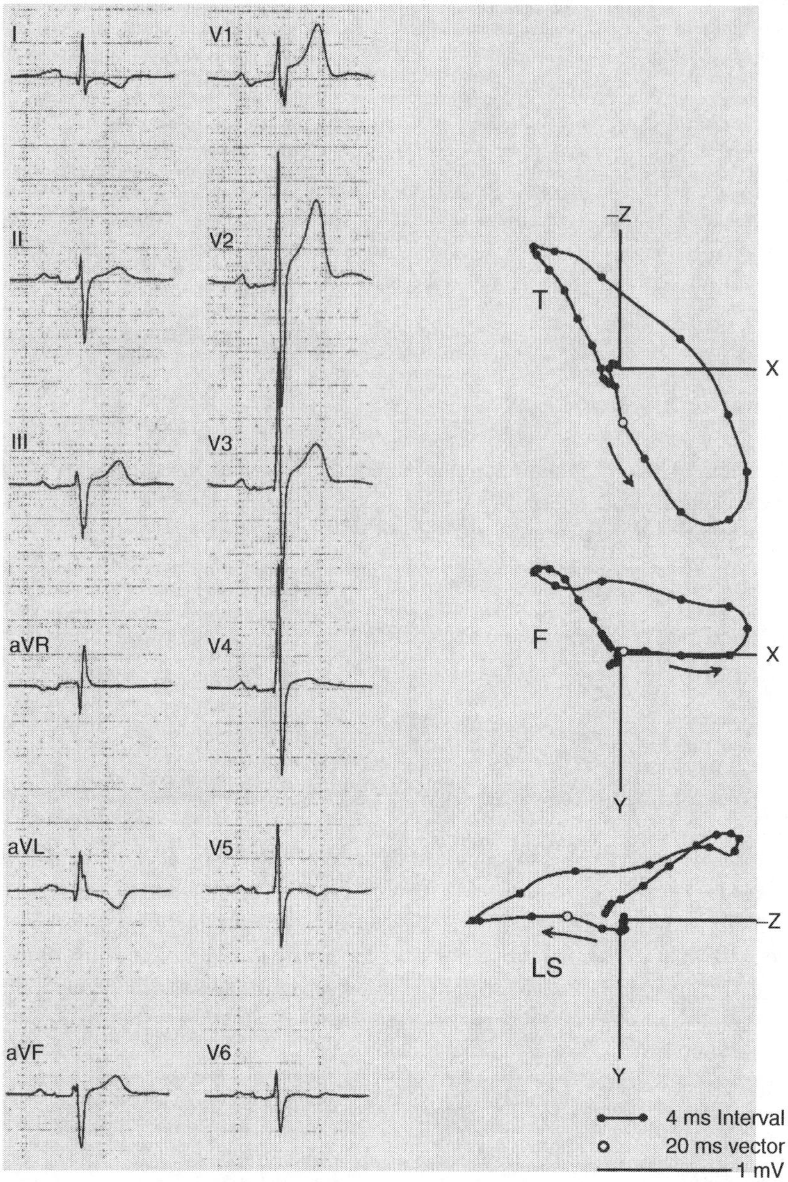

Figure 43.28
An example of right ventricular hypertrophy (RVH) Type A where there is a marked increase in anteriorly directed forces as seen in the transverse plane.

initially has normal counterclockwise inscription, but the second part of the loop is deviated anteriorly so that the net effect is to have a loop with clockwise or figure-of-eight inscription in the transverse plane (Fig. 43.29).

43.3.4.3 RVH: Type C

The third type of RVH is manifested as an abnormal transverse plane vectorcardiographic loop that is normally counterclockwise inscribed with the maximum QRS vector being oriented posteriorly and to the right (Fig. 43.30).

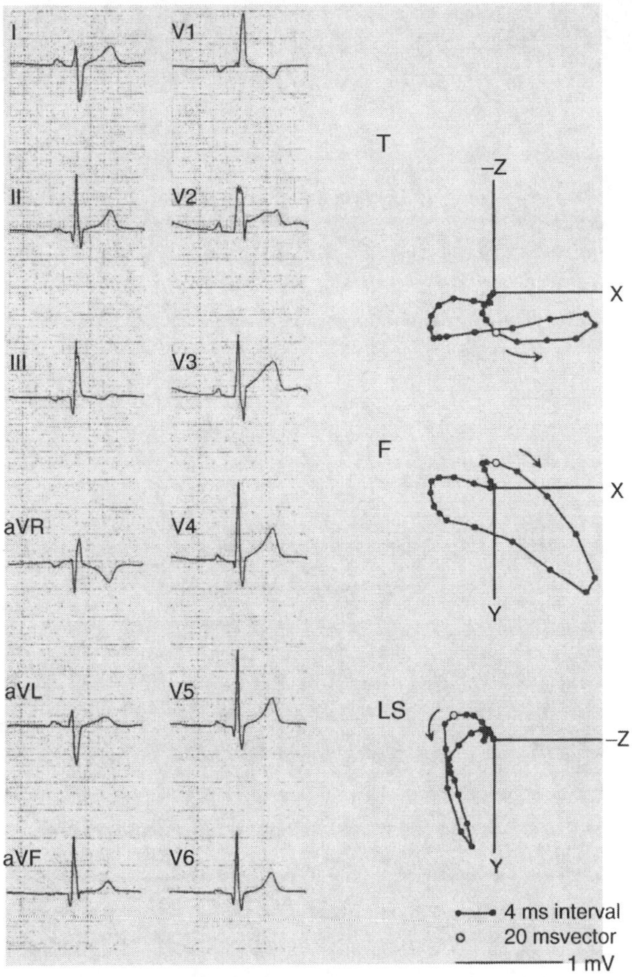

Figure 43.29
An example of RVH Type B where there is a figure-of-eight loop in the transverse plane in which all forces are seen to be anteriorly oriented. Note the late rightward-directed forces corresponding to the deep S wave in lead I.

This pattern is most often found in patients with chronic respiratory disease and therefore may be accompanied by P pulmonale. However, some patients with mitral stenosis also exhibit such findings. It is thought that the posterior rightward displacement of the QRS vector is due to hypertrophy of the basal portion of the right ventricle. Secondary ST-T changes can also be found in this pattern. The abnormal rightward forces also translate into right axis deviation in the frontal plane.

43.3.4.4 RVH: Type D

In the more severe forms of RVH, as may occur in certain forms of congenital heart disease the main QRS vector may be directed abnormally not only to the anterior but also to the right with clockwise inscription of QRS in the transverse plane (Fig. 43.31).

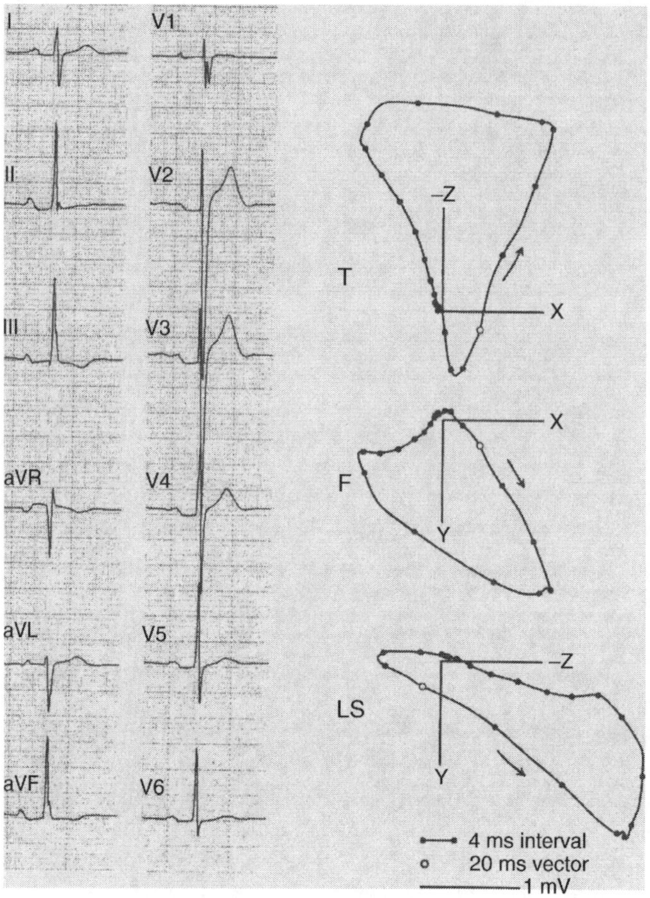

Figure 43.30
An example of RVH Type C. In this case, there is a large posterior and rightward-directed force in the later part of QRS.

43.3.5 Combined Ventricular Hypertrophy

It has been claimed that in most cases of LVH there is an accompanying RVH [23]. The vectorcardiographic diagnosis of biventricular hypertrophy is mainly based on finding features of both LVH and RVH separately. For example, if the transverse QRS loop is directed posteriorly and has an increased magnitude with the T loop oppositely directed, as in secondary changes of LVH, and the frontal QRS vector is oriented around 90°, then combined ventricular hypertrophy should be considered.

43.3.6 Pediatric Vectorcardiography

The various forms of congenital heart disease can produce a variety of vectorcardiographic patterns that can be extremely difficult to interpret. A few patterns are pathognomonic of rare forms of congenital heart disease but in any event, the report must remain an ECG interpretation. In other words, no attempt should be made to infer the anatomy of the congenital lesion in the majority of cases. Brohet [24] has reviewed the advantages of vectorcardiography in congenital heart disease.

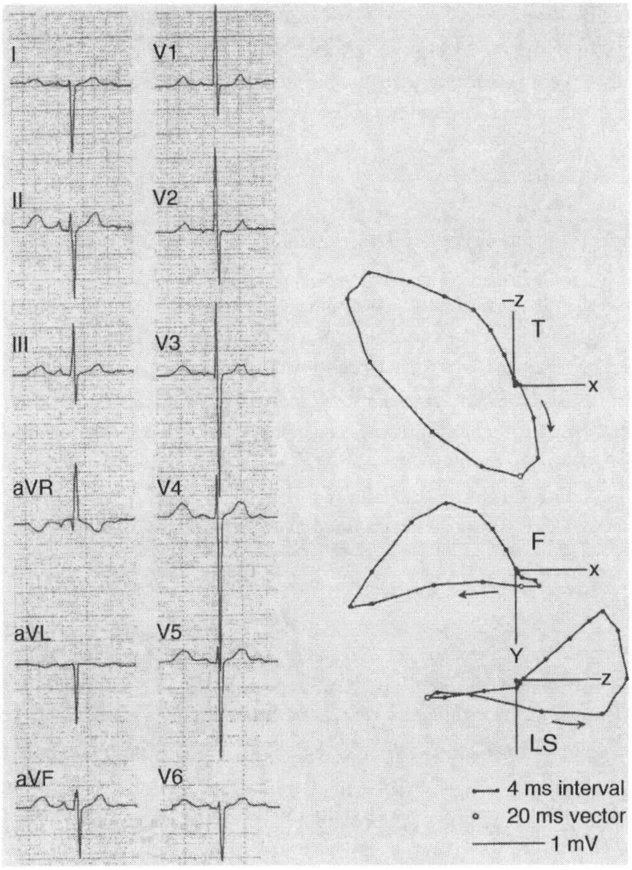

Figure 43.31
An example of RVH Type D where the QRS loop in the transverse plane is clearly abnormal, being inscribed in a clockwise direction and lying almost entirely to the right. Marked right axis deviation can also be seen in the frontal plane loop.

43.4 Myocardial Infarction

43.4.1 Introduction

43.4.1.1 Theoretical Considerations

A myocardial infarction is death of myocardial tissue as a result of insufficient blood supply, that is, due to a stenosis or occlusion of a coronary artery. An infarcted area is electrically inert and distorts the normal spread of excitation. The net effect is that the electrical forces, which are influenced by the "dead zone" or infarct vector, are directed away from the area of myocardial infarction. This is in contrast with hypertrophy that produces electrical forces directed toward the area of increased ventricular mass.

43.4.1.2 Anatomical Definitions

There is a lack of unanimity in describing myocardial infarction. In 1978, the American College of Cardiology Conference on Optimal Electrocardiography described myocardial infarction in terms of Q waves seen in the various leads of the

12-lead ECG [25]. This has resulted in terms such as septal, anteroseptal, and anterior infarction being used. However, there is still no consensus regarding the localization of an infarction among electrocardiographers.

Radiological or echocardiographic techniques can also be used to detect and localize a myocardial infarction. In coronary angiography, it is not the infarcted area per se that is detected but the stenosis or occlusion of the coronary artery that caused the myocardial infarction. An infarcted area causes abnormalities in the ventricular contraction pattern, which can usually be detected in a left ventriculogram or in an echocardiogram. The techniques described above show different aspects of the same disease. Differences in the position and orientation of the heart and a variation in the anatomy of the coronary arteries have varying influence on these techniques. Therefore, the correlations between electrocardiographic changes, abnormalities of ventricular contraction, and the degree of stenosis in relevant coronary arteries are not excellent. This should be borne in mind when comparing different techniques for the diagnosis of myocardial infarction.

Notwithstanding any of the above, in very general terms, occlusion of a left anterior descending coronary artery or a left main stem artery is likely to produce an infarct predominantly in the anterior (anteroseptal/anterosuperior) wall of the heart. An occlusion in the right coronary artery will generally produce an inferior myocardial infarction while occlusion of the left circumflex produces a lateral or posterolateral/inferior myocardial infarction.

43.4.2 The 12-Lead Vectorcardiogram in Myocardial Infarction

43.4.2.1 Anterior Infarction

In keeping with the concept of the dead zone or infarct vector, an anterior myocardial infarction will result in slightly increased posteriorly directed electromotive forces and a reduction, if not an absence, of anteriorly directed forces in the early part of ventricular depolarization. In 12-lead ECG terminology, this corresponds to a QS complex in V_2. The corresponding vectorcardiographic appearances are shown in ➲ Fig. 43.32. Here it can be seen that there is no electrical activity in the left anterior quadrant of the transverse plane vectorcardiogram. This type of clear-cut myocardial infarction, from an electrocardiographic point of view, is generally well delineated on the 12-lead ECG. Of more interest is the situation where the 12-lead ECG may be somewhat equivocal with low R waves in the anterior leads but the vectorcardiogram shows other features that are suggestive of myocardial infarction. In the transverse plane, common vectorcardiographic criteria for anterior myocardial infarction are listed in ➲ Table 43.4.

➲ Figure 43.33 gives an example of a QRS loop where there are initial anteriorly directed forces but the 30 ms QRS vector is posterior to 300°. ➲ Figure 43.34 shows a different form of vectorcardiographic change where there is an initial counterclockwise inscription leading to a bite in the vectorcardiographic loop (see ➲ Sect. 43.4.2.8). ➲ Figure 43.35 illustrates a case of anterior myocardial infarction where the area of the QRS loop in the left anterior quadrant <1%.

43.4.2.2 Anterior Myocardial Infarction Versus LVH

In cases of severe LVH, the vectorcardiographic loop may also resemble anterior myocardial infarction. In many cases, the difference between the two can clearly be separated on non-electrocardiographic considerations, clearly including the clinical history. If the QRS loop and T loop are oppositely directed in the transverse plane with lack of QRS electrical activity anteriorly, the higher probability is that appearances are due to LVH (➲ Fig. 43.27).

Of course, both abnormalities can be present simultaneously in an individual and this is more likely to be the case if the QRS–T angle is approximately 90°. A T vector directed posteriorly to the right suggests an ischemic component to the abnormality as isolated LVH would rarely produce a T vector so oriented.

43.4.2.3 Inferior Myocardial Infarction

The dead zone or infarct vector concept suggests that in inferior myocardial infarction there is an increase of electrical force superiorly. In turn, this produces a loss of inferiorly directed forces resulting in an initial superiorly directed vectorcardiographic loop in the frontal plane. However, as this is not altogether uncommon in normal individuals, the point

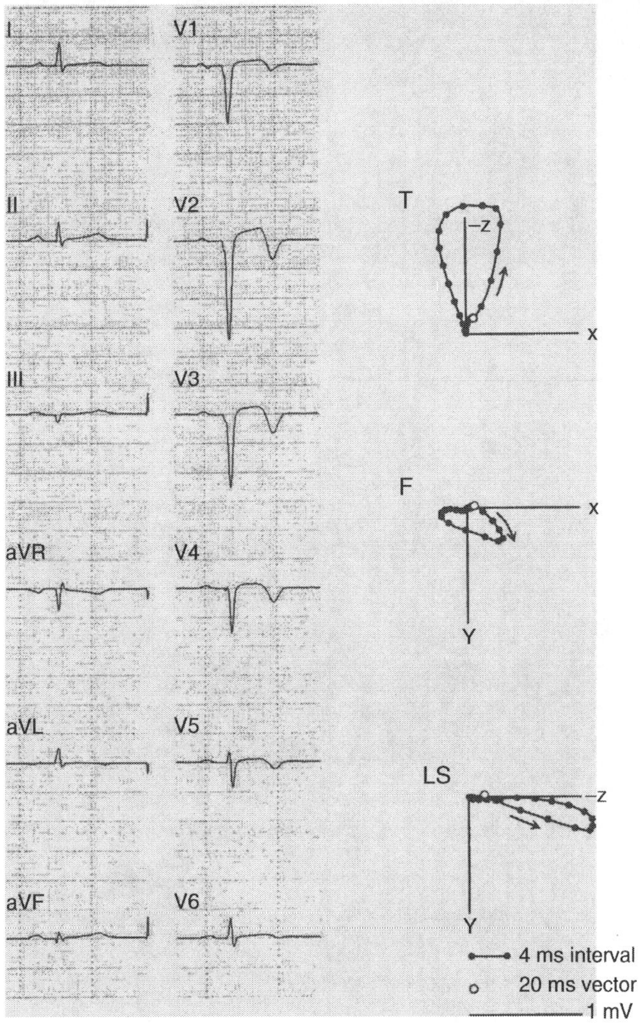

Figure 43.32
The classical features of anterior myocardial infarction with a complete absence of anteriorly directed QRS vectors. Note that the 30 ms vector is also posterior to 300° in the transverse plane.

Table 43.4
Criteria for anterior myocardial infarction

Any of the following in the transverse QRS loop
Direction of 30 ms vector 225–300°
Loop area in left anterior quadrant <1% of the total loop area
An early bite, that is, a clockwise inscription of the loop, with an amplitude >0.05 mV (see Sect. 43.4.2.8)

of importance is the length of time for which the loop persists in the superior quadrants and the ratio of superiorly to inferiorly directed forces. Common vectorcardiographic criteria for inferior myocardial infarction are presented in Table 43.5.

In addition to the superiorly directed forces having a duration >20 ms, the concept of X-intercept needs to be introduced. This is illustrated in Fig. 43.36. The X-intercept is the point at which a clockwise-inscribed frontal plane loop

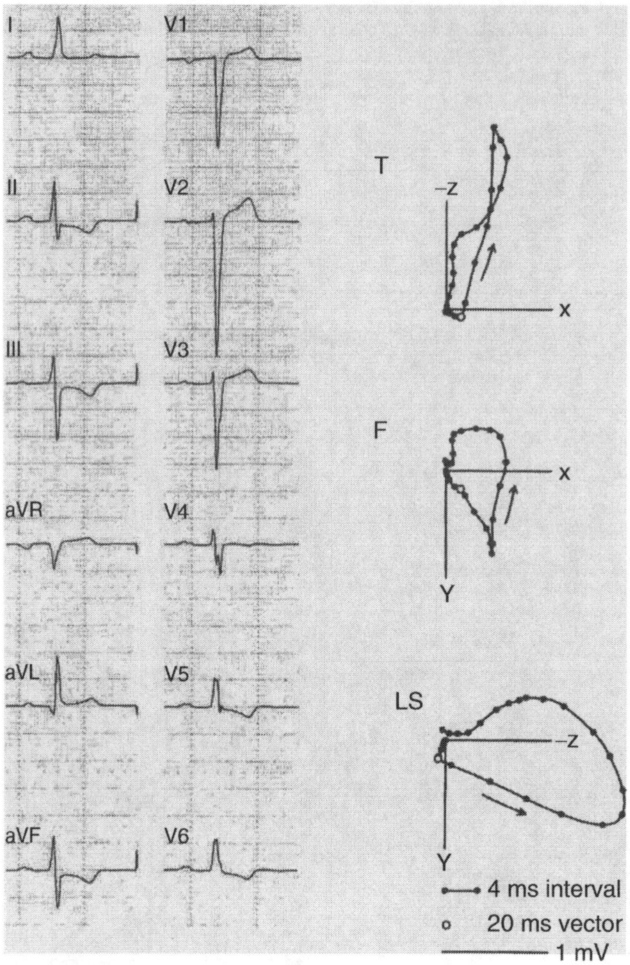

Figure 43.33

An example of anterior myocardial infarction where the initial QRS forces are anteriorly directed for a little over 20 ms but where the 30 ms vector is just posterior to 300°.

first crosses the X-axis. In a sense, it is clear that the longer the QRS loop remains superior to the X-axis, the greater the probability that the X-intercept will exceed 0.3 mV. Thus, in some ways, there is a correlation between this parameter and a superior duration >20 ms. Figure 43.37 gives an example of a vectorcardiographic frontal plane loop where all the criteria are met.

Analogous to the situation of low R waves in anterior leads consistent with myocardial infarction, it is possible that there can be very low amplitude R waves in the inferior lead Y in the presence of inferior infarction, that is, there is a very short inferiorly directed initial activation in the frontal plane. If the initial vector of activation is directed inferiorly and rightward followed by clockwise inscription, then the criteria of X-intercept >0.3 mV and a superior/inferior amplitude ratio >0.15 can often be met.

43.4.2.4 Inferior Myocardial Infarction Versus Left Anterior Fascicular Block

Often a common feature of the inferior myocardial infarction and left anterior fascicular block is the superior orientation of the frontal plane QRS vector loop. However, the left anterior fascicular block is always accompanied by a

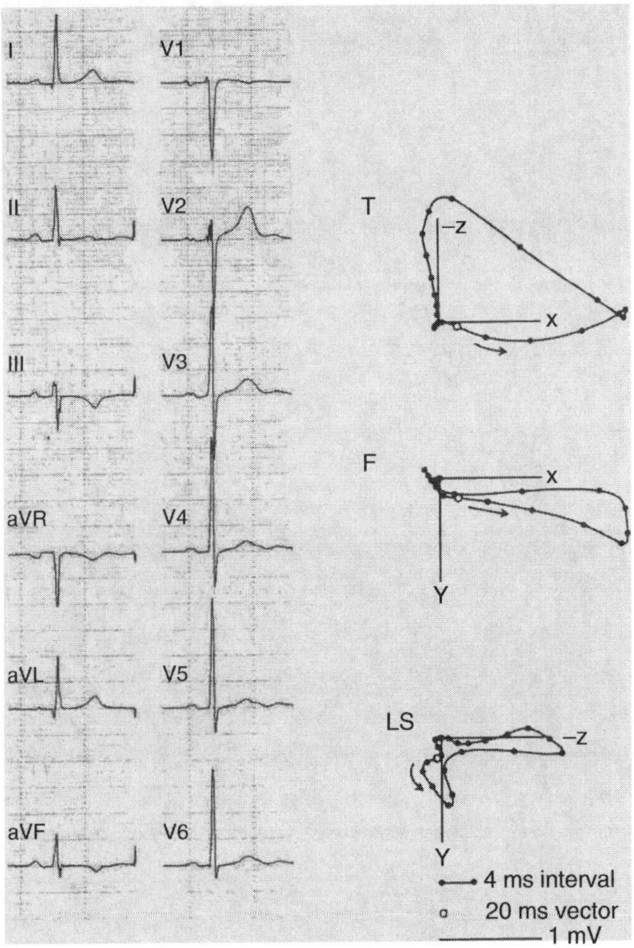

Figure 43.34
Anterior myocardial infarction proven by diagnostic cardiac catheterization in a 62-year-old male. Note the early bite in the QRS loop in the transverse plane.

counterclockwise inscription of the QRS loop whereas the inferior myocardial infarction invariably has a clockwise inscription of the QRS loop. A comparative example is shown in Fig. 43.38. It is also the case that in left anterior fascicular block, the QRS axis is generally more superiorly directed than in inferior myocardial infarction.

43.4.2.5 Posterior Myocardial Infarction

The theory of the dead zone or infarct vector applied to an infarction of the posterior wall of the heart indicates that there will be an increase in anteriorly directed electrical forces. The diagnosis of posterior myocardial infarction is again difficult, purely from an electrocardiographic standpoint. There may, of course, be other clinical factors that suggest a myocardial infarction which, taken together with the relevant vectorcardiographic changes, could point to infarction of the posterior wall.

The increase in anterior forces is reflected in an increased duration of the anteriorly directed forces in the transverse plane. In addition, the amplitude ratio of the anteriorly/posteriorly directed forces exceeds 1. Finally, the area

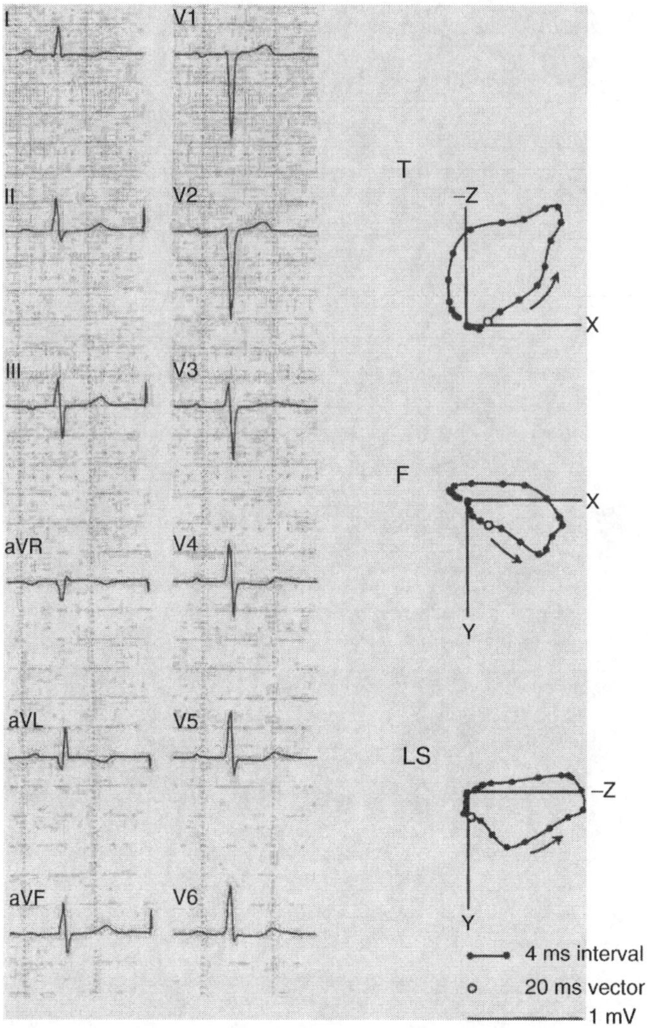

Figure 43.35
Anterior myocardial infarction proven by diagnostic cardiac catheterization in a 58-year-old male. Note the QRS loop area in the left anterior quadrant is almost 0% of the total QRS loop area.

Table 43.5
Criteria for inferior myocardial infarction

All of the following in the frontal QRS loop
An initial superior inscription >20 ms
A superior amplitude >0.1 mV
A superior/inferior amplitude ratio >0.15
An *X*-axis intercept >0.3 mV

in the left anterior quadrant of the transverse plane exceeds 50% of the total QRS loop area (Fig. 43.39). Common vectorcardiographic criteria for posterior myocardial infarction are presented in Table 43.6.

The other typical feature of posterior myocardial infarction is an increase in T vector amplitude. This may well correspond to a T vector oriented in the direction of 70–80° in the transverse plane.

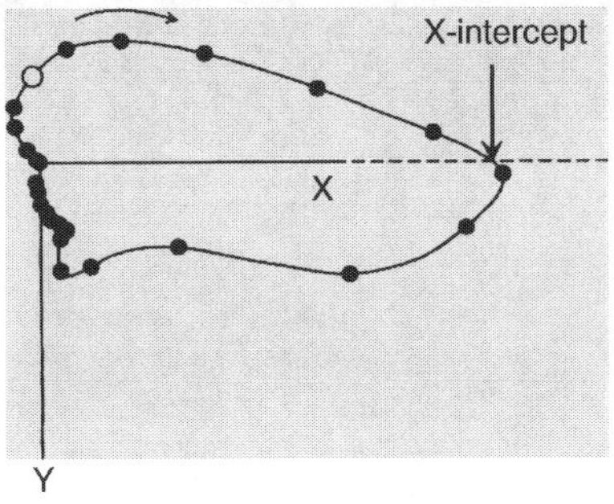

Figure 43.36
Frontal plane QRS loop showing the *X*-intercept.

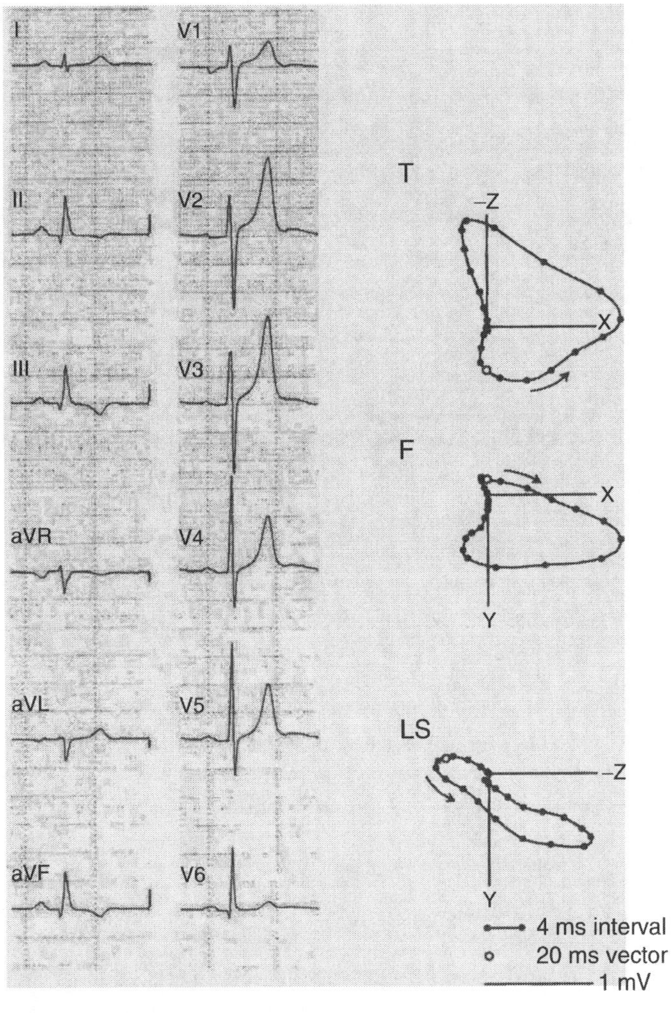

Figure 43.37
Vectorcardiogram of inferior myocardial infarction meeting the criteria of Table 43.5.

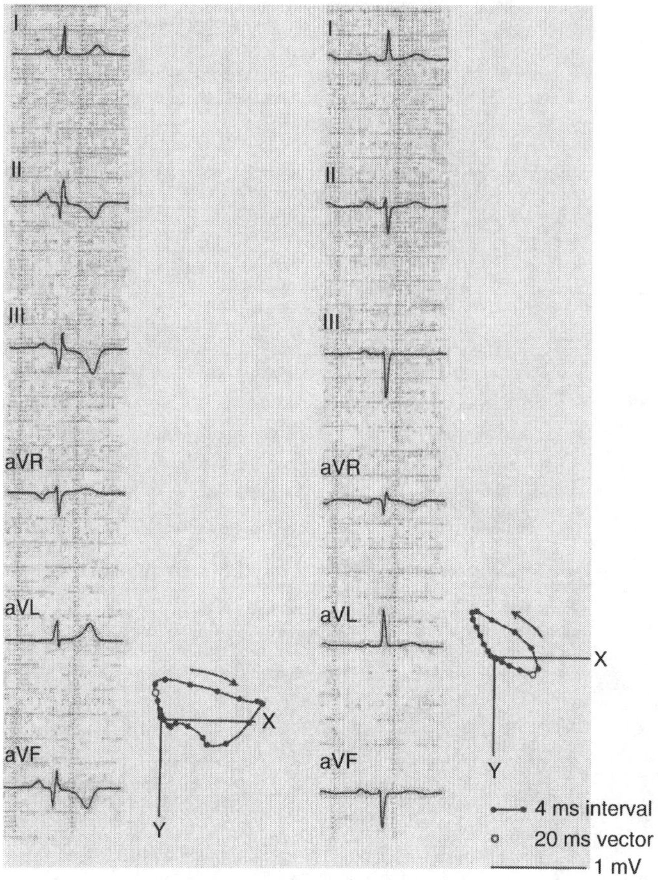

Figure 43.38
Left **panel shows inferior myocardial infarction;** ***right*** **panel shows left anterior fascicular block.**

43.4.2.6 Posterior Myocardial Infarction Versus RVH

The features of RVH have been discussed in Sect. 43.3. The Type A vectorcardiographic loop is not too dissimilar from that of posterior myocardial infarction and therefore the differentiation between the two is difficult. However, the criteria of anteriorly directed forces exceeding 50 ms is more likely to be met in a case of posterior myocardial infarction than in RVH. Also, the T vector loop is more likely to be anteriorly directed in posterior infarction than in RVH.

Another factor that influences the situation is that an inferior or even lateral myocardial infarction, which extends toward the posterior wall of the heart, may produce, in addition, an increase in anteriorly directed forces. In that situation, a report of "increased anterior forces probably reflecting inferior/posterior myocardial infarction" is more likely to be correct than one which suggests that there is RVH in addition to inferior infarction, for example.

43.4.2.7 Anterolateral Myocardial Infarction

The effect of a lateral wall myocardial infarction is to produce an initial electrical force which is directed in the range 90–270° in the transverse plane. This is the principal requirement for the diagnosis of anterolateral myocardial infarction, an example of which is shown in Fig. 43.40. Isolated anterolateral infarction is uncommon and more often than not, the changes are associated with an anterior rather than a purely anterolateral myocardial infarction.

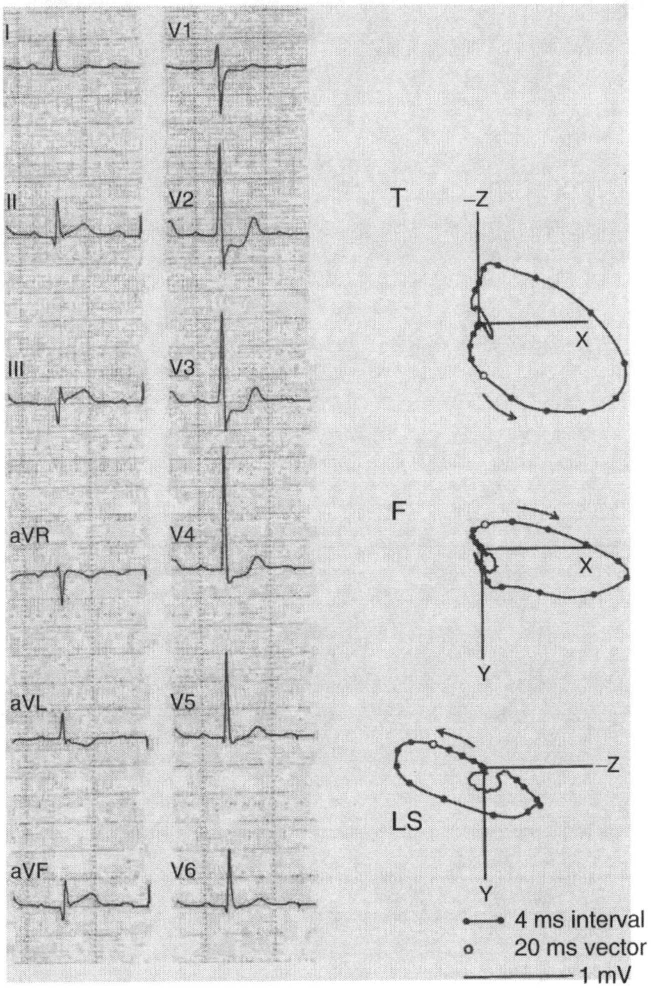

Figure 43.39
Inferior myocardial infarction with posterior wall involvement. Note that the area of the QRS loop in the left anterior quadrant exceeds 50% of the total QRS loop area. The *X*-intercept also greatly exceeds 0.3 mV.

Table 43.6
Criteria for posterior myocardial infarction

The QRS loop in the transverse plane shows:
an initial anteriorly directed loop >50 ms
an anterior/posterior amplitude ratio >1
a loop area in left anterior quadrant >50% of the total loop area

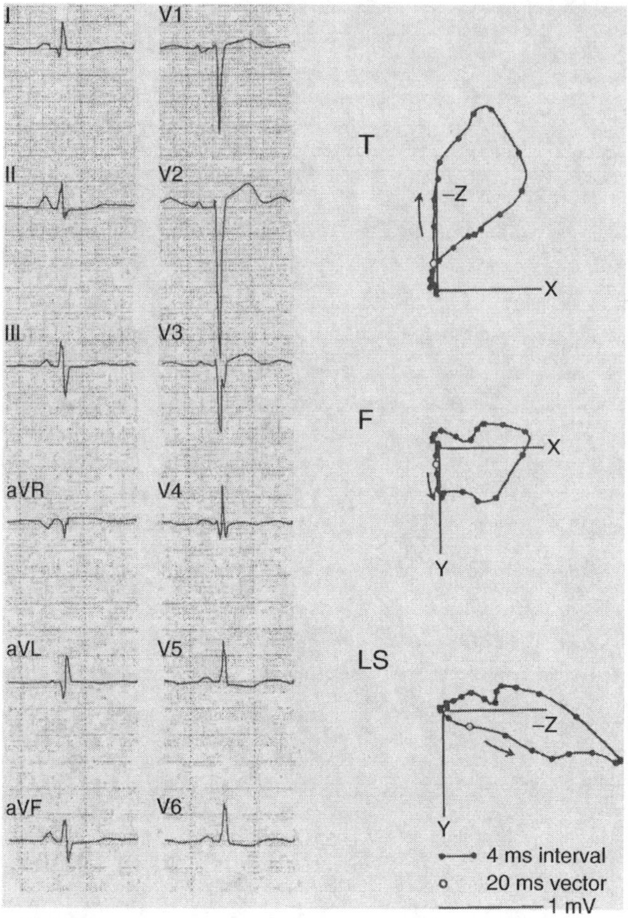

Figure 43.40
Vectorcardiographic example of anterolateral myocardial infarction. Note that in the transverse plane, the initial electrical forces are directed posteriorly to the right and that the inscription of the main body of this loop is clockwise. It should also be noted that there are no Q waves in the precordial leads of the 12-lead ECG.

43.4.2.8 Bites

The concept of a vectorcardiographic bite was mentioned earlier in this chapter. This is best explained by reference to Fig. 43.41, where a comparison is made between a normal transverse plane vectorcardiogram and one exhibiting a bite, that is, an indentation of the QRS loop, which often amounts to a reversal of the direction of inscription of the loop. For example, the normal QRS loop in the transverse plane has a constant counterclockwise inscription whereas the loop with a bite has an inscription that is initially counterclockwise, then changes to clockwise before returning to counterclockwise inscription. It is suggested that the amount of deviation of the bite from the normal loop gives an indication of the size of infarcted area. Considerable work was done in this area by Selvester and Sanmarco [26] with modeling studies and, indeed, a nomogram was produced, which linked the duration of the bite with the magnitude of the difference between the normal vectorcardiographic loop and the abnormal vectorcardiographic loop (Fig. 43.42). The nomogram allows calculation of an estimate of the percentage of myocardium that is damaged and in turn, an estimate of the ejection fraction of the left ventricle. Selvester et al. [27] also provided data on the occurrence of bites in diabetic patients. Edenbrandt et al. recently [28] developed a computer-assisted method for measuring the size of vectorcardiographic bites.

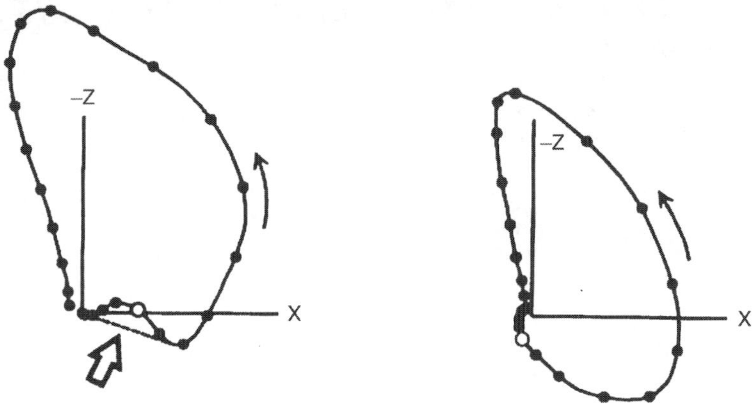

Figure 43.41
Vectorcardiographic loops in the transverse plane. The left loop shows an early bite whereas the right loop has a normal inscription.

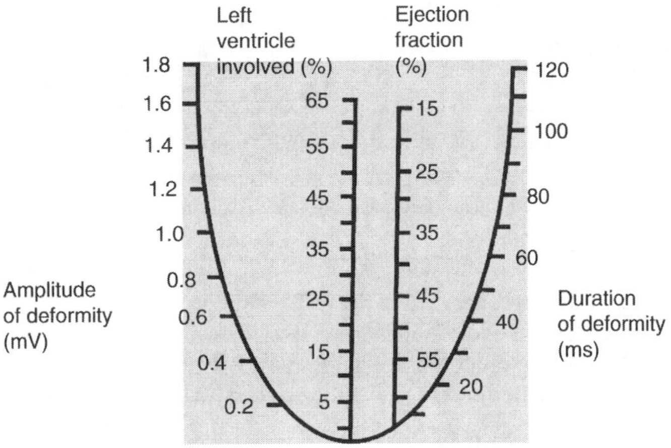

Figure 43.42
Nomogram for predicting infarct size from deformities in the Frank vectorcardiogram. The amplitude and duration of the deformity are transferred to the left side and right side of the nomogram, respectively. A line is then constructed through these two points and the points where it crosses the lines in the center are used to determine the angiographic percentage of left ventricle involved and the ejection fraction.

43.5 Conduction Defects

43.5.1 Introduction

It is certainly the case that conduction defects may be thought rather straightforward to diagnose from the 12-lead ECG. On the other hand, there are borderline situations where there can be doubt as to whether a QRS duration is prolonged. Indeed, it has gone almost unrecognized for many years that the normal mean QRS duration for males is almost 8 ms longer than for females in the 18–29 year age group (96.4 versus 87.7 ms) although as age increases the difference tends to decrease (92.7 versus 87.1 ms at 50 years and over) [29]. Notwithstanding, there are few if any diagnostic criteria that take cognizance of this fact.

One of the advantages of vectorcardiography is that conduction defects can be seen on the vectorcardiographic loop display as a slowing of the inscription of the loop, that is, the loop markers appear closer together. There are other features sometimes pathognomonic of particular defects, as will be seen later in this chapter.

43.5.1.1 Conduction System

The heart muscle consists of three types of tissue – automatic, specialized conducting, and contractile tissue. The process of depolarization is common to all three tissues but only the automatic and specialized conducting tissues have the ability to depolarize spontaneously.

In the human heart (➲ Fig. 43.43), the automatic tissue is concentrated mainly at the sinoatrial (SA) node. In the normal sequence of electrical events, an electrical impulse arising in the SA node, travels through the right atrium to the atrioventricular (AV) node and thereafter spreads into the ventricles via the specialized conducting tissue in the bundle of His.

Invasive recording of the signals in the atria and in particular in the area of the AV node has led to a much greater understanding of certain types of conduction defects but in particular, this approach has been of most value in the assessment of cardiac arrhythmias. On the other hand, this chapter is concerned more with abnormalities of conduction in the bundle of His and its various branches.

As seen in ➲ Fig. 43.43, the bundle of His divides at the base of the septum into the right bundle branch and the left bundle branch. The latter has been shown to divide into a variety of different forms, common to all of which are the left anterior and left posterior fascicles. Demoulin and Kulbertus [30] showed many years ago that there was often a third fascicle, which they called "the centroseptal fascicle."

Under normal circumstances, ventricular depolarization occurs first in the left ventricle, particularly on the left side of the septum, and then spreads to the free wall of the left ventricle. At the same time, shortly after left ventricular activation commences, the right ventricular excitation also starts. However, it is important to appreciate that the normal sequence of depolarization in the ventricles is from the left to the right side of the septum.

For completeness, it should be said that excitation in general terms spreads from the endocardium to the epicardium and from the apex to the base. In addition, ventricular repolarization takes place from the epicardium to the endocardium, giving rise to a normal upright T wave in the majority of precordial leads as well as most frontal plane leads with the exception of aVR. In the 12-lead vectorcardiogram, the T wave is normally upright in leads X, Y, and Z.

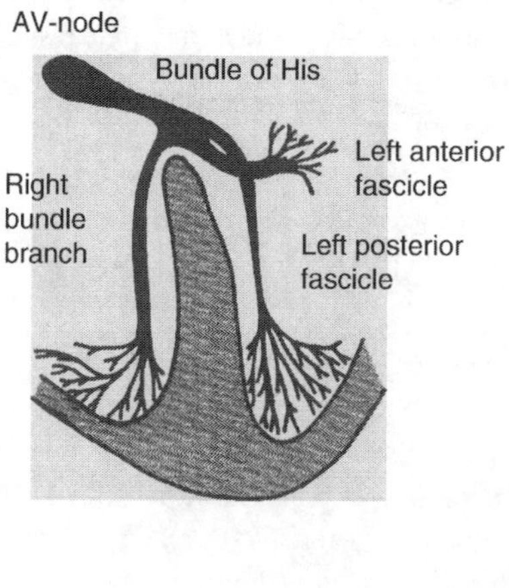

Figure 43.43
Conduction system of the human heart.

43.5.2 Bundle Branch Block

43.5.2.1 Left Bundle Branch Block

When the normal conduction to the left ventricle is prevented by a block in the left bundle branch, the excitation wavefront has to find an alternative pathway via normal cardiac muscle. A major consequence of this is naturally an increase in the time taken to depolarize the ventricles and hence, there is an increase in the QRS duration.

Because of the block in the left bundle branch above its division into the various fascicles, septal activation begins on the right side and progresses from the right to left. Thus, in the 12-lead vectorcardiogram in left bundle branch block (LBBB), there is frequently no primary R wave in lead Z and almost always, an absence of a Q wave in the anterolateral lead X. Right ventricular depolarization is virtually complete before the excitation waves later spread unopposed round both anterior and posterior walls of the left ventricle to meet at the lateral wall [31].

From a vectorcardiographic point of view, the initial QRS forces are therefore directed posteriorly and to the left. Commonly, the maximum QRS vector is also similarly directed posteriorly to the left and inferiorly. The vectorcardiographic appearances of LBBB are as shown in ➋ Fig. 43.44. These are quite distinctive with generally a long narrow QRS

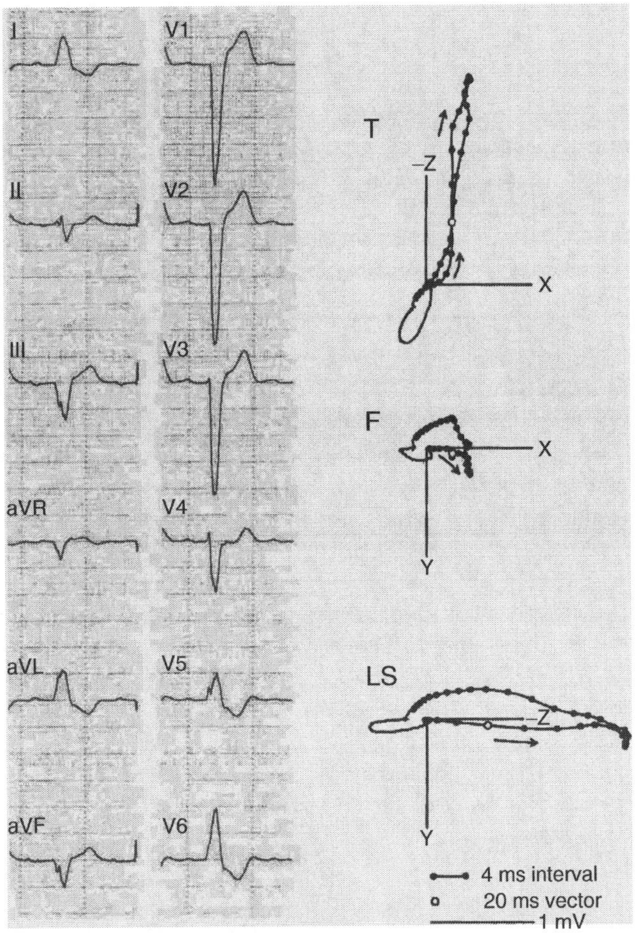

Figure 43.44
An example of left bundle branch block (LBBB). Note the closely spaced time-markers and the narrowness of the QRS loop particularly in the transverse plane. Also, it should be noted that the maximum QRS and T vectors are oppositely directed.

Table 43.7
Vectorcardiographic criteria for left bundle branch block (LBBB)

A maximum QRS vector directed oppositely to the T vector
An elongated or occasionally figure-of-eight loop in the transverse plane with minimal breadth
The main body of the loop is usually inscribed in a clockwise direction in the transverse plane
The timing markers are much closer together than in the normal vectorcardiographic loop

loop in the transverse plane with the T loop being oppositely directed. This corresponds to T wave inversion in lead X. It does also suggest that ventricular repolarization takes place in a sequence parallel to that of depolarization in view of the abnormally slow spread of the ventricular activation wave fronts. The characteristic vectorcardiographic features of LBBB are shown in Table 43.7.

43.5.2.2 Incomplete Left Bundle Branch Block

The concept of incomplete LBBB is perhaps more difficult to explain. It has been suggested that the progression of excitation via the left bundle branch is slower than normal although there is not a complete block. Initial septal activation will, therefore, be on the right side of the septum. From the vectorcardiographic point of view, the QRS loop is again elongated and narrow as in complete LBBB but the overall QRS duration is of the order of 120–140 ms. The remaining markers may also be a little closer than normal. An illustration is given in Fig. 43.45.

Often this pattern is associated with a slightly increased voltage and T wave abnormalities in the lateral leads, that is, a T vector loop almost oppositely directed to the QRS loop. This has given rise to the expression "incomplete LBBB/LVH pattern." It is possible that patients with hypertensive heart disease, for example, and LVH, exhibit fibrosis, which progresses to the extent that the conduction system becomes impaired resulting in the above described abnormalities.

43.5.2.3 Right Bundle Branch Block

In complete right bundle branch block (RBBB), there is a block in the conduction system in the right bundle branch below its junction with the bundle of His. Thus, left ventricular activation commences normally so that RBBB always manifests itself as an abnormality in the terminal part of the QRS complex. This allows other abnormalities such as myocardial infarction to be reported with reasonable confidence in the presence of RBBB.

Because of the normal left ventricular activation, appearances in RBBB show normal initial depolarization. After the greater part of the left ventricle has been depolarized, the delayed excitation wave has spread to the free wall of the right ventricle so that there is an electrically unopposed (but delayed) right ventricular depolarization. This action results in the terminal portion of the QRS complex exhibiting electrical forces oriented to the right, anteriorly. In addition, the QRS duration is abnormally prolonged beyond 120 ms. In the scalar presentation of lead Z there is a broad R′ wave.

The vectorcardiographic loop is shown in Fig. 43.46. It can be seen that the terminal portion of the QRS loop shows a marked slowing of the rate of inscription as evidenced by the closeness of the timing markers. The terminal loop is directed anteriorly, rightward, with quite characteristic features being demonstrated. The major part of the loop is inscribed in a counterclockwise direction.

43.5.2.4 Incomplete Right Bundle Branch Block

There can often be dispute as to whether on a scalar presentation there is, indeed, incomplete RBBB evidenced by the secondary r′ wave in V_1, V_2, or lead Z. It has been suggested that this late QRS activity is due to hypertrophy of the basal

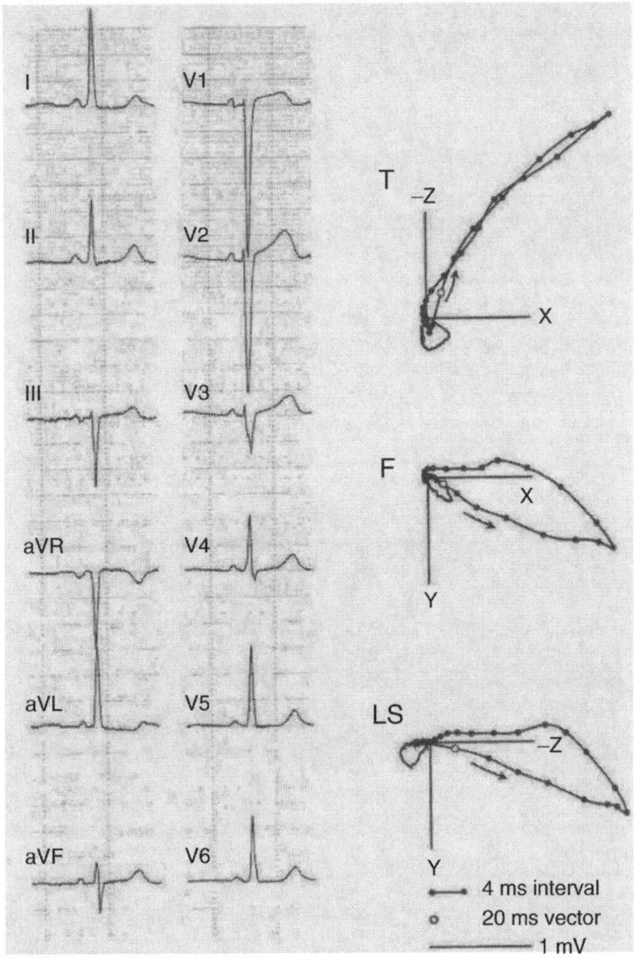

Figure 43.45
An example of incomplete LBBB where the QRS loop in the transverse plane is extremely narrow. The time-markers are also more closely spaced than normal.

portion of the right ventricle although, as it is often seen in younger individuals, it could indeed on occasions be considered a normal variant.

From the vectorcardiographic point of view, an incomplete RBBB is manifested as some slowing in the terminal inscription of the QRS loop seen best in the transverse plane. This portion of the loop will also be oriented anteriorly, rightward. In this situation, the presentation of a vectorcardiogram has an advantage over the scalar presentation in that terminal slowing can be seen.

It could be argued, since the normal QRS duration ranges from 60 to 110 ms, that an incomplete bundle branch block occurring in a patient who previously had a QRS duration of the order of 80 ms could well result in a relative prolongation of the QRS duration to 110 ms. Therefore, it should not be essential for conduction defects involving the right bundle branch in particular to be dependent on criteria which provide discrete time thresholds such as 110 ms for incomplete RBBB. On the other hand, it is unlikely in the adult that a QRS duration <90 ms would be consistent with even an incomplete RBBB. An illustration of the vectorcardiogram in incomplete RBBB is given in Fig. 43.47.

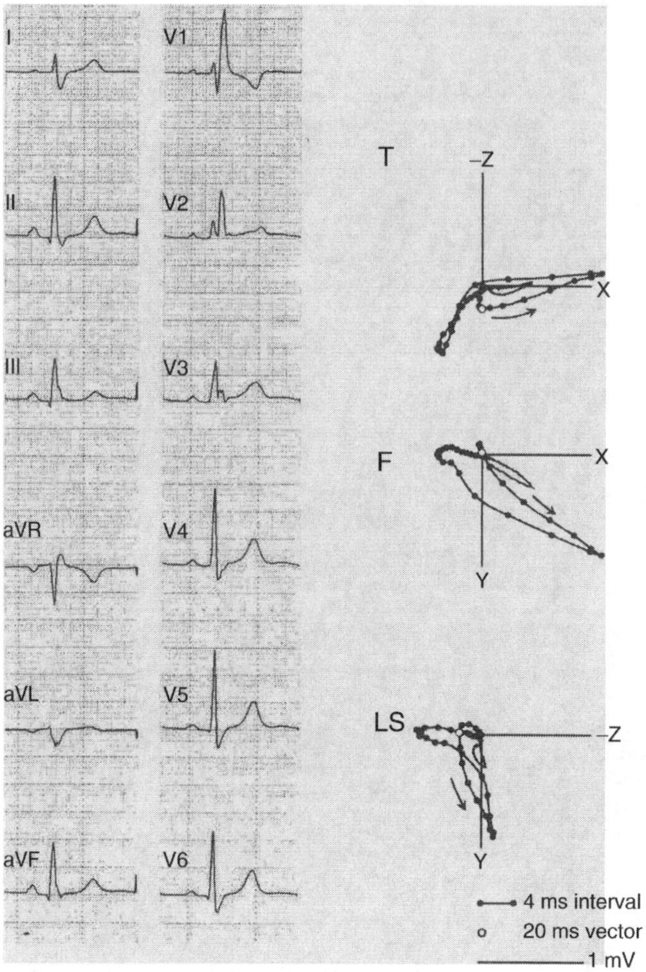

Figure 43.46
An example of right bundle branch block (RBBB) where the late QRS vectors are directed anteriorly to the right with very closely spaced time-markers in the loops corresponding to the slowed conduction. Note the terminal slowing of inscription of the QRS loop.

43.5.3 Fascicular Block

In 1966, Pryor and Blount [32] suggested that the pure left axis deviation might be due to a block in the superior division of the left bundle. They termed this abnormality the "Left Superior Intraventricular Block." They also suggested that right axis deviation could be caused by a block in the inferior division of the left bundle. It was surmised that the direction of the initial excitation depended on the nature of the lesion that produced the block, for example, fibrosis or necrosis. Subsequently, Rosenbaum [33] set out criteria for conduction defects arising from different parts of the conducting system. He postulated that localized abnormalities may occur in isolation, intermittently or in association with defects in other branches of the specialized conducting system. He also introduced the term "hemiblock" to replace "superior" and "inferior" intraventricular block but since there are often more than two specialized conducting fascicles in the left ventricle, the term is perhaps a misnomer. For this reason, it has been suggested [34] that the term "fascicular block" be used. This terminology will be used in the present discussion.

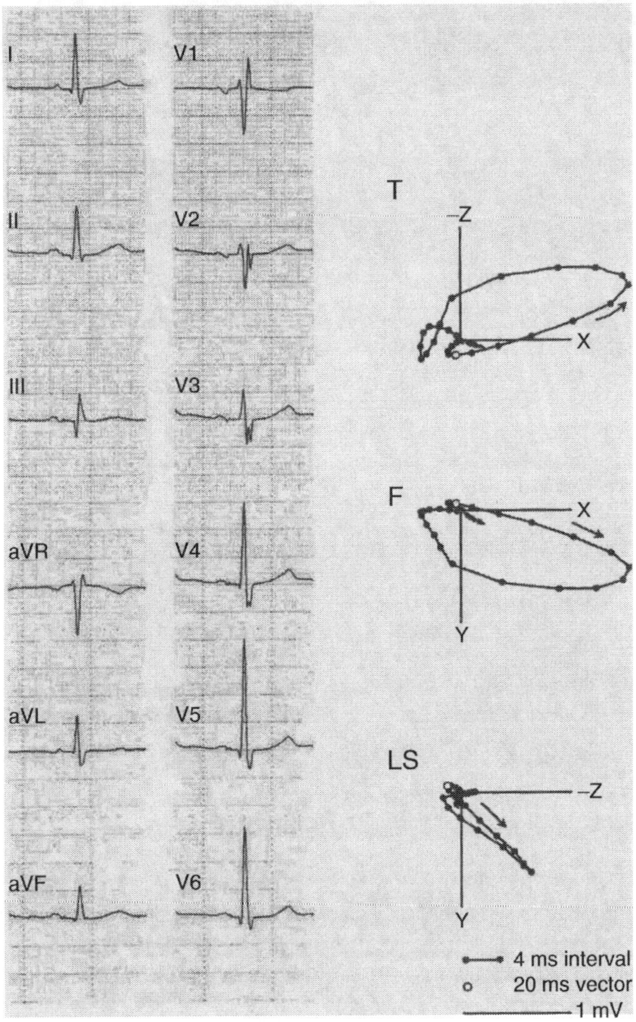

Figure 43.47
Example of incomplete RBBB.

43.5.3.1 Left Anterior Fascicular Block

In left anterior fascicular block, it is postulated that a block of the conduction of excitation arises in the anterior fascicle of the left bundle branch. In this case, ventricular depolarization probably takes place via the other intact fascicles. Excitation initially spreads inferiorly from the left posterior fascicle and it is feasible that there may also be some septal activation from another left-sided conducting fascicle such as the centroseptal fascicle. The net result is an initial, inferiorly directed and sometimes rightward spread of activation. Thereafter, the leftward upward spread of activation from the region of the left posterior fascicle becomes dominant, causing the ventricular resultant electrical force to be orientated posteriorly and superiorly producing a prominent S wave in lead Y.

The frontal plane vectorcardiographic loop shows left axis deviation in most cases, that is, the projection of the maximum QRS vector is superior to 0° and always has counterclockwise inscription [35]. This enables left anterior fascicular block to be differentiated from other forms of conduction abnormality such as due to inferior myocardial infarction.

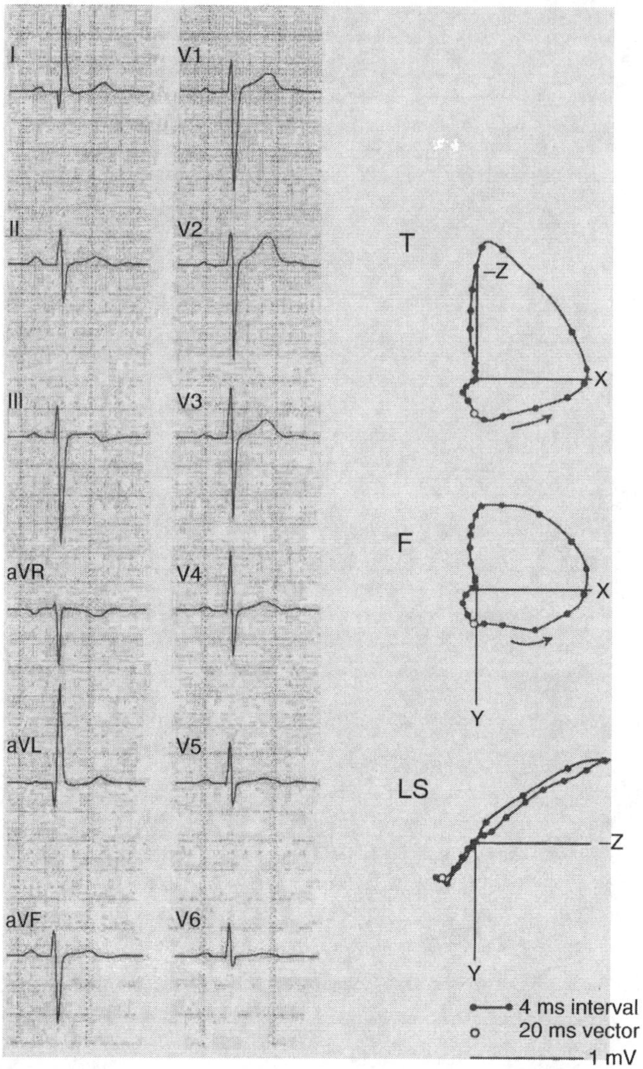

Figure 43.48
An example of left anterior fascicular block with counterclockwise inscription of the QRS loop in the frontal plane.

Table 43.8
Vectorcardiographic criteria for left anterior fascicular block

Initial QRS vectors directed inferiorly and rightward
Counterclockwise inscription in the frontal plane
QRS axis in the frontal plane superior to –30°

Figure 43.48 gives an example of left anterior fascicular block while Fig. 43.38 shows how inferior myocardial infarction can be separated from left anterior fascicular block by the fact that there is clockwise inscription in the frontal plane in the former. Lopes [36] has proposed that the vectorcardiographic criteria for left anterior fascicular block should include, those listed in Table 43.8.

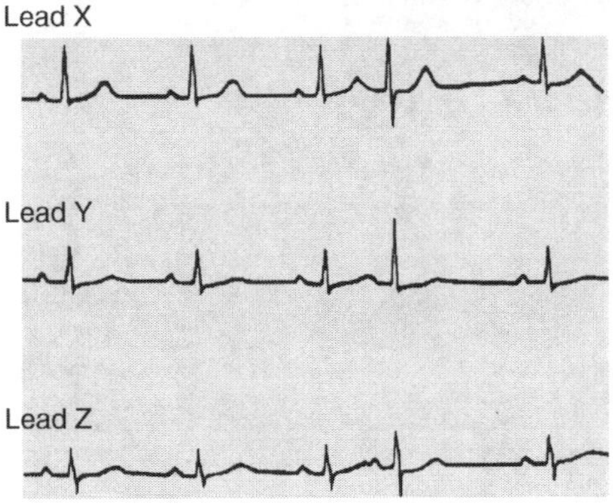

Figure 43.49
Example of intermittent left posterior fascicular block in the fourth beat.

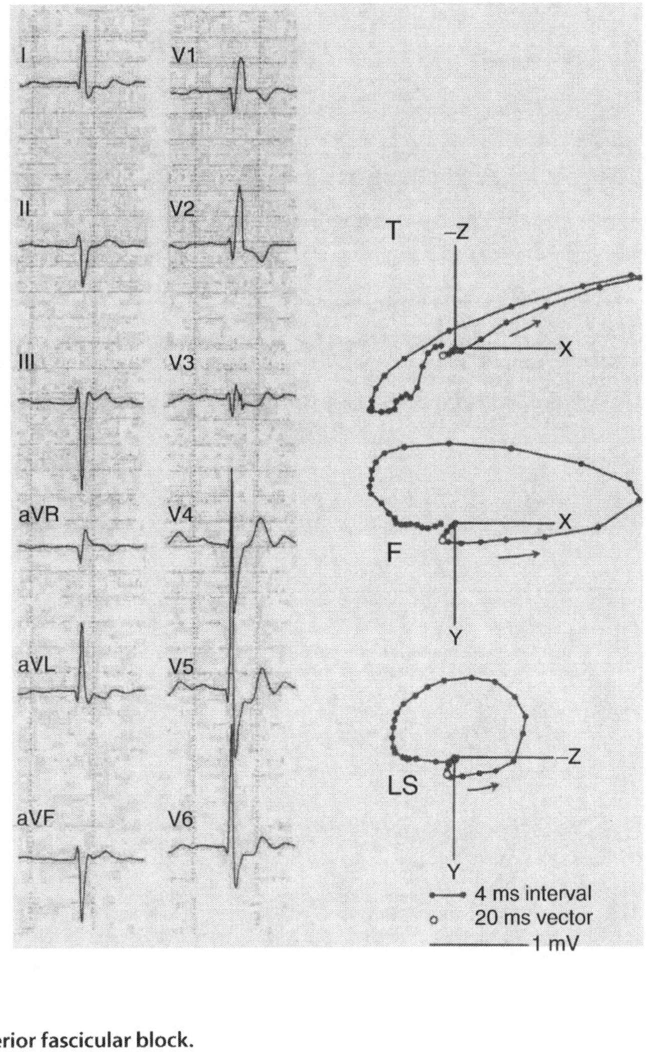

Figure 43.50
Example of RBBB + left anterior fascicular block.

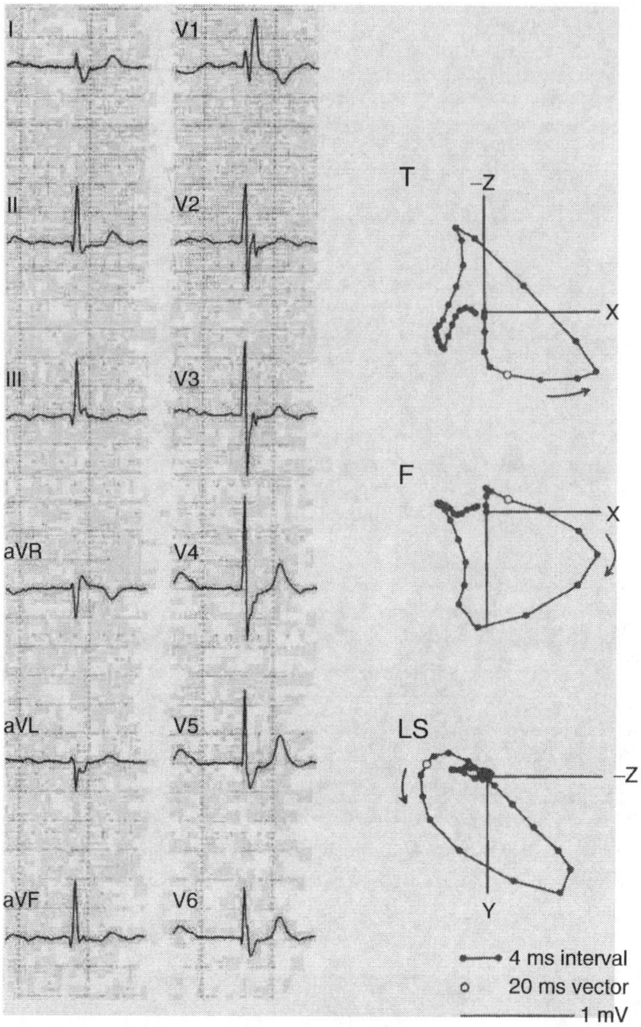

Figure 43.51
An example of RBBB plus left posterior fascicular block. Note that the QRS vector in the frontal plane exceeds 75° in this 39-year-old male.

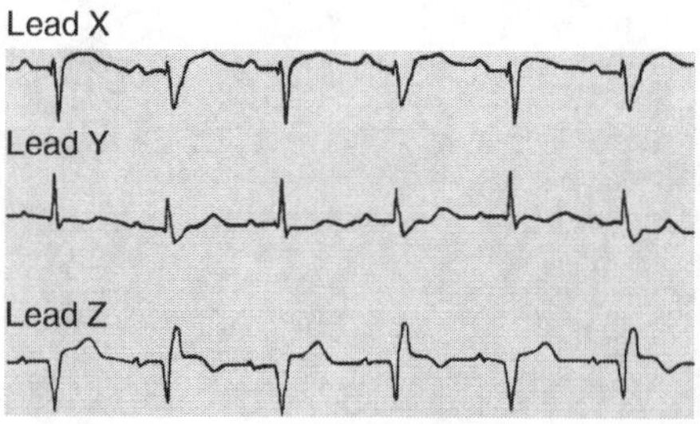

Figure 43.52
Example of alternans where every second beat shows RBBB plus left posterior fascicular block.

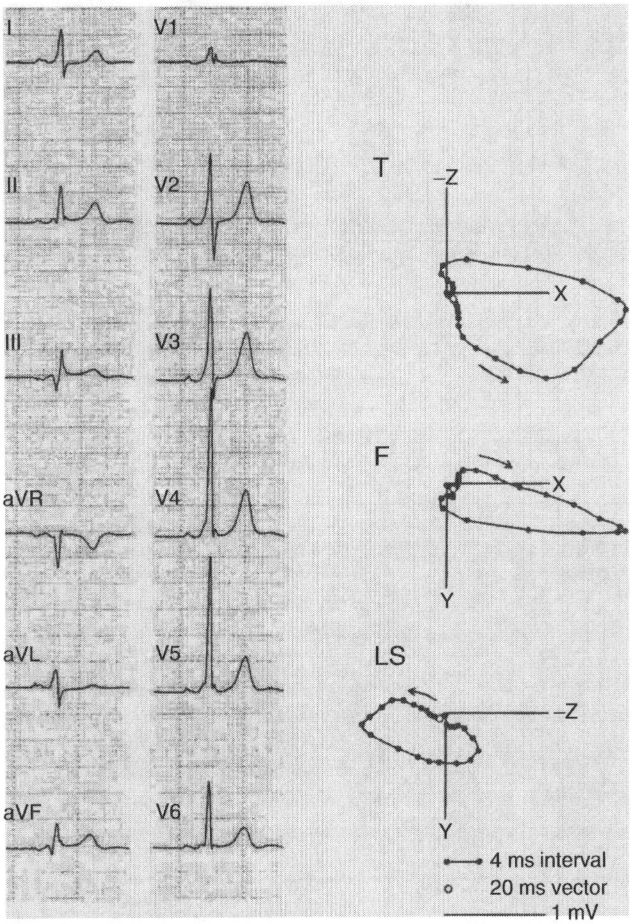

Figure 43.53
An example of Wolff–Parkinson–White (WPW) Type A with initial slowing of the inscription of the QRS loop and early QRS vectors directed anteriorly.

He also suggested that if the major axis of the QRS loop is not directed superiorly, but the late part of the loop lies in the left superior quadrant, possible left anterior fascicular block may be diagnosed.

43.5.3.2 Left Posterior Fascicular Block

It is postulated that if there is a block in the posterior fascicle of the left bundle branch, ventricular activation will proceed from the anterior wall of the left ventricle, spreading inferiorly. Right ventricular, and possibly septal activation are normal. However, the posterior wall of the left ventricle will be depolarized somewhat later than usual so that the excitation waves spreading anteriorly and inferiorly assume an increased importance in the development of ECG appearances. The QRS axis, therefore, shifts to the right, producing appearances suggestive of RVH. Often these appearances are best seen in an intermittent conduction defect in a rhythm strip, for example (Fig. 43.49). In this example, there is an atrial extrasystole that shows a change in QRS configuration compared to the dominant complex, namely, an increase in the amplitude of the R wave in the anteroseptal and inferior leads and a deepening of the S wave in the lateral lead. These appearances are in keeping with the concept of left posterior fascicular block, which, in this case, is intermittent, presumably being produced by a refractory left posterior fascicle at the time of the occurrence of the supraventricular extrasystole.

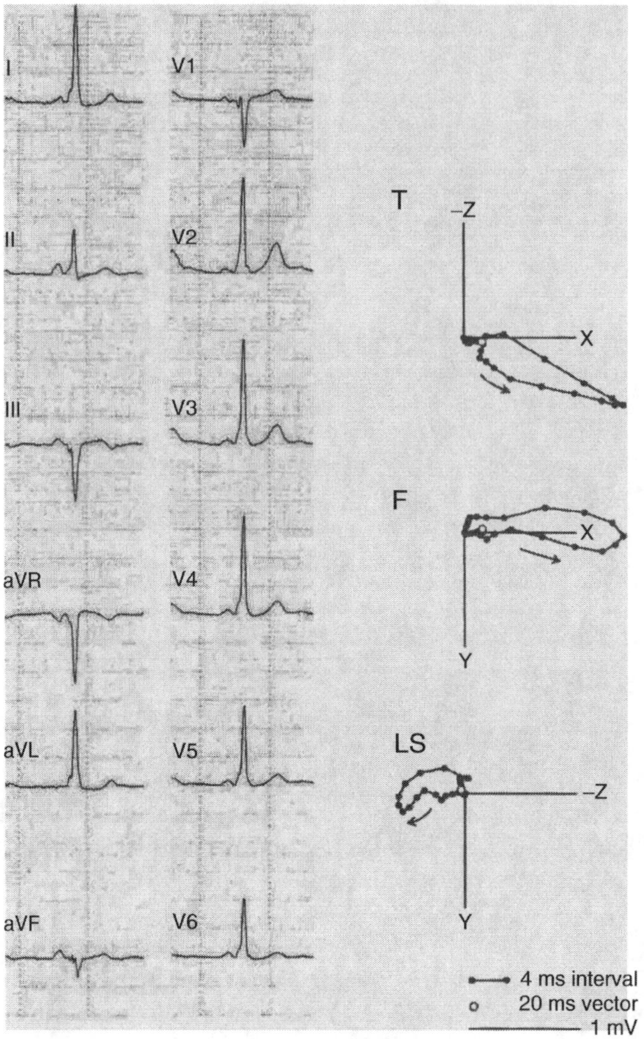

Figure 43.54
An example of WPW Type B where there is initial slowing of the early QRS vectors that are directed leftward for the first 20 ms.

The differentiation of pure left posterior fascicular block from RVH can be difficult. The former is best diagnosed only if there is a lack of clinical evidence to support RVH. It should be clear from Fig. 43.49 that the diagnosis of pure left posterior fascicular block from a single cardiac cycle is virtually impossible.

43.5.4 Bifascicular Block

Combinations of conduction abnormalities in the right bundle branch and different fascicles of the left bundle branch lead to bifascicular block. An example of bifascicular block is RBBB in association with left anterior fascicular block (Fig. 43.50). In this case, there is a marked superior displacement of the vectorcardiographic loop in the frontal plane while the terminal slowing of conduction is again apparent in the transverse plane where the terminal forces are directed anteriorly and to the right.

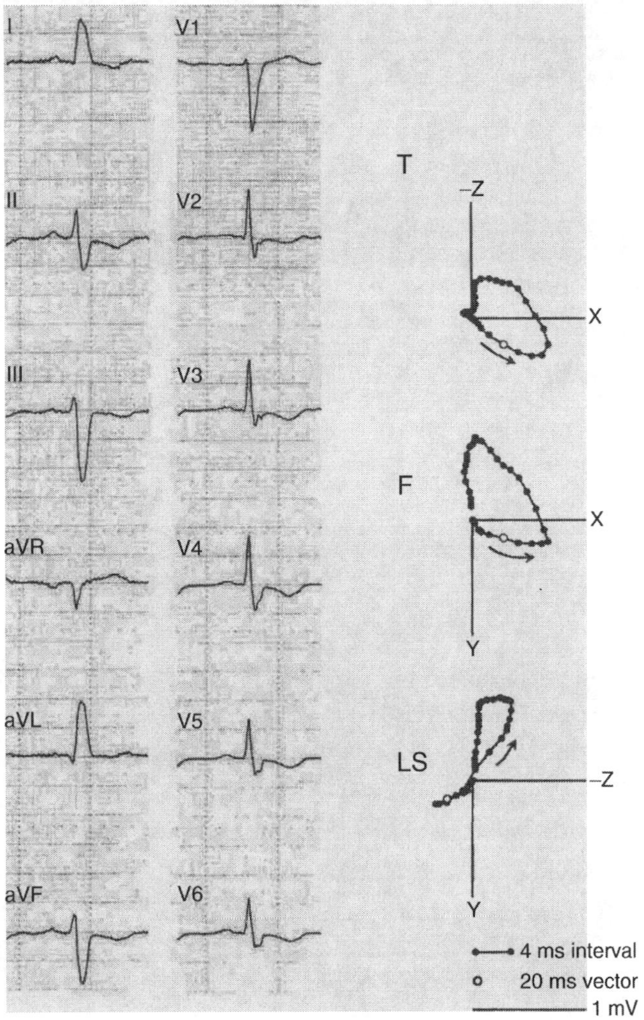

Figure 43.55
An intraventricular conduction defect exemplified by the closely spaced time-markers in all the QRS loops. The transverse loop is also open as opposed to being narrow in LBBB and does not have the late QRS vectors directed to the right as in RBBB.

On the other hand, RBBB with a left posterior fascicular block may be suspected when there is right axis deviation with other typical features of RBBB (Fig. 43.51). Again, the typical RBBB feature of terminal slowing of the QRS loop is apparent but the right axis deviation is abnormal. This feature can perhaps be seen more clearly in another rare example (Fig. 43.52) where there is electrical alternans. The narrow QRS complex suggests myocardial infarction with possible left posterior fascicular block while the alternate beats indicate RBBB in addition.

43.5.5 Wolff–Parkinson–White Pattern

The classical features of the Wolff–Parkinson–White (WPW) [37] pattern are those of an initial slurring of the QRS complex due to the spread of activation from atria to ventricles via an accessory pathway. Note that these features describe the WPW pattern, which, if associated with episodes of paroxysmal tachycardia, gives rise to the WPW syndrome. From the vectorcardiographic point of view, the main distinguishing feature is obviously the slowing of inscription in the initial

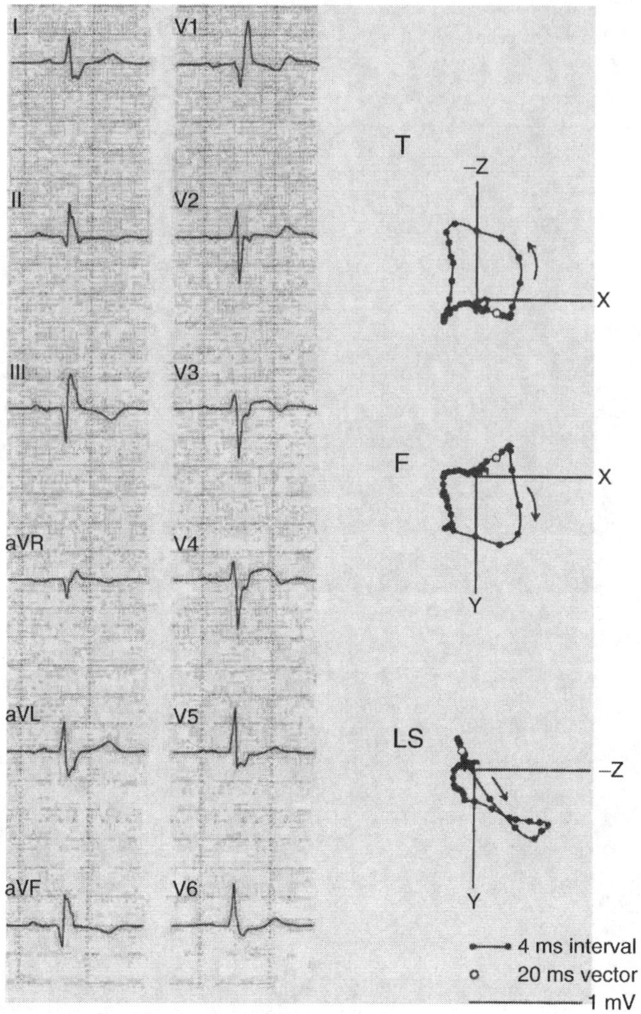

Figure 43.56
Example of RBBB and inferior myocardial infarction.

part of the vectorcardiographic loop. This may facilitate the diagnosis of WPW pattern in borderline cases where the initial slurring in the scalar presentation may be questioned as being technical in origin. There is a variety of accessory pathways such that there is no typical feature of the WPW pattern on the vectorcardiogram other than the initial slowing of inscription. In other words, initial vectors may be orientated in a whole spectrum of different directions depending on the location of the accessory pathway.

In broad terms, using the older classification of Type A and Type B, the initial vector orientation will be anterior in Type A and posterior/to the left in Type B. Figures 43.53 and 43.54 give examples of the vectorcardiographic loops in the WPW pattern.

43.5.6 Intraventricular Conduction Defects

There remains a class of conduction defects that does not fit any of the categories mentioned so far. In general terms, the QRS duration is prolonged in excess of 120 ms but the features of RBBB or LBBB are not apparent. On occasions, such defects manifest themselves as an open loop in the transverse plane, which permits the diagnosis of intraventricular conduction defect as opposed to LBBB, for example, even when the QRS duration is the order of 140–160 ms (➋ Fig. 43.55). This is another area where the vectorcardiographic loop can be of considerable diagnostic value.

43.5.7 Combined Conduction Defects and Myocardial Infarction

The diagnosis of myocardial infarction in the presence of RBBB is relatively straightforward. Because RBBB affects the terminal portion of the QRS complex and classical criteria for myocardial infarction affect the initial portion of the QRS complex, it is possible for myocardial infarction to be reported in the presence of RBBB. ➋ Figure 43.56 shows an example of inferior myocardial infarction and RBBB.

On the other hand, there is still controversy over the diagnosis of myocardial infarction in the presence of LBBB. In 1982, Havelda et al. suggested that inferior myocardial infarction could be diagnosed with 100% specificity in the presence of LBBB when there were QS complexes in the inferior leads [38]. Others have subsequently suggested that anterior infarction can be diagnosed in the presence of LBBB when, for example, there is a reversed R wave progression from V_3 to V_5 or there are Q waves in the lateral leads. On the other hand, left anterior fascicular block may mask or mimic inferior infarction [39].

43.5.8 ECG Versus 12-Lead Vectorcardiogram in Conduction Defects

This chapter has shown how there are certain situations where the vectorcardiographic display can exhibit the presence of conduction defects that may be less obvious on the scalar ECG. In particular, the slowing of timing marks on the vectorcardiogram is invariably an illustration of a conduction defect, whether it is at the beginning of the QRS complex as in the WPW pattern or at the end as in RBBB. If the slowing is throughout the QRS, then most likely the diagnosis is LBBB. It has also been shown how left anterior fascicular block and inferior infarction can be separated by the direction of inscription of the frontal plane vectorcardiographic loop. All of these points lend further weight to the conclusion that the vectorcardiogram has an important role to play in the diagnosis of ventricular conduction defects.

References

1. Einthoven, W., G. Fahr, and A. de Waart, Über die Richtung und die manifeste Grosse der Potentialschwankungen im menschlichen Herzen und über den Einfluss der Herzlage auf die Form des Elektrokardiogramms. *Pflügers Arch.*, 1913;**150**: 175–315. (Translation: Hoff, H.E. and P. Sekelj, *Am. Heart J.*, 1957;**40**: 163–194.)
2. Grishman, A. and L. Scherlis, *Spatial Vectorcardiography*, Philadelphia, PA: Saunders, 1952.
3. Frank, E., An accurate, clinically practical system for spatial vectorcardiography. *Circulation*, 1956;**13**: 737–749.
4. Mann, H., A method for analyzing the electrocardiogram. *Arch. Int. Med.*, 1920;**25**: 283–294.
5. American Heart Association Committee on Electrocardiography (Pipberger, H.V., R.C. Arzbaecher, and A.S. Berson, et al.), Recommendations for standardization of leads and of specifications for instruments in electrocardiography and vectorcardiography. *Circulation*, 1975;**52**(Suppl.): 11–31.
6. Watts, M.P. and D.B. Shoat, Trends in electrocardiograph design. *J. Inst. Electron. Radio Eng.*, 1987;**57**: 140–150.
7. Macfarlane, P.W., B. Devine, S. Latif, S. McLaughlin, D.B. Shoat, and M.P. Watts, Methodology of ECG interpretation in the Glasgow Program. *Meth. Inform. Med.*, 1990;**29**: 354–361.
8. Macfarlane, P.W., E.N. Coleman, B. Devine, et al., A new 12-lead pediatric ECG interpretation program. *J. Electrocardiol.*, 1990;**23**(Suppl.): 76–81.
9. Macfarlane, P.W., E.N. Coleman, E.O. Pomphrey, S. McLaughlin, and A. Houston, Normal limits of the high-fidelity pediatric ECG. *J. Electrocardiol.*, 1989;**22**(Suppl.): 162–168.
10. Lundh, B., On the normal scalar ECG. A new classification system considering age, sex and heart position. *Acta Med. Scand.*, 1984;691(Suppl.).
11. Chen, C.Y., B.N. Chiang, and P.W. Macfarlane, Normal limits of the electrocardiogram in a Chinese population. *J. Electrocardiol.*, 1989;**22**: 1–15.
12. Yang, T.F., C.Y. Chen, B.N. Chiang, and P.W. Macfarlane, Normal limits of derived vectorcardiogram in Chinese. *J. Electrocardiol.*, 1993;**26**: 97–106.
13. Downs, T.D., J. Liebman, R. Agusti, and H.C. Romberg, The statistical treatment of angular data in vectorcardiography, in

Proceedings of Long Island Jewish Hospital Symposium on Vectorcardiography, 1965, I. Hoffman and R.C. Traymore, Editors. Amsterdam: North Holland, 1966, pp. 272–278.
14. Yang, T.F. and P.W. Macfarlane, Comparison of the derived vectorcardiogram in apparently healthy Caucasians and Chinese. *Chest*, 1994;**106**: 1014–1020.
15. Draper, H.W., C.J. Peffer, F.W. Stallmann, D. Littmann, and H.V. Pipberger, The corrected orthogonal electrocardiogram and vectorcardiogram in 510 normal men (Frank lead system). *Circulation*, 1964;**30**: 853–864.
16. Nemati, M., J.T. Doyle, D. McCaughan, R.A. Dunn, and H.V. Pipberger, The orthogonal electrocardiogram in normal women. Implications of sex differences in diagnostic electrocardiography. *Am. Heart J.*, 1978;**95**: 12–21.
17. Namin, E.P., R.A. Arcilla, I.A. D'Cruz, and B.M. Gasul, Evaluation of the Frank vectorcardiogram in normal infants. *Am. J. Cardiol.*, 1964;**13**: 757.
18. Huwez, F.U., S.D. Pringle, and P.W. Macfarlane, A new classification of left ventricular geometry in patients with cardiac disease based on M-mode echocardiography. *Am. J. Cardiol.*, 1992;**70**: 681–688.
19. Kannel, W.B., Prevalence and natural history of electrocardiographic left ventricular hypertrophy. *Am. J. Med.*, 1983;**75**(Suppl. 3A): 4–11.
20. Macfarlane, P.W., British Regional Heart Study: The electrocardiogram and risk of myocardial infarction on follow-up. *J. Electrocardiol.*, 1987;**20**(Suppl.): 53–56.
21. Chou, T. and R.A. Helm, The pseudo P pulmonale. *Circulation*, 1965;**32**: 96–105.
22. Huwez, F.U., *Electrocardiography of the Left Ventricle in Coronary Artery Disease and Hypertrophy*, Ph.D. thesis. University of Glasgow, 1990.
23. Gottdiener, J.S., J.A. Gay, B.J. Maron, and R.D. Fletcher, Increased right ventricular wall thickness in left ventricular pressure overload: Echocardiographic determination of hypertrophic response of the 'non-stressed' ventricle. *J. Am. Coll. Cardiol.*, 1985;**6**: 550–555.
24. Brohet, C.R., Special value of the vectorcardiogram in pediatric cardiology. *J. Electrocardiol.*, 1990;**23**(Suppl.): 58–62.
25. American College of Cardiology, Tenth Bethesda conference report on optimal electrocardiology. *Am. J. Cardiol.*, 1978;**41**: 111–191.
26. Selvester, R.H. and M.E. Sanmarco, Infarct size in hi-gain hi-fidelity VCG's and serial ventriculograms in patients withproven coronary artery disease, in *Modern Electrocardiology*, Z Antoloczy, Editor. Amsterdam: Excerpta Medica, 1978, pp. 523–528.
27. Selvester, R.H., H.B. Rubin, J.A. Hamlin, and W.W. Pote, New quantitative vectorcardiographic criteria for the detection of unsuspected myocardial infarction in diabetics. *Am. Heart J.*, 1968;**75**: 335–348.
28. Edenbrandt, L., A. Ek, B. Lundh, and O. Pahlm, Vectorcardiographic bites. A method for detection and quantification applied on a normal material. *J. Electrocardiol.*, 1989;**22**: 325–331.
29. Macfarlane, P.W. and T.D.V. Lawrie, The normal electrocardiogram and vectorcardiogram, in *Comprehensive Electrocardiology*, Vol. 1, P.W. Macfarlane and T.D.V. Lawrie, Editors. Oxford: Pergamon Press, 1989, pp. 407–457.
30. Demoulin, J.C. and H.E. Kulbertus, Histopathological examination of the concept of left hemiblock. *Br. Heart J.*, 1972;**34**: 807–814.
31. van Dam, R.Th., Ventricular activation in human and canine bundle branch block, in *The Conduction System of the Heart*, H.J.J. Wellens, K.I. Lie, and M.J. Janse, Editors. Leiden: Stenfert Kroese, 1976, pp. 377–392.
32. Pryor, R. and S.G. Blount, The clinical significance of true left axis deviation. *Am. Heart J.*, 1966;**72**: 391–413.
33. Rosenbaum, B.M., The hemiblocks: Diagnostic criteria and clinical significance. *Mod. Concepts Cardiovasc. Dis.*, 1970;**39**: 141–146.
34. Pryor, R., Fascicular blocks and the bilateral bundle branch block syndrome. *Am. Heart J.*, 1972;**83**: 441.
35. Kulbertus, H.E., P. Collignon, and L. Humblet, Vectorcardiographic study of the QRS loop in patients with left anterior focal block. *Am. Heart J.*, 1970;**79**: 293–304.
36. Lopes, M.G., *Seminar in Vectorcardiography*, Stanford, CA: Stanford University Press, 1974.
37. Wolff, L., J. Parkinson, P.D. White, Bundle-branch block with short P–R interval in healthy young people prone to paroxysmal tachycardia. *Am. Heart J.*, 1930;**5**: 685–704.
38. Havelda, C.L.J., G.S. Sohi, N.C. Flowers, and L.G. Horan, The pathologic correlates of the electrocardiogram: Complete left bundle branch block. *Circulation*, 1982;**65**: 445–451.
39. Milliken, J.A., Isolated and complicated left anterior fascicular block: A review of suggested electrocardiographic criteria. *J. Electrocardiol.*, 1983;**16**: 199–211.

44 Magnetocardiography

Markku Mäkijärvi · Petri Korhonen · Raija Jurkko · Heikki Väänänen ·
Pentti Siltanen · Helena Hänninen

P. W. Macfarlane, A. van Oosterom, O. Pahlm, P. Kligfield, M. Janse, J. Camm (eds.), *Comprehensive Electrocardiology*, DOI 10.1007/978-1-84882-046-3_44,

44.1 Introduction

Although magnetocardiography (MCG) was first introduced in the early 1960s, it mainly remained as an experimental method practiced by engineers in research laboratories until the 1990s [1]. Today, it has developed into one of the new technologies in cardiology employed by medical doctors in several clinical laboratories. MCG still poses technical challenges, such as the instrumentation based on the use of liquid helium and the need for magnetically shielded rooms (MSR), but the clinical application of the method has significantly benefited from the availability of modern multichannel instrumentation in hospitals at the patient's bedside. In addition, the signal-to-noise ratio in routine MCG recordings is comparable to the best of electrical measurements. MCG studies provide online results quickly, and many groups are collecting libraries of reference data. Profound efforts in the direction of standardization and data comparability are also on the way.

There are several applications in which MCG has already provided clinically useful results. For example, an MCG can diagnose and localize acute myocardial infarction, separate myocardial infarction patients with and without susceptibility to malignant ventricular arrhythmias, detect ventricular hypertrophy and rejection after heart transplantation, localize the site of ventricular preexcitation and many types of cardiac arrhythmia, and can also reveal fetal arrhythmias and conduction disturbances [2].

In addition, several other clinical applications of MCG have recently been studied: detection and risk stratification of cardiomyopathies (dilated, hypertrophic, arrhythmogenic, and diabetic), risk stratification after idiopathic ventricular fibrillation, detection and localization of myocardial viability, and the follow-up of fetal growth and neural integrity. Some studies have clearly indicated that MCG is very sensitive to the changes of repolarization, for example, after myocardial infarction or in a hereditary long-QT syndrome [3].

In this review, we briefly overview the development in instrumentation, measurement techniques, data analysis, and the clinical aspects of MCG during the last few years and present a perspective for the near future.

44.2 Sources of MCG

44.2.1 Origin of Measured Magnetic Field

The biomagnetic fields are generated by the same bioelectric activity that generates electric potentials as discussed in Vol. 1. The ion pumps on the cell membrane and the diffusion gradients impress the flow of the charged ions (such as Na^+, K^+, and Ca^{2+}) in the myocardium. These ionic currents inside and in the vicinity of excited cells are called primary current density J_P [4]. The primary current causes changes of the electric potential ϕ, which in turn creates ohmic volume currents $J_V = -\sigma\nabla\phi$, where σ is the electric conductivity. The magnetic field of total current $J_T = J_P + J_V$ can then be solved by Maxwell's equations or in quasistatic approximation from the Biot–Savart law $dB = \frac{\mu_0}{4\pi}\frac{dJ\times r}{r^3}$, where the dB is the magnetic field from the differential current dJ in relative position r. By integrating the overall current density, we obtain the magnetic field B. The resulting magnetic field from all the cardiac activity is in the range of pT, below one millionth of the earth's magnetic field (30–60 μT).

44.2.2 Forward Problem

The calculation of the external magnetic field from the known cardiac currents, the *forward problem*, has similar properties to those of the forward problem in ECG. Analytically, the forward problem in a homogeneous volume conductor can be solved with only some simple geometry. Therefore, numerical methods are used in the calculations. In the boundary-element method (BEM), an analytical equation is discretized to linear matrix equations [5, 6], which can be extended to the torso models. In the finite element model (FEM), the anisotropic properties of the material can also be included in the calculations. As with the electric forward problem, realistic torso models are needed for accurate results. However, the magnetic field is not as sensitive for material properties of the torso as the electric potential, so the requirement for the model is not quite as strict. Interestingly, preliminary results for estimating the heart outline based only on MCG mapping have also been presented [7].

44.2.3 Inverse Problem

The biomagnetic inverse problem is ill-posed and has no unique solution even if the body surface potential mapping is combined with magnetic measurements. Therefore, different equivalent source models have to be used. The equivalent current dipole (ECD) is the most elementary source of magnetic fields and can be defined as $q = \int_V J_p(r)\mathrm{d}V$. Higher order equivalent generators, such as quadrupoles and octupoles, have also been presented by using multipole expansions [8]. Accuracies of 5–25 mm have been reported for best-fitting ECDs [9].

44.2.4 Inverse Problem with Distributed Source Model

To solve the inverse problem with a distributed source model like equivalent current density or uniform double-layer, a number of individual current dipoles are usually derived with the help of the lead field theory. The ill-posed problem of the lead field matrices requires the use of different regularization techniques for stabilizing the result. In addition to single-layer as well as double-layer sources, lead fields from each point on both the endocardial and epicardial surfaces, together with bidomain models for estimating the propagating wave front, are used [10, 11].

44.2.5 MCG vs ECG

Since the magnetocardiogram is generated by the same activity that generates the ECG, the signal waveforms corresponding to the P-, QRS-, and T-waves of the ECG are also seen in the magnetocardiogram. However, the differences in the information content between the MCG and ECG are still quite controversial; it can be shown that in an infinite and homogeneous volume conductor, J_V does not contribute to the magnetic field or electric potential which are independent of each other. In a homogeneous, semi-infinite volume conductor, all the magnetic field outside the conductor arises from the tangential current sources, while the ECG in general is more sensitive to the radial currents. In practice, due to torso inhomogeneities, conductivity differences, and anisotropy, the difference between the MCG and ECG is not that apparent. However, the sensitivity of MCG to the vortex currents seems also in practice to be remarkably different; an ideal vortex current does not produce any electric field outside the body [12]. The MCG has also been reported to be more sensitive to repolarization currents since it is not disturbed by electrode-skin potential, for example, distinguishing a real ST shift from an artifactual ST shift is possible [13]. As already discussed, the MCG is less affected by the torso inhomogeneities than the ECG, which is an advance, especially in inverse modeling. The measurement instrumentation where all the sensor localizations are always in the same relative positions to each other is also an advance, especially if the combination of measurement and torso coordinate systems is done correctly.

44.3 Measurement Technique and Instrumentation

44.3.1 History of MCG

The cardiac magnetic field was first measured by Baule and McFee [14] using a coil magnetometer. In 1970, D. Cohen et al. [15] introduced the superconducting quantum interference device (SQUID) for MCG, and 1 year after that, Zimmerman and Frederic [16] introduced a gradiometer that made MCG recording in an unshielded environment possible. In 1973 came the first modeling study [5]. Multichannel whole-thorax systems were introduced in the early 1990s, and high-Tc (high-temperature) superconductors appeared a few years later.

44.3.2 General

The DC-SQUID sensors offer the best sensitivity for MCG measurements. The strong environmental magnetic noise, unavoidable at urban hospitals and laboratories, makes the detection of biomagnetic signals impossible without special techniques for environmental interference suppression: The environmental magnetic noise is reduced by MSRs, which typically consist of a combination of μ-metal and eddy current shields. In addition, gradiometer coils are used to diminish residual magnetic noise within the shields. Alternatively, high-order gradiometers can be utilized if no magnetic shielding is employed.

44.3.3 Measurement System – Torso Position

The position of the subject's thorax with respect to the sensor array can be determined, for example, by using special marker coils attached to the skin. The positions of these coils are determined by a 3D digitizer before the measurement (in torso coordinates), and from the MCG recordings (in device coordinates) when electric current is fed to the coils. A typical measurement setup is seen in ➲ Fig. 44.1.

44.3.4 Low-Temperature and High-Temperature Sensors

The SQUID sensors are commonly classified as low-temperature SQUID sensors (LTS), which operate at the temperature of liquid helium or as high-temperature SQUID sensors (HTS) operating at liquid nitrogen temperatures. LTS are more commonly used. They are easier to manufacture and have less noise. On the other hand, cooling with liquid nitrogen is much less expensive than with liquid helium, and the cryogenic dewars for HTS are easier to manufacture.

44.3.5 Different Systems: Sensors

Most of the MCG systems have planar sensor loops, which detect the Z-component of the magnetic field, that is, the component directed toward the measurement system. Instead of measuring the magnitude of the magnetic field, the axial and planar field gradients are measured in some systems. Magnetometer sensors have the highest sensitivity for nearby and far-field sources. Thus, selecting a gradiometer with proper distance between the SQUID sensors reduces the amount of unwanted external noise components. The gradients are realized with wire-wound or bonded two or three gradiometers or by electronically combining several reference coils to one measurement coil. Therefore, it should be noted

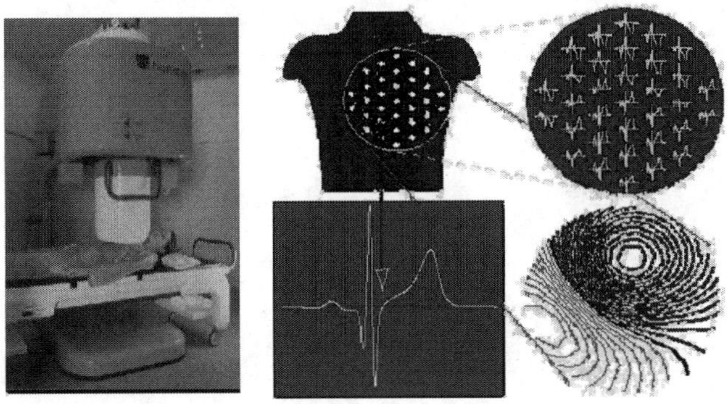

Figure 44.1
***Left*: MCG recording with 99-channel magnetocardiometer at 33- location (BioMag Laboratory, Helsinki, Finland). *Right*: The sensor arrangement in the cardiomagnetometer and an example of signal measured during 1 s and magnetic isofield integral map over atrial complex.**

that even one channel system may well contain nine SQUID sensors, and the 304-SQUID system may sample only at 57 sites. Hence, the measured number of channels does not necessarily tell everything about the effective dimension of the mapping. In some systems, gradients are also implemented with software from the separate magnetometer signals, and for multi-lead mappings also, many complex software noise cancellations are used (see below).

In addition to systems measuring only the Z-component of the field, there are also systems sensing only tangential fields or all three orthogonal vector component magnetometers. In quasistatic approximation ($\nabla \times B = 0$), it is directly seen that $\partial B_z/\partial x = \partial B_x/\partial z$, and so the tangential field gradients can be derived from the tangential gradients of the B_z.

44.3.6 Different Systems: Two Trends

Nowadays, the development of new MCG systems can be divided into two main trends: The "high-end," multichannel and high sensitivity systems mainly for research work and the "budget-priced" systems without MSRs for clinical use [17]. The lower cost systems bring MCG closer to the everyday clinical work. With efficient signal processing techniques, they can be used for single channel studies, but for mapping studies with these single or "quasi single" channel systems (seven or nine channel systems) several different measurements over certain sampling points are needed. Problems include the fact that scanning of the grid is very time-consuming, and the resulting measurement is not truly simultaneous although after averaging, signals should be time invariant. In high-end systems, more than 300 sensors are read simultaneously with a very high signal-to-noise ratio. ➲ Table 44.1 shows examples of different MCG systems with and without an MSR.

44.3.7 Different Systems: Need for Standardization

One of the problems with MCG studies is related to the number of systems with different sensor geometries, different sensor types, and different signal-to-noise ratios, which make multicenter studies difficult. In addition, the digital preprocessing of the measured signals varies. There are also problems with comparing the results obtained with different systems. Some standardization has been proposed several times including a proposal that was presented in the previous version of this book [1], but due to the huge variation of the systems in use, standardization has not been fully implemented.

Table 44.1
Examples of different MCG systems

Name	Type	Number of channels	Dewar diameter (cm)
Siemens Krenikon	Shielded, multichannel	37 axial gradiometers	19
Philips	Shielded, multichannel	2*31 axial gradiometers	13.5 + 13.5
PTB Berlin	Shielded, multichannel	63 electronical gradiometers + 20 reference	21
4-D Neuroimaging (Bti) Magnes	Shielded, multichannel	61 magnetometers	32.4
(Neuromag) VectorView	Shielded, multichannel	33 magnetometers, 66 planar gradiometers	30
AtB Argos	Shielded, multichannel	Vector-magnetometers	23
Hitachi	Shielded, multichannel	64 planar gradiometers	20
Cardiomag Imaging	Unshielded	9 + second-order axial gradiometers + three vectorial reference channels	
SQUID AG	Unshielded	4 + second-order axial gradiometers + 3 reference	
Jena (FSU)	HTS, unshielded	Two planar gradiometers	
Hitachi	HTS, open-ended shielding	16	

44.4 Digital Signal Processing

44.4.1 Preprocessing

44.4.1.1 Mapping Versus Multi-Lead Analysis

All signal processing meant for improving the signal quality, removing noise, and artifacts is hereinafter called "preprocessing." The preprocessing part, just as in the actual signal analysis and the calculation of the different measures, is further divided into mapping and multichannel methods. Mapping includes methods that utilize the spatial information on measurement channels, as the multichannel methods process one channel at a time and utilize the temporal information.

44.4.1.2 Order of Signal Processing

First, noise suppression preprocessing, can be divided into different phases. Selected map analysis may utilize methods like signal-space projection (SSP), signal-space separation (SSS), independent component analysis (ICA), filtering, baseline removal, and perhaps averaging. Sometimes filtering can be undertaken before the map-operator. The data conversion is usually made on the corrected signal. Then after possible conversion, the measures (markers) are calculated.

44.4.1.3 Signal-Space Projection

There are several different methods for noise suppression: The signal-space projection (SSP) method is based on defining a noise subspace from an empty room measurement and projecting the measured multichannel signal to the signal space orthogonal to the noise subspace [18]. A similar principle is also utilized in the "eigenvector-based spatial filtering of fetal biomagnetic signals," for example [19], where the fetal signal is extracted on the basis of projecting the data to the space that maximizes the fetal and maternal eigenvector ratio.

44.4.1.4 Signal-Space Separation

Signal-space separation (SSS) is based on mathematical separation of the signal sources to the cardiac sources that are inside a predefined subspace in normal space (a ball around the heart) and outside sources. After separating sources, the signal consisting only of cardiac activity can be formed. This method can also be used together with signal conversion to the different measurement geometry or for interpolating the poor quality channels [20].

44.4.1.5 Independent Component Analysis

Independent component analysis (ICA) is a method for the blind separation of linearly mixed signals. The assumptions in the method are that the number of mixed signals is, at a maximum, the number of measured signals. Mixed signals are non-Gaussian and linearly mixed, which in practice means that their sources are spatially separable. The most commonly used algorithm was presented by Hyvärinen [21]. The actual ICA algorithm is only a part of the solution as the algorithm separates the signals blindly, which means that the wanted and the unwanted signal components have to be somehow selected. The algorithm also loses the amplitude scaling of the signal, which has to be corrected afterwards. An automatic use of ICA for separating the fetal signal from the maternal signal was presented by Comani et al. [22].

44.4.1.6 One-Channel Digital Noise Suppression

In addition to signal processing in the mapping methods described above, preprocessing methods for individual MCG channels are used. The algorithms are similar to those used with ECG channels, where signals are band-pass filtered with finite (FIR) and infinite response filters (IIR). The 50 or 60 Hz powerline interference is filtered with different adaptive notch filters. Wavelet transforms are also used [23]. Low-frequency noise, such as baseline wander, is corrected with polynomial fits to the isofield (isopotential in ECG) intervals and signals are averaged. In fetal MCG, the maternal signal average is also used for separating the fetal and maternal signals.

44.4.1.7 Conversion Between Sensor Arrays

A separate part of preprocessing is the conversion between sensor arrays. There are several different MCG systems. Sensor locations and types vary. However, since there are many different systems, the comparison of the results is difficult. For that reason, different signal conversions based on multipole and minimum normal estimates have been proposed [24, 25]. In one of these studies [24], where the conversion from measurements was done with one multichannel MCG system to another, the reconstruction succeeded with 93–95% agreement. However, the overall accepted standard system for conversion is still missing.

44.5 Signal Analysis

44.5.1 MCG Morphology Analysis

Signal processing of individual MCG channels resembles that of an ECG lead. Often only scaling constants in algorithms for waveform detection are different. Markers derived from the MCG signal morphology are also very similar to those from the ECG, namely signal amplitudes (e.g., ST level), filtered signal amplitudes (e.g., late field/potential analysis), time intervals (e.g., QT interval), spectral characteristics (e.g., fragmentation index), etc.

44.5.2 MCG Mapping Analysis

The MCG mapping markers include mapping orientations (min-max, +−, centers of gravity, and maximum field gradient) of different time instants or time intervals of the signal (QT orientation, T-wave orientation, and ST orientation) and different markers representing spatial heterogeneity (nondipolar content of QRS, STT, QRST, QT-dispersion). Also, different spatiotemporal measures (ST, T-wave orientation, and heart rate dependency) and different source modeling parameterizations have been used.

44.5.3 Visualizations

Typically, the MCG data is visualized either as temporal time traces, like an ECG trace, from one or all the measurement channels, or with different isovalue maps from one time instant or from some defined marker (e.g., integral) from each channel (see ➋ Fig. 44.1). In these maps, the signal amplitude is displayed with color coding or with isocontours.

Another commonly used way for displaying the mapping information is the current arrow map, where pseudo current at each measurement location, derived from $\mathrm{d}B_Z/\mathrm{d}y\mathbf{x} + \mathrm{d}B_Z/\mathrm{d}x\mathbf{y}$ (**x** and **y** being unit vectors) is represented by a vector showing the amplitude (length of the arrow) and direction of the current at that location (➋ Fig. 44.1). This current arrow map can also be projected onto a 3D model of the heart surface.

44.5.4 Experimental MCG

Experimental work utilizing MCG techniques began more than 30 years ago [26]. As a fully contactless method, MCG is ideal for noninvasive cardiac mapping of small experimental animals, for example.

MCG has also been used to detect reentry currents in cardiac flutter and fibrillation. The magnetic field produced by induced atrial flutter has been measured in isolated rabbit hearts. A moving dipole model was proposed to analyze the experimental data and to locate the reentry path [27].

In another newer study, acute myocardial infarction was induced by ligation of the left anterior descending coronary artery in a dog model. Magnetic field maps of early reperfused myocardium showed spatio-temporal field distributions consistent with anterior myocardial infarction. The use of super-paramagnetic contrast agents increased the sensitivity of standard MCG and may have an important implication for MCG in the assessment of regional myocardial ischemia, infarction, and perfusion [28].

A recent study was able to show that MCG mapping can detect age-related changes in cardiac intervals and in maps having significantly longer QTe, JTe, and T peak-Te intervals (Te = T end) in older Wistar rats. As compared to ECG recordings, MCG methodology simplified reproducible multisite noninvasive cardiac mapping of ventricular repolarization in an experimental setting [29].

44.6 Localization of Preexcitation and Cardiac Arrhythmias by Magnetocardiographic Mapping

The clinical accuracy of the MCG method has been tested by localizing the site of earliest ventricular activation during preexcitation in Wolff–Parkinson–White syndrome patients. The clinical reference has been obtained either during endocardial catheter mapping, or from successful catheter or surgical ablation of the accessory pathway. The results of the earlier studies showed that the MCG method was accurate enough for localization of cardiac electric sources for clinical purposes [30, 31]. The methodological localization accuracy of the MCG mapping has also been confirmed to be quite good [32].

In a newer study, 28 patients with Wolff–Parkinson–White syndrome were examined by MCG mapping and imaging techniques and with the five most recent accessory pathway localization ECG algorithms. ECD, effective magnetic dipole, and distributed-current imaging models were used for the inverse solution. MCG classification of preexcitation was found to be more accurate than that of ECG algorithms, and also provided additional information for the identification of paraseptal pathways and the existence of multiple accessory pathways [33].

By applying the completely noninvasive techniques of MCG and magnetic resonance imaging (MRI), it has been shown to be possible to define different origins of right ventricular ectopic beats in a complex heart model of nonischemic cardiomyopathy of 84 patients with surgically repaired Tetralogy of Fallot [34].

Postmyocardial infarction patients were investigated with cardiac MRI and signal-averaged 62-lead MCG. Three of six patients were suffering from sustained ventricular tachycardia (VT). A close matching of the low current density areas based on the QRS complexes and the high current density areas based on the late field signals were found. In three patients, the premature ventricular complexes morphologically resembling the clinical VT were localized close to the exit sites of these arrhythmias. The authors concluded that the MCG was useful in steering catheter ablation and coronary revascularization therapies [35].

An MCG was recorded pre- and post interventional therapy in three patients with atrial flutter and four patients with atrial fibrillation (AF), and in 20 healthy volunteers. The serial conduction pathway of the QRS segment was superimposed on a 3D heart outline generated by a magnetic field and verified by the silhouette on the magnetic resonance (MR) images. The MCG revealed a counterclockwise rotation of the atrial conduction in patients with atrial flutter, and random microreentry in the cases of AF [7]. An example of MCG localization in a patient suffering from continuous atrial tachycardia is presented in ➋ Fig. 44.2.

In conclusion, MCG has been used for localization of other sources of cardiac arrhythmias, such as the site of origin of tachycardias and extrasystoles, as well as for localization of a cardiac pacing catheter. The localization accuracies for a pacing catheter have been reported to be <1.0 cm, for extrasystoles 0.5–2.0 cm, and for tachycardias accuracy is more variable at 1.6–4.0 cm. In particular, the results in localizing a pacing catheter and the accessory pathway in preexcitation

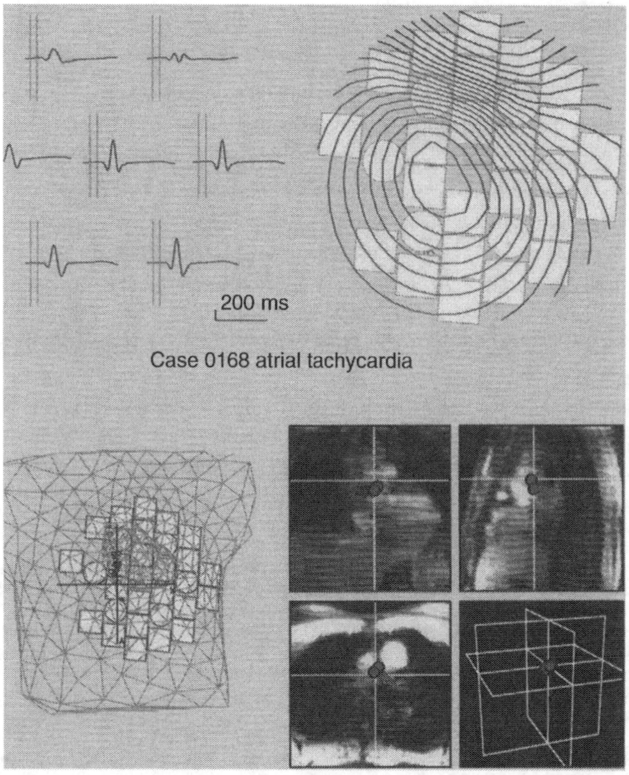

Figure 44.2

An example of MCG localization in a patient suffering from incessant atrial tachycardia. The integration of the MCG and MRI results was performed using electromagnetically localized MR positive fish-oil capsules. Invasive electrophysiologic study and successful catheter ablation confirmed the localization of the tachycardia focus in the intra-atrial septum. *Upper left panel*: time interval during the P-wave used for analysis. *Upper right panel*: magnetic field morphology relative to the sensors. *Lower left panel*: location of the tachycardia (*red dots*) relative to torso and sensors. *Lower right panel*: MCG localization of the tachycardia focus (*blue dots*) integrated to MR images.

confirm the good localization capability of the MCG method. The main obstacles in clinical cardiac source localization by MCG are the inherent problems concerning the clinical and imaging reference data integration.

44.7 Fetal Magnetocardiography

The fetal magnetocardiogram (FMCG) can be reliably recorded from approximately the 15th week of gestation onwards. The FMCG has the ability to accurately record cardiac time intervals, and to provide a real-time recording that reflects cardiac electrical activity. The FMCG has been demonstrated to be able to diagnose different types of atrioventricular conduction defects, paroxysmal supraventricular tachycardia, and repolarization abnormalities especially after 20 weeks of gestation. Standardization of the recording, signal processing, and measurement techniques has resulted in data that is reproducible, and when combined with normal values for different gestational ages, they can be of clinical value [36, 37].

In a study of 102 low-risk pregnant women, at the 20th to 42th gestational week, the fetal MCG provided a significant advantage via the technology available for recording the antenatal fetal heart. In the unaveraged recordings, the QRS complex was successfully detected in 68 cases (67%). Of those 68 traces, a P-wave was detected in 51 (77%) and a T-wave in 49 (72%) of the traces, using off-line signal-averaging techniques. Although good quality traces were obtained throughout the range of gestational ages, in general it was more difficult below 28 weeks. The QRS duration was found to increase significantly with increasing gestation [38]. In another study, various parameters concerning the electrical excitation of the heart, such as AV conduction, repolarization period, and morphology of the QRS complex, could be determined, leading to a more profound analysis of fetal arrhythmias [39].

Distinct patterns of initiation and termination of paroxysmal SVT have been detected using the MCG [40] as well such phenomena as Wolff–Parkinson–White syndrome, QRS aberrancy, and multiple reentrant pathways. A strong association between fetal trunk movement and the initiation and termination of SVT, suggesting autonomic influences, was also found.

The ability of the FMCG to reveal conduction system disease was evaluated in 702 fetuses initially referred on account of an arrhythmia. Altogether 306 had an irregular rhythm confirmed using either FMCG or postnatal 12-lead ECG. Eight (2.6%) had intermittent first- or second-degree AV block confirmed by FMCG and/or postnatal 12-lead ECG. Cases with conduction block may benefit from transplacental therapy with dexamethasone [41]. This method provides additional information concerning the effect of congenital heart disease (CHD) on the cardiac conduction system. As in the neonate, the FMCG changes do not reflect the severity of CHD. Indeed, the FMCG cannot serve as a primary diagnostic tool in the case of CHD for which echocardiography is more helpful.

Repolarization abnormalities in fetal heart can also be detected with the MCG during gestation. In one study, two patients were evaluated because of sustained fetal bradycardia, the diagnosis of QT-prolongation was made at the 29th and at the 35th gestational week. In both newborns, the QT-prolongation was confirmed by the postpartal ECG [42].

In general, good agreement of the magnetocardiograms and cardiac findings in newborns has been documented in a study investigating 189 fetal MCGs in 63 pregnant women between the 13th and the 42th week of pregnancy. In 16 recordings before the 20th gestational week, the signal strength was too weak to permit evaluation. Brief episodes of bradycardia, isolated supraventricular and ventricular premature beats, bigeminy and trigeminy, sinoatrial block, and atrioventricular conduction delays were found [43].

It has been suggested that the MCG can be used for noninvasive evaluation of hypertrophy of the fetal heart. In uncomplicated pregnancies, the magnitude of the MCG current dipole correlates with the gestational age, whereas in fetuses with cardiomegaly, the magnitude of the current dipole has been higher, reflecting the increased myocardial mass confirmed by ultrasound [44].

FMCG has shown that fetal P-wave and QRS complex durations increase with gestational age, reflecting a change in the cardiac muscle mass. A total of 230 FMCGs were obtained in 47 healthy fetuses between the 15th and 42nd week of gestation. The authors concluded that, from approximately the 18th week to term, fetal cardiac time intervals, which quantify depolarization times, can be reliably determined using MCG. The P-wave and QRS complex duration show a high dependence on age which to a large part reflects fetal growth. Gender instead, plays a role in QRS complex duration in the third trimester. ECD strength reflected gestational age slightly more reliably ($r^2 = 0.93$) than signal amplitude values (mean, median, maximum: $r^2 = 089, 0.88, 0.85$, respectively). The overall correlation of the amplitude to gestational age compared favorably with that of QRS complex duration. Fetal development is thus in part reflected in the fetal MCG and may be useful in the identification of intrauterine growth retardation [45, 46].

In conclusion, FMCG allows an insight into the electrophysiological aspects of the fetal heart, is accurate in the classification of fetal arrhythmias, and shows potential as a tool in defining a population at risk of congenital heart defects. FMCG offers unique capabilities for assessment of fetal heart rate (FHR) and fetal behavior, which are fundamental aspects of neurodevelopment. FMCG actograms are specific for fetal trunk movements, which are thought to be more important than isolated extremity movements and other small fetal movements. The ability to assess FHR, FHRV, and fetal trunk movement simultaneously makes fMCG a valuable tool for neurodevelopment research. In clinical reality, fMCG devices are still rare and a fetal arrhythmia or a congenital heart defect is in general discovered during prenatal evaluation by ultrasonography.

44.8 Arrhythmia Risk Assessment

MCG is sensitive to cardiac electrical activity of a very small amplitude and has therefore made it an interesting tool in the assessment of the risk of serious ventricular arrhythmias. The rapid development in instrumentation from single to multichannel devices capable of mapping large precordial areas has made clinical risk assessment studies possible.

In almost any structural heart disease, abnormalities in both depolarization and repolarization periods may lead to ventricular arrhythmias and sudden cardiac death. MCG studies considering ventricular arrhythmia risk assessment are still relatively few in number, with most of the data coming from postinfarction populations.

Postmortem studies in patients with postinfarction VT have revealed thin layers of surviving myocardial tissue, which might show mostly tangential currents and thus be detectable more readily by MCG than by ECG [47]. Late potentials in ECGs have been shown to be markers of delayed conduction serving as a substrate for reentrant ventricular arrhythmias. Late fields in MCG comparable to late potentials in ECGs have also been described [48]. Subsequently, late fields have shown a capability similar to that of late potentials in the discrimination of patients with and without ventricular arrhythmias after myocardial infarction [49, 50]. A few studies have applied magnetic source imaging in localization of the sources of late fields, which might be of value prior to catheter ablation or arrhythmia surgery [35].

In addition to late fields that may be detected at the end of the QRS, the abnormal delayed depolarization displays increased fragmentation of the whole magnetocardiographic QRS. The intra-QRS fragmentation method is based on finding the number of signal extrema during filtered QRS (➲ Fig. 44.3) and by computing the sum of the amplitude differences between neighboring extrema, thus yielding the intra-QRS fragmentation score (FRA) [51]. Recently, increased intra-QRS fragmentation in the MCG registered soon after acute MI has shown promise in the identification of patients at high risk of arrhythmic events in a prognostic study of 158 patients with acute MI and left ventricular ejection fraction (LVEF) < 50%. During follow-up of 50 +/− 15 months, 32 (20%) patients died and 18 (11%) had an arrhythmic event. Increased FRA in the MCG and LVEF < 30% yielded positive and negative predictive accuracies of 50 and 91% for arrhythmic events. The ECG predicted all-cause mortality ($P < 0.05$) but not arrhythmic events [52].

The dispersion of repolarization is increasingly recognized as a major factor in the genesis of malignant ventricular arrhythmias. Preliminary data from patients with hypertension and ischemic heart disease suggest that MCG might be especially sensitive to subtle changes in the repolarization period [53, 54]. Transmural repolarization in MCG, displayed as the terminal part of the T-wave, has been the subject of arrhythmia risk assessment studies in patients with ischemic

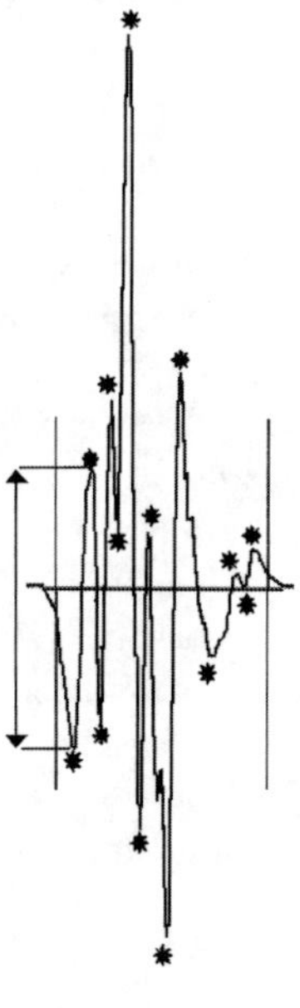

Figure 44.3

The principle of the intra-QRS fragmentation score analysis in MCG. After binomial filtering, the number of polarity changes or extrema () is computed. Next, the differences of each adjacent extrema are computed, and the differences are summed. As an example, the difference between the first and second extrema is shown with a thin arrow. Finally, the difference between the first and the last extrema is added to this sum, yielding the intra-QRS fragmentation score. The horizontal bars indicate the onset and offset of the filtered QRS.

heart disease and dilated cardiomyopathy. Patients with sustained ventricular arrhythmias showed prolonged T-wave peak to T-wave end interval as a sign of dispersion of transmural repolarization [55, 56]. In another study, nondipolar isointegral maps during the repolarization phase in postinfarction patients prone to ventricular arrhythmias were found [57]. It was suggested that the non-dipolarity might serve as a marker of increased repolarization heterogeneity.

Repolarization disparities related to idiopathic long-QT syndrome also seem to be detectable with MCG. Children with long-QT syndrome, both with and without symptoms were investigated [58]. All patients showed beat-to-beat variability in T-wave morphology. When the isofield maps during the T-wave were analyzed with eigenvectors for data reduction, the symptomatic patients displayed more disparity in their maps, suggesting more heterogeneous repolarization.

Modern multichannel devices seem to allow quick MCG registrations in the hospital environment thereby rendering it a potential tool in arrhythmia risk assessment. However, a few issues need more elucidation before the role of MCG in clinical decision making can be fully assessed. First, is there really essential information in MCG not available from the ECG? In order to evaluate this, we need risk assessment studies comparing MCG not to 12-lead ECGs but to signal-averaged ECGs (SAECGs) and body surface potential mapping equally covering large precordial areas. In addition, the MCG methodology needs to be subjected to well-designed large-scale prospective studies. Until this has been done, the role of the MCG in the assessment of the risk of ventricular arrhythmias and sudden cardiac death remains to be established.

44.8.1 AF

AF is the most prevalent clinically important rhythm disturbance. The initiating and perpetuating factors of AF may vary, and prolonged AF may lead to atrial electrical and mechanical remodeling [59–61]. The effective application of new treatments for AF calls for improved diagnostics and has also led to growing interest in noninvasive methods capable of detecting and characterizing abnormalities in atrial signals.

Subtle abnormalities in atrial electric and magnetic signals could serve as markers of foci or substrates for AF. The SAECG detects abnormalities in atrial signals in patients prone to AF. These include prolongation of atrial depolarization, abnormal frequency content, and increased spatial dispersion of atrial signal duration [62–64]. Recently, magnetocardiographic techniques have also been applied to the investigation of atrial electrophysiology and pathogenesis in AF.

44.8.1.1 Magnetocardiographic P-Wave in Patients with AF, Analyses of Non-Filter Signal and Application of High-Pass Filtering Techniques

Specific changes of the P-wave in MCG were found in a study in a total of 35 subjects, including 15 AF patients (50–70 years) with persistent AF converted to sinus rhythm and 20 healthy young men. The multichannel MCG over the anterior and posterior chest and 12-lead ECG were recorded simultaneously. Sum channels from all MCG channels and separately sums from anterior and posterior channels as well as ECG channels were created. The P-wave duration was manually measured in each sum channel. Also the homogeneity and fragmentation of MCG maps were evaluated. The P-wave was divided into four segments and the correlation between maps was calculated. All segments were compared with the first segment, and the average of these three correlations, the p-score, was used as the homogeneity factor. The fragmentation index of the P-wave was calculated as the sum of amplitude differences between two amplitude peaks multiplied by the total number of amplitude peaks in each sum channel. By MCG, the P-wave duration was longer and both correlation factor and fragmentation index were lower in patients compared to normals. Similar differences were not seen in the ECG. Differences were clearest in the sum channel of all MCG channels, for example, P-wave duration was 133 ms on average in patients versus 100 ms in controls [65].

High-pass filtering techniques have also been applied to the analysis of the atrial MCG signal. Multichannel MCG over the anterior chest and orthogonal three-lead ECGs were recorded in nine patients who had paroxysmal lone AF and in ten healthy subjects in duplicate at least 1 week apart. Data were averaged using an atrial wave template and high-pass filtered at 25, 40, and 60 Hz. Atrial signal duration with automatic detection of onset and offset and root mean square amplitudes

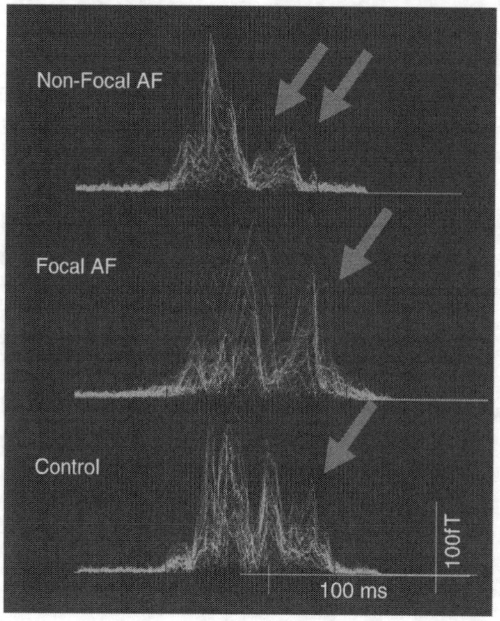

Figure 44.4
Examples of typical 40 Hz high-pass filtered atrial complex in a patient with non-focal AF, in a patient with focal AF and in a healthy subject. In the patients with focal AF, the atrial signal strength was normal also during late phase of atrial complex where the left atrium is depolarized, but in a patients not defined focal triggers late phase amplitudes were reduced. Data is expressed as superimposed display of 33 magnetometer channels.

(RMS) of the last portion of the atrial signal were determined. Reproducibility was best using a 40 Hz filter, somewhat better in MCG than in ECG and similar in patients and controls. For example, the difference between two measurements of atrial signal duration was 3.5 ms on average (coefficient of variation 3.3%) by MCG and 6.9 ms (coefficient of variation 6.1%) by ECG [66].

The atrial signal durations in MCG and SAECG were correlated, with $r = 0.64$ ($p < 0.01$). In addition, there was a correlation between RMS amplitudes of the first portion of the atrial complex in the MCG and the SAECG ($r = 0.49$, $p < 0.01$) but not in the last atrial portion ($r = 0.25$–0.30, $p =$ NS).

In this small population comprising patients with lone paroxysmal AF, the atrial wave duration was not longer but RMS amplitudes of the last 40 ms of the atrial complex were lower in patients.

Lately, the methods have been applied in a study of 81 patients with paroxysmal lone AF and 81 matched controls. In this study, the atrial depolarization complex was slightly prolonged in patients, namely, 109 ms on average versus 106 ms. The late RMS amplitudes were reduced in a subgroup of patients with non-focal AF but were otherwise normal (➲ Fig. 44.4). The findings are consistent with invasive measurements in focally triggered AF patients, in which only a minority of patients have shown conduction delay between or within the atria, or reduction in left atrial amplitude [67–69].

44.8.1.2 Spatial MCG Maps, Field Polarity, and Orientation during Atrial Activation and Application of Surface Gradient Methods

In another study, spatial MCG maps of 26 WPW patients were investigated [30]. Patients with AF attacks were found to have more dispersed atrial depolarization distributions compared to patients without AF. Altogether, 11/20 (55%) of the patients with AF problems were found to have more than two extrema in the atrial depolarization maps. On the other hand, 4/6 (67%) patients without AF had bipolar MCG maps. In an earlier study, multipolar MCG fields during the late phase of atrial activation have related to left atrial overloading [70]. In a study comprising patients with lone paroxysmal AF, for example, with normal left atria, multipolarity in MCG maps (integral maps over last 20 and 50 ms) was related to lone AF [71].

Magnetic field patterns can be visualized and parameterized using pseudo-current (90 degrees rotated field gradient) amplitude and direction distributions. This method is applied in some ongoing studies focused on evaluating atrial signal propagation during sinus rhythm in patients with focally triggered paroxysmal AF and in healthy subjects [71, 72]. In a study comprising 28 patients with focally triggered paroxysmal AF and 23 controls, the magnetic field orientation during the early part of atrial depolarization was mainly to the left and downward, and was similar in patients and controls. On the other hand, during the late phase of depolarization, field orientation was more variable in both groups and differed between groups as illustrated in ➋ Fig. 44.5. The degree of maximum gradient rotation over the atrial wave was larger, and the upward orientation was more common in patients. Results showed diversity and inhomogeneity in the propagation of atrial signals especially at the late phase of activation when the left atrium was depolarizing. Findings were more pronounced in patients but some seemed to occur also in healthy atria. It was concluded that the altered depolarization front may represent a conduction defect in the left atrium, and that this may be a normal variant facilitating the manifestation of AF in the presence of focal triggers.

In another study, MCG field patterns derived by the pseudo-current method were compared to invasive electroanatomic activation maps (EAM) obtained from patients undergoing catheter ablation treatment with prior AF. The orientation of the magnetic fields during the first 30 ms of atrial depolarization representing RA activation, and during early (40-70 ms from P onset) and later part (last 50%) of LA depolarization was determined. The mean of the angles of the top 30% of the strongest pseudocurrents was used, zero angle direction pointing from subject's right to left and positive clockwise. Breakthrough of electrical activation to LA occurred through Bachmann bundle (BB) in 14, margin of fossa ovalis (FO) in 3, coronary sinus ostial region (CS) in 2, and their combinations in 10 cases by invasive reference in total of 29 different P-waves. The pseudocurrent direction in MCG maps over the first 30 ms of atrial complex was mostly leftward down, with a mean angle of 43° (CSD 28°). Over the time interval of 40–70 ms from the onset of the atrial complex, the mean angle was 39° (CSD 30°). Over the time interval of last 50% of atrial complex, the mean angle was 3° (CSD 51°).

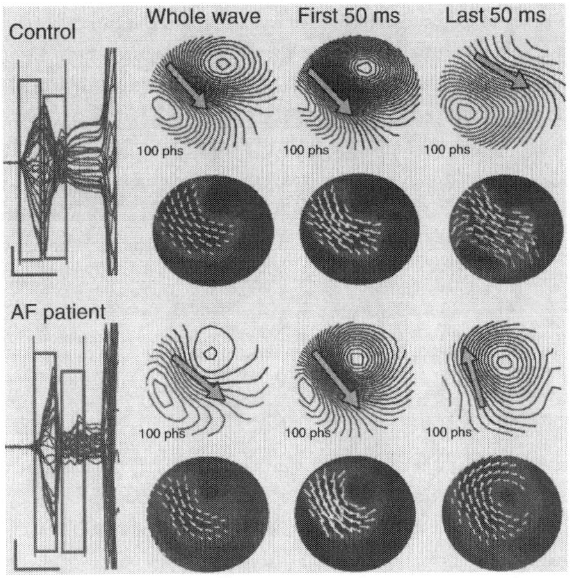

Figure 44.5
An example of orientation of magnetic fields during atrial depolarization from one healthy control (*upper rows*) and one patient with focally triggered lone AF (*lower rows*) performed as isofield integral maps (*rows 1 and 3*) and as pseudo-current maps (*rows 2 and 4*).

In the applied time interval the MCG pseudo-current angle corresponded the direction of propagation in EAM. When both the early and late LA MCG maps were viewed together, three types of combinations emerged: Type 1 with both maps showing pseudocurrent orientation leftward down, Type 2 with the map over the 40–70 ms orienting leftward down and the map over last 50% of atrial signal orienting leftward up, and Type 3 with both maps orienting leftward up. The three MCG atrial wave types correctly identified the LA breakthrough sites to BB, CS, FO, or their combinations in 27 of 29 (93%) cases [73].

Later on the method was applied in a population comprising 107 patients with lone paroxysmal AF (age 45 ± 12 years) and 94 controls. The Pd was longer in AF patients than in controls (112 ± 13 vs. 104 ± 13; $p < 0.001$), which was most obvious in Type 1 wave (109 ± 12 vs. 102 ± 11 ms, $p = 0.003$). The distribution of the atrial wave types differed between AF patients and controls: Type 1 occurred in 67% and Type 2 in 20% of controls whereas Type 1 occurred in 54% and Type 2 in 42% of AF patients, $p < 0.01$ for difference. Accordingly susceptibility to paroxysmal lone AF is associated with propagation of atrial signal to LA via margin of fossa ovalis or multiple pathways. When conduction occurs via Bachmann bundle, it is related with prolonged atrial activation. Thus altered and alternative conduction pathways may contribute to pathogenesis of lone AF [74].

44.8.1.3 Analyses of Atrial Signals during AF

The adaptation of QRS subtraction techniques and time frequency analysis has revealed the diagnosis of partial atrial standstill in an adult [75] and diagnosis of atrial flutter and AF in fetuses [76]. Now there are also ongoing studies designed to test the diagnostic performance of MCG mapping to separate clinical subgroups of AF, such as focal AF, by analyzing the atrial signal and its time domain as well as its spatial distribution during AF.

In conclusion, the effective application of new treatments for AF calls for improved diagnostics and has also led to a growing interest in noninvasive methods capable of detecting and characterizing abnormalities in atrial signals. The first applications of magnetocardiographic techniques to investigate atrial electrophysiology and pathogenesis in AF have presented extra dipoles in magnetic field maps during the last portion of the atrial depolarization signal in WPW patients with paroxysmal AF and in lone paroxysmal AF. Other abnormal features are altered field orientation during the late phase of atrial activation particularly in focally triggered lone AF, and reduced atrial signal amplitudes in AF patients without demonstrable focal triggers. In patients with persistent AF converted to sinus rhythm, prolongation and increased fragmentation of the atrial signal have been detected. Overall, abnormalities seem to be more pronounced in the late phase of the atrial complex corresponding to the depolarization of the left atrium. The diversity and inhomogeneity of propagation throughout the atrial signal seem to vary in different AF cohorts.

Currently, the patient series are still small, and further investigation is warranted before the real value of MCG in the clinical assessment of AF can be evaluated. However, MCG mapping techniques seem to be capable of noninvasively detecting abnormalities in the atrial activation sequences common in AF, even when the standard ECG is normal. It also seems that the MCG method may be applied to the study of pathophysiologic processes in AF and may identify subsets of patients with different underlying mechanisms for AF. Improved diagnostics could guide the selection of new treatment modalities, for example, catheter ablation or pharmacological treatment of AF.

44.9 Myocardial Ischemia and viability

One of the most interesting areas in clinical MCG is the detection and characterization of myocardial ischemia and viability. So far, few studies have been reported in this field, but the results are very encouraging. A new and accurate noninvasive method for recognition of acute and chronic ischemia could have important clinical applications, especially because therapeutic interventions for rapid and effective revascularization, such as percutaneous coronary interventions, are widely available.

It has been stated that some changes of the ST-segment could be explained by the ability of the MCG to detect DC currents, thus being more sensitive to ST changes than ECG or body surface mapping [53, 77–79]. In addition, the results of the equivalent dipole calculation during cardiac depolarization and repolarization has enabled the separation of patients with coronary artery disease from healthy controls [80].

The magnetic field orientation at rest has been proposed to separate coronary artery disease patients from healthy controls, with more prominent changes in the field orientation in severe disease. The magnetic field orientation is the angle between the line joining the field extrema and the right-left line of the torso [81]. In some studies, a semiautomatic surface gradient method helped to define the magnetic field orientation as the orientation of the maximum spatial field gradient, and this has been shown to separate both single-vessel and triple-vessel coronary artery disease patients from healthy controls during exercise-induced ischemia [82, 83]. Later, a method for heart rate adjustment of the magnetic field orientation during the recovery phase of exercise stress testing, which further improved ischemia detection, was developed [84].

The ST-segment depression and ST-segment slope, used in 12-lead ECG as ischemia parameters, can also detect ischemia in MCG. The ischemia-induced ST-depression takes place over the lower middle anterior thorax, and the reciprocal ST-elevation over the left anterior shoulder, locations orthogonal to those found in body surface potential mapping. The most prominent T-wave changes were found in patients with inferior ischemia and in patients with a history of myocardial infarction (➋ Fig. 44.6) [83, 85]. The ratio of the ST-T and QRS isointegral maxima has been reported to be reduced in coronary artery disease patients compared to healthy controls [86].

A few smaller studies have been designed to test the diagnostic performance of MCG mapping in detecting and localizing areas of hibernating myocardium. The viability of the myocardium has been first confirmed by a full set of other diagnostic tests: exercise ECG, thallium stress test, dobutamine MRI, and positron emission tomography (PET). Preliminary results look promising, but there are still some problems concerning the modeling of chronic ischemia. Both the MCG and the ECG showed a significant elevation or depression of the ST-segment during exercise-induced ischemia when they were investigated using a nonmagnetic bicycle ergometer in a MSR. The injury currents were in most cases directed from the ischemic area to the nonischemic area. For example, anterior ischemia caused an injury current directed from the apex to the base of the heart. The injury currents induced by transient ischemia and infarction of the same anatomical region were found to flow in the opposite direction. This change of direction was reflected as either ST-depression or ST-elevation in morphological signals. The anatomical location of the injury currents was in topographical agreement with the results of the reference methods (coronary angiography and myocardial scintigraphy) [87].

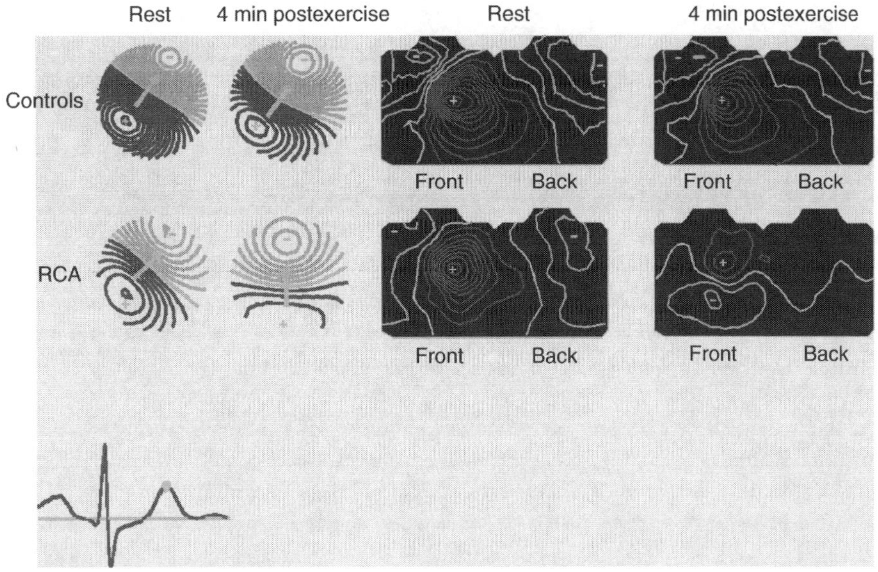

Figure 44.6
Exercise MCG in a coronary artery disease patient with a significant stenosis of the right coronary artery (RCA). The patient had the most extensive map rotation of the T-wave 4 min postexercise. Compared to the control group, the map rotation was more extensive ($45 \pm 39°$ vs $9 \pm 8°$, $p < 0.005$). Sensitivity for RCA was 75% at the specificity of 82% (Hänninen et al., 2000).

Corresponding to the absence of electrically active myocardial tissue, reduced current magnitudes were observed for the regions of infarcted myocardium (less than 1 $\mu A/mm^2$) in a series of 40 subjects, including normals and patients with a history of myocardial infarction. The current density distributions correlated well with other cardiological investigations (e.g., left ventricle catheterization, scintigraphy, echocardiography) [88]. Ischemia localization was attempted in a small study of four patients with single-vessel coronary artery disease with no previous myocardial infarction. Current density estimation (CDE) was focused on in the ischemic regions confirmed by positron emission tomography in all four patients and injury currents were restricted to the region of the stenosed coronary artery [89]. A recent large clinical study included 417 subjects: 177 patients with angiographically documented CAD (stenoses ≥ 50%), 123 symptomatic patients without hemodynamically relevant stenosis, and 117 healthy subjects [90]. Contrary to the ECG, unshielded MCG revealed significant differences between normals and symptomatic patients with and without relevant stenoses using current density reconstruction during repolarization at rest. The discrimination between normals and CAD patients was achieved with a sensitivity of 73.3%, and specificity of 70.1%. In another series of 50 patients with CAD (62 +/ − 10 years; EF = 76 + / − 11%; registration: before, 24 h, and 1 month ($n = 25$) after PCI) and 57 normals (51 +/ − 9 years), current density vector (CDV) maps were reconstructed within the ST-T interval and classified from category 0 (normal) to category 4 (grossly abnormal) [91]. Twenty-four hours after PCI, more maps were classified as category 2 ($P < 0.05$) and less as category 4 ($P < 0.005$). One month after PCI, the MCG results further improved: more maps were classified as category 1 ($P < 0.05$) and 2 ($P < 0.005$) and fewer maps as category 4 ($P < 0.0001$). The ECG remained unchanged in the course of PCI.

A few clinically interesting studies on acute myocardial ischemia in the emergency room setting have been published. A 6-min resting MCG scan has been shown to have sensitivity, specificity, positive and negative predictive values of 76.4, 74.3, 70.0, and 80.0%, respectively, for the detection of ischemia ($p < 0.0001$) in 125 patients with presumed ischemic chest pain [92]. In another new series of 264 acute chest pain patients without ST-elevation, T vector changes in the MCG were found to have a specificity and positive predictive value > 90% for angiographically confirmed diagnosis of significant coronary artery disease. It is noteable that 25% patients had to be excluded from the study for poor signal quality [93].

In conclusion, magnetocardiographic mapping seems to be an accurate method for ischemia and viability detection. Moreover, it is able to localize ischemic regions in the heart in a comparable fashion to other methods. Recent larger clinical series stress the potential of this noninvasive method for current clinical decision making. However, it is the task of future large-scale clinical studies to establish a clear mandate for the MCG method in every day patient work.

44.10 Discussion

The clinical application of any method means that it has to contribute at least in one of the following fields of clinical medicine: diagnosis, therapy, or prognosis. In particular, noninvasive methods improving the accuracy of diagnosis of a cardiac disease are valuable. However, most important new methods are able to deliver current palliative or curative therapy, either by improving the results of existing therapies or by providing new modes of therapy. Methods helping to assess better the prognosis of certain heart diseases in individual patients are also desired.

MCG has many advantages, and it may become a clinically accepted method. First of all, it is totally noninvasive. It is not necessary to attach electrodes or other sensors requiring direct contact to the patient to record an MCG. Currently operating multichannel magnetometers are feasible to use and enable fast MCG recordings. A full measurement can be carried out in few minutes. Such advantages of MCG recordings also mean high patient comfort. In addition, multiple temporal and spatial parameters can be extracted from a single heartbeat for complete electromagnetic characterization of the function of the patient's heart. The spatiotemporal resolution of MCG mapping is much higher than that of conventional ECG methods. Moreover, the MCG methods still have unexplored potential for extracting new physiological and pathophysiological information about the electrical activation in the myocardium.

However, the MCG has some disadvantages. The equipment is currently expensive, and requires the use of liquid helium and an MSR. Such technical demands so far exclude the wider applicability of MCG as a quick bedside test and during catheter interventions. The development of new supra-conducting materials, operating in higher temperatures achieved using liquid nitrogen, has not yet been able to solve this problem. Low-cost multichannel devices operating in an unshielded environment have recently been introduced with promising clinical results. This way of development might be one of the ways to bring this method closer to clinicians. One profound obstacle still remains: MCG systems are generally sensitive to moving magnetic objects, which excludes some patients from the studies.

44.11 Conclusions

From all magnetocardiographic studies published so far, it can be concluded that the diagnostic performance of the MCG is superior in several applications when compared to the conventional ECG or even body surface mapping. However, this advantage has not yet encouraged clinicians to widely accept and utilize the method, mostly because of the high cost, low availability, and lack of standardization. Nevertheless, clinical applications of MCG in the fields of therapy and prognosis are currently of growing interest. The localization of arrhythmias and the detection of arrhythmia risk are already established therapy-related applications. Non-pharmacologic antiarrhythmic and antirejection medical therapies guided by MCG have been proposed, but such suggestions need more comprehensive studies. The detection and localization of acute and chronic ischemia and assessment of myocardial viability would be applicable to a large number of cardiac patients, if the MCG method proves to be successful in larger patient series. Studies on the prognostic value of MCG after myocardial infarction, in long-QT syndrome, congenital heart disease, and cardiomyopathy, may offer new applications, taking into account the promising results of recent studies on the prophylactic use of implantable defibrillators in high-risk patient groups. The increasing availability of high-performance multichannel MCG systems in the clinical environment will strongly contribute to these efforts.

References

1. Siltanen, P.P., Magnetography, in *Comprehensive Eletrocardiology, Theory and Practice in Health and Disease*, vol. 2, M.W. Lawrie and T.D. Veitch, Editors. New York: Pergamon, 1989, pp. 1405–1438.
2. Stroink, G., W. Moshage, and S. Achenbach, Cardiomagnetism, in *Magnetism in* Medicine, W. Andrä and H. Nowak, Editors. Berlin: Wiley, 1998, pp. 136–189.
3. Mäkijärvi, M., K. Brockmeier, and U. Leder, et al.,, New trends in clinical magnetocardiography, in C. Aine*Biomag96: Proceedings of the 10th International Conference on Biomagnetism*, C. Aine, et al., Editors. New York: Springer, 2000, pp. 410–417.
4. Trip, J.H., Physical concepts and mathematical models, in *Biomagnetism: An Interdisciplinary Approach*, S.J. Williamson, Editor. New York: Plenum, 1982, pp. 101–149.
5. Horacek, B. M., Digital model for studies in magnetocardiography. *IEEE Trans. Magn.*, 1973;**9**: 440–444.
6. Nenonen, J., C. Purcell, B.M. Horacek, G. Stroink, et al.,, Magnetocardiographic functional localization using a current dipole in a realistic torso. *IEEE Trans. Biomed. Eng.*, 1991;**38**: 658–664.
7. Nakai, K., K. Kawazoe, H. Izumoto, J. Tsuboi, Y. Oshima, T. Oka, K. Yoshioka, M. Shozushima, A. Suwabe, M. Itoh, K. Kobayashi, T. Shimizu, and M. Yoshizawa, Construction of a three-dimensional outline of the heart and conduction pathway by means of a 64-channel magnetocardiogram in patients with atrial flutter and fibrillation. *Int. J. Cardiovasc. Imag.*, 2005;**21**: 555–561; discussion 563–564.
8. Katila, T. and P. Karp, Magnetocardiography: morphology and multipole presentations, in *Biomagnetism, An Interdisciplinary Approach*, S.J. Williamson, G.L. Romani, L. Kaufmann, and I. Modena, Editors. New York: Plenum, 1983, pp. 237–263.
9. Nenonen, J., Magnetocardiography, in *SQUID Handbook*, Vol. 2, Sect. 9.3, J. Clarke, and A. Braginski, Editors. Berlin: Wiley, 2005.
10. van Oosterom, A., T.F. Oostendorp, G.J. Huiskamp, and M. ter Brake, The magnetocardiogram as derived from electrocardiographic data. *Circ. Res.*, 1990;**67**: 1503–1509.
11. Sepulveda, N.G., B.J. Roth, and J. Wikswo Jr., Current injection into a two-dimensional anistropic bidomain. *Biophysical. J.*, 1989;**55**: 987–999.
12. Liehr, M., J. Haueisen, M. Goernig, P. Seidel, J. Nenonen, and T. Katila, Vortex shaped current sources in a physical torso phantom. *Ann. Biomed. Eng.*, 2005;**33**: 240–247.
13. Cohen, D. and L.A. Kaufman, Magnetic determination of the relationship between the S-T segment shift and the injury current produced by coronary artery occlusion. *Circ. Res.*, 1975;**36**: 414–424.
14. Baule, G. and R. McFee, Detection of the Magnetic Field of the Heart. *Am. Heart. J.*, 1963;**66**: 95–96.
15. Cohen, D. and E. Edelsack, J. Zimmerman, Magnetocardiograms taken inside a shielded room with a superconducting point-contact magnetometer *Appl. Phys. Lett.*, 1970;**16**: 278–280.
16. Zimmerman, J.E. and N.V. Frederick, Miniature ultrasensitive superconducting magnetic gradiometer and its use in cardiography and other applications. *Appl. Phys. Lett.*, 1971;**19**: 16–19.
17. Koch, H., Recent advances in magnetocardiography. *J. Electrocardiol.*, 2004;**37**: 117–122.
18. Uusitalo, M.A. and R.J. Ilmoniemi, Signal-space projection method for separating MEG or EEG into components. *Med. Biol. Eng. Comput.*, 1997;**35**: 135–140.
19. Chen, M. and R.T. Wakai, and B. Van Veen, Eigenvector based spatial filtering of fetal biomagnetic signals. *J. Perinat. Med.*, 2001;**29**(6): 486–496.
20. Taulu, S., M. Kajola, and J. Simola, Suppression of interference and artifacts by the signal space separation method. *Brain Topogr.*, 2004;**16**: 269–275.
21. Hyvärinen, A., Fast and robust fixed-point algorithms for independent component analysis. *IEEE Trans. Neural Networks*, 1999;**10**(3): 626–634.
22. Comani, S., D. Mantini, G. Alleva, S. Di Luzio, G.L.F. Romani, et al., Magnetocardiographic mapping using independent component analysis. *Physiol. Meas.*, 2004;**25**: 1459–1472.
23. Sternickel, K., A. Effern, K. Lehnertz, T. Schreiber, et al., Nonlinear noise reduction using reference data. *Phys. Rev.*, 2001;**E63**: 036209.
24. Burghoff, M., J. Nenonen, L. Trahms, and T. Katila, Conversion of magnetocardiographic recordings between two

different multichannel SQUID devices. *IEEE Trans. Biomed. Eng.*, 2000;**47**: 869–875.

25. Numminen, J., S. Ahlfors, R. Ilmoniemi, J. Montonen, et al., Transformation of multichannel magnetocardiographic signals to standard grid form. *IEEE Trans. Biomed. Eng.*, 1995;**42**: 72–78.
26. Cohen, D., Ferromagnetic contamination in the lungs and other organs of the human body. *Science*, 1973 May 18;**180**(87): 745–8.
27. Ribeiro, P.C., A.C. Bruno, P.L. Saboia e Silva, C.R. Barbosa, E.P. Ribeiro, E.C. Monteiro, and A.F. Costa, Detection of reentry currents in atrial flutter by magnetocardiography. *IEEE Trans. Biomed. Eng.*, 1999;**39**(8): 818–824.
28. Brazdeikis, A., C.W. Chu, P. Cherukuri, S. Litovsky, and M. Naghavi, Changes in magnetocardiogram patterns of infarcted-reperfused myocardium after injection of superparamagnetic contrast media. *Neurol. Clin. Neurophysiol.*, 2004;**30**: 16.
29. Brisinda, D., M.E. Caristo, and R. Fenici, Contactless magnetocardiographic mapping in anaesthetized Wistar rats: evidence of age-related changes of cardiac electrical activity. *Am. J. Physiol. Heart Circ. Physiol.*, 2005;Dec 22.
30. Mäkijärvi, M., J. Nenonen, L. Toivonen, J. Montonen, T. Katila, and P. Siltanen, Magnetocardiography: supraventricular arrhythmias and preexcitation syndromes. *Eur. Heart J.*, 1993;**14**(Suppl E): 46–52.
31. Moshage, W., S. Achenbach, K. Göhl, et al., Evaluation of the noninvasive localization accuracy of the cardiac arrhythmias attainable by multichannel magnetocardiography (MCG). *Int. J. Cardiac. Imag.*, 1996;**12**: 47–59.
32. Fenici, R., K. Pesola, P. Korhonen, M. Makijarvi, J. Nenonen, L. Toivonen, P. Fenici, and T. Katila, Magnetocardiographic pacemapping for nonfluoroscopic localization of intracardiac electrophysiology catheters. *PACE*, 1998,Nov;**21**(11 Pt 2): 2492–2499.
33. Fenici, R., D. Brisinda, J. Nenonen, and P. Fenici, Noninvasive study of ventricular preexcitation using multichannel magnetocardiography. *PACE*, 2003,Jan;**26**(1 Pt 2): 431–435.
34. Agren, P.L., H. Goranson, H. Jonsson, and L. Bergfeldt, Magnetocardiographic and magnetic resonance imaging for noninvasive localization of ventricular arrhythmia origin in a model of nonischemic cardiomyopathy. *PACE*, 2002,Feb;**25**(2): 161–166.
35. Leder, U., J. Haueisen, P. Pohl, R. Surber, J.P. Heyne, H. Nowak, H. R. Figulla, Localization of late potential sources in myocardial infarction. *Int. J. Cardiovasc. Imag.*, 2001,Aug;**17**(4): 315–325.
36. Grimm, B., J. Haueisen, M. Huotilainen, S. Lange, P. Van Leeuwen, T. Menendez, M. J. Peters, E. Schleussner, and U. Schneider, Recommended standards for fetal magnetocardiography. *PACE*, 2003 Nov;**26**(11): 2121–2126.
37. Stinstra, J., E. Golbach, P. van Leeuwen, S. Lange, T. Menendez, W. Moshage, E. Schleussner, C. Kaehler, H. Horigome, S. Shigemitsu, and M. J. Peters, Multicentre study of fetal cardiac time intervals using magnetocardiography. *BJOG*, 2002,Nov;**109**(11): 1235–1243.
38. Quinn, A., A. Weir, U. Shahani, et al., Antenatal fetal magnetocardiography: a new method for fetal surveillance? *Br. J. Obstet. Gynaecol.*, 1994;**101**: 866–870.
39. Menéndez, T., S. Achenbach, E. Beinder, et al., Usefulness of magnetocardiography for the investigation of fetal arrhythmias. *Am. J. Cardiol.*, 2001;**88**: 334–336.
40. Wakai, R. T., J. F. Strasburger, Z. Li, B. J. Deal, N. L. Gotteiner, Magnetocardiographic rhythm patterns at initiation and termination of fetal supraventricular tachycardia. *Circulation*, 2003 Jan 21;**107**(2): 307–12.
41. Cuneo, B.F., J.F. Strasburger, R.T. Wakai, and M. Ovadia, *Fetal Diagn. Ther.*, 2006;**21**(3): 307–313.
42. Menéndez, T., S. Achenbach, E. Beinder, et al., Prenatal diagnosis of QT prolongation by magnetocardiography. *PACE*, 2000;**23**: 1305–1307.
43. Van Leeuwen, P., B. Hailer, W. Bader, et al., Magnetocardiography in the diagnosis of fetal arrhythmia. *Br. J. Obstet. Gynaecol.*, 1999;**106**: 1200–1208.
44. Horigome, H., J. Shiono, S. Shigemitsu, et al., Detection of cardiac hypertrophy in the fetus by approximation of the current dipole using magnetocardiography. *Pediatr. Res.*, 2001;**50**: 242–245.
45. Van Leeuwen, P., S. Lange, A. Klein, D. Geue, and D.H. Gronemeyer, Dependency of magnetocardiographically determined fetal cardiac time intervals on gestational age, gender and postnatal biometrics in healthy pregnancies. *BMC Pregnancy Childbirth*, 2004,Apr 2;**4**(1): 6.
46. Van Leeuwen, P., Y. Beuvink, S. Lange, A. Klein, D. Geue, and D. Gronemeyer, Assessment of fetal growth on the basis of signal strength in fetal magnetocardiography. *Neurol. Clin. Neurophysiol.*, 2004, Nov 30;**47**(B).
47. Bolick, D.R., D.B. Hackel, K.A. Reimer, and R.E. Ideker, Quantitative analysis of myocardial infarct structure in patients with ventricular tachycardia. *Circulation*, 1986;**74**: 1266–1279.
48. Erne, S.N., R.R. Fenici, H.D. Hahlbom, W. Jaszczuk, H.P. Lehmann, and M. Masselli, High resolution magnetocardiographic recordings of the ST segment in patient with electrical late potentials. *Nuovo. Cimento.*, 1983;**2d**: 340–345.
49. Mäkijärvi, M., J. Montonen, L. Toivonen, P. Siltanen, M.S. Nieminen, M. Leiniö, and T. Katila, Identification of patients with ventricular tachycardia after myocardial infarction by high-resolution magnetocardography and electrocardiography. *J. Electrocardiol.*, 1993;**26**: 117–124.
50. Korhonen, P., J. Montonen, M. Mäkijärvi, T. Katila, M.S. Nieminen, and L. Toivonen, Late fields of the magnetocardiographic QRS complex as indicators of propensity to sustained ventricular tachycardia after myocardial infarction. *J. Cardiovasc. Electrophysiol.*, 2000;**11**: 413–420.
51. Müller, H.P., P. Gödde, K. Czerski, M. Oeff, R. Agrawal, P. Endt, W. Kruse, U. Steinhoff, and L. Trahms, Magnetocardiographic analysis of the two-dimensional distribution of intra-QRS fractionated activation. *Phys. Med. Biol.*, 1999;**44**: 105–120.
52. Korhonen, P., T. Husa, I. Tierala, H. Väänänen, M. Mäkijärvi, T. Katila, and L. Toivonen, Increased intra-QRS fragmentation in magnetocardiography as a predictor of arrhythmic events and mortality in patients with cardiac dysfunction after myocardial infarction. *J. Cardiovasc. Electrophysiol.*, 2006;**17**: 396–401.
53. Lant, J., G. Stroink, B. Ten Voorde, B. M. Horacek, and T. J. Montague, Complementary nature of electrocardiographic and magnetocardiographic data in patients with ischemic heart disease. *J. Electrocardiol.*, 1990;**23**: 315–322.
54. Nomura, M., K. Fujino, M. Katayama, A. Takeuchi, Y. Fukuda, M. Sumi, M. Murakami, Y. Nakaya, and H. Mori, Analysis of the T wave of the magnetocardiogram in patients with essential hypertensionby means of isomagnetic and vector array maps. *J. Electrocardiol.*, 1988;**21**: 174–182.
55. Oikarinen, L., M. Viitasalo, P. Korhonen, H. Väänänen, H. Hänninen, J. Montonen, M. Mäkijärvi, T. Katila, and L. Toivonen, Postmyocardial infarction patients susceptible to ventricular tachycardia show increased T wave dispersion independent

of delayed ventricular conduction. *J. Cardiovasc. Electrophysiol.*, 2001;**12**: 1115–1120.
56. Korhonen, P., H. Väänänen, M. Mäkijärvi, T. Katila, and L. Toivonen, Repolarization abnormalities detected by magnetocardiography in patients with dilated cardiomyopathy and ventricular arrhythmias. *J. Cardiovasc. Electrophysiol.*, 2001;**12**: 772–777.
57. Stroink, G., J. Lant, P. Elliot, P. Charlebois, and M.J. Gardner, Discrimination between myocardial infarct and ventricular tachycardia patients using magnetocardiographic trajectory plots and iso-integral maps. *J. Electrocardiol.*, 1992;**25**: 129–142.
58. Rovamo, L., M. Paavola, J. Montonen, M. Mäkijärvi, J. Nenonen, and T. Katila, Magnetocardiorapic repolarization maps in children with long QT syndrome, in *Biomagnetism: Fundamental Research and Clinical Applications*, C. Baumgartner, L. Deecke, G. Stroink, and S.J. Williamson, Editors. Amsterdam Oxford, Tokyo: IOS, 1995, pp. 615–618.
59. Haissaguerre, M., P. Jais, D. C. Shah, A. Takahashi, M. Hocini, G. Quiniou, S. Garrigue, A. Le Mouroux, P. Le Metayer, and J. Clementy, Spontaneous initiation of atrial fibrillation by ectopic beats originating in the pulmonary veins. *N. Engl. J. Med.*, 1998;**339**(10): 659–666.
60. Ausma, J., N. Litjens, M.H. Lenders, H. Duimel, F. Mast, L. Wouters, F. Ramaekers, M. Allessie, and M. Borgers, Time course of atrial fibrillation-induced cellular structural remodeling in atria of the goat. *J. Mol. Cell. Cardiol.*, 2001;**33**(12): 2083–2094.
61. Schotten, U., J. Ausma, C. Stellbrink, I. Sabatschus, M. Vogel, D. Frechen, F. Schoendube, P. Hanrath, and M.A. Allessie, Cellular mechanisms of depressed atrial contractility in patients with chronic atrial fibrillation. *Circulation*, 2001;**103**(5): 691–698.
62. Dilaveris, P.E., J.E. Gialafos, P-wave dispersion: a novel predictor of paroxysmal atrial fibrillation. *Ann. Noninvasive. Electrocardiol.*, 2001;**6**(2): 159–165.
63. Fukunami, M., T. Yamada, M. Ohmori, K. Kumagai, K. Umemoto, A. Sakai, N. Kondoh, T. Minamino, and N. Hoki, Detection of patients at risk for paroxysmal atrial fibrillation during sinus rhythm by P wave-triggered signal-averaged electrocardiogram. *Circulation*, 1991;**83**(1): 162–169.
64. Steinbigler, P. and R. Haberl B. Konig G. Steinbeck, P-wave signal averaging identifies patients prone to alcohol-induced paroxysmal atrial fibrillation. *Am J Cardiol*, 2003;**91**(4): 491–494.
65. Winklmaier, M., C. Pohle, S. Achenbach, M. Kaltenhauser, W. Moshage, and W.G. Daniel, P-wave analysis in MCG and ECG after conversion of atrial fibrillation. *Biomed. Tech. (Berl.)*, 1998;**43**(Suppl): 250–251.
66. Koskinen, R., M. Lehto, H. Vaananen, J. Rantonen, L.M. Voipio-Pulkki, M. Makijarvi, L. Lehtonen, J. Montonen, and L. Toivonen, Measurement and reproducibility of magnetocardiographic filtered atrial signal in patients with lone atrial fibrillation and in healthy subjects. *J. Electrocardiol.*, 2005;**38**(4): 330–336.
67. Hertervig, E., S. Yuan, S. Liu, O. Kongstad, J. Luo, S.B. Olsson, Electroanatomic mapping of transseptal conduction during coronary sinus pacing in patients with paroxysmal atrial fibrillation. *Scand. Cardiovasc. J.*, 2003;**37**(6): 340–343.
68. Markides, V. and R.J. Schilling, Atrial fibrillation: classification, pathophysiology, mechanisms and drug treatment. (Review) (5 refs) *Heart (British Cardiac. Soc.)*, 2003,Aug.; **89**(8): 939–943.
69. Verma, A., O.M. Wazni, N.F. Marrouche, D.O. Martin, F. Kilicaslan, S. Minor, R.A. Schweikert, W. Saliba, J. Cummings, J.D. Burkhardt, M. Bhargava, W.A. Belden, A. Abdul-Karim, and A. Natale, Pre-existent left atrial scarring in patients undergoing pulmonary vein antrum isolation: an independent predictor of procedural failure. *J. Am. Coll. Cardiol.*, 2005,Jan 18;**45**(2): 285–292.
70. Sumi, M., A. Takeuchi, M. Katayama, Y. Fukuda, M. Nomura, K. Fujino, M. Murakami, Y. Nakaya, H. Mori and P. Magnetocardiographic, Waves in normal subjects and patients with mitral stenosis. *Jpn. Heart J.*, 1986;**27**(5): 621–633.
71. Koskinen, R., H. Väänänen, V. Mäntynen, J. Montonen, J. Nenonen, L. Lehtonen, M. Mäkijärvi, and L. Toivonen, Field heterogeneity in magnetocardiographic atrial signals in patients with focally-triggered lone atrial fibrillation (abstract). *Heart Rhythm.*, 2004;**1**(1S): 226.
72. Kuusisto, J., R. Koskinen, V. Mäntynen, J. Nenonen, M. Mäkijärvi, J. Montonen, and L. Toivonen, Diversity in Activation of Healthy Atria by Magnetocardiographic Gradient Analysis (abstract), in *Proceedings of International Conference on Biomagnetism*, Boston, 2004: 407–408
73. Jurkko, R., V. Mäntynen, J. Tapanainen, J. Montonen, H. Väänänen, H. Parikka, L. Toivonen, Non-invasive detection of conduction pathways to left atrium using magnetocardiography: Validation by intracardiac electroanatomic mapping. *Europace* 2009;**11**: 169–177.
74. Jurkko, R., V. Mäntynen, M. Lehto, J.M. Tapanainen, J. Montonen, H. Parikka, L. Toivonen, Interatrial conduction in patients with paroxysmal atrial fibrillation and in healthy subjects. *Int. J. Cardiol.*, 2009, Jun 20.
75. Yamada, S., K. Tsukada, T. Miyashita, Y. Oyake, K. Kuga, I. Yamaguchi, Noninvasive diagnosis of partial atrial standstill using magnetocardiograms. *Circ. J.*, 2002;**66**(12): 1178–1180.
76. Kandori, A., T. Hosono, T. Kanagawa, S. Miyashita, Y. Chiba, M. Murakami, T. Miyashita, and K. Tsukada, Detection of atrial-flutter and atrial-fibrillation waveforms by fetal magnetocardiogram. *Med. Biol. Eng. Comput.*, 2002;**40**(2): 213–217.
77. Brockmeier, K., S. Comani, S. Erne, et al., Magnetocardiography and exercise testing. *J. Electrocardiol.*, 1994;**27**: 137–142.
78. Brockmeier, K., L. Schmitz, J.D. Bobadilla Chavez, et al., Magnetocardiography and 32-lead potential mapping: repolarization in normal subjects during pharmacologically induced stress. *J. Cardiovasc. Electrophysiol.*, 1997;**8**: 615–626.
79. Takala, P., H. Hänninen, J. Montonen, et al., Magnetocardiographic and electrocardiographic exercise mapping in healthy subjects. *Ann. Biomed. Eng.*, 2001;**46**: 975–982.
80. Van Leeuwen, P., B. Hailer, and M. Wehr, Changes in current dipole parameters in patients with coronary artery disease with and without myocardial infarction. *Biomed. Tech. (Berl.)*, 1997;**42**: 132–136.
81. Van Leeuwen, P., B. Hailer, S. Lange, et al., Spatial and temporal changes during the QT-interval in the magnetic field of patients with coronary artery disease. *Biomed. Tech. (Berl.)*, 1999;**44**: 139–142.
82. Hänninen, H., P. Takala, M. Mäkijärvi, et al., Detection of exercise induced myocardial ischemia by multichannel magnetocardiography in single vessel coronary artery disease. *Ann. Noninvas. Electrocardiol.*, 2000;**5**: 147–157.
83. Hänninen, H., P. Takala, P. Korhonen, et al., Features of ST segment and T-wave in exercise-induced myocardial ischemia evaluated with multichannel magnetocardiography. *Ann. Med.*, 2002;**34**: 1–10.
84. Takala, P., H. Hänninen, J. Montonen, et al., Heart rate adjustment of magnetic field map rotation in detection of myocardial

ischemia in exercise magnetocardiography. *Basic Res. Cardiol.*, 2002;**97**: 88–96.

85. Hänninen, H., P. Takala, M. Mäkijärvi, et al., Recording locations in multichannel magnetocardiography and body surface potential mapping sensitive for regional exercise-induced myocardial ischemia. *Basic Res. Cardiol.*, 2001;**96**: 405–414.
86. Tsukada, K., T. Miyashita, A. Kandori, et al., An iso-integral mapping technique using magnetocardiogram, and its possible use for diagnosis of ischemic heart disease. *Int. J. Card Imag.*, 2000;**16**: 55–66.
87. Seese, B., W. Moshage, S. Achenbach, et al., Magnetocardiographic (MCG) analysis of myocardial injury currents, in *Biomagnetism: Fundamental Research and Clinical Applications*, C. Baumgartner, et al., Editors. Amsterdam: Elsevier, 1995; 628–632.
88. Leder, U., H.P. Pohl, S. Michaelsen, et al., Non-invasive biomagnetic imaging in coronary artery disease based on individual current density. *Int. J. Cardiol.*, 1998;**64**: 83–92.
89. Pesola, K., H. Hänninen, K. Lauerma, J. Lötjönen, M. Mäkijärvi, J. Nenonen, P. Takala, L.M. Voipio-Pulkki, L. Toivonen, and T. Katila, Current densityestimation on the left ventricular epicardium: a potential method for ischemia localization. *Biomed. Tech. (Berl.)*, 1999b;**44**(suppl 2): 143–146.
90. Hailer, B., I. Chaikovsky, S. Auth-Eisernitz, H. Schafer, and P. Van Leeuwen, The value of magnetocardiography in patients with and without relevant stenoses of the coronary arteries using an unshielded system. *PACE*, 2005,Jan;**28**(1): 8–16.
91. Hailer, B., P. Van Leeuwen, I. Chaikovsky, S. Auth-Eisernitz, H. Schafer, and D. Gronemeyer, The value of magnetocardiography in the course of coronary intervention. *Ann. Noninvas. Electrocardiol.*, 2005,Apr;**10**(2): 188–196.
92. Tolstrup, K., B.E. Madsen, J.A. Ruiz, S.D. Greenwood, J. Camacho, R.J. Siegel, H.C. Gertzen, J.W. Park, and P.A. Smars, Non-invasive resting magnetocardiographic imaging for the rapid detection of ischemia in subjects presenting with chest pain. *Cardiology*, 2006;**106**(4): 270–276.
93. Park, J.W., P.M. Hill, N. Chung, P.G. Hugenholtz, and F. Jung, Magnetocardiography predicts coronary artery disease in patients with acute chest pain. *Ann. Noninvas. Electrocardiol.*, 2005,Jul;**10**(3): 312–323.

45 Polarcardiography

Gordon E. Dower

P. W. Macfarlane, A. van Oosterom, O. Pahlm, P. Kligfield, M. Janse, J. Camm (eds.), *Comprehensive Electrocardiology*, DOI 10.1007/978-1-84882-046-3_45,

45.1 Polarcardiography

Polarcardiography is a form of vectorcardiography. It provides a way of graphically representing the heart vector (➋ Chap. 7), but instead of doing so in the form of vectorcardiographic loops, which show only one beat and lack a time scale, the polarcardiograph separately plots the magnitude and direction of the heart vector against time. These are its polar coordinates, which are obtained from its rectangular, or *xyz*, coordinates by a simple mathematical transformation. Their use in electrocardiography goes back to Einthoven's manifest potential difference and electrical axis of the heart; that is, the magnitude and direction of the maximal heart vector in the frontal plane [1].

The choice of coordinate system can greatly affect the simplicity of certain mathematical problems. The rectangular coordinates of the heart vector are given directly by vectorcardiographic lead systems because of their simple relationship with potential differences on the body surface. The ideal vectorcardiographic lead system, however, would yield *xyz* signals as if they came from orthogonally placed surface electrodes on the body (assuming the body were a relatively large homogeneous sphere with the heart at its center). With this concept of the heart and body, it is natural to employ a spherical (polar) coordinate system. The most familiar application of such a system is to the terrestrial sphere. If altitude (radius) is ignored, only two coordinates, latitude and longitude, are required to fix a position on the earth's surface, whereas three rectangular coordinates would be needed.

It seemed intuitively obvious to Einthoven, and subsequently to several people [2], that polar coordinates, either in a plane or with respect to a sphere, should be helpful in studying the heart vector. For example, an increase in the peak magnitude of the heart vector, such as occurs in left ventricular hypertrophy (LVH), is an expected consequence of increase in the muscle mass of the left ventricle and is not affected by the orientation of the heart. In contrast, different orientations can produce various combinations of increased voltage in limb and precordial leads of the ECG, giving rise to a plethora of ECG voltage criteria.

Intriguing though the prospect was, the application of polarcardiography had to await the development of a device for continuously transforming rectangular into polar coordinates. The first clinically practical polarcardiograph was an analog computer completed in 1961 [3]. The polarcardiograms obtained from it provided the basis for developing diagnostic criteria covering a wide range of electrocardiographic conditions. It is important to appreciate that the transformation of coordinates does not increase information content, but it may display information in a manner that makes certain diagnostic features more obvious or discernible. Some polar criteria can be translated back into ECG or VCG criteria that had not previously been recognized, but others may be very difficult to discern in either of those displays.

The equipment now employed in computerized electrocardiography can execute the necessary transformations so that dedicated polarcardiographs are no longer needed; however, polar criteria and the polar approach play a useful role in the development of diagnostic programs. At the Woodward ECG computer system in the Vancouver General Hospital, polar coordinates form the basis of the measurement and analysis programs. These give the cardiologist information additional to that obtained from reading the ECG. However, the 12-lead ECG derived from the *xyz* signals (see ➋ Chap. 11) remains the primary graphic output of the system, with VCGs being generated when polar criteria for infarction are satisfied (see ➋ Fig. 45.18 and ➋ Table 45.3). Polarcardiograms can also be generated from any current or stored data, on request, although they are not used routinely for the following reasons:

(a) They are not really necessary, since the polar criteria have been satisfactorily programmed into the computer
(b) They are relatively voluminous, being written out at five times the paper speed of the conventional ECG and
(c) Most cardiologists are not familiar with them

However, polarcardiograms are used in problem cases, in research, and in the development of new criteria.

45.2 Spherical Coordinates Applied to the Body

➋ Figures 45.1 and ➋ 45.2 show how latitude and longitude are applied to the body to indicate the direction of the heart vector. The frontal plane of the body becomes the equatorial plane of the reference sphere. The angle in that plane α is the same as that used by Einthoven for the electrical axis of the heart. All points on a meridian of longitude have the same angle α. The angle of the heart vector with respect to the equatorial frontal plane (that is, anterior or posterior to it) is ψ. All points on a parallel of latitude have the same angle ψ.

Aitoff's equal-area projection (➋ Fig. 45.2) is a two-dimensional representation of the sphere that is useful in showing clustering of directions of corresponding heart vectors in groups of individuals (➋ Fig. 45.3). The orientation of the body in the Aitoff projection has been chosen so that the North–South axis corresponds to the posteroanterior (−z) axis of the transverse plane vector loop. Hence, anterior vectors are depicted downward in this form of Aitoff plot.

Elevation and azimuth are electrocardiographic synonyms for latitude and longitude, but they are less familiar terms to the general reader and have difficult dictionary definitions. Strictly, they are not appropriate because they define a direction with respect to an observer on the earth's surface. The celestial coordinates, declination, and right ascension would be conceptually more applicable, though not more so than latitude and longitude.

Although the position of a point on the surface of a sphere is defined by latitude and longitude, in order to fix that point in space, the radius of the sphere (R in ➋ Fig. 45.1) must also be specified. In polarcardiography, this third coordinate is known as magnitude, because it represents the magnitude of the heart vector.

45.3 Spatial Magnitude

For reasons of clarity, the magnitude coordinate will be discussed first. The spatial magnitude of the heart vector, when plotted against time, yields the M tracing. Magnitudes in the frontal, transverse, and sagittal planes (i.e., the distances from the origin in the corresponding VCG loops) are denoted by m_f, m_t, and m_s, respectively. The relationships between the magnitude tracings and their parent xyz signals are as follows:

$$M = \left(x^2 + y^2 + z^2\right)^{1/2}, M \geqslant 0$$

$$m_f = \left(x^2 + y^2\right)^{1/2}, m_f \geqslant 0$$

$$m_t = \left(x^2 + z^2\right)^{1/2}, m_t \geqslant 0$$

$$m_s = \left(y^2 + z^2\right)^{1/2}, m_s \geqslant 0$$

In the polarcardiographic examples in this chapter, the xyz signals have been derived from the Frank system (➋ Chap. 11).

At first sight, M tracings resemble ECGs and, so far as possible, they are similarly labeled. The P, QRS, and T events in the ECG have their counterparts in the M tracings (➋ Fig. 45.4). However, by definition, there cannot be Q or S waves because magnitudes are never negative. Although small initial peaks (e.g., ″r in ➋ Fig. 45.4) in the M tracing correspond approximately to the Q wave in the ECG, the degree of correspondence is variable and indefinite. It is useful, however, to retain the letters Q and S to denote the onset and offset points of the QRS complex. This is logical because it makes the time between the Q and S points equal to the QRS duration – a term desirable to retain, along with QRS complex and QT interval. Because of the multiplicity of peaks sometimes seen in QRS complexes in M tracings and the inapplicability of the terms Q wave and S wave, it is preferable not to refer to an R wave but rather to the individual R peaks in the manner shown in ➋ Fig. 45.4.

In left and right bundle branch block (LBBB and RBBB, respectively), the M tracing is very distinctive, the QRS complex being smooth in the former and notched in the latter (➋ Fig. 45.5). Nevertheless, angular data are included in the discrimination.

The M tracing is particularly valuable in determining QRS duration because of two useful properties. First, when the M tracing shows zero magnitude, all ECG leads must have zero signals, so problems relating to nonsimultaneity of the onset and offset of the QRS complex in various ECG leads (for example, an isoelectric Q wave) do not arise. Second, the M tracing returns, or almost returns, to the baseline to give a clear demarcation between QRS and ST events. (Exceptions to this are mentioned in ➋ Sect. 45.3.1.) The ECG does not give a good indication of this phenomenon and its QRS offset is often vague. The relative clarity of the Q and S points in the M tracing simplifies the programming of a computer to determine QRS duration [4]. This is a very important measurement in decision-tree logic leading to a diagnostic interpretation. Computer determinations of QRS durations based on the ECG show disappointing discrepancies from visual determinations. However, improving the agreement is difficult because visual determinations by different readers also

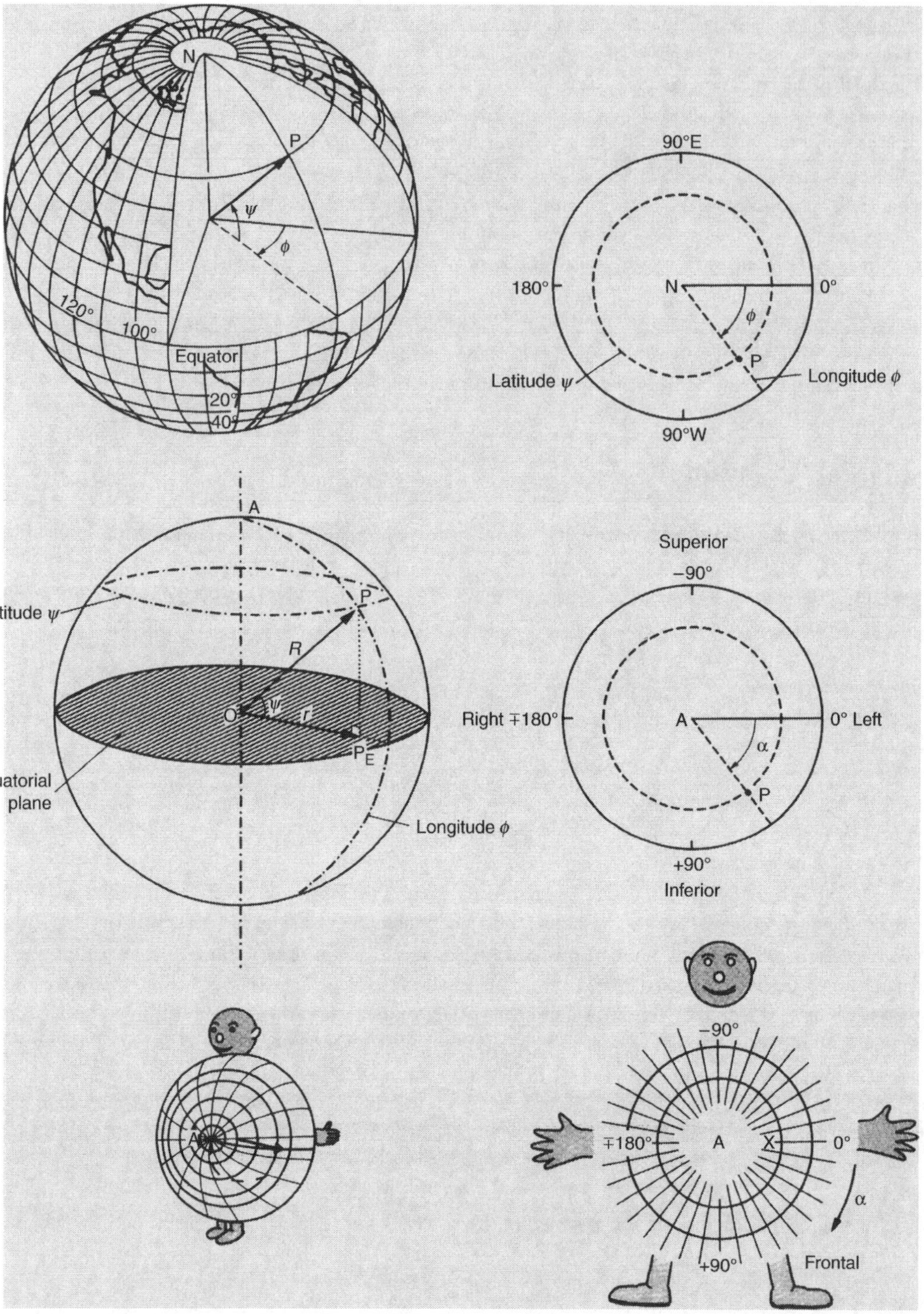

Figure 45.1

Definition of latitude and longitude. The vector from the center of the terrestrial sphere pierces the surface at P, the intersection of the meridian of longitude and the parallel of latitude. The angle Φ, between P meridian and reference (zero) meridian in the equatorial plane, is the longitude of P. The angle Ψ that the vector makes with respect to the equatorial plane is the latitude of P and is referred to as the posteroanterior (PA) latitude. Thus the latitude and longitude of P give spatial direction of the vector. The lower figures show spherical coordinates applied to the body: the North Pole becomes the anterior pole A, the equatorial plane becomes the frontal plane, and the angle Φ becomes α, following the tradition initiated by Einthoven.

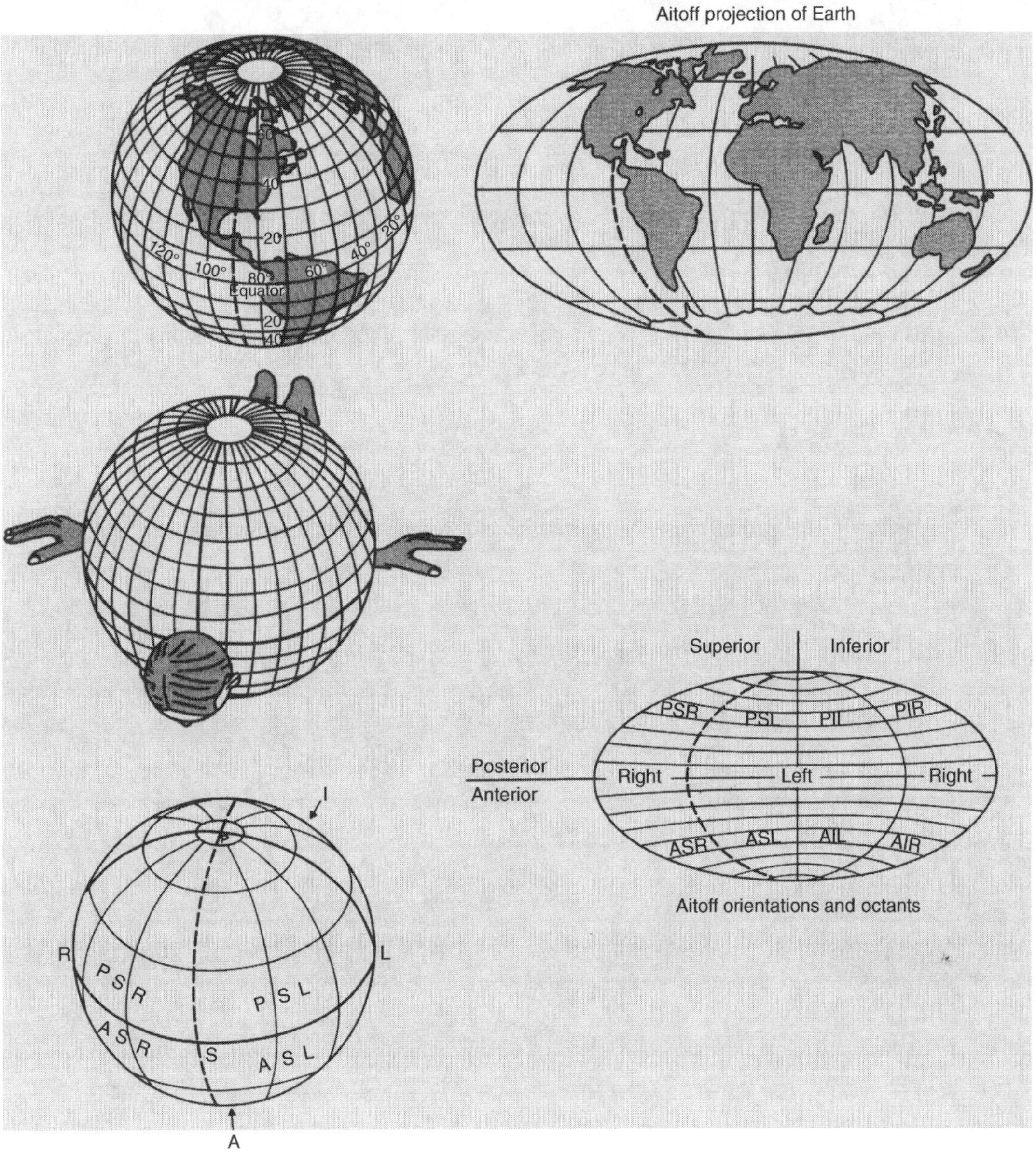

◘ Figure 45.2

The orientation of Aitoff equal-area projection for two-dimensional display of vector direction. The distribution of points on the sphere is uniform in this projection. PSR, posterior superior right; ASR, anterior superior right; PSL, posterior superior left; ASL, anterior superior left; PIL, posterior inferior left; AIL, anterior inferior left; PIR, posterior inferior right; AIR, anterior inferior right.

vary a good deal. On the other hand, there is good agreement between computer and visual determinations of QRS duration using the spatial magnitude. The normal limits of QRS duration so determined are given in ➲ Tables 45.1 and ➲ 45.2 as diagnostic cut-off points used by the Woodward computers.

The ST segment is that part of the M tracing between S and T_i, the onset of the T wave. Sometimes, for example, in LBBB (➲ Fig. 45.5), the ST segment initially rises steeply for a short distance then abruptly changes its slope to a more

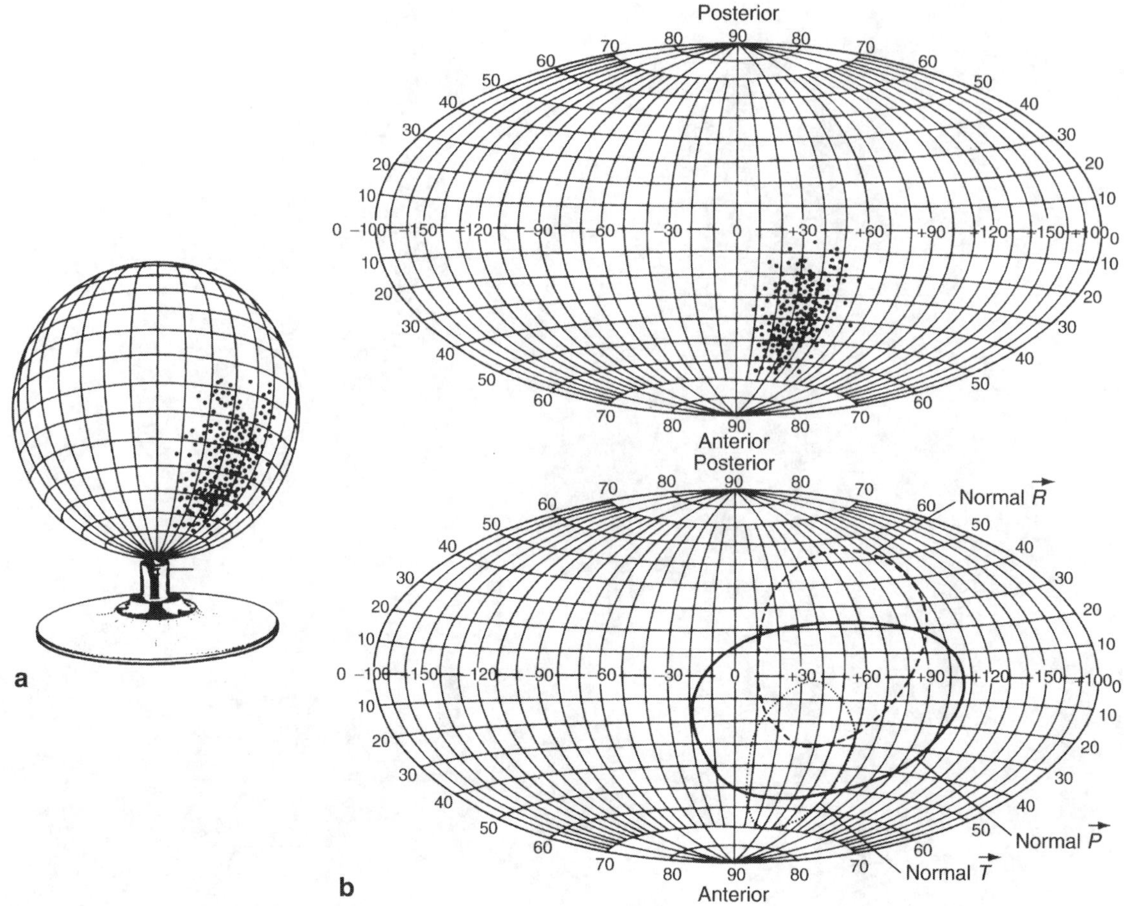

Figure 45.3
Direction of $\vec{T}$ in 195 young adults. Clustering in the sphere (a) is equally well shown in the Aitoff projection (*top* part of b). Normal "continents" of $\vec{P}$, $\vec{R}$, and $\vec{T}$ directions are thus defined (*bottom* part of b).

gradual one, forming a knee. This knee is labeled J because of its resemblance to the J point in the ECG (Fig. 45.4). Although, in the ECG, the J point is often used to mark the end of the QRS complex, the end of the QRS complex in the *M* tracing is defined as occurring at the S point because J is usually poorly defined.

M tracings reveal that there is not a clear-cut difference in identity between the ST segment and the T wave. However, it is useful to have some point in time that can be considered as representative of the ST segment so the ST point is taken as occurring midway between S and T, the peak of the T wave.

The points in the *M* tracing labeled P, R, ST, and T represent particular instants in time. A point U may often be identified also. These instants are not necessarily the same as those represented by similarly labeled points in the m_f, m_t, and m_s tracings. The vectors occurring at the points P, R, ST, and T in the *M* tracing are designated as $\vec{P}$, $\vec{R}$, $\vec{ST}$, and $\vec{T}$. The spatial magnitudes of these vectors are m_P, m_T, m_{ST}, and m_T: the heights of P, R, ST, and T above the baseline in the *M* tracing. By recording the *M*, m_f, m_t, and m_s tracings simultaneously, the magnitudes of these vectors in the frontal, transverse, and sagittal planes can be determined. These are designated as m_fP, m_tP, m_sP, and so on.

In vector loops, because of noise around the E point and lack of expansion on a time scale, the exact beginning of the QRS complex is difficult to determine, yet early QRS vectors, such as the 0.04 s vector, are often accorded considerable diagnostic importance. The timing of such vectors is clearly done more accurately from the *M* tracing. However, the greater detail afforded by the *M* tracing has disclosed that in many normal cases the development of early QRS vectors

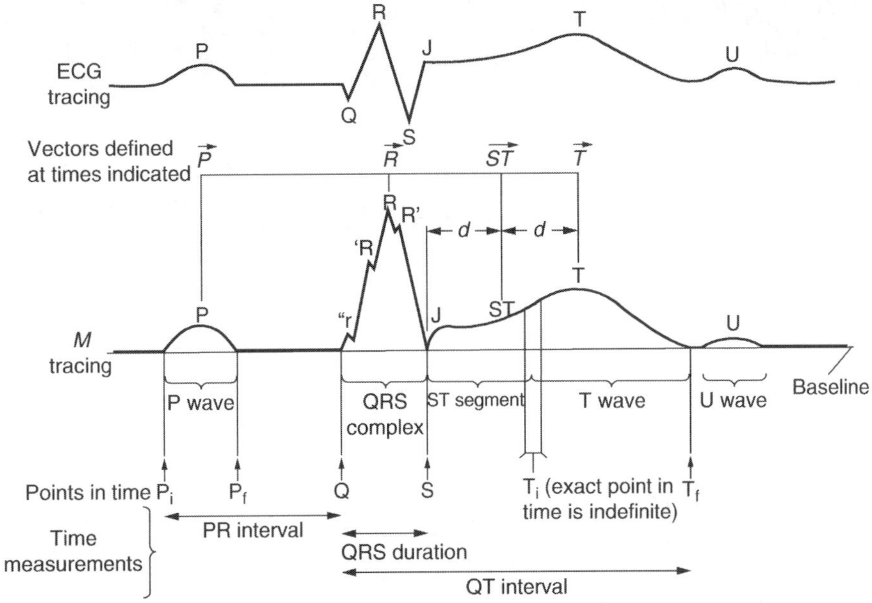

Figure 45.4
ECG and spatial vector magnitude (*M*) tracings are compared. All *M* deflections are above the baseline. The Q wave in the ECG corresponds to the so-called small foot "r in the *M* tracing. Q and S points in the *M* tracing mark the beginning and end of the QRS complex to conform with terms PR and QT intervals, and QRS duration. The J point is not a useful time reference because it is often absent, whereas the nadir S is usually clear. P_i and P_f are P-wave onset and end, respectively; T_i and T_f are T-wave onset and end, respectively.

can be very gradual: a slight initial deflection, or foot ("r in Fig. 45.4) may be clearly discernible, conjectural, or absent from case to case (Fig. 45.5) and sometimes from beat to beat. The presence or absence of this feature can make a difference of about 0.01 s to the QRS onset. This is something to be borne in mind when the QRS onset point is used as a time reference for defining QRS vectors. Its variability, however, still allows the *M* tracing to give a more consistent basis for determining QRS duration by computer than the QRS complex in the ECG [4].

The value of the *M* tracing in measuring QRS durations extends also to the determination of PR and QT intervals (Table 45.3) for the same reason; namely, a zero value of the *M* tracing indicates that ECG signals are "zero," so that different ECG leads do not have to be compared.

45.3.1 Baseline Clamping

An interesting requirement of magnitude tracings, not obvious from the formulae for deriving them, is that the baselines preceding the QRS complex in the parent *xyz* signals should be clamped (that is, set to zero) before the transformation of coordinates is carried out. If this is not done, the apparent baseline of the magnitude tracing will ride above the zero level and there will appear to be negative deflections from that baseline, which may meet, but cannot cross that level. Negative magnitudes are of course impossible. The appearance of these pseudo negative deflections is obvious (Fig. 45.6) and the consequent distortion is therefore avoidable. In fact, since the computer program clamps automatically and virtually without fail, this is no longer a problem, but the solution is not simple or obvious and, because of its crucial importance to polarcardiography, a brief account of it is merited.

The clamp point is located shortly before the Q point, at a time when signals are negligible in relation to an estimate of noise made in a 30 ms window in the region of the clamp point. The method used by the Woodward computers has not required alteration for several years and produces satisfactory *M* tracings in almost every case although rarely, P waves

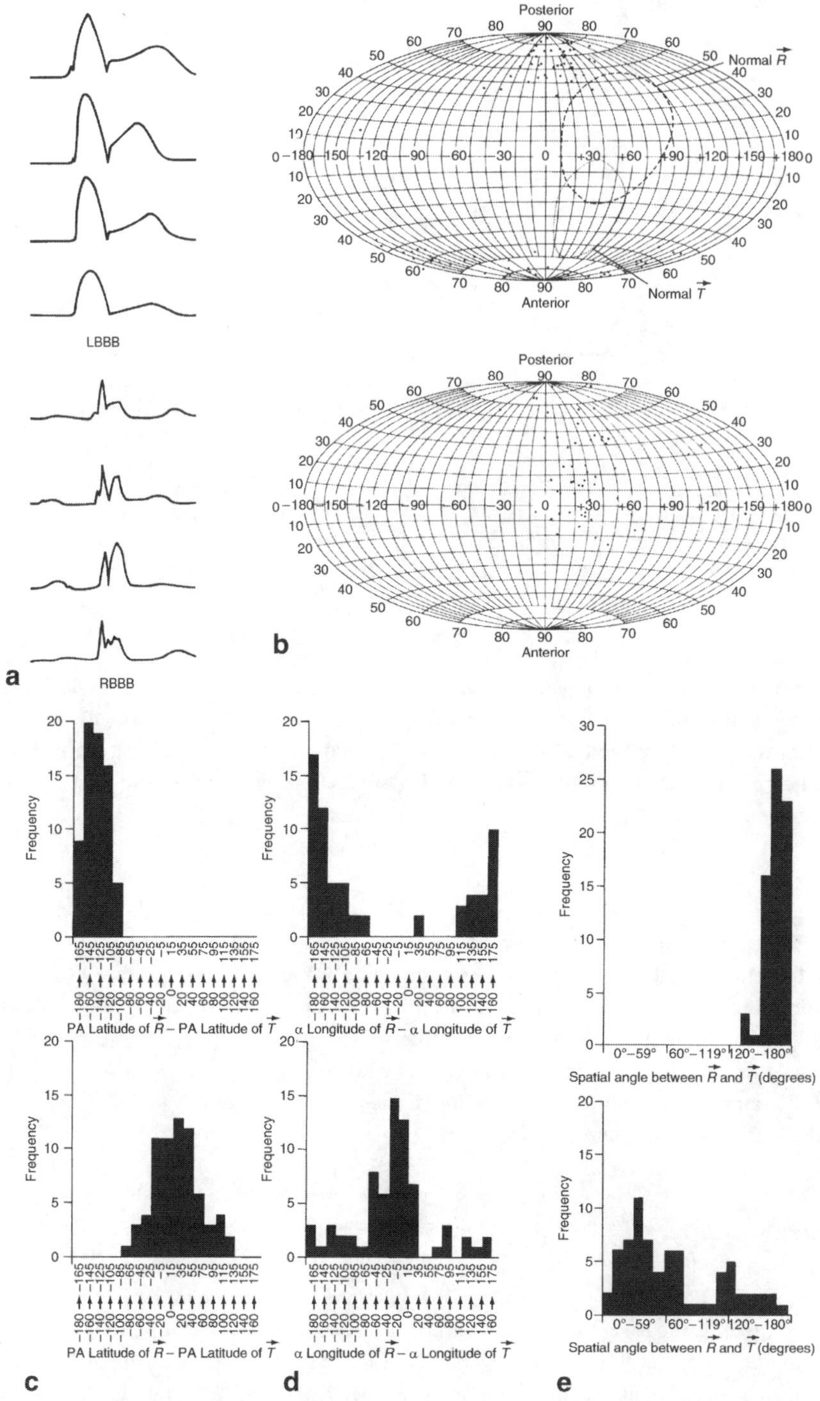

Figure 45.5

Left bundle branch block (LBBB) as compared to right bundle branch block (RBBB). *M* tracings in eight typical examples of LBBB and RBBB (a) show marked differences in the smoothness of QRS contour. Note the variability of the initial small deflection, or foot, in LBBB. Aitoff plots (b) show directions of $\vec{R}$ and $\vec{T}$ in 69 cases each of LBBB (*top*) and RBBB (*bottom*). In LBBB, these are at opposite poles. This difference is also shown in histograms of latitude differences (c), but less so in longitude differences (d) and spatial angles between $\vec{R}$ and $\vec{T}$ (e).

Table 45.1
Cutoff points for flagging interval measurements (ms): I, interpolate according to age; D, use default value

Interval	Flag	Default[a]	Age < 7	7 ⩽ age ⩽ 16	16 ⩽ age ⩽ 17
PR	>>	>210	> 180	I	>190 (rate > 100)
					>200 (rate ⩽ 100)
	<	<115	None	None	None
	<<	<106	<58	I	D
QRS	>>	>109	<88	I	D
	<<	<50	<46	I	D
QT_c	>>	>470	<473	I	D
	<<	<350	<300	I	D

[a] Age not otherwise specified

Table 45.2
Cutoff points for flagging vector magnitudes (mV), ratios, and ischemic index ($M_s\theta$): I, interpolate; D, use default value

Vector	Flag	Default	Age < 5	5 ⩽ age ⩽ 16	Age ⩾ 16 Female	16 ⩽ age ⩽ 28 Male
P	>>	> 0.240	> 0.290	I	D	D
	>	> 0.200	> 0.240	I	D	D
R	>>	> 2.200	> 3.320	I	> 2.000	> 2.400
	>	> 2.100	> 2.840	I	D	None
	<	None	< 0.920	I	None	D
	<<	< 0.850	< 0.440	I	< 0.750	D
m_t	>>	> 2.150	> 2.340	I	> 1.850	> 1.700
	>	> 1.950	> 2.030	I	> 1.700	None
ST/R	>>	> 0.25	D	D	D	> 0.20[a]
T/R	>>	> 0.75	> 0.52	I	D	> 0.60[a]
	<<	< 0.10	< 0.01	I	D	D
$M_s\theta$	*	⩾ 5	D	D	D	D
	*	⩾ 10	D	D	D	D

[a] For 16 ⩽ age ⩽ 30

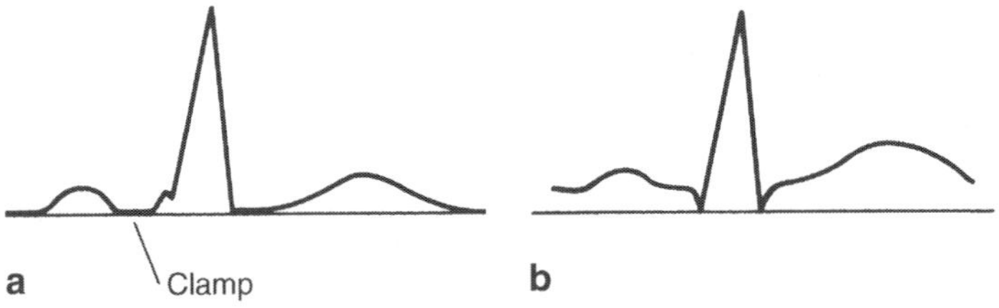

Figure 45.6
M tracings from clamped *xyz* signals (**a**). If the baselines of *xyz* signals are not centered to zero (clamped) prior to transformation to polar coordinates, the magnitude tracing will show an upward displacement of the baseline and downward deflections (**b**).

Table 45.3

Table of the information (with measurements made from polar coordinates) contained in computer outputs taken from a 99-year-old man with chest pain. Nine beats have been measured and analyzed. Measurements of time intervals are made from spatial magnitudes. Note the consistency of QRS durations, except for beat seven, which suggests aberrant conduction, probably owing to ectopy. P waves are not identified for every beat as they are rather small. R, ST, and T vectors all have abnormal longitudes (**); ($M_s\theta$) is increased (**). The extreme right-hand column gives baseline drift in 10 μV units. Under "computer comments," beats are ranked in order of increasing drift. Rows of asterisks indicate beats to which the comments apply, thus providing an indication of consistency. Criteria for anterior myocardial infarction (AMI) and inferior myocardial infarction (IMI) are present in each beat and are identified by letters in parentheses: B, a; T, b; d, i; y, g; G, c; where the upper case indicates a stronger positive criterion and the second lettered component refers to the lettered criterion in the text. For example, G, c indicates criterion c based on gamma downslope. "EQ vectors" refer to spherocardiographic criteria. The transverse QRS-vector loop indicates anterior infarction by having a notch in the efferent limb

Beat	R-R	PR	QRS	QT	QT_c	M	Latitude	Longitude	M	Latitude	Longitude	M	Latitude	Longitude	M	Latitude	Longitude	m_tR	M_S	$M_S\theta$	Drift
1			82	382	409				1.80	29P	−5**	0.18	40A	−171**	0.32	40A	−171**	1.79	0.12	11**	12
2	874		82	385	412				1.83	27P	−5**	0.16	29A	−180**	0.31	31A	−171**	1.82	0.12	12**	3
3	863		82	383	410				1.83	27P	−5**	0.17	85A	−171**	0.31	31A	−171**	1.82	0.15	14**	8
4	868	160	83	368	394	0.10	36P	−14	1.81	29P	−5**	0.14	44A	−168*	0.32	34A	−167**	1.81	0.10	8*	8
5	845	161	84	398	426	0.06	45P	0	1.80	25P	−4**	0.22	33A	−180**	0.37	29A	−176**	1.80	0.12	11**	4
6	854		85	362	388				1.78	28P	−4**	0.13	38A	−143*	0.25	34A	−168**	1.78	0.06	5*	5
7	871	154	99	386	413	0.09	43P	−18	2.0	86A	−171**	0.20	36A	−180**	0.33	32A	−180**	2.10	0.14	13**	4
8	844		85	382	409				1.84	27P	−6**	0.13	38A	−168*	0.25	40A	−161**	1.83	0.10	9*	8
9	887	162	81	375	402	0.12	30P	−36**	1.78	29P	−7**	0.22	33A	−180**	0.38	31A	−176**	1.77	0.14	14**	

Computer comments

2 7 5 6 3 4 8 1 9 (valid for beats shown, in order or drift)

* Abnormally directed P vector

*** * * ** ** Unusual R-vector direction

\+ LVH by voltage: possible *(+)/probable(*)

** * * ** *** QRS suggests (+)/indicates (*) AMI, (BT BT BT BT BT BT BT BT BT)

*** + ***** QRS suggests (+)/indicates (*) IMI (dy dy dy dy dy dy dy dyG dy)

I I I I I I I I I EQ vectors suggest apical (A), lateral (L) or inferior (I) MI.

*** * * ** ** Nonspecific T-wave abnormality

*** * ** Nonspecific ST abnormality

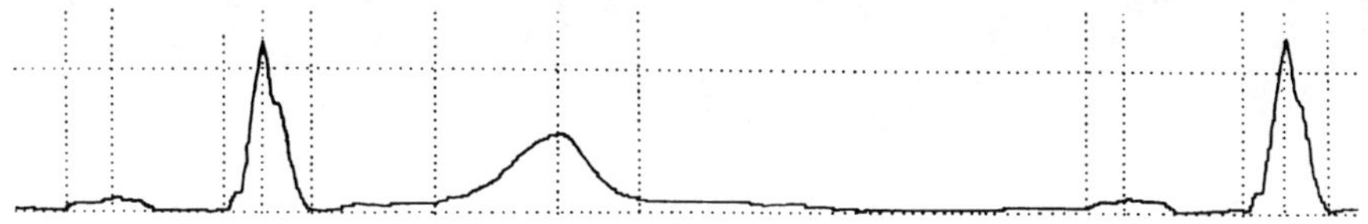

Figure 45.7
The normal spatial magnitude tracing in a computer-generated polarcardiogram. *Vertical dotted lines* show some computer-identified points in time. The time between dots horizontally is 10 ms. Note that the onset of each QRS in this *M* tracing is clamped to 0.

running into the QRS complex as is sometimes seen in the Wolff–Parkinson–White (WPW) syndrome, may give problems. The analysis program, however, bases its diagnosis of the WPW syndrome on the presence of delta waves in the ECG leads V_3–V_5, derived from the *xyz* signals. These signals are not distorted by clamping, since its effect is merely to straighten the baselines. Indeed, distortion resulting from drift tends to be reduced. The 12-lead ECGs derived from such clamped *xyz* signals (see Chap. 11) are free from baseline drift and are thereby much improved (Fig. 45.18b). A record of the drift between beats is printed out by the computer in its table of measurements (Table 45.3). The computer analyzes each beat separately and ranks its findings in order of increasing drift (Table 45.3). If beats show excessive drift they can be rejected by the program. Distortion might be expected from the so-called atrial T wave, occurring just before and during the QRS complex. This turns out not to be a problem, however, because atrial T waves are small and of low frequency, so they do not disturb early QRS vectors important in the diagnosis of infarction.

The designation of part of the ECG as baseline is arbitrary because the actual DC level (see Chap. 12) is not measurable with conventional equipment. Two possible choices for the baseline are the PR and TP segments. Early analog polarcardiographs used the TP segment, clamping being triggered by the preceding R wave, followed by a time delay adjusted by the operator. Fast heart rates and U waves gave problems, however. When tape-recorded *xyz* signals became available, the PR segment became the baseline of choice. Two reproducing heads allowed the R wave, detected through circuitry linked to the first head, to trigger clamping before the same QRS complex arrived at the second head; a short time delay allowed the clamp point to be positioned just before QRS onset. When baseline drift occurred between beats, clamping resulted in a step resetting the *M* tracing to zero (Fig. 45.8) prior to QRS onset. P vectors, preceding the clamp point, were thus affected by the total drift from the previous clamp. This problem was eliminated when digital signal-processing was adopted for polarcardiography, because baseline-drift removal could then be carried out forwards and backward between clamp points, so that drift corrections were linearly spread over the whole heart cycle. Digital processing, by allowing retrograde clamping, thus increased the accuracy of determinations of P-vector magnitudes and directions. It also permitted much refinement in the determination of the clamp points and eliminated operator adjustments.

It has been pointed out that no new information results from the transformation of rectangular to polar coordinates, although hidden features may be made obvious. It could be argued, however, that baseline clamping and straightening do add something to the tracings that is already known information but not contained in them, namely, that baseline wander or drift is non-cardiac in origin, and that the magnitude of the heart vector before the QRS complex may be taken as zero, to give a manifestly undistorted *M* tracing. The embodiment of this information in the polarcardiogram is, then, an addition. Clamping the *xyz* signals improves the quality and usefulness of the VCGs obtained from them because it provides accurate centering of the E point (Fig. 45.11). This often cannot be achieved with VCG loops, because the E point is a blurred spot resulting from the telescoping of noise and small signals around it.

45.3.2 Some Characteristics of Magnitude Tracings

45.3.2.1 P Waves

Generally, in magnitude (M) tracings P waves appear much as in the ECG, except that they are always positive. Sometimes there are two P-wave peaks, suggesting interatrial block. The so-called atrial T wave, sometimes seen in the ECG, will not

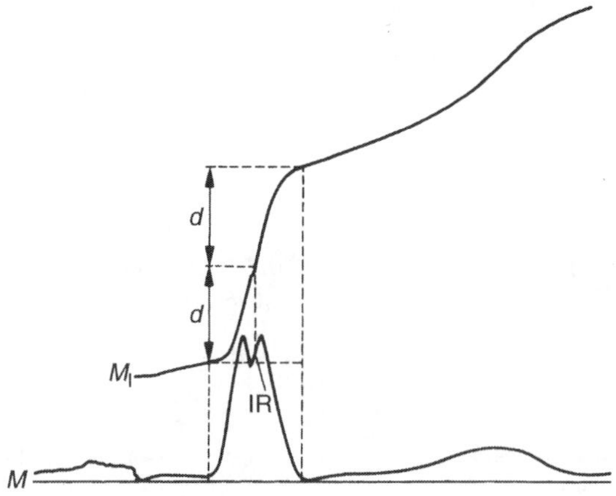

Figure 45.8
The definition of a representative QRS vector IR when there are two equal peaks in the *M* tracing. In these tracings, produced by an analog polarcardiograph, the *M* tracing is recorded together with its integral, *MI*. The height midway between Q and S points in the integral tracing defines the half-area point, IR. Note the step after P wave, caused by baseline clamping.

be observed directly in magnitude tracings because baselines are clamped during this period; however, its presence can be inferred because the baseline preceding the P wave will, in these cases, be displaced slightly upward. Atrial flutter and atrial fibrillation appear in magnitude tracings much as they do in the ECG. Large P waves in the M tracing, as in the ECG, suggest atrial hypertrophy. The cut-off point for the computer diagnosis of this condition is 0.22 mV.

45.3.2.2 QRS Complex

A typical normal QRS complex is shown in Fig. 45.7. It may begin with a decisive rise or show an initial foot. In myocardial infarction, the foot may be exaggerated and associated with diagnostic Q waves in the ECG. The foot may become diagnostic of infarction when it returns to the baseline before the main QRS deflection begins (Fig. 45.11). When this happens in the transverse magnitude, the criterion $m_t \rightarrow 0$ (see Sect. 45.5.1.1) is positive for anterior or anteroseptal infarction. In the VCG, $m_t \rightarrow 0$ may be discernible (Fig. 45.11 where the initial portion of the loop returns to the E point), but often it is lost in noise around the E point, in other parts of the QRS loop, or even in P and ST-T loops.

The peak of the QRS complex in the *M* tracing is usually distinct. The R vector whose magnitude is represented by this peak is then well defined. However, there may sometimes be two peaks of approximately equal magnitude and in this event the first or the second peak may alternately be the greater, from beat to beat. A useful definition of a characteristic QRS vector, in these cases, is the IR point, the point at which the area under the QRS complex is half its final value (Fig. 45.8). This is a more reliable time reference than the Q point, which can be displaced by the initial foot as described above. Such a foot would have little effect on the timing of IR.

Deep notching of the QRS complex in the *M* tracing is a feature of RBBB (Fig. 45.5); sometimes there appears to be a normal initial deflection followed by a delayed, slurred deflection, owing to the block. By contrast, in LBBB the QRS complex tends to be very smooth, but there is usually a small initial foot, corresponding to a small initial R wave in V_1 of the ECG. When magnitude and angle tracings are taken into account, the appearance of LBBB is so characteristic that myocardial infarction can sometimes be diagnosed in the presence of LBBB, which is much more difficult from the ECG (see also Chaps. 14 and 16). Fine notching of the *M* tracing near the peak is not normal and is associated with patchy fibrosis.

Delta waves in the *M* tracing are present in WPW syndrome, but they are not diagnostic. The computer program obtains better results from searching for delta waves in leads V 3–V 5 in the derived ECG. As would be expected, ventricular ectopy produces QRS complexes strikingly different from normally conducted beats, and various types may be classified.

45.3.2.3 ST Segment

Normally, the S point has a magnitude that is zero, or close to it. It is characteristically elevated in stress-induced ischemia (➲ Fig. 45.9) and in injury currents from infarction or pericarditis. This provides a sensitive measure that is easy to quantify by computer. After the S point, the ST segment, normally, gradually slopes upward to become the T wave. In ischemia it may, however, be downsloping. In this case, the S point is at the junction between the steeply descending QRS and the gradually descending ST segment. Fortunately, this point is readily identified by the computer program. There is not, however, any readily identifiable point to be designated as a representative of the ST segment. Therefore, a point midway between the S and T points is used as mentioned in ➲ Sect. 45.3 (➲ Fig. 45.4). Sometimes, in gross recent infarction, the ST segment may join with the QRS complex to form a monophasic pattern reminiscent of transmembrane potentials.

45.3.2.4 T Wave

In normal and most abnormal *M* tracings, the T wave is simple and smooth. Rarely, in recent infarction, double T-wave humps occur. Evolving inferior infarction often produces a marked increase in the amplitude of the T wave (➲ Fig. 45.15). Myocardial ischemia usually reduces the relative amplitude of the T wave (➲ Fig. 45.9). U waves, when they appear at all, are often attached to the terminal downslope of the T wave. T waves that are diphasic, or inverted, in the ECG arise from changes in vector direction, more than vector magnitude, and the corresponding abnormalities in the *M* tracing are often minimal though they are revealed by angle tracings.

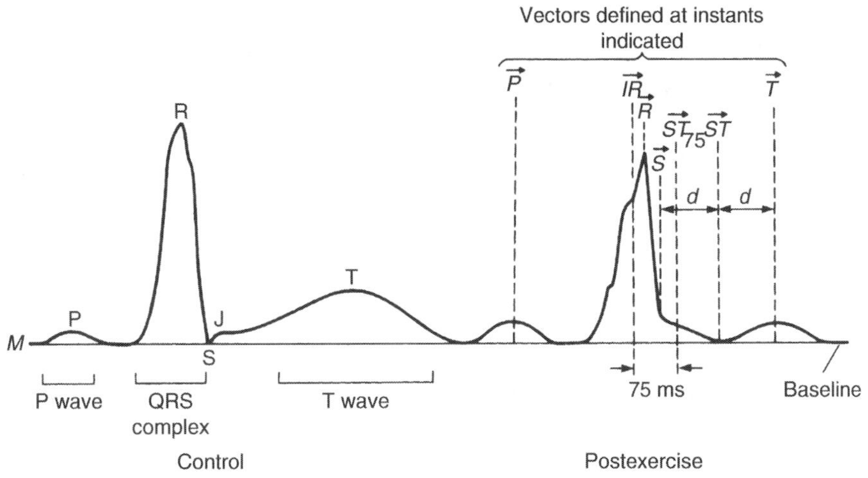

Figure 45.9
Spatial-magnitude changes in stress-induced ischemia, which causes the S-point to rise. The control *M* tracing is on the left and the post exercise *M* tracing on the right. The abnormal direction of ST_{75}, defined 75 ms after IR, is combined with the magnitude at S to give an ischemic index.

45.4 Angle Tracings

Directions of the heart vector in the frontal, transverse, and (left) sagittal planes of the VCG are denoted by angles α, β, and γ which, when plotted against time, give three of the angle tracings of the polarcardiogram (➋ Fig. 45.10). The polarity chosen is such that positive angles appear in the lower half of the scale (➋ Fig. 45.11) in conformity with the vectorcardiographic tradition initiated by Einthoven for describing the electrical axis. Consequently, upslopes in any of these angle tracings indicate counterclockwise rotations of the VCG loops about the E point. Just as the transverse QRS VCG loop in normal subjects consistently shows a counterclockwise rotation, the tracing shows an even upslope (➋ Fig. 45.12). Downslopes in the tracing occur in anterior infarction (➋ Figs. 45.11 and ➋ 45.18) and LBBB. There may be a reversal of direction, in normal individuals; that is, a small β downslope near the maximal radius of the loop which corresponds to the peak of the m_t tracing. Downslopes in the β tracing often appear as kinks in the VCG loop (➋ Fig. 45.11). Though less consistent than the β tracing, the γ tracing is useful in revealing inferior infarction by downslopes occurring before the peak of the m_s tracing. However, there must be an initial Q wave in the y-signal tracing, while γ remains in the range $-175^\circ < \gamma < -45^\circ$. The α tracing, like the frontal loop, is much more variable in normal subjects and has not given rise to any polar criteria for infarction.

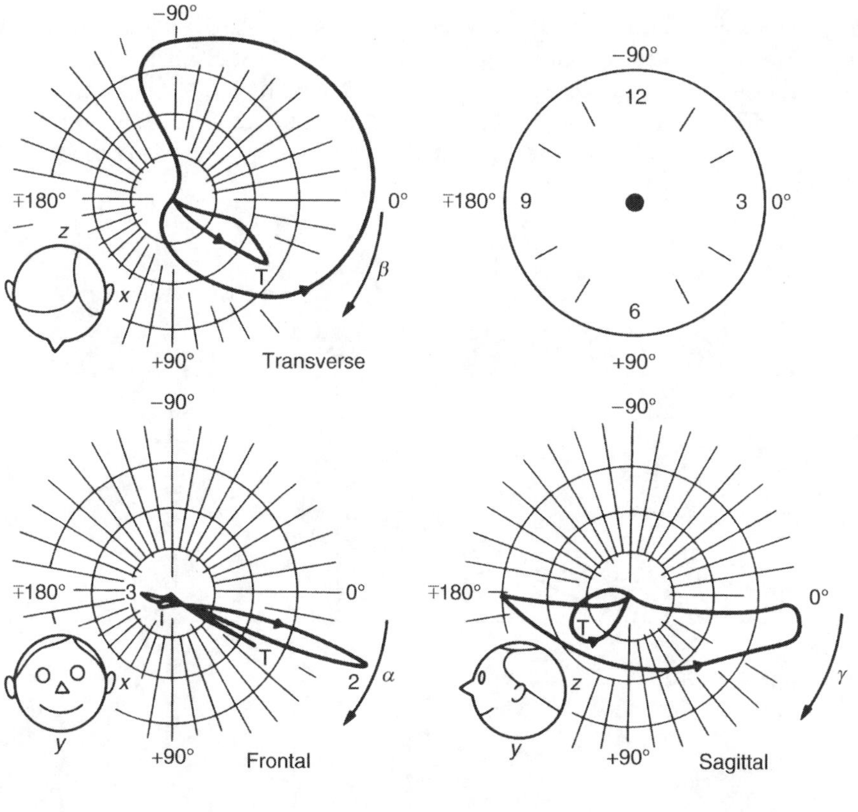

Figure 45.10
Using a clock face as reference for α, β, and γ angles in frontal, transverse, and left sagittal planes respectively, results in 0° at 3 o'clock, +90° at 6 o'clock, ±180° at 9 o'clock, and −90° at 12 o'clock.

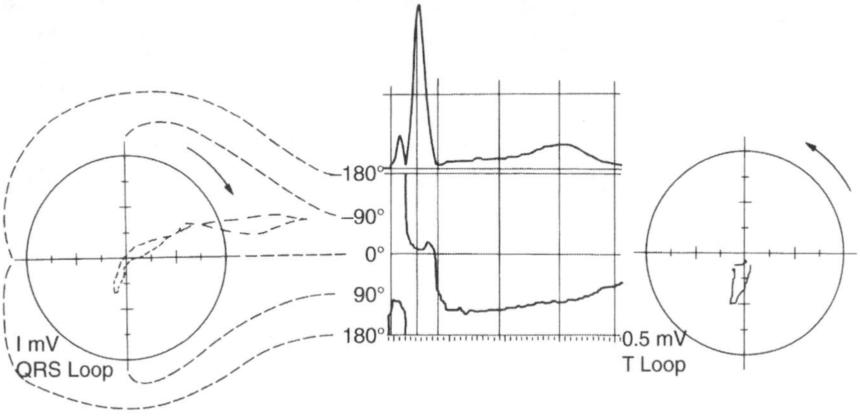

Figure 45.11
The relationship of the transverse QRS loop (*left*), T loop (*right*, amplification doubled), m_t magnitude, and angle tracings (*center*). Clockwise rotations of loops are associated with downsloping angle tracings. The example illustrates previous anterior infarction with $m_t \rightarrow 0$, indicated by the initial m_t deflection returning to the zero baseline and $\beta\downarrow$, indicated by the downsloping tracing, both of which criteria are present. That the initial lobe of the QRS loop returns exactly to the E point is indicated clearly by the m_t tracing; counterclockwise rotation of the T loop is indicated by the upsloping tracing opposite the T wave in the m_t tracing.

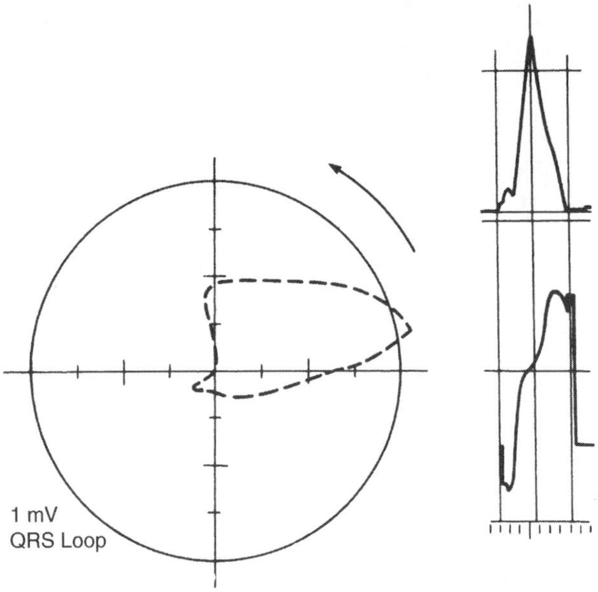

Figure 45.12
A normal transverse plane QRS loop with m_t and β tracings. Note the counterclockwise rotation of the loop and upsloping β tracing.

45.4.1 Latitudes and Longitudes

With the body conceptualized as a sphere, the frontal, transverse, and sagittal planes cut through the center to form equatorial planes of three different spherical polar coordinate systems. Each system has its respective polar axis: posteroanterior (PA), inferosuperior (IS), and right–left (RL). This polarity gives right-handed consistency of coordinates. The angles of the heart vector with respect to the planes give the PA-, IS- and RL-latitude tracings, respectively (➋ Fig. 45.13). Latitudes are expressed in degrees followed by one of the six letters to indicate the direction with respect to

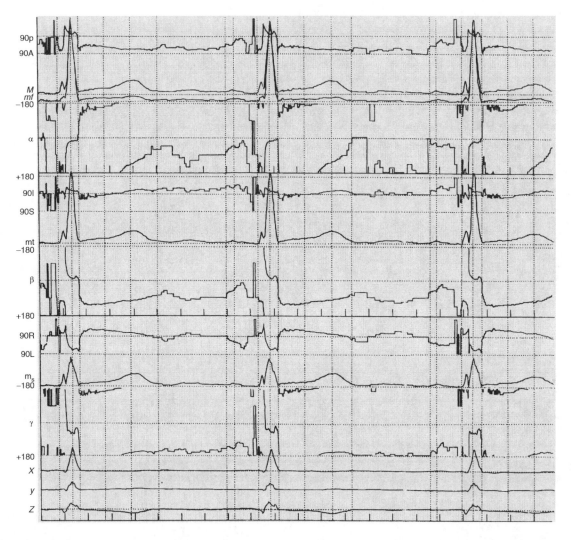

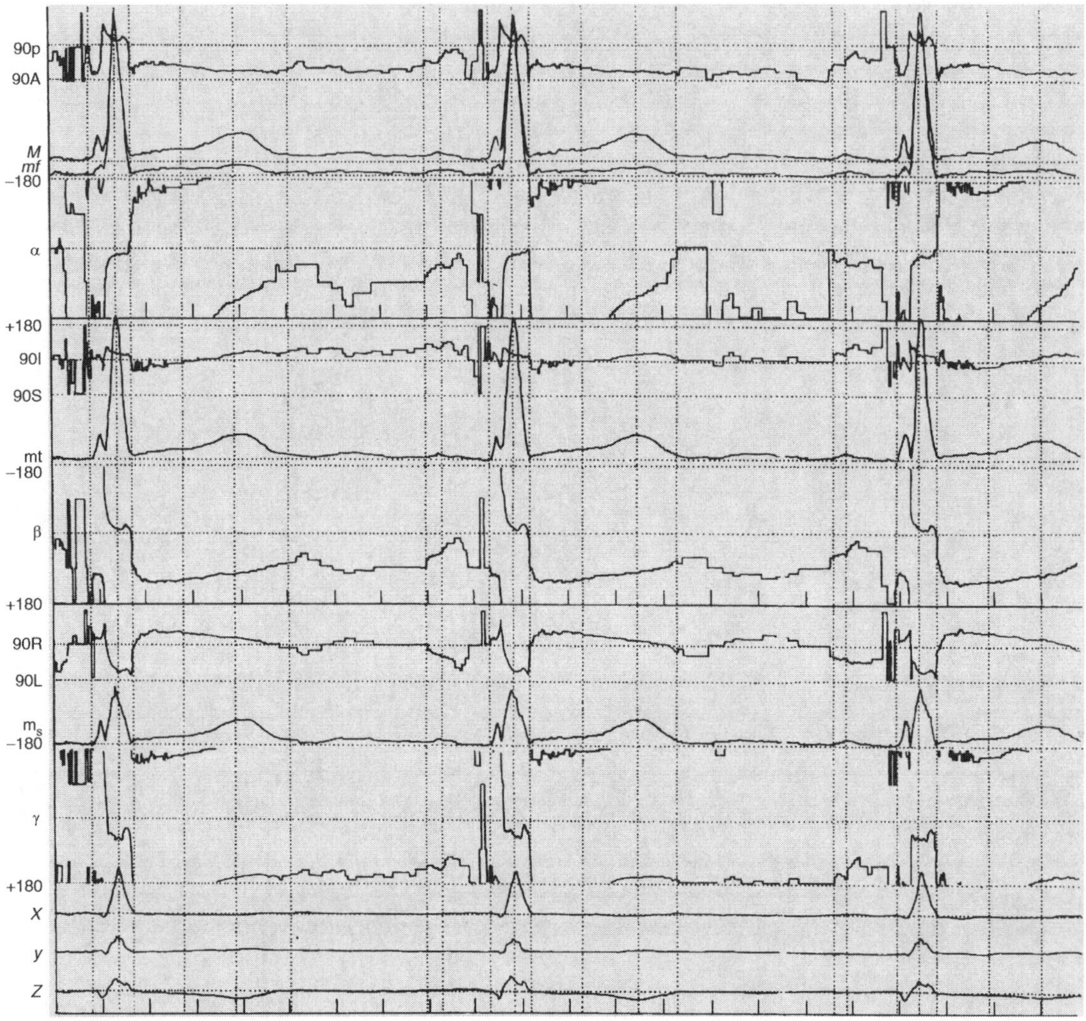

Figure 45.13
Polarcardiogram of the same patient as in ➋ Fig. 45.11 which shows the last of three beats displayed here. This beat had least baseline drift before clamping. Time marks occur every 10 ms. *XYZ* tracings (*bottom*) are of *xyz* signals which generate all other displays. *Vertical dotted lines* indicate computer-identified points (compare with ➋ Fig. 45.7) determined from the M tracing (second from *top*). Immediately below is the m_f tracing of frontal plane magnitude. Corresponding transverse and sagittal plane magnitudes appear lower down (m_t, m_s). Tracings of angles α, β, and γ appear below m_f, m_t, and m_s tracings, respectively. The remaining tracings are of angles anterior (A) and posterior (P) to the frontal plane, inferior (I) and superior (S) to the transverse plane, and right (R) and left (L) of the sagittal plane. Dots in vertical lines are 10° apart over angle ranges and 50 μV apart over magnitude ranges at standard recording sensitivity.

one of the three equatorial planes. The difference between the latitude of the *R* and *T* vectors differs distinctly in RBBB and LBBB – more so than the corresponding differences in longitudes or the spatial angle between these vectors (➋ Fig. 45.5). Latitude tracings have been less productive of criteria than longitude tracings, although Bruce [5] has defined a criterion for apical infarction based on the RL-latitude tracing: a downslope from 45R or more within the initial 10 ms of QRS, crossing the zero-degree line after the first 10 ms, associated with a Q wave in the *x*-signal tracing and between −45° and −175°. It is, of course, doubly redundant to use six angle coordinates to describe spatial directions and, in fact, this is not done. The directions of *P*, *R*, *ST*, and *T*, printed out by the measurement program, are given in terms of *α* longitude and PA latitude (of angle *ψ* in ➋ Fig. 45.1). Justification for displaying all six angle tracings in the polarcardiogram would be

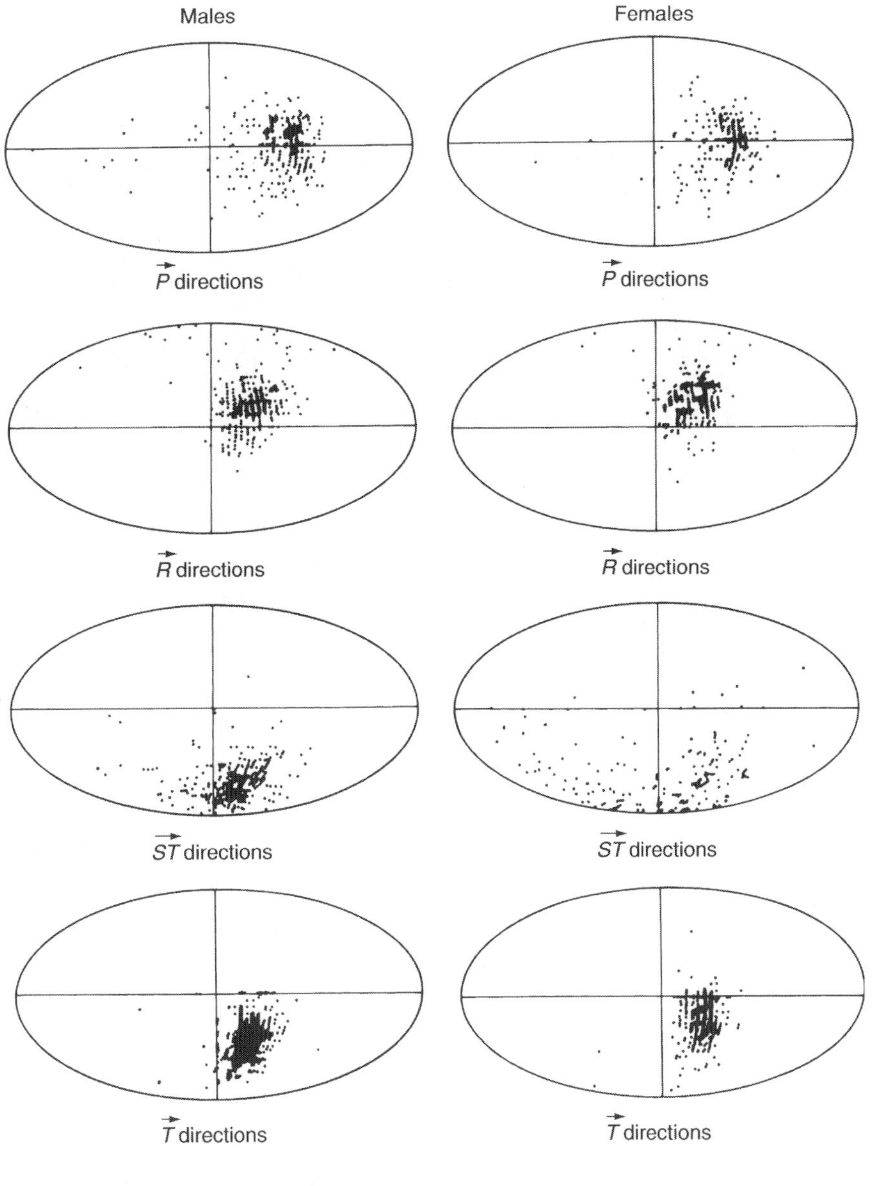

Figure 45.14
Directions of $\vec{P}$, $\vec{R}$, $\vec{ST}$ and $\vec{T}$ in 406 Cretan villagers. These findings should be compared with those of ➋ Fig. 45.3. Note that the T-vector distribution for women is slightly posterior to that for men.

the same as that for displaying six limb leads in the ECG or three VCG views when only two are sufficient: abnormalities may show better in some than in others.

45.4.2 Normal Directions of Representative Heart Vectors

The normal directions of $\vec{P}$, $\vec{R}$, and $\vec{T}$ are shown in ➋ Fig. 45.3 for 195 young adults, and in ➋ Fig. 45.14 for 406 Cretan villagers. ➋ Figure 45.14 also shows the directions of the ST vectors. As vector magnitudes decrease toward zero, angular

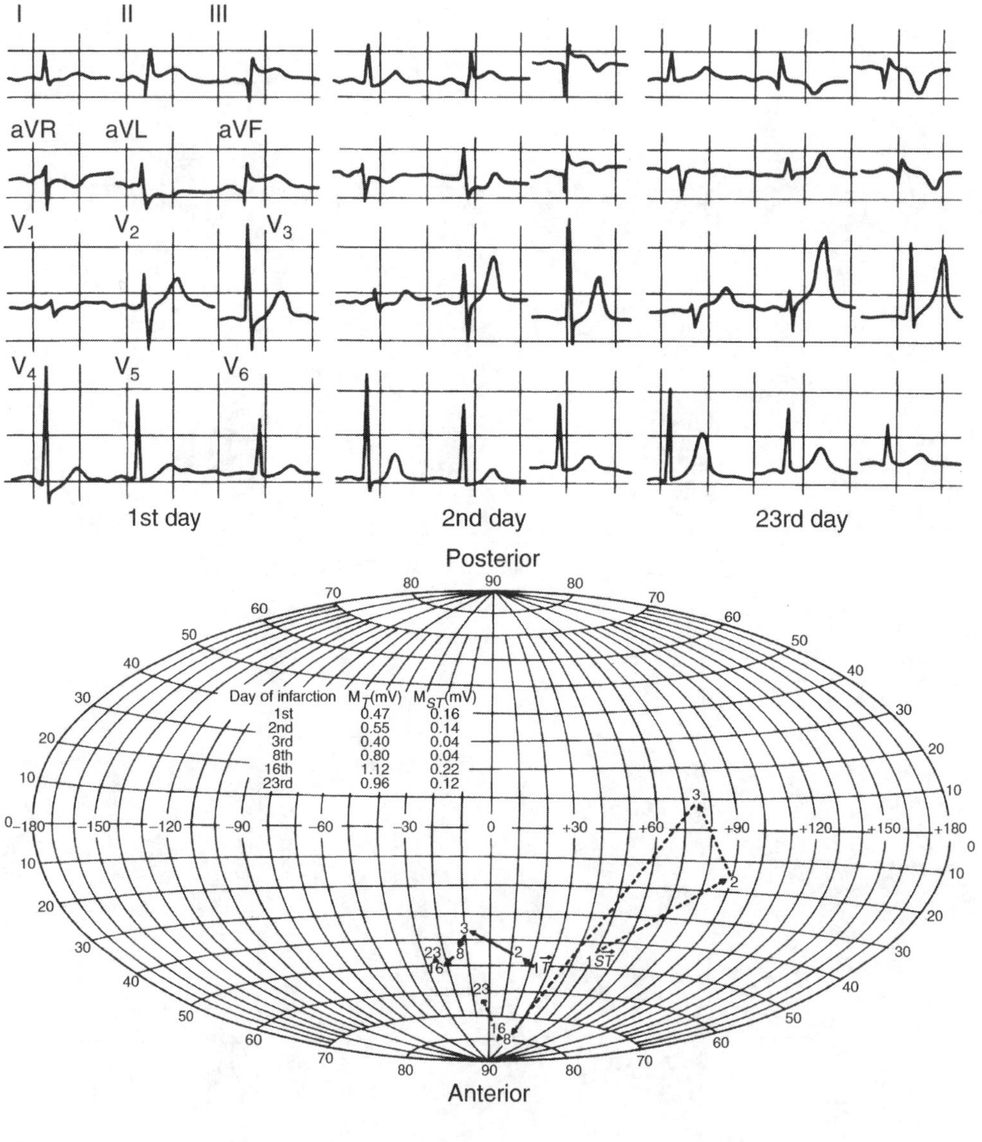

Figure 45.15
The typical migration of the T-wave vector in the Aitoff projection in evolving inferior infarction, with the movement toward "South America." The ST vector ultimately moves toward the same continent. ECGs (*above*) show evolving ST-T-wave changes. The spatial magnitudes of maximum $\vec{T}$ on first, second, third, eighth, 16th, and 23rd days are shown in the table (insert) together with the magnitudes of $\vec{ST}$. Note that on the 16th day, the magnitude of $\vec{T}$ is 2.4 times its initial value.

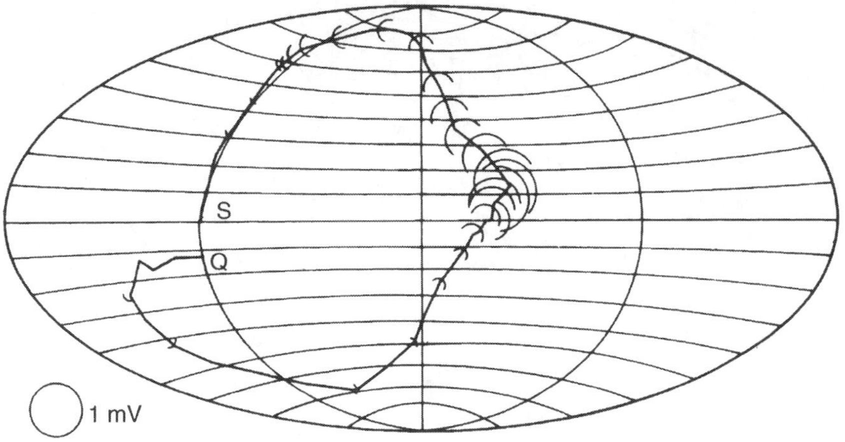

Figure 45.16
A normal spherocardiogram of the same subject as in Fig. 45.12. The snakelike pattern starts in the anterior (*lower*) hemisphere and crosses the equator "east" of the vertical, zero-degree meridian. The beginning and end of the snake are marked Q and S, respectively. Semicircles are drawn every 2 ms. Their diameters indicate spatial magnitude; the reference *circle* represents 1 mV.

directions have less resolution – at zero, directions are meaningless. This accounts for the wide scatter of $\vec{ST}$ directions among the women, whose $\vec{ST}$ magnitudes are often close to zero. In men, the $\vec{ST}$ tends to be larger and shows directions similar to those of the $\vec{T}$. The directions of $\vec{P}$ show considerable scatter, but it is the same in both sexes. $\vec{R}$ directions are well defined, but often change very quickly so that a few milliseconds difference in the timing of the R peak in the *M* tracing can result in a large difference in direction. This is responsible for some of the scatter seen for the $\vec{R}$ directions. $\vec{T}$, on the other hand, tends to change direction very gradually, if at all. It will be recalled that, in the VCG, the normal T loop is often a straight line. For this reason, and because magnitudes are large enough for accurate definition of direction, $\vec{T}$ directions show the least scatter. They form a well-defined "continent" that occupies only about 5% of the area of the globe (Fig. 45.3).

Abnormal $\vec{P}$ directions are useful in detecting abnormalities that, in the ECG, might be interpreted as a result of a low atrial focus. They have not, however, proven of much value in detecting atrial hypertrophy, which is more reliably done from the magnitude of $\vec{P}$.

Unusual $\vec{R}$ directions are helpful but tend to occur too frequently to justify calling them abnormal, if that is the only finding. It will now be appreciated by the reader that the longitude of $\vec{R}$ is a measure of Einthoven's electrical axis, which is not to be confused with the mean electrical axis (see Sect. 45.5.4).

The most useful vector direction is that of $\vec{T}$. Directions of $\vec{T}$ that fail to lie within the normal continent (Fig. 45.3) will give rise to abnormal T waves in the ECG. This efficient device for defining normality of T waves, presenting data in a way that immediately strikes the eye, should be compared with the prospect of defining T-wave normality on the basis of the 195 12-lead ECGs from the same subjects! It is an excellent example of the value of both the heart-vector concept and the use of spherical polar coordinates. It can also show subtle differences between populations. Figure 45.14 shows $\vec{T}$ distributions in male and female Cretan villagers: the center of the female distribution is slightly posterior to that of the male. A similar sex difference was observed in children aged 3–4 years [6] which is surprising in view of the similar body build at this age.

In evolving inferior infarction, there is a migration of $\vec{T}$ directions that is most characteristic (Fig. 45.15). If the normal T continent is analogous to the continent of Africa, $\vec{T}$ migrates westward to lie in South America. The concomitant ECG change is the development of inverted T waves in inferior leads.

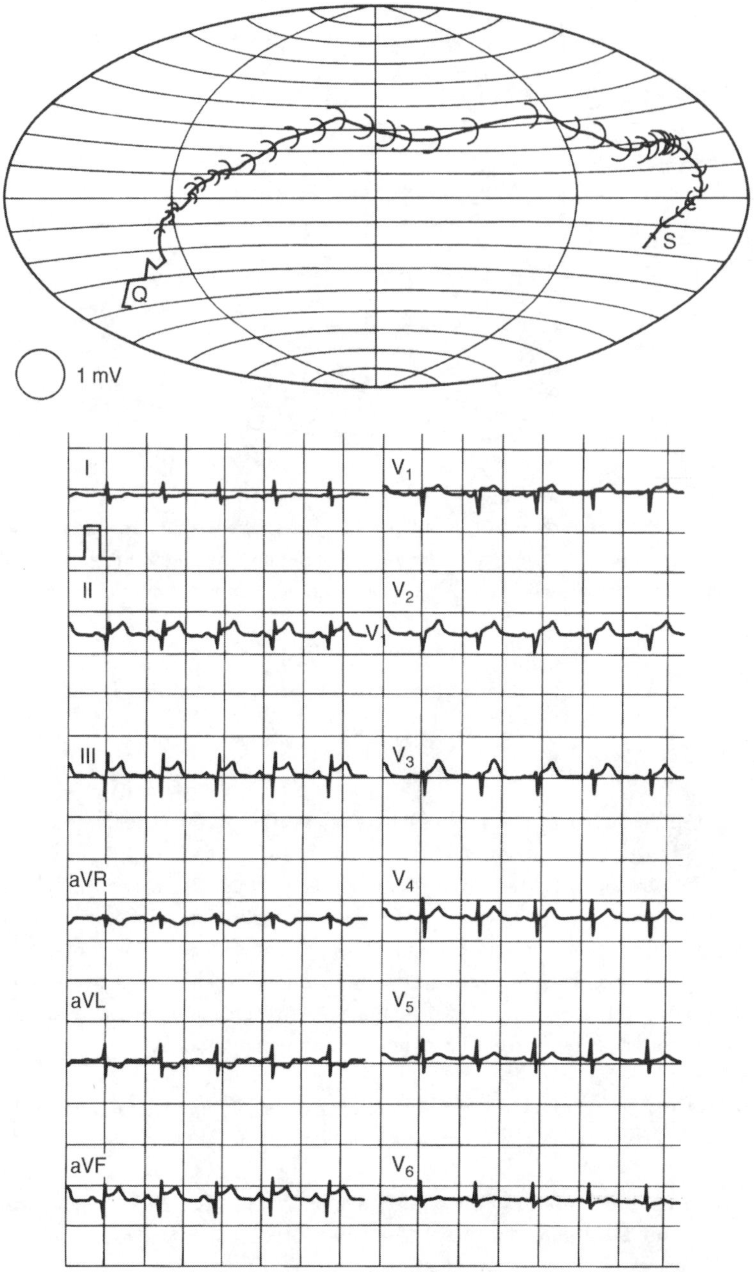

Figure 45.17
The spherocardiogram in inferior infarction. The "snake" begins in the anterior (*lower*) hemisphere and crosses the equator at longitude −80°. This should be compared with Fig. 45.16. The derived 12-1ead ECG below shows recent inferior infarction.

45.4.3 Spherocardiogram

Instead of plotting only a few vectors during the cardiac cycle, the Aitoff projection may be used to plot a full sequence of vectors, every 2 ms throughout the QRS complex (Fig. 45.16). The spherocardiogram [7] is such a display: vector magnitudes are represented by the radii of semicircles whose centers indicate direction and whose convexities indicate the next plot in the sequence. The result resembles a snake which, for normal subjects, has the thickest part of its body in the region of the "continent" of the direction of the normal $\vec{R}$ (➋ Fig. 45.3). The head of the snake lies in the "southern" hemisphere and its body crosses the equator slightly to the "east" of the zero meridian. Absent initial anterior forces, as in anteroseptal infarction, result in the head lying on the equator or in the "northern" hemisphere. When the body of the snake crosses the equator to the west, inferior infarction is indicated (➋ Fig. 45.17). Although the spherocardiogram has been useful in developing computer criteria (see ➋ Sect. 45.5), like the polarcardiogram, it is not routinely generated. Despite being based on spherical polar coordinates, the spherocardiogram is more like a three-dimensional VCG: it suffers from the same shortcoming that phenomena are not displayed against a time coordinate.

45.5 Diagnostic Criteria

Rather than present the various criteria developed for visual interpretation of polarcardiograms, it is thought more useful to outline the criteria now used in the computer program for further analysis. These differ from the older analog criteria [8, 9] in several important ways: they are, of necessity, totally explicit; they are designed to give higher specificity, at the expense of some sensitivity; and they include some new criteria [10].

45.5.1 Criteria for Myocardial Infarction

45.5.1.1 Criteria Relating to the QRS Complex

(a) Polarcardiographic criteria
  (i) $\beta\downarrow$: a downslope in the tracing of the angle in the transverse plane (➋ Figs. 45.11 and ➋ 45.13), occurring 10 ms before the peak $m_t\mathbf{R}$ of the transverse plane magnitude tracing (downslopes in the first 20 ms may be discounted). This is a criterion of anteroseptal or anterior infarction and corresponds to a region of clockwise rotation in the transverse plane vector loop (➋ Fig. 45.18a).

It must be noted that for the computer program, the definition of a downslope must be quite explicit. A downslope is an overall drop of 5° over a total downsloping time of at least 5 ms. A downslope begins at the onset of the first drop and ends at the beginning of the first rise following the fulfillment of its definition requirements of minimum time and overall drop. Small rises are allowed within a defined downslope provided that first, they are subtracted from the drop which has occurred up to that point and second, they do not thereby reduce the overall drop below 1°. Horizontal segments are also allowed, but neither they nor the segments containing rises contribute to the total downsloping time. Downslopes terminating with a net drop $\geqslant 10°$ are rated very marked, otherwise they are rated as significant. An upslope is similarly defined except that drops become rises.

  (ii) $m_t \to 0$: a return to the baseline of the transverse magnitude tracing immediately following an initial deflection (➋ Fig. 45.11). This is a criterion for anterior infarction that is often not apparent in the vector loop because of ambiguity of the E point. It may not be present in every beat (➋ Fig. 45.13).

In the above criterion, "initial" means between 10 ms after the Q point and the time of the peak of the m_t tracing. Return to the baseline means falling to 0.01 mV or less, or to 0.02 mV if the estimate for baseline noise is at least 0.02 mY. Between Q and the return to the baseline there must be a deflection of the m_t tracing of more than 0.05 mV and this amount must be greater than twice the value of the noise estimate.

  (iii) $\gamma\downarrow$: a downslope in the tracing of the angle in the left sagittal plane occurring 10 ms after the Q point and 10 ms before the peak m_sR of the sagittal plane magnitude tracing, provided that there is a Q wave in the Y tracing and longitude $< -45°$. This is a criterion of the inferior infarction; it corresponds to clockwise rotation in the initial anterosuperior segment of the left sagittal loop.

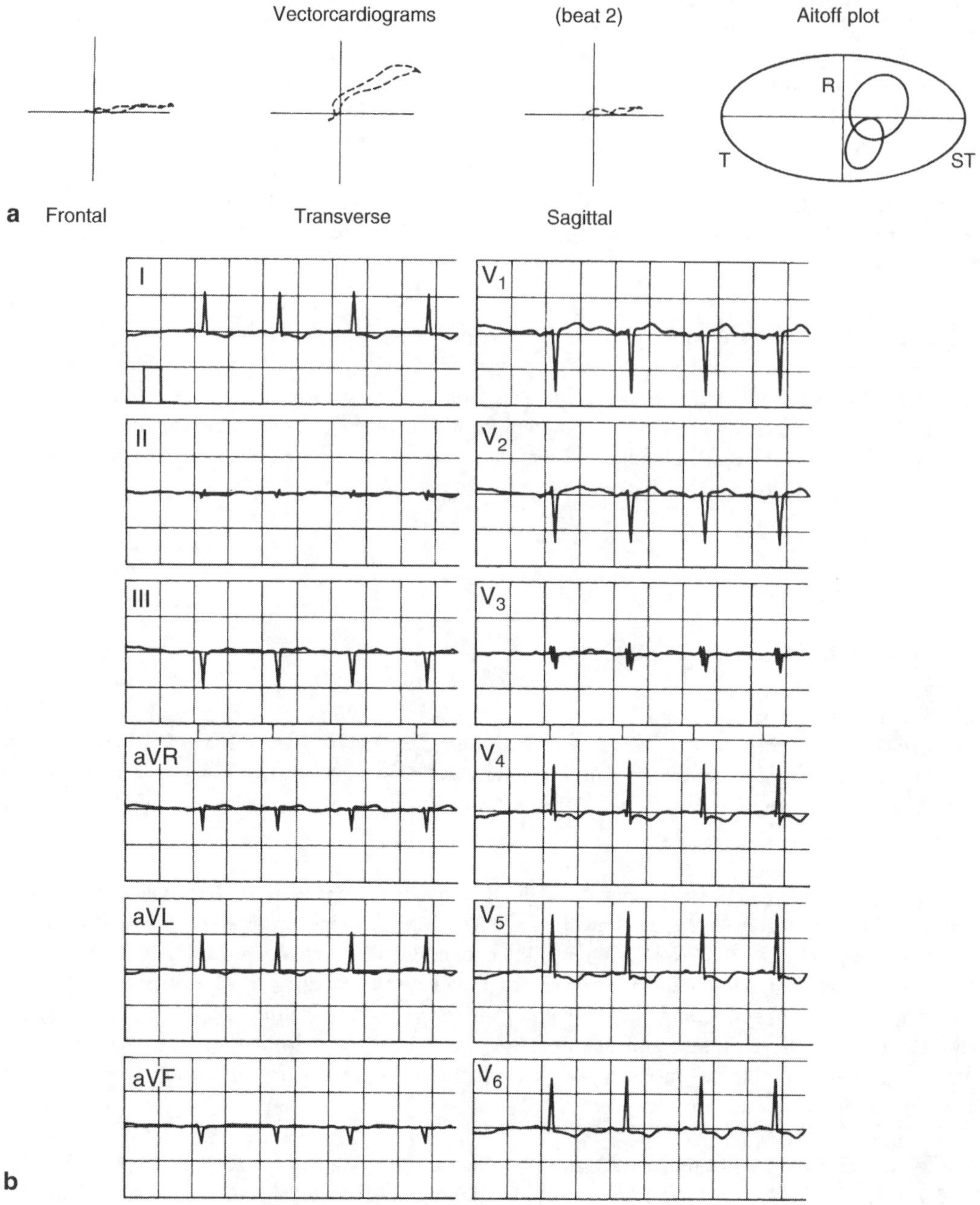

Figure 45.18

Part (a) shows VCGs and the Aitoff plot derived from the information for beat seven given in Table 45.3. The Aitoff plot illustrates that the directions of the R, ST, and T vectors are all outside the normal boundaries. Part (b) shows the ECG derived from three *xyz* signals of beats 1–4 (from Table 45.3); the ECG shows inferior infarction, but less-definite anterior infarction. Note the baseline drift prior to clamping of the first beat. Clamp points are indicated by short vertical lines between leads III and aVR, and V_3 and V_4.

(b) Quasi-electrocardiographic criteria

(iv) Absence of initial anterior forces, indicating by an initial Q wave in the Z tracing. This is a criterion of anteroseptal infarction; it corresponds to Q waves in V_1 and V_2 of the ECG.

(v) Initial anterior forces do not exceed 6% of posterior forces, as indicated in the Z tracing. This is a weaker criterion of anteroseptal infraction than (iv).

(vi) Posterior forces in the Z tracing do not exceed 9% of M_R, the maximum spatial QRS vector. This is a criterion of true posterior infarction; it is invalidated by complete or partial RBBB. It has no ECG equivalent but corresponds approximately to the finding of a small S wave in V_2.

(vii) Initial superior forces exceed 35% of inferior forces, indicated in the *Y* tracing. M_R is substituted for the inferior forces if these become less than 20% of M_R. This is a criterion of inferior infarction and corresponds to a relatively large Q wave in aVF.

(viii) Initial rightward forces exceed 23% of leftward forces, indicated in the *X* tracing. This suggests apical or lateral infarction; it corresponds to a relatively large Q wave in leads I, aVL or V_6.

(ix) Duration of the Q wave in the Y tracing exceeds 40% of the QRS complex in the spatial magnitude tracing. This criterion of inferior infarction roughly corresponds to a wide Q wave in aVF.

(c) Spherocardiographic criteria

(x) The locus of QRS directions on the sphere crosses the equator outside the normal longitude range of -15° to $+100^\circ$. This suggests inferior, lateral, or apical infarction (I, L, or A) according to the following ranges: $-130^\circ <$ longitude $< -15^\circ$ for inferior infarction; $+100^\circ <$ longitude $< +175^\circ$ for lateral infarction; and $+175^\circ <$ longitude $< -130^\circ$ for apical infarction (range passes through 180°). These criteria do not correspond to any ECG criteria.

Note that, near the crossing of the equator, the latitude must change from 5A to 5P within 10 ms. During this period, the vector magnitude must exceed 5% of M_R and the crossing must have occurred by 15 ms after the spatial magnitude tracing reaches its peak (M_R).

45.5.1.2 Non-QRS Criteria

(xi) $\vec{ST}$ suggests a current of injury if $M_S > 0.167M_R$, where M_S is the magnitude of the spatial vector at the end of QRS, that is, the S point. This criterion has no ECG counterpart. There are refinements of this criterion which indicate the region of the injury such as anteroseptal, lateral apical, according to the direction of $\vec{ST}$.

(xii) $\vec{T}$ suggests an evolving or old inferior infarction at first, $M_T > 0.2M_R$, where M_T is the spatial magnitude of the maximum $\vec{T}$; and second, this vector has a longitude between -10° and -60° and a latitude posterior to 70A and anterior to 20P.

(xiii) $\vec{T}$ suggests anterior injury if they have the following characteristics. First, $M_T > 0.2M_R$; second, the longitude of the maximum $\vec{T}$ is between -120° and $+165^\circ$; third, M_T has a latitude posterior to 20P, but lies outside a region defined between longitudes 0° and $+45^\circ$ and latitudes 19P and 45P; and finally, RBBB is not present. This corresponds to T-wave inversion in anterior precordial leads not of the type seen as a normal variant in women.

45.5.2 Evaluation of Performance of Infarction Criteria

The above criteria were evaluated with a test set of 369 subjects selected from a much larger population who were studied angiographically; 234 had 100% occlusion of a major coronary vessel and associated akinesia or dyskinesia (indicating infarction), and 135 had less than 50% occlusion, normal wall motion, normal hemodynamics, and no history of infarction [11]. The ECGs of the 369 subjects were separated into infarction and non-infarction groups by cardiologist ECG readers at two hospitals. A corresponding division was made by the computer using the above criteria. The computer's selection agreed slightly better with the angiographic evidence than the selections of the readers at the two hospitals: both its sensitivity and specificity were higher, though not significantly so. However, when the computer's selection of infarction cases was combined with the selections of either of the two groups of readers, the improvements in sensitivity were significant, while the corresponding reductions in specificity were not. This was satisfactory because the computer

Table 45.4
Apparent (worst case) specificity of polar computer program with respect to myocardial infarction (MI) tested on 473 presumed healthy Cretan villagers in various age-groups

	Men			Specificity
Age-group	Na	%	*MI + veb*	%
19–29	44	68	1	98
30–39	100	9	2	98
40–49	118	61	4	97
50–59	95	55	5	95
60–69	80	63	10	88
70–89	36	64	8	78
All	473	60	30	94

[a]Median ages for the men and women were both 46 years
[b]Positive diagnosis of previous infarction made by computer program

output was designed to be used as an aid to the cardiologist interpreting the ECG, rather than as an automated simulation of his interpretation. It is important to note, as an aside, that although the readers did not have VCGs for this evaluation, the computer generates these whenever it makes a diagnosis of infarction based on QRS evidence. This usually allows the reader to corroborate a positive diagnosis of infarction by the computer despite the lack of corroboration in the derived 12-lead ECG, that is, the ECGD (see Chap. 11). Sensitivity and specificity for the computer, alone, were 78% and 92%. However, the 92% specificity should be regarded as probably lower than it would be with a truly normal population because the 135 subjects who were negative angiographically could have had undetected lesions, since they all had sufficient chest pain to justify cardiac catheterization. When the computer criteria were applied to a test set of 118 Cretan villagers aged 40–49 years, the specificity was 97%. This is still probably lower than the true specificity because the Cretans were not examined – they were presumed healthy on the basis of the reported low prevalence of coronary artery disease among Cretan villagers [12]. Table 45.4 shows the specificity of the criteria in 473 Cretan villagers of various age-groups. Among the age-groups under 60 years, the specificity is 95% or better. There is a progressive decline in specificity with age, particularly after the age of 60 years. This decline is less than it appears, however, because 14 of the 30 positive cases had clear ECG evidence of previous myocardial infarction. Of course, it is possible to object that ECG evidence is inadmissible because it is not truly independent, but a cardiologist would find fault with failure of a computer program to detect "infarction" indicated by the ECG, so perhaps the program should not be faulted for those 14 cases. When this adjustment is made, the specificity for all subjects under 70% becomes 97%.

45.5.3 Criteria for Left Bundle Branch Block

This chapter would be unreasonably long if it contained every diagnostic criterion used by the polar analysis program. The myocardial infarction criteria have been given because of their importance and interest. The criteria for LBBB are also interesting and illustrate the polar approach when non-ECG evidence for a diagnosis is lacking.

Typical LBBB is recognized if:

(a) $\vec{R}$ is at least 70° posterior to $\vec{T}$
(b) The vector 30 ms before **IR** has transverse plane angle < 30°; that is, β_{IR}−30 ms < 30°
(c) The spatial angle between Rand $\vec{T}$ exceeds 120°
(d) β increases by less than 20° from **IR** to **IR** + 20 ms
(e) $\beta_{IR+10\,\mathrm{ms}} \leqslant \beta_{IR+10\,\mathrm{ms}} + 22$ and
(f) QRS duration is flagged in the table of measurements ">>" (see Table 45.1)

If condition (a) is false and $90^\circ \leqslant \vec{R} < -70^\circ$, or if $\vec{R}$ is posterior to 30P, then the LBBB is considered atypical.

If LBBB is otherwise typical but M_S exceeds 5% of M_R, there is the probability of injury or aneurysm.

Left ventricular hypertrophy is probably present in addition to LBBB if M_R exceeds 4 mV; it is also possible if M_R exceeds 3 mV.

45.5.3.1 Developing Criteria for LBBB

The marked differences between the above criteria for LBBB and those employed in conventional electrocardiography exemplify the polar approach used by the computer program. A view concerning the development of diagnostic criteria for computer analysis of the ECG that has enjoyed recent currency is that, to be valid, such criteria must be based on independent, that is, non-ECG disease classification. Independent means for selecting suitable populations are available for myocardial infarction and LVH (so-called type A diagnosis) but not for LBBB, a condition identified, and therefore defined by the ECG (a type B diagnosis [13]). Because no independent "gold standard" is available for LBBB, it might appear most reasonable to employ ECG criteria in the computer program. Instead of this, the ECG was used to select cases of LBBB whose polarcardiograms were then studied for clearly identifiable characteristics. Several polarcardiograms appeared that could not be simply stated in terms of the ECG. For example, striking differences in the contours of the M tracings separated LBBB from RBBB (➋ Fig. 45.5). The $\vec{R}$ and $\vec{T}$ of these conditions show marked differences in latitudes (➋ Fig. 45.5).

The approach followed in developing LBBB criteria was as follows:

(a) Identify, from ECGs, a population (training set) with well-recognized LBBB
(b) Compare polarcardiograms and vector loops of this population with those of a non-LBBB population and observe any clear-cut characteristics that separate the populations
(c) Organize these into an effective tool for identifying LBBB, and
(d) Observe how well this tool performs, in a separate test set, when the ECG is ambiguous, and whether it misidentifies LBBB in those cases where the ECG is definite as a gold standard

This approach can produce a better indicator even when there is no independent standard. It has historical parallels. Wenckebach's phenomenon, although nowadays an ECG diagnosis, was originally identified in polygraph tracings. Its description predated that of the electrocardiograph by 4 years. The polygraph, however, did not give satisfactory recordings in every case and was supplanted by the ECG as a gold standard for this condition.

An approach similar to the above was undertaken for RBBB and IV conduction delay. Once the various subtleties for discriminating these conditions had been programmed, approximately 100 examples of each were selected by the computer from stored data which was not used in developing the criteria, and the derived 12-lead ECGs were studied to check the accuracy of the computer identifications. There were no discrepancies beyond what would be considered acceptable differences of opinion. Since the ECG criteria for diagnosing LBBB have served so well over the years, it might be asked what the benefit is from introducing others, even if they seem less ambiguous. An example of the value of this is that the distinction between LBBB and LVH with IV conduction delay is sometimes difficult from a single ECG, whereas the computer tends not to confuse them. The ability of the program to identify injury or aneurysm in LBBB is particularly interesting: in LBBB, the S point in the M tracing returns to the baseline as it does in the normal subject. Departures from this are unusual and appear to connote injury or perhaps aneurysm. This feature cannot be detected from the ECG.

45.5.4 Left Anterior Fascicular Block

Like LBBB, left anterior fascicular block is an entity defined by the ECG but it seems to be less clearly defined. Its diagnosis hinges on the mean QRS axis being less than -45°. The mean QRS axis is defined from leads I and III using the Einthoven triangle, except that, instead of taking simultaneous points in time in each lead, the sums of the positive and negative deflections are normally used. Unfortunately, because simultaneity is abandoned, the triangle scheme no longer applies, since the "deflections" in lead II no longer fit the "vector" constructed from the other two leads. The result is not a vector at all and lacks a physical basis. The mean QRS axis was introduced in order to determine some indication of axis *from* the nonsimultaneous data provided by the then-standard single-channel electrocardiograph. The α tracing is a continuous display of the electrical axis throughout the heart cycle with the improvement provided by a scientifically based lead system. With this available, there would seem to be little merit in persisting with a technically inferior and theoretically questionable measure, such as the mean QRS axis, as a diagnostic discriminator. However, with a continuous range of axes available, which should be selected as most representative of QRS events?

The polarcardiographic criteria for left anterior fascicular block are that the longitude of $I\vec{R}$ (➜ Fig. 45.8) is between -110° and -45° with QRS duration <110 ms. The diagnosis is excluded if LBBB, RBBB, LVH, low voltage, inferior infarction, injury, current, or recent infarction are diagnosed. The use of $I\vec{R}$ rather than $\vec{R}$ gives a little more consistency in cases where there are two peaks in the M tracing. The directions of $I\vec{R}$ are generally the same as those of $\vec{R}$ but show slightly less scatter.

Polarcardiographic study of left anterior fascicular block has not revealed any clear-cut set of discriminators, such as those described for LBBB. Perhaps this gives support to the doubts expressed by some concerning the validity of left anterior fascicular block as a clinical entity [14]. Clearly, in this case, some independent "gold standard" would be useful.

Although the computer comment suggesting left anterior fascicular block tends to agree with the visual reading of the derived 12-lead ECG, occasional examples of disagreement should not be a surprise in the light of the foregoing. Of course, these examples could be reduced to zero by the expedient of adopting the mean QRS axis, despite its faults, as the discriminator. This could easily be done but it seems like a backward step.

45.5.5 Left Ventricular Hypertrophy

In the course of developing polarcardiographic criteria for LVH, electrocardiographic and autopsy criteria were closely examined. Problems were encountered with both. Nevertheless, some correlation exists between the mass of cardiac muscle and the strength of the electric field generated by the heart. This is exploited in the voltage criteria of the ECG, but amplitudes in several leads must be measured and correlated because individual variations in the orientation of the maximum heart vector produce maximal projections on different leads. This problem is solved in polarcardiography in which the *M* tracing gives a direct indication of field strength. Thus, one measurement, M_R, the spatial magnitude of the maximum QRS vector, $\vec{R}$, can replace measurements of several limbs and chest leads of the ECG. Another measurement, m_tR, the magnitude of the projection of $\vec{R}$ in the transverse plane, is also useful because there is a tendency for LVH to be associated with a 20° change in longitude of $\vec{R}$ toward the transverse plane, so that m_tR is increased by these two effects. A study of 168 autopsied cases, comparing the ratio of heart weight to body length with M_R and m_tR yielded correlation coefficients of 0.48 and 0.50, respectively [15]. These are not impressive, but fibrosis and scarring was present in many of the cases, and these tend to reduce voltage. (Pure LVH, observed in children, yielded much higher correlations [16].) Performance of ECG voltage criteria, if measurements are carefully made, on tracings obtained with good equipment turns out to be about the same, but not all cases of LVH recognized electrocardiographically meet polarcardiographic criteria for LVH, and vice versa.

An exact definition of LVH is difficult. Heart weights in normal subjects show such a wide variation that a patient's heart weight may undergo 50% hypertrophy, yet still not exceed the upper bounds *for* normal [17]. Taking body length into consideration reduces the spread somewhat but the coefficients of correlation are very low [17]. Body-surface area is commonly employed as a correlate of LVH diagnosed by echocardiography (see also ➜ Chap. 15). An echopolar study is overdue. Body build, athletic activity, and aging are normal factors that have a pronounced effect on body-surface voltages. If the effects of fibrosis, scarring, and myopathies are added to these, it is easy to appreciate that the diagnosis of LVH from body-surface potentials, whether these are studied as ECGs, VCGs, or vector magnitudes, is apt to be inexact. A further disquieting factor is that some patients show considerable day-to-day variations in voltage, beyond those attributable to differences in electrode placement or electrical activation.

The polarcardiographic criteria for LVH are indicated in ➜ Table 45.2. The appendix (Table Al.96) gives medians and upper 95 percentiles for 466 Cretan villagers for M_R and m_tR. Up to the age of 50 years, M_R is greater in males but gradually reduces with age in men, less clearly so in women. For the entire group, aged 19–82 years, for both sexes, the upper 95 percentiles for M_R and m_tR are 2.25 mV and 2.05 mV, respectively.

Of the ECG criteria of LVH – hypervoltage, left axis deviation, increased ventricular activation, and ST-T changes – only the first, and to some extent the second, have polarcardiographic equivalents (M_R and m_tR). The ST-T changes were considered secondary and nonspecific. When abnormal, ST-T-vector directions, as shown on the Aitoff plot, resemble those seen in ischemia. Ventricular activation time is based on the concept of intrinsicoid deflection, which is untenable vectorcardiographically since precordial-lead tracings can be simulated from derivations of the *xyz* signals.

45.5.6 Right Ventricular Hypertrophy

"Probable" right ventricular hypertrophy (RVH) is stated by the computer program if $\vec{R}$ is directed to the right, that is, if the longitude of $\vec{R}$ is not in the range $-90° <$ longitude of $R < +90°$, provided that:

(a) The ratio of maximum anterior forces to maximal posterior forces, during the QRS complex, exceeds 4
(b) $M_P > 0.22$ mV
(c) T is not posterior to 50P, and
(d) RBBB is not diagnosed

It is also stated that if the direction of $\vec{R}$ is more than 10° outside the normal bounds, $I\vec{R}$ is not in the range −I 10° < longitude of $I\vec{R}$ < +120°, and the angle between $\vec{R}$ and the direction of the lead I lead vector (59A, + 100) is less than 90°.

45.5.7 Ischemia

In myocardial ischemia, the **T** tend to be distributed around the 180° meridian (see Table 45.3), that is, approximately 180° from the normal direction of the $\vec{T}$. This corresponds to T-wave inversions in the ECG in ischemia. When ST segments of the ECG are depressed in ischemia, the directions of **ST** tend to parallel those of $\vec{T}$. During exercise testing, ischemia produces characteristic polarcardiographic changes which have been incorporated into an index of ischemia, $M_S\theta$. In the *M* tracing, the S point, instead of lying on or close to the zero baseline, becomes elevated (Fig. 45.9); M_S increases. The magnitude of the T wave, M_T in the *M* tracing tends to diminish, while that of the ST segment midway between the points S and T is variable. However, the directional change of *ST* is characteristic. The optimal time selected for sampling the directional *ST* changes is 75 ms after IR (Fig. 45.9) [18]. The directional change from normal is given by the spatial angle θ between the ST_{75} vector so defined and the normal direction of that vector, which is taken as (50A, +20) (latitude and longitude) after exercise, or (50A, +30) at rest. The product of M_S and θ gives the ischemic index $M_S\theta$. Because directions become meaningless as magnitudes reduce to zero, θ is considered to be zero if M_{ST75}, the spatial magnitude of the ST_{75} vector, is $\leqslant 0.04$ mV. For normal subjects, $M_S\theta < 7$ and the test is positive for ischemia if $M_S\theta$ exceeds 10.6; however, $M_S\theta$ may reach 60. Thus, ischemic changes seen in various leads of the ECG are reduced to a single number that has much more resolution than the microvolts of ST depression observed in various leads of the ECG. Furthermore, the ischemic index is easily programmed into the computer. It has shown statistically significant drug-induced changes that could not be demonstrated in the ECGs of the same patients [19].

45.6 Conclusion

The polar approach is but one of many means for studying and characterizing the electric field produced by the heart. Although it may strike the reader as strange, even alien, it often provides a relatively efficient means for categorizing abnormalities. In this regard, its very strangeness can be an asset since previously unrecognized features may be revealed. From a programming standpoint, the polar approach is attractive because it tends to be explicit.

References

1. Einthoven, W., G. Fahr, and A. De Waart, On the direction and manifest size of the variations of potential in the human heart and on the influence of the position of the heart on the form of the electrocardiogram. *Am. Heart J.*, 1950;**40**: 163–194.
2. Dower, G.E., *Polarcardiography*. Springfield, IL: Thomas, 1971, pp. 3–190.
3. Moore, A.D., P. Harding, and G.E. Dower, The polarcardiograph. An analogue computer that provides spherical coordinates of the heart vector. *Am. Heart J.*, 1962;**64**: 382–391.
4. Dower, G.E., D. Berghofer, and M. Kiely, Accuracy of computer measurements of QRS duration using spatial magnitude. Agreement with human readers. *Comput. Biomed. Res.*, 1983;**16**: 433–445.
5. Lerman, J., R.A. Bruce, and J.A. Murray, Correlation of polarcardiographic criteria for myocardial infarction with arteriographic and ventriculographic findings (substantiation of transmural and presentation of non-transmural criteria). *J. Electrocardiol.*, 1976;**9**: 219–226.
6. Niederberger, M., G.E. Dower, and H.B. Machado, Polarcardiographic and vectorcardiographic study of 70 normal children aged 3–4 years. *Clin. Cardiol.*, 1978;**1**: 142–151.
7. Niederberger, M. and G.A. Joskowicz, Global display of the heart vector (spherocardiogram). Applicability of vector and polarcardiographic infarct criteria. *J. Electrocardiol.*, 1977;**10**: 341–346.
8. Dower, G.E., H.E. Horn, and W.G. Ziegler, The polarcardiograph. Diagnosis of myocardial infarction. *Am. Heart J.*, 1965;**69**: 369–381.
9. Dower, G.E. and H.E. Horn, The polarcardiograph. Further studies of normal subjects, refinement of criteria for infarction, and a report on autopsied cases. *Am. Heart J.*, 1966;**72**: 451–462.
10. Dower, G.E., D. Berghofer, and M. Kiely, Polarcardiographic computer system vs cardiologists and derived ECG in diagnosis of infarction, in *Computers in Cardiology*, K.L. Ripley, Editor. New York: IEEE, 1982, pp. 75–80.
11. Dower, G.E., H. Bastos Machado, J.A. Osborne, D. Berghofer, and M. Kiely, Performance of computerized heart vector criteria in the diagnosis of myocardial infarction. *Rev. Port. Cardiol.*, 1984;**3**: 687–697.
12. Aravanis, C., A. Corcondilas, A.S. Dontas, D. Lekos, and A. Keys, IX The Greek islands of Crete and Corfu, in *Coronary Heart Disease in Seven Countries*, A. Keys, Editor. *Circulation*, 1910;**41**(Suppl I): 88–100.
13. Rautaharju, P.M., M. Ariet, T.A. Pryor, et al., Task Force III: computers in diagnostic electrocardiography. *Am. J. Cardiol.*, 1978;**41**: 158–170.
14. Rabkin, S.W., F.A.L. Mathewson, and R.B. Tate, Natural history of marked left axis deviation (left anterior hemiblock). *Am. J. Cardiol.*, 1979;**43**: 605–611.
15. Dower, G.E. and H.E. Horn, The polarcardiograph. Diagnosis of left ventricular hypertrophy. *Am. Heart J.*, 1967;**74**: 368–376.
16. Gamboa, R., P.G. Hugenholtz, and A.S. Nadas, Comparison of electrocardiograms and vectorcardiograms in congenital aortic stenosis. *Br. Heart J.*, 1965;**27**: 344–354.
17. Dower, G.E., H.E. Horn, and W.G. Ziegler, On electrocardiographic-autopsy correlations in left ventricular hypertrophy. A simple postmortem index of hypertrophy proposed. *Am. Heart J.*, 1967;**74**: 351–367.
18. Dower, G.E., R.A. Bruce, J. Pool, M.L. Simoons, M.W. Niederberger, and L.J. Meilink, Ischemic polarcardiographic changes induced by exercise: a new criterion. *Circulation*, 1973;**48**: 725–734.
19. Bruce, R.A., R. Alexander, Y.B. Li, et al., Electrocardiographic responses to maximal exercise in American and Chinese population samples, in *Measurement in Exercise Electrocardiography*, H. Blackburn, Editor. Springfield, IL: Thomas, 1969.

Appendix 1: Adult Normal Limits

P. W. Macfarlane, A. van Oosterom, O. Pahlm, P. Kligfield, M. Janse, J. Camm (eds.), *Comprehensive Electrocardiology*, DOI 10.1007/978-1-84882-046-3_A1

A1.1 Normal Limits of the 12-Lead ECG in White Caucasians

The data in the following tables have been derived from a study of over 1,450 apparently healthy Caucasians living in the west of Scotland. A few tables are based on smaller numbers. Further details can be found in ➋ Chap. 13. The results are presented as mean ± standard deviation together with the 96% range; that is, 2% of measurements have been excluded from each extreme of the range. P-, Q, S, T- wave amplitudes are presented as negative measurements (➋ Tables A1.1–A1.31).

Table A1.1

Interval and duration of measurements (in milliseconds) from Caucasian normals

Age-group	Sex	PR interval	QRS duration	QT interval	QT_c interval Hodges[a]	QT_c interval Bazett[b]	P-wave duration	Heart rate (bpm)
18→29	Male	152.5 ± 23.0	96.4 ± 8.6	385.5 ± 28.9	403.6 ± 19.0	413.9 ± 23.1	103.0 ± 14.2	70 ± 12
		112→208	80→114	336→442	368→444	370→463	72→128	48→98
		n = 265	n = 265	n = 265	n = 265	n = 265	n = 266	n = 266
	Female	145.9 ± 19.7	87.7 ± 7.8	380.0 ± 27.8	411.6 ± 18.0	429.7 ± 22.9	99.0 ± 12.7	76 ± 12
		114→194	72→104	322→440	378→451	386→477	70→122	55→108
		n = 317	n = 317	n = 317	n = 317	n = 317	n = 318	n = 318
30→39	Male	155.7 ± 21.4	95.4 ± 9.8	385.5 ± 29.5	404.8 ± 19.4	416.0 ± 22.9	105.0 ± 12.3	71 ± 12
		116→206	78→114	326→448	366→448	375→468	78→130	52→99
		n = 218	n = 218	n = 218	n = 218	n = 218	n = 221	n = 221
	Female	145.7 ± 18.6	88.6 ± 7.3	386.6 ± 27.7	415.2 ± 16.9	432.6 ± 20.9	99.0 ± 11.6	77 ± 13
		114→184	76→106	330→438	384→445	395→473	72→122	57→105
		n = 115	n = 115	n = 115	n = 115	n = 115	n = 118	n = 118
40→49	Male	157.2 ± 21.8	94.4 ± 9.9	390.8 ± 29.3	409.2 ± 17.9	420.0 ± 21.9	106.0 ± 11.2	70 ± 12
		116→210	78→114	340→450	377→450	377→464	84→128	49→96
		n = 119	n = 119	n = 119	n = 119	n = 119	n = 119	n = 119
	Female	154.9 ± 20.4	89.4 ± 7.9	386.1 ± 27.0	415.2 ± 22.5	433.7 ± 28.4	104.0 ± 12.9	77 ± 12
		108→200	74→108	328→434	347→457	350→483	78→128	59→106
		n = 72	n = 72	n = 72	n = 72	n = 72	n = 73	n = 73
50+	Male	161.5 ± 18.9	92.7 ± 9.3	385.5 ± 26.0	407.4 ± 17.5	420.9 ± 22.7	110.0 ± 10.5	73 ± 12
		120→196	74→112	320→434	374→444	380→475	86→134	54→100
		n = 123	n = 123	n = 123	n = 123	n = 123	n = 125	n = 125
	Female	155.6 ± 6.9	87.1 ± 8.7	390.7 ± 31.5	419.5 ± 22.7	438.2 ± 24.8	106.0 ± 9.5	77 ± 11
		122→196	68→104	336→488	376→486	392→506	88→126	59→104
		n = 79	n = 79	n = 79	n = 79	n = 79	n = 80	n = 80

[a] $QT_C = QT + 1.75\,(rate - 60)$ – Ref. [107] in ➋ Chap. 13.
[b] $QT_C = QT\,(rate/60)^{1/2}$ – Ref. [106] in ➋ Chap. 13.

Table A1.2

Interval measurements (in milliseconds) from Caucasian normals versus heart rate

Heart rate (bpm)	PR interval	QT interval	QT_c interval Hodges[a]	QT_c interval Bazett[b]
<50	145.067 ± 18.19	443.6 ± 21.62	420.5 ± 20.49	391.537 ± 17.70
	116→176	396→482	376.75→459.25	357.864→426.6
	n = 15	n = 15	n = 15	n = 15
50→59	154.753 ± 25.81	417.425 ± 21.27	410.197 ± 21.08	402.633 ± 21.28
	110→220	378→468	370→458.5	358.078→454.235
	N = 146	N = 146	n = 146	n = 146

Table A1.2 (Continued)

Heart rate (bpm)	PR interval	QT interval	QT_c interval Hodges[a]	QT_c interval Bazett[b]
60→69	154.769 ± 22.62	399.339 ± 21.51	408.173 ± 20.54	415.549 ± 21.38
	116→208	354→444	368.75→450.25	372.64→462.2
	n = 372	*n* = 372	*n* = 372	*n* = 372
70→79	151.196 ± 20.28	382.44 ± 19.82	407.516 ± 19.18	425.481 ± 21.51
	114→196	344→426	372.25→452	388.32→477.234
	n = 408	*n* = 408	*n* = 408	*n* = 408
80→89	150.51 ± 20.19	367.522 ± 17.36	409.076 ± 16.93	434.074 ± 20.06
	114→194	332→406	376.5→444.5	395.044→474.053
	n = 247	*n* = 247	*n* = 247	*n* = 247
90→99	151.368 ± 17.52	356.842 ± 17.59	415.958 ± 17.34	446.019 ± 21.80
	114→182	320→388	383.5→449.25	405.54→488.23
	n = 114	*n* = 114	*n* = 114	*n* = 114
≥ 100	151.459 ± 14.14	328.108 ± 16.25	408.372 ± 18.63	435.705 ± 23.93
	130→180	298→352	377.25→439.5	395.685→476.246
	n = 37	*n* = 37	*n* = 37	*n* = 37

[a] $QT_c = QT + 1.75(\text{rate} - 60)$
[b] $QT_c = QT\,(\text{rate}/60)^{1/2}$.

Table A1.3
Durations (in milliseconds) in Caucasian adults: lead I

Age-group	Sex	Q duration	R duration	S duration
18→29	Male	16 ± 5	45 ± 11	29 ± 11
		9 → 28	27 → 76	9 → 52
		n = 126	*n* = 224	*n* = 168
	Female	15 ± 4	46 ± 10	26 ± 9
		7 23	29 → 70	12 → 47
		n = 127	*n* = 284	*n* = 155
30→39	Male	17 ± 4	47 ± 12	28 ± 11
		8 → 26	29 → 80	12 → 53
		n = 121	*n* = 212	*n* = 142
	Female	16 ± 4	47 ± 11	25 ± 9
		8 → 26	29 → 68	11 → 46
		n = 53	*n* = 125	*n* = 61
40→49	Male	17 ± 4	48 ± 12	28 ± 11
		9 → 26	30 → 77	7 → 51
		n = 115	*n* = 187	*n* = 110
	Female	15 ± 4	49 ± 10	26 ± 8
		8 → 22	33 → 73	16 → 42
		n = 52	*n* = 82	*n* = 31
50+	Male	17 ± 4	50 ± 14	29 ± 10
		9 → 27	31 → 88	13 → 51
		n = 122	*n* = 197	*n* = 99
	Female	15 ± 4	50 ± 12	23 ± 7
		7 → 27	27 → 74	12 → 42
		n = 74	*n* = 124	*n* = 43

Table A1.4

Durations (in milliseconds) in Caucasian adults: lead II

Age-group	Sex	Q duration	R duration	S duration
18–29	Male	17 ± 5	52 ± 14	26 ± 11
		8 → 26	35 → 84	6 → 49
		n = 162	*n* = 230	*n* = 148
	Female	15 ± 4	47 ± 11	25 ± 9
		8 → 26	30 → 74	9 → 44
		n = 182	*n* = 287	*n* = 175
30–39	Male	17 ± 5	55 ± 15	27 ± 13
		9 → 28	31 → 88	6 → 56
		n = 118	*n* = 207	*n* = 115
	Female	16 ± 4	48 ± 10	24 ± 10
		10 → 23	33 → 72	6 → 46
		n = 95	*n* = 124	*n = 65*
40–49	Male	16 ± 4	54 ± 15	29 ± 14
		7 → 24	33 → 90	9 → 57
		n = 94	*n* = 182	*n = 97*
	Female	15 ± 4	51 ± 13	25 ± 10
		7 → 24	33 → 78	10 → 45
		n = 45	*n = 80*	*n = 40*
50+	Male	16 ± 4	53 ± 14	28 ± 13
		9 → 26	32 → 86	6 → 71
		n = 79	*n* = 195	*n* = 122
	Female	15 ± 4	52 ± 13	26 ± 11
		8 → 24	32 → 76	8 → 56
		n = 64	*n* = 129	*n = 57*

Table A1.5

Durations (in milliseconds) in Caucasian adults: lead III

Age-group	Sex	Q duration	R duration	S duration
18–29	Male	23 ± 10	47 ± 20	30 ± 17
		10 → 50	11 → 87	6 → 84
		n = 128	*n* = 221	*n* = 116
	Female	22 ± 11	40 ± 16	29 ± 14
		9 → 80	10 → 74	7 → 68
		n = 166	*n* = 278	*n* = 160
30–39	Male	24 ± 12	42 ± 22	33 ± 20
		9 → 48	8 → 86	7 → 86
		n = 106	*n* = 205	*n* = 114
	Female	20 ± 7	41 ± 18	27 ± 16
		10 → 48	9 → 79	7 → 78
		n = 86	*n* = 123	*n = 64*

Table A1.5 (Continued)

Age-group	Sex	Q duration	R duration	S duration
40–49	Male	23 ± 11	39 ± 21	37 ± 20
		8 → 58	11 → 86	6 → 79
		n = 72	*n* = 181	*n* = 114
	Female	22 ± 12	35 ± 18	36 ± 17
		12 → 78	12 → 74	9 → 70
		n = 33	*n* = 78	*n* = 48
50+	Male	26 ± 15	35 ± 20	41 ± 21
		9 → 100	10 → 83	10 → 90
		n = 68	*n* = 195	*n* = 137
	Female	22 ± 11	34 ± 18	35 ± 18
		9 → 62	8 → 76	6 → 70
		n = 49	*n* = 123	*n* = 82

Table A1.6
Durations (in milliseconds) in Caucasian adults: lead aVR

Age-group	Sex	Q duration	R duration	S duration
18–29	Male	50 ± 13	19 ± 8	46 ± 11
		32 → 92	8 → 44	30 → 75
		n = 74	*n* = 208	*n* = 160
	Female	47 ± 9	18 ± 8	45 ± 9
		34 → 74	9 → 38	30 → 65
		n = 130	*n* = 248	*n* = 159
30–39	Male	51 ± 13	20 ± 10	50 ± 12
		34 → 84	9 → 50	32 → 78
		n = 91	*n* = 181	*n* = 125
	Female	46 ± 10	17 ± 6	45 ± 8
		36 → 78	8 → 39	34 → 65
		n = 41	*n* = 108	*n* = 85
40–49	Male	52 ± 12	20 ± 10	50 ± 12
		34 → 78	10 → 49	34 → 75
		n = 76	*n* = 157	*n* = 110
	Female	51 ± 13	17 ± 6	48 ± 9
		35 → 84	10 → 36	31 → 66
		n = 32	*n* = 67	*n* = 51
50+	Male	52 ± 13	20 ± 9	48 ± 13
		34 → 84	9 → 42	31 → 78
		n = 104	*n* = 159	*n* = 95
	Female	53 ± 13	18 ± 7	48 ± 12
		29 → 80	8 → 42	25 → 70
		n = 69	*n* = 92	*n* = 61

Table A1.7

Durations (in milliseconds) in Caucasian adults: lead aVL

Age-group	Sex	Q duration	R duration	S duration
18–29	Male	21 ± 13	35 ± 16	39 ± 18
		8 → 74	9 → 77	7 → 75
		n = 100	*n* = 219	*n* = 172
	Female	18 ± 6	38 ± 16	30 ± 14
		8 → 35	13 → 74	7 → 62
		n = 107	*n* = 284	*n* = 187
30–39	Male	20 ± 10	39 ± 17	37 ± 16
		9 → 68	11 → 79	7 → 71
		n = 114	*n* = 209	*n* = 158
	Female	18 ± 5	37 ± 16	33 ± 16
		9 → 29	10 → 76	8 → 66
		n = 40	*n* = 125	*n = 82*
40–49	Male	19 ± 7	41 ± 16	33 ± 15
		10 → 45	12 → 75	7 → 67
		n = 113	*n* = 186	*n* = 118
	Female	17 ± 4	45 ± 14	29 ± 13
		11 → 25	21 → 72	12 → 56
		n = 46	*n = 82*	*n = 38*
50+	Male	20 ± 10	46 ± 17	34 ± 16
		10 → 52	19 → 89	10 → 72
		n = 125	*n* = 196	*n = 98*
	Female	17 ± 6	45 ± 14	27 ± 13
		8 → 38	15 → 69	5 → 61
		n = 71	*n* = 124	*n = 56*

Table A1.8

Durations (in milliseconds) in Caucasian adults: lead aVF

Age-group	Sex	Q duration	R duration	S duration
18–29	Male	19 ± 5	53 ± 17	28 ± 13
		9 → 27	15 → 90	8 → 69
		n = 143	*n* = 231	*n* = 129
	Female	17 ± 5	47 ± 12	27 ± 9
		8 → 28	25 → 74	12 → 49
		n = 174	*n* = 288	*n* = 162
30–39	Male	18 ± 5	55 ± 17	31 ± 16
		8 → 32	23 → 90	7 → 74
		n = 110	*n* = 208	*n = 95*
	Female	17 ± 4	47 ± 12	24 ± 10
		8 → 23	29 → 74	6 → 48
		n = 89	*n* = 124	*n = 66*

Table A1.8 (Continued)

Age-group	Sex	Q duration	R duration	S duration
40–49	Male	17 ± 5	51 ± *19*	32 ± *18*
		9 → 28	13 → 92	6 → 80
		n = 76	*n* = 183	*n* = 102
	Female	16 ± 4	48 ± 14	27 ± 11
		9 → 24	20 → 76	11 → 53
		n = 36	*n* = 81	*n* = 43
50+	Male	18 ± 7	49 ± 18	33 ± 18
		9 → 52	15 → 88	8 → 79
		n = 73	*n* = 198	*n* = 125
	Female	16 ± 5	48 ± 15	31 ± 14
		8 → 28	18 → 78	9 → 69
		n = 55	*n* = 129	*n* = *62*

Table A1.9
Durations (in milliseconds) in Caucasian adults: lead V_1

Age-group	Sex	Q duration	R duration	S duration
18–29	Male	80	31 ± 6	55 ± 10
		80 → 80	20 → 46	30 → 76
		n = 1	*n* = 228	*n* = 229
	Female	57 ± 31	26 ± 5	54 ± 9
		3 → 78	14 → 37	32 → 71
		n = 5	*n* = 282	*n* = 282
30–39	Male	46 ± 31	28 ± 7	57 ± 9
		11 → 74	11 → 42	34 → 75
		n = 4	*n* = 210	*n* = 210
	Female		26 ± 5	55 ± 8
			16 → 36	37 → 68
			n = 126	*n* = 126
40–49	Male	68 ± 5	28 ± 7	57 ± 10
		58 → 72	15 → 45	29 → 75
		n = 6	*n* = 179	*n* = 179
	Female	60 ± 8	24 ± 5	57 ± 8
		54 → 66	12 → 32	45 → 78
		n = 2	*n* = *80*	*n* = *80*
50+	Male	38 ± 34	27 ± 6	56 ± 10
		14 → 62	15 → 45	34 → 79
		n = 2	*n* = 198	*n* = 198
	Female	9	24 ± 7	55 ± 9
		9 → 9	10 → 39	28 → 71
		n = 1	*n* = 125	*n* = 125

Table A1.10

Durations (in milliseconds) in Caucasian adults: lead V_2

Age-group	Sex	Q duration	R duration	S duration
18–29	Male	74	33 ± 7	52 ± 9
		74 → 74	21 → 48	36 → 72
		n = 1	*n* = 228	*n* = 228
	Female	9 ± 6	30 ± 7	52 ± 9
		5 → 13	17 → 44	31 → 70
		n = 2	*n* = 281	*n* = 280
30–39	Male	41 ± 35	33 ± 9	54 ± 10
		16 → 66	20 → 47	31 → 73
		n = 2	*n* = 208	*n* = 207
	Female		31 ± 6	52 ± 9
			19 → 45	29 → 67
			n = 126	*n* = 126
40–49	Male	72 ± 5	34 ± 8	51 ± 11
		66 → 76	19 → 52	31 → 75
		n = 3	*n* = 180	*n* = 180
	Female	36 ± 40	30 ± 7	51 ± 10
		8 → 64	18 → 47	29 → 71
		n = 2	*n* = 82	*n* = 82
50+	Male	30 ± 35	35 ± 8	49 ± 11
		8 → 82	20 → 55	26 → 73
		n = 4	*n* = 199	*n* = 199
	Female	44 ± 42	30 ± 7	50 ± 9
		14 → 74	16 → 45	27 → 67
		n = 2	*n* = 126	*n* = 126

Table A1.11

Durations (in milliseconds) in Caucasian adults: lead V_3

Age-group	Sex	Q duration	R duration	S duration
18–29	Male	12 ± 5	43 ± 10	40 ± 13
		8 → 17	25 → 62	16 → 63
		n = 3	*n* = 229	*n* = 227
	Female	23 ± 27	42 ± 9	37 ± 12
		5 → 62	23 → 59	12 → 60
		n = 4	*n* = 284	*n* = 275
30–39	Male	14 ± 4	43 ± 10	42 ± 12
		6 → 18	26 → 74	15 → 62
		n = 8	*n* = 208	*n* = 204
	Female		43 ± 10	38 ± 13
			23 → 74	15 → 63
			n = 124	*n* = 118

Table A1.11 (Continued)

Age-group	Sex	Q duration	R duration	S duration
40–49	Male	14 ± 4	43 ± 9	41 ± 13
		9 → 18	26 → 62	14 → 65
		n = 5	*n* = 185	*n* = 182
	Female	13 ± 2	41 ± 8	39 ± 11
		11 → 14	22 → 60	18 → 58
		n = 2	*n* = 81	*n* = 79
50+	Male	12 ± 5	43 ± 10	41 ± 13
		5 → 21	27 → 66	18 → 68
		n = 13	*n* = 192	*n* = 190
	Female	12 ± 5	40 ± 8	39 ± 11
		8 → 20	22 → 58	18 → 60
		n = 5	*n* = 128	*n* = 125

Table A1.12
Durations (in milliseconds) in Caucasian adults: lead V_4

Age-group	Sex	Q duration	R duration	S duration
18–29	Male	16 ± 6	46 ± 10	28 ± 11
		7 → 29	32 → 78	7 → 53
		n = 86	*n* = 227	*n* = 209
	Female	14 ± 4	44 ± 9	29 ± 9
		8 → 25	31 → 72	12 → 46
		n = 72	*n* = 284	*n* = 242
30–39	Male	14 ± 4	46 ± 10	32 ± 11
		8 → 27	33 → 86	11 → 57
		n = 62	*n* = 210	*n* = 193
	Female	16 ± 4	45 ± 9	32 ± 10
		8 → 24	33 → 73	15 → 53
		n = 35	*n* = 125	*n* = 97
40–49	Male	13 ± 4	46 ± 9	33 ± 12
		7 → 25	34 → 76	12 → 61
		n = 49	*n* = 184	*n* = 169
	Female	12 ± 4	44 ± 8	32 ± 11
		5 → 19	34 → 66	12 → 54
		n = 15	*n* = 80	*n* = 70
50+	Male	14 ± 4	44 ± 9	35 ± 13
		8 → 24	31 → 72	12 → 57
		n = 54	*n* = 199	*n* = 188
	Female	13 ± 4	42 ± 7	33 ± 10
		9 → 23	29 → 60	13 → 58
		n = 28	*n* = 128	*n* = 122

Table A1.13

Durations (in milliseconds) in Caucasian adults: lead V_5

Age-group	Sex	Q duration	R duration	S duration
18–29	Male	17 ± 5	45 ± 12	27 ± 11
		9 → 27	30 → 80	6 → 49
		n = 171	*n* = 230	*n* = 181
	Female	14 ± 4	42 ± 9	27 ± 9
		8 → 24	29 → 69	10 → 46
		n = 148	*n* = 288	*n* = 214
30–39	Male	16 ± 5	46 ± 11	28 ± 12
		8 → 25	31 → 86	10 → 58
		n = 127	*n* = 211	*n* = 179
	Female	15 ± 5	43 ± 10	27 ± 9
		8 → 26	31 → 71	12 → 45
		n = 79	*n* = 124	*n = 88*
40–49	Male	15 ± 4	45 ± 10	29 ± 12
		7 → 24	33 → 79	8 → 55
		n = 103	*n* = 186	*n* = 149
	Female	13 ± 4	44 ± 9	29 ± 10
		8 → 20	33 → 67	12 → 51
		n = 35	*n = 81*	*n = 64*
50+	Male	15 ± 4	45 ± 11	33 ± 12
		8 → 26	31 → 78	11 → 68
		n = 101	*n* = 198	*n* = 160
	Female	14 ± 4	43 ± 9	28 ± 10
		8 → 25	28 → 66	13 → 49
		n = 62	*n* = 128	*n = 98*

Table A1.14

Durations (in milliseconds) in Caucasian adults: lead V_6

Age group	Sex	Q duration	R duration	S duration
18–29	Male	18 ± 5	47 ± 12	25 ± 11
		10 → 28	31 → 76	6 → 46
		n = 202	*n* = 228	*n* = 137
	Female	16 ± 4	44 ± 9	24 ± 8
		9 → 25	30 → 70	12 → 42
		n = 215	*n* = 289	*n* = 161
30–39	Male	17 ± 4	49 ± 13	26 ± 12
		8 → 26	31 → 82	8 → 58
		n = 159	*n* = 208	*n* = 129
	Female	17 ± 5	45 ± 9	25 ± 8
		10 → 28	34 → 71	11 → 45
		n = 104	*n* = 124	*n = 59*
40–49	Male	16 ± 4	50 ± 11	29 ± 11
		8 → 24	34 → 76	12 → 52
		n = 139	*n* = 186	*n = 96*
	Female	15 ± 4	47 ± 10	26 ± 8
		8 → 23	33 → 68	14 → 42
		n = 51	*n = 82*	*n = 42*
50+	Male	16 ± 4	48 ± 12	30 ± 11
		9 → 25	31 → 78	12 → 57
		n = 130	*n* = 195	*n* = 124
	Female	15 ± 4	48 ± 11	25 ± 9
		8 → 23	30 → 72	12 → 47
		n = 80	*n* = 128	*n = 60*

Table A1.15
Amplitudes (in millivolts) in Caucasian adults: lead I

Age group	Sex	P+	P–	Q	R	S	STj	T+	T–
18–29	Male	0.099 ± 0.029	−0.020 ± 0.000	−0.074 ± 0.051	0.756 ± 0.333	−0.227 ± 0.139	0.031 ± 0.022	0.343 ± 0.116	
		0.055–0.174	−0.020–0.000	−0.238–0.000	0.202–1.624	−0.568–0.000	−0.006–0.077	0.116–0.641	
		n = 224	*n* = 1	*n* = 126	*n* = 224	*n* = 168	*n* = 224	*n* = 224	
	Female	0.101 ± 0.026	−0.058 ± 0.011	−0.070 ± 0.046	0.699 ± 0.283	−0.163 ± 0.096	0.011 ± 0.015	0.283 ± 0.093	
		0.056–0.161	−0.066–0.000	−0.225–0.000	0.226–1.411	−0.468–0.000	−0.022–0.038	0.112–0.506	
		n = 283	*n* = 2	*n* = 127	*n* = 284	*n* = 155	*n* = 284	*n* = 284	
30–39	Male	0.102 ± 0.027	−0.019 ± 0.001	−0.070 ± 0.046	0.807 ± 0.353	−0.189 ± 0.122	0.023 ± 0.019	0.300 ± 0.099	
		0.056–0.163	−0.019–0.000	−0.223–0.000	0.251–1.685	−0.536–0.000	−0.011–0.061	0.119–0.522	
		n = 212	*n* = 2	*n* = 121	*n* = 212	*n* = 142	*n* = 212	*n* = 212	
	Female	0.102 ± 0.032	−0.023 ± 0.000	−0.070 ± 0.044	0.666 ± 0.290	−0.145 ± 0.103	0.008 ± 0.017	0.257 ± 0.093	
		0.051–0.202	−0.023–0.000	−0.208–0.000	0.198–1.557	−0.622–0.000	−0.028–0.042	0.096–0.478	
		n = 124	*n* = 1	*n* = 52	*n* = 124	*n* =61	*n* = 124	*n* = 124	
40–49	Male	0.098 ± 0.032	−0.024 ± 0.016	−0.069 ± 0.041	0.791 ± 0.331	−0.159 ± 0.100	0.019 ± 0.020	0.271 ± 0.105	−0.087 ± 0.000
		0.045–0.184	−0.038–0.000	−0.173–0.000	0.286–1.509	−0.465–0.000	−0.017–0.074	0.105–0.547	−0.087–0.000
		n = 207	*n* = 4	*n* = 125	*n* = 207	*n* = 121	*n* = 207	*n* = 207	*n* =1
	Female	0.103 ± 0.026	−0.038 ± 0.000	−0.068 ± 0.042	0.773 ± 0.256	−0.129 ± 0.078	0.010 ± 0.018	0.245 ± 0.090	−0.021 ± 0.000
		0.055–0.184	−0.038–0.000	−0.191–0.000	0.374–1.414	−0.351–0.000	−0.027–0.046	0.095–0.461	−0.021–0.000
		n = 89	*n* = 1	*n* = 56	*n* = 89	*n* = 32	*n* = 89	*n* = 89	*n* = 1
50+	Male	0.097 ± 0.030	−0.019 ± 0.008	−0.072 ± 0.045	0.800 ± 0.272	−0.145 ± 0.087	0.015 ± 0.020	0.234 ± 0.097	
		0.043–0.166	−0.024–0.000	−0.195–0.000	0.292–1.416	−0.404–0.000	−0.028–0.060	0.079–0.502	
		n = 200	*n* = 2	*n* = 125	*n* = 200	*n* =99	*n* = 200	*n* = 200	
	Female	0.102 ± 0.029	−0.029 ± 0.000	−0.067 ± 0.043	0.807 ± 0.299	−0.119 ± 0.081	0.004 ± 0.018	0.218–0.098	−0.036 ± 0.036
		0.052–0.179	−0.029–0.000	−0.277–0.000	0.231–1.544	−0.339–0.000	−0.039–0.032	0.060–0.452	−0.087–0.000
		n = 125	*n* = 1	*n* = 54	*n* = 125	*n* = 44	*n* = 125	*n* = 124	*n* = 4

■ Table A1.16
Amplitudes (in millivolts) in Caucasian adults: lead II

Age group	Sex	P+	P−	Q	R	S	STj	T+	T−
18–29	Male	0.148 ± 0.057	−0.028–0.027	−0.089 ± 0.057	1.364 ± 0.428	0.212 ± 0.126	0.041 ± 0.033	0.419 ± 0.141	−0.022 ± 0.001
		0.040–0.294	−0.124–0.000	−0.251–0.000	0.566–2.357	−0.538–0.000	−0.015–0.117	0.150–0.780	−0.023–0.000
		n = 229	*n* = 16	*n* = 162	*n* = 230	*n* = 148	*n* = 230	*n* = 230	*n* = 2
	Female	0.149 ± 0.055	−0.029 ± 0.024	−0.075 ± 0.052	1.133 ± 0.340	−0.169 ± 0.097	0.012 ± 0.024	0.334 ± 0.110	
		0.052–0.305	−0.086–0.000	−0.202–0.000	0.523–1.828	−0.437–0.000	−0.037–0.058	0.147–0.617	
		n = 287	*n* = 8	*n* = 182	*n* = 287	*n* = 175	*n* = 287	*n* = 287	
30–39	Male	0.150 ± 0.052	−0.020 ± 0.004	−0.089 ± 0.061	1.118 ± 0.424	−0.198 ± 0.143	0.033 ± 0.029	0.363 ± 0.131	
		0.043–0.281	−0.025–0.000	−0.343–0.000	0.412–2.230	−0.617–0.000	−0.025–0.101	0.142–0.710	
		n = 207	*n* = 7	*n* = 118	*n* = 207	*n* = 115	*n* = 207	*n* = 207	
	Female	0.158 ± 0.056		−0.079 ± 0.044	1.080 ± 0.335	−0.158 ± 0.110	0.007 ± 0.027	0.302 ± 0.106	−0.064 ± 0.000
		0.055–0.274		−0.178–0.000	0.450–1.793	−0.579–0.000	−0.069–0.074	0.118–0.554	−0.064–0.000
		n = 123		*n* = 95	*n* = 123	*n* = 65	*n* = 123	*n* = 123	*n* = 1
40–49	Male	0.145 ± 0.047	−0.041 ± 0.013	−0.063 ± 0.039	0.911 ± 0.370	−0.197 ± 0.147	0.026 ± 0.030	0.336 ± 0.125	
		0.062–0.274	−0.052–0.000	−0.187–0.000	0.267–1.731	−0.731–0.000	−0.042–0.088	0.136–0.656	
		n = 200	*n* = 4	*n* = 103	*n* = 201	*n* = 108	*n* = 201	*n* = 201	
	Female	0.144 ± 0.051	−0.022 ± 0.008	−0.060 ± 0.037	0.886 ± 0.280	−0.159 ± 0.111	0.014 ± 0.022	0.282 ± 0.095	−0.027 ± 0.000
		0.058–0.258	−0.031–0.000	−0.192–0.000	0.491–1.438	−0.622–0.000	−0.029–0.053	0.100–0.523	−0.027–0.000
		n = 87	*n* = 3	*n* = 48	*n* = 87	*n* = 43	*n* = 87	*n* = 87	*n* = 1
50+	Male	0.144 ± 0.047	−0.018 ± 0.003	−0.072 ± 0.051	0.835 ± 0.368	−0.188 ± 0.156	0.022 ± 0.026	0.304 ± 0.115	−0.057 ± 0.000
		0.047–0.242	−0.023–0.000	−0.200–0.000	0.258–1.980	−0.719–0.000	−0.030–0.079	0.114–0.590	−0.057–0.000
		n = 198	*n* = 6	*n* = 79	*n* = 198	*n* = 124	*n* = 198	*n* = 198	*n* = 1
	Female	0.153 ± 0.054	−0.021 ± 0.013	−0.063 ± 0.037	0.865 ± 0.296	−0.179 ± 0.106	0.005 ± 0.025	0.270 ± 0.101	−0.067 ± 0.038
		0.066–0.270	−0.038–0.000	−0.187–0.000	0.401–1.534	−0.447–0.000	−0.066–0.053	0.079–0.498	−0.131–0.000
		n = 130	*n* = 5	*n* = 64	*n* = 130	*n* = 57	*n* = 130	*n* = 130	*n* = 7

Table A1.17
Amplitudes (in millivolts) in Caucasian adults: lead III

Age group	Sex	P+	P–	Q	R	S	STj	T+	T–
18–29	Male	0.086 ± 0.049	–0.049 ± 0.030	–0.134 ± 0.111	0.777 ± 0.562	–0.224 ± 0.230	0.009 ± 0.025	0.142 ± 0.095	–0.101 ± 0.081
		0.000–0.224	–0.165–0.000	–0.531–0.000	0.000–2.115	–1.140–0.000	–0.053–0.062	0.000–0.381	–0.304–0.000
		n = 197	*n* = 104	*n* = 128	*n* = 221	*n* = 116	*n* = 225	*n* = 187	*n* = 77
	Female	0.082 ± 0.048	–0.038 ± 0.024	–0.110 ± 0.091	0.560 ± 0.391	–0.209 ± 0.180	0.001 ± 0.019	0.093 ± 0.061	–0.058 ± 0.044
		0.000–0.220	–0.104–0.000	–0.562–0.000	0.000–1.475	–0.912–0.000	–0.034–0.039	0.000–0.276	–0.207–0.000
		n = 252	*n* = 125	*n* = 166	*n* = 278	*n* = 160	*n* = 282	*n* = 245	*n* = 109
30–39	Male	0.084 ± 0.045	–0.044 ± 0.021	–0.130 ± 0.095	0.524 ± 0.477	–0.328 ± 0.290	0.010 ± 0.021	0.123 ± 0.092	–0.076 ± 0.052
		0.000–0.206	–0.105–0.000	–0.429–0.000	0.000–1.813	–1.261–0.000	–0.031–0.055	0.000–0.399	–0.248–0.000
		n = 188	*n* = 90	*n* = 105	*n* = 205	*n* = 114	*n* = 206	*n* = 172	*n* =78
	Female	0.088 ± 0.050	–0.041 ± 0.022	–0.110 ± 0.085	0.542 ± 0.356	–0.204 ± 0.198	–0.001 ± 0.021	0.089 ± 0.061	–0.055 ± 0.043
		0.000–0.216	–0.109–0.000	–0.399–0.000	0.000–1.407	–1.002–0.000	–0.051–0.043	0.000–0.284	–0.175–0.000
		n = 113	*n* = 51	*n* = 86	*n* = 122	*n* = 63	*n* = 124	*n* = 106	*n* = 59
40–49	Male	0.082 ± 0.042	–0.044 ± 0.028	–0.116 ± 0.106	0.381 ± 0.307	–0.348 ± 0.307	0.007 ± 0.025	0.123 ± 0.091	–0.074 ± 0.057
		0.000–0.190	–0.134–0.000	–0.478–0.000	0.000–1.483	–1.120–0.000	–0.033–0.050	0.000–0.362	–0.228–0.000
		n = 189	*n* = 90	*n* = 81	*n* = 200	*n* = 129	*n* = 204	*n* = 164	*n* = 80
	Female	0.072 ± 0.041	–0.042 ± 0.024	–0.079 ± 0.058	0.295 ± 0.253	–0.292 ± 0.245	0.004 ± 0.016	0.080 ± 0.066	–0.056 ± 0.040
		0.000–0.187	–0.100–0.000	–0.319–0.000	0.000–1.033	0.000–1.269	–0.028–0.037	0.000–0.298	–0.161–0.000
		n = 80	*n* = 38	*n* = 35	*n* = 85	*n* = 54	*n* = 86	*n* = 73	*n* = 35
50+	Male	0.081 ± 0.043	–0.044 ± 0.023	–0.140 ± 0.143	0.322 ± 0.329	–0.382 ± 0.315	0.007 ± 0.022	0.117 ± 0.094	–0.068 ± 0.054
		0.000–0.184	–0.113–0.000	–0.847–0.000	0.000–1.422	–1.235–0.000	–0.038–0.064	0.000–0.423	–0.236–0.000
		n = 190	*n* = 90	*n* = 68	*n* = 198	*n* = 140	*n* = 200	*n* = 169	*n* = 72
	Female	0.083 ± 0.045	–0.039 ± 0.024	–0.103 ± 0.123	0.290 ± 0.249	–0.336 ± 0.272	0.001 ± 0.019	0.091 ± 0.063	–0.059 ± 0.050
		0.000–0.209	–0.126–0.000	–0.849–0.000	0.000–0.944	–1.080–0.000	–0.048–0.038	0.000–0.274	–0.196–0.000
		n = 120	*n* = 49	*n* = 50	*n* = 124	*n* = 83	*n* = 126	*n* = 109	*n* = 47

■ Table A1.18

Amplitudes (in millivolts) in Caucasian adults: lead aVR

Age group	Sex	P+	P–	Q	R	S	STj	T+	T–
18–29	Male	0.024 ± 0.012	–0.119 ± 0.034	–0.942 ± 0.237	0.093 ± 0.083	–1.060 ± 0.261	–0.036 ± 0.025		–0.379 ± 0.109
		0.000–0.041	–0.194 to –0.058	–1.408–0.000	0.000–0.360	–1.583–0.000	–0.086–0.011		–0.631 to –0.179
		n = 4	*n* = 231	*n* = 74	*n* = 208	*n* = 160	*n* = 231		*n* = 231
	Female	0.023 ± 0.008	–0.120 ± 0.033	–0.811 ± 0.226	0.085 ± 0.065	–0.959 ± 0.230	0.012 ± 0.017		–0.307 ± 0.092
		0.000–0.032	–0.205 to –0.059	–1.379–0.000	0.000–0.250	–1.443–0.000	–0.051–0.020		–0.504 to –0.153
		n = 3	*n* = 289	*n* = 130	*n* = 248	*n* = 160	*n* = 289		*n* = 289
30–39	Male		–0.121 ± 0.031	–0.871 ± 0.248	0.104 ± 0.093	–0.990 ± 0.275	–0.029 ± 0.022		–0.332 ± 0.098
			–0.197 to –0.062	–1.564–0.000	0.000–0.380	–1.557–0.000	–0.096–0.014		–0.573 to –0.140
			n = 215	*n* = 91	*n* = 181	*n* = 125	*n* = 215		*n* = 215
	Female		–0.127 ± 0.035	–0.723 ± 0.212	0.071 ± 0.052	–0.909 ± 0.242	–0.008 ± 0.020		–0.279 ± 0.090
			–0.220 to –0.063	–1.247–0.000	0.000–0.269	–1.447–0.000	–0.060–0.032		–0.497 to –0.121
			n = 124	*n* = 41	*n* = 107	*n* = 84	*n* = 124		*n* = 124
40–49	Male	0.032 ± 0.007	–0.115 ± 0.031	–0.764 ± 0.218	0.087 ± 0.084	–0.893 ± 0.233	–0.022 ± 0.022		–0.302 ± 0.098
		0.000–0.026	–0.196 to –0.060	–1.210–0.000	0.000–0.310	–1.403–0.000	–0.080–0.019		–0.539 to –0.146
		n = 3	*n* = 206	*n* = 86	*n* = 175	*n* = 120	*n* = 206		*n* = 206
	Female	0.033 ± 0.000	–0.121 ± 0.032	–0.719 ± 0.208	0.065 ± 0.043	–0.872 ± 0.203	–0.011 ± 0.019	0.024 ± 0.000	–0.263 ± 0.082
		0.000–0.033	–0.194 to –0.062	–1.357–0.000	0.000–0.239	–1.310–0.000	–0.049–0.024	0.000–0.024	–0.413 to –0.096
		n = 1	*n* = 90	*n* = 35	*n* = 74	*n* = 56	*n* = 90	*n* = 1	*n* = 90
50+	Male	0.025 ± 0.000	–0.115 ± 0.030	–0.748 ± 0.202	0.086 ± 0.074	–0.840 ± 0.246	–0.018 ± 0.020		–0.268 ± 0.088
		0.000–0.025	–0.176 to –0.057	–1.352–0.000	0.000–0.372	–1.404–0.000	–0.063–0.024		–0.468 to –0.115
		n = 1	*n* = 200	*n* = 104	*n* = 162	*n* = 98	*n* = 200		*n* = 200
	Female	0.034 ± 0.000	–0.122 ± 0.035	–0.766 ± 0.193	0.077 ± 0.056	–0.877 ± 0.252	–0.005 ± 0.019	0.041 ± 0.023	0.246 ± 0.086
		0.000–0.034	–0.195 to –0.061	–1.238–0.000	0.000–0.215	–1.407–0.000	–0.037–0.040	0.000–0.072	–0.452 to –0.092
		n = 1	*n* = 130	*n* = 70	*n* = 92	*n* = 61	*n* = 130	*n* = 5	*n* = 128

■ Table A1.19
Amplitudes (in millivolts) in Caucasian adults: lead aVL

Age group	Sex	P+	P–	Q	R	S	STj	T+	T–
18–29	Male	0.055 ± 0.028	−0.036 ± 0.021	−0.105 ± 0.113	0.311 ± 0.279	−0.387 ± 0.261	0.011 ± 0.017	0.154 ± 0.096	−0.043 ± 0.029
		0.000–0.120	−0.092–0.000	−0.492–0.000	0.027–1.254	−1.098–0.000	−0.019–0.045	0.020–0.412	−0.124–0.000
		n = 197	*n* = 106	*n* = 100	*n* = 219	*n* = 172	*n* = 225	*n* = 215	*n* = 28
	Female	0.052 ± 0.023	−0.032 ± 0.023	−0.082 ± 0.053	0.283 ± 0.223	−0.253 ± 0.206	0.005 ± 0.012	0.124 ± 0.067	−0.023 ± 0.018
		0.000–0.111	−0.100–0.000	−0.254–0.000	0.040–0.980	−0.871–0.000	−0.021–0.030	0.014–0.284	−0.066–0.000
		n = 253	*n* = 132	*n* = 107	*n* = 284	*n* = 187	*n* = 285	*n* = 280	*n* = 22
30–39	Male	0.054 ± 0.024	−0.033 ± 0.018	−0.086 ± 0.059	0.414 ± 0.310	−0.288 ± 0.219	0.006 ± 0.013	0.136 ± 0.075	−0.051 ± 0.034
		0.000–0.114	−0.097–0.000	−0.287–0.000	0.040–1.199	−0.845–0.000	−0.021–0.034	0.022–0.334	−0.122–0.000
		n = 192	*n* = 108	*n* = 114	*n* = 209	*n* = 158	*n* = 212	*n* = 197	*n* = 24
	Female	0.053 ± 0.027	−0.035 ± 0.022	−0.077 ± 0.047	0.256 ± 0.241	−0.217 ± 0.147	0.005 ± 0.014	0.117 ± 0.068	−0.030 ± 0.020
		0.000–0.123	−0.106–0.000	−0.203–0.000	0.027–1.163	−0.614–0.000	−0.030–0.030	0.010–0.307	−0.071–0.000
		n = 103	*n* = 66	*n* =3 9	*n* = 124	*n* = 82	*n* = 124	*n* = 123	*n* = 12
40–49	Male	0.055 ± 0.028	−0.033 ± 0.017	−0.082–0.048	0.430 ± 0.326	−0.220 ± 0.166	0.006 ± 0.017	0.128 ± 0.081	−0.051 ± 0.034
		0.000–0.114	−0.083–0.000	−0.221–0.000	0.032–1.161	−0.799–0.000	−0.023–0.039	0.015–0.348	−0.155–0.000
		n = 176	*n* = 107	*n* = 123	*n* = 206	*n* = 130	*n* = 207	*n* = 186	*n* = 32
	Female	0.052 ± 0.024	−0.029 ± 0.017	−0.081 ± 0.055	0.397 ± 0.255	−0.173 ± 0.115	0.003 ± 0.013	0.114 ± 0.067	−0.031 ± 0.015
		0.000–0.113	−0.088–0.000	−0.298–0.000	0.040–1.070	−0.456–0.000	−0.022–0.023	0.017–0.285	−0.058–0.000
		n = 81	*n* = 40	*n* = 50	*n* = 89	*n* = 40	*n* = 89	*n* = 85	*n* = 9
50+	Male	0.055 ± 0.026	−0.034 ± 0.017	−0.093 ± 0.054	0.478 ± 0.298	−0.212 ± 0.156	0.004 ± 0.017	0.109 ± 0.077	−0.050 ± 0.039
		0.000–0.114	−0.085–0.000	−0.243–0.000	0.037–1.191	−0.648–0.000	−0.032–0.050	0.010–0.324	−0.181–0.000
		n = 171	*n* = 117	*n* = 128	*n* = 199	*n* = 98	*n* = 202	*n* = 179	*n* = 45
	Female	0.051 ± 0.024	−0.033 ± 0.018	−0.081 ± 0.052	0.441 ± 0.293	−0.159 ± 0.117	0.001 ± 0.013	0.096 ± 0.072	−0.035 ± 0.029
		0.000–0.121	−0.097–0.000	−0.302–0.000	0.028–1.209	−0.468–0.000	−0.027–0.025	0.011–0.316	−0.128–0.000
		n = 110	*n* = 65	*n* = 71	*n* = 125	*n* = 57	*n* = 126	*n* = 119	*n* = 26

Table A1.20

Amplitudes (in millivolts) in Caucasian adults: lead aVF

Age group	Sex	P+	P–	Q	R	S	STj	T+	T–
18–29	Male	0.111 ± 0.054	−0.034 ± 0.025	−0.095 ± 0.058	1.038 ± 0.510	−0.186 ± 0.121	0.025 ± 0.027	0.258 ± 0.122	−0.056 ± 0.052
		0.000–0.257	−0.144–0.000	−0.257–0.000	0.038–2.117	−0.489–0.000	−0.037–0.085	0.034–0.576	−0.164–0.000
		n = 226	*n* = 39	*n* = 143	*n* = 231	*n* = 129	*n* = 231	*n* = 229	*n* = 10
	Female	0.109 ± 0.052	−0.026 ± 0.022	−0.079 ± 0.049	0.822 ± 0.363	−0.163 ± 0.106	0.007 ± 0.020	0.198 ± 0.085	−0.041 ± 0.012
		0.000–0.237	−0.119–0.000	−0.203–0.000	0.174–1.652	−0.500–0.000	−0.037–0.045	0.055–0.411	−0.062–0.000
		n = 283	*n* = 30	*n* = 174	*n* = 288	*n* = 162	*n* = 288	*n* = 287	*n* = 9
30–39	Male	0.112 ± 0.047	−0.028 ± 0.014	−0.085 ± 0.059	0.772 ± 0.458	−0.217 ± 0.167	0.022 ± 0.023	0.222 ± 0.114	−0.029 ± 0.010
		0.000–0.237	−0.062–0.000	−0.320–0.000	0.072–2.015	−0.770–0.000	−0.023–0.070	0.033–0.546	−0.036–0.000
		n = 202	*n* = 26	*n* = 110	*n* = 208	*n* = 95	*n* = 208	*n* = 208	*n* = 6
	Female	0.118 ± 0.053	−0.021 ± 0.013	−0.078 ± 0.048	0.781 ± 0.348	−0.143 ± 0.104	0.003 ± 0.023	0.180 ± 0.083	−0.038 ± 0.024
		0.000–0.243	−0.052–0.000	−0.182–0.000	0.085–1.533	−0.532–0.000	−0.056–0.058	0.031–0.366	−0.069–0.000
		n = 121	*n* = 11	*n* = 89	*n* = 123	*n* = 66	*n* = 123	*n* = 123	*n* = 6
40–49	Male	0.107 ± 0.045	−0.033 ± 0.021	−0.065 ± 0.040	0.575 ± 0.383	−0.194 ± 0.171	0.016 ± 0.026	0.207 ± 0.109	−0.031 ± 0.013
		0.000–0.224	−0.075–0.000	−0.179–0.000	0.047–1.466	−0.955–0.000	−0.032–0.067	0.040–0.501	−0.047–0.000
		n = 201	*n* = 20	*n* = 85	*n* = 203	*n* = 115	*n* = 203	*n* = 202	*n* = 9
	Female	0.013 ± 0.046	−0.027 ± 0.011	−0.059 ± 0.031	0.529 ± 0.284	−0.174 ± 0.145	0.009 ± 0.017	0.164 ± 0.078	−0.028 ± 0.013
		0.000–0.223	−0.049–0.000	−0.147–0.000	0.093–1.231	−0.910–0.000	−0.029–0.047	0.055–0.404	−0.046–0.000
		n = 87	*n* = 10	*n* = 38	*n* = 88	*n* = 46	*n* = 88	*n* = 88	*n* = 4
50+	Male	0.108 ± 0.044	−0.026 ± 0.015	−0.072 ± 0.050	0.499 ± 0.373	−0.222 ± 0.205	0.014 ± 0.022	0.191 ± 0.105	−0.031 ± 0.021
		0.000–0.211	−0.069–0.000	−0.233–0.000	0.032–1.778	−0.977–0.000	−0.037–0.065	0.033–0.486	−0.062–0.000
		n = 200	*n* = 27	*n* = 73	*n* = 202	*n* = 127	*n* = 202	*n* = 202	*n* = 7
	Female	0.113 ± 0.050	−0.028 ± 0.015	−0.057 ± 0.029	0.509 ± 0.297	−0.216 ± 0.149	0.003 ± 0.021	0.165 ± 0.078	−0.049 ± 0.041
		0.000–0.238	−0.067–0.000	−0.154–0.000	0.100–1.198	−0.709–0.000	−0.058–0.042	0.040–0.339	−0.156–0.000
		n = 129	*n* = 14	*n* = 55	*n* = 130	*n* = 62	*n* = 130	*n* = 130	*n* = 13

Table A1.21
Amplitudes (in millivolts) in Caucasian adults: lead V_1

Age group	Sex	P+	P–	Q	R	S	STj	T+	T–
18–29	Male	0.073 ± 0.031	–0.039 ± 0.017	–1.317 ± 0.000	0.392 ± 0.200	–1.302 ± 0.455	0.055 ± 0.041	0.194 ± 0.147	–0.115 ± 0.082
		0.000–0.140	–0.091–0.000	–1.317–0.000	0.102–0.906	–2.246 to –0.503	–0.029–0.135	0.000–0.593	–0.290–0.000
		n = 229	*n* = 151	*n* = 1	*n* = 229	*n* = 229	*n* = 230	*n* = 201	*n* = 84
	Female	0.069 ± 0.030	–0.033 ± 0.017	–1.059 ± 0.672	0.262 ± 0.148	–1.075 ± 0.415	0.019 ± 0.028	0.086 ± 0.072	–0.102 ± 0.063
		0.000–0.135	–0.079–0.000	–1.544–0.000	0.050–0.648	–1.995 to –0.369	–0.076–0.070	0.000–0.321	–0.252–0.000
		n = 278	*n* = 155	*n* = 4	*n* = 282	*n* = 282	*n* = 285	*n* = 192	*n* = 193
30–39	Male	0.069 ± 0.028	–0.046 ± 0.020	–0.860 ± 0.848	0.295 ± 0.207	–1.158 ± 0.436	0.051 ± 0.033	0.179 ± 0.122	–0.095 ± 0.075
		0.000–0.143	–0.101–0.000	–1.594–0.000	0.040–1.030	–2.328 to –0.265	–0.023–0.119	0.000–0.519	–0.354–0.000
		n = 203	*n* = 162	*n* = 4	*n* = 211	*n* = 210	*n* = 213	*n* = 194	*n* = 54
	Female	0.064 ± 0.027	–0.038 ± 0.020		0.251 ± 0.146	–1.068 ± 0.384	0.020 ± 0.028	0.096 ± 0.073	–0.101 ± 0.068
		0.000–0.144	–0.095–0.000		0.039–0.629	–2.105 to –0.371	–0.076–0.056	0.000–0.316	–0.272–0.000
		n = 121	*n* = 80		*n* = 125	*n* = 125	*n* = 125	*n* = 88	*n* = 77
40–49	Male	0.061 ± 0.024	–0.046 ± 0.023	–1.208 ± 0.318	0.258 ± 0.163	–1.052 ± 0.408	0.045 ± 0.035	0.191 ± 0.123	–0.087 ± 0.068
		0.000–0.120	–0.117–0.000	–1.463–0.000	0.056–0.612	–2.110 to –0.364	–0.019–0.119	0.000–0.495	–0.305–0.000
		n = 194	*n* = 165	*n* = 6	*n* = 199	*n* = 199	*n* = 205	*n* = 192	*n* = 51
	Female	0.057 ± 0.022	–0.040 ± 0.019	–1.007 ± 0.188	0.214 ± 0.131	–0.951 ± 0.307	0.025 ± 0.018	0.079 ± 0.053	–0.073 ± 0.059
		0.000–0.104	–0.097–0.000	–1.140–0.000	0.029–0.634	–1.518 to –0.505	–0.010–0.067	0.000–0.255	–0.309–0.000
		n = 83	*n* = 73	*n* = 2	*n* = 87	*n* = 87	*n* = 89	*n* = 73	*n* = 50
50+	Male	0.057 ± 0.025	–0.046 ± 0.021	–0.443 ± 0.576	0.229 ± 0.123	–0.930 ± 0.365	0.047 ± 0.031	0.198 ± 0.117	–0.097 ± 0.083
		0.000–0.115	–0.110–0.000	–0.850–0.000	0.052–0.489	–1.904 to –0.319	–0.005–0.113	0.000–0.513	–0.337–0.000
		n = 193	*n* = 190	*n* = 2	*n* = 201	*n* = 201	*n* = 202	*n* = 187	*n* = 31
	Female	0.055 ± 0.024	–0.047 ± 0.021	–0.029	0.201 ± 0.130	–0.934 ± 0.340	0.030 ± 0.022	0.109 ± 0.065	–0.077 ± 0.058
		0.000–0.114	–0.105–0.000	–0.029–0.000	0.040–0.587	–1.630 to –0.349	–0.006–0.083	0.000–0.291	–0.319–0.000
		n = 116	*n* = 110	*n* = 1	*n* = 126	*n* = 126	*n* = 126	*n* = 108	*n* = 53

■ Table A1.22

Amplitudes (in millivolts) in Caucasian adults: lead V_2

Age group	Sex	P+	P−	Q	R	S	STj	T+	T−
18–29	Male	0.094 ± 0.033	−0.027 ± 0.011	−2.532 ± 0.000	0.833 ± 0.366	−2.164 ± 0.711	0.164 ± 0.077	0.887 ± 0.307	
		0.042–0.170	−0.064–0.000	−2.532–0.000	0.265–1.894	−3.562 to −0.826	0.016–0.320	0.342–1.634	
		n = 228	*n* = 39	*n* = 1	*n* = 228	*n* = 228	*n* = 229	*n* = 229	
	Female	0.090 ± 0.030	−0.021 ± 0.011	−0.026 ± 0.006	0.555 ± 0.259	−1.458 ± 0.554	0.055 ± 0.046	0.483 ± 0.213	−0.069 ± 0.041
		0.040–0.157	−0.050–0.000	−0.030–0.000	0.145–1.206	−2.889 to −0.550	−0.063–0.128	0.093–1.070	−0.116–0.000
		n = 281	*n* = 13	*n* = 2	*n* = 281	*n* = 280	*n* = 281	*n* = 281	*n* = 3
30–39	Male	0.089 ± 0.032	−0.030 ± 0.019	−0.969 ± 1.271	0.662 ± 0.355	−1.863 ± 0.671	0.136 ± 0.069	0.784 ± 0.276	−0.146 ± 0.110
		0.033–0.176	−0.118–0.000	−1.868–0.000	0.139–1.640	−3.518 to −0.664	−0.018–0.294	0.239–1.444	−0.267–0.000
		n = 209	*n* = 50	*n* = 2	*n* = 208	*n* = 207	*n* = 209	*n* = 208	*n* = 3
	Female	0.085 ± 0.033	−0.030 ± 0.019		0.505 ± 0.241	−1.375 ± 0.610	0.050 ± 0.045	0.462 ± 0.182	−0.146 ± 0.026
		0.027–0.169	−0.075–0.000		0.143–1.094	−3.126 to −0.485	−0.030–0.154	0.095–0.874	−0.174–0.000
		n = 125	*n* = 16		*n* = 125	*n* = 125	*n* = 125	*n* = 125	*n* = 3
40–49	Male	0.084 ± 0.032	−0.027 ± 0.014	−2.284 ± 0.215	0.604 ± 0.310	−1.502 ± 0.559	0.114 ± 0.061	0.708 ± 0.243	
		0.030–0.154	−0.094–0.000	−2.532–0.000	0.085–1.496	−2.724 to −0.548	−0.021–0.265	0.240–1.256	
		n = 201	*n* = 64	*n* = 3	*n* = 199	*n* = 199	*n* = 202	*n* = 202	
	Female	0.078 ± 0.028	−0.023 ± 0.008	−0.713 ± 0.969	0.430 ± 0.202	−1.175 ± 0.456	0.053 ± 0.030	0.418 ± 0.171	
		0.036–0.141	−0.043–0.000	−1.398–0.000	0.134–0.877	−2.470 to −0.390	−0.003 ± 0.110	0.161–0.823	
		n = 90	*n* = 15	*n* = 2	*n* = 89	*n* = 89	*n* = 90	*n* = 90	
50+	Male	0.079 ± 0.031	−0.025 ± 0.009	−0.428 ± 0.770	0.595 ± 0.331	−1.285 ± 0.528	0.103 ± 0.061	0.659 ± 0.249	−0.420 ± 0.000
		0.028–0.155	−0.047–0.000	−1.583–0.000	0.148–1.497	−2.425 to −0.271	−0.013–0.233	0.173–1.165	−0.420–0.000
		n = 203	*n* = 71	*n* = 4	*n* = 202	*n* = 202	*n* = 203	*n* = 202	*n* = 1
	Female	0.074 ± 0.027	−0.025 ± 0.012	−0.802 ± 1.076	0.453 ± 0.251	−1.126 ± 0.397	0.053 ± 0.036	0.383 ± 0.154	−0.066 ± 0.000
		0.019–0.134	−0.073–0.000	−1.563–0.000	0.148–1.081	−1.985 to −0.472	−0.033–0.132	0.073–0.672	−0.066–0.000
		n = 128	*n* = 48	*n* = 2	*n* = 127	*n* = 127	*n* = 128	*n* = 128	*n* = 1

■ Table A1.23

Amplitudes (in millivolts) in Caucasian adults: lead V_3

Age group	Sex	P+	P−	Q	R	S	STj	T+	T−
18–29	Male	0.093 ± 0.030	−0.024 ± 0.010	−0.061 ± 0.038	1.073 ± 0.506	−1.294 ± 0.655	0.151 ± 0.069	0.873 ± 0.280	
		0.038–0.160	−0.049–0.000	−0.103–0.000	0.326–2.344	−2.840 to −0.213	0.033–0.305	0.380–1.474	
		n = 228	*n* = 39	*n* = 3	*n* = 228	*n* = 227	*n* = 228	*n* = 228	
	Female	0.085 ± 0.027	−0.026 ± 0.036	−0.043 ± 0.037	0.761 ± 0.374	−0.753 ± 0.374	0.041 ± 0.036	0.517 ± 0.204	−0.130 ± 0.096
		0.043–0.151	−0.169–0.000	−0.086–0.000	0.177–1.800	−1.530 to −0.087	−0.019–0.108	0.160–1.012	−0.198–0.000
		n = 283	*n* = 18	*n* = 3	*n* = 284	*n* = 275	*n* = 284	*n* = 284	*n* = 2
30–39	Male	0.092 ± 0.028	−0.026 ± 0.011	−0.087 ± 0.068	0.983 ± 0.501	−1.188 ± 0.515	0.119 ± 0.065	0.779 ± 0.277	
		0.044–0.165	−0.067–0.000	−0.211–0.000	0.259–2.257	−2.338 to −0.186	0.004–0.269	0.253–1.472	
		n = 208	*n* = 29	*n* = 8	*n* = 208	*n* = 204	*n* = 208	*n* = 208	
	Female	0.086 ± 0.028	−0.028 ± 0.029		0.730 ± 0.367	−0.765 ± 0.439	0.038 ± 0.037	0.499 ± 0.200	−0.251 ± 0.208
		0.031–0.174	−0.116–0.000		0.180–2.174	−1.870 to −0.134	−0.035–0.132	0.135–0.987	−0.398–0.000
		n = 123	*n* = 12		*n* = 123	*n* = 117	*n* = 123	*n* = 122	*n* = 2
40–49	Male	0.088 ± 0.028	−0.022 ± 0.007	−0.066 ± 0.049	1.020 ± 0.506	−1.015 ± 0.490	0.096 ± 0.059	0.731 ± 0.257	
		0.039–0.152	−0.044–0.000	−0.150–0.000	0.250–2.221	−2.037 to −0.217	−0.014–0.203	0.311–1.322	
		n = 205	*n* = 33	*n* = 5	*n* = 205	*n* = 202	*n* = 205	*n* = 205	
	Female	0.082 ± 0.025	−0.022 ± 0.013	−0.062 ± 0.057	0.675 ± 0.329	−0.719 ± 0.366	0.033 ± 0.032	0.457 ± 0.184	
		0.038–0.139	−0.047–0.000	−0.128–0.000	0.187–1.633	−1.657 to −0.193	−0.035–0.099	0.132–0.917	
		n = 88	*n* = 10	*n* = 3	*n* = 88	*n* = 86	*n* = 88	*n* = 88	
50+	Male	0.088 ± 0.027	−0.023 ± 0.009	−0.066 ± 0.058	0.998 ± 0.506	−0.994 ± 0.494	0.086 ± 0.060	0.687 ± 0.270	
		0.040–0.167	−0.052–0.000	−0.219–0.000	0.200–2.235	−2.080 to −0.178	−0.017–0.230	0.229–1.307	
		n = 195	*n* = 23	*n* = 13	*n* = 195	*n* = 193	*n* = 195	*n* = 194	
	Female	0.084 ± 0.026	−0.021 ± 0.008	−0.051 ± 0.035	0.752 ± 0.409	−0.768 ± 0.372	0.030 ± 0.035	0.436 ± 0.169	−0.068 ± 0.059
		0.037–0.148	−0.039–0.000	−0.108–0.000	0.183–1.811	−1.696 to −0.214	−0.049–0.097	0.086–0.793	−0.110–0.000
		n = 128	*n* = 18	*n* = 5	*n* = 128	*n* = 126	*n* = 128	*n* = 128	*n* = 2

Table A1.24

Amplitudes (in millivolts) in Caucasian adults: lead V_4

Age group	Sex	P+	P–	Q	R	S	STj	T+	T–
18–29	Male	0.088 ± 0.028	–0.018 ± 0.005	–0.090 ± 0.075	1.973 ± 0.702	–0.584 ± 0.376	0.089 ± 0.054	0.705 ± 0.269	
		0.033–0.158	–0.025–0.000	–0.307–0.000	0.792–3.558	–1.660 to –0.074	–0.006–0.201	0.261–1.427	
		n = 227	*n* = 20	*n* = 85	*n* = 227	*n* = 209	*n* = 227	*n* = 227	
	Female	0.082 ± 0.027	–0.022 ± 0.020	–0.064 ± 0.049	1.247 ± 0.453	–0.372 ± 0.205	0.021 ± 0.027	0.413 ± 0.172	–0.060 ± 0.000
		0.038–0.150	–0.089–0.000	–0.271–0.000	0.452–2.264	–0.832 to –0.075	–0.037–0.088	0.141–0.847	–0.060–0.000
		n = 283	*n* = 14	*n* = 72	*n* = 284	*n* = 242	*n* = 284	*n* = 284	*n* = 1
30–39	Male	0.089 ± 0.027	–0.021 ± 0.009	–0.076 ± 0.063	1.717 ± 0.601	–0.627 ± 0.347	0.069 ± 0.050	0.625 ± 0.238	
		0.042–0.152	–0.041–0.000	–0.335–0.000	0.658–3.257	–1.571 to –0.113	–0.035–0.208	0.215–1.268	
		n = 208	*n* = 17	*n* = 62	*n* = 208	*n* = 193	*n* = 208	*n* = 208	
	Female	0.088 ± 0.028	–0.020 ± 0.011	–0.073 ± 0.039	1.277 ± 0.509	–0.433 ± 0.259	0.015 ± 0.031	0.418 ± 0.181	–0.030 ± 0.000
		0.032–0.155	–0.032–0.000	–0.197–0.000	0.285–2.705	–1.050 to –0.097	–0.056–0.095	0.081–0.961	–0.030–0.000
		n = 124	*n* = 3	*n* = 35	*n* = 124	*n* = 97	*n* = 124	*n* = 124	*n* = 1
40–49	Male	0.085 ± 0.024	–0.022 ± 0.008	–0.060 ± 0.047	1.620 ± 0.629	–0.571 ± 0.346	0.054 ± 0.046	0.593 ± 0.245	–0.077 ± 0.000
		0.044–0.142	–0.041–0.000	–0.306–0.000	0.516–3.222	–1.425 to –0.074	–0.017–0.166	0.213–1.185	–0.077–0.000
		n = 204	*n* = 18	*n* = 52	*n* = 204	*n* = 204	*n* = 204	*n* = 204	*n* = 1
	Female	0.085 ± 0.025	–0.021 ± 0.012	–0.058 ± 0.048	1.110 ± 0.454	–0.415 ± 0.257	0.010 ± 0.026	0.357 ± 0.159	
		0.035–0.142	–0.044–0.000	–0.175–0.000	0.384–2.200	–1.196 to –0.070	–0.046–0.075	0.094–0.787	
		n = 87	*n* = 6	*n* = 87	*n* = 87	*n* = 76	*n* = 87	*n* = 87	
50+	Male	0.086 ± 0.024	–0.018 ± 0.004	–0.079 ± 0.055	1.592 ± 0.546	–0.619 ± 0.377	0.045 ± 0.046	0.568 ± 0.248	–0.135 ± 0.000
		0.047–0.146	–0.025–0.000	–0.282–0.000	0.582–2.683	–1.500 to –0.099	–0.044–0.159	0.166–1.249	–0.135–0.000
		n = 202	*n* = 10	*n* = 54	*n* = 202	*n* = 191	*n* = 202	*n* = 201	*n* = 1
	Female	0.089 ± 0.027	–0.019 ± 0.009	–0.048 ± 0.030	1.150 ± 0.454	–0.444 ± 0.293	0.006 ± 0.030	0.352 ± 0.156	–0.080 ± 0.045
		0.041–0.151	–0.036–0.000	–0.148–0.000	0.435–2.326	–1.473 to –0.072	–0.075–0.074	0.074–0.690	–0.140–0.000
		n = 129	*n* = 12	*n* = 28	*n* = 129	*n* = 123	*n* = 129	*n* = 129	*n* = 4

Table A1.25
Amplitudes (in millivolts) in Caucasian adults: lead V_5

Age group	Sex	P+	P−	Q	R	S	STj	T+	T−
18–29	Male	0.082 ± 0.025	−0.020 ± 0.011	−0.101 ± 0.072	1.910 ± 0.598	−0.316 ± 0.228	0.051 ± 0.039	0.553 ± 0.226	
		0.028–0.136	−0.061–0.000	−0.324–0.000	0.958–3.530	−0.946 to −0.045	−0.010–0.155	0.194–1.152	
		n = 230	*n* = 20	*n* = 171	*n* = 230	*n* = 181	*n* = 230	*n* = 230	
	Female	0.079 ± 0.026	−0.016 ± 0.005	−0.070 ± 0.050	1.289 ± 0.383	−0.230 ± 0.131	0.013 ± 0.023	0.386 ± 0.141	
		0.037–0.143	−0.029–0.000	−0.233–0.000	0.574–2.112	−0.577 to −0.048	−0.032–0.061	0.149–0.720	
		n = 287	*n* = 14	*n* = 148	*n* = 288	*n* = 214	*n* = 287	*n* = 287	
30–39	Male	0.084 ± 0.023	−0.021 ± 0.008	−0.088 ± 0.069	1.761 ± 0.502	−0.330 ± 0.235	0.041 ± 0.039	0.497 ± 0.197	−0.370 ± 0.000
		0.035–0.135	−0.034–0.000	−0.283–0.000	0.756–3.012	−0.986 to −0.069	−0.029–0.154	0.181–1.028	−0.370–0.000
		n = 211	*n* = 11	*n* = 127	*n* = 211	*n* = 179	*n* = 211	*n* = 211	*n* = 1
40–49	Female	0.084 ± 0.026	−0.011 ± 0.005	−0.074 ± 0.048	1.299 ± 0.411	−0.239 ± 0.162	0.010 ± 0.027	0.369 ± 0.151	
		0.030–0.146	−0.014–0.000	−0.197–0.000	0.502–2.372	−0.689 to −0.057	−0.040–0.087	0.126–0.827	
		n = 123	*n* = 2	*n* = 78	*n* = 123	*n* = 88	*n* = 123	*n* = 123	
	Male	0.080 ± 0.022	−0.020 ± 0.010	−0.072 ± 0.049	1.576 ± 0.506	−0.306 ± 0.227	0.032 ± 0.034	0.458 ± 0.185	
		0.040–0.142	−0.044–0.000	−0.294–0.000	0.737–2.869	−0.952 to −0.049	−0.031–0.133	0.161–0.879	
		n = 206	*n* = 12	*n* = 111	*n* = 206	*n* = 165	*n* = 206	*n* = 206	
	Female	0.083 ± 0.026	−0.022 ± 0.012	−0.062 ± 0.046	1.190 v 0.393	−0.247 ± 0.190	0.004 ± 0.024	0.323 ± 0.128	−0.077 ± 0.000
		0.044–0.138	−0.043–0.000	−0.202–0.000	0.625–2.092	−1.034 to −0.046	−0.056–0.052	0.085–0.616	−0.077–0.000
		n = 88	*n* = 8	*n* = 39	*n* = 88	*n* = 68	*n* = 88	*n* = 87	*n* = 1
50+	Male	0.081 ± 0.022	−0.020 ± 0.011	−0.080 ± 0.054	1.581 ± 0.470	−0.366 ± 0.237	0.023 ± 0.036	0.427 ± 0.190	−0.067 ± 0.000
		0.043–0.134	−0.047–0.000	−0.241–0.000	0.745–2.488	−0.967 to −0.053	−0.047–0.104	0.114–0.916	−0.067–0.000
		n = 201	*n* = 8	*n* = 102	*n* = 201	*n* = 162	*n* = 201	*n* = 201	*n* = 1
	Female	0.084 ± 0.026	−0.016 ± 0.012	−0.063 ± 0.037	1.241 ± 0.379	−0.261 ± 0.189	−0.001 ± 0.025	0.310 ± 0.139	−0.098 ± 0.044
		0.029–0.145	−0.033–0.000	−0.229–0.000	0.570–2.295	−0.837 to −0.049	−0.050–0.052	0.088–0.636	−0.158–0.000
		n = 129	*n* = 5	*n* = 62	*n* = 129	*n* = 99	*n* = 129	*n* = 128	*n* = 4

Table A1.26

Amplitudes (in millivolts) in Caucasian adults: lead V_6

Age group	Sex	P+	P–	Q	R	S	STj	T+	T–
18–29	Male	0.075 ± 0.023	–0.011 ± 0.004	–0.106 ± 0.066	1.583 ± 0.463	–0.197 ± 0.142	0.032 ± 0.028	0.424 ± 0.166	
		0.032–0.125	–0.020–0.000	–0.262–0.000	0.766–2.806	–0.572 to –0.039	–0.017–0.091	0.153–0.897	
		n = 228	*n* = 11	*n* = 202	*n* = 228	*n* = 137	*n* = 228	*n* = 228	
	Female	0.074 ± 0.025	–0.020 ± 0.023	–0.076 ± 0.054	1.177 ± 0.319	–0.143 ± 0.083	0.009 ± 0.023	0.327 ± 0.109	–0.224 ± 0.168
		0.031–0.137	–0.075–0.000	–0.257–0.000	0.611–1.966	–0.359 to –0.046	–0.035–0.049	0.143–0.573	–0.343–0.000
		n = 289	*n* = 9	*n* = 215	*n* = 289	*n* = 161	*n* = 289	*n* = 289	*n* = 2
30–39	Male	0.077 ± 0.022	–0.019 ± 0.006	–0.090 ± 0.064	1.461 ± 0.456	–0.187 ± 0.141	0.026 ± 0.028	0.390 ± 0.150	
		0.040–0.131	–0.026–0.000	–0.271–0.000	0.623–2.540	–0.564 to –0.025	–0.026–0.088	0.149–0.782	
		n = 205	*n* = 5	*n* = 159	*n* = 207	*n* = 129	*n* = 207	*n* = 207	
	Female	0.079 ± 0.024	–0.013 ± 0.004	–0.083 ± 0.049	1.147 ± 0.323	–0.153 ± 0.125	0.006 ± 0.023	0.302 ± 0.119	
		0.032–0.140	–0.018–0.000	–0.229–0.000	0.605–1.978	–0.724 to –0.032	–0.043–0.074	0.081–0.649	
		n = 123	*n* =3	*n* = 102	*n* = 123	*n* = 59	*n* = 123	*n* = 123	
40–49	Male	0.075 ± 0.020	–0.024 ± 0.012	–0.071 ± 0.043	1.291 ± 0.399	–0.200 ± 0.137	0.021 ± 0.026	0.350 ± 0.140	
		0.040–0.132	–0.044–0.000	–0.202–0.000	0.664–2.290	–0.565 to –0.037	–0.031–0.094	0.137–0.680	
		n = 206	*n* = 8	*n* = 152	*n* = 206	*n* = 109	*n* = 206	*n* = 206	
	Female	0.079 ± 0.024	–0.036 ± 0.000	–0.065 ± 0.042	1.068 ± 0.315	–0.161 ± 0.127	0.004 ± 0.020	0.273 ± 0.104	–0.057 ± 0.000
		0.040–0.128	–0.036–0.000	–0.215–0.000	0.475–1.721	–0.728 to –0.025	–0.037–0.049	0.091–0.500	–0.057–0.000
		n = 89	*n* = 1	*n* = 55	*n* = 89	*n* = 45	*n* = 89	*n* = 88	*n* = 1
50+	Male	0.076 ± 0.020	–0.014 ± 0.004	–0.078 ± 0.050	1.304 ± 0.409	–0.200 ± 0.132	0.015 ± 0.029	0.318 ± 0.142	–0.049 ± 0.028
		0.038–0.117	–0.016–0.000	–0.233–0.000	0.649–2.287	–0.591 to –0.046	–0.045–0.067	0.102–0.680	–0.069–0.000
		n = 198	*n* = 2	*n* = 133	*n* = 198	*n* = 124	*n* = 198	*n* = 198	*n* = 2
	Female	0.080 ± 0.024	–0.015 ± 0.008	–0.067 ± 0.040	1.100 ± 0.335	–0.154 ± 0.084	–0.002 ± 0.021	0.259 ± 0.118	–0.063 ± 0.042
		0.033–0.137	–0.024–0.000	–0.185–0.000	0.523–1.859	–0.384 to –0.055	–0.051–0.037	0.078–0.561	–0.137–0.000
		n = 129	*n* = 6	*n* = 80	*n* = 129	*n* = 61	*n* = 129	*n* = 128	*n* = 6

Table A1.27
R/S ratio in Caucasians: leads $V_1 - V_6$

Age group	Sex	V_1	V_2	V_3	V_4	V_5	V_6
18–29	Male	0.333 ± 0.21	0.429 ± 0.25	1.345 ± 2.03	5.649 ± 7.15	10.327 ± 10.11	13.883 ± 12.47
		0.053–0.913	0.089–1.191	0.180–6.82	0.699–36.66	1.188–44.53	2.211–62.125
		n = 267	n = 267	n = 268	n = 243	n = 216	n = 169
	Female	0.289 ± 0.19	0.468 ± 0.31	1.548 ± 1.84	5.88 ± 8.07	7.879 ± 7.42	10.891 ± 8.81
		0.042–0.803	0.104–1.55	0.211–8.31	0.831–36.74	1.30–27.13	2.417–40.64
		n = 315	n = 318	n = 314	n = 284	n = 247	n = 188
30–39	Male	0.30 ± 0.25	0.420 ± 0.30	1.218 ± 1.62	4.623 ± 7.91	8.298 ± 7.87	13.093 ± 12.73
		0.050–0.940	0.090–1.214	0.194–4.04	0.63–19.14	1.33–23.07	2.305–62.07
		n = 214	n = 215	n = 214	n = 202	n = 181	n = 137
	Female	0.272 ± 0.17	0.480 ± 0.42	2.188 ± 6.17	4.482 ± 3.64	8.197 ± 6.07	13.757 ± 11.11
		0.035–0.62	0.102–1.728	0.241–11.48	0.781–14.684	1.104–25.228	1.916–47.47
		n = 116	n = 114	n = 113	n = 96	n = 88	n = 63
40–49	Male	0.269 ± 0.17	0.487 ± 0.48	1.478 ± 2.73	4.753 ± 6.47	8.818 ± 9.93	10.471 ± 9.22
		0.031–0.711	0.087–1.40	0.184–5.265	0.481–34.40	1.097–47.43	1.877–31.14
		n = 116	n = 115	n = 118	n = 114	n = 98	n = 68
	Female	0.239 ± 0.14	0.429 ± 0.26	1.354 ± 1.39	5.316 ± 7.79	7.99 ± 5.85	9.760 ± 6.18
		0.034–0.519	0.092–1.311	0.125–4.044	0.572–43.119	1.554–26.21	1.993–22.516
		n = 73	n = 73	n = 73	n = 71	n = 65	n = 48
50+	Male	0.263 ± 0.17	0.507 ± 0.35	1.340 ± 1.42	3.935 ± 3.71	7.29 ± 7.48	10.703 ± 10.62
		0.046–0.75	0.133–1.428	0.313–5.767	0.789–13.33	1.371–28.674	1.575–45.51
		n = 122	n = 121	n = 122	n = 117	n = 101	n = 80
	Female	0.227 ± 0.14	0.443 ± 0.28	1.369 ± 1.41	3.409 ± 3.07	8.274 ± 7.90	8.96 ± 6.93
		0.044–0.626	0.049–1.082	0.203–6.578	0.564–14.904	1.568–34.44	2.37–17.65
		n = 77	n = 79	n = 79	n = 77	n = 66	n = 40

Table A1.28

Q/R ratios (%) in Caucasians: leads I, II, III, aVL, aVF, V_5, V_6

Age group	Sex	I	II	III	aVL	aVF	V_5	V_6
18–29	Male	9.0 ± 5.5	6.0 ± 3.6	18 ± 51.8	37 ± 78.5	8.0 ± 6.6	5.0 ± 3.7	7.0 ± 4.1
		0.0–23.4	0.0–14.8	0.0–81.5	0.0–188	0.0–19.3	0.0–14.9	0.0–17.8
		n = 143	*n* = 189	*n* = 154	*n* = 119	*n* = 166	*n* = 199	*n* = 230
	Female	8.0 ± 4.5	6.0 ± 3.6	23 ± 38.9	27 ± 41.9	8.0 ± 4.8	5.0 ± 3.1	6.0 ± 3.7
		0.0–22.3	0.0–14.9	0.0–139	0.0–75.0	0.0–19.5	0.0–12.6	0.0–16.0
		n = 132	*n* = 204	*n* = 192	*n* = 115	*n* = 202	*n* = 168	*n* = 236
30–39	Male	8.0 ± 4.9	7.0 ± 4.0	35 ± 83.6	25 ± 33.6	9.0 ± 5.0	5.0 ± 3.2	6.0 ± 3.8
		0.0–21.6	0.0–19.3	0.0–236	0.0–138	0.0–22.2	*0.0–12.4*	0.0–14.7
		n = 122	*n* = 129	*n* = 115	*n* = 110	*n* = 120	*n* = 137	*n* = 171
	Female	9.0 ± 4.5	7.0 ± 3.3	21 ± 33.6	29 ± 25.2	9.0 ± 4.3	5.0 ± 3.1	7.0 ± 3.8
		0.0–18.9	0.0–14.6	0.0–85.3	0.0–92.0	0.0–17.5	0.0–12.2	0.0–15.0
		n = 54	*n* = 90	*n* = 76	*n* = 40	*n* = 83	*n* = 78	*n* = 99
40–49	Male	8.0 ± 3.6	6.0 ± 2.4	56 ± 11.0	24 ± 43.1	13 ± 24.2	4.0 ± 1.9	6.0 ± 2.6
		0.0–15.7	0.0–11.0	0.0–154	0.0–160	0.0–86.0	0.0–8.2	0.0–10.8
		n = 72	*n* = 65	*n* = 47	*n* = 69	*n* = 51	*n* = 70	*n* = 91
	Female	7.0 ± 4.7	6.0 ± 3.2	24 ± 31.5	19 ± 16.9	10 ± 8.60	5.0 ± 2.8	6.0 ± 3.0
		0.0–18.8	0.0–13.4	0.0–92.8	0.0–38.2	0.0–29.7	0.0–11.7	0.0–13.2
		n = 47	*n* = 41	*n* = 31	*n* = 44	*n* = 38	*n* = 38	*n* = 50
50+	Male	8.0 ± 5.5	7.0 ± 4.7	80 ± 192.7	26 ± 42.2	17 ± 31.2	5.0 ± 4.1	6.0 ± 4.6
		0.0–31.3	0.0–20.5	0.0–740	0.0–238	0.0–64.0	0.0–21.5	0.0–19.8
		n = 79	*n* = 52	*n* = 45	*n* = 79	*n* = 43	*n* = 69	*n* = 84
	Female	8.0 ± 4.3	7.0 ± 4.7	52 ± 109.4	14 ± 7.70	11 ± 8.90	5.0 ± 3.9	6.0 ± 4.8
		0.0–16.7	0.0–14.6	0.0–335	0.0–27.4	0.0–36.7	0.0–13.0	0.0–13.9
		n = 46	*n* = 39	*n* = 38	*n* = 43	*n* = 35	*n* = 39	*n* = 51

Table A1.29
Miscellaneous measures

Age group	Sex	QRS axis (°)	T axis (°)	QRS-T angle (frontal plane) (°)	Lewis index[a] (mV)	SV_1+RV_5 (mV)	SV_1+R V_6(mV)	SV_1+ RV_6 (mV)
18–29	Male	57.5 ± 25.6	40.9 ± 16.7	16.6 ± 23.9	−0.117 ± 0.956	3.331 ± 0.881	3.905 ± 0.995	2.99 ± 0.79
		−10–91	2–69	−39–71	−1.831–2.048	1.685–5.252	2.049–6.372	1.59–4.85
		n = 265	*n* = 265	*n* = 265	*n* = 266	*n* = 264	*n* = 265	*n* = 265
	Female	51.2 ± 24.9	38.7 ± 15.0	12.5 ± 25.1	0.110 ± 0.736	2.442 ± 0.690	2.741 ± 0.727	2.32 ± 0.64
		−9–91	6–65	−46–59	−1.288–1.574	1.112–3.816	1.430–4.635	1.20–3.68
		n = 317	*n* = 317	*n* = 317	*n* = 317	*n* = 316	*n* = 317	*n* = 318
30–39	Male	46.5 ± 29.2	39.3 ± 17.2	7.2 ± 27.3	0.266 ± 0.899	3.007 ± 0.775	3.430 ± 0.875	2.69 ± 0.72
		−22–92	−1–70	−61–59	−1.476–1.962	1.549–4.680	1.846–5.257	1.31–4.14
		n = 220	*n* = 220	*n* = 220	*n* = 221	*n* = 220	*n* = 219	*n* = 220
	Female	49.6 ± 24.2	39.0 ± 15.4	10.6 ± 21.7	0.129 ± 0.626	2.454 ± 0.644	2.662 ± 0.834	2.29 ± 0.60
		−14–81	1–66	−47–61	−1.011–1.293	0.483–4.287	1.565–4.803	1.43–3.59
		n = 118	*n* = 118	*n* = 118	*n* = 118	*n* = 118	*n* = 115	*n* = 118
40–49	Male	37.7 ± 31.6	42.5 ± 17.3	−4.8 ± 26.3	0.491 ± 0.892	2.702 ± 0.763	2.842 ± 0.673	2.39 ± 0.69
		−37–85	5–71	−67–37	−1.060–2.395	1.420–4.227	1.634–4.170	1.22–3.77
		n = 117	*n* = 117	*n* = 117	*n* = 118	*n* = 118	*n* = 117	*n* = 118
	Female	36.2 ± 29.1	38.0 ± 15.7	−1.8 ± 27.9	0.521 ± 0.698	2.291 ± 0.634	2.418 ± 0.566	2.13 ± 0.57
		−53–85	5–67	−86–40	0.762–1.818	1.087–3.411	1.429–3.433	1.07–3.16
		n = 73	*n* = 73	*n* = 73	*n* = 73	*n* = 73	*n* = 73	*n* = 73
50+	Male	31.4 ± 30.6	44.4 ± 18.7	−13.0 ± 27.1	0.627 ± 0.742	2.709 ± 0.736	2.746 ± 0.703	2.35 ± 0.63
		−33–77	4–74	−82–40	0.703–2.120	1.612–4.558	1.546–4.144	1.34–3.78
		n = 125	*n* = 125	*n* = 125	*n* = 125	*n* = 125	*n* = 124	*n* = 125
	Female	26.9 ± 29.2	41.9 ± 15.9	−15.0 ± 27.8	0.714 ± 0.691	2.259 ± 0.567	2.316 ± 0.574	2.10 ± 0.55
		−36–73	4–72	−89-26	0.578–2.165	1.086–3.635	1.298–3.419	0.94–3.14
		n = 80	*n* = 80	*n* = 80	*n* = 80	*n* = 79	*n* = 80	*n* = 80

[a]Lewis index = (RI + SIII)–(SI + RIII).

Table A1.30

ST slope (°) in selected leads in Caucasians

Age group	Sex	I	aVF	V_2	V_5	V_6
18–29	Male	19 ± 7	12 ± 8	50 ± 12	28 ± 10	20 ± 8
		6–33	−2–31	21–69	11–51	6–38
		n = 265	n = 265	n = 265	n = 265	n = 265
	Female	12 ± 6	7 ± 5	31 ± 11	15 ± 7	11 ± 6
		3–24	−3–19	8–53	2–33	1–25
		n = 317	n = 318	n = 317	n = 318	n = 318
30–39	Male	17 ± 8	9 ± 7	46 ± 13	26 ± 10	18 ± 9
		3–36	−3–27	19–66	10–53	3–42
		n = 220	n = 220	n = 220	n = 220	n = 220
	Female	10 ± 5	6 ± 5	28 ± 10	14 ± 7	10 ± 6
		1–21	−4–16	10–50	0–33	0–25
		n = 118	n = 118	n = 116	n = 118	n = 117
40–49	Male	13 ± 6	9 ± 6	41 ± 13	22 ± 9	15 ± 7
		3–25	−1–22	13–63	7–43	2–31
		n = 118	n = 117	n = 117	n = 119	n = 119
	Female	10 ± 5	6 ± 6	25 ± 10	13 ± 7	10 ± 6
		2–20	−2–21	7–46	0–26	2–32
		n = 73	n = 73	n = 73	n = 73	n = 73
50+	Male	12 ± 7	9 ± 6	40 ± 12	21 ± 13	14 ± 8
		0–28	−1–24	15–61	2–43	0–31
		n = 125	n = 125	n = 125	n = 125	n = 124
	Female	7 ± 4	5 ± 5	24 ± 10	12 ± 7	7 ± 5
		0–14	−2–17	10–42	0–28	−1–19
		n = 80	n = 80	n = 80	n = 80	n = 80

Table A1.31

P terminal force in V_1, intrinsicoid deflection (ID) in V_5, V_6, and body-surface area (BSA) in the Caucasians studied as described in Chap. 13.

Age group	Sex	P terminal force in V_1 (mV ms)	ID V_5 (ms)	ID V_6 (ms)	BSA (m²)
18–29	Male	1.09 ± 1.33	41 ± 13.4	41 ± 12.3	1.86 ± 0.15
		0.0–4.62	26–85	22–87	1.55–2.19
		n = 265	n = 265	n = 265	n = 257
	Female	0.75 ± 1.03	33 ± 8.40	35 ± 7.70	1.62 ± 0.14
		0.0–3.54	22–66	22–59	1.37–1.95
		n = 316	n = 318	n = 318	n = 315
30–39	Male	1.16 ± 1.21	37 ± 11.5	38 ± 10.7	1.89 ± 0.15
		0.0–3.94	22–79	22–84	1.60–2.26
		n = 219	n = 220	n = 220	n = 215
	Female	1.02 ± 1.25	35 ± 7.10	36 ± 6.90	1.62 ± 0.16
		0.0–4.29	22–48	24–47	1.38–2.07
		n = 118	n = 118	n = 117	n = 117
40–49	Male	1.49 ± 1.44	34 ± 9.40	35 ± 8.10	1.88 ± 0.17
		0.0–4.42	20–67	24–64	1.54–2.26
		n = 117	n = 119	n = 117	n = 115
	Female	1.57 ± 1.40	32 ± 6.70	33 ± 5.90	1.68 ± 0.16
		0.0–5.5	22–44	20–42	1.47–2.03
		n = 73	n = 73	n = 73	n = 69
50±	Male	1.47 ± 1.44	35 ± 10.7	33 ± 5.90	1.85 ± 0.14
		0.0–5.45	22–72	22–46	1.53–2.14
		n = 125	n = 175	n = 174	n = 121
	Female	1.42 ± 1.37	31 ± 6.30	32 ± 6.30	1.65 ± 0.11
		0.0–4.05	22–41	22–44	1.43–1.92
		n = 79	n = 80	n = 80	n = 79

A1.2 Normal Limits of the 12-Lead ECG in Chinese

The data in Tables A1.32–A1.38 were obtained from a study of 503 Chinese people with normal cardiovascular systems (see Chap. 13). There were 255 males and 248 females. The most significant differences with the Caucasian data presented in Sect. A1.1 were in the QRS amplitudes, and for this reason, only some amplitude data are tabulated here. Fuller details are available in: Chen C Y, Chiang B N, Macfarlane P W. Normal limits of the electrocardiogram in a Chinese population. *J. Electrocardiol.* 1989;**22**(1): 1–15. Most of these tables are reproduced from that article with the permission of Churchill Livingstone, New York. Data are presented as mean ± standard deviation together with 96% ranges; that is, 2% of the measurements have been excluded from each extreme of the range. Where numbers of subjects studied in a subgroup is small, the full range is given. P-, Q, S, T- wave amplitudes are presented as positive measurements.

A1.3 Normal Limits of the 12-Lead ECG in Japanese

The data in Tables A1.39–A1.43 are taken from a study of 1,329 normal Japanese individuals. The ECGs were recorded on paper and measurements made by hand. For this reason, only selected amplitudes are presented. These data have been reproduced from: The normal value of electrocardiogram in the Japanese. *Jpn. Heart J.* 1963; **4**: 141–172 with the permission of the University of Tokyo Press, Tokyo. Note that the maximum and minimum values are presented and not the 96% range. Q, S wave amplitudes are treated as positive values.

Table A1.32

Q-wave amplitude (millivolts) in normal Chinese in various age-groups: leads I-aVF

	18–29		30–39		40–49		50–59		60+	
	Male *n* = 56	Female *n* = 47	Male *n* = 50	Female *n* = 59	Male *n* = 50	Female *n* = 50	Male *n* = 50	Female *n* = 48	Male *n* = 49	Female *n* = 44
I	0.07 ± 0.04	0.04 ± 0.02	0.05 ± 0.03	0.06 ± 0.03	0.07 ± 0.05	0.05 ± 0.02	0.07 ± 0.04	0.06 ± 0.03	0.07 ± 0.04	0.06 ± 0.02
	0.00–0.15	0.00–0.07	0.00–0.10	0.00–0.10	0.00–0.10	0.00–0.10	0.00–0.11	0.00–0.12	0.00–0.14	0.00–0.08
	n = 23	*n* = 14	*n* = 15	*n* = 15	*n* = 15	*n* = 15	*n* = 21	*n* = 17	*n* = 22	*n* = 16
II	0.10 ± 0.05	0.07 ± 0.06	0.07 ± 0.06	0.08 ± 0.05	0.07 ± 0.05	0.06 ± 0.03	0.07 ± 0.05	0.07 ± 0.05	0.07 ± 0.04	0.07 ± 0.06
	0.00–0.17	0.00–0.20	0.00–0.22	0.00–0.15	0.00–0.15	0.00–0.10	0.00–0.11	0.00–0.11	0.00–0.15	0.00–0.10
	n = 23	*n* = 25	*n* = 29	*n* = 25	*n* = 19	*n* = 17	*n* = 18	*n* = 15	*n* = 23	*n* = 10
III	0.11 ± 0.06	0.12 ± 0.08	0.13 ± 0.10	0.10 ± 0.08	0.15 ± 0.13	0.11 ± 0.12	0.11 ± 0.09	0.10 ± 0.08	0.14 ± 0.12	0.17 ± 0.13
	0.00–0.20	0.00–0.27	0.00–0.28	0.00–0.21	0.00–0.25	0.00–0.43	0.00–0.27	0.00–0.19	0.00–0.34	0.00–0.37
	n = 23	*n* = 28	*n* = 32	*n* = 34	*n* = 23	*n* = 24	*n* = 17	*n* = 19	*n* = 25	*n* = 18
aVR	0.79 ± 0.22	0.73 ± 0.14	0.79 ± 0.17	0.76 ± 0.18	0.71 ± 0.18	0.70 ± 0.20	0.70 ± 0.20	0.68 ± 0.18	0.69 ± 0.20	0.69 ± 0.21
	0.00–1.22	0.00–0.95	0.00–1.12	0.00–1.05	0.00–0.95	0.00–1.28	0.00–1.03	0.00–1.04	0.00–0.88	0.00–1.05
	n = 33	*n* = 33	*n* = 31	*n* = 35	*n* = 35	*n* = 36	*n* = 29	*n* = 38	*n* = 30	*n* = 36
aVL	0.10 ± 0.07	0.09 ± 0.08	0.11 ± 0.09	0.13 ± 0.15	0.12 ± 0.12	0.07 ± 0.04	0.10 ± 0.07	0.07 ± 0.05	0.09 ± 0.05	0.10 ± 0.08
	0.00–0.19	0.00–0.23	0.00–0.27	0.00–0.44	0.00–0.27	0.00–0.10	0.00–0.23	0.00–0.18	0.00–0.21	0.00–0.21
	n = 27	*n* = 16	*n* = 17	*n* = 22	*n* = 22	*n* = 20	*n* = 29	*n* = 24	*n* = 25	*n* = 23
aVF	0.10 ± 0.05	0.09 ± 0.06	0.08 ± 0.06	0.08 ± 0.05	0.08 ± 0.05	0.06 ± 0.02	0.07 ± 0.06	0.07 ± 0.04	0.08 ± 0.05	0.08 ± 0.07
	0.00–0.18	0.00–0.19	0.00–0.23	0.00–0.17	0.00–0.15	0.00–0.10	0.00–0.14	0.00–0.14	0.00–0.16	0.00–0.12
	n = 22	*n* = 26	*n* = 29	*n* = 30	*n* = 20	*n* = 18	*n* = 18	*n* = 12	*n* = 24	*n* = 13

Table A1.33

Q-wave amplitude (millivolts) in normal Chinese in various age-groups: leads $V_1 - V_6$

	18–29		30–39		40–49		50–59		60+	
	Male *n* = 56	Female *n* = 47	Male *n* = 50	Female *n* = 59	Male *n* = 50	Female *n* = 50	Male *n* = 50	Female *n* = 48	Male *n* = 49	Female *n* = 44
V_1		1.08 ± 0.93			0.76 ± 0.81	0.98 ± 0.00	1.58 ± 0.00		0.33 ± 0.00	0.41 ± 0.01
		0.00–2.29			0.00–1.33	0.00–0.98	0.00–1.58		0.00–0.33	0.00–0.41
		n = 4			*n* = 2	*n* = 1	*n* = 1		*n* = 1	*n* = 3
V_2							0.04 ± 0.00		0.05 ± 0.00	0.04 ± 0.01
							0.00–0.04		0.00–0.05	0.00–0.04
							n = 1		*n* = 1	*n* = 3
V_3	0.07 ± 0.03	0.13 ± 0.06		0.03 ± 0.01			0.03 ± 0.00	0.23 ± 0.00	0.06 ± 0.01	0.06 ± 0.03
	0.00–0.09	0.00–0.18		0.00–0.04			0.00–0.03	0.00–0.23	0.00–0.07	0.00–0.10
	n = 2	*n* = 2		*n* = 3			*n* = 1	*n* = 1	*n* = 2	*n* = 4
V_4	0.11 ± 0.08	0.11 ± 0.10	0.07 ± 0.07	0.05 ± 0.02	0.09 ± 0.05	0.03 ± 0.01	0.07 ± 0.06	0.07 ± 0.07	0.09 ± 0.06	0.05 ± 0.04
	0.00–0.25	0.00–0.26	0.00–0.24	0.00–0.08	0.00–0.14	0.00–0.04	0.00–0.13	0.00–0.23	0.00–0.19	0.00–0.11
	n = 18	*n* = 7	*n* = 9	*n* = 14	*n* = 5	*n* = 5	*n* = 12	*n* = 8	*n* = 7	*n* = 13
V_5	0.11 ± 0.08	0.08 ± 0.07	0.09 ± 0.07	0.06 ± 0.03	0.08 ± 0.07	0.06 ± 0.02	0.08 ± 0.05	0.07 ± 0.05	0.07 ± 0.05	0.05 ± 0.03
	0.00–0.25	0.00–0.22	0.00–0.22	0.00–0.12	0.00–0.21	0.00–0.09	0.00–0.16	0.00–0.13	0.00–0.16	0.00–0.12
	n = 31	*n* = 14	*n* = 24	*n* = 22	*n* = 15	*n* = 13	*n* = 23	*n* = 14	*n* = 23	*n* = 19
V_6	0.11 ± 0.08	0.08 ± 0.06	0.09 ± 0.07	0.07 ± 0.03	0.08 ± 0.06	0.06 ± 0.03	0.08 ± 0.04	0.07 ± 0.04	0.07 ± 0.05	0.05 ± 0.03
	0.00–0.29	0.00–0.18	0.00–0.25	0.00–0.13	0.00–0.23	0.00–0.10	0.00–0.16	0.00–0.12	0.00–0.17	0.00–0.10
	n = 37	*n* = 21	*n* = 32	*n* = 36	*n* = 23	*n* = 20	*n* = 29	*n* = 20	*n* = 33	*n* = 21

■ Table A1.34

R-wave amplitude (millivolts) in normal Chinese in various age-groups: leads I-aVF

	18–29		30–39		40–49		50–59		60+	
	Male *n* = 56	Female *n* = 47	Male *n* = 50	Female *n* = 59	Male *n* = 50	Female *n* = 50	Male *n* = 50	Female *n* = 48	Male *n* = 49	Female *n* = 44
I	0.51 ± 0.23	0.41 ± 0.20	0.56 ± 0.25	0.50 ± 0.20	0.61 ± 0.30	0.60 ± 0.26	0.63 ± 0.30	0.61 ± 0.22	0.66 ± 0.29	0.66 ± 0.25
	0.13–1.00	0.08–0.86	0.22–1.14	0.17–0.87	0.17–1.27	0.25–1.21	0.17–1.27	0.30–1.07	0.23–1.37	0.19–1.11
	n = 56	*n* = 47	*n* = 50	*n* = 59	*n* = 50	*n* = 50	*n* = 50	*n* = 48	*n* = 49	*n* = 44
II	1.24 ± 0.42	1.19 ± 0.31	1.18 ± 0.40	1.09 ± 0.30	0.95 ± 0.41	0.93 ± 0.29	0.96 ± 0.41	0.81 ± 0.30	0.91 ± 0.45	0.78 ± 0.34
	0.49–2.17	0.68–1.65	0.58–2.21	0.53–1.70	0.21–1.60	0.51–1.47	0.25–1.97	0.39–1.37	0.30–1.90	0.35–1.49
	n = 56	*n* = 47	*n* = 50	*n* = 59	*n* = 50	*n* = 50	*n* = 50	*n* = 48	*n* = 49	*n* = 44
III	0.73 ± 0.49	0.90 ± 0.44	0.74 ± 0.51	0.72 ± 0.40	0.53 ± 0.42	0.46 ± 0.33	0.54 ± 0.50	0.39 ± 0.31	0.52 ± 0.48	0.35 ± 0.29
	0.02–1.71	0.21–1.85	0.04–2.08	0.06–1.46	0.04–1.38	0.04–1.10	0.05–1.94	0.05–1.36	0.05–1.63	0.04–1.16
	n = 56	*n* = 47	*n* = 50	*n* = 59	*n* = 49	*n* = 49	*n* = 50	*n* = 48	*n* = 49	*n* = 44
aVR	0.13 ± 0.12	0.11 ± 0.09	0.11 ± 0.09	0.08 ± 0.06	0.12 ± 0.09	0.09 ± 0.06	0.09 ± 0.08	0.11 ± 0.08	0.10 ± 0.06	0.11 ± 0.08
	0.00–0.43	0.00–0.32	0.00–0.37	0.00–0.23	0.00–0.31	0.00–0.24	0.00–0.23	0.00–0.24	0.00–0.26	0.00–0.28
	n = 51	*n* = 34	*n* = 43	*n* = 48	*n* = 41	*n* = 36	*n* = 39	*n* = 35	*n* = 40	*n* = 28
aVL	0.19 ± 0.16	0.16 ± 0.13	0.24 ± 0.19	0.18 ± 0.14	0.35 ± 0.27	0.27 ± 0.19	0.35 ± 0.31	0.32 ± 0.22	0.38 ± 0.32	0.38 ± 0.22
	0.04–0.59	0.04–0.52	0.04–0.59	0.00–0.54	0.00–0.98	0.04–0.69	0.05–1.07	0.07–0.83	0.04–1.07	0.06–0.83
	n = 56	*n* = 47	*n* = 49	*n* = 57	*n* = 48	*n* = 50	*n* = 49	*n* = 48	*n* = 49	*n* = 43
aVF	0.98 ± 0.48	1.04 ± 0.36	0.94 ± 0.46	0.90 ± 0.34	0.68 ± 0.44	0.68 ± 0.29	0.70 ± 0.48	0.55 ± 0.31	0.67 ± 0.49	0.49 ± 0.34
	0.09–2.07	0.48–1.81	0.13–2.13	0.28–1.56	0.07–1.50	0.13–1.34	0.04–1.92	0.03–1.27	0.09–1.73	0.03–1.24
	n = 56	*n* = 47	*n* = 50	*n* = 59	*n* = 50	*n* = 50	*n* = 50	*n* = 48	*n* = 49	*n* = 44

Table A1.35

R-wave amplitude (millivolts) in normal Chinese in various age groups: leads $V_1 - V_6$

	18–29		30–39		40–49		50–59		60+	
	Male *n* = 56	Female *n* = 47	Male *n* = 50	Female *n* = 59	Male *n* = 50	Female *n* = 50	Male *n* = 50	Female *n* = 48	Male *n* = 49	Female *n* = 44
V_1	0.43 ± 0.27	0.33 ± 0.28	0.33 ± 0.18	0.26 ± 0.12	0.27 ± 0.23	0.21 ± 0.13	0.27 ± 0.14	0.21 ± 0.15	0.27 ± 0.17	0.19 ± 0.12
	0.07–1.04	0.00–0.86	0.09–0.76	0.06–0.52	0.02–0.92	0.02–0.46	0.03–0.63	0.04–0.66	0.08–0.65	0.00–0.49
	n = 56	*n* = 44	*n* = 50	*n* = 59	*n* = 49	*n* = 49	*n* = 50	*n* = 48	*n* = 49	*n* = 41
V_2	0.92 ± 0.41	0.71 ± 0.41	0.85 ± 0.38	0.65 ± 0.25	0.72 ± 0.42	0.57 ± 0.28	0.73 ± 0.32	0.64 ± 0.30	0.76 ± 0.33	0.61 ± 0.35
	0.25–1.78	0.24–2.09	0.31–1.76	0.24–1.20	0.09–1.56	0.23–1.23	0.24–1.33	0.20–1.27	0.26–1.36	0.16–1.43
	n = 56	*n* = 47	*n* = 50	*n* = 59	*n* = 50	*n* = 50	*n* = 50	*n* = 48	*n* = 49	*n* = 44
V_3	1.21 ± 0.58	1.09 ± 0.47	1.21 ± 0.57	1.09 ± 0.48	0.99 ± 0.49	0.88 ± 0.39	1.19 ± 0.45	1.07 ± 0.50	1.30 ± 0.55	1.12 ± 0.46
	0.30–2.97	0.49–1.95	0.54–2.55	0.31–2.35	0.22–2.06	0.38–2.01	0.38–1.92	0.32–2.31	0.40–2.28	0.45–2.16
	n = 56	*n* = 47	*n* = 50	*n* = 59	*n* = 50	*n* = 50	*n* = 50	*n* = 48	*n* = 49	*n* = 44
V_4	2.06 ± 0.71	1.40 ± 0.45	1.83 ± 0.60	1.49 ± 0.48	1.76 ± 0.67	1.31 ± 0.40	1.85 ± 0.76	1.56 ± 0.48	1.97 ± 0.72	1.54 ± 0.48
	1.01–3.82	0.61–2.49	0.58–2.90	0.73–2.51	0.77–3.21	0.65–2.60	0.74–4.24	0.69–2.49	0.93–3.64	0.88–2.47
	n = 56	*n* = 47	*n* = 50	*n* = 59	*n* = 50	*n* = 50	*n* = 50	*n* = 48	*n* = 49	*n* = 44
V_5	1.84 ± 0.46	1.25 ± 0.38	1.83 ± 0.50	1.38 ± 0.35	1.83 ± 0.58	1.35 ± 0.39	1.82 ± 0.65	1.56 ± 0.45	1.89 ± 0.65	1.58 ± 0.47
	1.20–3.10	0.65–2.06	0.99–2.74	0.77–2.10	0.55–2.83	0.69–2.42	0.83–3.78	0.98–2.31	0.82–3.43	0.97–2.64
	n = 56	*n* = 47	*n* = 50	*n* = 59	*n* = 50	*n* = 50	*n* = 50	*n* = 48	*n* = 49	*n* = 44
V_6	1.43 ± 0.35	1.09 ± 0.32	1.50 ± 0.39	1.16 ± 0.26	1.47 ± 0.52	1.14 ± 0.34	1.48 ± 0.45	1.30 ± 0.41	1.47 ± 0.47	1.36 ± 0.43
	0.87–2.16	0.61–1.80	0.86–2.31	0.63–1.83	0.31–2.31	0.54–1.86	0.57–2.35	0.74–2.12	0.81–2.57	0.71–2.14
	n = 56	*n* = 47	*n* = 50	*n* = 59	*n* = 50	*n* = 50	*n* = 50	*n* = 48	*n* = 49	*n* = 44

Table A1.36

S-wave amplitude (millivolts) in normal Chinese in various age-groups: leads I-aVF

	18–29		30–39		40–49		50–59		60+	
	Male *n* = 56	Female *n* = 47	Male *n* = 50	Female *n* = 59	Male *n* = 50	Female *n* = 50	Male *n* = 50	Female *n* = 48	Male *n* = 49	Female *n* = 44
I	0.19 ± 0.13	0.18 ± 0.13	0.17 ± 0.09	0.14 ± 0.10	0.16 ± 0.11	0.13 ± 0.07	0.17 ± 0.11	0.16 ± 0.09	0.17 ± 0.09	0.14 ± 0.10
	0.00–0.49	0.00–0.44	0.00–0.33	0.00–0.40	0.00–0.44	0.00–0.29	0.00–0.41	0.00–0.36	0.00–0.30	0.00–0.34
	n = 47	*n* = 31	*n* = 42	*n* = 39	*n* = 40	*n* = 37	*n* = 30	*n* = 31	*n* = 39	*n* = 25
II	0.24 ± 0.16	0.17 ± 0.11	0.17 ± 0.13	0.17 ± 0.09	0.21 ± 0.12	0.15 ± 0.10	0.17 ± 0.13	0.20 ± 0.12	0.16 ± 0.07	0.18 ± 0.11
	0.00–0.63	0.00–0.43	0.00–0.48	0.00–0.48	0.00–0.48	0.00–0.33	0.00–0.35	0.00–0.44	0.00–0.26	0.00–0.37
	n = 39	*n* = 26	*n* = 28	*n* = 33	*n* = 34	*n* = 30	*n* = 30	*n* = 28	*n* = 31	*n* = 30
III	0.21 ± 0.16	0.17 ± 0.13	0.15 ± 0.14	0.15 ± 0.09	0.33 ± 0.28	0.17 ± 0.13	0.29 ± 0.36	0.27 ± 0.24	0.42 ± 0.31	0.32 ± 0.24
	0.00–0.55	0.00–0.42	0.00–0.48	0.00–0.33	0.00–0.90	0.00–0.41	0.00–1.36	0.00–0.79	0.00–0.91	0.00–0.81
	n = 43	*n* = 28	*n* = 27	*n* = 34	*n* = 33	*n* = 26	*n* = 36	*n* = 31	*n* = 26	*n* = 29
aVR	0.82 ± 0.36	0.80 ± 0.25	0.88 ± 0.30	0.76 ± 0.16	0.76 ± 0.40	0.85 ± 0.22	0.86 ± 0.18	0.62 ± 0.37	0.80 ± 0.29	0.70 ± 0.34
	0.00–1.43	0.00–1.33	0.00–1.39	0.00–1.16	0.00–1.72	0.00–1.19	0.00–1.22	0.00–1.07	0.00–1.30	0.00–1.15
	n = 26	*n* = 14	*n* = 20	*n* = 24	*n* = 17	*n* = 14	*n* = 21	*n* = 13	*n* = 21	*n* = 9
aVL	0.37 ± 0.21	0.41 ± 0.27	0.33 ± 0.24	0.32 ± 0.18	0.25 ± 0.16	0.20 ± 0.13	0.30 ± 0.28	0.22 ± 0.19	0.27 ± 0.24	0.17 ± 0.15
	0.00–0.89	0.00–0.98	0.00–0.92	0.00–0.72	0.00–0.55	0.00–0.45	0.00–0.99	0.00–0.75	0.00–0.71	0.00–0.52
	n = 50	*n* = 39	*n* = 45	*n* = 47	*n* = 38	*n* = 45	*n* = 34	*n* = 32	*n* = 41	*n* = 28
aVF	0.21 ± 0.14	0.15 ± 0.11	0.16 ± 0.10	0.16 ± 0.08	0.23 ± 0.16	0.14 ± 0.09	0.18 ± 0.23	0.19 ± 0.16	0.19 ± 0.14	0.18 ± 0.13
	0.00–0.51	0.00–0.34	0.00–0.32	0.00–0.31	0.00–0.54	0.00–0.33	0.00–0.98	0.00–0.44	0.00–0.49	0.00–0.39
	n = 40	*n* = 28	*n* = 22	*n* = 30	*n* = 34	*n* = 24	*n* = 31	*n* = 27	*n* = 30	*n* = 29

Table A1.37
S-wave amplitude (millivolts) in normal Chinese in various age-groups: leads V_1–V_6

	18–29		30–39		40–49		50–59		60+	
	Male *n* = 56	Female *n* = 47	Male *n* = 50	Female *n* = 59	Male *n* = 50	Female *n* = 50	Male *n* = 50	Female *n* = 48	Male *n* = 49	Female *n* = 44
V_1	1.05 ± 0.43	0.86 ± 0.35	0.90 ± 0.34	0.83 ± 0.35	0.84 ± 0.36	0.84 ± 0.37	0.83 ± 0.43	0.77 ± 0.36	0.76 ± 0.43	0.64 ± 0.28
	0.25–1.97	0.00–1.54	0.36–1.57	0.28–1.78	0.00–1.49	0.14–1.54	0.21–2.06	0.24–1.42	0.12–1.73	0.00–1.21
	n = 56	*n* = 44	*n* = 50	*n* = 59	*n* = 48	*n* = 49	*n* = 49	*n* = 48	*n* = 49	*n* = 41
V_2	2.07 ± 0.67	1.29 ± 0.55	1.68 ± 0.57	1.22 ± 0.46	1.49 ± 0.61	1.26 ± 0.48	1.26 ± 0.62	1.16 ± 0.45	1.25 ± 0.58	0.97 ± 0.39
	0.84–3.43	0.46–2.67	0.63–2.76	0.31–2.17	0.32–2.69	0.50–2.35	0.11–2.75	0.42–2.14	0.18–2.60	0.19–1.84
	n = 56	*n* = 47	*n* = 49	*n* = 59	*n* = 50	*n* = 50	*n* = 50	*n* = 48	*n* = 49	*n* = 44
V_3	1.14 ± 0.46	0.68 ± 0.41	1.04 ± 0.50	0.66 ± 0.32	1.04 ± 0.42	0.77 ± 0.32	0.98 ± 0.47	0.79 ± 0.37	1.05 ± 0.46	0.79 ± 0.36
	0.40–2.15	0.00–1.56	0.00–2.20	0.00–1.25	0.27–1.86	0.24–1.32	0.00–1.98	0.24–1.77	0.37–1.89	0.10–1.47
	n = 56	*n* = 42	*n* = 48	*n* = 57	*n* = 50	*n* = 49	*n* = 47	*n* = 48	*n* = 49	*n* = 43
V_4	0.61 ± 0.33	0.41 ± 0.29	0.61 ± 0.38	0.38 ± 0.21	0.63 ± 0.32	0.44 ± 0.24	0.59 ± 0.35	0.52 ± 0.28	0.74 ± 0.36	0.56 ± 0.27
	0.00–1.25	0.00–1.20	0.00–1.43	0.00–0.82	0.00–1.18	0.00–0.94	0.00–1.57	0.09–1.02	0.00–1.46	0.00–1.20
	n = 50	*n* = 38	*n* = 41	*n* = 51	*n* = 48	*n* = 48	*n* = 45	*n* = 47	*n* = 46	*n* = 41
V_5	0.38 ± 0.23	0.29 ± 0.20	0.40 ± 0.25	0.25 ± 0.16	0.43 ± 0.26	0.33 ± 0.17	0.37 ± 0.25	0.35 ± 0.21	0.49 ± 0.29	0.39 ± 0.21
	0.00–0.94	0.00–0.81	0.00–1.01	0.00–0.65	0.00–0.87	0.00–0.64	0.00–1.07	0.00–0.76	0.00–1.01	0.00–0.88
	n = 45	*n* = 34	*n* = 38	*n* = 49	*n* = 46	*n* = 41	*n* = 44	*n* = 45	*n* = 46	*n* = 40
V_6	0.22 ± 0.14	0.20 ± 0.13	0.21 ± 0.15	0.15 ± 0.11	0.26 ± 0.17	0.18 ± 0.10	0.21 ± 0.16	0.21 ± 0.16	0.26 ± 0.15	0.23 ± 0.14
	0.00–0.57	0.00–0.38	0.00–0.50	0.00–0.44	0.00–0.56	0.00–0.43	0.00–0.61	0.00–0.50	0.00–0.57	0.00–0.49
	n = 45	*n* = 24	*n* = 36	*n* = 42	*n* = 42	*n* = 37	*n* = 39	*n* = 42	*n* = 42	*n* = 36

Table A1.38

Maximal T-wave amplitude (millivolts) in normal Chinese in various age-groups

	18–29		30–39		40–49		50–59		60+	
	Male $n = 56$	Female $n = 47$	Male $n = 50$	Female $n = 59$	Male $n = 50$	Female $n = 50$	Male $n = 50$	Female $n = 48$	Male $n = 49$	Female $n = 44$
I	0.28 ± 0.08	0.21 ± 0.06	0.25 ± 0.10	0.22 ± 0.06	0.22 ± 0.09	0.20 ± 0.07	0.22 ± 0.09	0.18 ± 0.07	0.18 ± 0.08	0.17 ± 0.07
	0.12–0.44	0.11–0.29	0.10–0.47	0.11–0.33	0.08–0.41	0.10–0.33	0.08–0.41	0.07–0.29	0.08–0.35	0.06–0.33
II	0.37 ± 0.11	0.30 ± 0.12	0.35 ± 0.13	0.27 ± 0.10	0.28 ± 0.11	0.25 ± 0.08	0.29 ± 0.13	0.21 ± 0.09	0.27 ± 0.13	0.20 ± 0.10
	0.17–0.58	0.10–0.52	0.11–0.58	0.08–0.46	0.12–0.52	0.10–0.42	0.07–0.58	0.06–0.40	0.12–0.62	0.07–0.43
III	0.11 ± 0.13	0.11 ± 0.11	0.10 ± 0.14	0.06 ± 0.11	0.07 ± 0.13	0.06 ± 0.09	0.09 ± 0.14	0.03 ± 0.09	0.09 ± 0.15	0.04 ± 0.10
	−0.19–0.34	−0.14–0.33	−0.14–0.39	−0.12–0.26	−0.15–0.30	−0.10–0.23	−0.20–0.34	−0.11–0.18	−0.18–0.44	−0.14–0.17
aVR	−0.32 ± 0.08	−0.26 ± 0.07	−0.30 ± 0.09	−0.26 ± 0.07	−0.25 ± 0.08	−0.22 ± 0.07	−0.25 ± 0.09	−0.19 ± 0.07	−0.22 ± 0.08	−0.18 ± 0.07
	−0.50 to −0.20	−0.40 to −0.10	−0.46 to −0.10	−0.38 to −0.08	−0.42 to −0.12	−0.36 to −0.13	0.42 to −0.07	−0.35 to −0.07	−0.44 to −0.10	−0.35 to −0.08
aVL	0.09 ± 0.10	0.06 ± 0.08	0.08 ± 0.11	0.09 ± 0.07	0.08 ± 0.09	0.09 ± 0.06	0.07 ± 0.10	0.08 ± 0.06	0.05 ± 0.10	0.08 ± 0.07
	−0.08–0.26	−0.08–0.18	−0.10–0.30	−0.06–0.20	−0.07–0.27	−0.03–0.18	−0.10–0.27	−0.05–0.18	−0.15–0.27	−0.04–0.18
aVF	0.24 ± 0.11	0.20 ± 0.11	0.22 ± 0.12	0.17 ± 0.09	0.18 ± 0.10	0.15 ± 0.07	0.19 ± 0.12	0.12 ± 0.08	0.19 ± 0.12	0.12 ± 0.08
	0.04–0.45	−0.09–0.42	0.05–0.48	−0.05–0.37	0.03 ± 0.41	−0.03–0.29	−0.10–0.50	−0.06–0.26	0.04–0.51	−0.05–0.29
V_1	0.17 ± 0.15	−0.06 ± 0.14	0.12 ± 0.16	−0.05 ± 0.14	0.09 ± 0.15	−0.03 ± 0.09	0.12 ± 0.16	0.00 ± 0.11	0.09 ± 0.15	0.03 ± 0.09
	−0.14–0.50	−0.24–0.23	−0.25–0.35	−0.23–0.20	−0.18–0.36	−0.22–0.17	−0.21–0.39	−0.19–0.19	−0.15–0.33	0.18–0.10
V_2	0.82 ± 0.25	0.32 ± 0.21	0.79 ± 0.27	0.33 ± 0.22	0.62 ± 0.29	0.28 ± 0.19	0.63 ± 0.22	0.30 ± 0.16	0.51 ± 0.22	0.25 ± 0.17
	0.39–1.25	0.08–0.84	0.40–1.39	−0.12–0.71	0.09–1.18	−0.10–0.65	0.24–1.09	0.10–0.54	0.04–0.86	−0.11–0.55
V_3	0.90 ± 0.27	0.48 ± 0.20	0.81 ± 0.27	0.46 ± 0.25	0.68 ± 0.29	0.36 ± 0.23	0.69 ± 0.24	0.37 ± 0.20	0.55 ± 0.25	0.32 ± 0.13
	0.37–1.40	0.14–0.81	0.36–1.34	0.08–1.03	0.11–1.48	−0.13–0.83	0.27–1.18	0.10–0.83	−0.07–1.07	−0.04–0.80
V_4	0.77 ± 0.28	0.45 ± 0.16	0.69 ± 0.27	0.41 ± 0.21	0.60 ± 0.30	0.34 ± 0.19	0.60 ± 0.28	0.33 ± 0.20	0.50 ± 0.26	0.29 ± 0.19
	0.23–1.25	0.16–0.71	0.28–1.18	0.07–0.91	0.13–1.17	0.05–0.86	0.19–1.21	0.08–0.90	0.08–1.13	0.03–0.71
V_5	0.60 ± 0.20	0.41 ± 0.13	0.56 ± 0.23	0.38 ± 0.16	0.49 ± 0.20	0.33 ± 0.15	0.50 ± 0.26	0.31 ± 0.17	0.40 ± 0.21	0.28 ± 0.15
	0.22–0.93	0.20–0.65	0.16–1.08	0.09–0.73	0.14–1.01	0.06–0.59	0.09–0.92	0.07–0.71	−0.09–0.84	0.04–0.62
V_6	0.44 ± 0.14	0.34 ± 0.10	0.43 ± 0.17	0.32 ± 0.11	0.38 ± 0.15	0.29 ± 0.10	0.39 ± 0.20	0.26 ± 0.14	0.32 ± 0.14	0.25 ± 0.12
	0.18–0.72	0.19–0.57	0.13–0.80	0.10–0.57	0.14–0.74	0.12–0.47	0.08–0.75	0.07–0.52	0.08–0.59	0.05–0.50

Table A1.39

Amplitude of Q waves (in millimeters) in normal Japanese of various age-groups

		15–29		20–29		30–39		40–49		50–59		60–69		70+	
		Male $n = 61$	Female $n = 39$	Male $n = 165$	Female $n = 109$	Male $n = 119$	Female $n = 94$	Male $n = 153$	Female $n = 82$	Male $n = 135$	Female $n = 102$	Male $n = 101$	Female $n = 79$	Male $n = 48$	Female $n = 42$
I	Mn.	0.3	0.28	0.32	0.22	0.37	0.18	0.27	0.24	0.25	0.33	0.27	0.30	0.26	0.39
	Max.–Min.	1.9–0	1.5–0	1.4–0	1.5–0	2.47–0	1.8–0	2.1–0	2.0–0	1.7–0	2.0–0	2.3–0	2.3–0	2.0–0	1.7–0
II	Mn.	0.66	0.44	0.73	0.46	0.52	0.26	0.34	0.30	0.44	0.29	0.35	0.21	0.28	0.30
	Max.–Min.	3.2–0	2.5–0	9.2–0	2.7–0	3.2–0	1.9–0	2.4–0	2.1–0	5.6–0	1.1–0	2.0–0	1.7–0	1.1–0	1.7–0
III	Mn.	0.69	0.51	0.72	0.66	0.56	0.48	0.40	0.42	0.54	0.39	0.39	0.40	0.43	0.23
	Max.–Min.	3.0–0	3.2–0	8.8–0	3.2–0	2.9–0	2.11–0	2.9–0	5.6–0	11.0–0	4.4–0	2.0–0	2.3–0	2.6–0	2.9–0
aVR	Mn.	9.27	7.31	7.38	8.45	3.35	6.74	6.46	6.57	3.68	3.79	3.89	4.12	3.56	2.57
	Max.–Min.	2.0–0	10.1–5.5	13.5–0	14.3–2.8	12.9–0	11.5–0	18.2–0	14.5–0.8	14.0–0	10.9–0	12.9–0	9.9–0	16.9–0	1.39–0
		(30)[a]			(47)[a]										
aVL	Mn.	0.28	0.26	0.33	0.20	0.15	0.29	0.47	0.30	0.45	0.40	0.50	0.46	0.74	0.59
	Max.–Min.	2.0–0	2.1–0	4.0–0	3.0–0	3.2–0	4.4–0	5.2–0	1.8–0	3.8–0	2.6–0	4.1–0	4.1–0	5.3–0	3.2–0
aVF	Mn.	0.59	0.47	0.60	0.40	0.50	0.35	0.33	0.23	0.43	0.30	0.30	0.24	0.32	0.21
	Max.–Min.	3.8–0	2.4–0	5.8–0	2.5–0	3.0–0	3.4–0	3.1–0	2.2–0	5.3–0	2.6–0	2.0–0	1.7–0	1.1–0	1.4–0
V_4R	Mn.	0.3	0	0.15	0.03	0.11	0.01	0.15	0.18	0.23	0.09	0.07	0.18	0.01	0
	Max.–Min.	1.5–0	0	1.4–0	1.2–0	4.6–0	0.2–0	6.8–0	4.3–0	4.3–0	4.7–0	3.2–0	4.7–0	0.2–0	0
		(31)[a]		(117)[a]	(95)[a]		(75)[a]		(77)[a]	(96)[a]	(84)[a]	(74)[a]	(56)[a]	(23)[a]	
V_1	Mn.	0.03	0	0.01	0	0.01	0	0.11	0	0.09	0	0	0	0	0
	Max.–Min.	1.2–0	0	1.1–0	0	1.6–0	0	16.5–0	0	11.4–0	0	0	0.2–0	0	0.1–0
V_2	Mn.	0.04	0	0.01	0.01	0.01	0	0.10	0.01	0	0	0	0	0	0
	Max.–Min.	1.8–0	0	1.1–0	0.3–0	1.1–0		13.8–0	0.05–0	0	0.5–0	0.2–0	0	0.2–0	0.2–0
V_3	Mn.	0.03	0	0.01	0	0.01	0.01	0.04	0.02	0.02	0.01	0.03	0.01	0.04	0.04
	Max.–Min.	1.5–0	0	1.1–0	0	0.5–0	0.1–0	3.0–0	1.0–0	1.3–0	2.3–0	2.5–0	0.5–0	1.1–0	1.4–0
V_4	Mn.	0.35	0.16	0.31	0.13	0.4	0.01	0.27	0.70	0.24	0.25	0.26	0.23	0.35	0.50
	Max.–Min.	3.5–0	3.0–0	5.8–0	1.1–0	3.1–0	1.0–0	4.1–0	2.0–0	2.6–0	4.1–0	3.1–0	1.7–0	3.2–0	3.7–0
V_5	Mn.	1.01	0.39	0.87	0.33	0.70	0.04	0.52	0.39	0.51	0.34	0.53	0.34	0.36	0.59
	Max.–Min.	5.0–0	4.1–0	5.2–0	2.0–0	3.4–0	1.3–0	3.0–0	2.0–0	5.0–0	2.0–0	3.8–0	1.7–0	2.0–0	3.1–0
V_6	Mn.	0.95	0.56	0.92	0.45	0.58	0.39	0.59	0.48	0.31	0.49	0.52	0.36	0.45	0.52
	Max.–Min.	4.0–0	3.9–0	4.8–0	2.0–0	3.1–0	1.8–0	2.8–0	2.0–0	4.0–0	2.6–0	2.5–0	1.4–0	2.0–0	2.0–0

Mn., mean; Max., maximum; Min., minimum

[a]Number of cases in which measurement was made.

Table A1.40

Amplitude of R waves (in millimeters) in normal Japanese of various age-groups

		15–29		20–29		30–39		40–49		50–59		60–69		70+	
		Male *n* = 61	Female *n* =39	Male *n* = 165	Female *n* = 109	Male *n* = 119	Female *n* = 94	Male *n* = 153	Female *n* = 82	Male *n* = 135	Female *n* = 102	Male *n* = 101	Female *n* = 79	Male *n* = 48	Female *n* = 42
I	Mn., SD	5.68, 2.34	5.41, 2.13	5.14, 2.67	5.56, 2.59	5.46, 2.38	5.38, 2.38	5.56, 2.70	5.71, 2.87	5.30, 2.36	6.46, 2.75	5.27, 2.98	6.34, 25.1	5.05, 2.62	6.55, 3.18
	Max.–Min.	13.0–2.0	10.0–2.2	15.3–0.26	15.3–0.43	12.9–1.0	14.2–1.0	14.6–0.3	14.3–0	15.9–0	14.9–0.0	16.9–0	12.9–0	16.9–1.0	15.9–1.0
II	Mn., SD	14.25, 4.65	11.67, 2.99	13.73, 4.69	11.64, 3.97	12.99, 3.82	15.35, 3.16	10.74, 4.20	9.91, 3.66	10.78, 4.11	9.17, 3.24	9.35, 3.98	7.44, 4.06	7.89, 3.23	7.28, 2.98
	Max.–Min.	22.1–6.0	17.5–6.5	2.38–4.2	22.7–2.0	23.9–3.0	19.1–3.33	21.0–0.2	17.0–0	25.9–3.0	17.9–0.0	18.9–1.0	33.4–0	15.9–2.0	16,9–2.0
III	Mn., SD	10.0, 2.97	7.39, 3.95	9.44, 4.86	7.44, 4.26	8.15, 4.19	5.36, 3.34	6.24, 6.20	5.16, 4.15	6.53, 4.36	4.42, 3.15	5.51, 3.16	3.34, 2.66	6.62, 3.39	3.18, 2.41
	Max.–Min.	23.1–25	15.7–0.2	22.1–0.1	21.5–0	19.9–0	15.3–0.4	17.5–0.3	17.0–0.5	21.9–0.0	14.9–0.0	14.9–0	12.4–0	12.4–0	11.9–0.0
aVR	Mn., SD	0.48, 0.24	0.32, 0.73	0.49, 0.56	0.63, 0.74	0.45, 0.76	0.50, 0.72	0.74, 0.00	0.19, 0.00	0.36, 0.17	0.35, 0.54	0.37, 0.24	0.32, 0.43	0.30, 0.48	0.43, 0.48
	Max.–Min.	3.69–0.02	1.5–0	3.2–0	3.2–0	4.9–0	2.4–0	11.9–0	1.4–0	2.4–0	1.9–0.0	2.4–0	1.9–0	1.4–0	1.9–0.0
aVL	Mn., SD	1.62, 1.17	3.06, 1.69	1.93, 2.21	1.71, 2.09	1.89, 1.48	2.21, 2.01	2.37, 2.52	2.29, 2.09	2.16, 2.42	3.15, 2.02	2.53, 2.24	3.29, 2.28	3.22, 2.28	3.86, 0.63
	Max.–Min.	6.44–0.1	6.9–0	16.2–0	13.2–0	11.9–0	9.1–0	14.9–0.1	10.5–0.1	12.9–0	9.9–0	8.9–0	11.9–0	14.9–0	13.9–0.0
aVF	Mn., SD	11.80, 3.53	9.55, 3.70	11.07, 4.76	10.00, 3.88	9.57, 3.93	7.45, 4.71	8.01, 4.25	7.47, 4.71	8,19, 4.34	6.52, 3.36	6.84, 4.81	4.76, 2.68	5.97, 3.32	4.74, 2.30
	Max.–Min.	21.8–1.2	17.5–3.0	25.9–0.5	20.9–0.4	19.9–0	15.4–0.2	17.2–0.4	16.5–0.49	22.9–0.0	16.9–0.0	15.9–0	13.9–0	13.0–0	15.9–0.0
V_4R	Mn., SD	2.16, 3.26	1.43, 1.04	2.42, 1.65	1.51, 1.02	1.80, 1.05	0.98, 0.75	1.61, 0.00	5.50, 0.00	1.31, 0.74	1.04, 0.73	1.30, 0.96	1.09, 0.40	0.97	1.27, 0.85
	Max.–Min.	6.2–0.1	3.3–0	13.6–0	18.0–0	16.9–0	3.2–0.1	8.4–0.2	2.8–0.1	4.9–0	4.9–0.0	4.9–0	4.4–0	2.4–0	3.9–0.0
		(52)[a]	(34)[a]	(117)[a]	(97)[a]	(86)[a]	(76)[a]	(106)[a]	(76)[a]	(97)[a]	(80)[a]	(74)[a]			
V_1	Mn., SD	5.54, 2.56	3.95, 2.44	5.05, 3.19	3.53, 2.55	3.67, 2.06	2.81, 1.66	3.17, 3.64	2.44, 1.62	3.40, 2.28	2.59, 1.67	3.22, 2.57	2.35, 1.89	2.14, 1.57	2.57, 1.47
	Max.–Min.	15.5–1.6	10.8–0	15.8–0.54	10.0–0.23	10.9–0	7.5–0.2	13.9–0.2	6.1–0.3	13.9–0.0	8.9–0.0	13.9–0	9.9–0	8.9–0	8.9–0
V_2	Mn., SD	10.78, 4.16	7.32, 3.62	9.76, 4.69	7.10, 3.36	7.66, 3.95	6.48, 3.26	7.0, 4.12	5.52, 3.20	7,41, 4.27	5.95, 3.62	6.32, 3.38	6.38, 3.87	5.74, 3.87	6.28, 3.43
	Max.–Min.	15.7–3.5	18.0–0	22.4–0.8	23.7–1.0	17.9–0	12.5–1.2	18.6–1.0	17.2–0.7	21.9–0.0	15.9–0.0	17.9–0	23.9–0	14.4–0	13.9–0.0
V_3	Mn., SD	13.66, 5.80	9.05, 3.22	11.56, 5.73	9.83, 4.50	10.10, 5.37	4.04, 5.34	9.74, 5.38	9.33, 5.66	11.13, 3.25	9.95, 5.77	10.75, 5.69	11.65, 4.98	11.58, 7.40	13.43, 6.10
	Max.–Min.	33.3–3.1	18.5–0.2	3.70–1.17	24.6–3.5	29.9–0	26.1–0.4	24.3–0.6	27.0–1.3	29.9–0.0	35.9–0.0	27.9–0	21.9–0	35.9–0	31.9–2.0
V_4	Mn., SD	20.67, 7.05	13.49, 4.90	24.68, 7.53	14.60, 4.90	17.49, 7.35	14.65, 5.16	9.31, 6.51	15.86, 5.92	17.48, 6.04	15.99, 5.72	17.50, 8.41	17.29, 5.37	19.87, 7.94	21.09, 7.78
	Max.–Min.	35.8–2.3	24.8–6.8	53.2–5.0	30.6–2.87	41.9–2.0	34.0–7.1	33.8–3.4	26.8–5.5	35.9–0.0	37.9–0.0	49.9–2.0	37.4–2.0	37.9–0	39.9–6.0
V_5	Mn., SD	18.87, 5.31	13.51, 3.44	17.45, 5.84	13.4, 4.46	15.85, 6.29	14.10, 4.48	16.45, 6.09	15.50, 5.0	16.2, 5.52	15.20, 5.29	16.87, 8.23	14.43, 4.88	16.74, 5.81	17.33, 6.05
	Max.–Min.	55.0–8.0	24.0–9.0	43.0–3.98	27.4–2.01	33.9–2.0	26.4–1.7	35.2–1.8	28.5–5.6	35.9–2.0	35.9–0.0	55.9–4.0	24.4–0	24.4–0	35.9–6.0
V_6	Mn., SD	14.31, 4.31	11.56, 3.82	13.54, 5.11	11.07, 3.56	11.81, 4.54	11.65, 3.72	13.00, 5.06	11.88, 4.97	12.20, 4.27	11.85, 4.44	12.25, 5.99	10.30, 4.22	11.28, 4.05	10.0, 3.45
	Max.–Min.	28.2–7.0	21.8–7.0	29.0–2.4	23.7–1.2	23.9–2.0	24.9–1.1	24.7–0.9	26.1–1.7	25.9–2.0	23.9–0	37.9–2.0	23.9–0	23.9–20	17.9–4.0

Mn., mean; SD, standard deviation; Max., maximum; Min., minimum

[a]Number of cases in which measurement was made.

Table A1.41

Amplitude of S wave (in millimeters) in normal Japanese of various age-groups

		15–19		20–29		30–39		40–49		50–59		60–69		70+	
		Male $n = 61$	Female $n = 39$	Male $n = 165$	Female $n = 109$	Male $n = 119$	Female $n = 94$	Male $n = 153$	Female $n = 82$	Male $n = 135$	Female $n = 102$	Male $n = 101$	Female $n = 79$	Male $n = 48$	Female $n = 42$
I	Mn.	1.40	0.54	1.19	0.10	0.86	0.63	0.79	0.40	0.76	0.66	0.56	0.54	0.68	0.56
	Max.–Min.	5.7–0	3.4–0	5.35–0	4.0–0	3.9–0	2.7–0	5.0–0	3.1–0	3.9–0	3.4–0	3.4–0	3.4–0	3.4–0	2.4–0
II	Mn.	2.02	0.76	1.31	0.66	1.30	0.65	1.14	0.48	1.14	0.64	1.37	1.07	1.18	1.04
	Max.–Min.	5.8–0	2.9–0	6.2–0	6.2–0	6.9–0	4.6–0	9.1–0	3.7–0	7.9–0	4.9–0	8.4–0	8.9–0	5.9–0.0	4.9–0
III	Mn.	1.8	0.76	0.94	0.53	1.22	0.77	1.42	0.83	1.28	1.52	1.91	2.08	2.37	2.18
	Max.–Min.	4.9–0	4.8–0	8.9–0	6.2–0	11.9–0	4.1–0	14.0–0	5.5–0	11.9–0	8.9–0	9.9–0	12.9–0	11.9–0.0	10.9–0
aVR	Mn.	9.60	8.83	9.05	8.92	5.87	7.36	7.62	0.80	4.50	4.52	4.06	3.67	2.70	4.66
	Max.–Min.	14.4–0	11.2–5.6	21.7–0.17	21.0–4.0	15.9–0	11.6–3.6	14.0–0.7	13.8–0.1	17.9–0	15.9–0	19.9–0	13.9–0	11.9–0	15.9–0
aVL	Mn.	3.27	1.82	2.98	2.12	2.39	1.43	2.44	1.26	1.83	1.50	1.76	1.21	1.12	0.88
	Max.–Min.	9.5–0	7.1–0	14.2–0	12.0–0	15.9–0	7.2–0	8.8–0	8.2–0	8.9–0	5.9–0	7.9–0	5.9–0	5.9–0.0	3.4–0
aVF	Mn.	1.42	0.62	1.20	0.53	1.27	0.53	1.04	0.54	1.03	0.90	1.66	1.45	1.70	1.35
	Max.–Min.	5.84–0	2.2–0	12.0–0	3.2–0	12.9–0	2.0–0	7.1–0	3.8–0	8.9–0	4.9–0	22.9–0	10.9–0	6.9–0.0	5.9–0
V_4R	Mn.	2.99	3.3	3.99	2.73	3.47	3.35	4.15	2.70	3.82	3.43	4.05	3.68	4.45	3.16
	Max.–Min.	8.05–0	8.4–0	11.0–0	14.0–0	8.9–0	10.2–0	18.6–0	9.4–0	13.9–0	10.9–0	10.9–0	20.9–0	9.9–0.0	6.9–0
		(49)[a]													
V_1	Mn.	10.85	9.2	9.7	9.59	9.71	8.90	9.42	9.62	9.58	9.07	8.83	8.46	8.87	8.47
	Max.–Min.	22.7–0	22.0–0	27.4–0	24.5–2.3	1.9–0	15.5–0	24.2–0	23.3–1.1	21.9–0	21.9–0	19.9–0	21.9–0	17.9–0	17.9–0
V_2	Mn.	18.04	15.8	20.05	14.06	15.37	13.38	13.97	12.26	14.14	13.25	12.55	13.08	13.03	12.38
	Max.–Min.	42.5–0	26.5–6.5	55.1–0	34.2–3.6	37.9–0	28.8–2.1	27.0–0	30.7–1.2	31.9–0	27.9–0	29.9–2.0	33.9–0	29.9–0.0	27.9–2.0
V_3	Mn.	13.37	8.8	12.62	7.71	10.26	7.42	11.31	8.24	11.40	9.03	10.35	10.85	11.95	12.05
	Max.–Min.	25.0–0	17.0–0	34.0–0	16.0–0	27.9–0	16.9–0	29.5–0	19.4–0	27.9–2.0	27.9–0	27.9–0	37.9–0	27.9–0	29.9–0
V_4	Mn.	7.22	4.5	6.46	6.42	5.54	4.00	6.43	3.91	6.82	5.85	7.01	6.36	7.66	7.38
	Max.–Min.	45.0–0	11.5–0	22.6–0	11.1–0	23.9–0	12.3–0	20.0–0	12.7–0	21.9–0	21.9–0	27.9–0	19.9–0	27.9–0	21.9–0
V_5	Mn.	2.92	2.7	2.28	1.86	2.02	1.73	2.46	1.46	2.53	2.06	2.75	2.80	2.26	2.50
	Max.–Min.	40.0–0	5.1–0	10.0–0	6.4–0	14.9–0	5.55–0	11.5–0	6.3–0	18.9–0	10.9–0	16.9–0	23.9–0	7.9–0	10.9–0
V_6	Mn.	1.32	0.62	0.99	0.54	0.85	0.51	0.64	0.39	0.91	0.72	0.99	0.94	0.80	0.84
	Max.–Min.	17.8–0	3.3–0	10.7–0	4.5–0	6.9–0	3.64–0	6.9–0	2.6–0.4	7.9–0	3.9–0	9.9–0	8.9–0	5.9–0	4.9–0

Mn., mean; Max., maximum; Min., minimum

[a]Number of cases in which measurement was made.

Table A1.42

ST segment "J" point (in millimeters) in normal Japanese of various age-groups

		15–29		20–29		30–39		40–49		50–59		60–69		70+	
		Male $n = 60$	Female $n = 39$	Male $n = 165$	Female $n = 109$	Male $n = 119$	Female $n = 94$	Male $n = 153$	Female $n = 82$	Male $n = 135$	Female $n = 102$	Male $n = 101$	Female $n = 79$	Male $n = 48$	Female $n = 42$
I	Mn., SD	0.03, 0.21	0.1, 0.21	0.01, 0.25	−0.06, 0.23	0.08, 0.17	−0.01, 0.22	0.03, 0.17	0.08, 0.24	0.00, 0.22	−0.00, 0.18	0.06, 0.22	0.03, 0.10	0.00, 0.27	0.00, 0.09
	Max.–Min.	0.9 to −0.8	0.7 to −0.8	0.9 to −1.0	0.05 to −0.8	0.7 to −0.8	0.9 to −0.6	0.5 to −0.8	1.3 to −0.8	1.0 to −1.2	0.5 to −0.8	1.1 to −1.0	0.7 to −0.4	0.9 to −1.0	0.3 to −0.4
II	Mn., SD	0.02, 0.55	0.03, 0.45	0.13, 0.39	0.77, 0.42	0.16, 0.38	0.00, 0.23	0.03, 0.36	0.12, 0.46	0.07, 0.33	0.03, 0.26	0.05, 0.41	0.06, 0.08	0.00, 0.27	0.09, 0.41
	Max.–Min.	2.9 to −1.01	2.4 to −0.6	1.7 to −1.0	2.1 to −0.2	1.4 to −1.0	0.7 to −0.8	1.1 to −1.6	2.5 to −0.8	1.7 to −1.0	1.1 to −1.0	1.5 to −1.6	1.1 to −0.6	0.7 to −1.0	1.9 to −0.5
III	Mn., SD	0.05, 0.28	0.01, 0.30	0.07, 0.35	0.03, 0.40	0.15, 0.35	−0.01, 0.30	0.00, 0.41	0.07, 0.33	0.03, 0.28	0.06, 0.23	0.05, 0.38	0.09, 0.18	0.04, 0.17	0.03, 0.21
	Max.–Min.	1.1 to −1.2	1.5 to −0.6	1.9 to −0.8	1.3 to −0.8	1.9 to −0.8	1.3 to −1.0	1.1 to −1.2	1.1 to −1.0	1.0 to −1.1	0.9 to −1.0	1.1 to −2.0	1.1 to −0.6	0.5 to −0.4	1.9–0.5
aVR	Mn., SD	0.06, 0.41	−0.05, 0.20	0.00, 0.38	−0.05, 0.52	0.04, 0.34	0.01, 0.28	0.06, 0.29	0.04, 0.31	0.11, 0.28	0.04, 0.26	0.03, 0.29	0.04, 0.19	0.00, 0.23	−0.08, 0.19
	Max.–Min.	1.3 to −0.8	0.5 to −0.6	2.1 to −1.2	1.3 to −1.4	1.1 to −1.6	1.1 to −1.0	1.0 to −0.8	1.1 to −1.4	1.1 to −1.2	0.9 to −0.8	1.9 to −0.6	0.9 to −0.8	0.7 to −0.6	0.3 to −0.7
aVL	Mn., SD	−0.02, 0.23	0.03, 0.17	0.03, 0.31	0.20, 0.35	0.04, 0.19	−0.02, 0.27	0.06, 0.21	0.031, 0.22	−0.04, 0.21	0.05, 0.13	0.08, 0.17	0.02, 0.14	0.03, 0.13	0.02, 0.05
	Max.–Min.	0.9 to −0.8	0.5 to −0.6	1.3 to −1.2	1.5 to −0.6	0.9 to −2.0	1.0 to −1.8	1.1 to −1.0	0.7 to −1.2	0.7 to −1.0	0.7 to −0.6	1.3 to −0.6	0.3 to −0.6	0.3 to −0.6	0.2 to −1.0
aVF	Mn., SD	0.01, 0.35	0.3, 0.24	0.10, 0.41	0.53, 0.35	0.16, 0.35	0.00, 0.36	0.07, 0.37	0.09, 0.31	0.07, 0.33	0.04, 0.20	0.07, 0.37	0.09, 0.19	−0.03, 0.31	0.03, 0.30
	Max.–Min.	1.1 to −1.0	1.1 to −0.6	1.8 to −1.0	1.5 to −0.4	2.1 to −1.0	1.5 to −1.0	1.5 to −1.2	1.1 to −0.6	1.6 to −1.0	0.9 to −1.0	1.7 to −1.6	1.1 to −0.6	1.9 to −1.0	1.3 to −0.5
V_4R	Mn., SD	0.23, 0.36	0.00, 0.23	0.02, 0.24	0.17, 0.26	0.11, 0.10	−0.02, 0.21	0.08, 0.24	−0.02, 0.21	0.02, 0.19	0.07, 0.21	0.13, 0.14	0.09, 0.18	0.16, 0.10	0.08, 0.10
	Max.–Min.	1.1 to −0.5	1.1 to −0.6	1.1 to −0.6	1.1 to −0.6	0.9 to −0.2	1.1 to −0.6	1.1 to −1.4	0.9 to −0.8	1.7 to −0.8	1.1 to −1.0	0.9 to −0.4	0.9 to −2.0	0.7 to −0.4	0.9–0.1
		(32)[a]	(36)[a]	(114)[a]	(94)[a]	(87)[a]	(68)[a]	(107)[a]	(74)[a]	(93)[a]	(78)[a]	(74)[a]	(56)[a]	(23)[a]	(17)[a]
V_1	Mn., SD	0.28, 0.44	0.12, 0.35	0.44, 0.52	0.36, 0.29	0.38, 0.44	0.09, 0.25	0.39, 0.50	0.09, 0.38	0.39, 0.40	0.18, 0.25	0.08, 0.57	0.19, 0.30	0.46, 0.35	0.32, 0.39
	Max.–Min.	1.3 to −0.8	1.7 to −0.8	2.5 to −1.0	1.1 to −0.6	1.7 to −1.6	0.9 to −0.6	2.9 to −1.4	1.1 to −1.2	1.5 to −1.2	1.1 to −0.6	2.1 to −1.0	1.1 to −1.0	1.1 to −0.5	1.5 to −0.0
V_2	Mn., SD	0.94, 0.87	0.26, 0.48	0.93, 0.89	0.47, 0.31	0.70, 0.30	0.28, 0.47	0.88, 0.86	0.32, 0.50	0.69, 0.79	0.40, 0.47	0.75, 0.48	0.51, 0.54	0.83, 0.63	0.62, 0.43
	Max.–Min.	3.3–0	1.1 to −1.4	4.1 to −1.0	1.3 to −0.4	4.1 to −1.4	2.7 to −0.8	6.1 to −0.6	2.1 to −1.2	2.8 to −2.6	1.9 to −1.0	3.3 to −0.5	2.7 to −1.0	2.5 to −0.5	1.5 to −0.7
V_3	Mn., SD	1.00, 0.57	0.47, 0.57	0.87, 0.99	0.46, 0.29	0.67, 0.61	0.18, 0.44	0.76, 1.22	0.11, 0.93	0.56, 1.25	0.30, 0.62	0.60, 0.37	0.22, 0.40	0.59, 0.66	0.43, 0.57
	Max.–Min.	2.7 to −1.5	2.3 to −0.8	5.5 to −0.6	1.3 to −0.4	2.7 to −1.2	1.9 to −1.0	9.0 to −6.8	2.5 to −6.3	2.2 to −1.5	2.1 to −2.0	2.1 to −1.2	1.5 to −1.2	2.3 to −0.8	2.0 to −0.7
V_4	Mn., SD	0.51, 0.72	0.10, 0.47	0.57, 0.90	0.37, 0.23	0.59, 0.79	0.04, 0.44	0.37, 0.59	0.01, 0.38	0.23, 0.50	0.04, 0.51	0.29, 0.55	0.03, 0.43	0.09, 0.57	0.06, 0.46
	Max.–Min.	2.9 to −1.0	2.3 to −0.8	6.4 to −1.0	1.1 to −0.4	5.1 to −1.6	1.7 to −0.8	3.3 to −2.4	2.1 to −1.0	1.8 to −1.4	2.5 to −1.2	1.9 to −2.0	1.5 to −1.6	1.7 to −1.6	1.0 to −1.0
V_5	Mn., SD	0.35, 0.50	0.03, 0.31	0.26, 0.57	0.34, 0.22	0.25, 0.49	0.09, 0.33	0.19, 0.41	0.09, 0.54	0.05, 0.38	0.02, 0.35	0.09, 0.37	−0.02, 0.32	−0.02, 0.39	0.03, 0.25
	Max.–Min.	2.3 to −0.4	1.1 to −0.8	3.0 to −1.0	1.1 to −0.2	2.9 to −1.2	1.3 to −0.8	1.9 to −2.2	2.5 to −1.0	1.5 to −1.1	1.5 to −1.0	1.5 to −1.0	1.1 to −1.0	1.1 to −1.2	0.6 to −0.9
V_6	Mn., SD	0.24, 0.50	0.0, 0.14	0.24, 0.50	0.34, 0.25	0.21, 0.41	0.13, 0.37	0.14, 0.26	0.20, 0.54	0.10, 0.32	0.06, 0.26	0.09, 0.33	0.09, 0.19	0.15, 0.30	0.05, 0.27
	Max.–Min.	2.2–0.4	0.7 to −0.4	5.1 to −0.6	1.3 to −0.2	2.5 to −2.1	1.5 to −1.0	1.1 to −1.2	2.5 to −0.6	1.5 to −1.0	1.9 to −0.6	1.5 to −1.0	1.1 to −0.6	0.7 to −1.0	1.0 to −0.7

Mn., SD, standard deviation; Max., maximum; Min., minimum

[a]Number of cases in which measurement was made.

Table A1.43

Amplitude of T wave (in millimeters) in normal Japanese of various age-groups

		15–19		20–29		30–39		40–49		50–59		60–69		70+	
		Male $n = 61$	Female $n = 38$	Male $n = 165$	Female $n = 109$	Male $n = 119$	Female $n = 94$	Male $n = 153$	Female $n = 82$	Male $n = 135$	Female $n = 102$	Male $n = 101$	Female $n = 79$	Male $n = 48$	Female $n = 42$
I	Mn., SD	3.21, 1.13	2.50, 0.88	2.80, 1.11	2.42, 0.98	2.42, 0.93	2.30, 0.79	2.42, 0.95	1.87, 0.89	2.23, 0.93	2.08, 0.81	2.07, 0.96	1.83, 0.96	1.54, 1.74	1.83, 0.78
	Max.–Min.	5.9–1.0	4.9–0.0	7.9–0.0	5.9–0.0	5.9–0.0	5.9 to −1.0	6.9–0.0	3.9–0.0	5.9–0.0	4.9–0.0	4.9–0.0	4.9–0.0	2.9 to −1.0	4.9 to −0.0
II	Mn., SD	4.65, 1.34	2.88, 1.04	4.26, 1.56	2.98, 1.40	3.79, 1.47	2.91, 1.15	3.63, 1.39	2.62, 1.02	3.40, 1.44	2.47, 1.22	3.18, 1.45	2.27, 1.02	2.56, 1.17	2.12, 0.87
	Max.–Min.	7.9–1.0	4.9–0.0	9.9–0.0	7.9 to −3.0	9.9–0.0	6.1–0.0	7.9–0.0	5.9–0.0	8.9 to −1.0	5.9 to −1.0	7.9 to −2.0	7.9–0.0	5.9–0.0	3.9–0.0
III	Mn., SD	1.93, 0.99	0.42, 0.95	1.78, 1.30	0.58, 1.10	1.65, 1.33	0.70, 0.76	1.45, 1.54	0.75, 0.99	1.40, 1.10	0.56, 0.90	1.46, 1.48	0.54, 1.07	1.33, 1.14	0.62, 0.78
	Max.–Min.	4.9 to −1.0	3.9 to −3.0	5.9 to −2.0	4.9 to −2.0	5.9 to −2.0	2.9 to −2.0	6.9 to −3.0	4.9 to −2.0	5.9 to −2.0	2.9 to −2.0	5.9 to −3.0	3.9 to −2.0	4.9 to −1.0	2.9 to −2.0
aVR	Mn., SD	−3.68, 1.58	2.42, 0.83	−2.37, 1.38	−0.3, 0.95	−2.74, 1.46	−2.50, 0.87	−2.00, 2.28	−1.55, 1.53	−2.31, 1.42	−1.88, 1.10	−2.16, 1.43	−1.58, 1.16	−0.16, 1.17	−1.12, 1.48
	Max.–Min.	−0.1 to −70	−0.1 to −5.0	−0.1 to −12.0	−0.1 to −5.0	3.9 to −7.6	−0.1 to −5.0	4.9 to −6	3.9 to −4.0	4.9 to −6.0	2.9 to −5.0	2.9 to −5.0	2.9 to −4.0	1.9 to −5.0	2.9 to −4.0
aVL	Mn., SD	1.05, 0.83	1.26, 0.83	0.89, 1.03	1.34, 1.02	0.70, 1.00	1.10, 0.72	0.87, 1.14	0.85, 0.66	0.73, 0.89	0.96, 0.72	0.52, 1.15	0.27, 0.84	0.52, 0.71	0.90, 0.85
	Max.–Min.	3.0 to −1.0	3.9 to −1.0	5.9 to −2.0	6.9 to −2.0	3.9 to −2.0	3.9 to −1.0	7.9 to −2.0	2.9 to −2.0	3.9 to −2.0	2.9 to −1.0	3.9 to −3.0	3.9 to −2.0	1.9 to −2.0	3.9 to −2.0
aVF	Mn., SD	3.18, 1.28	1.73, 0.95	2.98, 1.56	1.88, 1.20	2.62, 1.46	1.86, 0.93	2.45, 1.58	1.72, 0.90	2.21, 1.26	1.47–0.86	2.43, 1.37	1.41, 0.75	1.75, 1.23	1.23, 1.06
	Max.–Min.	5.9–0.0	4.9 to −1.0	7.9 to −6.0	6.9 to −2.0	6.9 to −4.0	4.9 to −1.0	12.0 to −1.0	4.9 to −0.2	6.9 to −2.0	4.9 to −1.0	6.9 to −2.0	3.9 to −1.0	4.9 to −1.0	3.9 to −4.0
V_4R	Mn., SD	−0.26, 1.10	−0.55, 0.78	0.17, 1.25	−0.68, 0.95	0.35, 1.02	0.67, 1.07	−0.01, 1.26	−0.49, 0.79	−0.02, 0.96	−0.42, 0.81	−0.26, 0.99	−0.55, 0.84	0.00, 1.16	−0.32, 0.57
	Max.–Min.	1.9 to −4.0	1.9 to −2.0	3.9 to −4.0	1.9 to −3.0	2.9 to −3.0	11.9 to −3.0	3.9 to −5.0	2.9 to −2.0	3.9 to −3.0	1.9 to −2.0	1.9 to −3.0	2.9 to −3.0	2.9 to −2.0	0.9 to −2.0
		(50)[a]	(37)[a]	(117)[a]	(96)[a]	(86)[a]	(67)[a]	(108)[a]	(74)[a]	(94)[a]	(78)[a]	(74)[a]	(56)[a]	(23)[a]	(17)[a]
V_1	Mn., SD	1.42, 2.55	−0.15, 1.38	1.99, 1.81	−0.45, 1.65	1.77, 2.02	−0.19, 1.34	1.52, 4.21	0.53, 1.28	1.35, 2.10	−0.13, 1.45	1.20, 1.95	0.08, 1.45	1.63, 1.67	0.29, 1.86
	Max.–Min.	7.9 to −5.0	2.9 to −4.0	6.9 to −3.0	4.9–4.0	8.9 to −2.0	2.9 to −3.0	10.9–5.0	2.9 to −4.0	7.9 to −5.0	4.9 to −4.0	5.9 to −3.0	4.9 to −3.0	5.9 to −3.0	6.9 to −4.0
V_2	Mn., SD	8.38, 3.67	3.20, 2.90	8.14, 3.65	2.96, 2.16	6.90, 3.57	3.17, 2.03	6.48, 3.04	2.78, 1.85	6.43, 3.01	3.35, 2.25	5.38, 2.90	3.08, 2.19	5.04, 2.80	3.24, 2.42
	Max.–Min.	19.9 to −1.0	10.9 to −3.0	23.9–0.0	8.9 to −4.0	18.9 to −2.0	8.9 to −2.0	13.9 to −1.0	7.9 to −2.0	13.9 to −3.0	9.9 to −3.0	11.9 to −5.0	10.9 to −2.0	11.9–0.0	9.9 to −3.0
V_3	Mn., SD	9.92, 3.52	4.50, 2.24	8.86, 3.63	4.05, 2.16	7.58, 3.37	4.21, 2.17	7.92, 3.03	3.78, 2.14	7.95, 2.97	4.43, 2.52	7.27, 2.98	4.36, 2.37	7.08, 2.98	4.62, 2.41
	Max.–Min.	20.9–0.0	10.9–0.0	22.9–2.0	9.9 to −0.0	20.9–1.0	10.9–0.0	16.9–0.0	9.9 to −2.0	18.9–2.0	12.9 to −3.0	15.9 to −2.0	11.9 to −1.0	14.9–2.0	8.9 to −1.0
V_4	Mn., SD	9.38, 3.75	5.15, 2.08	8,72, 3.93	4.80, 1.84	7.58, 3.89	4.87, 2.46	7.61, 3.35	4.36, 2.14	7.78, 2.93	4.32, 2.74	7.07, 3.27	4.22, 1.99	6.19, 2.63	4.33, 2.58
	Max.–Min.	16.9–1.0	10.9–0.0	24.9–1.0	10.9–0.0	40.9–1.0	13.0–0.0	26.9–1.0	11.9 to −1.0	15.9 to −2.0	13.9 to −2.0	17.4 to −5.0	12.9–0.0	13.9 to −4.0	11.9–0.0
V_5	Mn., SD	8.53, 3.58	4.45, 1.70	5.30, 2.82	4.25, 1.65	5.44, 2.86	4.45, 2.01	5.54, 2.43	4.03, 1.95	5.50, 2.43	3.11–1.66	5.48, 2.51	3.32, 1.77	4.42, 2.49	3.41, 1.66
	Max.–Min.	15.9–0.0	8.9–0.0	16.9–1.0	10.9 to −2.0	14.9–0.0	11.0–0.0	14.9–1.0	11.9–6	14.9 to −2.0	9.9–0.0	12.9–0.0	9.9–0.0	10.9 to −2.0	8.9–0.0
V_6	Mn., SD	4.96, 2.21	3.69, 1.37	4.48, 2.15	3.43, 2.41	3.80, 1.85	3.55, 1.45	3.87, 1.75	3.12, 1.42	3.27, 1.60	2.98, 1.33	3.96, 2.06	2.45, 1.00	2.92, 1.54	2.34, 1.11
	Max.–Min.	11.9–0.0	6.9–0.0	14.9–0.0	8.9–0.0	8.9–0.0	7.9–1.0	9.9–0.0	7.9–0.0	8.9 to −1.0	7.9–0.0	13.9–0.0	4.9–0.0	6.9 to −1.0	4.9–0.0
															(38)[a]

Mn., mean; SD, standard deviation; Max., maximum; Min., minimum

[a]Number of cases in which measurement was made.

A1.4 Normal Limits of Right-Sided Chest Leads in Caucasians

The tables in this section were derived from 109 subjects with no evidence of heart disease (➋ Tables A1.44–A1.51b). The data is reproduced from: Andersen HR, Nielsen D, Hansen LG. The normal right chest electrocardiogram. *J. Electrocardiol.* 1987;**20**: 27–32 with the permission of Churchill Livingstone, New York.

Table A1.44
R-wave amplitudes (millimeters)

	Median	95% fractile values	Mean $\bar{x}$	Range	Number *n*
V_3R	1.5	0.4–3.9	1.7	0.1–4.3	107
V_4R	1.0	0.3–3.3	1.2	0.2–4.2	101
V_5R	1.0	0.3–2.4	1.0	0.2–5.0	79
V_6R	0.7	0.2–2.6	0.9	0.2–5.1	59
V_7R	0.5	0.3–3.6	1.0	0.3–5.6	37

Table A1.45
S-wave amplitudes (millimeters)

	Median	95% fractile values	Mean $\bar{x}$	Range	Number *n*
V_3R	5.1	1.7–11.4	5.7	0.9–13.5	108
V_4R	3.6	1.1–8.0	3.9	0.9–9.7	104
V_5R	3.6	1.0–7.7	3.5	0.5–8.8	89
V_6R	2.5	0.7–6.4	2.9	0.4–7.3	70
V_7R	2.7	0.8–5.4	2.8	0.7–6.8	45

Table A1.46
Secondary r-wave (qr, rSr′) amplitudes (millimeters)

	Median	95% fractile values	Mean $\bar{x}$	Range	Number *n*
V_3R	0.8	0.5–1.5	0.9	0.4–1.5	6
V_4R	0.7	0.2–1.8	0.8	0.2–1.8	11
V_5R	0.8	0.3–2.5	1.0	0.2–2.5	36
V_6R	0.9	0.3–3.6	1.3	0.2–4.2	50
V_7R	0.0	0.3–3.6	1.2	0.1–4.8	71

Table A1.47
Q-wave amplitudes (millimeters)

	Median	95% fractile values	Mean $\bar{x}$	Range	Number *n*
V_3R					2
V_4R	2.8	1.5–8.1	4.4	1.5–8.1	8
V_5R	2.5	0.8–5.3	2.7	0.7–5.9	30
V_6R	1.9	1.0–4.7	2.2	0.9–4.9	47
V_7R	1.8	0.9–3.9	2.0	0.6–4.1	71

Table A1.48

r + r′ amplitudes (millimeters)

	Median	95% fractile values	Mean $\bar{x}$	Range	Number *n*
V_3R	1.5	0.3–3.9	1.7	0.0–4.3	109
$\sum_{V_3R}^{V_4R} r + r'$	2.5	0.5–6.8	2.8	0.3–8.1	109
$\sum_{V_3R}^{V_5R} r + r'$	3.5	0.5–9.1	3.8	0.3–13.3	109
$\sum_{V_3R}^{V_6R} r + r'$	4.0	0.8–11.3	4.9	0.4–15.7	109
$\sum_{V_3R}^{V_7R} r + r'$	4.9	1.4–14.8	6.0	0.4–20.5	109

Table A1.49

J-point deviation (millimeters)

	Median	95% fractile values	Mean $\bar{x}$	Range	Number *n*
V_3R	−0.1	0.4 to −0.5	−0.1	−0.6 to +0.5	109
V_4R	−0.1	0.3 to −0.6	−0.1	−0.9 to +0.5	104
V_5R	0.0	0.3 to −0.7	−0.1	−0.8 to +0.5	107
V_6R	0.0	0.5 to −0.6	0.0	−0.8 to +0.5	108
V_7R	0.0	0.5 to −0.5	0.0	−0.9 to +0.6	103

Table A1.50

ST-segment deviation (millimeters) 40 ms after last QRS deflection

	Median	95% fractile values	Mean $\bar{x}$	Range	Number *n*
V_3R	0.1	0.6 to −0.5	0.1	−1.2 to +0.8	109
V_4R	0.1	0.5 to −0.4	0.0	−1.4 to +0.6	109
V_5R	0.1	0.5 to −0.4	0.1	−0.8 to +0.6	109
V_6R	0.0	0.5 to −0.3	0.1	−0.5 to +1.0	109
V_7R	0.1	0.4 to −0.4	0.1	−0.5 to +0.9	109

Table A1.51(a)

ST-segment deviation (millimeters) 80 ms after last QRS deflection

	Median	95% fractile values	Mean $\bar{x}$	Range	Number *n*
V_3R	0.3	0.9 to −0.3	0.3	−0.4 to +1.0	109
V_4R	0.2	0.6 to −0.3	0.2	−0.5 to +0.9	109
V_5R	0.1	0.5 to −0.3	0.1	−0.7 to +1.0	109
V_6R	0.0	0.5 to −0.3	0.0	−0.6 to +0.6	109
V_7R	0.0	0.4 to −0.3	0.0	−0.8 to +0.5	109

Table A1.51(b)

Q-wave duration (ms)

	Median	95% fractile values	Mean $\bar{x}$	Range	Number *n*
V_3R					2
V_4R	53	20–80	50	20–81	8
V_5R	34	20–64	40	17–108	30
V_6R	34	17–74	36	17–81	47
V_7R	34	17–58	34	14–64	71

A1.5 Normal Limits of the Three-Orthogonal-Lead ECG

A1.5.1 Normal Limits in Males

Some of the earliest work on the three-orthogonal-lead ECG was undertaken in the laboratory of Pipberger. His group studied 510 men, including whites and blacks, and with the use of computer methods, analyzed the ECGs automatically. ➋ Tables A1.52–A1.61 are reproduced from Draper H W et al. The corrected orthogonal electrocardiogram and vectorcardiogram in 510 normal men (Frank-lead system). *Circulation* 1964;**30**: 853–864 with the permission of the American Heart Association, Dallas, Texas.

Table A1.52

Measurements of P, QRS and T in scalar orthogonal leads *X*, *Y* and *Z*. The mean and standard deviation of each item is shown on the upper line. The second line indicates the limits of a 96 percentile range. A similar pattern is used for A1.52 to A1.61. Figures in parentheses in A1.52 show the actual number of measurements taken (e.g., Q waves in lead *X* were present in 306 cases only). Results not followed by a number in parentheses were obtained from the total series. All wave durations are based on the total QRS duration derived from the three leads. The earliest or last deflection in any one of the simultaneously recorded leads indicates onset or end of this complex

Item	*X*		*Y*		*Z*
P amplitude (mV)	0.06 ± 0.03		0.11 ± 0.07		0.03 ± 0.0
	0.03–0.12		0.05–0.23		−0.06–0.10
Q amplitude (mV)	0.10 ± 0.05	(306)	0.10 ± 0.07	(333)	0.41 ± 0.21
	0.03–0.25		0.01–0.29		0.09–0.93
Q duration (s)	0.019 ± 0.004	(306)	0.021 ± 0.005	(333)	0.033 ± 0.007
	0.012–0.028		0.008–0.032		0.020–0.048
R amplitude (mV)	1.17 ± 0.37		1.03 ± 0.41		0.93 ± 0.35
	0.51–1.97		0.35–1.95		0.36–1.79
R duration (s)	0.051 ± 0.016		0.061 ± 0.019		0.059 ± 0.010
	0.028–0.88		0.028–0.100		0.032–0.080
S amplitude (mV)	0.27 ± 0.15	(407)	0.18 ± 0.12	(274)	
	0.06–0.68		0.03–0.49		
S duration (s)	0.039 ± 0.008	(407)	0.035 ± 0.010	(274)	
	0.024–0.056		0.020–0.056		
T amplitude (mV)	0.27 ± 0.13		0.22 ± 0.13		−0.28± 0.13
	0.06–0.56		−0.11–0.48		−0.58 to −0.06
Q/R amplitude ratio	0.08 ± 0.04	(306)	0.10 ± 0.05	(333)	0.49 ± 0.35
	0.02–0.21		0.01–0.22		0.10–1.21
R/S amplitude ratio	5.74 ± 4.62	(407)	9.07 ± 9.39	(274)	
	1.40–19.25		1.11–38.51		
R/T amplitude ratio	5.44 ± 4.01		5.18 ± 3.25		4.02 ± 2.66
	1.63–20.16		1.67–13.79		1.12–12.42
Time from beginning of QRS to largest R peak (s)	0.037 ± 0.005		0.039 ± 0.005		0.049 ± 0.006
	0.028–0.048		0.028–0.052		0.036–0.064

Table A1.53

Time intervals obtained from three simultaneously recorded leads. The method of measurement is the same as indicated for Table A1.52

Item	Measurement (s)
P duration	0.102+0.016
	0.068–0.140
PR interval	0.153 ± 0.023
	0.112–0.204
PR segment	0.051 ± 0.019
	0.021–0.096
QRS duration	0.093 ± 0.009
	0.076–0.112
QT interval (uncorrected)	0.367 ± 0.034
	0.312–0.448

Table A1.54

Maximal P, QRS and T vectors in the frontal, sagittal and horizontal planes together with spatial amplitude and orientation. The vectors in the plane projections were obtained from *XY*, *YZ* and *XZ* leads, respectively, and do not therefore represent projections of the spatial maximal vectors onto these planes

Item	Maximal P vector	Maximal QRS vector	Maximal T vector
Frontal plane			
Amplitude (mV)	0.18 ± 0.06	1.57 ± 0.42	0.36 ± 0.14
	0.08–0.31	0.81–2.53	0.12–0.69
Direction (°)	67 ± 18	41 ± 14	40 ± 20
	22–91	14–71	4–74
Sagittal plane			
Amplitude (mV)	0.17 ± 0.06	1.32 ± 0.45	0.36 ± 0.13
	0.06–0.31	0.60–2.42	0.13–0.67
Direction (°)	87 ± 23	48 ± 30	142 ± 23
	54–129	343–114	93–180
Horizontal plane			
Amplitude (mV)	0.09 ± 0.03	1.39 ± 0.36	0.40 ± 0.14
	0.04–0.14	0.74–2.19	0.15–0.72
Direction (°)	349 ± 41	327 ± 34	46 ± 19
	285–91	245–29	8–83
Spatial amplitude (m)V	0.18 ± 0.06	1.73 ± 0.44	0.46 ± 0.16
	0.09–0.32	0.92–2.75	0.18–0.82
Spatial orientation			
Azimuth (°)	342 ± 38	331 ± 27	44 ± 19
	277–75	263–23	4–79
Elevation (°)	63 ± 17	35 ± 13	29 ± 13
	20–86	7–60	2–58

Table A1.55

Quantitative analysis of early QRS vectors. The upper row provides mean ±SD and the lower row gives a 96% range

Instantaneous vectors[a]	Scalar amplitude (mV)			Planar direction (°)			Spatial magnitude and orientation		
	X	*Y*	*Z*	Frontal	Sagittal	Horizontal	Amplitude (mV)	Azimuth (°)	Elevation (°)
0.01 s after	−0.04± 0.04	−0.03± 0.06	−0.11± 0.06	210 ± 61	189 ± 36	110 ± 34	0.14 ± 0.06	110 ± 34	−10± 27
QRS onset	−0.14–0.07	−0.13–0.08	−0.25–0.01	59–330	79–242	17–168	0.05–0.29	17–168	−59–58
0.02 s after	0.05 ± 0.14	−0.01± 0.12	−0.31± 0.15	325 ± 87	180 ± 25	81 ± 26	0.37 ± 0.15	81 ± 26	−1± 20
QRS onset	−0.19–0.38	−0.25–0.29	0.68 to −0.06	162–136	117–220	23–124	0.12–0.75	23–124	−39–42
0.03 s after	0.56 ± 0.27	0.35 ± 0.25	−0.20± 0.32	29 ± 23	120 ± 40	21 ± 30	0.79 ± 0.30	21 ± 30	26 ± 15
QRS onset	0.06–1.19	−0.06–0.97	−0.89–0.48	350–59	41–186	319–79	0.29–1.52	319–79	−6–53
0.04 s after	1.05 ± 0.37	0.86 ± 0.37	0.34 ± 0.46	40 ± 16	72 ± 27	343 ± 25	1.51 ± 0.42	343 ± 25	35 ± 11
QRS onset	0.33–1.79	−0.24–1.79	−0.59–1.26	14–66	28–139	294–35	0.78–2.53	294 ± 35	13–59
0.05 s after	0.65 ± 0.51	0.74 ± 0.47	0.77 ± 0.40	52 ± 36	42 ± 24	307 ± 28	1.40 ± 0.50	307 ± 28	31 ± 16
QRS onset	−0.27–1.76	−0.09–1.74	−0.01–1.67	343–147	351–89	248–1	0.56–2.57	248–1	−5–59

[a] Scalar components, plane projections, spatial magnitude and orientation of five initial instantaneous QRS vectors taken 0.01 s intervals after the onset of QRS. As in all other tables the beginning of QRS is taken at the earliest deflection in any one of the simultaneously recorded scalar leads. In this as in all consecutive tables the ranges for azimuth are to be followed in clockwise direction.

Table A1.56

Quantitative analysis of late QRS vectors. The upper row provides mean ±SD and the lower row gives a 96% range

Instantaneous vectors[a]	Scalar amplitude (mV)			Planar direction (°)			Spatial magnitude and orientation		
	X	*Y*	*Z*	Frontal	Sagittal	Horizontal	Amplitude (mV)	Azimuth (°)	Elevation (°)
End of QRS or	0.01 ± 0.03	0.03 ± 0.04	−0.07 ± 0.04	79 ± 74	160 ± 30	83 ± 34	0.09 ± 0.04	83 ± 34	19 ± 27
J point	−0.06–0.08	−0.06–0.10	−0.17–0.00	279–241	95–227	9–153	0.03–0.19	9–153	−42–70
0.01 S before	−0.01 ± 0.06	0.01 ± 0.08	0.01 ± 0.08	156 ± 89	77 ± 97	197 ± 94	0.11 ± 0.05	197 ± 94	6 ± 45
End of QRS	−0.13–0.10	−0.13–0.15	−0.15–0.16	4–315	267–250	26–1	0.03–0.24	26–1	−72–79
0.02 S before	−0.09 ± 0.12	0.00 ± 0.16	0.24 ± 0.19	173 ± 72	358 ± 53	247 ± 52	0.34 ± 0.17	247 ± 52	−2 ± 34
End of QRS	−0.36–0.12	−0.30–0.33	−0.10–0.67	31–305	242–122	96–30	0.07–0.74	96–30	−74–63
0.03 S before	−0.03 ± 0.32	0.18 ± 0.33	0.60 ± 0.28	128 ± 74	12 ± 32	263 ± 32	0.76 ± 0.31	263 ± 32	10 ± 26
End of QRS	−0.51–0.95	−0.39–1.1	0.05–1.14	353–268	290–68	202–338	0.25–1.51	202–338	−54–55
0.04 S before	0.46 ± 0.59	0.61 ± 0.49	0.77 ± 0.37	69 ± 55	35 ± 29	295 ± 36	1.30 ± 0.48	295 ± 36	26 ± 20
End of QRS	−0.54–1.70	−0.31–1.66	−0.04–1.52	341–214	335–93	230–1	0.43–2.27	230–1	−26–58
0.05 S before	0.87 ± 0.48	0.81 ± 0.41	0.49 ± 0.50	46 ± 27	65 ± 32	331 ± 33	1.44 ± 0.48	331 ± 33	34 ± 13
End of QRS	−0.28–1.75	0.05–1.80	0.47–1.56	13–131	7–148	251–28	0.57–2.53	251–28	2–58

[a]Scalar components, planar projections, spatial magnitude and orientation of five terminal instantaneous QRS vectors taken in retrograde fashion from the end of QRS at 0.01 s intervals.

Table A1.57

Quantitative analysis of ST segment. The upper row provides mean ±SD and the lower row gives a 96% range

Instantaneous vectors	Scalar amplitude (mV)			Spatial magnitude and orientation		
	X	*Y*	*Z*	Amplitude (mV)	Azimuth (°)	Elevation (°)
0.02 s after J point	0.01 ± 0.03 −0.06–0.08	0.02 ± 0.04 −0.06–0.10	−0.10 ± 0.04 −0.19 to −0.02	0.11 ± 0.04 0.04–0.21	82 ± 22 37–127	12 ± 20 −35–55
0.04 s after J point	0.03 ± 0.03 −0.04–0.10	0.03 ± 0.04 −0.06–0.11	−0.11 ± 0.05 −0.22–0.03	0.13 ± 0.05 0.05–0.24	77 ± 20 35–117	13 ± 19 −32–56
0.06 s after J point	0.04 ± 0.04 −0.03–0.13	0.04 ± 0.04 −0.05–0.14	−0.12 ± 0.05 −0.25 to −0.03	0.15 ± 0.06 0.05–0.29	72 ± 19 35–109	16 ± 17 −22–54

Table A1.58

Quantitative analysis of eight instantaneous QRS vectors

Instantaneous vectors [a]	Scalar amplitude (mV)			Spatial magnitude and orientation		
	X	Y	Z	Amplitude (mV)	Azimuth (°)	Elevation (°)
1/8 QRS	−0.04± 0.05	−0.03± 0.07	−0.14± 0.08	0.18 ± 0.07	108 ± 27	−10± 26
	−0.16–0.09	−0.18–0.10	−0.31–0.00	0.06–0.36	51–156	−59–58
2/8 QRS	0.17 ± 0.17	0.05 ± 0.13	−0.34± 0.19	0.44 ± 0.19	63 ± 27	7 ± 19
	−0.11–0.61	−0.21–0.34	−0.82–0.00	0.14–0.93	359–110	−30–50
3/8 QRS	0.87 ± 0.31	0.63 ± 0.30	0.03 ± 0.40	1.17 ± 0.37	359 ± 25	33 ± 11
	0.33–1.63	0.17–1.44	−0.74–0.90	0.59–2.19	308–47	12–55
4/8 QRS	0.89 ± 0.49	0.89 ± 0.44	0.70 ± 0.44	1.58 ± 0.48	320 ± 26	34 ± 14
	−0.15–1.84	0.02–1.85	−0.05–1.74	0.70–2.62	261–3	2–61
5/8 QRS	0.09 ± 0.40	0.34 ± 0.39	0.77 ± 0.29	1.00 ± 0.34	275 ± 30	18 ± 21
	−0.55–1.13	−0.36–1.28	0.10–1.38	0.39–1.73	224–341	−30–56
6/8 QRS	−0.13± 0.14	0.01 ± 0.19	0.34 ± 0.20	0.44 ± 0.18	248 ± 35	1 ± 29
	−0.41–0.17	−0.34–0.40	−0.04–0.76	0.12–0.89	153–328	−68–57
7/8 QRS	−0.02± 0.06	0.01 ± 0.09	0.02 ± 0.09	0.13 ± 0.05	288 ± 96	3 ± 45
	−0.15–0.11	−0.15–0.16	−0.15–0.20	0.04–0.27	116–101	−75–76
8/8 QRS	0.01 ± 0.03	0.03 ± 0.04	−0.07± 0.04	0.09 ± 0.04	83 ± 34	19 ± 27
(J point)	−0.06–0.08	−0.06–0.10	−0.17–0.00	0.03–0.19	9–153	−42–70

[a] Obtained after each eighth of the QRS duration.

Table A1.59

Quantitative analysis of ST and T vectors

Instantaneous vectors [a]	Scalar amplitude (mV)			Spatial magnitude and orientation		
	X	Y	Z	Amplitude (mV)	Azimuth (°)	Elevation (°)
1/8 (ST-T)	0.02 ± 0.02	0.00 ± 0.02	−0.04 ± 0.03	0.06 ± 0.02	67 ± 36	1 ± 29
	−0.02–0.06	−0.05–0.05	−0.10–0.02	0.02–0.10	317–133	−49–58
2/8 (ST-T)	0.04 ± 0.03	0.02 ± 0.03	−0.07 ± 0.04	0.10 ± 0.04	59 ± 23	11 ± 21
	−0.01–0.11	−0.04–0.08	−0.16–0.01	0.03–0.19	353–98	−33–52
3/8 (ST-T)	0.08 ± 0.05	0.05 ± 0.04	−0.12 ± 0.06	0.16 ± 0.07	55 ± 19	17 ± 15
	0.00–0.19	−0.03–0.15	−0.27 to −0.02	0.05–0.33	13–87	−16–48
4/8 (ST-T)	0.15 ± 0.08	0.10 ± 0.07	−0.19 ± 0.09	0.28 ± 0.11	52 ± 17	21 ± 12
	0.02–0.35	−0.01–0.29	−0.42 to −0.03	0.09–0.56	14–85	−3–47
5/8 (ST-T)	0.23 ± 0.11	0.18 ± 0.10	−0.25 ± 0.12	0.40 ± 0.15	46 ± 19	27 ± 12
	0.04–0.51	0.02–0.43	−0.51 to −0.02	0.14–0.73	6–81	5–55
6/8 (ST-T)	0.22 ± 0.12	0.20 ± 0.10	−0.17 ± 0.11	0.37 ± 0.14	37 ± 28	35 ± 15
	0.00–0.49	0.02–0.43	−0.38–0.02	0.06–0.68	352–81	8–68
7/8 (ST-T)	0.07 ± 0.05	0.08 ± 0.04	−0.04 ± 0.04	0.12 ± 0.06	28 ± 47	40 ± 22
	−0.02–0.17	−0.01–0.18	−0.11–0.03	0.02–0.24	244–163	−29–78

[a] Determined through time normalization of the ST-T segment.

Table A1.60

Time integrals of QRS and T derived from scalar leads *X*, *Y* and *Z*. Combinations of two scalar components lead to planar projections in the frontal, sagittal and horizontal planes. Spatial magnitude and orientation are obtained by vectorial addition of all three scalar components. Addition of the time integrals of QRS and T results in the ventricular gradient ($V\bar{G}$). The P wave was included in the QRS time integral as described in the text. The QRS-T angle is derived from the two vectors of the respective time integrals

	Planar direction (°)			Scalar amplitude (μVs)			Spatial magnitude, orientation and spatial angles		
Item	Frontal	Sagittal	Horizontal	*X*	*Y*	*Z*	Amplitude (μVs)	Azimuth (°)	Elevation (°)
SÂQRS	44 ± 23	53 ± 27	323 ± 25	22.20 ± 11.18	23.41 ± 13.36	17.11 ± 12.35	40.08 ± 13.51	323 ± 25	36 ± 17
	351–85	355–110	274–14	1.23–46.97	−2.83–51.87	−6.11–44.28	15.16–71.32	274–14	−5–71
SÂT	33 ± 24	152 ± 21	52 ± 18	29.09 ± 15.78	20.12 ± 14.73	−37.79 ± 17.50	54.97 ± 20.72	52 ± 18	21 ± 14
	338–78	106–197	13–86	2.33–64.86	−6.86–56.06	−76.45 to −6.65	20.29–106.33	13–86	−11–49
SVĜ	40 ± 14	114 ± 23	21 ± 19	51.29 ± 20.79	43.54 ± 21.28	−20.67 ± 19.99	74.35 ± 26.63	21 ± 19	36 ± 13
	9–67	71–159	340–58	15.30–102.55	5.51–94.79	−66.46–13.35	27.17–138.42	340–58	7–63
QRS-T	12 ± 33	97 ± 39	89 ± 31						78 ± 26
angle	0–88	24–163	23–148						26–134

Table A1.61

Eigenvectors $\vec{A}$, $\vec{B}$ and $\vec{C}$ of the P, QRS, and T loops. The polar vector is identical with eigenvector $\vec{C}$. The ratios between eigenvectors give an estimate of the planarity of the loop and its configuration. The magnitude of the polar vector is based on the area of the loop in its 'broadside' projection. A multiplication constant was used for P and T because of their small magnitude

Item	P loop	QRS loop	T loop
Eigenvector $\vec{C}$ (polar vector)			
Azimuth (°)	35 ± 34	339 ± 30	337 ± 24
	304–95	285–46	292–36
Elevation (°)	−14 ± 15	−50 ± 12	−31 ± 17
	−50–11	−73 to −25	−63–8
Eigenvector $\vec{B}$			
Azimuth (°)	122 ± 31	65 ± 37	97 ± 33
	64–183	329–143	27–170
Elevation (°)	11 ± 15	3 ± 23	−37 ± 18
	−26–43	−45–49	−69–6
Eigenvector $\vec{A}$			
Azimuth (°)	352 ± 35	332 ± 42	39 ± 21
	293–79	244–56	354–79
Elevation (°)	62 ± 20	30 ± 15	31 ± 15
	0–84	−6–57	1–61
Eigenvector ratios			
C/B	0.20 ± 0.16	0.02 ± 0.04	0.05 ± 0.09
	0.00–0.65	0.00–0.10	0.00–0.38
B/A	0.21 ± 0.17	0.41 ± 0.20	0.17 ± 0.13
	0.00–0.67	0.06–0.85	0.01–0.51
C/A	0.03 ± 0.03	0.01 ± 0.01	0.01 ± 0.01
	0.00–0.14	0.00–0.03	0.00–0.03
Spatial amplitude of the Polar vector (area of "broadside" projection of the spatial loop) (mV^2)	$6.16 \pm 3.43 \times 10^{-3}$	1.62 ± 0.85	$5.03 \pm 3.48 \times 10^{-2}$
	$1.69–15.76 \times 10^{-3}$	0.45–3.89	$0.66–14.76 \times 10^{-2}$

A1.5.2 Normal Limits in Females

The group of Pipberger, in a later paper, published data on the normal limits of the Frank orthogonal-lead ECG in 450 women, including both whites and blacks. ➋ Tables A1.62–A1.68 are reproduced from Nemati M et al. The orthogonal electrocardiogram in normal women. Implications of sex differences in diagnostic electrocardiography. *Am. Heart J.* 1978;**95**: 12–21, with the permission of Mosby, St. Louis, Missouri.

Table A1.62
Measurements of P, QRS and T in orthogonal *X*, *Y* and *Z* leads

Item	*X*		*Y*		*Z*	
P amplitude[a]	0.05 ± 0.02[c]		0.09 ± 0.04		Positive component	
	0.02–0.09[d]		0.02–0.17		0.03 ± 0.02 0–0.08 Negative component −0.03 ± 0.02 −0.07–0	
Q amplitude	−0.08 ± 0.05	(237)[e]	−0.08 ± 0.05	(238)	−0.31 ± 0.17	(446)
	−0.22 to −0.01		−0.23 to −0.01		−0.77 to −0.07	
R amplitude	0.94 ± 0.35		0.81 ± 0.33		0.68 ± 0.25	
	0.35–1.75		0.27–1.55		0.22–1.25	
S amplitude	−0.17 ± 0.11	(319)	−0.17 ± 0.12	(252)	−0.19 ± 0.17	(9)
	−0.47 to −0.01		−0.52 to −0.01		−0.60 to −0.04	
T amplitude	0.22 ± 0.10		0.16 ± 0.09		−0.08 ± 0.10	
	0.05–0.44		−0.03–0.39		−0.31–0.10	
Q/R amplitude ratio	0.07 ± 0.04	(237)	0.09 ± 0.06	(238)	0.54 ± 0.42	(446)
	0.01–0.18		0.01–0.22		0.10–1.73	
R/S amplitude ratio	11.36 ± 18.00	(319)	9.20 ± 13.00	(252)	4.00 ± 3.30	(9)
	1.12–71.70		0.64–56.00		0.70–12.00	
Q duration[b]	0.016 ± 0.003	(237)	0.017 ± 0.004	(238)	0.030 ± 0.008	(446)
	0.010–0.024		0.009–0.028		0.014–0.047	
R duration	0.045 ± 0.012		0.050 ± 0.014		0.048 ± 0.010	
	0.028–0.071		0.028–0.081		0.028–0.069	
S duration	0.029 ± 0.009	(319)	0.028 ± 0.090	(252)	0.028 ± 0.009	(9)
	0.012–0.048		0.012–0.049		0.018–0.044	
R peak time	0.037 ± 0.005		0.039 ± 0.005		0.048 ± 0.006	
(instrinsicoid deflection)	0.028–0.048		0.027–0.050		0.036–0.062	
RX + RZ	1.60 ± 0.42					
	0.87–2.50					

[a]Amplitudes are in millivolts [b]Durations are in seconds [c]Mean ± standard deviation of each item [d]Limits of a 96 percentile range [e]Figures in parentheses show the actual number of observation: results not followed by a number in parentheses were obtained from total series.

A1.5.3 Effect of Age, Build and Race on the Orthogonal-Lead ECG

The correlation of various ECG parameters with age and race (black and white) was investigated by Pipberger *et al.* In general terms, voltages were higher in black men than in white men. ➲ Tables A1.69–A1.73 are reproduced from Pipberger HV et al. Correlations of the orthogonal electrocardiogram and vectorcardiogram with constitutional variables in 518 normal men. *Circulation* 1967;**35**: 536–551 with the permission of the American Heart Association, Dallas, Texas. ➲ Table A1.69 sets out the distribution of ages of subjects in ➲ Tables A1.70–A1.73.

Table A1.63

Measurements of initial, terminal and instantaneous vectors in orthogonal *X*, *Y* and *Z* leads

	X	*Y*	*Z*
Initial vectors from the onset of QRS			
0.01 s	−0.01 ± 0.05	−0.00 ± 0.06	−0.11 ± 0.06
	−0.11–0.10	−0.13–0.12	−0.23–0.01
0.02 s	0.16 ± 0.16	0.10 ± 0.14	−0.22 ± 0.16
	−0.13–0.56	−0.13–0.45	−0.51–0.14
0.03 s	0.62 ± 0.27	0.44 ± 0.26	−0.13 ± 0.31
	0.10–1.25	−0.01–1.07	−0.66–0.65
0.04 s	0.77 ± 0.41	0.63 ± 0.38	0.39 ± 0.36
	0.00–1.70	−0.08–1.46	−0.53–1.10
Terminal vectors from the end of QRS			
J point	0.00 ± 0.03	0.00 ± 0.03	0.01 ± 0.03
	−0.06–0.05	−0.06–0.06	−0.07–0.04
0.01 s	−0.01 ± 0.05	0.00 ± 0.08	0.09 ± 0.07
	−0.12–0.11	−0.15–0.14	−0.07–0.22
0.02 s	−0.03 ± 0.15	0.00 ± 0.17	0.31 ± 0.17
	−0.27–0.36	−0.28–0.38	−0.04–0.62
0.03 s	0.19 ± 0.38	0.21 ± 0.32	0.52 ± 0.26
	−0.29–1.18	−0.35–0.97	−0.05–1.04
0.04 s	0.61 ± 0.48	0.53 ± 0.40	0.44 ± 0.36
	−0.25–1.61	−0.29–1.32	−0.31–1.16
Instantaneous ST vectors			
0.02 s after J point	−0.02 ± 0.07	−0.03 ± 0.04	−0.01 ± 0.05
	−0.10–0.06	−0.12–0.07	−0.09–0.07
0.04 s after J point	−0.01 ± 0.07	−0.03 ± 0.05	−0.02 ± 0.04
	−0.09–0.07	−0.12–0.07	−0.09–0.05
0.06 s after J point	0.00 ± 0.08	−0.02 ± 0.07	−0.02 ± 0.04
	−0.08–0.07	−0.11–0.07	−0.10–0.05

Table A1.64

Measurements of P, PR, QRS and QT

Item	Measurement (s)
P duration	0.106 ± 0.019
	0.064–0.142
PR interval	0.154 ± 0.022
	0.112–0.208
PR segment	0.048 ± 0.018
	0.020–0.096
QRS duration	0.084 ± 0.008
	0.068–0.104
QT interval (uncorrected)	0.372 ± 0.026
	0.319–0.428

Table A1.65
Measurement of P, QRS and T vectors in frontal, left sagittal and horizontal plane

Item	Frontal		Left sagittal		Horizontal	
	Amplitude	Angle	Amplitude	Angle	Amplitude	Angle
Maximal P vector	0.01 ± 0.04[a]	61[b]	0.10 ± 0.04	96	0.06 ± 0.02	14
	0.04–0.19	0–93[b]	0.03–0.18	178–44	0.03–0.12	115 to −98
Maximal QRS vector	1.24 ± 0.38	41	1.00 ± 0.31	125	1.10 ± 0.32	29
	0.63–2.19	10–88	0.49–1.70	−156–48	0.55–1.84	114 to −35
Maximal T vector	0.28 ± 0.11	32	0.20 ± 0.08	62	0.25 ± 0.10	−17
	0.10–0.57	2–68	0.07–0.42	170 to −88	0.09–0.49	42 to −70
Half-area QRS vector	1.12 ± 0.45	42	0.93 ± 0.36	120	0.97 ± 0.33	29
	0.22–2.20	−1–82	0.24–1.71	−174–52	0.36–1.70	78 to −22
Initial QRS vectors						
0.01 s	0.07 ± 0.04	−137[c]	0.13 ± 0.05	−4	0.13 ± 0.05	−95
	0.01–0.18	20 to −10	0.03–0.24	80 to −58	0.03–0.24	−5 to −146
0.02 s	0.23 ± 0.17	28	0.29 ± 0.14	24	0.32 ± 0.14	−54
	0.02–0.69	−129–174	0.04–0.62	144 to −30	0.06–0.69	38 to −115
0.03 s	0.78 ± 0.32	34	0.53 ± 0.27	88	0.69 ± 0.27	−2
	0.17–1.51	1–64	0.04–1.16	156–10	0.12–1.32	54 to −57
0.04 s	1.06 ± 0.45	39	0.84 ± 0.35	124	0.96 ± 0.36	30
	0.10–2.00	−4–84	0.10–1.59	−168–42	0.20–1.75	88 to −26
Terminal QRS vectors						
0.01 s	0.08 ± 0.05	−150[c]	0.12 ± 0.06	−172	0.11 ± 0.06	96[c]
	0.01–0.20	16 to −18	0.01–0.24	−32–82	0.01–0.23	−98 to −36
0.02 s	0.18 ± 0.14	−169[c]	0.36 ± 0.16	−177	0.35 ± 0.17	99
	0.03–0.55	15 to −30	0.05–0.69	−82–108	0.03–0.72	−150–16
0.03 s	0.45 ± 0.35	63[c]	0.64 ± 0.27	163	0.66 ± 0.29	78
	0.04–1.44	−20 to −80	0.12–1.24	−124–84	0.13–1.34	150 to −5
0.04 s	0.90 ± 0.49	43	0.81 ± 0.34	132	0.89 ± 0.36	41
	0.06–2.00	−25 to −160	0.19–1.56	−140–54	0.18–1.75	122 to −26

[a]Amplitudes are in millivolts [b]Angles are in °. All angular ranges should be read in a clockwise sequence [c]Angles show no significant clustering as evidenced by wide 96 percentile ranges.

Table A1.66
Spatial magnitude and orientation of maximal P, QRS and T vectors

Item	Magnitude (mV)	Azimuth (°)	Elevation (°)
Maximal P*XYZ* vector	0.11 ± 0.04	14	48
	0.05–0.20	115 to −98[a]	−22–80
Maximal QRS*XYZ* vector	1.35 ± 0.36	29	34
	0.75–2.25	114 to −35	−5–61
Maximal T*XYZ* vector	0.33 ± 0.13	−17	31
	0.12–0.64	42 to −70	−50–64

[a]Angular ranges should be read in a clockwise sequence.

Table A1.67

Direction of inscription of QRS loops in the frontal, left sagittal and horizontal planes in 450 normal women: CW, clockwise; CW/CCW, figure-of-eight, clockwise then counterclockwise; CCW/CW, figure-of-eight, counterclockwise then clockwise; CCW, counterclockwise

	Frontal	Left sagittal	Horizontal
CW	230 (51%)	8 (2%)	0 (0%)
CW/CCW	37 (8%)	12 (3%)	0 (0%)
CCW/CW	61 (14%)	27 (6%)	8 (2%)
CCW	122 (27%)	403 (89%)	442 (98%)

Table A1.68

Sex-specific limits. Data are derived from records of 510 normal men and 450 normal women, unless indicated otherwise in parentheses ($p < 0.001$ for all measurements)

Item	Men		Women	
Scalar measurements				
QRS duration[a]	0.093 ± 0.009		0.084 ± 0.008	
	0.076–0.112		0.068–0.104	
QZ amplitude[b]	−0.41 ± 0.21		−0.31 ± 0.17	(446)
	−0.93 to −0.09		−0.77 to −0.07	
RX amplitude	1.17 ± 0.37		0.94 ± 0.35	
	0.51–1.97		0.35–1.75	
RY amplitude	1.03 ± 0.41		0.81 ± 0.33	
	0.35–1.95		0.27–1.55	
RZ amplitude	0.93 ± 0.35		0.68 ± 0.25	
	0.36–1.79		0.22–1.25	
SX amplitude	−0.27 ± 0.15	(407)	−0.17 ± 0.11	(319)
	−0.68 to −0.06		−0.47 to −0.01	
TZ amplitude	−0.28 ± 0.13		−0.08 ± 0.10	
	−0.58 to −06		−0.31–0.10	
Q/RZ amplitude ratio	0.50 ± 0.35		0.54 ± 0.42	(446)
	0.10–1.20		0.10–1.73	
RX + RZ amplitude	2.00 ± 0.52		1.60 ± 0.42	
	1.06–3.10		0.87–2.50	
J point in lead Z	−0.07 ± 0.04		−0.01 ± 0.03	
	−0.17–0.00		−0.07–0.04	

A1.6 Normal Limits of Polarcardiographic Data

The technique of polarcardiography is explained in Chap. 45. As an adjunct, some tables of normal limits derived from the work of Dower are presented in Tables A1.74 to A1.77. The terminology is also discussed in the chapter. In addition, data from exercise testing of normals is included.

Table A1.68 (Continued)

Item	Men		Women	
Planar measurements				
Max. QRSXY amplitude	1.57 ± 0.42		1.24 ± 0.38	
	0.81–2.53		0.63–2.19	
Max. QRSZY amplitude	1.32 ± 0.45		1.00 ± 0.31	
	0.60–2.42		0.49–1.71	
Max. QRSXZ amplitude	1.39 ± 0.36		1.10 ± 0.31	
	0.74–2.19		0.55–1.83	
Max. TZY amplitude	0.36 ± 0.13		0.20 ± 0.08	
	0.13–0.67		0.07–0.42	
Max. TXZ amplitude	0.40 ± 0.14		0.25 ± 0.10	
	0.15–0.72		0.09–0.49	
Max. TZY angle[c]	38		62	
	87–0		170 to −88	
Max. TXZ angle	−46		−17	
	−8 to −83		42 to −70	
Spatial measurements				
Max. QRSXYZ magnitude	1,073 ± 0.44		1.35 ± 0.36	
	0.92–2,075		0.75–2.25	
Max. TXYZ magnitude	0.46 ± 0.16		0.30 ± 0.12	
	0.18–0.82		0.09–0.59	

[a]Duration are in seconds [b]Amplitudes are in millivolts [c]Angles are in degrees.

Table A1.69
Age distribution of subjects included in the study

Age-group	20–29	30–39	40–49	50–59	60–78	total
Number of cases	78	179	151	56	54	518

Note that in Table A1.74, the median values were obtained from grouped data. Prominent sex differences can be seen by comparing adjacent columns marked*. Progressive change with age can be seen by comparing corresponding columns at the same level; for example, systolic blood pressure (BP) increases from 120 to 145 mmHg for male and female (M + F) with increasing age.

With increasing age, the R vector decreases in magnitude and tends toward the transverse and frontal planes (longitude and latitude of R); the ST vector tends to be greater in the men than in the women; the longitude difference between the T and R vectors (long. T–long. R) also tends to decrease. Offered for comparative purposes the systolic blood pressure shows an increase with age, especially in the women.

With respect to sex differences the T vector is more anterior (latitude of T) in the men, and its magnitude is greater, especially in the 20–29 age-group. Relative to the R vector, the T vector is more strongly anterior (lat. T–lat. R) in the men, and of greater magnitude (M_T/M_R) (Tables A1.75–A1.79).

Table A1.70

Correlations between age-groups and ECG measurements. The means and standard deviations in this and the following tables are shown in the upper line of each measurement. The lower line indicates the limits of 96 percentile ranges. In the last column results of *t*-tests are given, comparing the youngest and oldest age-groups. Vector magnitude is denoted vc

	Correlation coefficient	Age-group (years)					*p* (youngest vs. oldest group)
		20–29	30–39	40–49	50–59	60–78	
Wave durations							
P duration	0.223	0.096 ± 0.012	0.098 ± 0.012	0.099 ± 0.011	0.104 ± 0.011	0.104 ± 0.013	<0.01
(s)		0.072–0.116	0.072–0.116	0.076–0.124	0.088–0.124	0.080–0.128	
Scalar QRS measurements							
RY	−0.338	1.25 ± 0.42	1.06 ± 0.39	1.02 ± 0.42	0.88 ± 0.38	0.72 ± 0.37	<0.001
(mV)		0.53–2.08	0.47–1.95	0.34–1.99	0.26–1.69	0.24–1.63	
RZ	−0.277	1.03 ± 0.37	0.98 ± 0.35	0.89 ± 0.31	0.78 ± 0.26	0.74 ± 0.33	<0.001
(mV)		(78)	(179)	(151)	(56)	(54)	
		0.51–1.71	0.42–1.76	0.38–1.54	0.33–1.24	0.28–1.74	
SX	−0.214	−0.27 ± 0.21	−0.26 ± 0.16	−0.25 ± 0.15	−0.24 ± 0.16	−0.15 ± 0.10	<0.01
(mV)		(37)	(100)	(86)	(42)	(41)	
		−0.04 to −0.79	−0.01 to −0.64	−0.03 to −0.57	−0.01 to −0.55	−0.01–0.34	
Initial 0.02 s lead *Z*	0.268	−0.30 ± 0.19	−0.26 ± 0.16	−0.24 ± 0.13	−0.18 ± 0.10	−0.17 ± 0.12	<0.001
(mV)		(78)	(179)	(151)	(56)	(54)	
		−0.63 to −0.13	−0.62 to −0.05	−0.63 to −0.03	−0.47 to −0.02	−0.45–0.03	
Initial 0.02 s lead *Z*	0.192	0.29 ± 0.47	0.34 ± 0.49	0.21 ± 0.41	0.15 ± 0.38	0.07 ± 0.42	<0.01
(mV)		−0.50–1.22	−0.51–1.23	−0.54–1.11	−0.61–0.77	−0.81–0.91	
Initial 0.05 s lead *Z*	−0.257	0.83 ± 0.44	0.76 ± 0.42	0.65 ± 0.39	0.53 ± 0.37	0.50 ± 0.40	<0.001
(mV)		0.10–1.56	−0.07–1.67	−0.12–1.38	−0.12–1.16	−0.26–1.43	
Planar QRS measurements							
Maximal vc. sagittal plane	−0.371	1.56 ± 0.46	1.37 ± 0.45	1.27 ± 0.41	1.12 ± 0.36	0.97 ± 0.42	<0.001
(mV)		0.73–2.46	0.72–2.39	0.59–2.25	0.43–2.00	0.48–2.03	
Maximal vc. direction, frontal plane	−0.288	50 ± 15	41 ± 13	42 ± 15	38 ± 18	32 ± 16	<0.001
(°)		17–77	17–66	13–73	6–74	3–56	
Maximal vc. direction, horizontal plane	−0.217	312 ± 30	325 ± 32	326 ± 36	331 ± 36	340 ± 36	<0.001
(°)		254–354	250–26	243–19	241–20	264–46	

Table A1.70 (Continued)

	Correlation coefficient	Age-group (years)					*p* (youngest vs. oldest group)
		20–29	30–39	40–49	50–59	60–78	
Spatial QRS measurements							
Maximal vc.	−0.367	1.88 ± 0.42	1.80 ± 0.43	1.65 ± 0.45	1.45 ± 0.33	1.36 ± 0.48	<0.001
(mV)		1.13–2.82	0.96–2.76	0.94–2.74	0.91–2.19	0.77–2.47	
Initial 0.02 s vc.	−0.269	0.36 ± 0.16	0.31 ± 0.17	0.27 ± 0.14	0.24 ± 0.11	0.22 ± 0.11	<0.001
(mV)		0.15–0.69	0.11–0.68	0.10–0.68	0.08–0.51	0.05–0.48	
Initial 0.05 s vc.	−0.359	1.65 ± 0.44	1.49 ± 0.47	1.36 ± 0.45	1.19 ± 0.41	1.07 ± 0.46	<0.001
(mV)		0.91–2.54	0.77–2.48	0.55–2.42	0.49–1.91	0.40–2.00	
Spatial ST-T measurements							
Maximal T vc.	−0.312	0.50 ± 0.17	0.45 ± 0.15	0.42 ± 0.14	0.40 ± 0.13	0.31 ± 0.12	<0.001
(mV)		0.26–0.82	0.20–0.80	0.16–0.75	0.20–0.71	0.10–0.61	
0.04 s ST-T vc.	−0.213	0.12 ± 0.05	0.11 ± 0.05	0.11 ± 0.04	0.09 ± 0.04	0.09 ± 0.04	<0.01
(mV)		0.02–0.22	0.04–0.23	0.03–0.20	0.02–0.17	0.03–0.20	
3/8 ST-T vc.	−0.361	0.19 ± 0.08	0.17 ± 0.07	0.15 ± 0.06	0.13 ± 0.06	0.10 ± 0.05	<0.001
(mV)		0.08–0.36	0.05–0.32	0.04–0.32	0.04–0.25	0.03–0.24	
4/8 ST-T vc.	−0.376	0.30 ± 0.11	0.26 ± 0.10	0.24 ± 0.10	0.21 ± 0.09	0.15 ± 0.07	<0.001
(mV)		0.11–0.54	0.10–0.48	0.06–0.46	0.09–0.42	0.05–0.32	
Time integrals							
SÂQRS	−0.243	42.0 ± 13.0	42.0 ± 14.5	39.5 ± 13.0	34.5 ± 11.4	32.4 ± 13.4	<0.001
(μVs)		18.7–65.6	15.0–72.1	18.9–74.0	16.9–56.2	12.3–62.7	
SÂT	−0.352	57.8 ± 22.8	52.8 ± 20.0	47.0 ± 17.7	44.0 ± 17.3	31.9 ± 14.9	<0.001
(μVs)		26.5–100.0	17.1–95.9	13.4–90.4	19.3–92.5	6.3–58.7	
SVĜ	−0.343	78.9 ± 27.5	73.3 ± 25.7	66.3 ± 24.4	59.4 ± 24.4	46.2 ± 20.8	<0.001
(μVs)		36.2–140.0	31.7–133.0	20.8–120.0	15.9–115.0	16.1–92.2	

Table A1.71
Correlations between subjects of the white and black race

	Correlation coefficient	Race				*p*
		White		Black		(W versus B)
Scalar QRS measurements						
QRS duration (s)	−0.320	0.098 ± 0.009		0.091 ± 0.009		<0.001
		0.080–0.116		0.076–0.108		
Q*X* duration (s)	−0.160	0.019 ± 0.006	(173)	0.017 ± 0.004	(96)	<0.02
		0.008–0.028		0.008–0.024		
Q*Y* duration (s)	−0.377	0.021 ± 0.007	(133)	0.015 ± 0.006	(72)	<0.001
		0.008–0.032		0.008–0.028		
Q*Y* (mV)	−0.267	0.09 ± 0.07	(133)	0.06 ± 0.05	(72)	<0.001
		0.01–0.27		0.01–0.23		
R*X* (mV)	0.264	1.06 ± 0.35		1.25 ± 0.36		<0.001
		0.49–1.84		0.63–1.94		
R*Y* (mV)	0.197	1.96 ± 0.42		1.13 ± 0.41		<0.001
		0.28–1.99		0.41–1.99		
R*Z* (mV)	0.275	1.84 ± 0.30		1.04 ± 0.37		<0.001
		0.33–1.53		0.44–1.98		
Q/R*X* ratio	−0.128	0.06 ± 0.05	(173)	0.05 ± 0.03	(96)	<0.05
		0.00–0.20		0.00–0.11		
Q/R*Y* ratio	−0.392	0.09 ± 0.06	(133)	0.05 ± 0.03	(72)	<0.001
		0.01–0.23		0.00–0.14		
Initial 0.04 s, lead *X* (mV)	0.280	1.88 ± 0.32		1.08 ± 0.38		<0.001
		0.31–1.59		0.44–1.79		
Initial 0.02 s, lead *Y* (mV)	0.279	−0.02± 0.09		0.04 ± 0.13		<0.001
		−0.22–0.15		−0.20–0.31		
Initial 0.03 s, lead *Y* (mV)	0.315	0.19 ± 0.20		0.33 ± 0.23		<0.001
		−0.09–0.73		−0.02–0.90		
Initial 0.04 s, lead *Y* (mV)	0.297	0.67 ± 0.36		0.91 ± 0.41		<0.001
		0.10–1.62		0.23–1.76		
Terminal 0.03 s, lead *Z* (mV)	0.284	0.51 ± 0.27		0.69 ± 0.30		<0.001
		0.06–1.12		0.13–1.37		
Planar QRS measurements						
Maximal vc. frontal (mV)	0.299	1.44 ± 0.41		1.70 ± 0.42		<0.001
		0.73–2.38		0.96–2.65		
Maximal vc. sagittal plane (mV)	0.238	1.22 ± 0.44		1.45 ± 0.46		<0.001
		0.52–2.12		0.73–2.46		
Maximal vc. horizontal plane (mV)	0.312	1.26 ± 0.34		1.49 ± 0.36		<0.001
		0.65–2.05		0.89–2.24		
Spatial QRS measurements						
Maximal vc. (mV)	0.305	1.58 ± 0.43		1.87 ± 0.45		<0.001
		0.83–2.52		1.06–2.94		
Initial 0.02 s vc., elevation angle (°)	0.256	−3 ± 22		9 ± 22		<0.001
		−44–47		−25–60		
Initial 0.03 s vc. (mV)	0.348	0.56 ± 0.25		0.77 ± 0.31		<0.001
		0.18–1.16		0.28–1.50		
Initial 0.04 s vc. (mV)	0.386	1.21 ± 0.40		1.58 ± 0.46		<0.001
		0.52–2.17		0.74–2.58		
Terminal 0.05 s vc., azimuth angle (°)	0.243	324 ± 37		342 ± 34		<0.001
		240–28		266–42		

Table A1.72

Correlations between chest configuration and ECG measurements. The ratio SD/TD was derived from sagittal and transverse chest diameters. Note the small increases of the mean Q/*RY* ratio accompanied by a marked increase of 136% of the upper limit of the 96 percentile range. Mean results failed frequently to reflect the magnitude of changes in range limits

	Correlation coefficient	Chest configuration				*p* (first versus fourth subgroup)
		0.69 (SD/TD)	0.70–0.74 (SD/TD)	0.75–0.79 (SD/TD)	0.80 (SD/TD)	
Scalar QRS measurements						
RY (mV)	−0.227	1.16 ± 0.45	1.09 ± 0.40	0.96 ± 0.42	0.91 ± 0.39	<0.001
		0.40–2.1	0.34–2.02	0.28–1.86	0.27–1.72	
RZ (mV)	−0.237	1.02 ± 0.33	0.94 ± 0.32	0.93 ± 0.36	0.80 ± 0.33	<0.001
		0.46–1.67	0.42–1.74	0.33–1.98	0.33–1.41	
Initial 0.04 s, lead *Y* (mV)	−0.220	0.85 ± 0.43	0.81 ± 0.39	0.72 ± 0.39	0.65 ± 0.36	<0.001
		0.23–1.85	0.20–1.71	0.08–1.75	0.03–1.40	
Initial 0.04 s, lead *Z* (mV)	−0.235	0.37 ± 0.49	0.29 ± 0.45	0.26 ± 0.47	0.10 ± 0.38	<0.001
		−0.50–1.23	−0.51–1.14	−0.55–1.27	−0.58–0.96	
Q/*RY* ratio	0.147	0.06 ± 0.04	0.07 ± 0.05	0.08 ± 0.06	0.09 ± 0.07	<0.05
		(38)	(67)	(41)	(59)	
		0.01–0.14	0.00–0.18	0.01–0.21	0.01–0.33	
Planar QRS measurements						
Maximal vc. direction, frontal plane (°)	−0.218	46 ± 16	44 ± 14	40 ± 14	37 ± 17	<0.001
		16–79	16–71	10–72	5–66	
Maximal vc. direction, horizontal plane (°)	0.195	320 ± 32	320 ± 34	326 ± 34	335 ± 36	<0.001
		252–13	239–19	247–24	248–36	
Spatial QRS measurements						
Maximal vc. (mV)	−0.200	1.81 ± 0.42	1.74 ± 0.44	1.64 ± 0.48	1.57 ± 0.46	<0.001
		1.01–2.69	0.94–2.76	0.84–2.81	0.81–2.64	
Initial 0.04 s vc., azimuth angle (°)	0.206	341 ± 28	345 ± 28	347 ± 25	355 ± 22	<0.001
		296–40	298–38	302–34	312–40	
Initial 0.04 s vc, elevation angle (°)	−0.190	37 ± 12	35 ± 12	33 ± 12	31 ± 14	<0.01
		15–60	14–58	5–55	2–54	
Spatial ST-T measurements						
0.04 s ST-T vc. (mV)	−0.196	0.12 ± 0.05	0.11 ± 0.04	0.10 ± 0.05	0.09 ± 0.04	<0.001
		0.05–0.23	0.03–0.19	0.03–0.26	0.03–0.19	

Table A1.73

Correlation between body weight and ECG measurements. When relative body weight or deviations from ideal weight were used instead of absolute weight, ECG items exhibiting significant relationships were almost identical but most correlation coefficients decreased noticeably

	Correlation coefficient	Weight				p (first versus fourth subgroup)
		≤134 (lb)	135–159 (lb)	160–184 (lb)	≥185 (lb)	
Scalar QRS measurements						
RZ (mV)	−0.171	0.99 ± 0.39	0.94 ± 0.35	0.89 ± 0.32	0.82 ± 0.29	<0.01
		0.41–1.94	0.36–1.74	0.41–1.65	0.38–1.44	
Initial 0.05 s, lead *X* (mV)	0.233	0.51 ± 0.49	0.70 ± 0.49	0.80 ± 0.50	0.86 ± 0.48	
		−0.35–1.49	−0.18–1.56	−0.28–1.77	−0.06–1.72	<0.001
Initial 0.04 s, lead *Z* (mV)	−0.316	0.43 ± 0.43	0.33 ± 0.45	0.16 ± 0.43	0.04 ± 0.41	
		−0.34–1.23	−0.44–1.22	−0.58–0.106	−0.61–0.95	<0.001
Initial 0.05 s, lead *Z* (mV)	−0.270	0.82 ± 0.41	0.75 ± 0.41	0.62 ± 0.40	0.52 ± 0.42	
		0.13–1.64	−0.10–1.67	−0.30–1.38	−0.34–1.29	
Planar QRS measurements						
Maximal vc. direction, frontal plane (°)	−0.262	47 ± 15	43 ± 15	40 ± 14	35 ± 17	<0.001
		19–74	13–69	11–70	6–73	
Maximal vc. Direction, horizontal plane (°)	0.261	310 ± 35	352 ± 34	326 ± 33	339 ± 32	<0.001
		243–4	248–32	248–19	263–38	
Spatial QRS measurements						
Maximal vc., azimuth angle (°)	0.232	318 ± 27	329 ± 24	330 ± 28	339 ± 30	<0.001
		265–5	271–15	252–18	274–24	
Initial 0.04 s vc., azimuth angle (°)	0.313	336 ± 26	344 ± 27	353 ± 25	358 ± 22	<0.001
		296–27	298–36	302–42	308–35	
Initial 0.04 s vc., elevation angle (°)	−0.263	38 ± 11	36 ± 12	32 ± 13	29 ± 12	<0.001
		17–52	13–58	9–53	6–56	
Initial 0.05 s vc., azimuth angle (°)	0.295	301 ± 29	310 ± 27	320 ± 29	325 ± 29	<0.001
		250–352	259–5	249–19	265–15	
Spatial ST-T measurements						
7/8 ST-T vc., elevation angle (°)	−0.190	42 ± 14	41 ± 14	37 ± 13	35 ± 14	<0.01
		13–68	16–71	9–66	6–58	
Polar vectors						
P, elevation angle (°)	−0.260	−5 ± 19	−9 ± 20	−13 ± 19	−21± 21	<0.001
		−41–29	−54–32	−62–20	−63–17	
QRS, elevation angle (°)	−0.229	−46 ± 12	−48 ± 12	−51± 12	−55 ± 13	<0.001
		−67 to −24	−75 to −25	−76 to −32	−81 to −28	
T, elevation angle (°)	−0.176	−27 ± 19	−29 ± 20	−35± 17	−36 ± 19	<0.01
		−58–5	−64–8	−70–5	−63–11	

Table A1.74

Polarcardiographic median normal values of Cretan villagers. See Chapter 45 for definitions

	M	F	M+F	M	F	M+F	M	F	M+F	M	F	M+F	M	F	M+F	M	F	M+F
Age-groups	20–82			20–29			30–39			40–49			50–59			60–82		
Number of cases	275	191	466	30	16	46	53	42	95	70	46	116	51	43	94	71	44	115
Median age (years)	46	48	48	26	26	26	36	36	36	44	46	44	54	54	54	68	68	68
Intervals (ms): PR	135	135	135	130	125	130	135	135	135	140	135	135	130	135	135	145	135	140
QRS	85	80	80	80	80	80	85	80	80	85	80	80	85	85	85	85	75	80
QT_c	370	360	370	350	360	360	370	360	360	360	370	360	370	360	360	390	370	380
Latitudes: P	10A	10A	10A	10A	10A	10A	10A	10A	10A	10A	10A	10A	10A	0	0	0	0	0
R	10P	20P	20P	20P	20P	20P	10P	20P	20P	10P	20P	20P	20P	10P	10P	10P	10P	10P
ST	60A	50A	60A	60A	60A	60A	60A	60A	60A	60A	60A	60A	60A	40A	60A	60A	50A	60A
T	40A*	20A	30A	40A	20A	30A	30A	30A	30A	40A	20A	30A	40A	10A	30A	30A	30A	30A
Lat. T–Lat. R	50	40	40	50	40	50	50	40	50	50	40	40	50	20	40	40	40	40
Longitudes: P	+60	+70	+70	+70	+70	+70	+70	+70	+70	+70	+70	+70	+70	+60	+70	+70	+60	+70
R	+30	+40	+30	+40	+50	+50	+40	+40	+40	+30	+30	+30	+30	+30	+30	+20	+30	+30
ST	+30	+50	+40	+50	−90	+40	+30	+50	+30	+30	+20	+30	+20	+70	+40	+50	+65	+50
T	+30	+40	+30	+30	+40	+40	+30	+40	+30	+30	+30	+30	+30	+40	+40	+30	+40	+30
Long. T–Long. R	−10	0	0	−10	−10	−10	0	−10	−10	0	−10	0	0	0	0	10	10	10
Magnitudes (mV): P	0.14	0.14	0.14	0.17	0.13	0.16	0.14	0.15	0.15	0.15	0.14	0.14	0.14	0.15	0.15	0.13	0.14	0.13
R	1.55	1.50	1.50	1.75	1.65	1.75	1.60	1.55	1.55	1.60	1.45	1.50	1.45	1.45	1.45	1.45	1.45	1.45
ST	0.14*	0.08	0.11	0.21	0.08	0.17	0.17	0.07	0.12	0.15	0.07	0.10	0.14	0.06	0.10	0.12	0.08	0.10
T	0.500*	0.350	0.450	0.625	0.400	0.575	0.525	0.400	0.475	0.500	0.325	0.450	0.500	0.325	0.400	0.475	0.350	0.425
m_1R	1.30	1.25	1.30	1.55	1.25	1.35	1.25	1.20	1.25	1.35	1.15	1.30	1.20	1.25	1.25	1.30	1.25	1.25
S	0.045	0.040	0.045	0.045	0.045	0.045	0.060	0.040	0.050	0.045	0.030	0.040	0.045	0.030	0.040	0.040	0.040	0.040
$M_S.\theta$	1	1	1	1	1	1	1	1	1	1	1	1	1	1	1	1	1	1
M_T/M_R	0.32*	0.24	0.30	0.36	0.24	0.30	0.34	0.28	0.30	0.30	0.26	0.28	0.34	0.22	0.30	0.32	0.22	0.30
M_P/M_R	0.08	0.10	0.08	0.08	0.08	0.08	0.10	0.10	0.10	0.08	0.10	0.10	0.10	0.12	0.10	0.08	0.10	0.08
Systolic BP (mmHg)	130	130	130	125	120	120	125	125	125	130	130	130	130	135	135	140	150	145

Table A1.75

Magnitudes in space (M_R) and transverse plane (m_tR) in Cretan villagers. Up to age 50, M_R is greater in males, but not m_tR. In males, M_R reduces with age, also in FEMALEs but only up to age 50. In females m_tR tends to increase with age, but not in males. No overall sex difference for M_R and m_tR, but younger males show greater M_R. Medians of M_R are always greater than m_tR, also 95 percentiles. 95 percentiles for R or T in either sex do not show change with age, but numbers are too small to draw a reliable conclusion

Age-group	Sex	*n*	M_R(mV) median	M_R (mV) upper 95%	m_t R (mV) median	m_t R (mV) upper 95%
20–82	M+F	466	1.50	2.25	1.30	2.05
	M	275	1.55	2.30	1.30	2.10
	F	191	1.50	2.10	1.25	1.85
20–29	M+F	46	1.75	2.50	1.35	2.15
	M	30	1.75	2.40	1.55	2.15
	F	16	1.65	2.20	1.25	1.50
30–39	M+F	95	1.55	2.15	1.25	1.75
	M	53	1.65	2.15	1.25	1.80
	F	42	1.50	2.10	1.20	1.75
40–49	M+F	116	1.50	2.30	1.30	1.95
	M	70	1.60	2.25	1.35	2.00
	F	46	1.45	1.95	1.15	1.65
50–59	M+F	94	1.45	2.10	1.25	1.95
	M	51	1.45	2.10	1.20	1.95
	F	43	1.45	2.05	1.25	1.90
60–82	M+F	115	1.40	2.35	1.25	2.10
	M	71	1.45	2.35	1.30	2.00
	F	44	1.40	2.10	1.25	2.05

Table A1.76

Normal lower and upper 2½ percentiles (After Dower G. Polarcardiography. 1961. © Thomas: Springfield, Illinois. Reproduced with permission)

	74 young women on university campus		121 young men on university campus		192 elderly men with normal ECGs	
	Lower	Upper	Lower	Upper	Lower	Upper
PR interval (s)	0.111	0.202	0.113	0.207	0.111	0.200
QRS duration (ms)	62.0	105.0	70.0	109.0	61.0	106.0
QT_C interval (s)	0.369	0.451	0.357	0.442		
QT interval (ms)					0.302	0.426
M_P (mV)	0.028	0.182	0.042	0.198		
M_R(Mv)	0.751	2.172	0.951	2.111	0.695	2.028
M_{ST}(mV)	0.042	0.193	0.088	0.332	0.000	0.222
M_T(mV)	0.209	0.654	0.300	0.875	0.193	0.742
M_{ST}/M_R	0.035	0.133	0.064	0.240	0.000	0.151
M_T/M_R	0.147	0.536	0.207	0.700	0.141	0.604
Latitude difference (°)	−78.0	−9.50	−105.0	−11.7		
Longitude difference (°)	−7.88	43.8	−18.7	73.3		
Angle between $\hat{R}$ and $\hat{T}$ (°)	14.7	80.9	14.9	114.5	18.2	149.6

Table A1.77

Lower 2½ percentiles, medians, and upper 2½ percentiles for various polarcardiographic quantities (measured by computer in apparently healthy hospital staff). M_P, spatial magnitude of maximum P vector, $\vec{P}$; M_R, spatial magnitude of maximum QRS vector, $\vec{R}$; M_T spatial magnitude of maximum T vector, $\vec{T}$; m_tR, magnitude of maximum transverse plane QRS vector; M_{ST}, spatial magnitude of vector $\vec{ST}$ occurring midway in time between end of QRS complex and $\vec{T}$; lat., latitude posterior or anterior to frontal plane; long., longitude, or angle α in the frontal plane (After Dower G E, Osborne JA. Polarcardiographic study of hospital staff-abnormalities found in smokers. *J. Electrocardiol.* 1972; 5: 273–80. © Churchill Livingstone, New York. Reproduced with permission)

	137 women, median age 47			117 men, median age 51		
	Lower 2½ percentile	Median	Upper 2½ percentile	Lower 2½ percentile	Median	Upper 2½ percentile
PR interval (s)	0.10 (0.11)[a]	0.14	0.19 (0.20)[a]	0.11 (0.11)	0.16	0.19 (0.21)[a]
QRS duration (ms)	60 (62)	80	100 (105)	70 (70)	80	100 (109)
QT_C interval (s)	0.36 (0.37)	0.41	0.45 (0.45)	0.35 (0.36)	0.40	0.44 (0.44)
M_P (mV)	0.06 (0.03)	0.14	0.26 (0.18)	0.06 (0.04)	0.14	0.24 (0.20)
			0.22 (0.17)[b]			0.22 (0.19)[b]
M_R (mV)	0.82 (0.75)	1.41	2.24 (2.17)	1.02 (0.95)	1.50	2.20 (2.11)
			1.98 (1.82)[b]			2.14 (1.90)[b]
m_tR (mV)	0.72 (0.56)	1.16	1.84 (1.40)	0.77 (0.77)	1.31	2.13 (1.68)
			1.71 (1.28)[b]			1.95 (1.59)[b]
M_{ST} (mV)	0.04 (0.04)	0.12	0.22 (0.19)	0.09 (0.09)	0.21	0.35 (0.33)
M_T (mV)	0.21 (0.21)	0.44	0.82 (0.65)	0.28 (0.30)	0.60	0.92 (0.88)
M_{ST}/M_R	0.02 (0.04)	0.08	0.20 (0.13)	0.06 (0.06)	0.13	0.27 (0.24)
M_T/M_R	0.13 (0.15)	0.31	0.68 (0.54)	0.19 (0.21)	0.39	0.68 (0.70)
Lat. difference between $\hat{R}$ and $\hat{T}$ (°)	−115 (−80)	−44	+3 (−10)	−109 (−105)	−47	−2 (−10)
Long. difference between $\hat{R}$ and $\hat{T}$ (°)	−41 (−10)	+5	+139 (+45)	−45 (−20)	0	157 (+75)
Angle between $\hat{R}$ and $\hat{T}$ (°)	9 (15)	46	134 (81)	11 (15)	50	125 (115)

[a]Corresponding values from 74 young women and 121 young men.
[b]Upper 5 percentiles.

A1.7 Linear and Directional Statistics

A1.7.1 Acknowledgment

The work in this section has been compiled by Dr. Jerome Liebman of the Rainbow Babies and Children's Hospital, Cleveland, Ohio.

A1.7.2 Linear Statistics

Linear statistics are utilized for all linear measurements such as durations and magnitudes. Standard statistical methods are readily available, although there are many issues which must be understood in order to ensure appropriate use.

First, most standard ECGs are recorded with a maximal frequency response of 100–125 Hz, as are some orthogonal ECGs. However, for most orthogonal ECGs, the frequency response has been at least 250 Hz, often 500 Hz. In the author's laboratory, for recording the standard ECG, an upper frequency response limit above 250 Hz is preferred while

Table A1.78

PCG variables, at rest and after maximal exercise, in 30 healthy middle-aged men and 32 healthy young men compared. (After Bruce R A, Li Y B, Dower G E, Nilson K. Polarcardiographic responses to maximal exercise and to changes in posture in healthy middle aged men. J. Electrocardiol. 1973; 6: 91–6. I Churchill Livingstone, New York. Reproduced with permission)

	At rest					After exercise					
	Middle-aged		Young			Middle-aged		Young			Middle-aged rest versus exercise
Variables	$\bar{X}$	σ	$\bar{X}$	σ	p	$\bar{X}$	Σ	$\bar{X}$	σ	p	p
Heart rate (bmp)	67	11	72	13	NS	155	8	197	8	<0.001	<0.0001
PR interval (ms)	158	24	128	28	<0.001	121	15	110	15	<0.01	<0.0001
QRS duration (ms)	68	2	80	8	<0.05	75	6	79	8	<0.05	NS
RT interval (ms)	237	40				145	15				<0.0001
QT interval (ms)	360	30	341	40	<0.05	228	22	214	15	<0.01	<0.0001
$\hat{P}$: M (mV)	0.14	0.05	0.14	0.05	NS	0.20	0.05	0.27	0.08	< 0.001	<0.0001
α long. (°)	+45	28	+57	49	NS	+57	42	+63	41	NS	NS
PA lat. (°)	6P	23	2P	22	NS	7A	21	13A	23	NS	0.03
$\hat{R}$: M (mV)	1.48	0.36	1.76	0.46	NS	1.30	0.38	1.42	0.55	NS	NS
α long. (°)	+27	35	+42	15	<0.05	+38	57	+34	81	NS	NS
PA lat. (°)	26P	16	19P	22	NS	31P	28	43	20	NS	NS
$\hat{ST}$: M (mV)	0.15	0.05	0.20	0.06	<0.002	0.11	0.05	0.22	0.08	<0.001	<0.01
α long. (°)	+28	19	+33	18	NS	−8	94	+20	48	NS	NS
PA lat. (°)	41A	25	39A	24	NS	49A	26	43A	27	NS	NS
$\hat{T}$: M (mV)	0.56	0.17	0.59	0.18	NS	0.60	0.15	0.67	0.17	NS	NS
α long. (°)	+24	11	+33	12	<0.01	+26	20	+30	21	NS	NS
PA lat. (°)	29A	19	29A	16	NS	39A	20	35A	21	NS	

Table A1.79

Polarcardiographic responses in 72 apparently healthy middle-aged women and 40 young women before and after maximal exercise. M, spatial magnitude (After Bruce RA, Dower GE, Whitkanack S, Voigt AE. Polar-cardiographic responses to maximal exercise in middle-aged women. *J. Electrocardiol*. 1974; 7: J15–22. © Churchill Livingstone, New York. Reproduced with permission)

Variables	Before maximal exercise			After maximal exercise			Before versus after exercise middle aged *p*
	Young	(Mean ± SD) Middle-aged	*p*	Young	(Mean ± SD) Middle-aged	p	
Heart-rate (bpm)	81.0 ± 15.0	74.7 ± 10.6	0.01	192.2 ± 9.0	151.6 ± 17.1	0.00001	0.00001
PR interval (ms)	138 ± 26	141 ± 17	NS	108 ± 14	129 ± 16	0.00001	0.001
QRS duration (ms)	73 ± 6	77 ± 8	0.01	72 ± 10	76 ± 7	0.05	NS
QT interval (ms)	339 ± 43	351 ± 25	NS	215 ± 15	248 ± 23	0.00001	0.00001
$\hat{P}$: *M* (mV)	0.15 ± 0.07	0.14 ± 0.06	NS	0.22 ± 0.06	0.23 ± 0.09	NS	0.00001
α longitude (°)	62.0 ± 44.0	30.5 ± 38.4	0.001	69.0 ± 37.0	32.8 ± 48.0	0.0001	NS
PA latitude (°)	A4.6 ± 29.0	A1.5 ± 27.3	NS	A10.0 ± 31.0	A11.1 ± 31.1	NS	0.05
$\hat{R}$: *M* (mV)	1.40 ± 0.34	1.49 ± 0.37	NS	1.23 ± 34.0	1.32 ± 32.3	NS	NS
α longitude (°)	46.0 ± 20.0	32.7 ± 22.6	0.01	41.0 ± 50.0	39.2 ± 30.6	NS	NS
PA latitude (°)	P27.0 ± 18.0	P26.2 ± 22.7	NS	P40.0 ± 21.0	P32.9 ± 22.6	NS	NS
ST: *M* (mV)	0.10 ± 0.05	0.10 ± 0.04	NS	0.16 ± 0.07	0.10 ± 0.04	0.00001	NS
α longitude (°)	29.0 ± 34.0	8.7 ± 48.5	0.05	16.0 ± 60.0	−9.5± 72.4	NS	NS
PA latitude (°)	A36.0 ± 23.0	A57.7 ± 212	0.00001	A41.0 ± 30.0	A53.7 ± 29.2	0.05	NS
$\hat{T}$: *M* (mV)	0.37 ± 0.15	0.42 ± 0.14	NS	0.53 ± 0.16	0.45 ± 0.12	0.01	NS
α longitude (°)	33.0 ± 13.0	21.4 ± 13.7	0.0001	33.0 ± 20.0	17.9 ± 33.3	0.05	NS
PA latitude (°)	A15.0 ± 20.0	A27.0 ± 13.2	0.001	A35.0 ± 18.0	A40.0 ± 17.2	NS	0.00001

the machine which is regularly used has an upper limit listed at 700 Hz. Although it is not well known by most electrocardiographers, the higher-frequency response electrocardiograph allows detection of higher-frequency content in the ECG. It has been shown that in order to detect a frequency *content* of at least 100 Hz, it is necessary that the instrument have a frequency *response* of at least 200 Hz [1]. It should be appreciated that, in the *same patient*, with the electrodes attached and an appropriate electrocardiograph running, merely switching from the normal high-frequency response to the low-frequency response usually to avoid ac interference produces a significant decrease in QRS voltage. Therefore, separate tables of normal ranges should be available [2, 3] for both low-frequency response machines and high-frequency response machines. This is particularly true for the pediatric ECG. Manufacturers have a responsibility to produce equipment that meets the required standards. Nowadays, equipment has to meet strict standards but technical staff often opt for a low frequency setting to reduce noise, thereby running the risk of distorting the ECG.

A second issue is that from puberty, different age/sex groups have different normal voltage ranges. Pediatric cardiologists are aware that the voltages in pubescent females are lower than those of males. The same is true in virtually all age-groups through adulthood. In addition, for each decade throughout adulthood, total voltage decreases in both males and females, so that standards should be available for each, throughout the decades [4].

Prior to puberty, there is no difference in normal ECG ranges between males and females.

However, of great interest is the fact that the newborn baby has lower voltages than older infants and children, and these increase dramatically from the newborn period till about the age of 3 months. It is also striking that the prematurely-born infant has even lower voltages than does the full-term baby [5, 6]. Catch-up occurs at varying ages for each individual baby, and usually at about 3 months.

Racial differences are also important, but, again, probably not until puberty. Blacks may have much higher voltages than whites, and, occasionally, individual normal black teenagers may have extremely high voltages, with no available explanation. Unfortunately, separate large tables for normal blacks are not available.

A fourth major issue is that electrocardiographic amplitude data is skewed. The mean and the median are rarely the same, the distribution in the various age ranges not being Gaussian [2, 3]. Almost always the 50th percentile is below the mean, in addition to which the mean plus twice standard deviation may be higher than the maximal value recorded. Therefore, the use of means and standard deviations for magnitude is not appropriate, which is why the use of percentiles to describe the distribution is necessary (see ➋ Chap. 13). In describing the complete distribution, as is frequently done, a series of percentiles is optimal. In using the percentiles to describe low and high limits of the normal range, the $2\frac{1}{2}$ and $97\frac{1}{2}$ percentiles are recommended. However, because of known inaccuracies of measurement, the 5th and 95th percentiles denoted p_5 and p_{95} are considered as being satisfactory by the author.

A1.7.3 Directional Statistics

For directional (circular and spherical) data, linear statistics are not appropriate. The fallacies in employing linear methods for directional statistics are easily understood, although, because of the lack of available methodology, such incorrect measurements were almost invariably used [7]. For example, if two vectors, 1° and 359° are grouped, the mean of 360°/2 = 180° is obviously incorrect, since it is 180° away from the true average direction of 0°. Why not, therefore, change the terminology to +1° and −1°? The mean of 0°/2 = 0° would be correct. However, suppose two vectors at 179° and 181° were then averaged. According to the new terminology, the electrocardiographer would then have to average +179° and −179° in order to be consistent. The average of +179° and −179° = 0°/2 = 0°, which is again 180° away from the true average direction of 180°. Some of these rotational adjustments have been reported [8], but in using linear statistics, the adjustments have a minimal error only when there is an intense clustering of the data. Every linear treatment of directional data has inherent error. Whether the angles are, measured from −180° to +180°, or from 0° to 360°, and so on, some of the observations (actually close together) *must inevitably be treated as being far apart*. Therefore, appropriate statistical analyses of planar data as well as spherical data were developed by Downs et al. [9, 10], based upon the Von Mises distribution [11–13]. This new set of statistics was termed the "center of gravity method." In linear analysis, there is an arithmetic mean. In circular analysis, there is a prevalent direction.

For details of the methods, the reader is referred to pertinent literature, including utilization with tabular data. However, some explanations and descriptions will be given here. ➋ Table A1.80 is a segment of a table from 50 premature infants where the ECG was recorded with the Frank system.

Table A1.80
Segment of a table from 50 premature infants where the ECG was recorded with the Frank-lead system. n_f, n_s, n_h are the number of angles under study; d_f, d_s, d_h are the distances to the center of gravity; $\widehat{\alpha}_f$, $\widehat{\alpha}_s$, $\widehat{\alpha}_h$ are the prevalent directions of the vector. All parameters represent measurements taken in the frontal, sagittal and horizontal planes, respectively (denoted by the subscripts f, s, h)

Age (h)	Times vector (ms)	n_f	d_f	$\hat{\alpha}_f$	χ_f^2	n_s	d_s	$\hat{\alpha}_s$	χs^2	n_h	d_h	$\hat{\alpha}_h$	χ_h^2
24	10	50	0.41	7	16.4	50	0.95	178	90.6	50	0.96	83	91.6

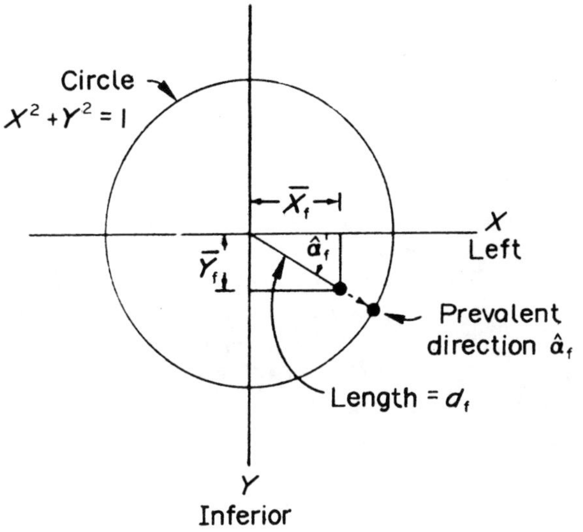

Figure A1.1
The center of gravity (X_f, Y_f) or, in polar coordinates (d_f, α_f) of points (x_f, y_f) on the circle $X^2 + Y^2 = 1$ in the frontal plane.

Imagine the data for the 10 ms vector of the 50 premature infants to be on a circular disk, balanced on a fulcrum at its center. If all the observations were in approximately the same direction, then they would all cluster together and the disk would tilt maximally. The direction of tilt would be toward the center of gravity of the points: this is termed the prevalent direction. If the points were scattered equally around the disk, there would be no tilt and there would be no prevalent direction.

In calculating the prevalent direction $\widehat{\alpha}_f$ the "average" of the 50 10 ms angles (each α_f) in the frontal plane, reference is made to Fig. A1.1.

For each angle α_f, X_f and Y_f are computed as follows.

$$X_f = \cos \alpha_f, \qquad Y_f = \sin \alpha_f \tag{A1.1}$$

The means are then computed from:

$$\overline{X}_f = \sum X_f / n_f \qquad \overline{Y}_f = \sum Y_f / n_f \tag{A1.2}$$

where n_f is number of angles under study in the frontal plane.

The center of gravity corresponds to the point $(\overline{X}_f, \overline{Y}_f)$ and the distance d_f to the center of gravity is then

$$d_f = (\overline{X}_f + \overline{Y}_f) \tag{A1.3}$$

The direction α_f toward the center of gravity is calculated from

$$\cos \hat{\alpha}_f = \overline{X}_f / d_f \quad \text{or} \quad \sin \hat{\alpha}_f = \overline{Y}_f / d_f \tag{A1.4}$$

wnere d_f is termed the precision. If it is zero, then there is no prevalent direction. If it is 1.0, then *all* the individual measurements are the same. The higher the precision *d*, the more the clustering; the lower the precision d, the more the scatter.

In order to determine whether a calculated prevalent direction can be trusted, that is, whether a true prevalent direction actually exists, a χ^2 value can be calculated where:

$$\chi_f^2 = 2n_f d_f^2 \tag{A1.5}$$

Values of χ^2 greater than 5.99 are significant at the 5% level, and values of χ^2 greater than 9.21 are significant at the 1% level.

For Table A1.80, d_f is not high (0.41), so that there is little cluster of the $\hat{\alpha}_f$. However, the prevalent direction of 7° can be trusted, since the $\chi_f^2 = 16.4$. For the sagittal and horizontal planes, the values of χ^2 are extremely high at 90.6 and 91.6 so that the high precisions $d_s = 0.95$ and $d_h = 0.96$ (and thus high cluster) can be very reliably trusted. (To determine whether a particular measured angle α is likely to be within the accepted normal range, it is only necessary to determine whether the angle is between the p_5 and p_{95} for the normal population.)

For spatial data, a methodology has been developed similar to that of the above planar data. In this case, consider that there exists a sample of spatial vectors with coordinates (X, Y, Z) and that it is desired to determine an "average" spatial direction for n measurements. Imagine a sphere placed in the surface of a liquid. If the *n* measurements cluster about some direction, then the sphere will rotate in the liquid until the center of gravity of the *n* measurements points downward. This is the spatial prevalent direction (Fig. A1.2).

As before, the distance from the center of the sphere to the center of gravity $(\overline{X}, \overline{Y}, \overline{Z})$ is a measure of how much the observations cluster, and is called the *spatial precision*. The center of gravity $(\overline{X}, \overline{Y}, \overline{Z})$ can also be expressed in spherical coordinates $(d, \hat{\alpha}, \widehat{\beta})$ where $\hat{\alpha}$ is the longitude or angular deviation from the left in the horizontal plane (0–360°) and $\widehat{\beta}$ is the colatitude or angular deviation from the superior (0–180°). (The spherical α is identical to the planar α_h.) The spatial prevalent direction is $(\hat{\alpha}, \widehat{\beta})$ and can be determined uniquely.

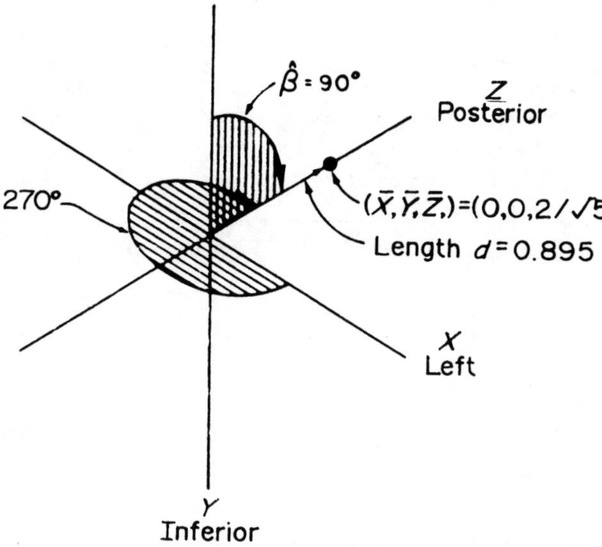

Figure A1.2
An example of the spatial prevalent direction expressed in spherical and Cartesian coordinates.

Table A1.81
Segment of spatial data for the same 50 premature babies tabulated in Table A1.80

Age (h)	Timed vector (ms)	X	Y	Z	d	$\hat{\alpha}$ (°)	$\hat{\beta}$ (°)	χ^2	n
24	10	0.122	0.045	-0.910	0.920	82	93	127.1	50

A χ^2 can be calculated to test whether the clustering about the spatial prevalent direction can be trusted. Values of χ^2 greater than 7.81 are significant at the 5% level and values of χ^2 larger than 11.34 are significant at the 1% level. For further details and the methodology (with equations) for calculation of the above, the reader is referred to an original paper and a recent summary [14].

Table A1.81 is a segment of spatial data for the same 50 premature babies given in Table A1.80.

The d and χ^2 are both very high, consistent with the high d for the horizontal and sagittal planes, despite the low d for the frontal plane. The α of 82° is actually the $\hat{\alpha}_h$. The $\widehat{\beta}$ of 93° has no corollary in the planar directions. However, from these spatial data, all the planar data can be calculated.

For detailed research (as well as descriptive) purposes, this methodology has been very satisfactory. As with linear statistics, for example, it is frequently necessary to determine whether two different samples of directional measurements are from the same or different populations. The necessary methodology is detailed in the appendix to [9].

References

1. Thomas, C., Electrocardiographic measurement system response, in *Pediatric Electrocardiography*, J. Liebman, R. Plonsey, and P. Gillette, Editors. Baltimore: Williams and Wilkins, 1982, Chap. 5, pp. 40–59.
2. Liebman, J. and R. Plonsey, Electrocardiography, in *Moss' Heart Disease in Infants, Children and Adolescents*, H. Adams and G.C. Emmanouillides, Editors. Baltimore: Williams and Wilkins, 1983, Chap. 3.
3. Liebman, J., Tables of normal standards, in *Pediatric Cardiology*, J. Liebman, R. Plonsey, and P. Gillette, Editors. Baltimore: Williams and Wilkins, 1982, Chap. 8, pp. 82–133.
4. Pipberger, H.V., M.J. Goldman, D. Littman, F.P. Murphy, J. Cosma, and J.R. Snyder, Correlations of the orthogonal electrocardiogram and vectorcardiogram with constitutional variables in 518 normal men. *Circulation*, 1967; **35**: 536.
5. Sreenivasan, V.V, B.J. Fisher, J. Liebman, and T.D. Downs, A longitudinal study of the standard electrocardiogram in the healthy premature infant during the first year of life. *Am. J. Cardiol.*, 1973;**31**: 57.
6. Liebman, J., H.C. Romberg, T.D. Downs, and R. Agusti, The Frank QRS vectorcardiogram in the premature infant, in *Vectorcardiography, 1965*, I. Hoffman and R.C. Taymore, Editors. Amsterdam: North-Holland, 1966.
7. Hugenholtz, P.C. and J. Liebman, The orthogonal electrocardiogram in 100 normal children (Frank system), with some comparative data recorded by the cube system. *Circulation*, 1962;**26**: 891.
8. Liebman, J., C. Doershuk, C. Rapp, and L. Matthews, The vectorcardiogram in cycstic fibrosis: diagnostic significance and correlation with pulmonary function tests. *Circulation*, 1967;**32**: 552.
9. Downs, T.D., J. Liebman, R. Agusti, and H.C. Romberg, The statistical treatment of angular data in vectocardiography, in *Vectorcardiography, 1965*, I. Hoffman and R.C. Traymore, Editors. Amsterdam: North-Holland, 1966, p. 272.
10. Downs, T.D. and J. Liebman, Statistical methods for vectorcardiographic directions. *IEEE Trans. Biomed. Eng.*, 1969;**16**: 87.
11. Stephens, M.A., *The Statistics of Directions. The Von Mises and Fisher Distributions*, PhD thesis. Tornto: University of Toronto, 1962.
12. Downs, T.D., *The Von Mises Distribution: Derived Distributions Regression Theory, and Some Applications to Meterological Problems*, PhD thesis. Ann Arbor: University of Michigan, 1965.
13. Downs, T.D., Some relationships amont the von Mises distributions of different dimensions. *Biometrika*, 1966;**53**: 269.
14. Liebman, J., Statistics related to electrocardiographic interpretation, in *Pediatric Electrocardiography*, J. Liebman, R. Plonsey, and P. Gillete, Editors. Baltimore: Williams and Wilkins, 1982, Chap. 7, pp. 76–81.

Appendix 2: Paediatric Normal Limits

P. W. Macfarlane, A. van Oosterom, O. Pahlm, P. Kligfield, M. Janse, J. Camm (eds.), *Comprehensive Electrocardiology*, DOI 10.1007/978-1-84882-046-3_A2

A2.1 Normal Limits of the Paediatric 12-Lead ECG

This appendix is based on a series of 1,784 ECGs collected from neonates, infants and children in Glasgow, Scotland in the late 1980s. Some outline information was previously published [1, 2] but the detailed normal limits as presented in this chapter have never previously been published. Lead V_{4R} has been recorded to the exclusion of lead V_3. Precordial leads are therefore presented in the sequence V_1, V_2, V_{4R}, V_4, V_5, V_6.

One of the significant aspects of this data is the availability of ECGs from over 500 neonates from birth to 7 days of life. Nowadays, new mothers tend to be discharged from hospital within 24–48 h and so the difficulty of collecting such a database is significantly increased.

The standard presentation of mean together with standard deviation and 96 percentile range has been used throughout. (➋ Tables A2.1–A2.24) With small numbers in some groups this leads to an irregular upper limit of normal for some measures but in practice, continuous equations for normal limits can be developed (see ➋ Chap. 13).

Table A2.1
Durations (in milliseconds) in Caucasian children: lead I

Age-group	Q duration	S duration	R duration	R′ duration
<24 h	36 ± 17	16 ± 4	29 ± 4	
	6 → 62	10 → 23	22 → 40	
	n = 10	n = 35	n = 35	
<1 day	33 ± 15	17 ± 7	29 ± 4	
	9 → 52	6 → 25	22 → 40	
	n = 36	n = 112	n = 110	
<2 days	29 ± 17	17 ± 5	28 ± 5	24
	8 → 56	6 → 26	19 → 38	24 → 24
	n = 34	n = 99	n = 99	n = 1
<3 days	30 ± 16	17 ± 5	28 ± 5	30
	8 → 56	7 → 28	17 → 38	30 → 30
	n = 41	n = 80	n = 79	n = 1
<1 week	23 ± 17	16 ± 5	29 ± 6	17
	8 → 56	5 → 27	20 → 43	17 → 17
	n = 38	n = 104	n = 104	n = 1
<1 month	16 ± 15	18 ± 5	26 ± 6	
	5 → 52	11 → 32	17 → 43	
	n = 14	n = 43	n = 43	
<3 months	11 ± 5	22 ± 6	24 ± 6	24
	6 → 30	10 → 34	13 → 34	24 → 24
	n = 34	n = 66	n = 65	n = 1
<6 months	12 ± 4	24 ± 5	25 ± 6	
	5 → 18	17 → 41	14 → 35	
	n = 21	n = 50	n = 46	
<1 year	12 ± 3	24 ± 5	24 ± 7	33
	8 → 19	14 → 38	14 → 42	33 → 33
	n = 40	n = 73	n = 70	n = 1
1–2 years	13 ± 4	27 ± 7	24 ± 7	24 ± 2
	5 → 25	16 → 46	10 → 39	22 → 25
	n = 69	n = 111	n = 95	n = 2
3–4 years	14 ± 4	30 ± 7	28 ± 8	
	7 → 21	20 → 51	14 → 45	
	n = 77	n = 141	n = 109	

Table A2.1 (Continued)

Age-group	Q duration	S duration	R duration	R′ duration
5–6 years	14 ± 5	31 ± 6	27 ± 8	
	6 → 40	21 → 46	13 → 44	
	n = 71	*n* = 155	*n* = 136	
7–8 years	13 ± 5	32 ± 8	27 ± 8	10
	7 → 21	21 → 54	14 → 45	10 → 10
	n = 80	*n* = 141	*n* = 114	*n* = 1
9–10 years	13 ± 4	36 ± 9	27 ± 9	17 ± 5
	7 → 21	24 → 64	6 → 45	13 → 22
	n = 49	*n* = 115	*n* = 92	*n* = 3
11–12 years	14 ± 4	37 ± 7	29 ± 9	
	8 → 29	24 → 55	14 → 50	
	n = 62	*n* = 119	*n* = 90	
13–14 years	15 ± 4	39 ± 8	28 ± 10	23 ± 6
	7 → 27	27 → 63	12 → 55	19 → 27
	n = 62	*n* = 141	*n* = 113	*n* = 2
15–16 years	15 ± 5	39 ± 9	30 ± 11	
	8 → 29	27 → 63	11 → 61	
	n = 46	*n* = 92	*n* = 71	
17–18 years	32			
	32 → 32			
	n = 1			

Table A2.2
Durations (in milliseconds) in Caucasian children: lead II

Age-group	Q duration	S duration	R duration	R′ duration
<24 h	18 ± 11	21 ± 8	18 ± 5	
	9 → 68	9 → 39	12 → 35	
	n = 41	*n* = 41	*n* = 22	
<1 day	16 ± 5	21 ± 9	19 ± 6	19 ± 13
	7 → 24	7 → 41	10 → 39	5 → 38
	n = 132	*n* = 136	*n* = 71	*n* = 5
<2 days	16 ± 5	21 ± 9	19 ± 7	18 ± 9
	9 → 38	7 → 44	10 → 45	12 → 28
	n = 106	*n* = 116	*n* = 69	*n* = 3
<3 days	16 ± 6	22 ± 12	19 ± 6	23 ± 12
	10 → 38	7 → 43	7 → 32	12 → 36
	n = 102	*n* = 105	*n* = 56	*n* = 3
<1 week	17 ± 6	19 ± 6	20 ± 7	15 ± 0
	9 → 44	7 → 35	6 → 37	15 → 15
	n = 115	*n* = 114	*n* = 75	*n* = 2

Table A2.2 (Continued)

Age-group	Q duration	S duration	R duration	R′ duration
<1 month	16 ± 3	21 ± 6	19 ± 6	24
	11 → 23	9 → 39	12 → 32	24 → 24
	n = 41	*n* = 45	*n* = 28	*n* = 1
<3 months	16 ± 3	25 ± 7	17 ± 4	26 ± 0
	12 → 24	11 → 45	12 → 26	26 → 26
	n = 58	*n* = 67	*n* = 46	*n* = 2
<6 months	16 ± 5	27 ± 6	19 ± 6	19 ± 6
	6 → 24	19 → 38	13 → 34	15 → 23
	n = 40	*n* = 49	*n* = 32	*n* = 2
<1 year	14 ± 5	29 ± 7	21 ± 6	36
	6 → 26	18 → 54	12 → 34	36 → 36
	n = 54	*n* = 73	*n* = 52	*n* = 1
1–2 years	16 ± 4	32 ± 9	23 ± 7	21 ± 7
	8 → 26	21 → 56	12 → 37	13 → 26
	n = 79	*n* = 111	*n* = 74	*n* = 3
3–4 years	16 ± 4	34 ± 10	24 ± 9	17 ± 4
	7 → 23	23 → 57	8 → 41	14 → 21
	n = 106	*n* = 141	*n* = 98	*n* = 3
5–6 years	15 ± 4	36 ± 9	23 ± 8	24
	7 → 22	24 → 60	12 → 43	24 → 24
	n = 118	*n* = 155	*n* = 114	*n* = 1
7–8 years	15 ± 6	37 ± 8	25 ± 9	18
	7 → 24	26 → 61	13 → 46	18 → 18
	n = 94	*n* = 141	*n* = 115	*n* = 1
9–10 years	15 ± 4	40 ± 9	23 ± 10	17 ± 5
	7 → 21	26 → 64	7 → 47	14 → 23
	n = 86	*n* = 115	*n* = 84	*n* = 3
11–12 years	14 ± 5	43 ± 11	25 ± 9	17 ± 1
	7 → 26	29 → 78	7 → 45	16 → 18
	n = 83	*n* = 119	*n* = 87	*n* = 3
13–14 years	14 ± 4	43 ± 11	26 ± 10	13 ± 4
	7 → 25	30 → 75	9 → 50	7 → 17
	n = 100	*n* = 141	*n* = 111	*n* = 5
15–16 years	16 ± 6	45 ± 12	28 ± 11	16
	7 → 37	29 → 76	11 → 53	16 → 16
	n = 60	*n* = 92	*n* = 67	*n* = 1
17–18 years	15	22		
	15 → 15	22 → 22		
	n = 1	*n* = 1		

Table A2.3

Durations (in milliseconds) in Caucasian children: lead III

Age-group	Q duration	S duration	R duration	R′ duration
<24 h	17 ± 3	28 ± 7	18 ± 4	12
	12 → 23	15 → 41	12 → 21	12 → 12
	n = 41	*n* = 43	*n* = 4	*n* = 1
<1 day	18 ± 3	28 ± 8	15 ± 3	32 ± 6
	13 → 23	12 → 44	11 → 21	27 → 39
	n = 131	*n* = 138	*n* = 32	*n* = 3
<2 days	17 ± 3	27 ± 8	15 ± 5	16 ± 7
	9 → 24	7 → 44	7 → 32	9 → 25
	n = 110	*n* = 119	*n* = 32	*n* = 5
<3 days	17 ± 3	28 ± 12	15 ± 3	20 ± 3
	10 → 25	8 → 56	8 → 22	15 → 23
	n = 101	*n* = 106	*n* = 22	*n* = 6
<1 week	17 ± 3	27 ± 8	16 ± 4	14 ± 6
	10 → 24	8 → 43	12 → 31	9 → 20
	n = 115	*n* = 117	*n* = 32	*n* = 3
<1 month	19 ± 4	28 ± 9	15 ± 4	26 ± 7
	13 → 28	9 → 54	8 → 22	21 → 31
	n = 41	*n* = 45	*n* = 11	*n* = 2
<3 months	20 ± 3	27 ± 8	17 ± 5	24 ± 5
	13 → 28	8 → 43	10 → 32	18 → 30
	n = 59	*n* = 67	*n* = 17	*n* = 7
<6 months	22 ± 4	25 ± 9	17 ± 7	23 ± 5
	14 → 30	6 → 42	8 → 41	14 → 31
	n = 38	*n* = 50	*n* = 20	*n* = 9
<1 year	21 ± 4	23 ± 9	16 ± 6	22 ± 10
	14 → 30	5 → 38	8 → 39	5 → 53
	n = 47	*n* = 73	*n* = 35	*n* = 23
1–2 years	21 ± 6	25 ± 11	20 ± 11	20 ± 8
	9 → 54	7 → 54	8 → 63	8 → 39
	n = 68	*n* = 110	*n* = 63	*n* = 33
3–4 years	21 ± 6	31 ± 13	22 ± 12	24 ± 9
	6 → 32	6 → 64	6 → 58	9 → 44
	n = 94	*n* = 140	*n* = 79	*n* = 23
5–6 years	18 ± 5	33 ± 13	19 ± 10	26 ± 7
	5 → 27	6 → 60	6 → 49	13 → 43
	n = 106	*n* = 155	*n* = 84	*n* = 22
7–8 years	19 ± 6	35 ± 13	24 ± 11	28 ± 14
	8 → 28	7 → 62	7 → 53	8 → 48
	n = 82	*n* = 141	*n* = 94	*n* = 19
9–10 years	19 ± 4	39 ± 14	21 ± 11	23 ± 12
	9 → 28	13 → 72	5 → 60	5 → 47
	n = 78	*n* = 115	*n* = 70	*n* = 15

Table A2.3 (Continued)

Age-group	Q duration	S duration	R duration	R′ duration
11–12 years	18 ± 6	40 ± 16	24 ± 12	23 ± 12
	7 → 32	7 → 80	5 → 51	5 → 45
	n = 75	*n* = 119	*n* = 74	*n* = 14
13–14 years	19 ± 6	41 ± 17	26 ± 13	26 ± 17
	8 → 31	9 → 79	6 → 57	5 → 71
	n = 88	*n* = 141	*n* = 93	*n* = 21
15–16 years	19 ± 6	44 ± 17	28 ± 14	28 ± 14
	9 → 37	8 → 82	6 → 67	11 → 48
	n = 55	*n* = 92	*n* = 55	*n* = 8
17–18 years	14	17		
	14 → 14	17 → 17		
	n = 1	*n* = 1		

Table A2.4

Durations (in milliseconds) in Caucasian children: lead aVR

Age-group	Q duration	S duration	R duration	R′ duration
<24 h	17 ± 7	22 ± 13	16 ± 6	21 ± 7
	8 → 30	8 → 68	6 → 31	11 → 31
	n = 15	*n* = 42	*n* = 24	*n* = 15
<1 day	20 ± 6	22 ± 12	15 ± 7	19 ± 6
	8 → 35	8 → 50	6 → 40	0 → 33
	n = 46	*n* = 135	*n* = 73	*n* = 57
<2 days	21 ± 6	19 ± 10	14 ± 5	22 ± 6
	13 → 38	7 → 46	5 → 30	11 → 42
	n = 35	*n* = 117	*n* = 72	*n* = 62
<3 days	20 ± 6	20 ± 11	15 ± 6	20 ± 7
	8 → 33	7 → 48	6 → 35	11 → 32
	n = 26	*n* = 105	*n* = 67	*n* = 51
<1 week	18 ± 7	19 ± 11	14 ± 4	21 ± 5
	10 → 34	8 → 56	5 → 22	12 → 33
	n = 20	*n* = 116	*n* = 86	*n* = 80
<1 month	22 ± 3	15 ± 7	16 ± 4	21 ± 6
	17 → 27	8 → 42	8 → 26	13 → 37
	n = 8	*n* = 45	*n* = 36	*n* = 33
<3 months	26 ± 7	14 ± 5	21 ± 4	20 ± 5
	11 → 41	5 → 29	14 → 34	12 → 28
	n = 13	*n* = 67	*n* = 55	*n* = 48
<6 months	26 ± 4	16 ± 6	24 ± 3	21 ± 5
	18 → 31	6 → 33	20 → 34	13 → 32
	n = 16	*n* = 50	*n* = 34	*n* = 27
<1 year	28 ± 5	16 ± 8	26 ± 6	21 ± 6
	11 → 37	6 → 36	17 → 46	12 → 38
	n = 25	*n* = 72	*n* = 49	*n* = 37

Table A2.4 (Continued)

Age-group	Q duration	S duration	R duration	R′ duration
1–2 years	29 ± 7	17 ± 7	28 ± 6	22 ± 8
	11 → 46	7 → 35	19 → 45	8 → 38
	n = 33	n = 107	n = 80	n = 56
3–4 years	35 ± 8	17 ± 7	31 ± 6	24 ± 8
	24 → 72	8 → 38	21 → 51	12 → 43
	n = 39	n = 133	n = 102	n = 78
5–6 years	35 ± 7	17 ± 7	32 ± 5	23 ± 8
	27 → 68	7 → 36	24 → 48	12 → 44
	n = 52	n = 148	n = 103	n = 79
7–8 years	35 ± 4	18 ± 8	34 ± 7	25 ± 8
	29 → 46	8 → 44	24 → 51	12 → 40
	n = 45	n = 140	n = 96	n = 75
9–10 years	37 ± 5	17 ± 8	36 ± 7	23 ± 9
	29 → 48	7 → 41	27 → 62	7 → 49
	n = 43	n = 108	n = 72	n = 55
11–12 years	39 ± 7	19 ± 9	39 ± 8	25 ± 8
	28 → 60	8 → 46	28 → 67	13 → 44
	n = 48	n = 112	n = 71	n = 50
13–14 years	41 ± 8	19 ± 9	39 ± 7	25 ± 9
	33 → 74	8 → 43	29 → 62	12 → 52
	n = 59	n = 135	n = 83	n = 69
15–16 years	44 ± 9	22 ± 10	39 ± 6	27 ± 9
	33 → 68	9 → 48	30 → 60	15 → 50
	n = 41	n = 82	n = 50	n = 37
17–18 years		18	7	
		18 → 18	7 → 7	
		n = 1	n = 1	

Table A2.5

Durations (in milliseconds) in Caucasian children: lead aVL

Age-group	Q duration	S duration	R duration	R′ duration
<24 h	24 ± 17	16 ± 4	31 ± 4	
	9 → 40	10 → 22	23 → 42	
	n = 4	n = 41	n = 41	
<1 day	25 ± 18	17 ± 4	30 ± 5	13 ± 1
	7 → 54	11 → 23	21 → 41	0 → 13
	n = 8	n = 136	n = 134	n = 2
<2 days	23 ± 16	17 ± 4	30 ± 5	15 ± 4
	7 → 46	9 → 25	19 → 39	12 → 17
	n = 11	n = 116	n = 114	n = 2

Table A2.5 (Continued)

Age-group	Q duration	S duration	R duration	R′ duration
<3 days	20 ± 15	16 ± 4	30 ± 5	21 ± 7
	7 → 52	8 → 25	20 → 44	15 → 28
	n = 12	*n* =102	*n* = 102	*n* = 3
<1 week	16 ± 17	17 ± 4	30 ± 6	13 ± 5
	6 → 56	8 → 26	18 → 42	8 → 22
	n = 9	*n* = 115	*n* = 115	*n* = 6
<1 month	39 ± 27	19 ± 4	28 ± 6	12 ± 0
	9 → 62	11 → 29	16 → 47	12 → 12
	n = 5	*n* = 42	*n* = 42	*n* = 2
<3 months	12 ± 7	21 ± 4	27 ± 6	11 ± 2
	6 → 32	12 → 32	13 → 37	9 → 12
	n = 13	*n* = 66	*n* = 66	*n* = 2
<6 months	11 ± 4	24 ± 6	26 ± 6	20
	6 → 19	13 → 50	17 → 35	20 → 20
	n = 13	*n* = 50	*n* = 48	*n* = 1
<1 year	12 ± 5	22 ± 5	26 ± 7	25 ± 18
	6 → 26	12 → 37	13 → 41	12 → 37
	n = 30	*n* = 73	*n* = 71	*n* = 2
1–2 years	13 ± 5	24 ± 7	25 ± 8	17 ± 5
	6 → 28	14 → 52	8 → 43	13 → 26
	n = 54	*n* = 111	*n* = 98	*n* = 6
3–4 years	15 ± 8	25 ± 9	30 ± 10	19 ± 6
	6 → 40	10 → 49	6 → 48	6 → 33
	n = 60	*n* = 140	*n* = 118	*n* = 15
5–6 years	15 ± 10	23 ± 8	31 ± 11	18 ± 5
	5 → 62	8 → 44	9 → 51	12 → 28
	n = 58	*n* = 154	*n* = 140	*n* = 26
7–8 years	18 ± 10	25 ± 12	30 ± 12	21 ± 6
	6 → 54	10 → 60	6 → 53	12 → 35
	n = 71	*n* = 140	*n* = 115	*n* = 33
9–10 years	21 ± 17	25 ± 10	34 ± 12	21 ± 8
	7 → 76	10 → 57	10 → 60	9 → 44
	n = 37	*n* = 112	*n* = 99	*n* = 24
11–12 years	20 ± 10	27 ± 13	32 ± 14	23 ± 8
	10 → 57	6 → 66	7 → 59	7 → 39
	n = 49	*n* = 117	*n* = 100	*n* = 23
13–14 years	19 ± 11	29 ± 14	34 ± 16	22 ± 9
	6 → 74	9 → 69	6 → 72	7 → 42
	n = 55	*n* = 139	*n* = 118	*n* = 37
15–16 years	24 ± 17	30 ± 14	34 ± 14	22 ± 10
	8 → 78	10 → 63	6 → 65	7 → 45
	n = 41	*n* = 88	*n* = 70	*n* = 19
17–18 years		13	22	
		13 → 13	22 → 22	
		n = 1	*n* = 1	

Table A2.6
Durations (in milliseconds) in Caucasian children: lead aVF

Age-group	Q duration	S duration	R duration	R′ duration
<24 h	16 ± 4	24 ± 8	17 ± 8	
	11 → 24	10 → 41	6 → 37	
	n = 42	n = 43	n = 14	
<1 day	17 ± 4	25 ± 9	17 ± 5	21 ± 14
	10 → 24	11 → 45	11 → 36	9 → 37
	n = 133	n = 137	n = 44	n = 3
<2 days	17 ± 3	24 ± 8	17 ± 7	15 ± 3
	8 → 23	7 → 39	7 → 46	13 → 19
	n = 110	n = 119	n = 50	n = 3
<3 days	17 ± 4	25 ± 12	18 ± 5	22 ± 10
	10 → 27	5 → 51	10 → 31	15 → 29
	n = 101	n = 106	n = 36	n = 2
<1 week	17 ± 3	22 ± 7	18 ± 7	
	11 → 26	8 → 37	10 → 54	
	n = 116	n = 117	n = 56	
<1 month	17 ± 3	25 ± 7	17 ± 4	
	8 → 25	13 → 43	12 → 24	
	n = 43	n = 45	n = 19	
<3 months	17 ± 3	27 ± 8	15 ± 5	25 ± 8
	12 → 23	11 → 50	8 → 26	16 → 36
	n = 57	n = 67	n = 28	n = 6
<6 months	19 ± 4	26 ± 7	17 ± 7	20 ± 3
	11 → 27	8 → 42	8 → 38	16 → 24
	n = 40	n = 50	n = 26	n = 5
< 1 year	18 ± 4	28 ± 9	19 ± 7	26 ± 11
	11 → 28	5 → 58	7 → 36	5 → 36
	n = 50	n = 73	n = 40	n = 7
1–2 years	18 ± 5	30 ± 9	21 ± 9	23 ± 11
	8 → 27	9 → 52	5 → 46	5 → 38
	n = 73	n = 111	n = 66	n = 9
3–4 years	18 ± 4	35 ± 12	22 ± 10	23 ± 6
	11 → 26	8 → 62	6 → 41	10 → 31
	n = 95	n = 141	n = 85	n = 9
5–6 years	16 ± 4	36 ± 10	22 ± 8	29 ± 2
	7 → 24	13 → 62	12 → 43	27 → 31
	n = 114	n = 155	n = 99	n = 3
7–8 years	17 ± 6	38 ± 10	25 ± 10	18 ± 10
	7 → 24	24 → 68	7 → 47	6 → 33
	n = 87	n = 141	n = 97	n = 6
9–10 years	16 ± 4	42 ± 11	24 ± 10	16 ± 3
	7 → 22	25 → 72	9 → 53	13 → 19
	n = 79	n = 115	n = 74	n = 4
11–12 years	16 ± 5	44 ± 12	26 ± 10	10
	6 → 27	28 → 80	12 → 46	10 → 10
	n = 76	n = 119	n = 77	n = 1

Table A2.6 (Continued)

Age-group	Q duration	S duration	R duration	R′ duration
13–14 years	16 ± 5	44 ± 13	28 ± 10	19 ± 6
	8 → 26	29 → 80	9 → 48	13 → 27
	n = 94	*n* = 141	*n* = 96	*n* = 4
15–16 years	17 ± 6	45 ± 13	27 ± 12	44 ± 18
	6 → 38	24 → 76	6 → 53	24 → 57
	n = 57	*n* = 92	*n* = 63	*n* = 3
17–18 years	14	17		
	14 → 14	17 → 17		
	n = 1	*n* = 1		

Table A2.7

Durations (in milliseconds) in Caucasian children: lead V_1

Age-group	Q duration	S duration	R duration	R′ duration
<24 h		30 ± 7	23 ± 6	30
		19 → 44	14 → 36	30 → 30
		n = 43	*n* = 40	*n* = 1
<1 day	9 ± 1	32 ± 7	23 ± 6	
	8 → 10	21 → 52	12 → 35	
	n = 2	*n* = 137	*n* = 124	
<2 days	11 ± 8	33 ± 8	21 ± 6	
	5 → 17	20 → 54	13 → 37	
	n = 2	*n* = 117	*n* = 100	
<3 days	12 ± 6	32 ± 7	22 ± 10	11 ± 5
	8 → 16	20 → 48	12 → 42	7 → 17
	n = 2	*n* = 102	*n* = 93	*n* = 3
<1 week		35 ± 9	20 ± 5	
		23 → 60	12 → 34	
		n = 114	*n* = 99	
<1 month	10	35 ± 8	21 ± 6	
	10 → 10	24 → 62	13 → 36	
	n = 1	*n* = 41	*n* = 37	
<3 months	10 ± 3	33 ± 9	24 ± 8	7 ± 2
	8 → 12	20 → 56	7 → 55	6 → 10
	n = 2	*n* = 62	*n* = 56	*n* = 4
<6 months		31 ± 8	28 ± 9	12 ± 5
		20 → 60	9 → 43	9 → 19
		n = 49	*n* = 48	*n* = 4
<1 year		29 ± 8	29 ± 10	23 ± 11
		15 → 55	8 → 50	9 → 44
		n = 70	*n* = 69	*n* = 10
1–2 years	12 ± 6	28 ± 8	37 ± 9	14 ± 8
	8 → 16	14 → 58	12 → 51	6 → 32
	n = 2	*n* = 107	*n* = 103	*n* = 10
3–4 years	86	25 ± 5	39 ± 12	21 ± 9
	86 → 86	16 → 39	13 → 59	6 → 36
	n = 1	*n* = 135	*n* = 135	*n* = 27

Table A2.7 (Continued)

Age-group	Q duration	S duration	R duration	R′ duration
5–6 years	28 ± 28	25 ± 5	42 ± 11	18 ± 10
	8 → 48	13 → 40	15 → 55	4 → 43
	n = 2	n = 146	n = 146	n = 23
7–8 years		25 ± 4	43 ± 11	22 ± 9
		18 → 34	16 → 59	12 → 49
		n = 135	n = 135	n = 28
9–10 years		25 ± 4	42 ± 11	23 ± 11
		17 → 38	15 → 61	5 → 45
		n = 108	n = 108	n = 25
11–12 years		25 ± 5	44 ± 12	24 ± 9
		16 → 37	22 → 65	9 → 46
		n = 104	n = 104	n = 27
13–14 years	48 ± 36	25 ± 5	47 ± 13	23 ± 14
	8 → 76	13 → 36	18 → 82	6 → 59
	n = 3	n = 107	n = 107	n = 27
15–16 years	39 ± 38	26 ± 5	45 ± 13	28 ± 11
	12 → 66	15 → 37	11 → 67	11 → 49
	n = 2	n = 70	n = 70	n = 17
17–18 years		27	20	
		27 → 27	20 → 20	
		n = 1	n = 1	

Table A2.8

Durations (in milliseconds) in Caucasian children: lead V_2

Age-group	Q duration	S duration	R duration	R′ duration
<24 h		26 ± 5	29 ± 6	8!
		19 → 46	19 → 42	8 → 8
		n = 43	n = 42	n = 1
<1 day	13	26 ± 4	29 ± 5	10
	13 → 13	19 → 32	18 → 41	10 → 10
	n = 1	n = 137	n = 136	n = 1
<2 days	6	27 ± 5	28 ± 5	13
	6 → 6	19 → 45	19 → 48	13 → 13
	n = 1	n = 116	n = 114	n = 1
<3 days	10 ± 5	27 ± 5	27 ± 10	10 ± 4
	4 → 15	18 → 42	16 → 43	7 → 13
	n = 4	n = 104	n = 104	n = 2

Table A2.8 (Continued)

Age-group	Q duration	S duration	R duration	R′ duration
<1 week	10	29 ± 8	26 ± 5	10
	10 → 10	20 → 64	17 → 38	10 → 10
	n = 1	*n* = 114	*n* = 111	*n* = 1
<1 month	9	28 ± 6	26 ± 5	
	9 → 9	23 → 62	18 → 38	
	n = 1	*n* = 42	*n* = 41	
<3 months	8	30 ± 5	28 ± 5	
	8 → 8	22 → 46	19 → 50	
	n = 1	*n* = 63	*n* = 62	
<6 months	7 ± 1	29 ± 4	32 ± 5	
	6 → 8	22 → 42	23 → 40	
	n = 2	*n* = 48	*n* = 48	
<1 year	6	29 ± 6	35 ± 7	21 ± 8
	6 → 6	22 → 66	14 → 51	15 → 26
	n = 1	*n* = 73	*n* = 72	*n* = 2
1–2 years	11 ± 2	28 ± 6	39 ± 7	21 ± 16
	9 → 13	14 → 46	21 → 52	9 → 32
	n = 3	*n* = 106	*n* = 106	*n* = 2
3–4 years	21	27 ± 4	42 ± 8	21 ± 5
	21 → 21	19 → 39	19 → 59	14 → 30
	n = 1	*n* = 140	*n* = 140	*n* = 11
5–6 years	6 ± 3	26 ± 4	44 ± 7	17 ± 8
	4 → 8	18 → 36	26 → 57	5 → 29
	n = 2	*n* = 155	*n* = 155	*n* = 8
7–8 years		27 ± 4	45 ± 8	16 ± 6
		18 → 39	25 → 59	6 → 25
		n = 140	*n* = 140	*n* = 10
9–10 years		27 ± 4	46 ± 9	24 ± 9
		20 → 36	21 → 62	11 → 44
		n = 114	*n* = 114	*n* = 11
11–12 years	6	28 ± 6	47 ± 10	20 ± 10
	6 → 6	18 → 43	22 → 66	4 → 34
	n = 1	*n* = 119	*n* = 119	*n* = 10
13–14 years	43 ± 49	28 ± 6	46 ± 12	22 ± 9
	8 → 78	16 → 40	23 → 71	9 → 38
	n = 2	*n* = 138	*n* = 138	*n* = 29
15–16 years	9	29 ± 6	47 ± 11	23 ± 6
	9 → 9	17 → 40	28 → 69	12 → 34
	n = 1	*n* = 89	*n* = 89	*n* = 13
17–18 years		24	25	
		24 → 24	25 → 25	
		n = 1	*n* = 1	

Table A2.9

Durations (in milliseconds) in Caucasian children: lead V_{4R}

Age-group	Q duration	S duration	R duration	R′ duration
<24 h	11 ± 5	34 ± 6	17 ± 3	
	6 → 18	22 → 50	12 → 28	
	n = 7	*n* = 43	*n* = 29	
<1 day	13 ± 3	34 ± 7	18 ± 4	30
	9 → 18	21 → 50	12 → 29	0 → 30
	n = 14	*n* = 136	*n* = 93	*n* = 1
<2 days	10 ± 4	35 ± 7	17 ± 4	27
	6 → 20	23 → 52	12 → 30	27 → 27
	n = 19	*n* = 116	*n* = 72	*n* = 1
<3 days	10 ± 3	37 ± 8	16 ± 4	34
	6 → 15	24 → 56	12 → 28	34 → 34
	n = 15	*n* = 102	*n* = 52	*n* = 1
<1 week	10 ± 3	38 ± 8	16 ± 4	31
	6 → 16	23 → 56	12 → 35	31 → 31
	n = 15	*n* = 113	*n* = 62	*n* = 1
<1 month	12 ± 3	38 ± 8	17 ± 10	21 ± 3
	8 → 16	15 → 56	6 → 59	18 → 24
	n = 5	*n* = 44	*n* = 24	*n* = 3
<3 months	12 ± 8	36 ± 11	19 ± 7	19 ± 6
	5 → 25	14 → 66	5 → 52	9 → 27
	n = 5	*n* = 65	*n* = 47	*n* = 8
<6 months	18 ± 9	34 ± 11	19 ± 8	15 ± 6
	12 → 28	18 → 53	5 → 37	7 → 26
	n = 3	*n* = 49	*n* = 44	*n* = 9
<1 year	11 ± 3	30 ± 10	23 ± 11	18 ± 8
	6 → 14	16 → 70	6 → 44	7 → 45
	n = 4	*n* = 70	*n* = 66	*n* = 20
1–2 years	14 ± 9	27 ± 9	28 ± 13	15 ± 8
	8 → 33	13 → 54	5 → 53	6 → 34
	n = 7	*n* = 110	*n* = 106	*n* = 26
3–4 years	15 ± 12	25 ± 7	36 ± 13	17 ± 8
	8 → 40	14 → 44	9 → 54	4 → 35
	n = 6	*n* = 137	*n* = 136	*n* = 24
5–6 years	24 ± 28	25 ± 7	40 ± 12	14 ± 7
	5 → 56	15 → 48	11 → 54	5 → 28
	n = 3	*n* = 146	*n* = 145	*n* = 18
7–8 years	11	26 ± 6	43 ± 12	17 ± 10
	11 → 11	16 → 47	10 → 58	5 → 38
	n = 1	*n* = 136	*n* = 134	*n* = 13
9–10 years	12 ± 9	26 ± 8	43 ± 12	16 ± 10
	5 → 25	13 → 52	14 → 63	7 → 35
	n = 4	*n* = 109	*n* = 107	*n* = 10

Table A2.9 (Continued)

Age-group	Q duration	S duration	R duration	R′ duration
11–12 years	33 ± 27	25 ± 7	46 ± 11	21 ± 11
	7 → 64	13 → 43	20 → 64	7 → 40
	n = 5	*n* = 102	*n* = 102	*n* = 10
13–14 years	38 ± 31	26 ± 8	45 ± 14	21 ± 12
	10 → 72	13 → 59	8 → 66	4 → 44
	n = 4	*n* = 109	*n* = 107	*n* = 17
15–16 years	31 ± 33	27 ± 9	45 ± 14	19 ± 9
	7 → 54	12 → 64	9 → 68	5 → 32
	n = 2	*n* = 73	*n* = 71	*n* = 12
17–18 years		28	17	
		28 → 28	17 → 17	
		n = 1	*n* = 1	

Table A2.10

Durations (in milliseconds) in Caucasian children: lead V_4

Age-group	Q duration	S duration	R duration	R′ duration
<24 h	9 ± 2	26 ± 4	26 ± 5	
	6 → 11	18 → 34	18 → 43	
	n = 8	*n* = 43	*n* = 42	
<1 day	9 ± 3	25 ± 5	26 ± 5	
	4 → 13	16 → 35	16 → 38	
	n = 23	*n* = 138	*n* = 137	
<2 days	9 ± 3	26 ± 5	25 ± 6	
	4 → 15	14 → 36	13 → 38	
	n = 20	*n* = 119	*n* = 119	
<3 days	9 ± 3	25 ± 6	23 ± 6	22 ± 8
	5 → 15	13 → 37	8 → 34	17 → 32
	n = 25	*n* = 106	*n* = 104	*n* = 3
<1 week	10 ± 4	24 ± 6	23 ± 7	19
	4 → 33	13 → 39	12 → 40	19 → 19
	n = 57	*n* = 116	*n* = 115	*n* = 1
<1 month	10 ± 3	23 ± 5	24 ± 8	
	4 → 16	16 → 37	15 → 67	
	n = 20	*n* = 45	*n* = 45	
<3 months	11 ± 4	25 ± 7	24 ± 5	
	6 → 21	15 → 52	15 → 38	
	n = 35	*n* = 67	*n* = 65	
<6 months	12 ± 4	27 ± 6	26 ± 6	
	5 → 16	19 → 45	16 → 40	
	n = 20	*n* = 50	*n* = 50	

Table A2.10 (Continued)

Age-group	Q duration	S duration	R duration	R′ duration
<1 year	12 ± 4	27 ± 6	27 ± 8	22
	6 → 20	15 → 43	12 → 49	22 → 22
	n = 27	*n* = 72	*n* = 71	*n* = 1
1–2 years	13 ± 4	28 ± 7	27 ± 8	24 ± 6
	6 → 25	11 → 41	12 → 44	19 → 32
	n = 47	*n* = 110	*n* = 108	*n* = 4
3–4 years	13 ± 4	33 ± 7	28 ± 9	17 ± 5
	7 → 22	21 → 47	10 → 45	14 → 25
	n = 50	*n* = 141	*n* = 134	*n* = 4
5–6 years	12 ± 5	35 ± 7	28 ± 9	16 ± 7
	6 → 32	24 → 49	12 → 47	8 → 27
	n = 56	*n* = 155	*n* = 152	*n* = 5
7–8 years	13 ± 6	36 ± 7	30 ± 9	17 ± 7
	6 → 40	23 → 49	9 → 48	12 → 30
	n = 39	*n* = 141	*n* = 140	*n* = 5
9–10 years	11 ± 5	38 ± 6	29 ± 10	18 ± 7
	4 → 21	27 → 51	10 → 48	10 → 34
	n = 25	*n* = 115	*n* = 111	*n* = 7
11–12 years	13 ± 5	40 ± 8	31 ± 10	15 ± 2
	7 → 26	28 → 63	12 → 51	13 → 16
	n = 26	*n* = 119	*n* = 115	*n* = 2
13–14 years	12 ± 4	42 ± 7	32 ± 9	14
	6 → 18	31 → 52	14 → 54	14 → 14
	n = 26	*n* = 141	*n* = 138	*n* = 1
15–16 years	13 ± 6	42 ± 6	32 ± 11	
	6 → 25	30 → 57	10 → 56	
	n = 17	*n* = 91	*n* = 90	
17–18 years		25	22	
		25 → 25	22 → 22	
		n = 1	*n* = 1	
<24 h	11 ± 3	20 ± 5	24 ± 6	
	4 → 16	29 → 12	16 → 42	
	n = 22	*n* = 43	*n* = 41	
<1 day	11 ± 3	20 ± 6	23 ± 5	
	5 → 17	12 → 37	14 → 37	
	n = 77	*n* = 137	*n* = 134	
<2 days	12 ± 3	21 ± 6	24 ± 6	
	7 → 20	12 → 34	13 → 36	
	n = 59	*n* = 118	*n* = 116	

Table A2.11

Durations (in milliseconds) in Caucasian children: lead V_5

Age-group	Q duration	S duration	R duration	R′ duration
<3 days	11 ± 3	20 ± 6	22 ± 5	57 ± 24
	5 → 15	12 → 31	13 → 32	40 → 74
	n = 59	*n* = 106	*n* = 104	*n* = 2
<1 week	12 ± 4	18 ± 4	23 ± 7	
	6 → 18	11 → 35	12 → 39	
	n = 96	*n* = 116	*n* = 109	
<1 month	13 ± 3	19 ± 5	22 ± 6	30 ± 23
	8 → 23	14 → 38	13 → 40	14 → 46
	n = 36	*n* = 44	*n* = 42	*n* = 2
<3 months	13 ± 3	21 ± 5	22 ± 6	20
	5 → 23	14 → 36	12 → 33	20 → 20
	n = 56	*n* = 66	*n* = 65	*n* = 1
<6 months	14 ± 4	22 ± 4	23 ± 6	23
	8 → 24	15 → 31	12 → 38	23 → 23
	n = 38	*n* = 48	*n* = 46	*n* = 1
<1 year	14 ± 4	23 ± 6	24 ± 7	36 ± 18
	7 → 26	13 → 44	12 → 42	23 → 48
	n = 58	*n* = 73	*n* = 67	*n* = 2
1–2 years	15 ± 5	25 ± 7	24 ± 8	28 ± 15
	8 → 26	9 → 46	10 → 38	9 → 50
	n = 83	*n* = 111	*n* = 100	*n* = 5
3–4 years	15 ± 4	29 ± 7	25 ± 9	16 ± 2
	8 → 24	21 → 53	5 → 45	13 → 17
	n = 121	*n* = 140	*n* = 120	*n* = 4
5–6 years	14 ± 4	30 ± 6	25 ± 8	16 ± 9
	7 → 23	23 → 54	11 → 43	9 → 35
	n = 125	*n* = 155	*n* = 142	*n* = 7
7–8 years	15 ± 5	31 ± 6	27 ± 8	14 ± 3
	6 → 26	23 → 45	12 → 44	12 → 18
	n = 118	*n* = 141	*n* = 134	*n* = 3
9–10 years	15 ± 4	33 ± 8	25 ± 10	17 ± 5
	6 → 24	22 → 64	7 → 44	13 → 32
	n = 97	*n* = 115	*n* = 106	*n* = 10
11–12 years	15 ± 5	36 ± 8	26 ± 10	16 ± 2
	6 → 27	25 → 63	10 → 50	14 → 18
	n = 91	*n* = 118	*n* = 106	*n* = 4
13–14 years	16 ± 5	37 ± 7	28 ± 9	17 ± 6
	8 → 26	27 → 52	10 → 47	12 → 29
	n = 87	*n* = 140	*n* = 135	*n* = 6
15–16 years	16 ± 6	39 ± 8	29 ± 12	19 ± 4
	8 → 34	28 → 64	6 → 59	15 → 22
	n = 50	*n* = 92	*n* = 86	*n* = 3
17–18 years	12	15	17	
	12 → 12	15 → 15	17 → 17	
	n = 1	*n* = 1	*n* = 1	

Table A2.12

Durations (in milliseconds) in Caucasian children: lead V_6

Age-group	Q duration	S duration	R duration	R′ duration
<24 h	16 ± 10	18 ± 5	22 ± 8	
	8 → 60	12 → 31	12 → 45	
	n = 36	n = 40	n = 32	
<1 day	15 ± 8	18 ± 7	22 ± 6	18 ± 2
	6 → 50	7 → 39	9 → 40	0 → 19
	n = 100	n = 127	n = 104	n = 3
<2 days	14 ± 8	17 ± 6	22 ± 6	16 ± 4
	6 → 48	8 → 42	12 → 37	13 → 20
	n = 85	n = 111	n = 100	n = 3
<3 days	14 ± 8	18 ± 6	21 ± 6	21 ± 7
	6 → 48	9 → 34	10 → 32	14 → 27
	n = 81	n = 99	n = 79	n = 3
<1 week	15 ± 6	17 ± 5	21 ± 7	21 ± 4
	7 → 46	10 → 35	12 → 38	18 → 24
	n = 99	n = 110	n = 87	n = 2
<1 month	14 ± 3	18 ± 4	20 ± 6	22 ± 1
	9 → 23	10 → 27	12 → 37	21 → 22
	n = 38	n = 45	n = 40	n = 2
<3 months	14 ± 3	22 ± 5	20 ± 5	
	8 → 26	16 → 35	12 → 31	
	n = 61	n = 65	n = 53	
<6 months	15 ± 4	22 ± 4	20 ± 7	30 ± 1
	8 → 25	13 → 36	12 → 38	29 → 31
	n = 42	n = 49	n = 41	n = 2
<1 year	15 ± 4	24 ± 9	21 ± 7	24 ± 6
	7 → 23	8 → 64	10 → 38	18 → 33
	n = 59	n = 73	n = 60	n = 5
1–2 years	17 ± 5	27 ± 7	21 ± 7	30 ± 10
	7 → 27	11 → 50	11 → 34	18 → 43
	n = 91	n = 110	n = 76	n = 5
3–4 years	16 ± 4	29 ± 7	24 ± 8	29 ± 16
	8 → 25	19 → 52	12 → 44	16 → 48
	n = 131	n = 140	n = 99	n = 4
5–6 years	16 ± 4	31 ± 6	23 ± 7	13
	8 → 22	23 → 48	12 → 42	13 → 13
	n = 139	n = 154	n = 119	n = 1
7–8 years	16 ± 4	31 ± 6	24 ± 8	26 ± 16
	8 → 24	23 → 47	12 → 43	15 → 37
	n = 129	n = 140	n = 119	n = 2
9–10 years	16 ± 4	34 ± 8	23 ± 9	17 ± 1
	8 → 24	25 → 64	6 → 42	15 → 18
	n = 107	n = 115	n = 88	n = 4
11–12 years	16 ± 4	37 ± 8	25 ± 9	9 ± 2
	8 → 29	27 → 64	13 → 47	7 → 10
	n = 104	n = 117	n = 86	n = 2

Table A2.12 (Continued)

Age-group	Q duration	S duration	R duration	R′ duration
13–14 years	16 ± 5	37 ± 6	25 ± 9	15 ± 5
	8 → 29	29 → 58	5 → 44	12 → 21
	n = 124	*n* = 140	*n* = 122	*n* = 3
15–16 years	17 ± 5	40 ± 10	28 ± 10	38
	9 → 34	27 → 76	13 → 59	38 → 38
	n = 58	*n* = 88	*n* = 72	*n* = 1
17–18 years	15	16		
	15 → 15	16 → 16		
	n = 1	*n* = 1		

Table A2.13
Amplitudes (in μV) in Caucasian children: lead I

Age-group	P+	P−	Q	R	S	STj	T+	T−
<24 h	80 ± 27	−59 ± 0	−428 ± 311	208 ± 181	−742 ± 312	10 ± 18	84 ± 51	−70 ± 95
	37 → 136	−59 → −59	−903 → −36	30 → 906	−1,586 → −296	37 → −42	23 → 231	−378 → −12
	n = 43	*n* = 1	*n* = 10	*n* = 35	*n* = 35	*n* = 43	*n* = 41	*n* = 14
<1 day	81 ± 27	−42 ± 49	−427 ± 315	206 ± 161	−711 ± 227	18 ± 21	93 ± 50	−57 ± 38
	23 → 142	−137 → −11	−1,199 → −27	25 → 595	−1,392 → −347	−15 → 62	18 → 219	−158 → −12
	n = 138	*n* = 6	*n* = 36	*n* = 112	*n* = 110	*n* = 138	*n* = 133	*n* = 34
<2 days	79 ± 25	−26 ± 10	−333 ± 251	214 ± 145	−687 ± 202	22 ± 26	116 ± 57	−39 ± 25
	43 → 146	−37 → −17	−752 → −29	30 → 599	−1,171 → −335	−34 → 76	31 → 261	−95 → −12
	n = 119	*n* = 3	*n* = 34	*n* = 99	*n* = 99	*n* = 119	*n* = 119	*n* = 12
<3 days	82 ± 33	−22 ± 12	−303 ± 207	210 ± 141	−658 ± 239	28 ± 28	125 ± 54	−49 ± 30
	39 → 147	−41 → −8	−674 → −34	54 → 631	−1,193 → −240	−10 → 125	38 → 291	−101 → −7
	n = 106	*n* = 8	*n* = 41	*n* = 80	*n* = 79	*n* = 106	*n* = 106	*n* = 9
<1 week	83 ± 28	−25 ± 12	−244 ± 275	191 ± 138	−643 ± 198	37 ± 31	151 ± 58	−69 ± 49
	27 → 141	−43 → −10	−952 → −29	32 → 615	−1,185 → −351	−4 → 139	35 → 276	−138 → −22
	n = 116	*n* = 7	*n* = 38	*n* = 104	*n* = 104	*n* = 117	*n* = 117	*n* = 5
<1 month	89 ± 30	−37 ± 18	−141 ± 206	282 ± 199	−556 ± 197	37 ± 25	209 ± 76	−28 ± 3
	33 → 176	−48 → −16	−743 → −24	49 → 759	−1,043 → −215	−7 → 93	68 → 400	−30 → −26
	n = 45	*n* = 3	*n* = 14	*n* = 43	*n* = 43	*n* = 45	*n* = 45	*n* = 2
<3 months	96 ± 27		−78 ± 62	626 ± 291	−474 ± 185	28 ± 35	257 ± 89	−64 ± 47
	44 → 168		−350 → −23	56 → 1,270	−870 → −55	−98 → 101	64 → 457	−117 → −28
	n = 67		*n* = 34	*n* = 66	*n* = 65	*n* = 69	*n* = 66	*n* = 3
<6 months	102 ± 27	−23 ± 0	−107 ± 74	807 ± 230	−467 ± 206	34 ± 23	288 ± 85	
	48 → 197	−23 → −23	−336 → −22	321 → 1,307	−1,019 → −147	−12 → 85	159 → 539	
	n = 50	*n* = 1	*n* = 21	*n* = 50	*n* = 46	*n* = 50	*n* = 50	

Table A2.13 (Continued)

Age-group	P+	P–	Q	R	S	STj	T+	T–
<1 year	112 ± 31	–53 ± 24	–107 ± 71	870 ± 330	–374 ± 206	35 ± 31	310 ± 92	–47 ± 23
	41 → 188	–95 → –31	–334 → –33	339 → 1,714	–1, 054 → –94	–36 → 139	89→524	–63 → –31
	n = 73	*n* = 5	*n* = 40	*n* = 73	*n* = 70	*n* = 74	*n* = 73	*n* = 2
1–2 years	112 ± 28	–75 ± 49	–130 ± 96	821 ± 364	–311 ± 161	25 ± 33	330 ± 120	–50 ± 0
	61 → 197	–128 → –23	–501 → –22	238 → 1,838	–695 → –71	–25 → 146	112 → 680	–50 → –50
	n = 110	*n* = 5	*n* = 69	*n* = 111	*n* = 95	*n* = 116	*n* = 111	*n* = 1
3–4 years	99 ± 21	–28 ± 8	–103 ± 62	684 ± 261	–278 ± 141	14 ± 19	281 ± 87	–89 ± 50
	59 → 145	–34 → –19	–280 → –28	252 → 1,248	–614 → –67	–22 → 68	130 → 463	–136→–37
	n = 141	*n* = 3	*n* = 77	*n* = 141	*n* = 109	*n* = 141	*n* = 141	*n* = 3
5–6 years	97 ± 20	–59 ± 26	–89 ± 55	624 ± 238	–244 ± 127	11 ± 24	276 ± 86	–62 ± 40
	58 → 141	–77 → –40	–263 → –22	185 → 1,259	–581→–64	–60 →71	123 → 433	–94→–17
	n = 154	*n* = 2	*n* = 71	*n* = 155	*n* = 136	*n* = 155	*n* = 154	*n* = 3
7–8 years	95 ± 21	–45 ± 42	–80 ± 61	592 ± 278	–220 ± 140	11 ± 22	279 ± 88	–86 ± 64
	60 → 140	–74 → –15	–203 → –21	187 → 1,354	–781 → –62	–36 → 74	132 → 481	–131 → –40
	n = 140	*n* = 2	*n* = 80	*n* = 141	*n* = 114	*n* = 141	*n* = 141	*n* = 2
9–10 years	97 ± 26	–38 ± 0	–72 ± 53	541 ± 219	–202 ± 130	14 ± 24	285 ± 98	–57 ± 9
	50 → 162	–38 → –38	–259 → –21	190 → 1,042	–668 → –41	–19 → 73	103 → 519	–66 → –48
	n = 115	*n* = 1	*n* = 49	*n* = 115	*n* = 92	*n* = 115	*n* = 115	*n* = 3
11–12 years	101 ± 24	–102 ± 0	–80 ± 48	649 ± 245	–233 ± 121	10 ± 28	281 ± 86	–75 ± 42
	64 → 153	–102 → –102	–241 → –23	230 → 1,250	–563 → –70	–43 → 106	135 → 518	–135 → –41
	n = 118	*n* = 1	*n* = 62	*n* = 119	*n* = 90	*n* = 119	*n* = 119	*n* = 4
13–14 years	99 ± 25		–78 ± 57	630 ± 274	–221 ± 121	17 ± 32	297 ± 102	–146 ± 0
	58 → 153		–378 → –23	232 → 1,390	–491 → –46	–17 → 128	106 → 557	–146 → –146
	n = 141		*n* = 62	*n* = 141	*n* = 113	*n* = 141	*n* = 140	*n* = 1
15–16 years	92 ± 22	–37 ± 0	–71 ± 54	625 ± 308	–243 ± 153	13 ± 28	275 ± 107	–124 ± 40
	48 → 149	–37 → –37	–226 → –23	169 → 1,469	–761 → –46	–47 → 81	129 → 576	–152 → –95
	n = 92	*n* = 1	*n* = 46	*n* = 92	*n* = 71	*n* = 92	*n* = 91	*n* = 2
17–18 years	132 ± 0		–606 ± 0			24 ± 0	158 ± 0	
	132 → 132		–606 → –606			24 → 24	158 → 158	
	n = 1		*n* = 1			*n* = 1	*n* = 1	

Table A2.14

Amplitudes (in μV) in Caucasian children: lead II

Age-group	P+	P–	Q	R	S	STj	T+	T–
<24 h	146 ± 50		–208 ± 143	597 ± 303	–393 ± 197	46 ± 30	149 ± 65	–117 ± 81
	65 → 276		–722 → –47	111 → 1,398	–961 → –178	–3 → 131	55 → 314	–174 → –59
	n = 43		*n* = 41	*n* = 41	*n* = 22	*n* = 43	*n* = 42	*n* = 2
<1 day	157 ± 48	–16 ± 0	–208 ± 127	618 ± 365	–433 ± 193	48 ± 32	157 ± 65	–147 ± 204
	56 → 258	–16 → –16	–526 → –34	53 → 1,389	–966 → –55	1 → 115	51 → 299	–597 → –20
	n = 138	*n* = 1	*n* = 132	*n* = 136	*n* = 71	*n* = 138	*n* = 137	*n* = 7
<2 days	164 ± 53	–35 ± 19	–204 ± 107	579 ± 300	–460 ± 204	51 ± 31	182 ± 70	–99 ± 66
	68 → 298	–54 → –17	–510 → –44	116 → 1,348	–905 → –165	–9 → 121	47 → 404	–175 → –60
	n = 118	*n* = 3	*n* = 106	*n* = 116	*n* = 69	*n* = 119	*n* = 118	*n* = 3

Table A2.14 (Continued)

Age-group	P+	P−	Q	R	S	STj	T+	T−
<3 days	163 ± 61	−36 ± 24	−210 ± 126	542 ± 303	−425 ± 229	58 ± 35	199 ± 69	−32 ± 13
	60 → 322	−77 → −19	−609 → −42	86 → 1,399	−976 → −94	−9 → 169	70 → 394	−41 → −22
	n = 105	*n* = 5	*n* = 102	*n* = 105	*n* = 56	*n* = 106	*n* = 106	*n* = 2
<1 week	168 ± 53	−25 ± 0	−239 ± 107	662 ± 345	−453 ± 220	65 ± 37	222 ± 72	−58 ± 0
	73 → 305	−25 → −25	−542 → −27	104 → 1,485	−975 → −89	−14 → 173	100 → 398	−58 → −58
	n = 117	*n* = 1	*n* = 115	*n* = 114	*n* = 75	*n* = 117	*n* = 117	*n* = 1
<1 month	159 ± 46	−23 ± 8	−199 ± 103	695 ± 312	−308 ± 160	47 ± 43	260 ± 84	−29 ± 0
	88 → 250	−29 → −17	−567 → −70	86 → 1,392	−662 → −94	−1 → 274	135 → 607	−29 → −29
	n = 45	*n* = 2	*n* = 41	*n* = 45	*n* = 28	*n* = 45	*n* = 45	*n* = 1
<3 months	150 ± 32		−218 ± 108	1,029 ± 407	−226 ± 76	44 ± 35	287 ± 91	
	87 → 231		−470 → −43	99 → 2,177	−405 → −111	−23 → 162	123 → 568	
	n = 67		*n* = 58	*n* = 67	*n* = 46	*n* = 69	*n* = 67	
<6 months	150 ± 39		−209 ± 132	1,067 ± 360	−247 ± 129	47 ± 32	293 ± 84	
	72 → 274		−492 → −34	404 → 1,830	−732 → −108	−22 → 127	126 → 560	
	n = 50		*n* = 40	*n* = 49	*n* = 32	*n* = 50	*n* = 50	
<1 year	159 ± 35	−96 ± 23	−178 ± 137	1,112 ± 445	−260 ± 116	55 ± 35	327 ± 102	
	81 → 241	−113 → −70	−676 → −22	468 → 2,666	−510 → −87	−19 → 164	84 → 597	
	n = 73	*n* = 3	*n* = 54	*n* = 73	*n* = 52	*n* = 74	*n* = 73	
1–2 years	146 ± 43	−39 ± 44	−169 ± 123	1,068 ± 379	−257 ± 121	28 ± 35	299 ± 112	
	50 → 240	−100 → 0	−480 → −29	349 → 1,928	−561 → −43	−20 → 161	93 → 579	
	n = 111	*n* = 4	*n* = 79	*n* = 111	*n* = 74	*n* = 116	*n* = 111	
3–4 years	149 ± 42	−29 ± 16	−140 ± 87	1,252 ± 381	−233 ± 125	13 ± 27	361 ± 127	
	73 → 241	−40 → −18	−332 → −23	475 → 2,119	−507 → −52	−41 → 66	129 → 653	
	n = 140	*n* = 2	*n* = 106	*n* = 141	*n* = 98	*n* = 141	*n* = 141	
5–6 years	136 ± 53	−27 ± 11	−126 ± 87	1,342 ± 420	−236 ± 113	16 ± 30	370 ± 108	
	36 → 277	−47 → −12	−357 → −22	637 → 2,394	−531 → −58	−35 → 92	171 → 594	
	n = 155	*n* = 12	*n* = 118	*n* = 155	*n* = 114	*n* = 155	*n* = 155	
7–8 years	137 ± 49	−35 ± 26	−119 ± 96	1,317 ± 380	−249 ± 137	15 ± 30	398 ± 118	−57 ± 0
	45 → 242	−105 → −18	−329 → −22	708 → 2,205	−620 → −54	−29 → 93	196 → 685	−57 → −57
	n = 140	*n* = 10	*n* = 94	*n* = 141	*n* = 115	*n* = 141	*n* = 141	*n* = 1
9–10 years	129 ± 53	−22 ± 8	−109 ± 62	1,423 ± 378	−218 ± 120	21 ± 38	401 ± 134	
	38 → 253	−38 → −13	−260 → −24	619 → 2,180	−497 → −55	−39 → 200	173 → 705	
	n = 115	*n* = 8	*n* = 86	*n* = 115	*n* = 84	*n* = 115	*n* = 115	
11–12 years	137 ± 49	−23 ± 13	−89 ± 55	1,413 ± 379	−245 ± 137	16 ± 31	391 ± 125	−62 ± 9
	37 → 232	−45 → −10	−217 → −23	759 → 2,342	−617 → −54	−60 → 122	148 → 672	−68 → −55
	n = 119	*n* = 6	*n* = 83	*n* = 119	*n* = 87	*n* = 119	*n* = 119	*n* = 2
13–14 years	125 ± 61	−27 ± 12	−90 ± 64	1,368 ± 408	−271 ± 152	23 ± 45	393 ± 137	−231 ± 0
	21 → 250	−60 → −14	−303 → −23	584 → 2,387	−740 → −47	−33 → 139	172 → 736	−231 → −231
	n = 139	*n* = 16	*n* = 100	*n* = 141	*n* = 111	*n* = 141	*n* = 140	*n* = 1
15–16 years	141 ± 59	−36 ± 23	−94 ± 65	1,290 ± 407	−278 ± 183	16 ± 41	366 ± 123	−77 ± 9
	35 → 297	−82 → −18	−394 → −27	593 → 2,089	−732 → −54	−77 → 98	144 → 631	−87 → −67
	N = 91	*n* = 6	*n* = 60	*n* = 92	*n* = 67	*n* = 92	*n* = 92	*n* = 4
17–18 years	270 ± 0		−284 ± 0	792 ± 0		85 ± 0	246 ± 0	
	270 → 270		−284 → −284	792 → 792		85 → 85	246 → 246	
	n = 1		*n* = 1	*n* = 1		*n* = 1	*n* = 1	

Table A2.15
Amplitudes (in μV) in Caucasian children: lead III

Age-group	P+	P−	Q	R	S	STj	T+	T−
<24 h	83 ± 46	−26 ± 10	−289 ± 157	1,072 ± 443	−377 ± 244	35 ± 30	103 ± 72	−30 ± 15
	18 → 198	−41 → −6	−731 → −78	209 → 2,028	−725 → −162	−15 → 114	14 → 110	−55 → −16
	n = 41	*n* = 15	*n* = 41	*n* = 43	*n* = 4	*n* = 43	*n* = 43	*n* = 7
<1 day	89 ± 45	−26 ± 15	−311 ± 145	1,048 ± 419	−228 ± 95	30 ± 22	97 ± 56	−67 ± 131
	14 → 195	−64 → −12	−651 → −101	49 → 1,954	−428 → −108	−10 → 85	15 → 272	−612 → −7
	n = 135	*n* = 29	*n* = 131	*n* = 138	*n* = 32	*n* = 138	*n* = 135	*n* = 20
<2 days	101 ± 43	−42 ± 27	−290 ± 143	950 ± 413	−212 ± 121	28 ± 32	98 ± 51	−61 ± 43
	27 → 213	−104 → −15	−655 → −72	32 → 2,017	−497 → −74	−30 → 96	22 → 240	−174 → −14
	n = 112	*n* = 23	*n* = 110	*n* = 119	*n* = 32	*n* = 119	*n* = 109	*n* = 20
<3 days	98 ± 50	−39 ± 18	−271 ± 128	882 ± 374	−197 ± 107	30 ± 29	105 ± 57	−51 ± 27
	26 → 225	−86 → −16	−621 → −59	54 → 1,859	−445 → −52	−38 → 100	15-277	−108 → −13
	n = 99	*n* = 23	*n* = 101	*n* = 106	*n* = 22	*n* = 106	*n* = 97	*n* = 17
<1 week	102 ± 47	−28 ± 14	−304 ± 126	972 ± 412	−209 ± 105	27 ± 26	104 ± 57	−37 ± 23
	17 → 223	−69 → −5	−626 → −92	97 → 2,095	−506 → −91	−34 → 82	19-241	−102 → −10
	n = 113	*n* = 22	*n* = 115	*n* = 117	*n* = 32	*n* = 117	*n* = 112	*n* = 26
<1 month	86 ± 51	−29 ± 12	−263 ± 126	913 ± 375	−158 ± 69	9 ± 40	86 ± 55	−42 ± 26
	10 → 208	−44 → −13	−575 → −77	23 → 1,571	−277 → −57	−29 → 208	13 → 228	−88 → −11
	n = 44	*n* = 13	*n* = 41	*n* = 45	*n* = 11	*n* = 45	*n* = 37	*n* = 16
<3 months	73 ± 31	−30 ± 18	−348 ± 166	909 ± 456	−220 ± 163	15 ± 31	88 ± 64	−78 ± 51
	16 → 168	−78 → −12	−726 → −76	54 → 2,038	−704 → −77	−24 → 141	16 → 317	−210 → −16
	n = 65	*n* = 18	*n* = 59	*n* = 67	*n* = 17	*n* = 69	*n* = 57	*n* = 22
<6 months	64 ± 25	−24 ± 14	−372 ± 139	765 ± 450	−256 ± 203	11 ± 29	79 ± 52	−79 ± 67
	25 → 130	−67 → −10	−682 → −104	24 → 1,816	−876 → −72	−36 → 63	20 → 218	−315 → −21
	n = 50	*n* = 16	*n* = 38	*n* = 50	*n* = 20	*n* = 50	*n* = 36	*n* = 30
<1 year	66 ± 28	−38 ± 28	−373 ± 170	651 ± 482	−291 ± 206	19 ± 24	88 ± 58	−85 ± 45
	17 → 123	−162 → −16	−807 → −122	25 → 1,575	−824 → −60	−28 → 76	18 → 314	−230 → −29
	n = 70	*n* = 34	*n* = 47	*n* = 73	*n* = 35	*n* = 74	*n* = 57	*n* = 35
1–2 years	64 ± 36	−36 ± 22	−280 ± 140	563 ± 432	−289 ± 243	2 ± 19	69 ± 53	−98 ± 56
	12 → 158	−109 → −10	−751 → −61	25 → 1,739	−967 → −47	−35 → 39	13 → 262	−248 → −28
	n = 101	*n* = 56	*n* = 68	*n* = 110	*n* = 64	*n* = 116	*n* = 66	*n* = 77
3–4 years	76 ± 39	−34 ± 16	−232 ± 138	817 ± 524	−204 ± 147	−2 ± 22	131 ± 86	−73 ± 51
	15 → 177	−99 → −13	−648 → −27	35 → 1,893	−604 → −40	−37 → 43	14 → 352	−232 → −15
	n = 127	*n* = 57	*n* = 94	*n* = 140	*n* = 79	*n* = 141	*n* = 121	*n* = 46
5–6 years	74 ± 38	−41 ± 20	−191 ± 126	903 ± 550	−156 ± 112	4 ± 21	125 ± 78	−51 ± 30
	14 → 174	−91 → −13	−456 → −24	34 → 2,102	−335 → −41	−44 → 62	17 → 361	−153 → −14
	n = 126	*n* = 76	*n* = 106	*n* = 155	*n* = 84	*n* = 155	*n* = 146	*n* = 43
7–8 years	71 ± 37	−40 ± 23	−177 ± 118	882 ± 527	−217 ± 159	3 ± 23	140 ± 90	−45 ± 25
	16 → 169	−113 → −13	−528 → −25	28 → 1,972	−722 → −47	−33 → 96	26 → 366	−129 → −16
	n = 120	*n* = 72	*n* = 82	*n* = 141	*n* = 94	*n* = 141	*n* = 139	*n* = 31
9–10 years	67 ± 41	−44 ± 26	−162 ± 98	1,035 ± 518	−163 ± 110	6 ± 27	150 ± 96	−71 ± 58
	13 → 183	−132 → −9	−416 → −27	45 → 1,973	−589 → −36	−37 → 79	23 → 402	−265 → −16
	n = 92	*n* = 70	*n* = 78	*n* = 115	*n* = 70	*n* = 115	*n* = 107	*n* = 28
11–12 years	71 ± 35	−42 ± 20	−126 ± 78	908 ± 527	−193 ± 151	6 ± 22	142 ± 84	−62 ± 40
	14 → 149	−108 → −14	−2, 901 → −21	38 → 2,271	−808 → −28	−43 → 53	21 → 346	−163 → −13
	n = 100	*n* = 65	*n* = 75	*n* = 119	*n* = 74	*n* = 119	*n* = 111	*n* = 30

Table A2.15 (Continued)

Age-group	P+	P−	Q	R	S	STj	T+	T−
13–14 years	71 ± 44	−51 ± 25	−111 ± 66	878 ± 511	−227 ± 158	5 ± 26	133 ± 78	−63 ± 37
	13 → 220	−114 → −18	−303 → −24	34 → 1,985	−736 → −31	−46 → 71	21 → 341	−140 → −13
	n = 105	*n* = 90	*n* = 88	*n* = 141	*n* = 93	*n* = 141	*n* = 126	*n* = 35
15–16 years	82 ± 53	−38 ± 21	−123 ± 81	797 ± 482	−225 ± 158	3 ± 26	121 ± 78	−56 ± 28
	14 → 219	−137 → −14	−402 → −27	31 → 1,805	−751 → −41	−55 → 54	20 → 125	−156 → −23
	n = 77	*n* = 53	*n* = 55	*n* = 92	*n* = 55	*n* = 92	*n* = 89	*n* = 30
17–18 years	138 ± 0		−241 ± 0	1,399 ± 0		61 ± 0	95 ± 0	
	138 → 138		−241 → −241	1,399 → 1,399		61 → 61	95 → 95	
	n = 1		*n* = 1	*n* = 1		*n* = 1	*n* = 1	

Table A2.16
Amplitudes (in μV) in Caucasian children: lead aVR

Age-group	P+	P−	Q	R	S	STj	T+	T−
<24 h		−110 ± 32	−188 ± 180	273 ± 245	−228 ± 132	−28 ± 19	84 ± 87	−113 ± 49
		−181 → −60	−671 → −31	22 → 974	−533 → −64	−64 → 14	23 → 229	−231 → −46
		n = 43	*n* = 15	*n* = 42	*n* = 24	*n* = 43	*n* = 5	*n* = 42
<1 day	15 ± 2	−116 ± 32	−269 ± 155	283 ± 227	−263 ± 170	−32 ± 25	65 ± 72	−121 ± 50
	13 → 16	−181 → −46	−662 → −42	30 → 786	−778 → −46	−88 → 11	12 → 300	−237 → −42
	n = 2	*n* = 138	*n* = 46	*n* = 135	*n* = 73	*n* = 138	*n* = 18	*n* = 137
<2 days	9 ± 0	−118 ± 34	−286 ± 130	217 ± 197	−237 ± 141	−36 ± 24	51 ± 34	−147 ± 55
	9 → 9	−209 → −60	−571 → −42	22 → 809	−625 → −35	−96 → 11	14 → 114	−289 → −39
	n = 1	*n* = 119	*n* = 35	*n* = 117	*n* = 72	*n* = 119	*n* = 6	*n* = 117
<3 days	34 ± 33	−119 ± 42	−264 ± 124	218 ± 181	−248 ± 152	−43 ± 28	47 ± 22	−160 ± 51
	11 → 57	−203 → −42	−484 → −36	27 → 659	−723 → −42	−105 → 29	31 → 62	−289 → −68
	n = 2	*n* = 106	*n* = 26	*n* = 105	*n* = 67	*n* = 106	*n* = 2	*n* = 106
<1 week	22 ± 4	−122 ± 34	−240 ± 145	202 ± 183	−305 ± 173	−51 ± 31	59 ± 33	−183 ± 57
	19 → 24	−217 → −54	−638 → −77	29 → 892	−696 → −74	−130 → 0	36 → 97	−332 → −83
	n = 2	*n* = 117	*n* = 20	*n* = 116	*n* = 86	*n* = 117	*n* = 3	*n* = 117
<1 month	23 ± 10	−121 ± 29	−384 ± 121	151 ± 134	−389 ± 221	−42 ± 29	33 ± 8	−233 ± 69
	14 → 33	−193 → −62	−533 → −235	25 → 680	−1,050 → −114	−84 → 11	27 → 39	−430 → −112
	n = 3	*n* = 45	*n* = 8	*n* = 45	*n* = 36	*n* = 45	*n* = 2	*n* = 45
<3 months		−119 ± 24	−609 ± 260	138 ± 105	−786 ± 272	−37 ± 32		−269 ± 79
		−190 → −78	−1, 017 → −74	24 → 465	−1, 262 to −309	−114 → 80		−511 → −92
		n = 67	*n* = 13	*n* = 67	*n* = 55	*n* = 69		*n* = 67
<6 months		−122 ± 32	−792 ± 259	188 ± 152	−906 ± 190	−41 ± 24		−288 ± 70
		−233 → −56	−1, 246 to −79	26 → 730	−1, 264 to −615	−92 → 11		−523 → −154
		n = 50	*n* = 16	*n* = 50	*n* = 34	*n* = 50		*n* = 50
<1 year	62 ± 36	−132 ± 29	−852 ± 299	179 ± 153	−980 ± 303	−45 ± 31		−316 ± 86
	27 → 99	−208 → −72	−1, 478 → −81	20 → 729	−2, 188 to −616	−94 → 51		−454 → −117
	n = 3	*n* = 73	*n* = 25	*n* = 72	*n* = 49	*n* = 74		*n* = 73
1–2 years	89 ± 34	−125 ± 29	−711 ± 231	174 ± 119	−983 ± 270	−26 ± 32		−313 ± 106
	65 → 113	−187 → −74	−1, 118 → −97	22 → 501	−1, 426 to −499	−87 → 24		−646 → −126
	n = 2	*n* = 110	*n* = 33	*n* = 107	*n* = 80	*n* = 116		*n* = 111
3–4 years		−120 ± 26	−797 ± 185	122 ± 89	−948 ± 221	−13 ± 20		−319 ± 93
		−177 → −69	−1, 362 → −457	23 → 392	−1, 405 to −466	−55 → 28		−552 → −149
		n = 141	*n* = 39	*n* = 133	*n* = 102	*n* = 141		*n* = 141

Table A2.16 (Continued)

Age-group	P+	P−	Q	R	S	STj	T+	T−
5–6 years	27 ± 18	−112 ± 31	−824 ± 166	125 ± 91	−975 ± 259	−13 ± 25	38 ± 0	−321 ± 87
	15 → 58	−184 → −54	−1, 288 → −553	23 → 363	−1, 568 → −607	−67 → 67	38 → 38	−495 → −148
	n = 5	*n* = 155	*n* = 52	*n* = 148	*n* = 103	*n* = 155	*n* = 1	*n* = 155
7–8 years	86 ± 0	−113 ± 28	−842 ± 251	114 ± 96	−924 ± 241	−13 ± 23	127 ± 0	−337 ± 91
	86 → 86	−184 → −65	−1, 451 → −463	21 → 384	−1, 459 → −535	−70 → 27	127 → 127	−562 → −183
	n = 1	*n* = 140	*n* = 45	*n* = 140	*n* = 96	*n* = 141	*n* = 1	*n* = 141
9–10 years	21 ± 0	−110 ± 30	−837 ± 194	103 ± 83	−977 ± 218	−17 ± 28	35 ± 0	−341 ± 102
	21 → 21	−182 → −57	−1, 240 → −521	22 → 364	−1, 578 → −585	−73 → 27	35 → 35	−560 → −166
	n = 1	*n* = 115	*n* = 43	*n* = 108	*n* = 72	*n* = 115	*n* = 1	*n* = 115
11–12 years	67 ± 0	−115 ± 30	−908 ± 177	122 ± 107	−1, 041 ± 223	−12 ± 27	58 ± 0	−334 ± 92
	67 → 67	−181 → −59	−1, 290 → −610	23 → 479	−1, 662 → −650	−77 → 81	58 → 58	−570 → −179
	n = 1	*n* = 118	*n* = 48	*n* = 112	*n* = 71	*n* = 119	*n* = 1	*n* = 119
13–14 years	14 ± 3	−107 ± 37	−826 ± 214	126 ± 107	−1, 036 ± 265	−20 ± 36	182 ± 0	−343 ± 110
	11 → 18	−176 → −41	−1, 414 → −465	21 → 401	−1, 573 → −522	−113 → 29	182 → 182	−600 → −145
	n = 4	*n* = 141	*n* = 59	*n* = 135	*n* = 83	*n* = 141	*n* = 1	*n* = 140
15–16 years	17 ± 10	−111 ± 33	−805 ± 277	139 ± 129	−1, 035 ± 241	−13 ± 32	84 ± 10	−317 ± 107
	9 → 31	−196 → −40	−1, 675 → −352	23 → 521	−1, 737 → −582	−63 → 75	77 → 91	−584 → −125
	n = 4	*n* = 92	*n* = 41	*n* = 82	*n* = 50	*n* = 92	*n* = 2	*n* = 92
17–18 years		−200 ± 0		166 ± 0	−108 ± 0	−54 ± 0		−199 ± 0
		−200 → −200		166 → 166	−108 → −108	−54 → −54		−199 → −199
		n = 1		*n* = 1	*n* = 1	*n* = 1		*n* = 1

Table A2.17

Amplitudes (in μV) in Caucasian children: lead aVL

Age-group	P+	P−	Q	R	S	STj	T+	T−
<24 h	34 ± 17	−29 ± 20	−295 ± 283	224 ± 160	−865 ± 322	−12 ± 19	47 ± 28	−62 ± 62
	9 → 65	−90 → −9	−681 → −47	34 → 731	−1, 654 → −394	−49 → 27	11 → 119	−369 → −15
	n = 31	*n* = 28	*n* = 4	*n* = 41	*n* = 41	*n* = 43	*n* = 22	*n* = 34
<1 day	31 ± 18	−30 ± 26	−291 ± 286	223 ± 140	−832 ± 251	−5 ± 15	51 ± 45	−57 ± 37
	7 → 76	−96 → −7	−759 → −36	46 → 546	−1, 518 → −465	−33 → 31	11 → 147	−165 → −12
	n = 104	*n* = 90	*n* = 8	*n* = 136	*n* = 134	*n* = 138	*n* = 79	*n* = 88
<2 days	24 ± 22	−32 ± 18	−264 ± 261	211 ± 131	−780 ± 250	−3 ± 25	57 ± 42	−40 ± 24
	0 → 92	−78 → −9	−609 → −24	42 → 582	−1, 563 → −354	−49 → 112	13 → 164	−113 → −9
	n = 119	*n* = 81	*n* = 11	*n* = 116	*n* = 114	*n* = 119	*n* = 82	*n* = 67
<3 days	33 ± 25	−31 ± 26	−177 ± 210	191 ± 118	−718 ± 230	−1 ± 22	53 ± 42	−45 ± 31
	6 → 86	−182 → −6	−548 → −24	32 → 535	−1, 318 → −308	−44 → 59	10 → 182	−159 → −6
	n = 75	*n* = 69	*n* = 12	*n* = 102	*n* = 102	*n* = 106	*n* = 78	*n* = 52
<1 week	29 ± 20	−31 ± 21	−158 ± 240	206 ± 111	−750 ± 248	5 ± 21	64 ± 38	−51 ± 31
	6 → 81	−101 → −5	−659 → −22	36 → 488	−1, 490 → −373	−41 → 47	11 → 139	−161 → −11
	n = 82	*n* = 90	*n* = 9	*n* = 115	*n* = 115	*n* = 117	*n* = 94	*n* = 44

Table A2.17 (Continued)

Age-group	P+	P–	Q	R	S	STj	T+	T–
<1 month	40 ± 22	–33 ± 24	–353 ± 396	225 ± 123	–703 ± 232	14 ± 25	98 ± 52	–56 ± 26
	9 → 93	–88 → –9	–912 → –26	65 → 625	–1, 147 → –243	–69 → 55	12 → 233	–93 → –17
	n = 34	*n* = 26	*n* = 5	*n* = 42	*n* = 42	*n* = 45	*n* = 40	*n* = 8
<3 months	40 ± 22	–22 ± 14	–76 ± 65	397 ± 199	–647 ± 246	7 ± 28	125 ± 69	–54 ± 59
	10 → 123	–76 → –6	–236 → –20	58 → 988	–1, 205 → –174	–110 → 59	25 → 322	–216 → –14
	n = 61	*n* = 38	*n* = 13	*n* = 66	*n* = 66	*n* = 69	*n* = 64	*n* = 11
<6 months	40 ± 19	–18 ± 10	–88 ± 46	480 ± 188	–598 ± 232	11 ± 19	154 ± 77	–30 ± 33
	14 → 83	–41 → –4	–163 → –25	110 → 1,092	–1, 159 → –208	–21 → 61	15 → 427	–78 → –4
	n = 50	*n* = 24	*n* = 13	*n* = 50	*n* = 48	*n* = 50	*n* = 49	*n* = 4
<1 year	52 ± 24	–23 ± 14	–103 ± 63	522 ± 238	–474 ± 224	8 ± 21	160 ± 70	–42 ± 24
	9 → 108	–73 → –9	–245 → –28	136 → 1,199	–1, 089 → –112	–32 → 57	32 → 331	–71 → –5
	n = 69	*n* = 33	*n* = 30	*n* = 73	*n* = 71	*n* = 74	*n* = 71	*n* = 8
1–2 years	56 ± 25	–30 ± 26	–117 ± 89	475 ± 275	–386 ± 207	11 ± 20	188 ± 90	–48 ± 17
	12 → 133	–132 → –9	–460 → –26	93 → 1,287	–1, 035 → –50	–22 → 68	27 → 410	–72 → –26
	n = 106	*n* = 38	*n* = 54	*n* = 111	*n* = 98	*n* = 116	*n* = 110	*n* = 5
3–4 years	48 ± 20	–27 ± 17	–106 ± 84	336 ± 185	–447 ± 272	8 ± 16	121 ± 65	–53 ± 38
	13 → 84	–88 → –9	–407 → –23	66 → 756	–10–74 → –48	–21 → 41	19 → 269	–159 → –7
	n = 127	*n* = 64	*n* = 60	*n* = 140	*n* = 118	*n* = 141	*n* = 132	*n* = 21
5–6 years	49 ± 22	–25 ± 13	–86 ± 69	255 ± 156	–454 ± 259	3 ± 17	109 ± 60	–46 ± 31
	10 → 94	–70 → –6	–410 → –22	38 → 719	–1, 114 → –59	–41 → 40	15 → 245	–136 → –10
	n = 146	*n* = 58	*n* = 58	*n* = 154	*n* = 140	*n* = 155	*n* = 145	*n* = 28
7–8 years	48 ± 23	–24 ± 13	–104 ± 77	248 ± 183	–440 ± 279	4 ± 17	102 ± 61	–43 ± 26
	11 → 98	–63 → –7	–483 → –23	43 → 768	–1, 176 → –46	–32 → 38	14 → 240	–107 → –8
	n = 129	*n* = 58	*n* = 71	*n* = 140	*n* = 115	*n* = 141	*n* = 130	*n* = 34
9–10 years	53 ± 28	–26 ± 15	–130 ± 129	197 ± 131	–484 ± 260	4 ± 16	110 ± 72	–51 ± 38
	11 → 127	–72 → –8	–473 → –23	41 → 558	–1, 104 → –63	–31 → 33	14 → 347	–169 → –13
	n = 105	*n* = 42	*n* = 37	*n* = 112	*n* = 99	*n* = 115	*n* = 107	*n* = 23
11–12 years	53 ± 23	–26 ± 17	–116 ± 97	232 ± 190	–415 ± 260	2 ± 19	109 ± 62	–52 ± 33
	13 → 104	–88 → –8	–595 → –28	38 → 867	–976 → –38	–44 → 51	20 → 262	–145 → –15
	n = 112	*n* = 47	*n* = 49	*n* = 117	*n* = 100	*n* = 119	*n* = 107	*n* = 18
13–14 years	60 ± 24	–29 ± 20	–110 ± 111	228 ± 196	–389 ± 235	7 ± 18	114 ± 67	–32 ± 20
	17 → 117	–96 → –9	–699 → –25	27 → 790	–945 → –43	–26 → 56	18 → 261	–81 → –7
	n = 134	*n* = 43	*n* = 55	*n* = 139	*n* = 118	*n* = 141	*n* = 136	*n* = 20
15–16 years	48 ± 23	–36 ± 18	–125 ± 124	231 ± 191	–373 ± 261	6 ± 19	111 ± 69	–55 ± 42
	11 → 103	–81 → –11	–537 → –21	46 → 884	–1, 050 → –56	–24 → 49	8 → 281	–169 → –16
	n = 85	*n* = 40	*n* = 41	*n* = 88	*n* = 70	*n* = 92	*n* = 85	*n* = 18
17–18 years		–16 ± 0		113 ± 0	–1, 002 ± 0	–18 ± 0	39 ± 0	
		–16 → –16		113 → 113	–1, 002 → –1,002	–18 → –18	39 → 39	
		n = 1		*n* = 1	*n* = 1	*n* = 1	*n* = 1	

Table A2.18
Amplitudes (in μV) in Caucasian children: lead aVF

Age-group	P+	P−	Q	R	S	STj	T+	T−
<24 h	111 ± 48	−16 ± 9	−226 ± 114	798 ± 360	−260 ± 184	40 ± 28	120 ± 58	−34 ± 38
	221 → 21	−25 → −7	−541 → −47	82 → 1,492	−703 → −70	−1 → 117	34 → 296	−78 → −11
	n = 43	*n* = 4	*n* = 42	*n* = 43	*n* = 14	*n* = 43	*n* = 43	*n* = 3
<1 day	121 ± 45	−22 ± 14	−251 ± 113	811 ± 380	−309 ± 131	39 ± 26	121 ± 55	−150 ± 253
	36 → 227	−40 → −8	−428 → −74	73 → 1,725	−691 → −106	0 → 98	29 → 260	−601 → −20
	n = 138	*n* = 5	*n* = 133	*n* = 137	*n* = 44	*n* = 138	*n* = 137	*n* = 5
<2 days	129 ± 50	−39 ± 23	−233 ± 112	723 ± 353	−316 ± 156	40 ± 28	135 ± 58	−42 ± 53
	21 → 247	−79 → −12	−555 → −36	38 → 1,462	−657 → −90	95 → −19	31 → 293	−150 → −13
	n = 118	*n* = 9	*n* = 110	*n* = 119	*n* = 50	*n* = 119	*n* = 117	*n* = 6
<3 days	128 ± 54	−29 ± 17	−232 ± 106	675 ± 337	−314 ± 141	44 ± 29	142 ± 63	−14 ± 0
	37 → 268	−58 → −11	−541 → −58	27 → 1,571	−622 → −141	−15 → 134	28 → 322	−14 → −14
	n = 104	*n* = 9	*n* = 101	*n* = 106	*n* = 36	*n* = 106	*n* = 106	*n* = 1
<1 week	132 ± 50	−15 ± 8	−264 ± 104	783 ± 373	−316 ± 149	46 ± 28	155 ± 61	−15 ± 0
	34 → 241	−22 → −6	−517 → −71	139 → 1,801	−785 → −65	−7 → 110	53 → 312	−15 → −15
	n = 117	*n* = 4	*n* = 116	*n* = 117	*n* = 56	*n* = 117	*n* = 117	*n* = 1
<1 month	120 ± 47	−21 ± 2	−212 ± 104	780 ± 300	−221 ± 101	27 ± 40	160 ± 72	
	44 → 225	−22 → −19	−571 → −62	276 → 1,377	−454 → −85	−14 → 241	72 → 480	
	n = 45	*n* = 2	*n* = 43	*n* = 45	*n* = 19	*n* = 45	*n* = 45	
<3 months	109 ± 30		−255 ± 126	880 ± 427	−154 ± 74	30 ± 28	172 ± 72	−42 ± 1
	55 → 176		−579 → −52	34 → 2,049	−351 → −66	−22 → 111	67 → 358	−43 → −41
	n = 67		*n* = 58	*n* = 67	*n* = 28	*n* = 69	*n* = 65	*n* = 2
<6 months	105 ± 30	−31 ± 0	−254 ± 114	831 ± 380	−158 ± 86	29 ± 28	158 ± 74	−60 ± 2
	53 → 185	−31 → −31	−459 → −62	46 → 1,671	−421 → −59	−26 → 92	30 → 316	−61 → −58
	n = 50	*n* = 1	*n* = 40	*n* = 50	*n* = 26	*n* = 50	*n* = 50	*n* = 2
<1 year	108 ± 30	−45 ± 43	−235 ± 136	788 ± 474	−160 ± 76	37 ± 26	185 ± 78	−69 ± 66
	35 → 171	−136 → −16	−628 → −53	26 → 1,819	−384 → −53	−6 → 94	72 → 454	−116 → −22
	n = 73	*n* = 7	*n* = 50	*n* = 73	*n* = 40	*n* = 74	*n* = 72	*n* = 2
1–2 years	100 ± 38	−28 ± 15	−183 ± 122	746 ± 405	−178 ± 87	15 ± 23	146 ± 78	−33 ± 20
	32 → 177	−52 → −12	−553 → −27	24 → 1,769	−353 → −49	69 → −24	17 → 329	−77 → −16
	n = 109	*n* = 10	*n* = 73	*n* = 111	*n* = 66	*n* = 116	*n* = 110	*n* = 10
3–4 years	107 ± 42	−24 ± 12	−165 ± 92	998 ± 469	−182 ± 106	6 ± 23	230 ± 107	−68 ± 33
	24 → 191	−57 → −11	−413 → −30	42 → 1,955	−512 → −45	44 → −42	25 → 466	−116 → −46
	n =140	*n* = 11	*n* = 95	*n* = 141	*n* = 85	*n* = 141	*n* = 139	*n* = 4
5–6 years	96 ± 49	−25 ± 13	−135 ± 100	1,113 ± 466	−183 ± 89	10 ± 23	237 ± 90	−35 ± 0
	14 → 228	−56 → −8	−366 → −21	111 → 2,245	−405 → −59	−34 → 73	87 → 452	−35 → −35
	n = 152	*n* = 34	*n* = 114	*n* = 155	*n* = 99	*n* = 155	*n* = 155	*n* = 1
7–8 years	99 ± 44	−31 ± 21	−133 ± 96	1,093 ± 426	−219 ± 133	9 ± 25	262 ± 100	−64 ± 20
	25 → 206	−74 → −11	−409 → −23	329 → 2,033	−551 → −48	−29 → 70	104 → 509	−78 → −50
	n = 136	*n* = 24	*n* = 87	*n* = 141	*n* = 97	*n* = 141	*n* = 141	*n* = 2
9–10 years	92 ± 48	−26 ± 15	−128 ± 76	1,237 ± 410	−189 ± 99	13 ± 30	263 ± 114	−52 ± 32
	15 → 216	−63 → −8	−304 → −29	414 → 2,072	−516 → −42	−41 → 137	43 → 539	−83 → −20
	n = 109	*n* = 25	*n* = 79	*n* = 115	*n* = 74	*n* = 115	*n* = 115	*n* = 3

Table A2.18 (Continued)

Age-group	P+	P–	Q	R	S	STj	T+	T–
11–12 years	98 ± 42	–24 ± 15	–97 ± 58	1,161 ± 428	–202 ± 111	11 ± 23	254 ± 107	–47 ± 28
	20 → 185	–68 → –5	–233 → –273	404 → 2,170	–477 → –52	–39 → 66	45 → 494	–91 → –17
	n = 115	*n* = 22	*n* = 76	*n* = 119	*n* = 77	*n* = 119	*n* = 119	*n* = 5
13–14 years	94 ± 51	–32 ± 18	–93 ± 62	1,117 ± 433	–249 ± 137	14 ± 33	250 ± 106	–54 ± 48
	13 → 211	–94 → –12	–287 → –24	309 → 2,144	–683 → –57	–35 → 89	77 → 484	–138 → –23
	n = 125	*n* = 47	*n* = 94	*n* = 141	*n* = 96	*n* = 141	*n* = 140	*n* = 5
15–16 years	104 ± 57	–26 ± 22	–98 ± 67	1,020 ± 434	–228 ± 157	9 ± 32	237 ± 90	–53 ± 21
	21 → 259	–105 → –10	–295 → –21	201 → 1,861	–605 → –275	–69 → 74	96 → 417	–80 → –15
	n = 89	*n* = 20	*n* = 57	*n* = 92	*n* = 63	*n* = 92	*n* = 91	*n* = 7
17–18 years	204 ± 0		–263 ± 0	1,095 ± 0		73 ± 0	171 ± 0	
	204 → 204		–263 → –263	1,095 → 1,095		73 → 73	171 → 171	
	n = 1		*n* = 1	*n* = 1		*n* = 1	*n* = 1	

Table A2.19
Amplitudes (in μV) in Caucasian children: lead V_1

Age-group	P+	P–	Q	R	S	STj	T+	T–
<24 h	93 ± 44	–40 ± 14		1,248 ± 547	–764 ± 458	–18 ± 31	175 ± 120	–121 ± 68
	28 → 186	–74 → –19		481 → 2,291	–1,780 → –141	–75 → 47	15 → 590	–254 → –39
	n = 40	*n* = 25		*n* = 43	*n* = 40	*n* = 43	*n* = 33	*n* = 22
<1 day	90 ± 52	–39 ± 16	–36 ± 7	1,169 ± 487	–786 ± 503	–24 ± 37	148 ± 102	–111 ± 57
	13 → 201	–83 → –13	–41 → –31	329 → 2,370	–2,176 → –74	–112 → 53	31 → 432	–281 → –24
	n = 132	*n* = 65	*n* = 2	*n* = 137	*n* = 124	*n* = 137	*n* = 12	*n* = 99
<2 days	83 ± 50	–49 ± 24	–69 ± 59	1,102 ± 524	–603 ± 379	–28 ± 40	124 ± 83	–145 ± 78
	25 → 244	–133 → –15	–110 → –27	404 → 2,592	–1,530 → –137	–110 → 64	18 → 320	–355 → –34
	n = 106	*n* = 58	*n* = 2	*n* = 117	*n* = 100	*n* = 117	*n* = 76	*n* = 102
<3 days	85 ± 43	–44 ± 26	–87 ± 70	1,102 ± 536	–615 ± 385	–37 ± 69	98 ± 52	–182 ± 104
	18 → 199	–121 → –15	–136 → –37	363 → 2,832	–1,718 → –145	–136 → 62	35 → 261	–621 → –47
	n = 97	*n* = 40	*n* = 2	*n* = 102	*n* = 93	*n* = 102	*n* = 57	*n* = 97
<1 week	86 ± 49	–44 ± 31		1,085 ± 445	–494 ± 326	–47 ± 42	102 ± 67	–211 ± 79
	17 → 249	–229 → –18		2,078 → 354	–1,530 → –103	–137 → 15	25 → 332	–428 → –70
	n = 109	*n* = 51		*n* = 114	*n* = 99	*n* = 114	*n* = 28	*n* = 110
<1 month	83 ± 49	–40 ± 20	–28 ± 0	873 ± 391	–463 ± 346	–32 ± 42	102 ± 68	–192 ± 102
	29 → 237	–81 → –10	–28 → –28	425 → 2,303	–1,529 → –102	–105 → 53	30 → 224	–467 → –58
	n = 38	*n* = 16	*n* = 1	*n* = 41	*n* = 37	*n* = 41	*n* = 16	*n* = 38
<3 months	65 ± 34	–45 ± 21	–134 ± 117	772 ± 345	–496 ± 337	–20 ± 60	102 ± 75	–251 ± 143
	17 → 144	–98 → –14	–217 → –51	194 → 1,656	–1,735 → –120	–137 → 336	38 → 314	–465 → 0
	n = 52	*n* = 40	*n* = 2	*n* = 63	*n* = 56	*n* = 65	*n* = 12	*n* = 65
<6 months	59 ± 28	–45 ± 28		829 ± 413	–511 ± 358	–19 ± 45	80 ± 46	–319 ± 99
	16 → 131	–127 → –18		298 → 2,222	–1,660 → –96	–92 → 192	40 → 156	–493 → –139
	n = 41	*n* = 32		*n* = 49	*n* = 48	*n* = 49	*n* = 5	*n* = 48
<1 year	68 ± 29	–53 ± 35		779 ± 420	–538 ± 330	–22 ± 27	160 ± 83	–343 ± 97
	21 → 142	–194 → –15		47 → 2,037	–1,562 → –95	–74 → 106	77 → 243	–577 → –204
	n = 59	*n* = 43		*n* = 70	*n* = 69	*n* = 71	*n* = 3	*n* = 69

Table A2.19 (Continued)

Age-group	P+	P–	Q	R	S	STj	T+	T–
1–2 years	91 ± 33	–57 ± 36	–114 ± 59	807 ± 382	–771 ± 385	–3 ± 31	126 ± 64	–317 ± 105
	26 → 152	–204 → –21	–155 → –72	99 → 1,804	–1,701 → –183	–86 → 65	46 → 224	–578 → –139
	n = 100	*n* = 46	*n* = 2	*n* = 107	*n* = 103	*n* = 112	*n* = 7	*n* = 105
3–4 years	95 ± 36	–56 ± 32	–526 ± 0	610 ± 244	–895 ± 394	13 ± 30	85 ± 37	–270 ± 92
	30 → 168	–153 → –16	–526 → –526	106 → 1,112	–1,893 → –205	–45 → 91	43 → 163	–467 → –88
	n = 129	*n* = 57	*n* = 1	*n* = 135	*n* = 135	*n* = 136	*n* = 25	*n* = 136
5–6 years	98 ± 35	–55 ± 25	–342 ± 409	509 ± 232	–1,008 ± 398	22 ± 28	77 ± 31	–249 ± 96
	33 → 182	–132 → –24	–631 → –53	84 → 1,020	–2,011 → –395	–28 → 88	40 → 155	–448 → –51
	n = 141	*n* = 62	*n* = 2	*n* = 146	*n* = 146	*n* = 147	*n* = 37	*n* = 145
7–8 years	88 ± 35	–54 ± 30		483 ± 283	–969 ± 426	25 ± 26	78 ± 37	–251 ± 98
	28 → 166	–180 → –17		142 → 1,388	–1,921 → –264	–37 → 78	38 → 185	–431 → –76
	n = 133	*n* = 69		*n* = 135	*n* = 135	*n* = 135	*n* = 34	*n* = 129
9–10 years	76 ± 32	–49 ± 25		403 ± 191	–1,090 ± 462	31 ± 31	111 ± 98	–218 ± 99
	21 → 150	–115 → –15		131-919	–2,213 → –214	–25 → 145	32 → 468	–436 → –43
	n = 104	*n* = 73		*n* = 108	*n* = 108	*n* = 108	*n* = 35	*n* = 102
11–12 years	75 ± 31	–57 ± 37		383 ± 225	–1,130 ± 451	30 ± 33	129 ± 121	–206 ± 99
	22 → 148	–228 → –16		66 → 1,090	–2,366 → –363	–33 → 109	34 → 573	–488 → –35
	n = 97	*n* = 70		*n* = 104	*n* = 104	*n* = 104	*n* = 39	*n* = 87
13–14 years	70 ± 30	–58 ± 32	–788 ± 692	329 ± 196	–1,097 ± 418	33 ± 36	91 ± 65	–176 ± 92
	25 → 136	–151 → –18	–1,364 → –20	52 → 912	–2,265 → –366	–26 → 164	31 → 342	–450 → –32
	n = 99	*n* = 81	*n* = 3	*n* = 107	*n* = 107	*n* = 109	*n* = 44	*n* = 96
15–16 years	72 ± 34	–54 ± 25	–115 ± 99	348 ± 211	–1,194 ± 529	38 ± 36	79 ± 114	–156 ± 93
	22 → 232	–110 → –19	–185 → –45	71 → 1,186	–2,303 → –98	–43 → 140	492 → 0	–420 → –37
	n = 65	*n* = 50	*n* = 2	*n* = 70	*n* = 70	*n* = 71	*n* = 71	*n* = 48
17–18 years	56 ± 0	26 ± 0		1,028 ± 0	–466 ± 0	–17 ± 0		–139 ± 0
	56 → 56	–26 → –26		1,028 → 1,028	–466 → –466	–17 → –17		–139 → –139
	n = 1	*n* = 1		*n* = 1	*n* = 1	*n* = 1		*n* = 1

Table A2.20

Amplitudes (in μV) in Caucasian children: lead V_2

Age-group	P+	P–	Q	R	S	STj	T+	T–
<24 h	123 ± 55	–19 ± 8		1,642 ± 621	–1,758 ± 690	–17 ± 38	198 ± 151	–113 ± 57
	15 → 238	–33 → –13		709 → 3,673	–2,967 → –618	–77 → 77	21 → 487	–226 → –35
	n = 43	*n* = 5		*n* = 43	*n* = 42	*n* = 43	*n* = 35	*n* = 19
<1 day	114 ± 52	–51 ± 71	–36 ± 0	1,402 ± 500	–1,534 ± 674	–35 ± 48	169 ± 98	–133 ± 88
	36 → 229	–342 → –12	–36 → –36	494 → 2,393	–2,927 → –405	–119 → 70	25 → 417	–352 → –25
	n = 133	*n* = 21	*n* = 1	*n* = 137	*n* = 136	*n* = 137	*n* = 103	*n* = 91
<2 days	103 ± 47	–33 ± 17	–60 ± 0	1,402 ± 536	–1,346 ± 609	–27 ± 44	133 ± 91	–138 ± 75
	28 → 230	–65 → –8	–60 → –60	484 → 2,536	–2,887 → –414	–111 → 77	21 → 407	–313 → –28
	n = 116	*n* = 22	*n* = 1	*n* = 116	*n* = 114	*n* = 116	*n* = 84	*n* = 93
<3 days	111 ± 39	–33 ± 25	–83 ± 65	1,343 ± 504	–1,278 ± 556	–29 ± 72	125 ± 79	–179 ± 114
	32 → 192	–77 → –9	–166 → –21	501 → 2,689	–2,895 → –300	–161 → 76	26 → 376	–595 → –34
	n = 102	*n* = 9	*n* = 4	*n* = 104	*n* = 104	*n* = 104	*n* = 56	*n* = 89

Table A2.20 (Continued)

Age-group	P+	P−	Q	R	S	STj	T+	T−
<1 week	110 ± 53	−57 ± 55	−60 ± 0	1,332 ± 480	−1, 130 ± 499	−35 ± 63	116 ± 75	−210 ± 128
	17 → 248	−219 → −21	−60 → −60	583 → 2,542	−2, 070 → −238	−172 → 86	16 → 326	−574 → −37
	n = 112	n = 16	n = 1	n = 114	n = 111	n = 114	n = 43	n = 102
<1 month	108 ± 54	−36 ± 31	−52 ± 0	1,159 ± 519	−1, 000 ± 494	0 ± 63	118 ± 78	−145 ± 110
	22 → 235	−87 → −10	−52 → −52	380 → 3,156	−2, 431 → −276	−99 → 92	33 → 302	−500 → −20
	n = 42	n = 6	n = 1	n = 42	n = 41	n = 42	n = 32	n = 24
<3 months	102 ± 40	−49 ± 27	−56 ± 0	1,406 ± 411	−972 ± 433	11 ± 61	144 ± 94	−209 ± 134
	33 → 224	−96 → −20	−56 → −56	566 → 2,292	−2, 306 → −307	−87 → 343	42 → 474	−554 → −50
	n = 63	n = 6	n = 1	n = 63	n = 62	n = 65	n = 35	n = 47
<6 months	96 ± 43	−32 ± 14	−33 ± 11	1,449 ± 464	−1, 015 ± 489	0 ± 44	121 ± 91	−269 ± 142
	19 → 184	−56 → −5	−41 → −25	750 → 2,829	−2, 142 → −344	−86 → 101	24 → 397	−638 → −48
	n = 48	n = 10	n = 2	n = 48	n = 48	n = 48	n = 18	n = 42
<1 year	104 ± 42	−44 ± 25	−134 ± 0	1,520 ± 463	−1, 083 ± 466	1 ± 38	99 ± 40	−271 ± 132
	23 → 251	−82 → −15	−134 → −134	736 → 3,269	−2, 650 → −378	−74 → 155	28 → 187	−671 → −58
	n = 73	n = 12	n = 1	n = 73	n = 72	n = 74	n = 24	n = 71
1–2 years	114 ± 37	−77 ± 63	−70 ± 24	1,473 ± 593	−1, 478 ± 688	11 ± 46	134 ± 70	−307 ± 160
	31 → 193	−200 → −14	−84 → −43	275 → 2,720	−2, 722 → −185	−108 → 124	51 → 380	−695 → −36
	n = 101	n = 13	n = 3	n = 106	n = 106	n = 111	n = 46	n = 89
3–4 years	110 ± 36	−48 ± 20	−362 ± 0	1,285 ± 442	−1, 703 ± 613	38 ± 39	156 ± 83	−222 ± 112
	187 → 44	−95 → −21	−362 → −362	434 → 2,200	−3, 188 → −572	−50 → 142	43 → 396	−486 → −50
	n = 140	n = 11	n = 1	n = 140	n = 140	n = 140	n = 107	n = 96
5–6 years	110 ± 32	−34 ± 9	−28 ± 10	1,090 ± 385	−1, 819 ± 639	54 ± 46	184 ± 119	−154 ± 85
	60 → 214	−48 → −25	−35 → −21	399 → 2,004	−3, 161 → −647	−28 → 175	46 → 478	−339 → −38
	n = 155	n = 8	n = 2	n = 155	n = 155	n = 155	n = 133	n = 85
7–8 years	106 ± 36	−29 ± 16		1,030 ± 444	−1, 822 ± 638	58 ± 41	198 ± 128	−141 ± 93
	45→208	−57 → −15		353 → 2,050	−3, 224 → −713	−23 → 184	47 → 547	−540 → −34
	n = 140	n = 13		n = 140	n = 140	n = 140	n = 131	n = 58
9–10 years	102 ± 34	−25 ± 13		870 ± 344	−1, 891 ± 624	67 ± 46	280 ± 179	−129 ± 72
	41 → 211	−45 → −13		314 → 1,707	−3, 635 → −744	−18 → 198	52 → 865	−259 → −27
	n = 113	n = 9		n = 114	n = 114	n = 114	n = 105	n = 32
11–12 years	94 ± 33	−44 ± 26	−21 ± 0	718 ± 295	−1, 704 ± 590	69 ± 51	285 ± 184	−108 ± 61
	36 → 177	−109 → −18	−21 → −21	208 → 1,465	−3, 335 → −644	−19 → 212	50 → 809	−237 → −32
	n = 118	n = 18	n = 1	n = 119	n = 119	n = 119	n = 117	n = 23
13–14 years	83 ± 34	−42 ± 27	−1, 079±1, 484	650 ± 336	−1, 562 ± 611	66 ± 47	281 ± 186	−84 ± 54
	26 → 160	−125 → −12	−2, 128 → −30	170 → 1,493	−2, 910 → −387	−10 → 194	43 → 781	−244 → −21
	n = 134	n = 32	n = 2	n = 138	n = 138	n = 139	n = 134	n = 20
15–16 years	84 ± 29	−38 ± 17	−26 ± 0	632 ± 356	−1, 603 ± 584	78 ± 59	367 ± 238	−117 ± 68
	34 → 152	−69 → −10	−26 → −26	183 → 2,034	−2, 888 → −645	6-242	69 → 1,125	−228 → −49
	n = 86	n = 21	n = 1	n = 89	n = 89	n = 89	n = 87	n = 5
17–18 years	113 ± 0			1,481 ± 0	−1, 599 ± 0	−43 ± 0	75 ± 0	−131 ± 0
	113 → 113			1,481 → 1,481	−1, 599 → −1,599	−43 → −43	75-7-5	−131 → −131
	n = 1			n = 1	n = 1	n = 1	n = 1	n = 1

Table A2.21

Amplitudes (in μV) in Caucasian children: lead V_{4R}

Age-group	P+	P−	Q	R	S	STj	T+	T−
<24 h	75 ± 42	−40 ± 14	−64 ± 76	989 ± 365	−344 ± 169	−6 ± 27	116 ± 87	−82 ± 53
	17 → 178	−69 → −20	−225 → −22	307 → 1,747	−749 → −156	−55 → 74	17 → 440	−244 → −7
	n = 39	*n* = 24	*n* = 7	*n* = 43	*n* = 29	*n* = 43	*n* = 34	*n* = 23
<1 day	86 ± 43	−36 ± 20	−72 ± 31	992 ± 339	−359 ± 263	−13 ± 24	107 ± 71	−85 ± 49
	11 → 200	−142 → −13	−121 → −22	373 → 1,653	−1, 273 → −93	−66 → 36	22 → 292	−228 → −17
	n = 123	*n* = 69	*n* = 14	*n* = 136	*n* = 93	*n* = 136	*n* = 99	*n* = 103
<2 days	80 ± 39	−37 ± 22	−55 ± 34	980 ± 338	−298 ± 181	−15 ± 28	81 ± 52	−109 ± 65
	19 → 189	−149 → −12	−143 → −20	288 → 1,993	−804 → −36	−65 → 61	17 → 291	−285 → −24
	n = 110	*n* = 57	*n* = 19	*n* = 116	*n* = 72	*n* = 116	*n* = 70	*n* = 98
<3 days	74 ± 39	−35 ± 17	−39 ± 18	924 ± 321	−262 ± 161	−23 ± 48	67 ± 36	−129 ± 69
	16 → 166	−102 → −13	−73 → −20	405 → 1,713	−658 → −67	−84 → 30	25 → 199	−322 → −32
	n = 97	*n* = 47	*n* = 15	*n* = 102	*n* = 52	*n* = 102	*n* = 43	*n* = 99
<1 week	88 ± 44	−40 ± 30	−47 ± 22	962 ± 345	−299 ± 214	−34 ± 40	80 ± 59	−165 ± 62
	19 → 235	−169 → −12	−113 → −23	415 → 1,747	−1, 508 → −68	−170 → 29	26 → 328	−295 → −57
	n = 107	*n* = 34	*n* = 15	*n* = 113	*n* = 62	*n* = 113	*n* = 34	*n* = 109
<1 month	77 ± 44	−38 ± 22	−56 ± 57	758 ± 335	−242 ± 232	−28 ± 32	86 ± 43	−152 ± 71
	12 → 161	−110 → −9	−155 → −20	95 → 1,629	−1, 106 → −52	−91 → 82	13 → 167	−342 → −51
	n = 41	*n* = 20	*n* = 5	*n* = 44	*n* = 24	*n* = 44	*n* = 14	*n* = 42
<3 months	60 ± 26	−37 ± 25	−89 ± 91	551 ± 243	−241 ± 126	−26 ± 31	60 ± 46	−225 ± 88
	16 → 121	−105 → −11	−238 → −23	102 → 1,257	−553 → −65	−80 → 34	13 → 128	−464 → −51
	n = 60	*n* = 40	*n* = 5	*n* = 65	*n* = 47	*n* = 67	*n* = 7	*n* = 62
<6 months	53 ± 26	−35 ± 21	−232 ± 255	514 ± 237	−257 ± 210	−18 ± 26	56 ± 0	−236 ± 77
	15 → 144	−104 → −8	−526 → −76	110 → 1,189	−1, 088 → −54	−57 → 31	56 → 56	−466 → −110
	n = 46	*n* = 25	*n* = 3	*n* = 49	*n* = 44	*n* = 49	*n* = 1	*n* = 49
<1 year	58 ± 21	−35 ± 19	−38 ± 8	426 ± 221	−247 ± 148	−14 ± 28	61 ± 24	−234 ± 75
	22 → 123	−104 → −13	−47 → −29	95 → 1,248	−896 → −44	−79 → 116	48 → 97	−474 → −99
	n = 65	*n* = 38	*n* = 4	*n* = 70	*n* = 66	*n* = 71	*n* = 4	*n* = 69
1–2 years	63 ± 25	−36 ± 35	−92 ± 110	348 ± 173	−302 ± 201	−4 ± 37	83 ± 79	−192 ± 78
	16 → 131	−220 → −12	−341 → −35	89 → 835	−1, 026 → −42	−84 → 104	14 → 285	−407 → −52
	n = 102	*n* = 46	*n* = 7	*n* = 110	*n* = 106	*n* = 115	*n* = 10	*n* = 108
3–4 years	73 ± 31	−27 ± 12	−82 ± 68	349 ± 171	−419 ± 229	3 ± 24	46 ± 20	−175 ± 81
	22 → 135	−59 → −12	−202 → −29	77 → 745	−991 → −95	−40 → 62	18 → 100	−352 → −37
	n = 135	*n* = 41	*n* = 6	*n* = 137	*n* = 136	*n* = 137	*n* = 43	*n* = 136
5–6 years	68 ± 30	−28 ± 14	−117 ± 131	295 ± 140	−477 ± 236	10 ± 18	57 ± 60	−166 ± 75
	15 → 131	−75 → −10	−267 → −27	81 → 624	−1, 046 → −104	−24 → 45	26 → 419	−332 → 0
	n = 142	*n* = 65	*n* = 3	*n* = 146	*n* = 145	*n* = 147	*n* = 41	*n* = 147
7–8 years	66 ± 29	−28 ± 13	−70 ± 0	287 ± 151	−488 ± 249	11 ± 18	51 ± 29	−171 ± 74
	12 → 123	−63 → −10	−70 → −70	75 → 648	−1, 109 → −100	−36 → 49	20 → 164	−318 → −35
	n = 129	*n* = 63	*n* = 1	*n* = 136	*n* = 134	*n* = 136	*n* = 42	*n* = 128

Table A2.21 (Continued)

Age-group	P+	P–	Q	R	S	STj	T+	T–
9–10 years	59 ± 29	−31 ± 16	−27 ± 7	245 ± 122	−536 ± 304	13 ± 21	59 ± 39	−160 ± 75
	16 → 142	−115 → −12	−37 → −21	46 → 579	−1, 679 → −109	−42 → 63	26 → 204	−348 → −26
	n = 101	*n* = 67	*n* = 4	*n* = 109	*n* = 107	*n* = 109	*n* = 35	*n* = 104
11–12 years	59 ± 24	−32 ± 15	−270 ± 387	222 ± 131	−539 ± 287	9 ± 24	50 ± 26	−146 ± 80
	17 → 110	−93 → −11	−934 → −20	40 → 631	−1, 277 → −144	−57 → 47	21 → 172	−362 → −33
	n = 95	*n* = 71	*n* = 5	*n* = 102	*n* = 102	*n* = 104	*n* = 44	*n* = 95
13–14 years	46 ± 27	−34 ± 15	−282 ± 278	190 ± 115	−486 ± 274	13 ± 17	41 ± 22	−113 ± 63
	10 → 102	−76 → −12	−569 → −20	43 → 575	−1, 460 → −36	−26 → 51	13 → 94	−318 → −18
	n = 95	*n* = 90	*n* = 4	*n* = 109	*n* = 107	*n* = 111	*n* = 45	*n* = 103
15–16 years	51 ± 25	−29 ± 14	−166 ± 206	184 ± 135	−482 ± 320	13 ± 24	59 ± 49	−104 ± 66
	13 → 109	−76 → −11	−311 → −20	25 → 914	−1, 209 → −31	−36 → 83	15 → 219	−298 → −16
	n = 66	*n* = 46	*n* = 2	*n* = 73	*n* = 71	*n* = 74	*n* = 38	*n* = 55
17–18 years	24 ± 0	−48 ± 0		655 ± 0	−249 ± 0	−16 ± 0		−132 ± 0
	24 → 24	−48 → −48		655 → 655	−249 → −249	−16 → −16		−132 → −132
	n = 1	*n* = 1		*n* = 1	*n* = 1	*n* = 1		*n* = 1

Table A2.22
Amplitudes (in μV) in Caucasian children: lead V_4

Age-group	P+	P–	Q	R	S	STj	T+	T–
<24 h	139 ± 43	−78 ± 86	−83 ± 39	1,554 ± 484	−1, 573 ± 562	10 ± 51	187 ± 128	−92 ± 40
	66 → 250	−177 → −24	−156 → −32	700 → 2,427	−3, 213 → −592	−65 → 141	42 → 177	−175 → −51
	n = 42	*n* = 3	*n* = 8	*n* = 43	*n* = 42	*n* = 43	*n* = 38	*n* = 11
<1 day	141 ± 58	−104 ± 108	−72 ± 50	1,578 ± 483	−1, 565 ± 560	7 ± 53	178 ± 135	−134 ± 93
	57 → 279	−313 → −24	−233 → −20	2,513 → 536	−2, 834 → −530	−110 → 125	24 → 776	−446 → −24
	n = 138	*n* = 6	*n* = 23	*n* = 138	*n* = 137	*n* = 138	*n* = 113	*n* = 56
<2 days	134 ± 40	−62 ± 66	−107 ± 83	1,567 ± 510	−1, 439 ± 524	14 ± 52	144 ± 111	−115 ± 78
	62 → 236	−160 → −18	−367 → −30	540 → 3,072	−2, 608 → −424	−132 → 117	15 → 429	−350 → −250
	n = 119	*n* = 4	*n* = 20	*n* = 119	*n* = 119	*n* = 119	*n* = 89	*n* = 57
<3 days	144 ± 39	−39 ± 11	−91 ± 67	1,545 ± 534	−1, 254 ± 539	25 ± 69	139 ± 99	−166 ± 113
	69 → 229	−47 → −31	−218 → −20	67 → 2,847	−2, 502 → −132	−110 → 152	15 → 440	−580 → −43
	n = 106	*n* = 2	*n* = 25	*n* = 106	*n* = 104	*n* = 106	*n* = 75	*n* = 57
<1 week	143 ± 44	−71 ± 75	−116 ± 98	1,645 ± 542	−1, 173 ± 489	36 ± 63	178 ± 97	−187 ± 121
	56 → 251	−157 → −19	−481 → −20	695 → 3,126	−2, 392 → −261	−98 → 180	39 → 402	−644 → −54
	n = 116	*n* = 3	*n* = 57	*n* = 116	*n* = 115	*n* = 116	*n* = 79	*n* = 56
<1 month	142 ± 43	−72 ± 0	−140 ± 97	1,497 ± 463	−954 ± 369	32 ± 57	259 ± 127	−193 ± 129
	55 → 239	−72 → −72	−373 → −23	732 → 2,757	−2, 350 → −317	−76 → 157	41 → 561	−442 → −25
	n = 45	*n* = 1	*n* = 20	*n* = 45	*n* = 45	*n* = 45	*n* = 38	*n* = 11
<3 months	113 ± 37	−26 ± 8	−156 ± 123	1,983 ± 489	−863 ± 355	44 ± 61	273 ± 155	−162 ± 187
	48 → 208	−35 → −16	−540 → −26	1,094 → 3,084	−1, 649 → −150	−36 → 373	21 → 735	−647 → −40
	n = 67	*n* = 4	*n* = 35	*n* = 67	*n* = 65	*n* = 69	*n* = 63	*n* = 9
<6 months	104 ± 43	−29 ± 10	−185 ± 152	1,741 ± 549	−795 ± 331	48 ± 37	257 ± 180	−132 ± 65
	49 → 276	−38 → −18	−428 → −20	877 → 3,202	−1, 659 → −307	−11 → 162	58 → 910	−233 → −39
	n = 50	*n* = 3	*n* = 20	*n* = 50	*n* = 50	*n* = 50	*n* = 46	*n* = 11

Table A2.22 (Continued)

Age-group	P+	P–	Q	R	S	STj	T+	T–
<1 year	109 ± 40 49 → 257 *n* = 72	–31 ± 19 –52 → –16 *n* = 3	–190 ± 184 –777 → –25 *n* = 27	1,792 ± 675 541 → 3,768 *n* = 72	–794 ± 410 –2, 462 → –222 *n* = 71	44 ± 48 –42 → 263 *n* = 73	270 ± 148 52 → 663 *n* = 63	–223 ± 162 –713 → –63 *n* = 16
1–2 years	86 ± 27 25 → 155 *n* = 109	–37 ± 28 –89 → –15 *n* = 10	–179 ± 129 –549 → –25 *n* = 47	1,668 ± 668 50 → 3,095 *n* = 110	–741 ± 414 –1, 995 → –165 *n* = 108	36 ± 33 –23 → 114 *n* = 115	282 ± 187 24 → 719 *n* = 98	–147 ± 94 –348 → –27 *n* = 32
3–4 years	87 ± 27 41 → 150 *n* = 141	–17 ± 5 –26 → –11 *n* = 12	–143 ± 113 –523 → –22 *n* = 50	2,125 ± 724 886 → 3,760 *n* = 141	–861 ± 482 –2, 083 → –145 *n* = 134	34 ± 33 –39 → 105 *n* = 141	433 ± 259 49 → 993 *n* = 140	–113 ± 122 –443 → –31 *n* = 10
5–6 years	86 ± 25 43 → 154 *n* = 155	–23 ± 12 –49 → –12 *n* = 17	–134 ± 118 –613 → –21 *n* = 56	2,263 ± 828 929 → 4,305 *n* = 155	–963 ± 532 –2, 694 → –200 *n* = 152	44 ± 41 –20 → 153 *n* = 155	523 ± 223 138 → 987 *n* = 153	–206 ± 52 –261 → –158 *n* =3
7–8 years	84 ± 28 47 → 159 *n* = 141	–25 ± 12 –51 → –14 *n* = 13	–110 ± 93 –459 → –24 *n* = 39	2,063 ± 793 911 → 3,804 *n* = 141	–1, 037 ± 566 –2, 307 → –114 *n* = 140	44 ± 42 –36 → 122 *n* = 141	571 ± 246 150 → 1,232 *n* = 141	–129 ± 0 –129 → –129 *n* = 1
9–10 years	82 ± 24 44 → 151 *n* = 115	–22 ± 6 –37 → –14 *n* = 12	–100 ± 80 –312 → –22 *n* = 25	1,955 ± 771 878 → 4,040 *n* = 115	–972 ± 510 –2, 359 → –136 *n* = 111	56 ± 51 –19 → 198 *n* = 115	652 ± 264 168 → 1,328 *n* = 115	
11–12 years	87 ± 25 43 → 145 *n* = 118	–25 ± 22 –86 → –10 *n* = 10	–102 ± 82 –292 → –22 *n* = 26	1,864 ± 686 713 → 3,400 *n* = 119	–891 ± 495 –2, 475 → –149 *n* = 115	45 ± 51 –47 → 200 *n* = 119	589 ± 251 203 → 1,292 *n* = 119	
13–14 years	79 ± 28 27 → 137 *n* = 140	–23 ± 12 –47 → –7 *n* = 22	–55 ± 32 –174 → –24 *n* = 26	1,632 ± 624 607 → 3,010 *n* = 141	–1, 045 ± 547 –2, 342 → –206 *n* = 138	68 ± 64 –21 → 253 *n* = 141	602 ± 251 168 → 1,176 *n* = 141	–48 ± 0 –48 → –48 *n* = 1
15–16 years	83 ± 27 31 → 144 *n* = 91	–19 ± 7 –38 → –10 *n* = 11	–85 ± 97 –381 → –20 *n* = 17	1,638 ± 735 582 → 3,787 *n* = 91	–898 ± 554 –2, 332 → –168 *n* = 90	60 ± 67 –38 → 229 *n* = 91	587 ± 289 188 → 1,314 *n* = 91	–109 ± 58 –150 → –68 *n* = 2
17–18 years	199 ± 0 199 → 199 *n* = 1			1,687 ± 0 1,687 → 1,687 *n* = 1	–2, 201 ± 0 –2, 201 → –2,201 *n* = 1	72 ± 0 72 → 72 *n* = 1	143 ± 0 143 → 143 *n* = 1	

Table A2.23

Amplitudes (in μV) in Caucasian children: lead V_5

Age-group	P+	P–	Q	R	S	STj	T+	T–
<24 h	110 ± 40 45 → 218 *n* = 42	–99 ± 109 –176 → –22 *n* = 2	–132 ± 102 –381 → –22 *n* = 22	924 ± 419 292 → 1,890 *n* = 43	–893 ± 427 –1, 848 → –148 *n* = 41	29 ± 41 –53 → 157 *n* = 43	161 ± 104 18 → 440 *n* = 43	–55 ± 32 –98 → –11 *n* = 8
<1 day	112 ± 51 55 → 213 *n* = 135	–60 ± 32 –102 → –13 *n* = 7	–104 ± 64 –278 → –23 *n* = 77	999 ± 434 296 → 1,886 *n* = 137	–964 ± 470 –1, 958 → –206 *n* = 134	29 ± 34 –37 → 106 *n* = 137	154 ± 88 29 → 385 *n* = 131	–66 ± 39 –151 → –20 *n* = 33
<2 days	105 ± 33 43 → 205 *n* = 116	–24 ± 10 –40 → –10 *n* = 6	–127 ± 100 –543 → –27 *n* = 59	941 ± 419 221 → 1,840 *n* = 118	–924 ± 451 –2, 171 → –210 *n* = 116	33 ± 38 –65 → 117 *n* = 118	173 ± 90 27 → 458 *n* = 110	–64 ± 49 –204 → –11 *n* = 20

Table A2.23 (Continued)

Age-group	P+	P−	Q	R	S	STj	T+	T−
<3 days	115 ± 35	−34 ± 24	−124 ± 69	1,002 ± 427	−876 ± 477	52 ± 40	192 ± 91	−106 ± 65
	55 → 186	−51 → −17	−256 → −24	291 → 2,255	−2, 343 → −225	−30 → 138	51 → 436	−278 → −22
	n = 106	*n* = 2	*n* = 59	*n* = 106	*n* = 104	*n* = 106	*n* = 97	*n* = 18
<1 week	114 ± 36	−26 ± 15	−172 ± 101	1,090 ± 473	−794 ± 385	59 ± 50	243 ± 114	−170 ± 167
	38 → 191	−40 → −11	−466 → −28	273 → 2,412	−1, 685 → −154	−60 → 205	42 → 519	−510 → −18
	n = 116	*n* = 3	*n* = 96	*n* = 116	*n* = 109	*n* = 117	*n* = 109	*n* = 13
<1 month	119 ± 44	−45 ± 2	−184 ± 114	1,163 ± 538	−647 ± 375	42 ± 46	310 ± 150	−34 ± 16
	58 → 298	−46 → −43	−478 → −43	324 → 2,690	−1, 767 → −112	−32 → 172	90 → 696	−45 → −22
	n = 43	*n* = 2	*n* = 36	*n* = 44	*n* = 42	*n* = 44	*n* = 43	*n* = 2
<3 months	101 ± 33	−24 ± 10	−228 ± 159	1,849 ± 630	−578 ± 272	47 ± 43	358 ± 130	−93 ± 0
	39 → 220	−38 → −17	−959 → −32	913 → 3,271	−1, 367 → −177	−25 → 202	86 → 648	−93 → −93
	n = 66	*n* = 4	*n* = 56	*n* = 66	*n* = 65	*n* = 68	*n* = 66	*n* = 1
<6 months	96 ± 51	−31 ± 30	−195 ± 117	1,569 ± 563	−536 ± 329	60 ± 51	357 ± 202	−130 ± 0
	31 → 277	−75 → −9	−496 → −35	572 → 3,111	−1, 368 → −157	−11 → 203	107 → 928	−130 → −130
	n = 48	*n* = 4	*n* = 38	*n* = 48	*n* = 46	*n* = 48	*n* = 47	*n* = 1
<1 year	94 ± 36	−29 ± 13	−218 ± 159	1,758 ± 667	−481 ± 272	53 ± 42	364 ± 148	−72 ± 18
	34 → 293	−52 → −18	−891 → −38	59 → 3,238	−1, 277 → −95	−5 → 317	35 → 748	−88 → −45
	n = 73	*n* = 5	*n* = 58	*n* = 73	*n* = 67	*n* = 74	*n* = 73	*n* = 4
1–2 years	84 ± 48	−34 ± 27	−224 ± 155	1,569 ± 690	−379 ± 234	30 ± 48	333 ± 160	−109 ± 47
	29 → 391	−103 → −13	−626 → −35	35 → 3,080	−999 → −65	−25 → 120	63 → 724	−172 → −35
	n = 109	*n* = 13	*n* = 83	*n* = 111	*n* = 100	*n* = 116	*n* = 107	*n* = 6
3–4 years	78 ± 21	−18 ± 11	−189 ± 126	2,003 ± 661	−388 ± 234	15 ± 26	465 ± 204	−37 ± 26
	35 → 116	−45 → −6	−596 → −26	803 → 3,381	−1, 047 → −56	−38 → 71	62 → 877	−76 → −21
	n = 140	*n* = 10	*n* = 121	*n* = 140	*n* = 120	*n* = 140	*n* = 140	*n* = 4
5–6 years	78 ± 22	−22 ± 16	−185 ± 130	2,443 ± 684	−493 ± 294	20 ± 33	578 ± 196	
	36 → 120	−85 → −12	−534 → −20	1,227 → 3,889	−1, 475 → −105	−60 → 98	235 → 1,065	
	n = 155	*n* = 18	*n* = 125	*n* = 155	*n* = 142	*n* = 155	*n* = 155	
7–8 years	76 ± 21	−39 ± 62	−160 ± 114	2,392 ± 640	−488 ± 276	14 ± 32	627 ± 196	
	41 → 134	−267 → −11	−481 → −21	1,302 → 4,029	−1, 159 → −109	−48 → 87	262 → 1,131	
	n = 139	*n* = 16	*n* = 118	*n* = 141	*n* = 134	*n* = 141	*n* = 141	
9–10 years	74 ± 21	−17 ± 4	−154 ± 108	2,373 ± 606	−424 ± 278	22 ± 34	676 ± 226	
	39 → 136	−23 → −12	−411 → −23	1,209 → 3,918	−1, 200 → −86	−34 → 110	219 → 1,146	
	n = 115	*n* = 11	*n* = 97	*n* = 115	*n* = 106	*n* = 115	*n* = 115	
11–12 years	81 ± 23	−26 ± 29	−123 ± 91	2,317 ± 686	−430 ± 264	16 ± 41	604 ± 255	−110 ± 0
	43 → 153	−90 → −10	−346 → −20	933 → 3,796	−1, 169 → −77	−58 → 152	200 → 1,286	−110 → −110
	n = 117	*n* = 7	*n* = 91	*n* = 118	*n* = 106	*n* = 118	*n* = 118	*n* = 1
13–14 years	74 ± 27	−19 ± 9	−123 ± 108	2,272 ± 795	−519 ± 309	29 ± 46	598 ± 240	−51 ± 0
	18 → 130	−43 → −5	−441 → −25	973 → 4,521	−1, 213 → −126	−46 → 153	208 → 1,192	−51 → −51
	n = 138	*n* = 24	*n* = 87	*n* = 140	*n* = 135	*n* = 140	*n* = 139	*n* = 1
15–16 years	79 ± 24	−16 ± 4	−117 ± 94	1,901 ± 738	−508 ± 394	25 ± 39	515 ± 237	−101 ± 25
	35 → 150	−21 → −10	−383 → −23	665 → 3,687	−1, 691 → −70	−23 → 105	186 → 1,089	−129 → −81
	n = 92	*n* = 8	*n* = 50	*n* = 92	*n* = 86	*n* = 92	*n* = 92	*n* = 3
17–18 years	146 ± 0		−125 ± 0	969 ± 0	−426 ± 0	69 ± 0	241 ± 0	
	146 → 146		−125 → −125	969 → 969	−426 → −426	69 → 69	241 → 241	
	n = 1		*n* = 1	*n* = 1	*n* = 1	*n* = 1	*n* = 1	

Table A2.24

Amplitudes (in μV) in Caucasian children: lead V_6

Age-group	P+	P−	Q	R	S	STj	T+	T−
<24 h	83 ± 30	−97 ± 91	−140 ± 92	521 ± 258	−372 ± 198	26 ± 37	162 ± 124	−40 ± 23
	42 → 166	−199 → −24	−412 → −26	83 → 1,118	−867 → −139	−29 → 200	36 → 604	−88 → −21
	n = 41	*n* = 3	*n* = 36	*n* = 40	*n* = 32	*n* = 42	*n* = 42	*n* = 8
<1 day	83 ± 31	−41 ± 28	−133 ± 84	563 ± 389	−467 ± 353	22 ± 28	131 ± 69	−54 ± 36
	29 → 136	−76 → −10	−328 → −22	67 → 1,705	−1,708 → −55	−32 → 81	28 → 323	−137 → −10
	n = 127	*n* = 11	*n* = 100	*n* = 127	*n* = 104	*n* = 132	*n* = 128	*n* = 16
<2 days	82 ± 30	−39 ± 29	−116 ± 83	488 ± 314	−438 ± 324	32 ± 42	159 ± 81	−75 ± 38
	38 → 161	−85 → −2	−380 → −22	38 → 1,328	−1, 326 → −100	−24 → 130	37 → 464	−134 → −21
	n = 110	*n* = 7	*n* = 85	*n* = 111	*n* = 100	*n* = 115	*n* = 107	*n* = 10
<3 days	92 ± 34	−13 ± 4	−122 ± 82	545 ± 345	−493 ± 361	44 ± 38	173 ± 76	−83 ± 37
	30 → 177	−17 → −8	−377 → −24	68 → 1,562	−1, 558 → −60	−21 → 171	51 → 391	−129 → −20
	n = 101	*n* = 4	*n* = 81	*n* = 99	*n* = 79	*n* = 103	*n* = 100	*n* = 9
<1 week	86 ± 33	−30 ± 22	−163 ± 99	574 ± 368	−407 ± 244	53 ± 65	203 ± 105	−55 ± 44
	29 → 177	−73 → −3	−388 → −27	69 → 1,814	−998 → −66	−38 → 301	25 → 493	−148 → −6
	n = 109	*n* = 8	*n* = 99	*n* = 110	*n* = 87	*n* = 112	*n* = 109	*n* = 9
<1 month	100 ± 39	−42 ± 33	−183 ± 93	743 ± 456	−339 ± 230	41 ± 50	267 ± 122	−7 ± 0
	40 → 225	−93 → −23	−346 → −49	43 → 1,867	−1, 023 → −68	−12 → 237	90 → 666	−7 → −7
	n = 45	*n* = 4	*n* = 38	*n* = 45	*n* = 40	*n* = 45	*n* = 45	*n* = 1
<3 months	96 ± 30	−30 ± 21	−223 ± 127	1,359 ± 581	−361 ± 223	39 ± 36	330 ± 106	
	52 → 219	−55 → −17	−738 → −31	538 → 2,976	−1, 087 → −60	−25 → 133	126 → 603	
	n = 64	*n* = 3	*n* = 61	*n* = 65	*n* = 53	*n* = 67	*n* = 65	
<6 months	88 ± 37	−33 ± 26	−200 ± 110	1,193 ± 472	−326 ± 212	52 ± 35	326 ± 143	−167 ± 0
	39 → 277	−73 → −11	−479 → −35	142 → 2,321	−820 → −64	−7 → 178	140 → 843	−167 → −167
	n = 48	*n* = 5	*n* = 42	*n* = 49	*n* = 41	*n* = 49	*n* = 48	*n* = 1
<1 year	91 ± 53	−31 ± 21	−223 ± 135	1,230 ± 561	−274 ± 223	51 ± 54	349 ± 113	−183 ± 0
	34 → 411	−57 → −6	−740 → −26	30 → 2,539	−1, 600 → −42	−20 → 334	159 → 651	−183 → −183
	n = 72	*n* = 8	*n* = 59	*n* = 73	*n* = 60	*n* = 74	*n* = 72	*n* = 1
1–2 years	75 ± 29	−30 ± 22	−216 ± 134	1,196 ± 547	−227 ± 151	23 ± 30	282 ± 119	−55 ± 24
	28 → 146	−65 → −6	−539 → −23	54 → 2,324	−701 → −56	−25 → 120	54 → 550	−83 → −24
	n = 109	*n* = 9	*n* = 91	*n* = 110	*n* = 76	*n* = 115	*n* = 109	*n* = 4
3–4 years	74 ± 20	−28 ± 36	−179 ± 109	1,428 ± 511	−211 ± 109	7 ± 22	367 ± 135	−99 ± 104
	36 → 113	−135 → −6	−530 → −28	435 → 2,320	−442 → −66	−41 → 45	100 → 642	−255 → −37
	n = 138	*n* = 12	*n* = 131	*n* = 140	*n* = 99	*n* = 140	*n* = 137	*n* = 4
5–6 years	72 ± 21	−19 ± 8	−181 ± 112	1,747 ± 477	−240 ± 149	11 ± 29	434 ± 148	−63 ± 13
	26 → 115	−37 → −11	−511 → −25	803 → 2,974	−650 → −56	−51 → 85	168 → 801	−72 → −54
	n = 154	*n* = 10	*n* = 139	*n* = 154	*n* = 119	*n* = 154	*n* = 154	*n* = 2
7–8 years	72 ± 20	−27 ± 15	−162 ± 98	1,797 ± 538	−239 ± 153	9 ± 31	487 ± 150	
	40 → 120	−60 → −11	−417 → −35	968 → 3,137	−799 → −51	−51 → 110	234 → 908	
	n = 139	*n* = 13	*n* = 129	*n* = 140	*n* = 119	*n* = 140	*n* = 140	
9–10 years	70 ± 20	−14 ± 4	−153 ± 92	1,851 ± 453	−221 ± 140	13 ± 30	519 ± 173	
	30 → 125	−21 → −9	−404 → −28	1,030 → 3,135	−642 → −53	−37 → 136	193 → 946	
	n = 115	*n* = 12	*n* = 107	*n* = 115	*n* = 88	*n* = 115	*n* = 115	
11–12 years	76 ± 21	−22 ± 18	−130 ± 79	1,836 ± 478	−239 ± 134	7 ± 29	463 ± 173	
	42 → 121	−65 → −8	−306 → −23	907 → 3,022	−573 → −76	−55 → 121	892 → 163	
	n = 116	*n* = 9	*n* = 104	*n* = 117	*n* = 86	*n* = 117	*n* = 117	

Table A2.24 (Continued)

Age-group	P+	P–	Q	R	S	STj	T+	T–
13–14 years	70 ± 26	−21 ± 12	−117 ± 92	1,762 ± 515	−244 ± 139	15 ± 40	450 ± 177	−89 ± 49
	20 → 127	−45 → −6	−391 → −23	874 → 3,107	−384 → −36	−65 → 151	204 → 983	−124 → −54
	n = 136	*n* = 17	*n* = 124	*n* = 140	*n* = 122	*n* = 140	*n* = 139	*n* = 2
15–16 years	75 ± 23	−14 ± 6	−118 ± 74	1,574 ± 498	−267 ± 206	11 ± 29	395 ± 170	−127 ± 66
	28 → 144	−96 → −11	−377 → −26	656 → 2,691	−1,075 → −62	−34 → 64	135 → 806	−226 → −89
	n = 87	*n* = 6	*n* = 58	*n* = 88	*n* = 72	*n* = 88	*n* = 87	*n* = 4
17–18 years	136 ± 0		−177 ± 0	595 ± 0		44 ± 0	153 ± 0	
	136 → 136		−177 → → 177	595 → 595		44 → 44	153 → 153	
	n = 1		*n* = 1	*n* = 1		*n* = 1	*n* = 1	

A2.2 Percentile Charts

The following percentile charts showing normal 12-lead ECG limits were obtained from a study of 2,141 white children between birth and 16 years. Data were derived from computer-assisted methods where the sampling rate was 333 samples per second so that there may be some underestimation of upper limits of normal. The charts are reproduced from: Davignon A *et al.* Normal ECG standards for infants and children. *Pediatr. Cardiol.* 1979/1980; **1**: 133–52 with the permission of Springer, New York. Note that V_3R was included while V_3 was omitted in this study (Figs. A2.1–A2.39).

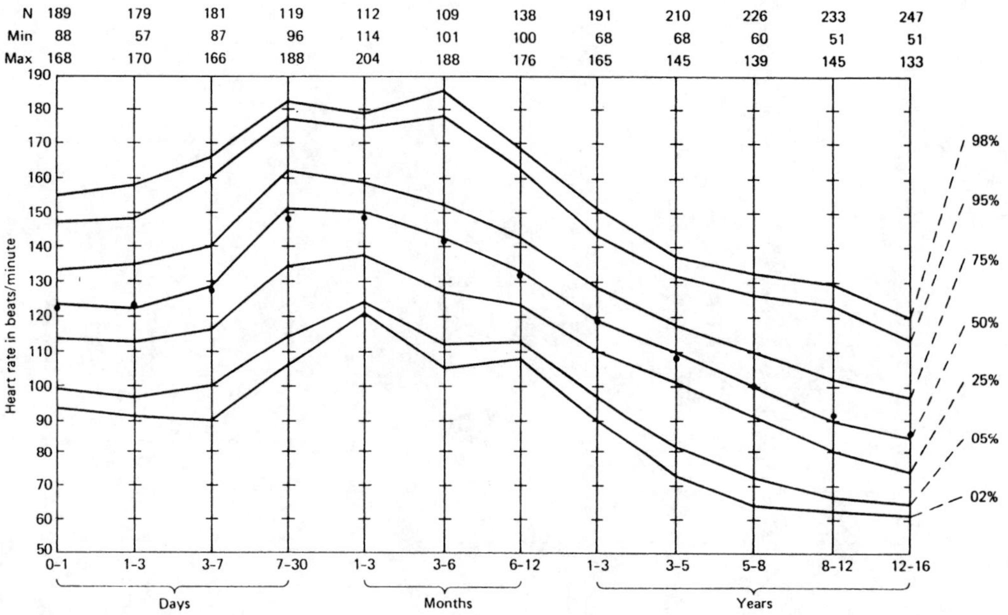

Figure A2.1
Heart rate versus age (•, mean).

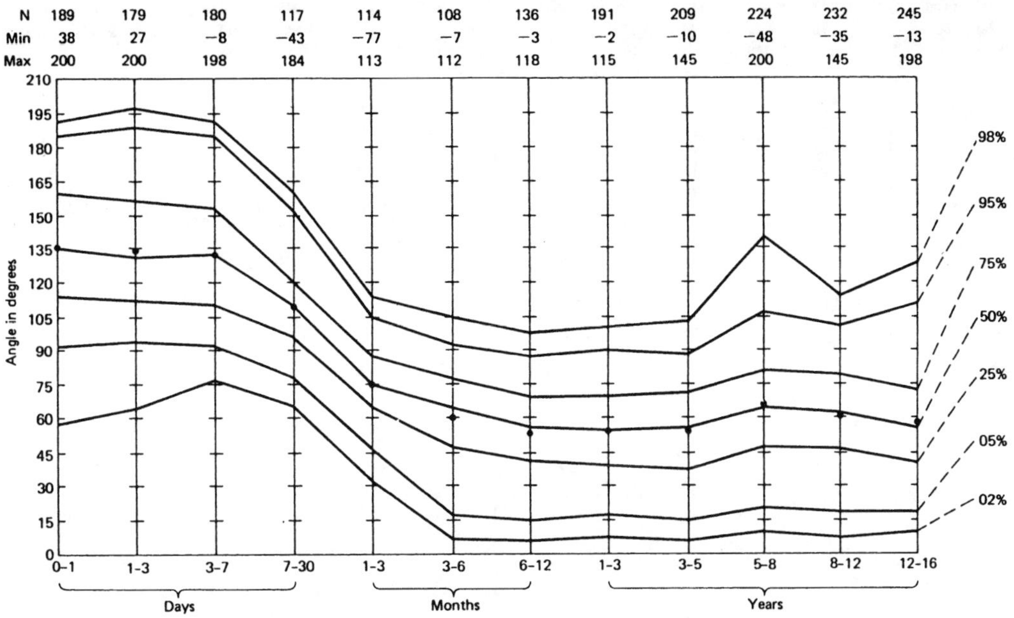

■ Figure A2.2
Frontal plane QRS angle versus age (•, mean).

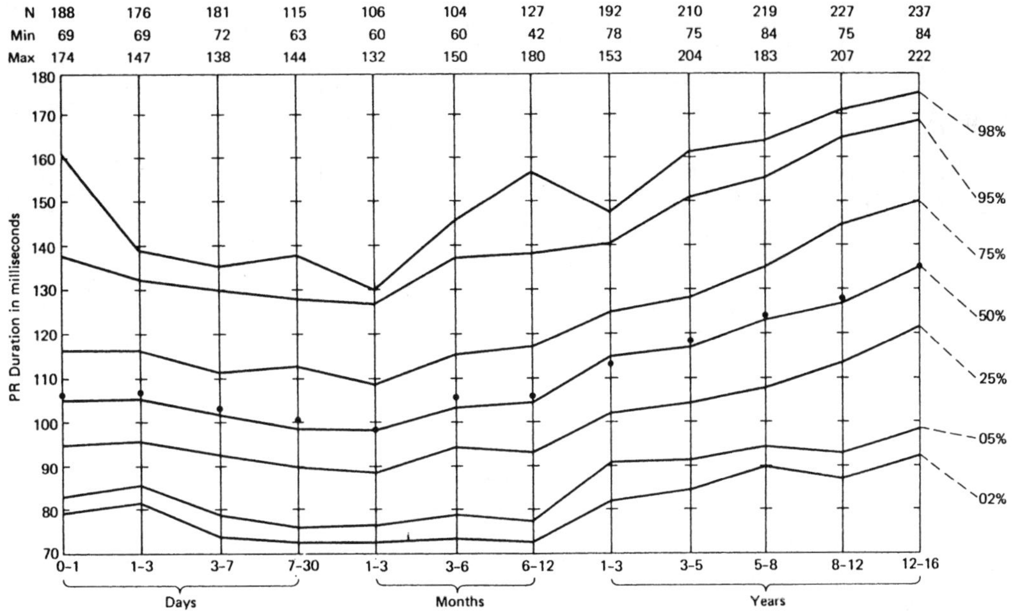

■ Figure A2.3
PR duration in lead II versus age (•, mean).

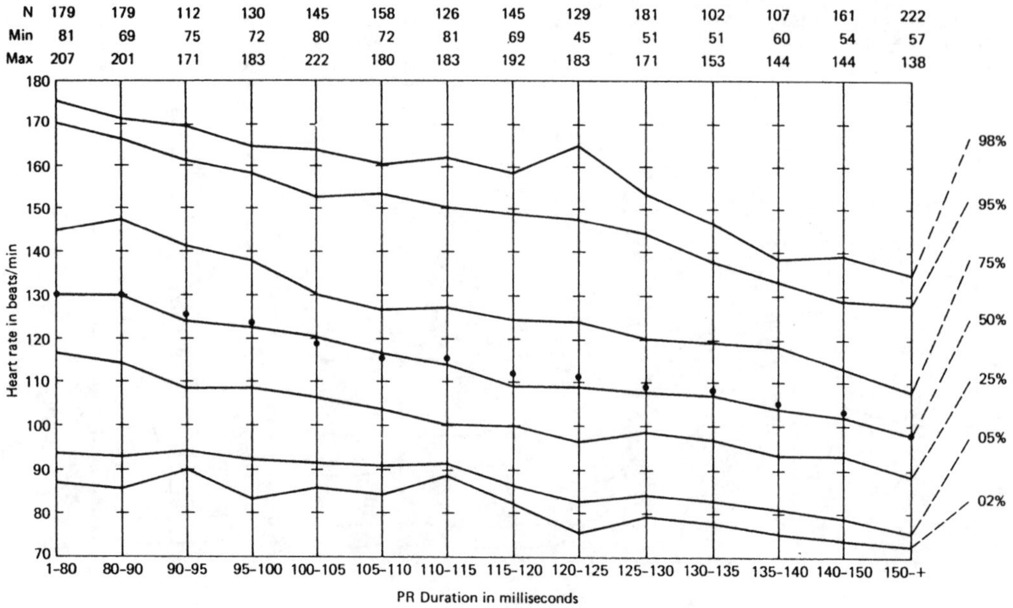

Figure A2.4
Heart rate versus PR duration in lead II (•, mean).

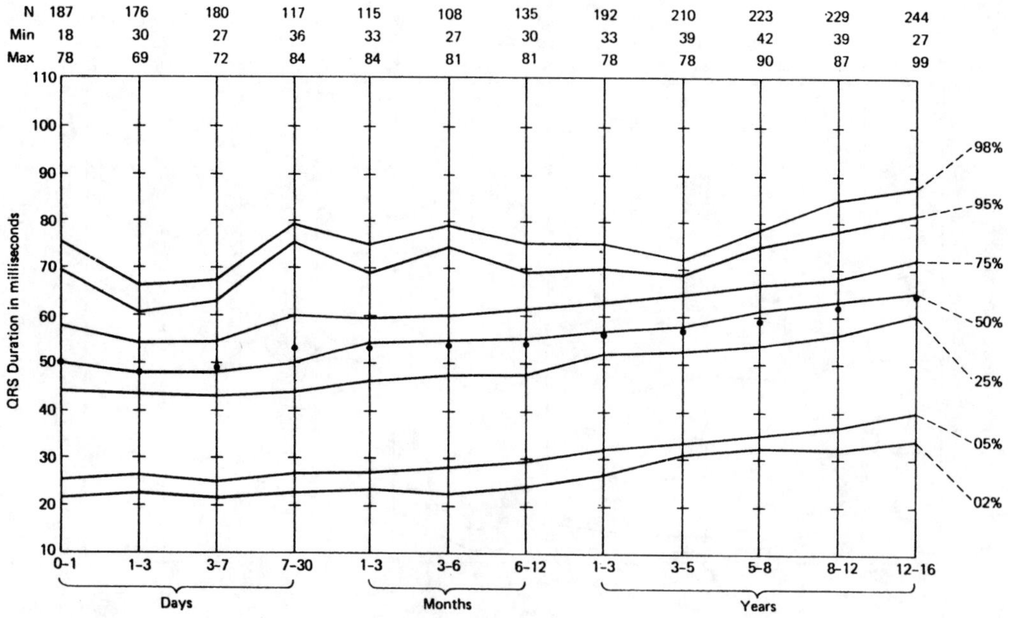

Figure A2.5
QRS duration in lead V_5 versus age (•, mean).

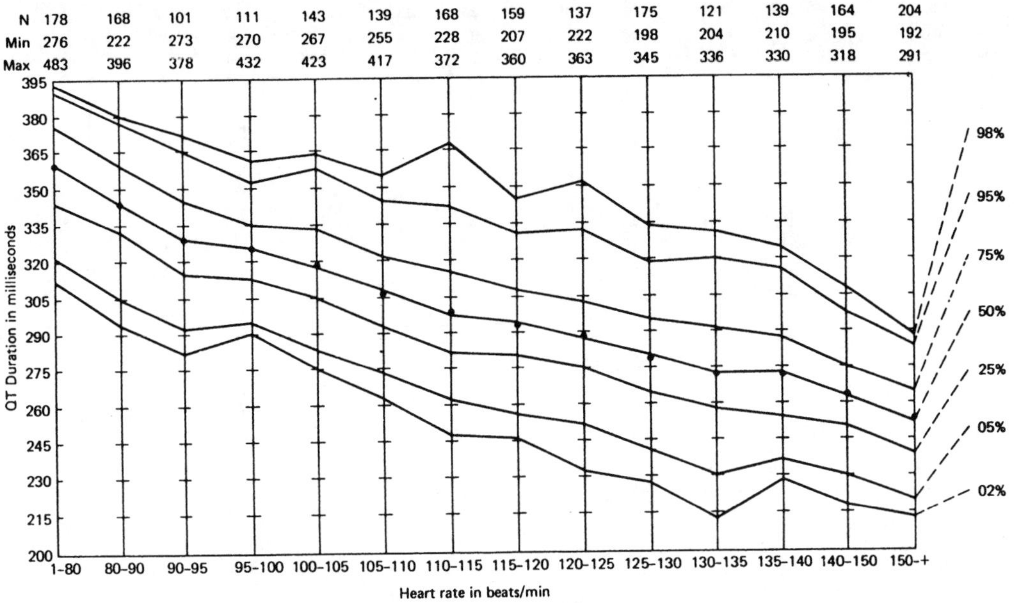

Figure A2.6
QT duration in lead V_5 versus heart rate (•, mean).

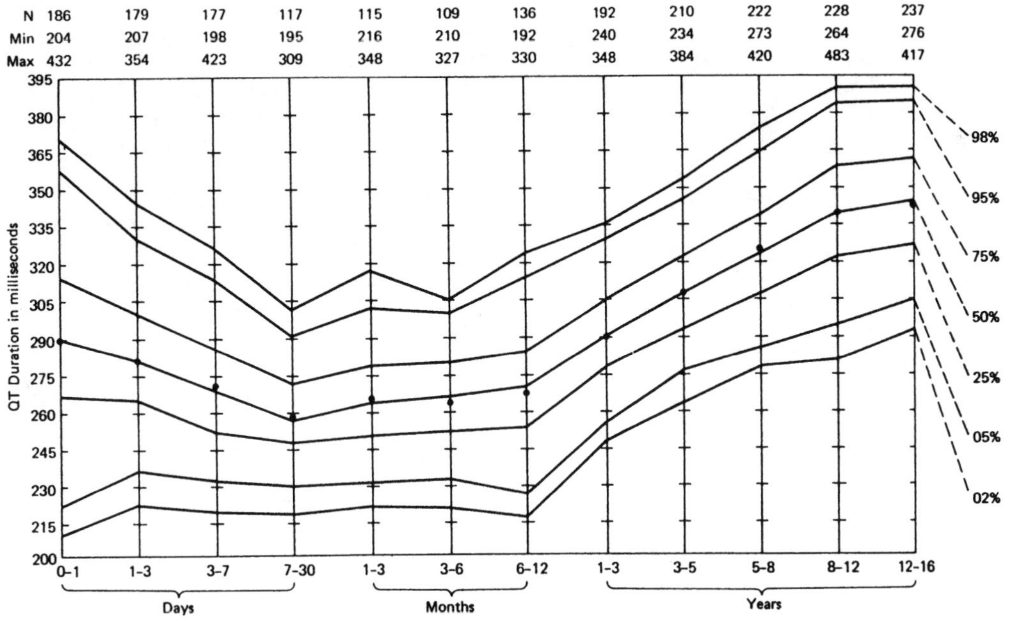

Figure A2.7
QT duration in lead V_5 versus age (•, mean).

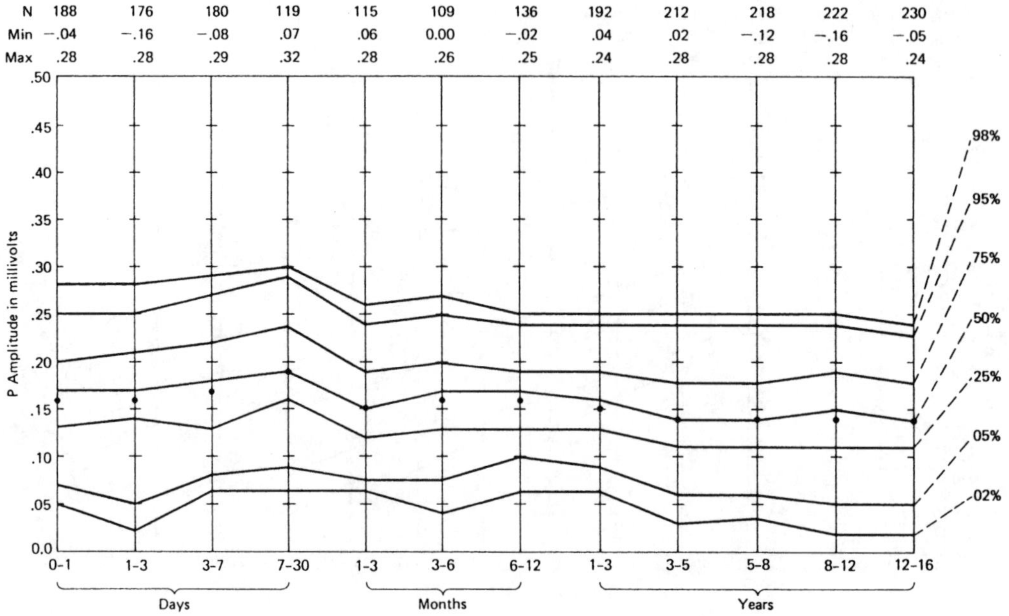

Figure A2.8
P amplitude in lead II versus age (•, mean).

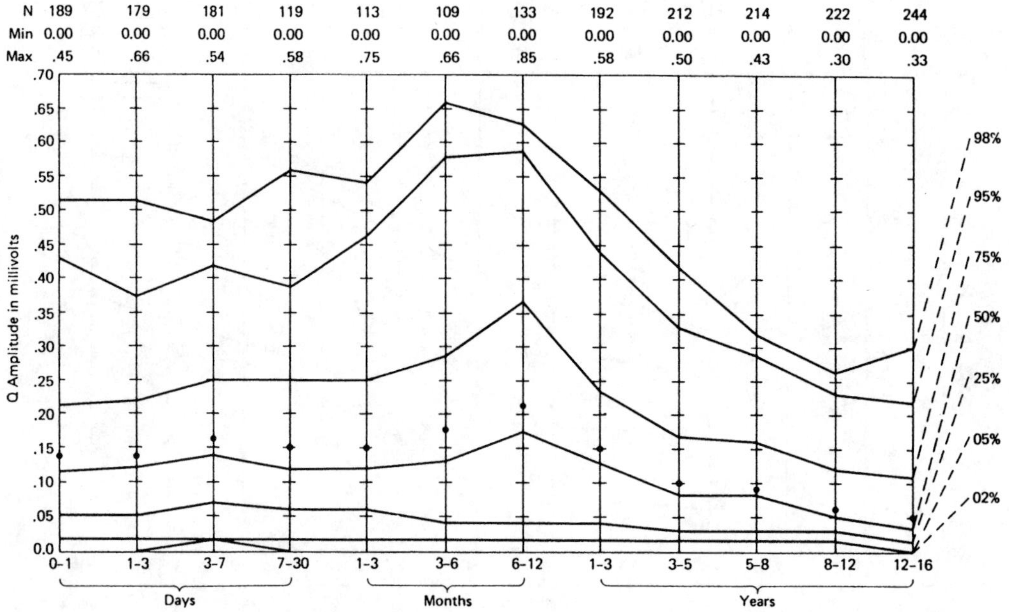

Figure A2.9
Q amplitude in lead III versus age (•, mean).

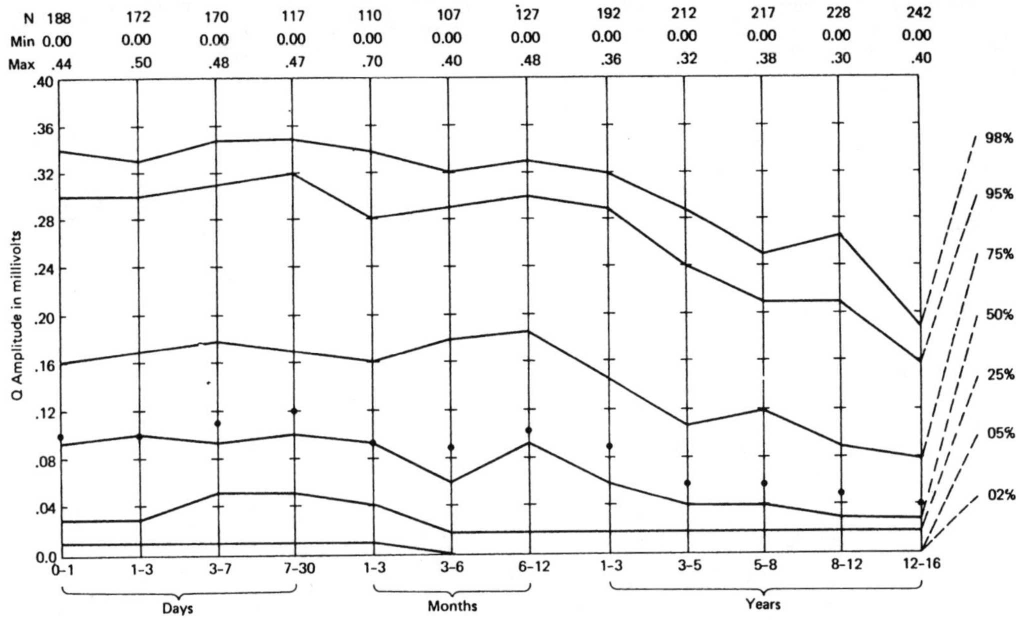

Figure A2.10
Q amplitude in lead a VF versus age (•, mean).

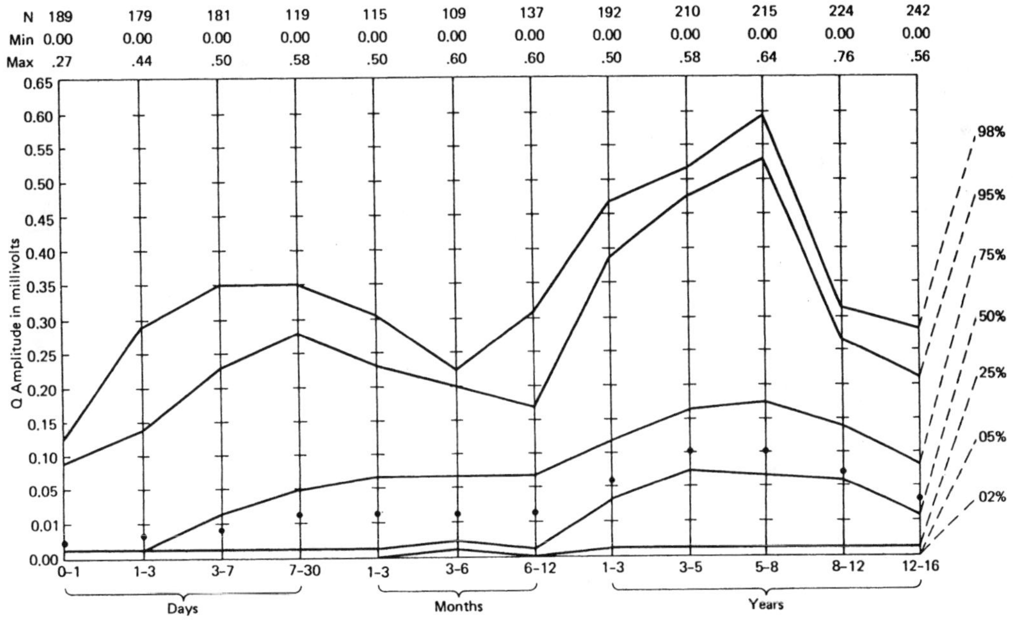

Figure A2.11
Q amplitude in lead V_5 versus age (•, mean).

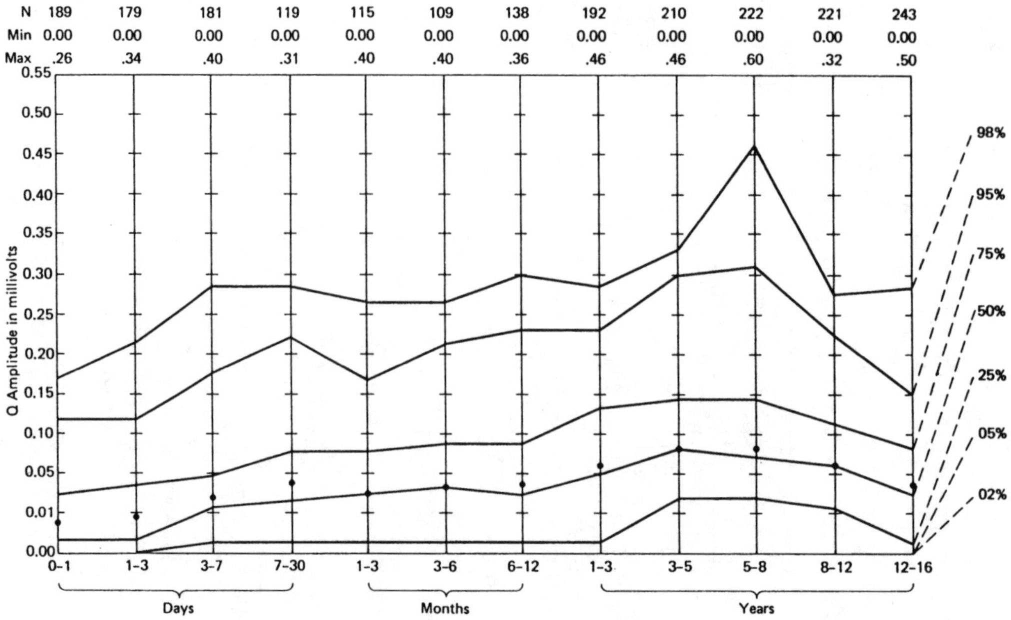

Figure A2.12
Q amplitude in lead V_6 versus age (•, mean).

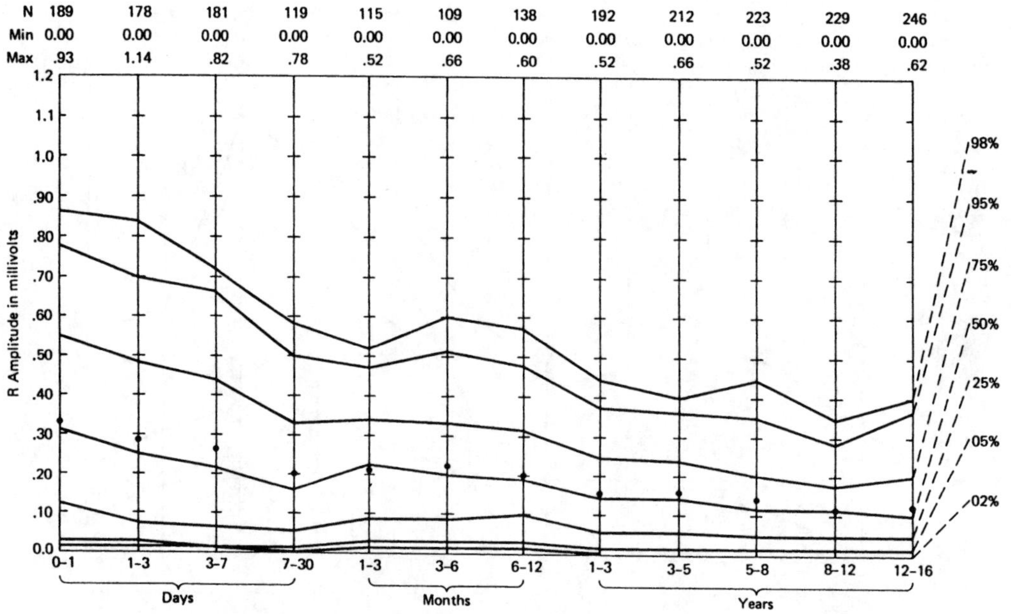

Figure A2.13
R amplitude in lead aVR versus age (•, mean).

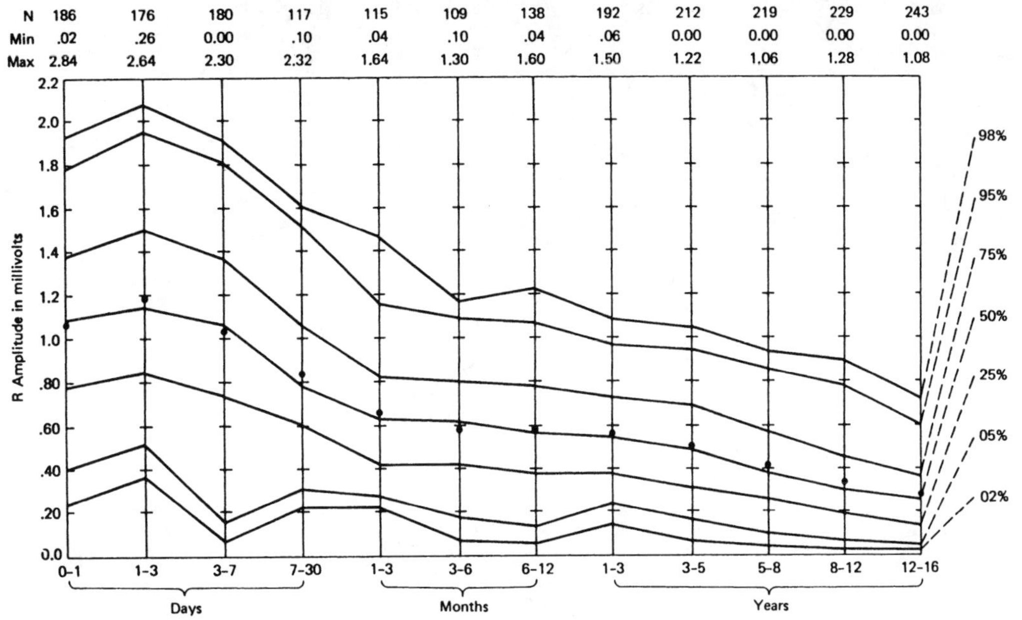

■ Figure A2.14
R amplitude in lead V_3R versus age (•, mean).

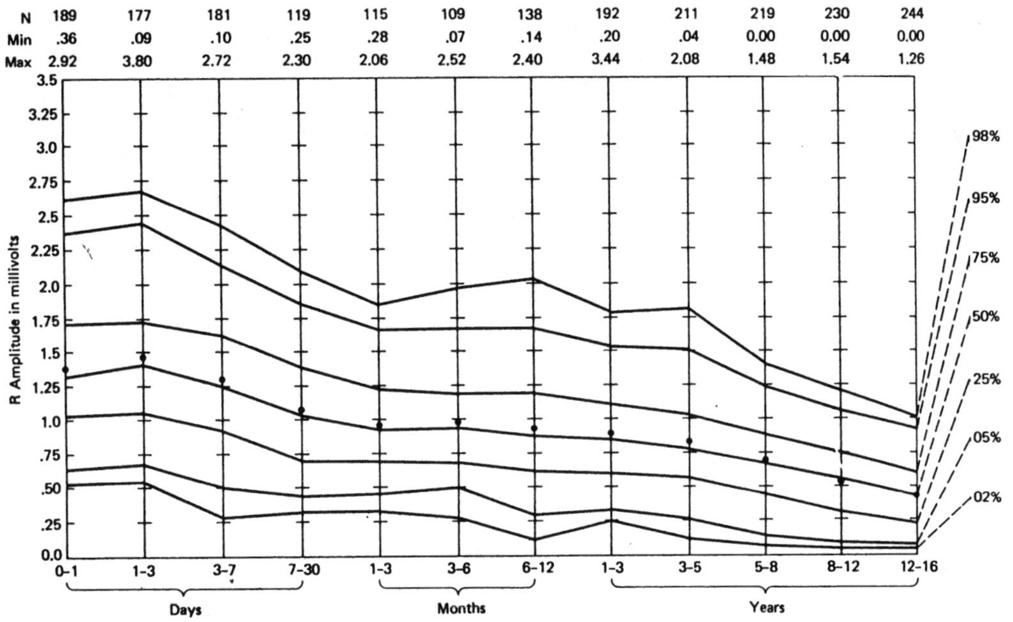

■ Figure A2.15
R amplitude in lead V_1 versus age (•, mean).

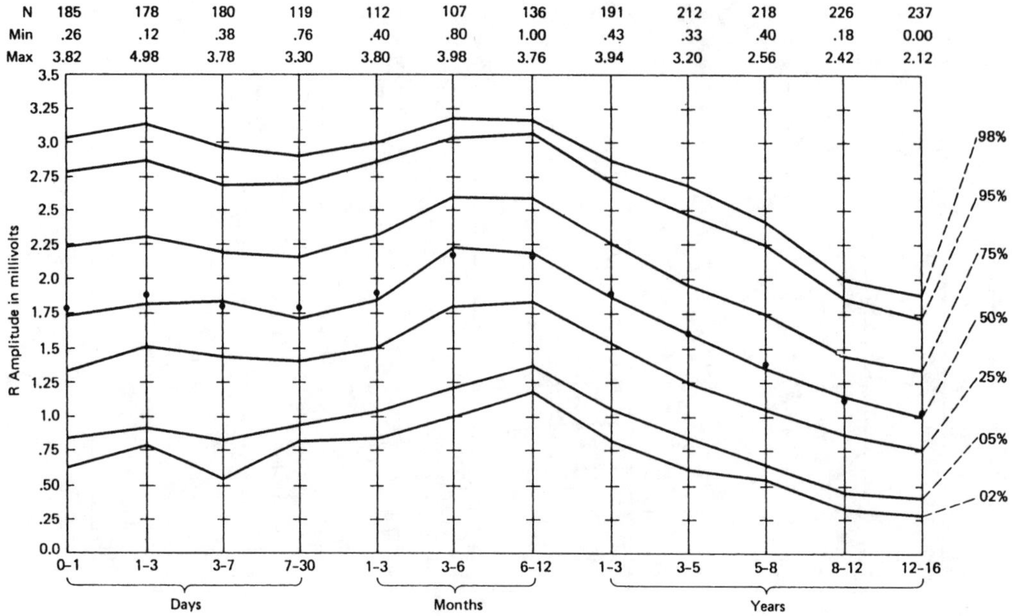

■ Figure A2.16
R amplitude in lead V_2 versus age (•, mean).

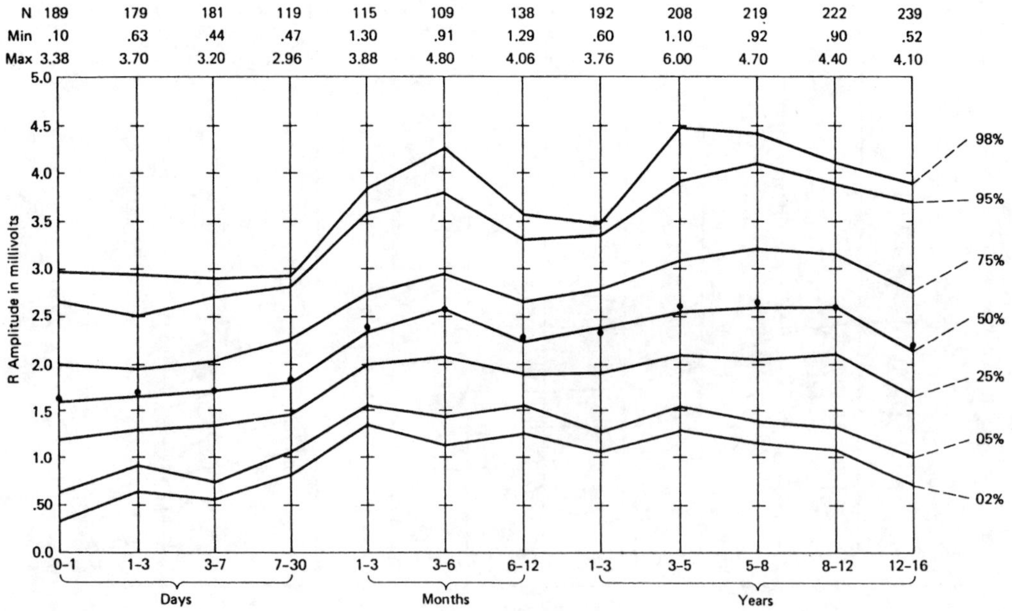

■ Figure A2.17
R amplitude in lead V_4 versus age (•, mean).

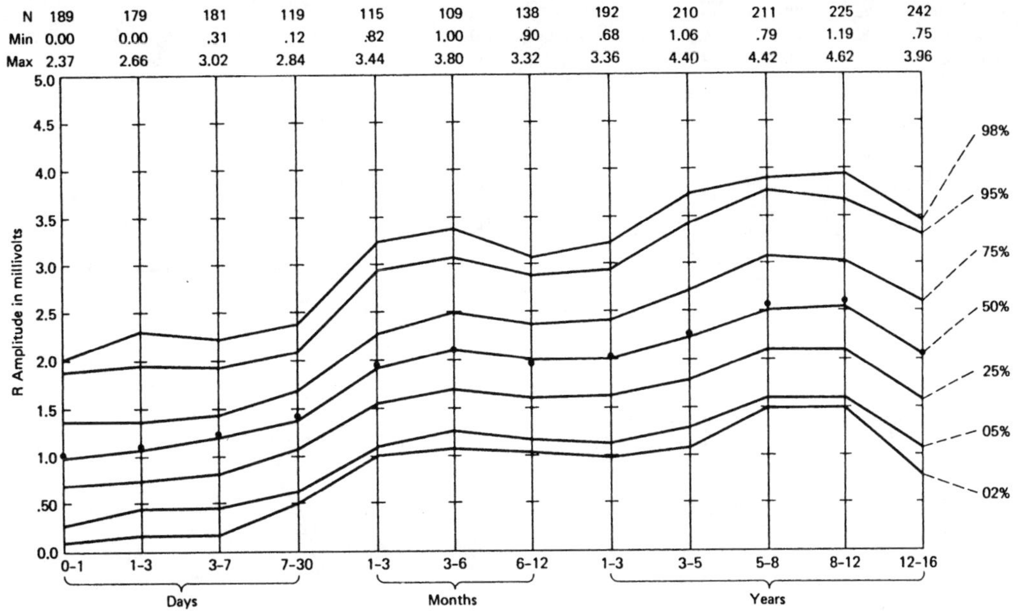

■ Figure A2.18
R amplitude in lead V_5 versus age (•, mean).

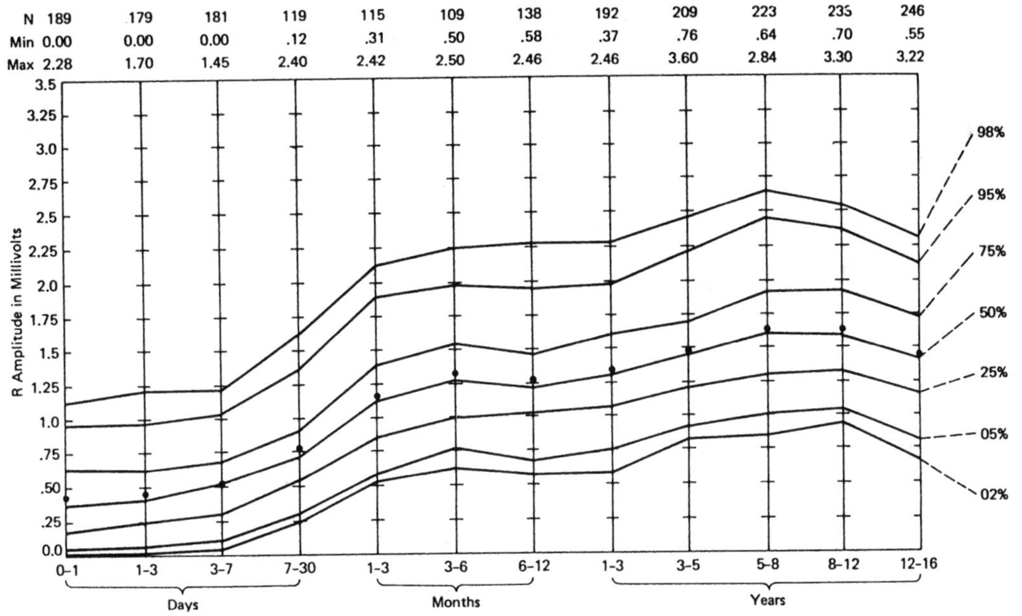

■ Figure A2.19
R amplitude in lead V_6 versus age (•, mean).

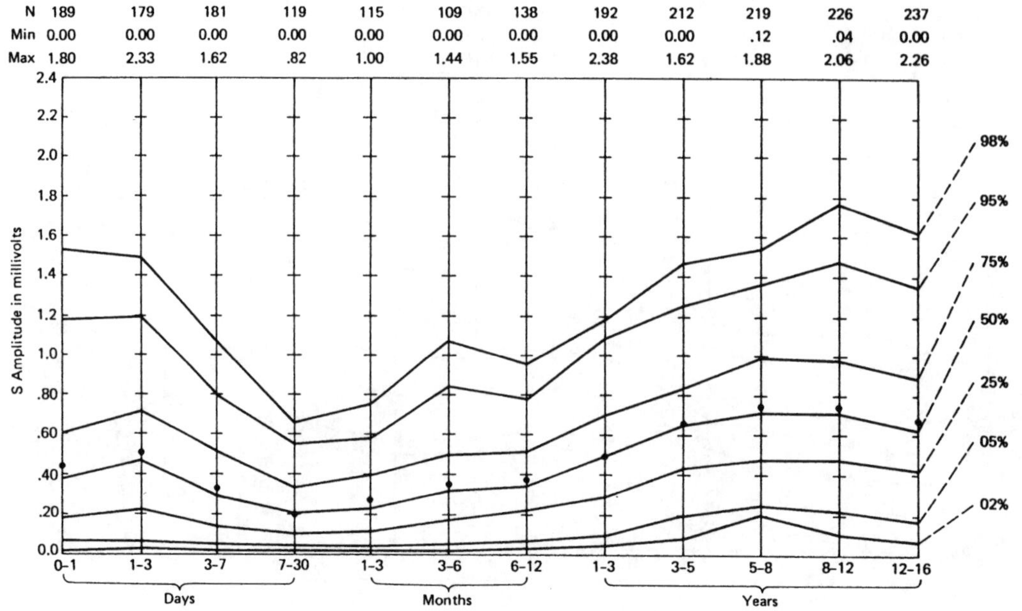

Figure A2.20
S amplitude in lead V_3R versus age (•, mean).

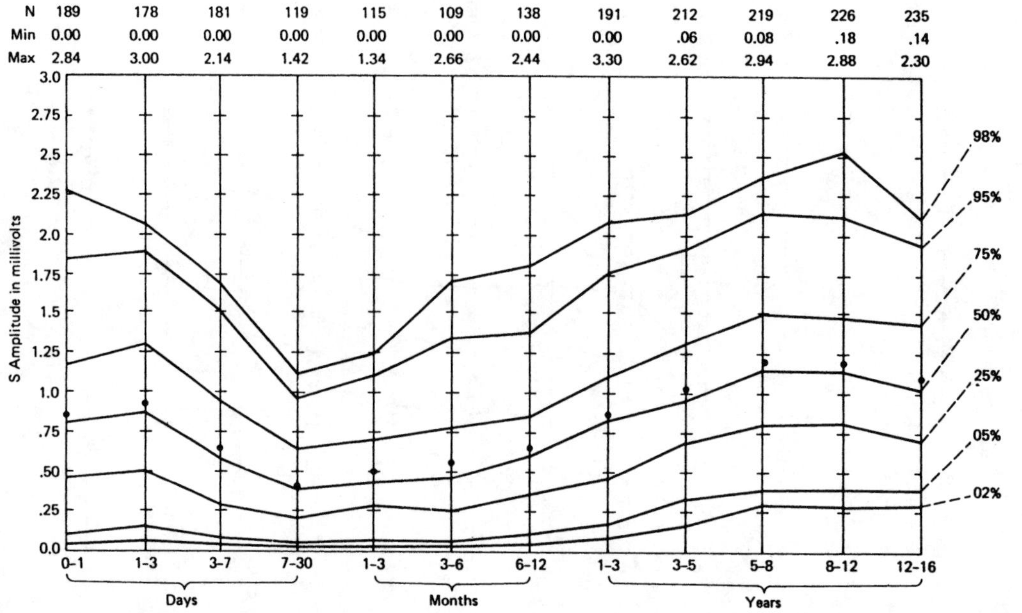

Figure A2.21
S amplitude in lead V_1 versus age (•, mean).

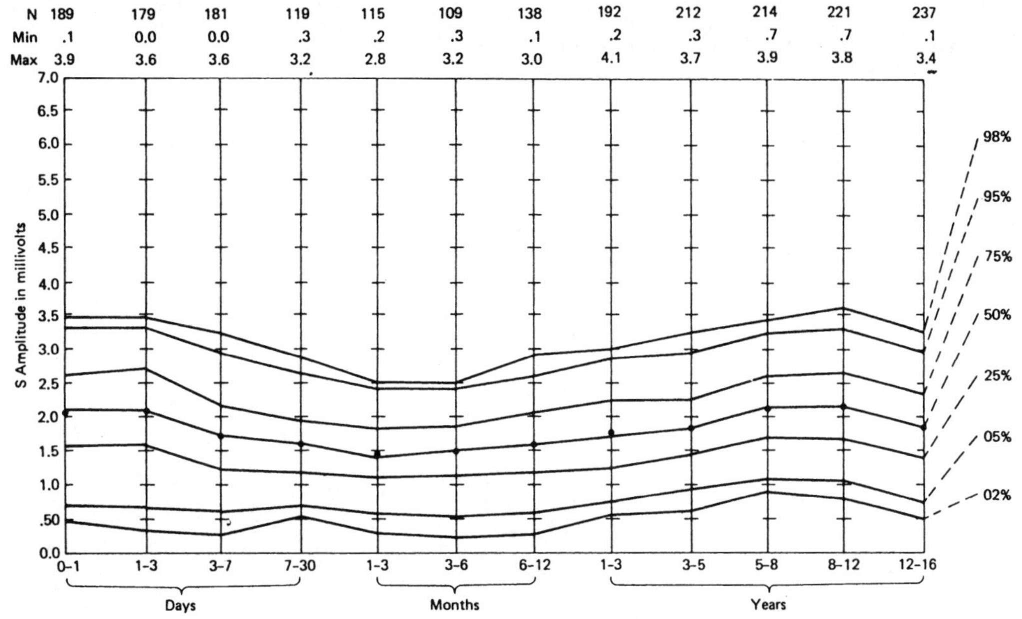

Figure A2.22
S amplitude in lead V_2 versus age (•, mean).

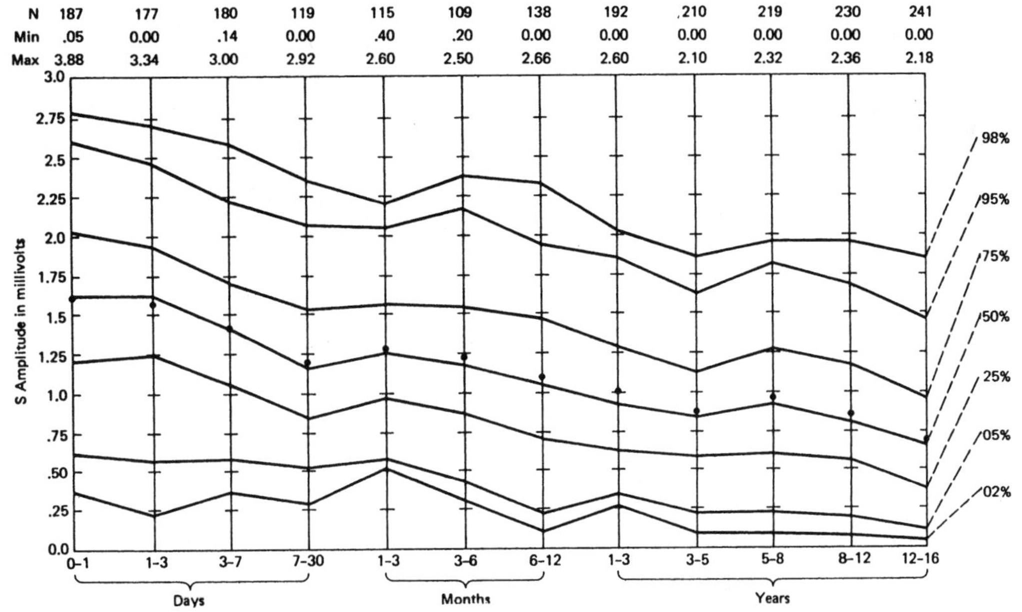

Figure A2.23
S amplitude in lead V_4 versus age (•, mean).

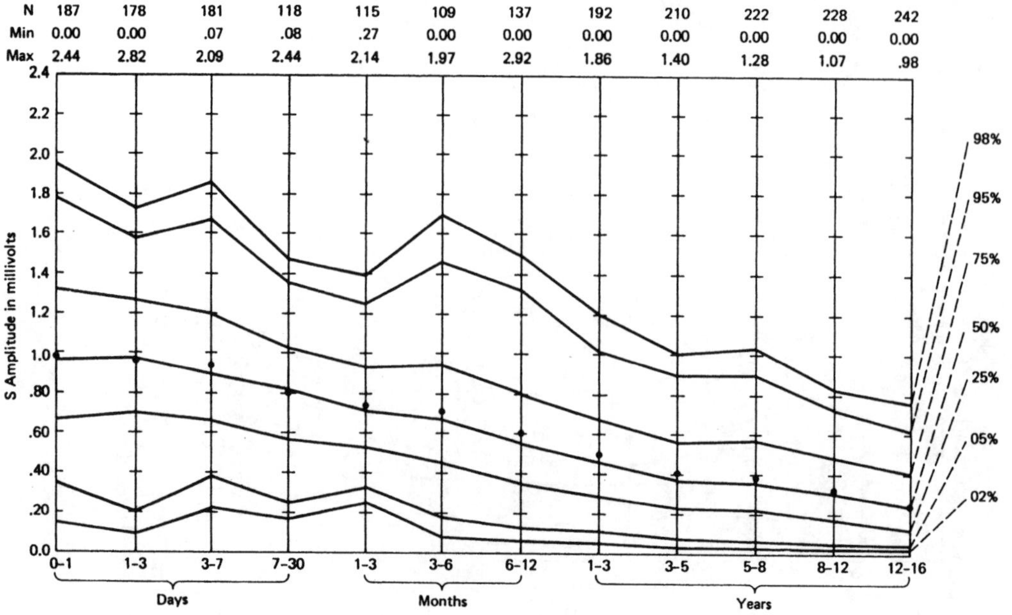

■ Figure A2.24
S amplitude in lead V_5 versus age (•, mean).

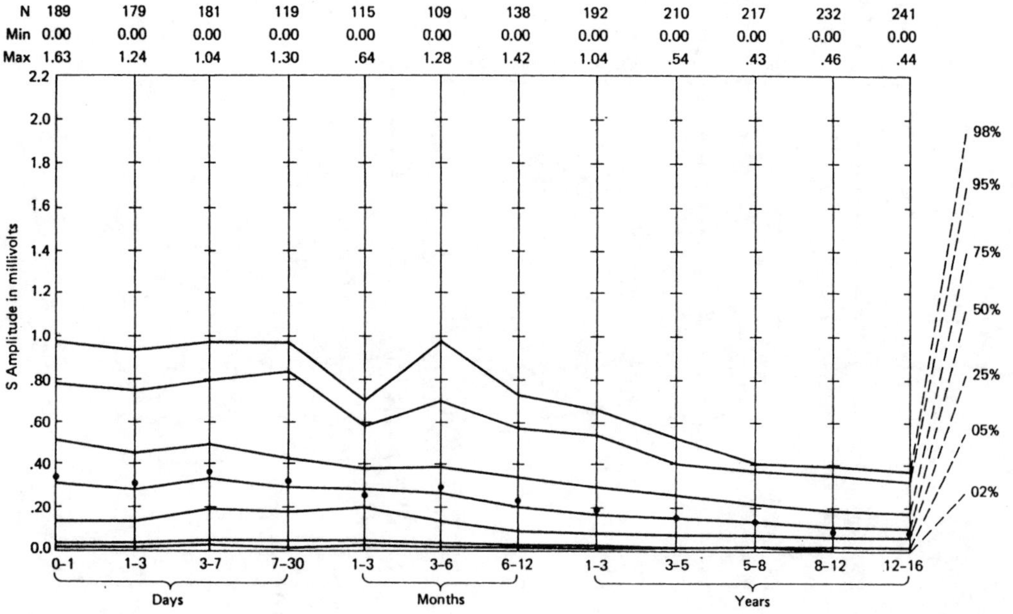

■ Figure A2.25
S amplitude in lead V_6 versus age (•, mean).

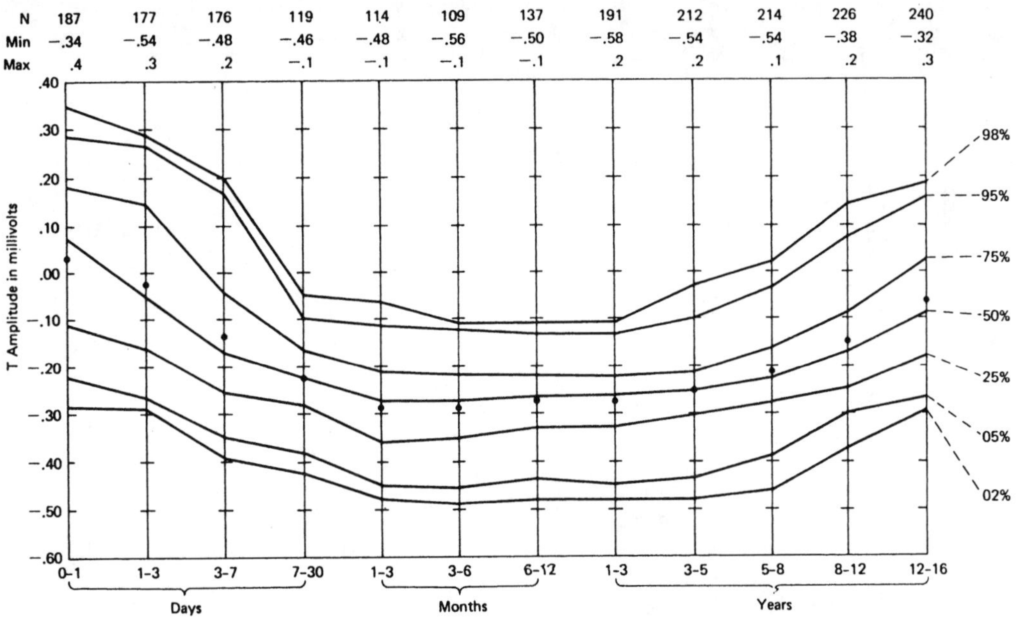

Figure A2.26
T amplitude in lead V_3R versus age (•, mean).

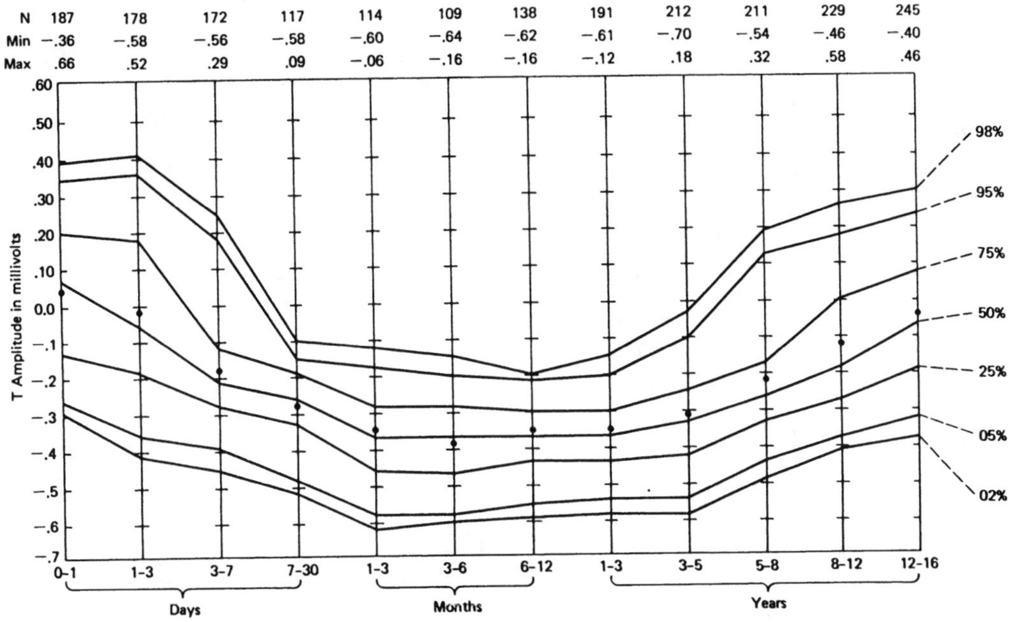

Figure A2.27
T amplitude in lead V_1 versus age (•, mean).

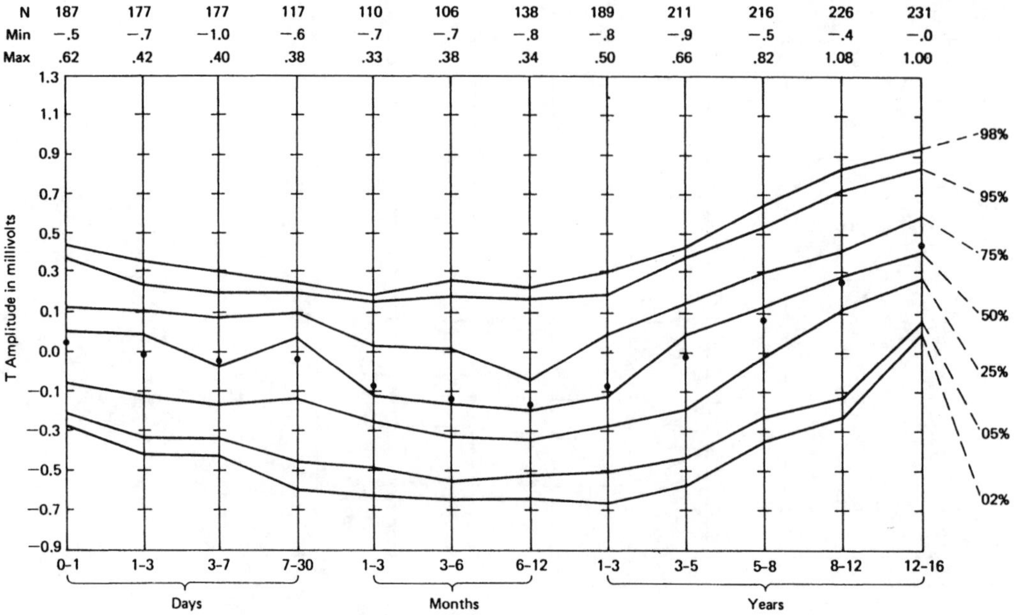

Figure A2.28
T amplitude in lead V_2 versus age (•, mean).

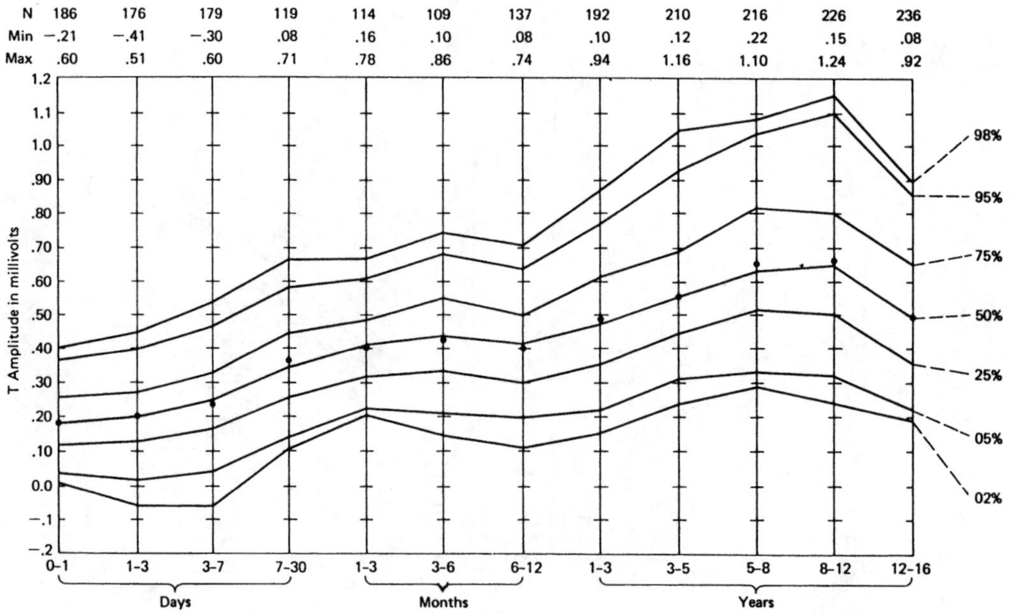

Figure A2.29
T amplitude in lead V_5 versus age (•, mean).

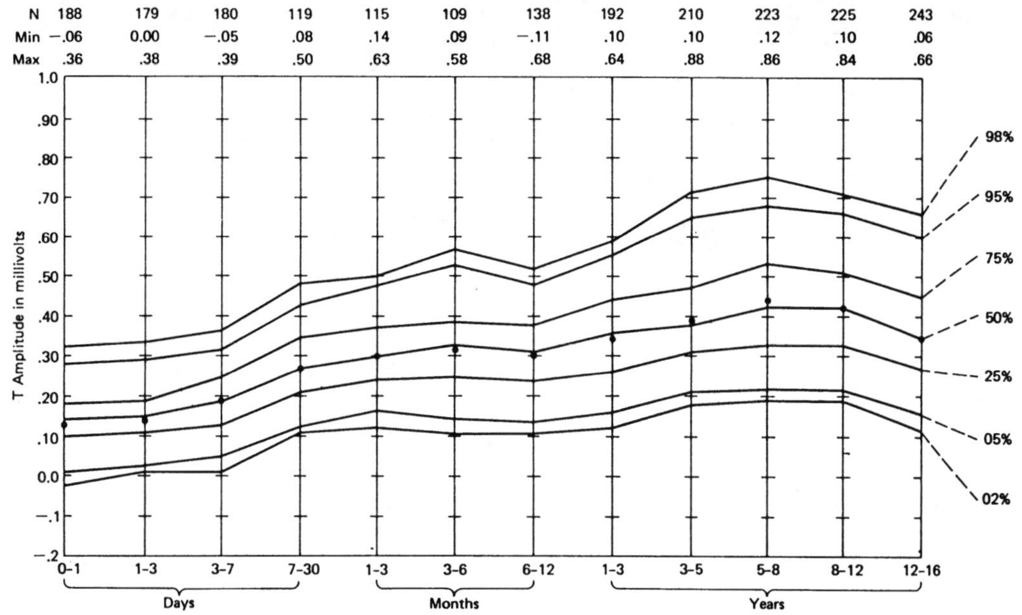

Figure A2.30
T amplitude in lead V_6 versus age (•, mean).

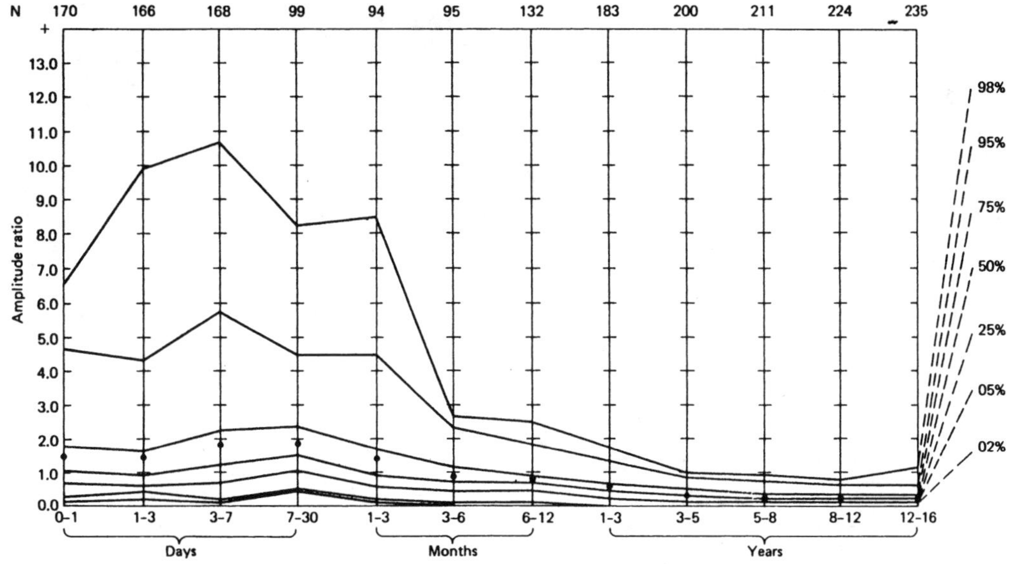

Figure A2.31
R/S amplitude ratio in lead V_3R versus age (•, mean).

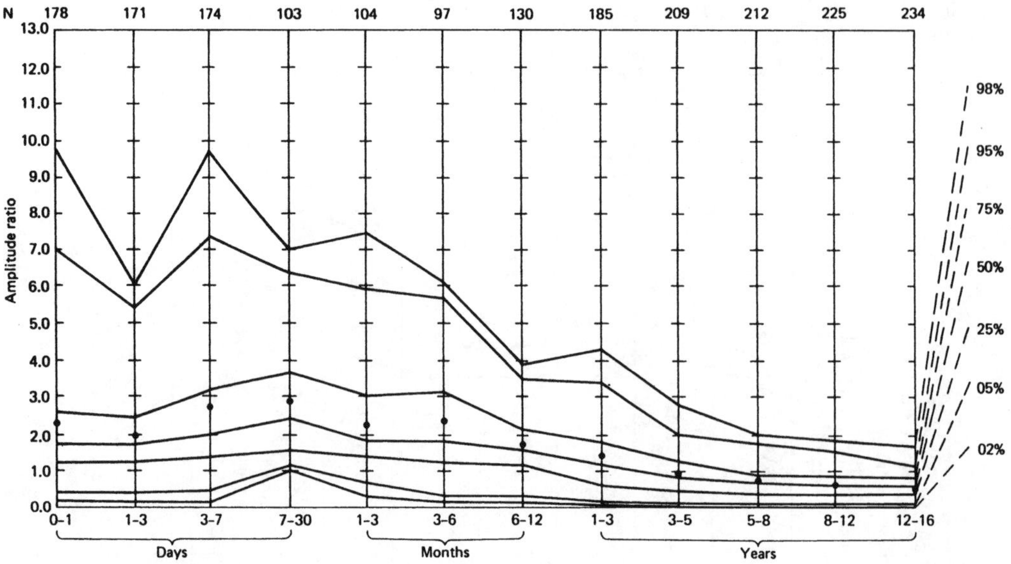

■ Figure A2.32
R/S amplitude ratio in lead V_1 versus age (•, mean).

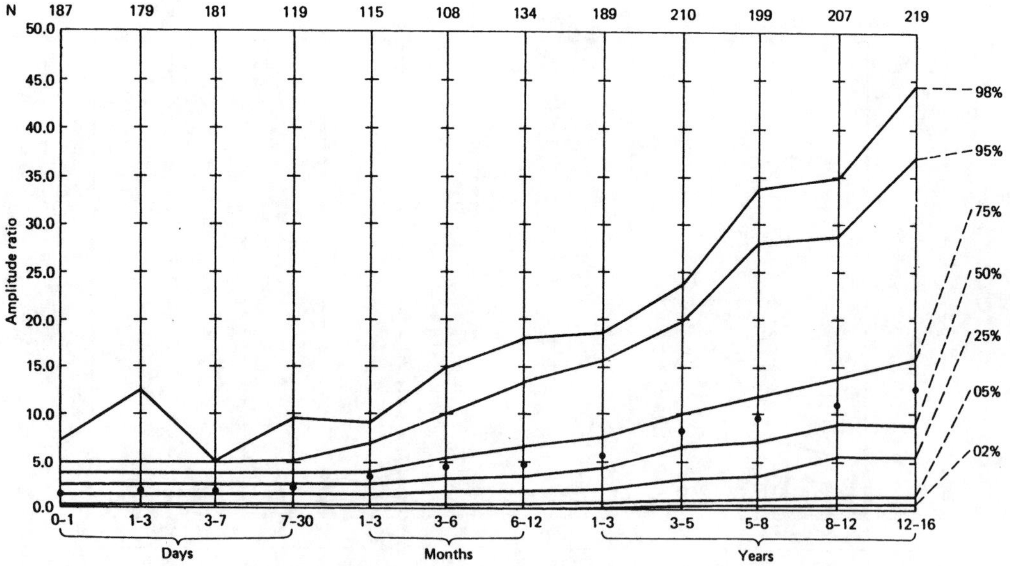

■ Figure A2.33
R/S amplitude ratio in lead V_5 versus age (•, mean).

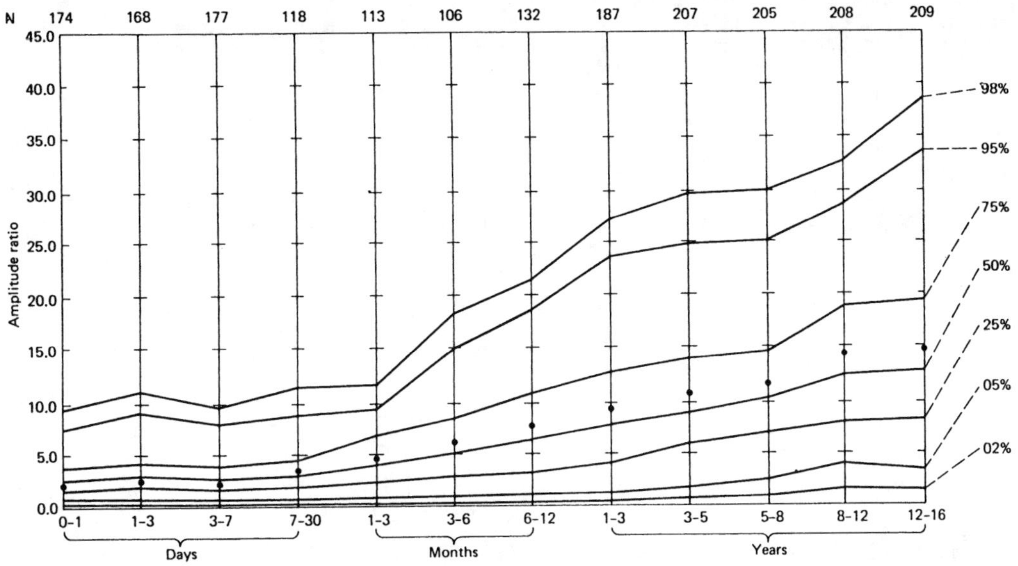

Figure A2.34
R/S amplitude ratio in lead V_6 versus age (•, mean).

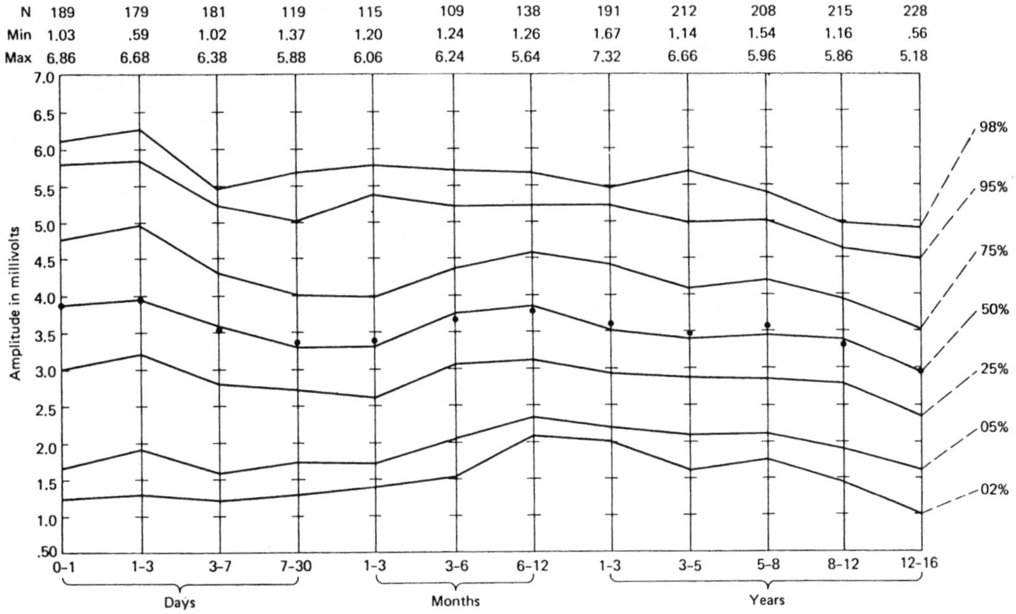

Figure A2.35
R + S amplitude in lead V_2 versus age (•, mean).

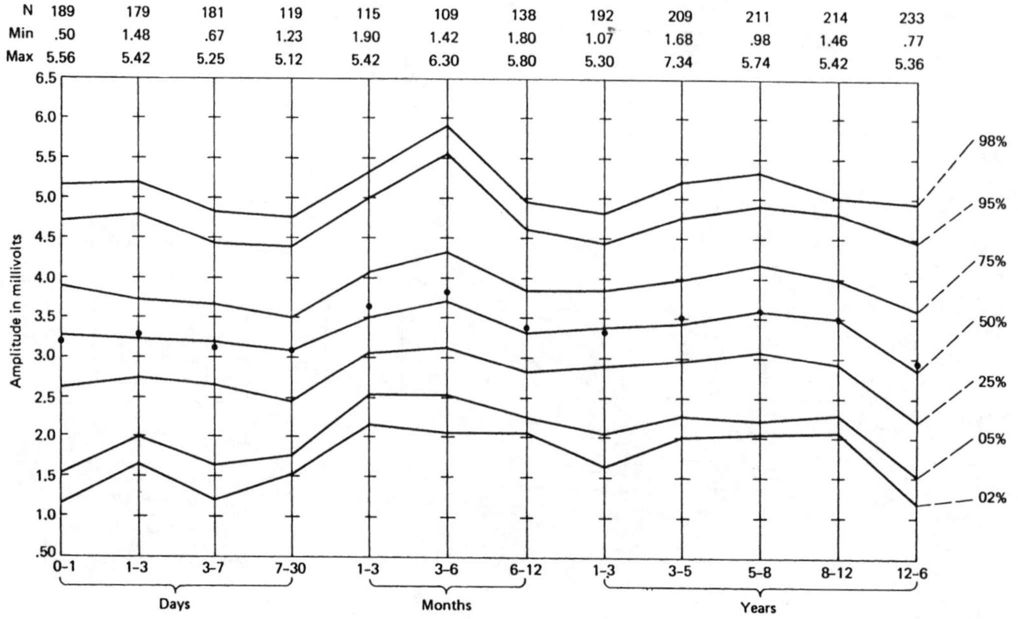

Figure A2.36
R + S amplitude in lead V_4 versus age (•, mean).

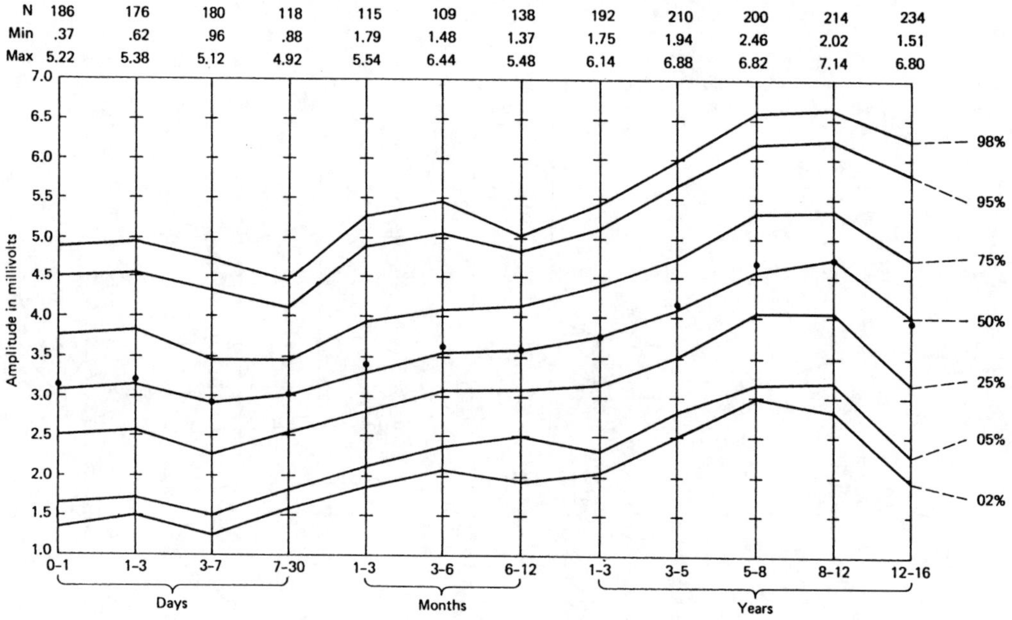

Figure A2.37
R amplitude in lead V_5 + S amplitude in lead V_2 versus age (•, mean).

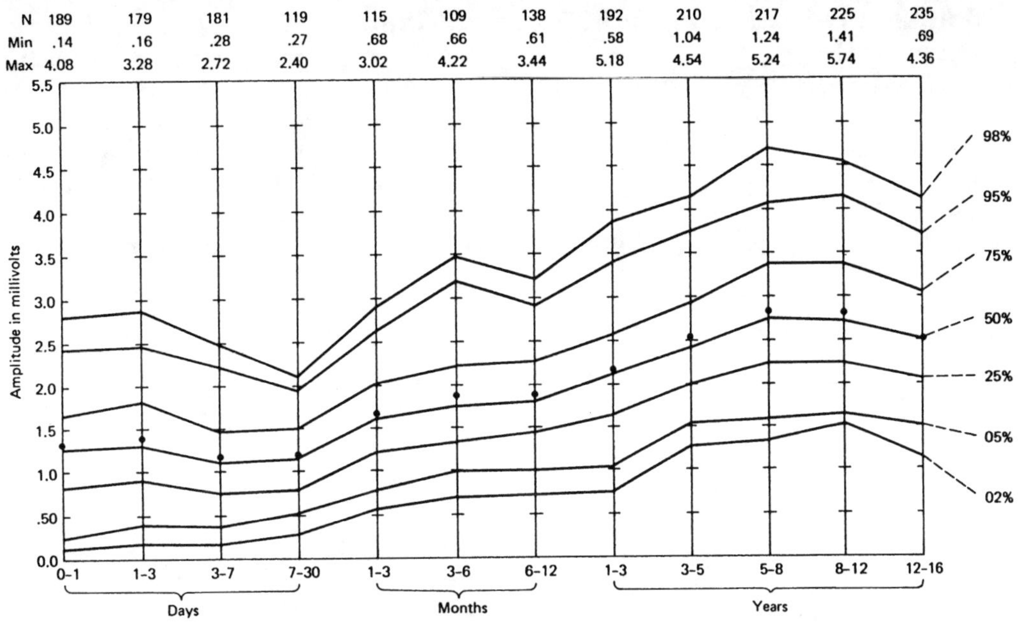

Figure A2.38
R amplitude in lead V_6 + S amplitude in lead V_1 versus age (•, mean).

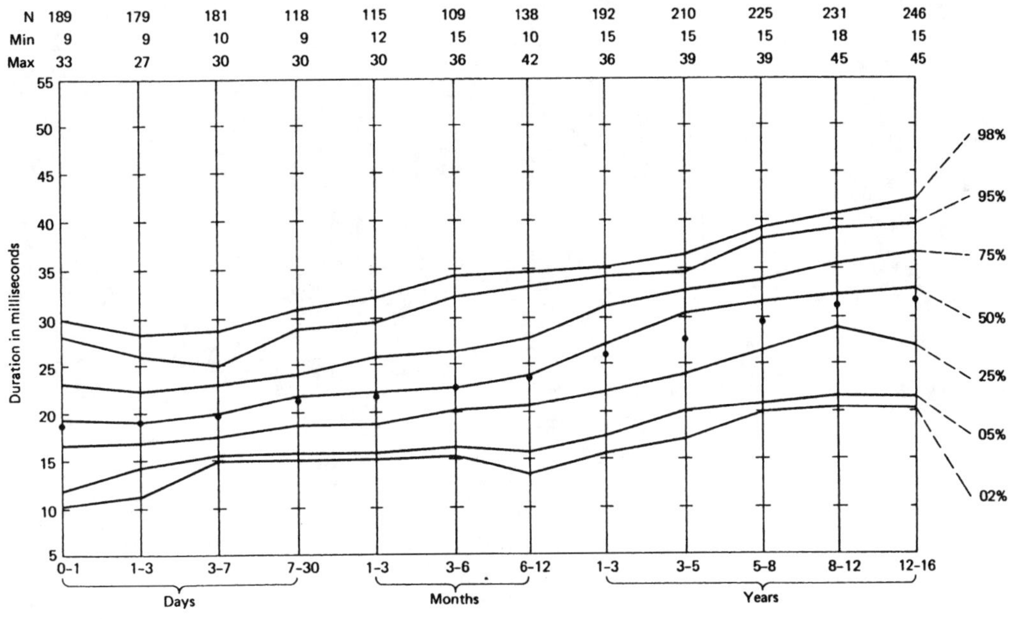

Figure A2.39
Ventricular activation time in lead V_5 versus age (•, mean).

A2.3 Additional 12-Lead Pediatric ECG Normal Limits

Liebman, who has contributed ➋ Chap. 21 as well as a short technical note on linear and directional statistics (see ➋ Appendix 1, Sect. A1.7), has published extensively on normal limits of the pediatric and adolescent ECG. He has stressed the difference between using equipment with high (adequate) and low (poor) frequency response. Some tables which are mostly complementary to ➋ Sect. A2.2 are reproduced from: Liebman J, Plonsey R, Gillette PC, eds. *Pediatric Cardiology*. 1982, with the permission of Williams and Wilkins, Baltimore, Maryland. These include a mixture of "high-" and "low-frequency data." The former are more correct for use with modern recording equipment (➋ Tables A2.25–A2.30).

Table A2.25
High-frequency data

	Heart rate (bpm)						PR interval (s)					
Age	Min	5%	Mean	95%	Max.	SD	Min	5%	Mean	95%	Max.	SD
0–24 h	85	94	119	145	145	16.1	0.07	0.07	0.10	0.12	0.13	0.012
1–7 days	100	100	133	175	175	22.3	0.5	0.07	0.09	0.12	0.13	0.014
8–30 days	115	115	163	190	190	19.9	0.07	0.07	0.09	0.11	0.13	0.010
1–3 months	115	124	154	190	205	18.6	0.07	0.07	0.10	0.13	0.17	0.017
3–6 months	115	111	140	179	205	21.0	0.07	0.07	0.10	0.13	0.13	0.014
6–12 months	115	112	140	177	175	18.7	0.07	0.08	0.10	0.13	0.15	0.013
1–3 years	100	98	126	163	190	19.8	0.07	0.08	0.11	0.15	0.17	0.019
3–5 years	55	65	98	132	145	18.0	0.09	0.09	0.12	0.15	0.17	0.015
5–8 years	70	70	98	115	145	16.1	0.09	0.10	0.13	0.16	0.19	0.017
8–12 years	55	55	79	107	115	15.0	0.09	0.10	0.14	0.17	0.27	0.022
12–16 years	55	55	75	102	115	13.5	0.09	0.11	0.14	0.16	0.21	0.018
	P-wave duration (s)						QRS duration (s)					
0–24 h	0.040	0.040	0.051	0.065	0.075	0.066	0.05	0.05	0.065	0.084	0.09	0.010
1–7 days	0.035	0.038	0.046	0.061	0.065	0.066	0.04	0.04	0.056	0.079	0.08	0.010
8–30 days	0.040	0.040	0.048	0.057	0.065	0.064	0.04	0.04	0.057	0.073	0.08	0.009
1–3 months	0.040	0.040	0.046	0.058	0.065	0.063	0.05	0.05	0.062	0.080	0.08	0.007
3–6 months	0.040	0.040	0.049	0.065	0.065	0.072	0.06	0.06	0.068	0.080	0.08	0.008
6–12 months	0.040	0.046	0.058	0.068	0.075	0.058	0.05	0.05	0.065	0.080	0.08	0.008
1–3 years	0.045	0.053	0.065	0.082	0.085	0.090	0.05	0.05	0.064	0.080	0.08	0.008
3–5 years	0.040	0.051	0.069	0.087	0.095	0.108	0.06	0.06	0.072	0.084	0.09	0.009
5–8 years	0.050	0.059	0.070	0.084	0.095	0.095	0.05	0.05	0.067	0.080	0.08	0.017
8–12 years	0.050	0.061	0.075	0.092	0.105	0.098	0.05	0.05	0.073	0.084	0.09	0.008
12–16 years	0.060	0.064	0.081	0.095	0.105	0.095	0.04	0.04	0.068	0.080	0.10	0.010

Table A2.26
Heart rate and durations: adolescents

	Sex	Mean	SD	5%	50%	95%
Heart rate	Male	69	11	52	70	90
	Female	73	12	57	70	92
	Total	71	12	55	70	90
P-wave duration	Male	0.09	0.02	0.06	0.08	0.11
	Female	0.09	0.02	0.06	0.09	0.12
	Total	0.09	0.02	0.06	0.09	0.11
PR interval	Male	0.15	0.03	0.10	0.15	0.20
	Female	0.15	0.02	0.12	0.15	0.19
	Total	0.15	0.02	0.11	0.15	0.20
QRS duration	Male	0.08	0.01	0.06	0.08	0.10
	Female	0.08	0.01	0.05	0.08	0.09
	Total	0.08	0.01	0.06	0.08	0.10
QT interval	Male	0.38	0.04	0.32	0.38	0.42
	Female	0.37	0.03	0.32	0.38	0.42
	Total	0.37	0.04	0.32	0.38	0.43
T-wave duration	Male	0.21	0.06	0.16	0.20	0.35
	Female	0.19	0.04	0.15	0.18	0.28
	Total	0.20	0.05	0.15	0.20	0.31

Table A2.27
Heart rate (bpm): prematures

Infants		Mean	Percentile	
Age	Number	Heart rate	5th	95th
24 h	66	141	109	173
72 h	69	150	127	182
1 week	62	164	134	200
1 month	42	170	133	200
2 months	30	171	128	203
3 months	24	159	130	202
6 months	16	145		
1 year	18	142		

Table A2.28
P amplitude in lead II ×10 (mV): prematures

Infants		Mean	Percentile	
Age	Number	P amplitude	5th	95th
24 h	65	1.1	0.5	2.0
72 h	69	1.3	0.5	2.0
1 week	62	1.3	0.5	2.6
1 month	40	0.8	0.3	1.5
2 months	30	0.9	0.4	1.5
3 months	24	1.0	0.5	1.9
6 months	16	1.1		
1 year	18	1.2		

Table A2.29
Amplitudes in V_3(high-frequency data) ×10 (mV)

	R wave						S wave					
Age	Min.	5%	Mean	95%	Max.	SD	Min.	5%	Mean	95%	Max.	SD
0–24 h	12.0	12.7	18.8	26.7	28.0	4.12	10.0	12.0	25.0	32.0	38.0	6.05
1–7 days	4.0	8.8	18.1	30.0	40.0	6.55	0.0	2.	17.1	33.0	38.0	8.37
8–30 days	0.0	8.3	18.8	33.8	36.0	7.50	2.0	4.2	12.4	20.0	26.0	5.47
1–3 months	12.0	13.5	21.8	29.1	32.0	4.50	2.0	4.9	14.0	21.3	22.0	4.95
3–6 months	12.0	14.5	20.1	30.0	32.0	4.56	2.0	4.5	14.7	23.5	26.0	6.14
6–12 months	8.0	9.7	17.8	23.5	24.0	4.32	0.0	4.9	14.2	22.6	26.0	5.94
1–3 years	8.0	8.8	15.4	27.0	32.0	4.92	2.0	4.5	13.4	23.5	26.0	5.24
3–5 years	4.0	6.0	15.0	22.7	24.0	4.49	2.0	3.3	12.2	20.0	26.0	6.00
5–8 years	0.0	4.5	13.1	23.2	32.0	5.81	0.0	3.7	15.0	24.8	30.0	6.10
8–12 years	0.0	4.4	10.8	21.0	32.0	5.22	0.0	2.4	13.2	24.0	30.0	6.80
12–16 years	0.0	3.8	10.0	19.6	28.0	5.34	0.0	2.2	10.8	25.4	34.0	6.82

Table A2.30
Amplitude in lead V_4R (low-frequency data) ×10 (mV)

	R wave					S wave				
Age	Min.	5%	Mean	95%	Max.	Min.	5%	Mean	95%	Max.
30 h	3.5	4.0	8.6	14.2	15.0	0.0	0.2	3.8	13.0	12.0
1 month	3.0	3.3	6.3	8.5	12.0	0.0	0.8	1.8	4.6	9.0
2–3 months	0.5	1.1	5.1	10.1	15.0	0.0	0.0	3.4	9.3	15.0
4–5 months	2.0	2.4	5.2	7.5	9.0	1.0	0.3	3.5	6.7	9.0
6–8 months	2.0	1.3	4.4	7.1	7.0	0.0	0.2	3.9	11.7	10.0
9 months–2 years	1.0	0.2	4.0	6.6	8.0	0.0	0.8	4.9	8.1	10.5
2–5 years	1.0	1.6	3.4	7.4	8.0	1.0	1.2	4.8	9.5	12.0
6–13 years	0.2	0.6	2.5	5.7	7.0	0.5	0.9	5.8	12.5	20.0

A2.4 Normal Limits of the Pediatric Orthogonal-Lead ECG

A number of tables are included to provide an indication of the normal limits of the orthogonal-lead ECG in infants and children. The terminology is discussed in Chap. 21 and in addition, the concept of prevalent direction is described in Appendix 1, Sect. A1.7 by Liebman. Tables have been reproduced from Liebman J, Plonsey R, Gillette PC. *Pediatric Electrocardiography*. 1982, with the permission of Williams and Wilkins, Baltimore, Maryland. However, some of these tables are from earlier publications and the list at the end of the section relates to the identifying superscript in the legend to each table (Tables A2.31–A2.51). Figures A2.40 and A2.41 are reproduced from the work of Davignon and Rautaharju with the permission of Springer (See A2.2).

Table A2.31
Direction of inscription of the QRS complex: adolescents[a]

	Sex	Number	Clockwise		Figure-of-eight		Counterclockwise	
			No.	%	No.	%	No.	%
Frontal plane	male	67	42	62.6	2	2.9	23	34.3
	female	47	29	61.7	3	6.3	15	31.9
	total	114	71	62.2	5	4.3	38	33.3
Horizontal plane	male	67	0	0	1	1.4	66	98.5
	female	47	0	0	2	4.2	45	95.7
	total	114	0	0	3	2.6	111	97.3

Table A2.32
Direction of inscription of the QRS complex: prematures C, clockwise; CC, counterclockwise[b]

Age of infants	Frontal plane		Horizontal plane			
	C (%)	CC (%)	C (%)	CC (%)	Figure-of-eight initially CC (%)	Narrow[a] loop (%)
24 h	97	3	1.5	11	34	54
72 h	98.5	1.5	1.5	16.5	43.5	39
1 week	97	3		21.5	54	24.5
1 month	97.5	2.5	2.5	17	51	29.5
2 months	93.5	6.5	3.5	30	53.5	13.5
3 months	100		4.5	29	62.5	4
6 months	100		6	37.5	50	6
1 year	100		5.5	61	28	5.5

[a]Presence of a QRS loop so narrow that the specific direction of inscription has no significance.

Table A2.33
Evolution of the Frank vectorcardiogam in normal infants. In frontal plane, initial QRS to left 66.6% from birth to 30 days; initial QRS to left 55.5% from 1 to 2 months; initial QRS to left 10.5% from to 2 to 3 months; initial QRS to left 6.2% by 6–8 months[c]

Groups	Frontal			Horizontal		
	CW	Figure-of-eight	CCW	CW	Figure-of-eight	CCW
I (birth–30 h)	91.3	7.0	1.7	61.5	19.2	19.3
II (1 month)	88.9	11.1	0	18.5	40.7	40.8
III (2–3 months)	97.4	0	2.6	2.6	39.5	57.9
IV (4–5 months)	87.0	6.5	6.5	0	35.5	64.5
V (6–8 months)	87.6	6.2	6.2	0	21.9	78.1
VI (12–18 months)	83.3	10.0	6.7	0	6.6	93.4

Table A2.34

Prevalent direction of the QRS in the frontal plane: adolescents[a]

Prevalent direction							
Sex	Age	d	χ^2	$\hat{\alpha}_f$	5%	50%	95%
Male	11 → 15	0.94	51.2	65	10	70	90
Male	16 → 19	0.92	63.9	63	9	70	91
Female	11 → 15	0.97	48.5	62	38	60	87
Female	16 → 19	0.93	36.0	56	11	65	80
Male	11 → 19	0.93	115.0	64	10	70	90
Female	11 → 19	0.95	84.2	59	15	60	80
Total	11 → 19	0.93	198.9	62	14	65	90

Table A2.35

Prevalent direction of the QRS in the horizontal plane: adolescents[a]

Prevalent direction							
Sex	Age	d	χ^2	$\hat{\alpha}_h$	5%	50%	95%
Male	11 → 15	0.94	51.6	330	292	335	8
Male	16 → 19	0.92	64.4	333	289	340	8
Female	11 → 15	0.96	48.0	334	290	340	354
Female	16 → 19	0.95	38.2	332	300	340	0
Male	11 → 19	0.93	116.0	332	292	340	5
Female	11 → 19	0.96	86.2	333	292	340	0
Total	11 → 19	0.94	202.1	332	293	340	0

Table A2.36

Prevalent direction of the T in the frontal plane: adolescents[a]

Prevalent direction							
Sex	Age	d	χ^2	$\hat{\alpha}_f$	5%	50%	95%
Male	11 → 15	0.96	53.8	42	15	45	60
Male	16 → 19	0.94	67.7	48	5	55	70
Female	11 → 15	0.98	49.6	42	17	45	60
Female	16 → 19	0.96	39.0	42	3	45	74
Male	11 → 19	0.95	121.2	45	7	45	70
Female	11 → 19	0.97	88.6	42	17	45	60
Total	11 → 19	0.96	209.6	44	10	45	63

Table A2.37

Prevalent direction of the T in the horizontal plane: adolescents[a]

Prevalent direction							
Sex	Age	d	$\chi2$	$\hat{\alpha}_h$	5%	50%	95%
Male	11 → 15	0.90	47.5	15	310	15	60
Male	16 → 19	0.88	58.3	24	338	15	74
Female	11 → 15	0.96	48.4	12	335	15	38
Female	16 → 19	0.97	39.2	12	310	15	25
Male	11 → 19	0.89	105.1	20	334	15	60
Female	11 → 19	0.97	87.6	12	335	15	33
Total	11 → 19	0.92	191.4	17	335	15	60

Table A2.38

Frontal angular deviation of T from QRS: adolescents[a]

Prevalent direction							
Sex	Age	d	χ^2	$\hat{\alpha}_f$	5%	50%	95%
Male	11 → 15	0.90	46.9	337	285	335	40
Male	16 → 19	0.94	66.6	345	305	345	17
Female	11 → 15	0.97	48.8	340	302	340	0
Female	16 → 19	0.93	36.0	346	316	340	42
Male	11 → 19	0.92	113.0	342	299	340	34
Female	11 → 19	0.95	84.5	343	309	340	15
Total	11 → 19	0.93	197.5	342	305	340	18

Table A2.39

Horizontal angular deviation of T from QRS: adolescents[a]

Prevalent direction							
Sex	Age	d	$\chi2$	$\hat{\alpha}_h$	5%	50%	95%
Male	11 → 15	0.85	41.8	45	330	35	98
Male	16 → 19	0.83	52.8	52	334	45	108
Female	11 → 15	0.95	46.4	39	4	35	80
Female	16 → 19	0.92	35.2	39	330	35	75
Male	11 → 19	0.84	94.3	49	339	45	103
Female	11 → 19	0.93	81.7	39	4	35	78
Total	11 → 19	0.87	174.3	45	356	40	93

Table A2.40

Prevalent direction of QRS: prematures[b]

	Frontal plane				First horizontal vector				Second horizontal vector			
Age of infants	No. of infants	Prevalent direction	5%	95%	No. of infants	Prevalent Direction	5%	95%	No. of Infants	Prevalent direction	5%	95%
24 h	60	127	75	194	58	74	338	340	16	239		
72 h	68	121	75	195	59	84	295	220	15	233		
1 week	61	117	75	165	54	69	332	216	21	231	9	332
1 month	42	80	17	171	35	58	340	115	13	232		
2 months	30	63	345	105	30	46	340	60	8	231		
3 months	24	59	352	105	23	51	346	108	2	223		
6 months	15	58			16	30			1	240		
1 year	18	46			18	12			2	253		

Table A2.41

Prevalent direction of T: prematures[b]

	Frontal plane				Horizontal plane			
Age of infants	No. of infants	Prevalent direction	5%	95%	No. of infants	Prevalent direction	5%	95%
24 h	52	41	319	84	56	314	240	111
72 h	65	28	315	60	61	318	241	101
1 week	60	31	345	75	59	335	290	15
1 month	42	53	30	75	36	346	281	52
2 months	30	43	353	67	30	334	259	15
3 months	24	47	19	74	24	335	295	14
6 months	14	41			15	333		
1 year	18	46			18	343		

Table A2.42

QRS magnitudes ×10 (mV): prematures[b]

Age of infants	No. of infants	R wave	5%	95%	No. of infants	S wave	5%	95%
Lead V_5 (*X* axis)								
24 h	64	6.5	2.0	12.6	61	6.8	0.06	17.6
72 h	65	7.4	2.6	14.9	64	6.5	1.00	16.0
1 week	61	8.7	3.8	16.8	56	6.8	0.00	15.0
1 month	38	13.0	6.2	21.6	38	6.2	1.20	14.0
2 months	30	18.3	12.1	31.5	29	7.0	0.96	15.0
3 months	24	21.0	14.6	31.5	24	6.7	1.30	21.4
6 months	16	20.3			16	6.8		
1 year	18	17.5			17	3.0		

Table A2.42 (Continued)

Age of infants	No. of infants	R wave	5%	95%	No. of infants	S wave	5%	95%
Lead aVF (*Y* axis)								
24 h	63	6.7	0.85	16.6	28	0.96	0.00	4.5
72 h	68	7.1	0.86	13.9	33	1.20	0.00	5.5
1 week	61	7.6	1.3	14.1	30	0.98	0.00	3.3
1 month	42	9.0	1.8	18.8	20	0.86	0.00	4.0
2 months	30	10.0	1.2	21.7	13	0.90	0.00	5.3
3 months	24	11.1	1.9	23.0	14	0.77	0.00	3.8
6 months	16	12.0			9	0.53		
1 year	18	9.1			9	0.56		
Lead V_2 (*Z* axis)								
24 h	65	11.4	3.5	21.3	65	15.0	2.5	26.5
72 h	66	11.9	5.0	20.8	66	13.5	2.6	26.0
1 week	60	12.3	4.0	20.5	60	14.0	3.0	25.0
1 month	41	15.0	8.3	21.0	41	14.0	5.1	26.3
2 months	30	19.0	8.6	32.0	30	17.1	8.0	34.5
3 months	23	20.1	13.3	30.0	23	16.1	6.0	37.6
6 months	16	20.6			16	18.5		
1 year	18	16.3			18	16.0		

Table A2.43

Spatial magnitude and orientation (MSVR in mV). M, male; F, female; T, total[d]

	n	$\hat{\alpha}$	β	D	χ^2	MSVR mean mag.	SD	$P_{2.5}$	P_5	P_{10}	P_{50}	P_{90}	P_{95}	$P_{97.5}$
Frank-lead system														
M 2-5	23	246	108	0.58	22.8	1.11	0.46		0.33	0.44	1.10	1.80	1.86	
M 6-10	60	254	107	0.73	96.6	1.11	0.41	0.45	0.49	0.59	1.02	1.78	1.99	214.
M 2-10	83	252	107	0.69	117.8	1.11	0.42	0.36	0.47	0.58	1.04	1.79	1.92	2.08
F 2-5	29	257	113	0.74	47.4	1.06	0.35		0.44	0.60	1.09	1.56	1.65	
F 6-10	63	253	109	0.69	90.3	1.01	0.39	0.20	0.34	0.51	0.95	1.46	1.80	2.05
F 2-10	92	254	111	0.71	137.5	1.02	0.38	0.25	0.39	0.56	0.95	1.52	1.67	1.97
T 2-5	52	252	111	0.66	68.6	1.08	0.40	0.33	0.34	0.58	1.10	1.69	1.80	1.85
T 6-10	123	254	108	0.71	186.7	1.06	0.40	0.33	0.47	0.56	0.99	1.62	1.91	2.08
T 2-10	175	253	109	0.70	254.9	1.07	0.40	0.33	0.46	0.58	1.02	1.65	1.81	2.02
T 2-19	341	252	112	0.63	406.5	0.94	0.43	0.22	0.30	0.40	0.92	1.51	1.71	1.94

Table A2.44

Spatial magnitude and orientation (MSVL). M, male; F, female; T, total[d]

	n		β	D	χ	MSVR mean mag.	SD	$P_{2.5}$	P_5	P_{10}	P_{50}	P_{90}	P_{95}	$P_{97.5}$
Frank-lead system														
M 2-5	23	347	129	0.85	50.2	1.73	0.60		0.83	0.87	1.80	2.59	2.74	
M 6-10	60	329	133	0.87	135.9	1.82	0.53	1.00	1.10	1.13	1.77	2.67	2.93	3.15
M 2-10	83	334	133	0.86	183.9	1.80	0.55	0.85	0.96	1.10	1.80	2.59	2.81	3.05
F 2-5	29	341	133	0.83	60.0	1.71	0.44		1.02	1.10	1.69	2.13	2.66	
F 6-10	63	340	138	0.93	162.5	1.77	0.44	1.03	1.10	1.13	1.76	2.36	2.60	2.80
F 2-10	92	340	136	0.90	221.7	1.75	0.44	1.02	1.09	1.13	1.75	2.31	2.58	2.81
T 2-5	52	344	131	0.84	109.9	1.72	0.51	0.93	0.88	1.01	1.76	2.40	2.70	2.97
T 6-10	123	334	136	0.90	296.5	1.80	0.49	1.05	1.10	1.13	1.76	2.43	2.77	2.93
T 2-10	175	337	135	0.88	404.5	1.77	0.49	0.95	1.05	1.12	1.76	2.42	2.75	2.89
T 2-19	341	336	136	0.88	801.1	1.71	0.48	0.88	1.00	1.11	1.69	2.31	2.59	2.82

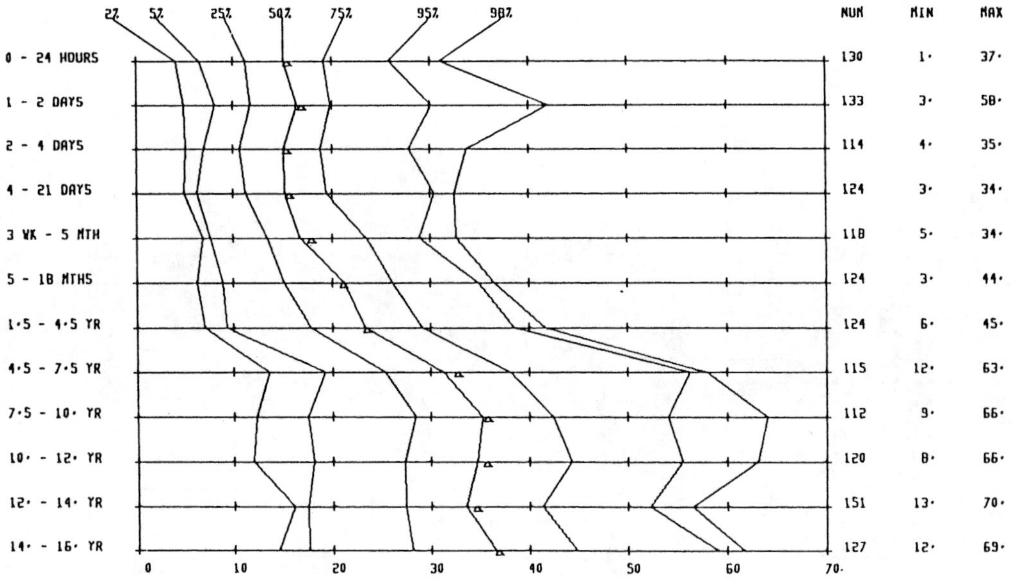

Figure A2.40

Percentile distributions for the spatial magnitude of the QRS integral sector in normal children from birth to aged 16 years.[9]

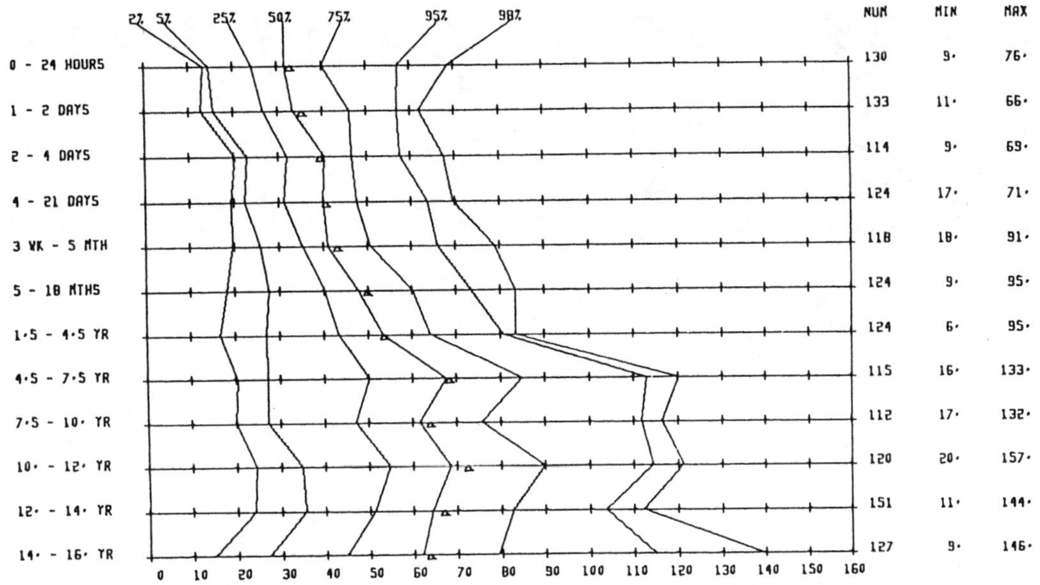

Figure A2.41
Percentile distribution for the spatial magnitude of the ST-T integral vector in normal children from birth to aged 16 years.[9]

Table A2.45
Maximal projections. M, male; F, female; T, total[d]: Magnitude is in mV

	n	Frank system Mean mag.	SD	$p_{2.5}$	p_5	p_{10}	p_{50}	p_{90}	p_{95}	$p_{97.5}$
X initial right										
M 2-5	23	0.18	0.16		0.00	0.00	0.15	0.41	0.51	
M 6-10	60	0.12	0.12	0.00	0.00	0.00	0.09	0.28	0.36	0.48
M 2-10	83	0.14	0.14	0.00	0.00	0.00	0.11	0.33	0.36	0.55
F 2-5	29	0.11	0.09		0.00	0.00	0.11	0.24	0.26	
F 6-10	63	0.10	0.09	0.00	0.00	0.00	0.09	0.21	0.25	0.39
F 2-10	92	0.10	0.08	0.00	0.00	0.00	0.09	0.22	0.25	0.35
T 2-5	52	0.14	0.13	0.00	0.00	0.00	0.11	0.32	0.38	0.50
T 6-10	123	0.11	0.11	0.00	0.00	0.00	0.09	0.25	0.35	0.38
T 2-10	175	0.12	0.11	0.00	0.00	0.00	0.09	0.25	0.35	0.40
T 2-19	341	0.10	0.10	0.00	0.00	0.00	0.08	0.23	0.30	0.36

Table A2.45 (Continued)

		Frank system								
	n	Mean mag.	SD	$p_{2.5}$	p_5	p_{10}	p_{50}	p_{90}	p_{95}	$p_{97.5}$
X left										
M 2-5	23	1.23	0.52		0.49	0.55	1.22	2.02	2.22	
M 6-10	60	1.17	0.47	0.37	0.41	0.68	1.09	1.80	2.23	2.47
M 2-10	83	1.19	0.48	0.40	0.49	0.60	1.14	1.81	2.22	2.34
F 2-5	29	1.15	0.47		0.55	0.60	1.16	1.60	2.21	
F 6-10	63	1.08	0.36	0.18	0.46	0.65	1.08	1.55	1.66	1.89
F 2-10	92	1.10	0.40	0.28	0.52	0.60	1.10	1.55	1.64	2.03
T 2-5	52	1.19	0.49	0.49	0.53	0.59	1.18	1.79	2.15	2.62
T 6-10	123	1.13	0.42	0.34	0.46	0.68	1.08	1.66	1.82	2.25
T 2-10	175	1.14	0.44	0.40	0.52	0.61	1.10	1.68	1.90	2.25
T 2-19	341	1.06	0.44	0.35	0.46	0.56	1.04	1.60	1.80	1.99
X terminal right										
M 2-5	23	0.32	0.23		0.00	0.01	0.28	0.66	0.69	
M 6-10	60	0.29	0.20	0.00	0.00	0.03	0.30	0.52	0.58	0.84
M 2-10	83	0.30	0.21	0.00	0.00	0.02	0.30	0.55	0.65	0.70
F 2-5	29	0.30	0.21		0.00	0.06	0.25	0.62	0.75	
F 6-10	63	0.23	0.17	0.00	0.00	0.02	0.24	0.48	0.52	0.78
F 2-10	92	0.25	0.19	0.00	0.00	0.02	0.24	0.50	0.63	0.79
T 2-5	52	0.31	0.22	0.00	0.00	0.03	0.26	0.64	0.68	0.80
T 6-10	123	0.26	0.19	0.00	0.00	0.02	0.26	0.50	0.55	0.75
T 2-10	175	0.27	0.20	0.00	0.00	0.02	0.26	0.52	0.63	0.74
T 2-19	341	0.23	0.19	0.00	0.00	0.00	0.20	0.50	0.55	0.65

Table A2.46

Maximal projections. M, male; F, female; T total[d]: Magnitude is in mV

		Frank-lead system								
	n	Mean Mag.	SD	$p_{2.5}$	p_5	p_{10}	p_{50}	p_{90}	p_{95}	$p_{97.5}$
X initial superior										
M 2-5	23	0.12	0.12		0.00	0.00	0.10	0.28	0.46	
M 6-10	60	0.11	0.09	0.00	0.00	0.00	0.08	0.26	0.30	0.32
M 2-10	83	0.11	0.10	0.00	0.00	0.00	0.10	0.26	0.30	0.35
F 2-5	29	0.10	0.10		0.00	0.00	0.08	0.24	0.37	
F 6-10	63	0.10	0.09	0.00	0.00	0.00	0.08	0.21	0.30	0.37
F 2-10	92	0.10	0.09	0.00	0.00	0.00	0.08	0.22	0.31	0.38
T 2-5	52	0.11	0.11	0.00	0.00	0.00	0.10	0.25	0.36	0.47
T 6-10	123	0.10	0.09	0.00	0.00	0.00	0.08	0.22	0.30	0.35
T 2-10	175	0.10	0.10	0.00	0.00	0.00	0.08	0.24	0.30	0.35
T 2-19	341	0.10	0.08	0.00	0.00	0.00	0.08	0.20	0.25	0.30

Table A2.46 (Continued)

		Frank-lead system								
	n	Mean Mag.	SD	$p_{2.5}$	p_{5}	p_{10}	p_{50}	p_{90}	p_{95}	$p_{97.5}$
Y inferior										
M 2-5	23	1.02	0.45		0.39	0.45	1.00	1.78	1.98	
M 6-10	60	1.17	0.41	0.51	0.56	0.64	1.09	1.76	1.97	2.15
M 2-10	83	1.13	0.43	0.45	0.47	0.58	1.04	1.76	1.96	2.05
F 2-5	29	1.09	0.32		0.46	0.71	1.12	1.50	1.54	
F 6-10	63	1.27	0.40	0.64	0.68	0.72	1.30	1.82	2.00	2.24
F 2-10	92	1.22	0.39	0.58	0.67	0.72	1.21	1.64	1.97	2.07
T 2-5	52	1.06	0.38	0.36	0.42	0.50	1.06	1.51	1.71	1.97
T 6-10	123	1.22	0.41	0.56	0.63	0.70	1.20	1.77	2.00	2.10
T 2-10	175	1.17	0.41	0.46	0.56	0.66	1.16	1.69	1.96	2.03
T 2-19	341		0.40	0.47	0.58	0.65	1.12	1.70	1.88	2.02
Y terminal superior										
M 2-5	23	0.11	0.11		0.00	0.00	0.07	0.24	0.40	
M 6-10	60	0.13	0.11	0.00	0.00	0.00	0.15	0.28	0.32	0.38
M 2-10	83	0.12	0.11	0.00	0.00	0.00	0.10	0.27	0.32	0.43
F 2-5	29	0.08	0.10		0.00	0.00	0.06	0.23	0.35	
F 6-10	63	0.11	0.15	0.00	0.00	0.00	0.04	0.28	0.40	0.61
F 2-10	92	0.10	0.14	0.00	0.00	0.00	0.04	0.26	0.39	0.47
T 2-5	52	0.09	0.10	0.00	0.00	0.00	0.07	0.23	0.32	0.45
T 6-10	123	0.12	0.13	0.00	0.00	0.00	0.08	0.28	0.32	0.44
T 2-10	175	0.11	0.12	0.00	0.00	0.00	0.08	0.26	0.32	0.44
T 2-19	341	0.10	0.12	0.00	0.00	0.00	0.07	0.26	0.32	0.38

Table A2.47
Maximal projections. M, male; F, female; T total[d]: Magnitude is in mV

		Frank-lead system								
	n	Mean mag.	SD	$p_{2.5}$	p_{5}	p_{10}	p_{50}	p_{90}	p_{95}	$p_{97.5}$
Z anterior										
M 2-5	23	0.86	0.41		0.36	0.42	0.78	1.52	2.02	
M 6-10	60	0.64	0.27	0.23	0.28	0.31	0.60	1.00	1.18	1.37
M 2-10	83	0.70	0.32	0.25	0.30	0.36	0.62	1.09	1.25	1.68
F 2-5	29	0.72	0.30		0.26	0.34	0.70	1.10	1.35	
F 6-10	63	0.62	0.21	0.17	0.22	0.36	0.62	0.90	1.01	1.15
F 2-10	92	0.65	0.24	0.19	0.24	0.36	0.65	0.95	1.09	1.24
T 2-5	52	0.78	0.35	0.25	0.31	0.37	0.74	1.10	1.64	1.97
T 6-10	123	0.63	0.24	0.20	0.26	0.33	0.62	0.95	1.05	1.24
T 2-10	175	0.67	0.29	0.22	0.28	0.36	0.64	1.01	1.16	1.42
T 2-19	341	0.57	0.29	0.12	0.16	0.23	0.55	0.94	1.08	1.25

Table A2.47 (Continued)

	n	Frank-lead system								
		Mean mag.	SD	$p_{2.5}$	p_5	p_{10}	p_{50}	p_{90}	p_{95}	$p_{97.5}$
Z posterior										
M 2-5	23	1.02	0.44		0.29	0.45	1.04	1.70	1.72	
M 6-10	60	1.15	0.37	0.64	0.68	0.70	1.12	1.72	1.86	2.10
M 2-10	83	1.11	0.39	0.41	0.62	0.65	1.08	1.70	1.79	2.03
F 2-5	29	1.09	0.26		0.65	0.75	1.10	1.44	1.53	
F 6-10	63	0.97	0.39	0.18	0.23	0.56	0.94	1.41	1.87	2.01
F 2-10	92	1.01	0.35	0.22	0.42	0.61	0.99	1.42	1.63	1.93
T 2-5	52	1.06	0.35	0.31	0.48	0.63	1.08	1.54	1.70	1.71
T 6-10	123	1.06	0.39	0.23	0.56	0.66	1.00	1.65	1.85	2.04
T 2-10	175	1.06	0.38	0.26	0.55	0.65	1.02	1.58	1.75	1.93
T 2-19	341	0.97	0.40	0.22	0.32	0.48	0.94	1.45	1.70	1.90

Table A2.48

Ratios of maximal projections. M, male; F, female; T, total[d]: Magnitude is in mV

	X terminal right/*X* left						*Y* terminal superior/*Y* interior					*Z* anterior/*Z* posterior							
	n	Mean mag.	SD	P_{50}	P_{90}	P_{95}	Mean mag.	SD	P_{50}	P_{90}	P_{95}	n	Mean mag.	SD	P_5	P_{10}	P_{50}	P_{90}	P_{95}
Frank-lead system																			
M 11-15	62	0.23	0.21	0.18	0.63	0.75	0.10	0.12	0.06	0.28	0.42	62	0.57	0.28	0.21	0.24	0.52	0.94	1.11
M 16-19	30	0.29	0.46	0.15	0.53	1.55	0.14	0.13	0.11	0.30	0.43	29	0.54	0.24	0.21	0.25	0.50	0.92	0.97
M 11-19	92	0.25	0.31	0.17	0.55	0.75	0.11	0.12	0.08	0.28	0.28	91	0.56	0.27	0.21	0.26	0.50	0.94	1.04
F 11-15	42	0.26	0.40	0.13	0.71	0.83	0.08	0.13	0.02	0.21	0.46	42	0.85	1.44	0.15	0.17	0.56	1.35	1.98
F 16-19	32	0.23	0.25	0.12	0.65	0.88	0.12	0.13	0.09	0.29	0.28	32	0.63	0.50	0.13	0.16	0.49	1.47	1.95
F 11-19	74	0.25	0.34	0.13	0.69	0.84	0.10	0.13	0.03	0.26	0.43	74	0.76	1.13	0.14	0.17	0.50	1.33	1.94
F 11-15	104	0.24	0.30	0.15	0.67	0.75	0.09	0.12	0.04	0.26	0.43	104	0.68	0.94	0.18	0.22	0.53	1.10	1.36
T 16-19	62	0.26	0.37	0.13	0.55	0.85	0.13	0.13	0.11	0.29	0.32	61	0.59	0.40	0.15	0.21	0.50	1.11	1.53
Total	166	0.25	0.32	0.15	0.59	0.75	0.10	0.12	0.06	0.27	0.40	165	0.65	0.79	0.18	0.22	0.50	1.10	1.36

Table A2.49

Maximal spatial angles and magnitudes- QRS + T. M, male; F, female; T, total[d]: Magnitude is in mV

	n		β	D	χ	Mean Mag.	SD	$P_{2.5}$	P_5	P_{10}	P_{50}	P_{90}	P_{96}	$P_{97.5}$
Frank-lead system														
QRS														
M 2-5	23	347	129	0.85	49.5	1.73	0.60		0.83	1.87	1.81	2.59	2.74	
M 6-10	60	327	133	0.85	130.7	1.83	0.53	1.00	1.11	1.14	1.77	2.67	2.93	3.15
M 2-10	83	332	133	0.84	177.6	1.80	0.55	0.85	0.96	1.11	1.80	2.59	2.81	3.05
F 2-5	29	341	133	0.83	59.4	1.72	0.44		1.02	1.15	1.69	2.13	2.66	
F 6-10	63	336	138	0.90	152.1	1.79	0.43	1.08	1.12	1.15	1.76	2.36	2.60	2.80
F 2-10	92	338	136	0.87	210.9	1.76	0.43	1.07	1.11	1.15	1.76	2.31	2.58	2.81
T 2-5	52	344	131	0.83	108.6	1.72	0.51	0.83	0.88	1.01	1.76	2.40	2.70	2.97
T 6-10	123	331	136	0.87	281.3	1.81	0.48	1.07	1.11	1.15	1.76	2.43	2.77	2.93
T 2-10	175	335	135	0.86	387.5	1.78	0.49	0.95	1.06	1.13	1.76	2.42	2.75	2.89
T 2-19	341	334	136	0.87	772.9	1.71	0.48	0.91	1.02	1.12	1.69	2.31	2.59	2.82

Table A2.49 (Continued)

	n		β	D	χ	Mean Mag.	SD	$P_{2.5}$	P_5	P_{10}	P_{50}	P_{90}	P_{96}	$P_{97.5}$
T wave														
M 2-5	22	33	123	0.89	51.7	0.48	0.15		0.18	0.25	0.53	0.68	0.72	
M 6-10	58	360	128	0.93	149.2	0.54	0.18	0.22	0.27	0.32	0.49	0.79	0.83	1.00
M 2-10	80	352	127	0.90	195.3	0.52	0.17	0.20	0.25	0.31	0.51	0.76	0.79	0.92
F 2-5	29	341	121	0.91	72.3	0.51	0.18		0.25	0.28	0.51	0.81	0.90	
F 6-10	62	0	132	0.91	153.4	0.51	0.20	0.18	0.23	0.28	0.51	0.77	0.93	1.11
F 2-10	91	354	129	0.90	220.6	0.51	0.19	0.21	0.23	0.28	0.51	0.78	0.89	1.03
T 2-5	51	338	122	0.90	123.6	0.50	0.17	0.18	0.23	0.28	0.53	0.68	0.82	0.93
T 6-10	120	360	130	0.92	302.2	0.52	0.19	0.21	0.25	0.29	0.51	0.78	0.83	1.05
T 2-10	171	353	128	0.90	415.9	0.52	0.18	0.21	0.24	0.29	0.51	0.77	0.82	0.96
T 2-19	335	5	130	0.88	781.6	0.48	0.18	0.18	0.21	0.28	0.48	0.72	0.79	0.94

Table A2.50

Normal values – magnitudes – Frank scalars (atrial)[f]

Age	Median	*n*	Mean	SD	p_5	p_{10}	p_{50}	P_{90}	p_{96}
X left (μV)									
Birth–6 months	1.8 months	27	102	37	44	61	100	167	188
6 months–5 years	2.5 years	22	78	26	38	43	76	127	139
5–12 years	8.1 years	42	76	17	47	54	76	95	117
12–19 years	15.6 years	24	69	17	42	46	66	93	97
Entire group	6.6 years	115	81	27	45	51	76	117	134
Y interior (μV)									
Birth–6 months	1.8 months	27	114	43	45	62	112	175	189
6 months–5 years	2.5 years	22	134	41	25	64	135	175	191
5–12 years	8.1 years	42	118	47	43	55	114	187	202
12–19 years	15.6 years	24	137	45	55	69	150	171	189
Entire group	6.6 years	115	124	45	48	68	124	179	193
Z interior (μV)									
Birth–6 months	1.8 months	27	41	24	6	11	40	72	92
6 months–5 years	2.5 years	22	45	15	18	22	44	65	76
5–12 years	8.1 years	42	44	16	20	24	43	62	80
12–19 years	15.6 years	24	40	17	18	20	36	67	76
Entire group	6.6 years	115	43	18	18	20	41	65	77
Z posterior (μV)									
birth–6 months	1.8 months	27	40	24	0	18	42	76	85
6 months–5 years	2.5 years	22	32	19	1	14	30	64	67
5–12 years	8.1 years	42	30	16	1	18	27	54	57
12–19 years	15.6 years	24	26	18	1	9	20	57	61
Entire group	6.6 years	115	32	18	0	15	29	58	65

Table A2.50 (Continued)

Age	Median	n	Mean	SD	p_5	p_{10}	p_{50}	p_{90}	p_{96}
Spatial voltage anterior (μV)									
Birth–6 months	1.8 months	27	117	42	52	65	113	182	201
6 months–5 years	2.5 years	22	94	31	50	57	91	145	150
5–12 years	8.1 years	42	93	30	41	53	93	134	150
12–19 years	15.6 years	24	109	37	42	65	104	168	173
Entire group	6.6 years	115	102	36	50	60	99	151	167
Spatial voltage posterior (μV)									
Birth–6 months	1.8 months	27	109	58	0	0	110	189	196
6 months–5 years	2.5 years	22	113	58	3	23	115	195	213
5–12 years	8.1 years	42	85	38	2	35	82	126	157
12–19 years	15.6 years	24	84	44	7	39	68	160	179
Entire group	6.6 years	115	36	50	0	30	92	166	185

Table A2.51

Normal values (timing) – Frank scalars (atrial)[f]

Age	Median	n	Mean	SD	p_5	p_{10}	p_{50}	p_{90}	p_{96}
P duration (ms)									
Birth–6 months	1.8 months	27	63	12	44	48	65	85	89
6 months–5 years	2.5 years	22	77	9	58	64	78	90	92
5–12 years	8.1 years	42	85	8	67	77	84	95	102
12–19 years	15.6 years	24	91	10	75	77	91	103	114
Entire group	6.6 years	115	80	14	55	60	82	95	100
PR interval (ms)									
Birth–6 months	1.8 months	27	107	13	84	87	108	124	126
6 months–5 years	2.5 years	22	125	15	97	108	125	146	146
5–12 years	8.1 years	42	137	20	107	112	133	161	161
12–19 years	15.6 years	24	144	17	108	126	140	165	165
Entire group	6.6 years	115	120	21	91	101	130	160	160
Macruz index									
birth–6 months	1.8 months	27	1.58	0.61	0.82	0.91	1.41	2.72	3.23
6 months–5 years	2.5 years	22	1.80	0.80	1.10	1.13	1.56	3.46	3.99
5–12 years	8.1 years	42	1.81	0.58	0.89	1.08	1.75	2.76	2.89
12–19 years	15.6 years	24	1.93	0.77	0.92	1.19	1.85	2.68	4.18
Entire group	6.6 years	115	1.78	0.68	0.91	1.13	1.61	2.67	3.05

Table A2.51 (Continued)

Age	Median	*n*	Mean	SD	p_5	p_{10}	p_{50}	p_{90}	p_{96}
Time maximal anterior (%)									
Birth–6 months	1.8 months	27	39	11	19	23	40	52	64
6 months–5 years	2.5 years	22	34	17	20	25	35	43	45
5–12 years	8.1 years	42	38	9	27	28	37	51	56
12–19 years	15.6 years	24	37	7	26	27	37	48	51
Entire group	6.6 years	115	37	9	24	27	37	48	53
Time ant.-post. shift (%)									
Birth–6 months	1.8 months	27	59	14	32	42	54	76	96
6 months–5 years	2.5 years	22	57	13	40	42	57	83	87
5–12 years	8.1 years	42	61	10	44	48	59	85	91
12–19 years	15.6 years	24	62	14	44	48	59	85	91
Entire group	6.6 years	115	60	12	43	47	58	78	84

(a) Strong, W.B., T.F. Downs, J. Liebman, and R. Liebowitz, The normal adolescent electrocardiogram. *Am. Heart J.*, 1972;**83**: 115. The tables are reproduced with the permission of Mosby, St Louis, Missouri.
(b) Sreenivasan, V.V., B.J. Fisher, J. Liebman, and T.D. Downs, A longitudinal study of the standard electrocardiogram in the healthy premature infant during the first year of life. *Am. J. Cardiol.*, 1973;**31**: 57. The tables are reproduced with the permission of Yorke Medical Group, New York.
(c) Namin, E.P., R.A. Arcilla, I.A. D'Cruz, and B.M. Gasul, Evolution of the Frank vectorcardiogram in normal infants. *Am. J. Cardiol.*, 1964;**13**: 757. The tables are reproduced with the permission of Yorke Medical Group, New York.
(d) Kan, J.S., J. Liebman, M.H. Lee, and A. Whitney, Quantification of the normal Frank and McFee-Parungao orthogonal electrocardiogram at ages two to ten years. *Circulation*, 1977;**55**: 31. The tables are reproduced with the permission of the American Heart Association, Dallas, Texas.
(e) Liebman, J., M.H. Lee, P.S. Rao, and W. McKay, Quantification of the normal and Frank-McFee-Parungao orthogonal electrocardiogram in the adolescent. *Circulation*, 1973;**48**: 735. The tables are reproduced with the permission of the American Heart Association, Dallas, Texas.
(f) Ferrer, P.L. and R.C. Ellison, The Frank scalar atrial vectorcardiogram in normal children. *Am. Heart J.*, 1974;**88**: 467. The tables are reproduced with the permission of Mosby, St Louis, Missouri.
(g) Rautaharju, P.M., A. Davignon, F. Soumis, E. Boiselle, and C. Hoquette, Evolution of QRS- T relationship from birth to adolescence in Frank-lead orthogonal electrocardiograms of 1492 normal children. *Circulation*, 1979;**60**: 196–204. The tables are reproduced with the permission of the American Heart Association, Dallas, Texas.

References

1. Macfarlane, P.W., E.N. Coleman, E.O. Pomphrey, S. McLaughlin, A. Houston, and T. Aitchison, Normal limits of the high-fidelity pediatric ECG. *J. Electrocardiol.*, 1989;**22**(Suppl): 162–168.
2. Macfarlane, P.W., S.C. McLaughlin, and J.C. Rodger, Influence of lead selection and population on automated measurement of QT dispersion. *Circulation*, 1998;**98**: 2160–2167.

Appendix 3: Instrumentation Standards and Recommendations

P. W. Macfarlane, A. van Oosterom, O. Pahlm, P. Kligfield, M. Janse, J. Camm (eds.), *Comprehensive Electrocardiology*, DOI 10.1007/978-1-84882-046-3_A3,

A3.1 Introduction

All ECG instruments such as direct writing electrocardiographs, cardiac monitors, ambulatory monitoring electrocardiographic devices, stress ECG machines, fetal monitors and other ECG devices that process signals from galvanic biopotential sensors share the same design principles. Modern ECG equipment utilizes integrated digital signal processors (DSP) to perform signal amplification, analog-to-digital conversion, digital filtering, formatting and communication.

This section lists several of the available standards and recommendations related to the design of such instruments. The concepts discussed and their terminology are introduced in ➋ Sect. 12.4 (Vol. 1).

A3.2 General Design Considerations

Electrocardiographs have been optimized to best suit their particular application. The function of the analog processor is to form an ECG lead signal, amplify it, eliminate the common mode noise, and minimize the external interference.

The front-end, which is responsible for processing of the body surface potentials must be able to work with low voltage alternating current (AC) signals ranging from 0.01 to 5.0 mV, combined with a direct current (DC) common-mode component of up to ±300 mV resulting from the electrode–skin interface and a common-mode noise of up to 1.5 V AC.

The bandwidth of the electrocardiograph depends on the application and can range from 50 Hz for monitoring applications to 1 kHz for late-potential measurements. The bandwidth is of great importance in ECG diagnosis since many interpretation criteria are based on exact measurements of small notches and slurs. The faithful reproduction of the lower frequency regions, such as the ST segment, is essential since these have a critical diagnostic value. Many studies have been carried out for determining the frequency content of the adult and pediatric ECGs. The American Heart Association (AHA) recommends 150 Hz as minimum bandwidth and 500 Hz as minimum sampling rate for recording adult and pediatric ECGs. The report also states that it is unknown how far the bandwidth of systems may need to be extended, due to limitations of previous studies [1].

The analog ECG signal is then sampled and converted into digital data (➋ Sect. 12.A). In modern electrocardiographs, the sampling rate is usually much higher than is required by the Nyquist criterion and may be as high as 50 kHz in some cases (➋ Sect. 12.A.1.2). The advantage of sampling at such a high rate is the elimination of the need for using anti-aliasing filters at the front-end and to facilitate the reduction of noise in the input signal by computing running averages.

Rijnbeek et al. [2] reported on normal ECG features observed when using the higher sampling rate of 1,200 sps. On the basis of this study, a minimum bandwidth of 250 Hz for recording pediatric ECGs was recommended. He showed that with a bandwidth of 150 Hz, 38% of the cases in the study had an error >25 μV in the maximum positive deflection in lead V4. For leads V2 and V6, these percentages were 25% and 23%, respectively. Furthermore, 15% of the positive deflections and 7% of the negative deflections in V4 have amplitude errors >50 μV when a 150 Hz filter is used. The effect of age on the frequency content of the ECG signals was also addressed. It was found that the frequency content gradually decreases from infancy to adulthood. The data for children aged 12 to 16 years indicate that the system bandwidth should be 150 Hz to yield amplitude errors less than 25μV in 95% of the cases in this age group. This is close to the 125 Hz recommendation of the AHA for the adult ECG [3]. In vectorcardiographic leads, Berson et al. [4] found amplitude errors >50 μV in 8% of the R-wave amplitudes and 5% of the S-wave amplitudes when using a 150 Hz filter.

The differences between the results of the two studies may, in part, be explained by the difference in the sampling rates used (500 versus 1,200 sps) and by the use of different lead systems. Furthermore, the analyses by Berson et al. [4] were not performed on separate leads but on leads X, Y, and Z combined, which is likely to underestimate the effect of filtering on individual lead signals. More importantly, it was concluded that a threshold of 25 μV instead of 50 μV is preferable for measuring the effect of a reduced bandwidth on signal amplitudes.

Rijnbeek recommended using the minimum bandwidth of 250 Hz for the entire pediatric population. The higher bandwidth demands the sampling rate to be at least twice the bandwidth of the signal (➋ Sect. 12.A.2). In order to facilitate a high quality data representation, the AHA recommends a sampling rate of at least two or three times this theoretical minimum. As a rule of thumb, the pediatric ECG should be sampled at least at 1,000 samples per second.

A3.2.1 FDA Performance Standards

All diagnostic pieces of ECG equipment are classified by the Federal Drug Administration as Class II devices. Hence their design is subject both to safety and performance standards. These standards have been developed by the American Association of Medical Instrumentation and adopted by the American National Standards Institute as ANSI/AAMI EC11:1991/(R)2001. The standards establish minimum safety and performance requirements for ECG systems with direct-writing devices that are intended for use under the operating conditions specified. Also included are standards for the analysis of rhythm and of detailed morphology of complex cardiac complexes. Subject to this standard are all parts of the ECG system necessary to (a) obtain the signal from the surface of the patient's body, (b) amplify this signal, and (c) display it in a form suitable for diagnosing the heart's electrical activity. Hence, this standard includes requirements for the entire electrocardiographic recording system, ranging from the input electrodes right up to the displayed output.

Included within the scope of this standard are:

1. Direct-writing electrocardiographs.
2. Electrocardiographs used in other medical devices (e.g., patient monitors, defibrillators, stress testing devices), when such devices are intended for use in obtaining diagnostic ECG signatures.
3. Electrocardiographs having a display that is remote from the patient (via cable, telephone, telemetry, or storage media), when such devices are intended for use in obtaining ECG signatures. These devices are subject to the functional performance requirements at the system output-input levels.

Excluded from the scope of this standard are:

1. Devices that collect ECG data from locations other than the external surface of the body
2. Devices for interpretation and pattern recognition (e.g., QRS detectors, alarm circuits, rate meters, diagnostic algorithms)
3. Fetal ECG monitors
4. Ambulatory monitoring electrocardiographic devices, including ECG recorders and associated scanning and read-out devices
5. Diagnostic electrocardiographic devices utilizing non-permanent displays
6. Vectorcardiographs, that is, a device for displaying loops derived from X,Y,Z leads (see ➋ Chap. 11)
7. Electrocardiographic devices intended for use under extreme or uncontrolled environmental conditions outside of a hospital environment or physician's office
8. Cardiac monitorsl

See FDA: http://www.fda.gov/cdrh/ode/ecgs.pdf for further details

A3.2.2 Guidance for Diagnostic ECG

The Industry Diagnostic ECG Guidance (including Non-Alarming ST Segment Measurement) was issued on November 5, 1998. This guidance applies to most of the diagnostic electrocardiographs covered by the ANSI/AAMI EC11-1991 standard for Electrocardiographs (EC-11 standard). Included in the EC-11 standard are ECG devices intended for diagnostic purposes.

The Guidance for Industry, Cardiac Monitors (including Cardiotachometer and Rate Alarm) was issued on: November 5, 1998. See

http://www.fda.gov/cdrh/ode/cmonitor.pdf

This guidance applies to most of the cardiac monitors covered by the ANSI/AAMI EC13-1992 standard for Cardiac Monitors, Heart Rate Meters, and Alarms (EC13 standard). Included in the EC13 standard are ECG devices intended for monitoring purposes.

A3.2.3 Typical Design Specifications for an Electrocardiograph

Sampling rate:

- Digital sampling rate: 4,000 sps/channel
- 10,000 samples/s/channel used for pacemaker spike detection
- ECG analysis frequency: 500 sps

Dynamic range:

- AC differential: ±10 mV DC offset: ±320 mV
- Resolution: 4.88 μV/LSB @ 500 sps
- Frequency response: –3 dB @ 0.01–150 Hz
- Common mode rejection: >140 dB (123 dB with AC filter disabled)
- Input impedance: >10 MΩ @ 10 Hz, defibrillator protected
- Patient leakage: <10 μA
- Pace detect: Orthogonal LA, LL and V6; 750 μV @ 50 μs

Communication:

- Modem and Fax transmission, WI-FI wireless 802.11X

Writers:

- Writer technology: Thermal dot array
- Writer speeds: 5, 12.5, 25 and 50 mm/s (same as displayed)
- Number of traces: 3, 6, 12 or 15, user selectable (same as displayed)
- Writer sensitivity/gain: 2.5, 5, 10, 20, 10/5 (split calibration) mm/mV (same as displayed)
- Writer speed accuracy: ±2%
- Writer amplitude accuracy: ±5%
- Writer resolution: Horizontal: 1,000 dpi @ 25 mm/s, Vertical: 200 dpi

Electrical power:

- Power supply: AC or battery operation
- Voltage: 100–240 VAC +10, –15%
- Current: 0.5A @ 115 VAC, 0.3 A @ 240 VAC, typical
- Mains Frequency: 50–60 Hz ± 10%
- Battery type: User replaceable, 18 V @ 3.5 AH ± 15%, rechargeable NiMH

A3.2.4 Typical Performance Requirements for Cardiac Monitors

Cardiac monitors, with or without heart rate meters and alarms, are intended primarily for detecting cardiac rhythm and are covered by the ANSI/AAMI standard for cardiac monitors, heart rate meters and alarms – ANSI/AAMI EC 13-2002. A separate AAMI standard EC 38-1998 covers the Ambulatory Electrocardiographs. The objective of these standards was to provide minimum labeling, performance, and safety requirements and to help ensure a reasonable level of clinical efficacy and patient safety in the use of cardiac monitors. With the few exceptions noted below, performance and disclosure requirements in that standard remain appropriate.

In one section, the standard defines performance requirements, specifying a minimum, a maximum or a range of values, as applicable, that must be met. A separate section lists several performance parameters without specifying values; the requirement is for disclosure of the achieved value to the consumer in a standardized manner. For example, a minimum heart rate meter accuracy for irregular rhythms is not specified, but the accuracy of detecting several types of defined ECG complexes must be disclosed. Designation of specifications as either performance requirements or disclosures is appropriate. The reader is referred to standard-10 for further details, including test procedures and rationale. Below is a partial listing of some important parameters in each category:

1. Protection from overload: Protection should be adequate (no damage) for 1 V (peak to peak), 60 Hz, applied for 10 s to any electrode connection. The device should recover within 8 s after defibrillation shocks of up to at least 5,000 V, delivered with energies up to 360 J.
2. Isolated patient connection: The system should include isolated patient connections to meet standards defined in the publication: American National Standard for Safe Current Limits for Electromedical Apparatus: 11.

The American Heart Association has also published its own set of instrumentation and practice standards for electrocardiographic monitoring in coronary care units, intensive care units, telemetry units, surgical suites, emergency rooms, and all other areas in which ECG monitoring functions are performed. These were directed primarily at the cardiac monitors that detect and diagnose arrhythmias and also at those that detect ST segment changes that suggest myocardial ischemia. The AHA felt that specific guidelines for ECG monitors are required, even though much of the technology is similar to other forms of ECG measurement. The clinical environment, including severity and acuteness of illness and immediacy of treatment, combined with limited time for over-reading and editing, and initiation of urgent therapy by non-physicians mandate more critical evaluation of automated arrhythmia detection and diagnostic systems.

A3.2.5 Typical Specifications for the ECG Monitoring Devices

Frequency response	–3 dB @ 0.01–150 Hz
Sampling Frequency	500 samples per second (sps)
Dynamic range	AC differential: ±10 mV
	DC offset: ±320 mV
Resolution	4.88 μV/LSB @ 500 sps
Common mode rejection	>140 dB
Input impedance	>10MΩ@10 Hz, defibrillator protected
Patient leakage current	<10μA
Pace detect	750 μV @ 50 μs
Front-end circuits	See below

To ensure correct operation with typical electrode-skin impedances and typical interference sources, the front-end of an electrocardiograph should meet the following demands:

- Very high common mode input impedance (>100 MΩ at 50/60 Hz)
- High differential mode input impedance (>10 MΩ at 50/60 Hz)
- Equal common mode input impedances for all inputs
- High common mode rejection ratio (>80 dB at 50/60 Hz)

A3.3 Patient Safety Standards

Power-line–operated electromedical equipment, connected to patients for monitoring, may permit accidental flow (leakage) of weak alternating current (AC) through a patient's body to ground. An intracardiac catheter may provide a low-resistance path to ground through the patient's heart and thereby place the patient at risk for electrically induced ventricular tachycardia (VT) or ventricular fibrillation (VF).

In addition to the performance standards, the Safe Current Limits for Electromedical Apparatus (ANSI/AAMI ES1-1993) standards apply to all electrocardiograph designs. It states that: "The electrocardiographic (ECG) or vectorcardiographic apparatus shall be designed so that no more than 50 μA root mean square, from direct current component to the tenth harmonic of the power line frequency shall flow through any patient-connected lead under either normal or single-fault conditions." This raised the limit from 10 to 50 μA [5], the value of the European standard since 1988 [6]. Both the 10-μA standard [7, 8] and the 50-μA standard were based on estimates of the risk of AC-induced VF. However, AC may cause cardiovascular collapse at levels that are below the VF threshold [9–13]. This adverse response to AC was not considered in the selection of either safety standard. Furthermore, safe levels of AC have not been determined in closed-chest humans.

The 10-μA standard was adopted in 1967 to ensure patient safety during cardiac catheterization [14] and pacemaker [15] procedures. The annual number of invasive cardiac procedures in the United States has increased from less than 60,000 when the 10-μA standard was adopted to more than 3 million today. The potential number of adverse outcomes from leakage current increased correspondingly.

Electromedical devices contain electrical isolation circuits and insulation to limit leakage current. Manufacturers continue to comply with the original 10-μA standard, but they may realize substantial cost savings by equipment designs that comply only with the newer 50-μA standard [16]. However, the American Heart Association continues to recommend the 10-μA standard [17–19].

A3.4 Recommendations for the Standardization and Interpretation of the Electrocardiogram

Around 2005, the American Heart Association, the American College of Cardiology Foundation and the Heart Rhythm Society agreed to collaborate on the establishment of a series of recommendations for the standardization and interpretation of the ECG. A number of working groups were set up and over the next few years, various publications emerged on the topic. These were all endorsed by the International Society for Computerized Electrocardiology. The six papers form an important contribution to the field and are recommended as essential reading for anyone interested in the development of diagnostic criteria, particularly for computer assisted interpretation of the ECG.

The titles of the different papers are self explanatory and there is no need to expand on the content. However, it is worth mentioning that the first paper [20] is wide ranging and includes many aspects of ECG recording through electrode positioning to requirements for digital filters. The other papers deal with terminology [21], intraventricular conduction disturbances [22], the ST Segment, T and U Waves, and the QT Interval [23], cardiac chamber hypertrophy [24] and acute ischemia/infarction [25].

It is also relevant to highlight again the guidelines published over 30 years ago in a paper [8] entitled "Recommendations for standardization of leads and their specifications for instruments in electrocardiography and vectorcardiography," which still contains many points that are of relevance in this area.

A3.5 Guidelines

A3.5.1 Heart Rate Variability

A Task Force of the European Society of Cardiology and the North American Society of Pacing and Electrophysiology produced a guideline paper relating to heart rate variability [26]. The paper dealt with standards of measurement, physiological interpretation and clinical use.

A3.5.2 Electrocardiographic Monitoring in Hospital Settings

A scientific statement for the American Heart Association Councils on Cardiovascular Nursing, Clinical Cardiology and Cardiovascular disease in the young, dealing with electrocardiographic monitoring in hospital settings was published in 2004 [27]. This sets out recommendations for the different types of monitoring that are necessary in the hospital environment, ranging from arrhythmia analysis through to ischemia monitoring and QT interval assessment. Different types of ECG lead systems for monitoring are described and recommendations relating to staffing, training and quality improvement are given.

A3.5.3 Recommendations for Ambulatory Electrocardiography

A paper on instrumentation and ambulatory electrocardiography published as a set of recommendations was published in 1985 [28]. Much of the information there is still of relevance although equipment is now predominantly based around digital recording techniques. A follow up paper was published in 1999 [29]. These more recent guidelines review equipment and also deal with assessment of symptoms that may be related to disturbances of rhythm and assessment of risk in patients without symptoms of arrhythmias.

A3.5.4 Exercise Testing

Guidelines for clinical exercise testing laboratories were published in 1995 [30]. These outlined the environment in which exercise testing should be undertaken and discussed equipment requirements, etc. Clinical guidelines were published in 1997 [31] and updated in 2002 [32]. These papers essentially deal with the conditions under which exercise testing is deemed to be appropriate.

A3.5.5 Clinical Competence

The ACC and AHA issued a statement on competence for reporting ECGs and ambulatory ECGs [33]. This publication outlines the diagnostic areas where physicians are expected to have a high degree of competence in reporting resting and ambulatory ECGs.

A3.5.6 Pacemakers/Electrophysiology Testing

One of the earliest guidelines on electrophysiology testing and pacemakers was published in 1984 [34]. This outlined the three position and the five position coding scheme for pacemakers. This was updated in 2002 [35]. A paper on guidelines for implantation of pacemakers, which was published in 2002 [36], contained earlier 1998 guidelines together with updated guidelines.

References

1. Kligfield, P., L.S. Gettes, J.J. Bailey, et al., Recommendations for the standardization and interpretation of the electrocardiogram: Part I: the electrocardiogram and its technology. A scientific statement from the American Heart Association Electrocardiography and Arrhythmias Committee, Council on Clinical Cardiology; the American College of Cardiology Foundation; and the Heart Rhythm Society Endorsed by the International Society for Computerized Electrocardiology. *Circulation*, 2007;**115**(10).
2. Rijnbeek, P.R., J.A. Kors, and M. Witsenburg, Minimum bandwidth requirements for recording of pediatric electrocardiograms. *Circulation*, 2001;**104**(25): 3087–3090.
3. Bailey, J.J., A.S. Berson, A. Garson, Jr., L.G. Horan, P.W. Macfarlane, D.W. Mortara, and C. Zywietz, Recommendations for standardization and specifications in automated electrocardiography: bandwidth and digital signal processing. *Circulation*, 1990;**81**: 730–739.

4. Berson, A.S., Y.K. Francis, B.A. Lau, J.M. Wojick, and H.V. Pipberger, Distortions in infant electrocardiograms caused by inadequate high-frequency response. *Am. Heart J.*, 1977;**93**: 730–734.
5. *Safe Current Limits for Electromedical Apparatus*. 1993, Association for the Advancement of Medical Instrumentation.
6. Medical Electrical Equipment- Part 1: General Requirements for Safety in Collateral Standard: Electromagnetic Compatibility – Requirements and Test. 1993.
7. Pipberger, H.V., R.C. Arzbaecher, A.S. Berson, et al., Amendment of recommendations for standardization of specifications for instruments in electrocardiography and vectorcardiography concerning safety and electrical shock hazards. 1972; Committee of Electrocardiography, American Heart Association. *Circulation*, 1972;**46**: 1–2.
8. Pipberger, H.V., R.C. Arzbaecher, A.S. Berson, et al., Recommendations for standardization of leads and of specifications for instruments in electrocardiography and vectorcardiography. 1975. Committee on Electrocardiography, American Heart Association. *Circulation*, 1975;**52**: 11–31.
9. Green, H., E.B. Raftery, and I.C. Gregory, Ventricular fibrillation threshold of healthy dogs to 50 Hz current in relation to earth leakage currents of electromedical equipment. *Biomed. Eng.*, 1972;**7**: 408–414.
10. Raftery, E., H. Green, and I. Gregory, Disturbances of heart rhythm produced by 50 Hz leakage currents in dogs. *Cardiovasc. Res.*, 1975;**9**: 256–262.
11. Raftery, E.B., H.L. Green, and M.H. Yacoub, Disturbances of heart rhythm produced by 50 Hz leakage currents in human subjects. *Cardiovasc. Res.*, 1975;**9**: 263–265.
12. Roy, O.Z., J.R. Scott, and G.C. Park, 60 Hz ventricular fibrillations and pump failure thresholds versus electrode area. *IEEE Eng. Med. Biol.*, 1976;**BME-23**: 45–48.
13. Graystone, P. and J. Ledsome, Microshock hazards in hospital: fibrillation thresholds: the wrong parameter, in *Digest of the 10th International Conference on Medical and Biologic Engineering*, Dresden, Germany, 1973.
14. Weinberg, D.I., J.L.D. Artley, R.E. Whalen, and H.D. Mcintosh, Electric shock hazards in cardiac catheterization. *Circ. Res.*, 1962;**11**: 1004–1009.
15. Whalen, R.E., C.F. Starmer, and H.D. McIntosh, Electrical hazards associated with cardiac pacemaking. *Ann. NY Acad. Sci.*, 1964;**111**: 922–931.
16. Bruner, J.M.R. and P.F. Leonard, Codes and standards: who makes the rules?, in *Electricity, Safety and the Patient*. Chicago: Year Book Medical Publishers, 1989, pp. 240–279.
17. Laks, M.M., R. Arzbaecher, D. Geselowitz, et al., Will relaxing safe current limits for electromedical equipment increase hazards to patients? *Circulation*, 2000;**102**: 823.
18. Laks, M.M., R. Arzbaecher, J.J. Bailey, et al., Recommendations for safe current limits for electrocardiographs: a statement for healthcare professionals from the Committee of Electrocardiography; American Heart Association. *Circulation*, 1996;**93**: 837–839.
19. Laks, M.M., R. Arzbaecher, J.J. Bailey, et al., Comments on "Special report: recommendations for safe current limits for electrocardiographs". *Circulation*, 1997;**95**: 277–278.
20. Kligfield, P., L.S. Gettes, J.J. Bailey, et al., AHA/ACC/HRS recommendations for the standardization and interpretation of the electrocardiogram. Part I: The electrocardiogram and its technology. A scientific statement From the American Heart Association Electrocardiography and Arrhythmias Committee, Council on Clinical Cardiology; the American College of Cardiology Foundation; and the Heart Rhythm Society (Endorsed by the International Society for Computerized Electrocardiology). *J. Am. Coll. Cardiol.*, 2007;49: 1109–1127, doi:10.1016/j.jacc.2007.01.024.
21. Mason, J.W., E.W. Hancock, and L.S. Gettes, AHA/ACC/HRS recommendations for the standardization and interpretation of the electrocardiogram. Part II: Electrocardiography diagnostic statement list. A scientific statement from the American Heart Association Electrocardiography and Arrhythmias Committee, Council on Clinical Cardiology; the American College of Cardiology Foundation; and the Heart Rhythm Society. (Endorsed by the International Society for Computerized Electrocardiology). *J. Am. Coll. Cardiol.*, 2007;**49**: 1128–1135, doi:10.1016/j.jacc.2007.01.025.
22. Surawicz, B., R. Childers, B.J. Deal, and L.S. Gettes, AHA/ACC/HRS recommendations for the standardization and interpretation of the electrocardiogram. Part III: Intraventricular conduction disturbances. A scientific statement from the American Heart Association Electrocardiography and Arrhythmias Committee, Council on Clinical Cardiology; the American College of Cardiology Foundation; and the Heart Rhythm Society. (Endorsed by the International Society for Computerized Electrocardiology). *J. Am. Coll. Cardiol.*, 2009;**53**: 976–981, doi:10.1016/j.jacc.2008.12.013.
23. Rautaharju, P.M., B. Surawicz, and L.S. Gettes, AHA/ACCF/HRS recommendations for the standardization and interpretation of the electrocardiogram. Part IV: The ST segment, T and U waves, and the QT interval. A scientific statement from the American Heart Association Electrocardiography and Arrhythmias Committee, Council on Clinical Cardiology; the American College of Cardiology Foundation; and the Heart Rhythm Society. (Endorsed by the International Society for Computerized Electrocardiology). *J. Am. Coll. Cardiol.*, 2009;**53**: 982–991, doi:10.1016/j.jacc.2008.12.014.
24. Hancock, E.W., B.J. Deal, D.M. Mirvis, et al., AHA/ACCF/HRS recommendations for the standardization and interpretation of the electrocardiogram. Part V: Electrocardiogram changes associated with cardiac chamber hypertrophy. A scientific statement from the American Heart Association Electrocardiography and Arrhythmias Committee, Council on Clinical Cardiology; the American College of Cardiology Foundation; and the Heart Rhythm Society. (Endorsed by the International Society for Computerized Electrocardiology). *J. Am. Coll. Cardiol.*, 2009;**53**: 992–1002, doi:10.1016/j.jacc.2008.12.015.
25. Wagner, G.S., P.W. Macfarlane, H. Wellens, et al., AHA/ACCF/HRS recommendations for the standardization and interpretation of the electrocardiogram. Part VI: Acute ischemia/infarction. A scientific statement from the American Heart Association Electrocardiography and Arrhythmias Committee, Council on Clinical Cardiology; the American College of Cardiology Foundation; and the Heart Rhythm Society Endorsed by the International Society for Computerized Electrocardiology. *J. Am. Coll. Cardiol.*, 2009;**53**: 1003–1011, doi:10.1016/j.jacc.2008.12.016.
26. Task Force of the European Society of Cardiology and the North American Society of Pacing Electrophysiology, Heart rate variability. Standards of measurement, physiological interpretation, and clinical use. *Circulation*, 1996;**93**: 1043–1065.

27. Drew, B., R.M. Califf, M. Funk, et al., AHA scientific statement. Practice standards for electrocardiographic monitoring in hospital settings. *Circulation*, 2004;**110**: 2721–2746.
28. Sheffield, L.T., A. Berson, and D. Bragg-Remschel, Recommendations for standards and instrumentation and practice in the use of ambulatory electrocardiography. The Task force of the Committee on Electrocardiography and Cardiac Electrophysiology of the Council on Clinical Cardiology. *Circulation*, 1985;**71**: 626A–636A.
29. Crawford, M.H., S.J. Bernstein, P.C. Deedwania, et al., ACC/AHA guidelines for ambulatory electrocardiography. Executive summary and recommendations: A report of the American College of Cardiology/American Heart Association Task Force on Practice Guidelines (Committee to Revise the Guidelines for Ambulatory Electrocardiography). *Circulation*, 1999;**100**: 886–893.
30. Pina, I.L., G.J. Balady, P. Hanson, A.J. Labovitz, D.W. Madonna, and J. Myers, Guidelines for clinical exercise testing laboratories. A statement for healthcare professionals from the Committee on Exercise and Cardiac Rehabilitation, American Heart Association. *Circulation*, 1995;**91**: 912–921.
31. Gibbons, R., G.J. Balady, J.W. Beasley, et al., ACC/AHA guidelines for exercise testing. A report of the American College of Cardiology/American Heart Association Task Force on Practice Guidelines (Committee on Exercise Testing). *J. Am. Coll. Cardiol.*, 1997;**30**: 260–311.
32. Gibbons, R., G.J. Balady, J.T. Bricker, et al., ACC/AHA 2002 guideline update for exercise testing: summary article. A report of the American College of Cardiology/American Heart Association Task Force on Practice Guidelines (Committee to Update the 1997 Exercise testing Guidelines). *Circulation*, 2002;**106**: 1883–1892.
33. Kadish, A.H., A.E. Buxton, H.L. Kennedy, B.P. Knight, C.D. Schuger, and C.M. Tracy, A report of the ACC/AHA/ACP-ASIM task force on clinical competence (ACC/AHA Committee to develop a clinical competence statement on electrocardiography and ambulatory electrocardiography). *Circulation*, 2001;**104**: 3169–3178.
34. Gettes, L.S., D.P. Zipes, P.C. Gillette, et al., Personnel and equipment required for electrophysiologic testing. Report of the committee on electrocardiography and cardiac electrophysiology, Council on Clinical Cardiology, the American Heart Association. *Circulation*, 1984;**69**: 1219A–1221A.
35. Bernstein, A.D., J.-C. Daubert, R.D. Fletcher, et al., The revised NASPE/BPEG generic code for antibradycardia, adaptive-rate, and multisite pacing. *PACE*, 2002;**25**: 260–264.
36. Gregoratis, G., J. Abrams, A.E. Epstein, et al., ACC/AHA/NASPE 2002 guideline update for implantation of cardiac pacemakers and antiarrhythmia devices: Summary article. A report of the American College of Cardiology/American Heart Association Task Force on Practice Guidelines (ACC/AHA/NASPE Committee to Update the 1998 Pacemaker Guidelines). *Circulation*, 2002;**106**: 2145–2161.

Appendix 4: Coding Schemes

P. W. Macfarlane, A. van Oosterom, O. Pahlm, P. Kligfield, M. Janse, J. Camm (eds.), *Comprehensive Electrocardiology*, DOI 10.1007/978-1-84882-046-3_A4,

A4.1 The Minnesota Code

The Minnesota code was initially developed and published in 1960 (see Ref. [20] in (➔ Chap. 13). It remains the most widely used ECG coding scheme in epidemiological practice and has recently been revised and extended. The following section has been reprinted from: Prineas RJ, Crow RS, Zhang Z-M. The Minnesota Code Manual of Electrocardiographic Findings. 2009, with the permission of Springer, New York.

A4.1.1 Minnesota Code 2009

A4.1.1.1 Q and QS Patterns

(Do not code in the presence of Wolff-Parkinson-White (WPW) code 6-4-1), or artificial pacemaker code 6-8 or code 6-1, 8-2-1, 8-2-2, or 8-4-1 with a heart rate ≥ 140. To qualify as a Q wave, the deflection should be at least 0.1 mV (1 mm in amplitude).

Anterolateral site (leads I, aVL, V_6)

- 1-1-1 Q/R amplitude ratio ≥1/3, plus Q duration ≥0.03 s in lead I or V_6.
- 1-1-2 Q duration ≥0.04 s in lead I or V_6.
- 1-1-3 Q duration ≥0.04 s, plus R amplitude ≥3 mm in lead aVL.
- 1-2-1 Q/R amplitude ratio ≥1/3, plus Q duration ≥0.02 s and <0.03 s in lead I or V_6.
- 1-2-2 Q duration ≥0.03 s and <0.04 s in lead I or V_6.
- 1-2-3 QS pattern in lead I. Do not code in the presence of 7-1-1.
- 1-2-8 Initial R amplitude decreasing to 2 mm or less in every beat (and absence of codes 3-2, 7-1-1, 7-2-1 or 7-3) between V_5 and V_6. (All beats in lead V_5 must have an initial R >2 mm.)
- 1-3-1 Q/R amplitude ratio ≥1/5 and <1/3, plus Q duration ≥0.02 s and <0.03 s in lead I or V_6.
- 1-3-3 Q duration ≥0.03 s and <0.04 s, plus R amplitude ≥3 mm in lead aVL.
- 1-3-8[1] Initial R amplitude decreasing to 2 mm or less in every beat (and absence of codes 3-2, 7-1-1, 7-2-1, or 7-3) between V_5 and V_6 (All beats in lead V_5 must have an initial R > 2 mm.)

Posterior (inferior) site (leads II, III, aVF).

- 1-1-1 Q/R amplitude ratio ≥1/3, plus Q duration ≥0.03 s in lead II.
- 1-1-2 Q duration ≥0.04 s in lead II.
- 1-1-4 Q duration ≥0.05 s in lead III, plus a Q-wave amplitude ≥1.0 mm in the majority of beats in lead aVF.
- 1-1-5 Q duration ≥0.05 s in lead aVF.
- 1-2-1 Q/R amplitude ratio ≥1/3, plus Q duration ≥0.02 s and<0.03 s in lead II.
- 1-2-2 Q duration ≥0.03 s and <0.04 s in lead II.
- 1-2-3 QS pattern in lead II. Do not code in the presence of 7-1-1.
- 1-2-4 Q duration ≥0.04 s and <0.05 s in lead III, plus a Q wave ≥1.0 mm amplitude in the majority of beats in aVF.
- 1-2-5 Q duration ≥0.04 s and <0.05 s in lead aVF.
- 1-3-1 Q/R amplitude ratio ≥1/5 and <1/3, plus Q duration ≥0.02 s and <0.03 s in lead II.
- 1-3-4 Q duration ≥0.03 s and <0.04 s in lead III, plus a Q wave ≥1.0 mm amplitude in the majority of beats in lead aVF.
- 1-3-5 Q duration ≥0.03 s and <0.04 s in lead aVF.
- 1-3-6 QS pattern in each of leads III and aVF. (Do not code in the presence of 7-1-1.)
- 1-3-7[2] QS pattern in lead a aVF only. (Do not code in the presence of 7-1-1)

Anterior site (leads V_1, V_2, V_3, V_4, V_5)

- 1-1-1 Q/R amplitude ratio ≥1/3 plus Q duration ≥0.03 s in any of leads V_2, V_3,V_4,V_5.
- 1-1-2 Q duration ≥0.04 s in any of leads V_1, V_2, V_3, V_4, V_5.
- 1-1-6 QS pattern when initial R wave is present in adjacent lead to the right on the chest, in any of leads V_2, V_3, V_4, V_5, V_6.
- 1-1-7 QS pattern in all of leads V_1–V_4 or V_1–V_5.
- 1-2-1 Q/R amplitude ratio ≥1/3, plus Q duration ≥0.02 s and <0.03 s, in any leads V_2, V_3, V_4, V_5.
- 1-2-2 Q duration ≥0.03 s and <0.04 s in any leads V_2, V_3, V_4, V_5.
- 1-2-7 QS pattern in all of lead V_1, V_2, and V_3. (Do not code in the presence of 7-1-1.)
- 1-3-1 Q/R amplitude ratio ≥1/5 and <1/3 plus Q duration ≥0.02 s and <0.03 s in any of leads V_2, V_3, V_4, V_5.
- 1-3-2 QS pattern in lead V_1 and V_2. (Do not code in the presence of 3-1 or 7-1-1.)
- 1-3-8[1] Initial R amplitude decreasing to 2.0 mm or less in every beat (and absence of codes 3-2, 7-1-1, 7-2-1, or 7-3) between any of leads V_2 and V_3, V_3, and V_4, or V_4 and V_5. (All beats in the lead immediately to the right on the chest must have an initial R > 2 mm.)

A4.1.1.2 QRS Axis Deviation

(Do not code in presence of low-voltage QRS code 9-1, WPW 6-4-1, artificial pacemaker code 6-8, ventricular conduction defects 7-1-1, 7-2-1, 7-4 or 7-8.)

- 2-1 Left. QRS axis from −30° through −90° in leads I, II, III. (The algebraic sum of major positive and major negative QRS waves must be 0 or positive in I, negative in III, and 0 or negative in II).
- 2-2 Right. QRS axis from +120° through −150° in leads I, II, III. (The algebraic sum of major positive and major negative QRS waves must be negative in I, and zero or positive in III, and in I must be one half or more of that in III.)
- 2-3 Right (optional code when 2-2 is not present). QRS axis from +90° through +119°in leads I, II, III. (The algebraic sum of major positive and major negative QRS waves must be zero or negative in I and positive in II and III.)
- 2-4 Extreme axis deviation (usually S1, S2, S3 pattern). QRS axis from −90° through −149° in leads I, II and III. (The algebraic sum of major positive and major negative QRS waves must be negative in each of leads I, II and III.)
- 2-5 Indeterminate axis. QRS axis approximately 90° from the frontal plane. (The algebraic sum of major positive and major negative QRS waves is zero in each of leads I, II and III, or the information from these three leads is incongruous.)

A4.1.1.3 High-Amplitude R Waves

Do not code in the presence of codes 6-4-1, 6-8, 7-1-1, 7-2-1, 7-4, or 7-8.

- 3-1 Left: R amplitude >26 mm in either V_5 or V_6, or R amplitude >20.0 mm in any of leads I, II, III, aVF, or R amplitude >12.0 mm in lead aVL measured only on second to last complete normal beat.

- 3-2 Right: R amplitude ≥5.0 mm and R amplitude ≥S amplitude in the majority of beats in lead V_1, when S amplitude is >R amplitude somewhere to the left on the chest of V_1 (codes 7-3 and 3-2, if criteria for both are present).
- 3-3 Left (optional code when 3-1 is not present): R amplitude >15.0 mm but ≤20.0 mm in lead I, or R amplitude in V_5 or V_6, plus S amplitude in V_1 > 35.0 mm.
- 3-4 Criteria for 3-1 and 3-2 both present.

A4.1.1.4 ST Junction (J) and Segment Depression

(Do not code in the presence of codes 6-4-1, 6-8, 7-1-1, 7-2-1, 7-4, or 7-8. When 4-1, 4-2, or 4-3 is coded, then a 5-code most often must also be assigned except in lead V_1.)

Anterolateral site (leads I, aVL, V_6)

- 4-1-1 STJ depression ≥2.0 mm and ST segment horizontal or downward sloping in any of leads I, aVL, or V_6
- 4-1-2 STJ depression ≥1.0 mm but <2.0 mm, and ST segment horizontal or downward sloping in any of leads I, aVL, or V_6
- 4-2 STJ depression ≥0.5 mm and <1.0 mm and ST segment horizontal or downward sloping in any of leads I, aVL, or V_6
- 4-3 No STJ depression as much as 0.5 mm but ST segment downward sloping and segment or T-wave nadir ≥0.5 mm below P-R baseline, in any of leads I, aVL, or V_6
- 4-4 STJ depression ≥1.0 mm and ST segment upward sloping or U-shaped, in any of leads I, aVL, or V_6

Posterior (inferior) site (leads II, III, aVF)

- 4-1-1 STJ depression ≥2.0 mm and ST segment horizontal or downward sloping in lead II or aVF
- 4-1-2 STJ depression ≥1.0 mm but <2.0 mm and ST segment horizontal or downward sloping in lead II or aVF
- 4-2 STJ depression ≥0.5 mm and <1.0 mm and ST segment horizontal or downward sloping in lead II or aVF
- 4-3 No STJ depression as much as 0.5 mm, but ST segment downward sloping and segment or T-wave nadir ≥0.5 mm below P-R baseline in lead II
- 4-4 STJ depression ≥1.0 mm and ST segment upward sloping, or U shaped, in lead II

Anterior site (leads V_1, V_2, V_3, V_4, V_5)

- 4-1-1 STJ depression ≥2.0 mm and ST segment horizontal or downward sloping in any of leads V_1, V_2, V_3, V_4, V_5
- 4-1-2 STJ depression ≥1.0 mm but <2.0 mm and ST segment horizontal or downward sloping in any of leads V_1, V_2, V_3, V_4, V_5
- 4-2 STJ depression ≥0.5 mm and <1.0 mm and ST segment horizontal or downward sloping in any of leads V_1, V_2, V_3, V_4, V_5
- 4-3 No STJ depression as much as 0.5 mm, but ST segment downward sloping and segment or T-wave nadir ≥0.5 mm below P-R baseline in any of leads V_2, V_3, V_4, V_5
- 4-4 STJ depression ≥1.0 mm and ST segment upward sloping or U-shaped in any of leads V_1, V_2, V_3, V_4, V_5

A4.1.1.5 T-Wave Items

(Do not code in the presence of codes 6–4–1, 6-8, 7–1–1, 7–2–1, 7-4, or 7–8.)

Anterolateral site (leads I, aVL, V_6)

- 5-1 T amplitude negative 5.0 mm or more in either of leads I, V_6, or in lead aVL when R amplitude is ≥5.0 mm
- 5-2 T amplitude negative or diphasic (positive–negative or negative–positive type) with negative phase at least 1.0 mm but not as deep as 5.0 mm in lead I or V_6, or in lead aVL when R amplitude is ≥5.0 mm
- 5-3 T amplitude zero (flat), or negative, or diphasic (negative-positive type only) with less than 1.0 mm negative phase in lead I or V_6, or in lead aVL when R amplitude is ≥5.0 mm
- 5-4 T amplitude positive and T/R amplitude ratio <1/20 in any of leads I, aVL, V_6; R-wave amplitude must be ≥10.0 mm

Posterior (inferior) site (leads II, III, aVF)

- 5-1 T amplitude negative 5.0 mm or more in lead II, or in lead aVF when QRS is mainly upright
- 5-2 T amplitude negative or diphasic with negative phase (negative-positive or positive–negative type) at least 1.0 mm but not as deep as 5.0 mm in lead II, or in lead aVF when QRS is mainly upright
- 5-3 T amplitude zero (flat), or negative, or diphasic (negative-positive type only) with less than 1.0 mm negative phase in lead II; not coded in lead aVF
- 5-4 T amplitude positive and T/R amplitude ratio <1/20 in lead II; R-wave amplitude must be ≥10.0 mm

Anterior site (leads V_2, V_3, V_4, V_5)

- 5-1 T amplitude negative 5.0 mm or more in any of leads V_2, V_3, V_4, V_5
- 5-2 T amplitude negative (flat) n any codes, or diphasic (negative–positive or positive–negative type) with negative phase at least 1.0 mm but not as deep as 5.0 mm, in any of leads V_2, V_3, V_4, V_5
- 5-3 T amplitude zero (flat), or negative, or diphasic (negative–positive type only) with less than 1.0 mm negative phase, in any of leads V_3, V_4, V_5
- 5-4 T amplitude positive and T/R amplitude ratio <1/20 in any of leads V_3, V_4, V_5; R-wave amplitude must be ≥10.0 mm

A4.1.1.6 AV Conduction Defect in Codes

- 6-1 Complete (third degree) AV block (permanent or intermittent) in any lead. Atrial and ventricular complexes independent, and atrial rate faster than ventricular rate, with ventricular rate <60.
- 6-2-1 Mobitz type II (occurrence of P wave on time with dropped QRS and T).
- 6-2-2 Partial (second degree) AV block in any lead (2:1 or 3:1 block).
- 6-2-3 Wenckebach's phenomenon (PR interval increasing from beat to beat until QRS and T dropped).
- 6-3 PR(PQ) interval ≥0.22 s in the majority of beats in any of leads I, II, III, aVL, aVF.
- 6-4-1 Wolff-Parkinson-White pattern (WPW), persistent. Sinus P wave. PR interval <0.12 s, plus QRS duration ≥0.12 s, plus R peak duration ≥0.06 s, coexisting in the same beat and present in the majority of beats in any of leads I, II, aVL, V_4, V_5, V_6. (6-4-1 suppresses 1-2-3, 1-2-7, 1-3-2, 1-3-6, 1-3-8, all 3, 4, 5, 7, 9-2, 9-4, 9-5 codes.)

- 6-4-2 WPW pattern, intermittent. WPW pattern in ≤50% of beats in appropriate leads.
- 6-5 Short PR interval. PR interval <0.12 s in all beats of any two of leads I, II, III, aVL, aVF.
- 6-6 Intermittent aberrant atrioventricular conductions. PR > 0.12 s (except in presence of 6-5 or heart rate greater than 100); wide QRS complex >0.12 s; normal P wave when most beats are sinus rhythm. (Do not code in the presence of 6-4-2.)
- 6-8 Artificial pacemaker.

A4.1.1.7 Ventricular Conduction Defect in Codes

- 7-1-1 Complete left bundle branch block (LBBB). (Do not code in presence of 6-1, 6-4-1, 6-8, 8-2-1 or 8-2-2,) QRS duration ≥0.12 s in a majority of beats (of the same QRS pattern) in any of leads I, II, III, aVL, aVF, ***plus*** R peak duration ≥0.06 s in a majority of beats (of the same QRS pattern) in any of leads I, II aVL, V_5, V_6 (7-1-1 suppresses 1-2-3, 1-2-7, 1-2-8, 1-3-2, 1-3-6, all 2, 3, 4, 5, 9-2, 9-4, 9-5 codes. If any other codable Q wave coexists with the LBBB pattern, code the Q and diminish the 7-1-1 code to a 7-4 code.)
- 7-1-2 Intermittent LBBB. Same as 7-1-1 but with presence of normally conducted QRS complexes of different shape than the LBBB pattern.
- 7-2-1 Complete right bundle branch block (RBBB). (Do not code in the presence of 6-1, 6-4-1, 6-8, 8-2-1 or 8-2-2.) QRS duration ≥0.12 s in a majority of beats (of the same QRS pattern) in any of leads I, II, III, aVL, aVF, ***plus***: R′ > R in V_1 or QRS mainly upright, ***plus*** R peak duration ≥0.06 s in V_1 or V_2; or S duration > R duration in all beats in lead I or II. (Suppresses 1-2-8 + 1-3-8, all 2-, 3-, 4- and 5- codes, 9-2, 9-4, 9-5.)
- 7-2-2 Intermittent RBBB. Same as 7-2-1 but with presence of normally conducted QRS complexes of different shape than the RBBB pattern.
- 7-3 Incomplete right bundle branch block. QRS duration <0.12 s in each of leads I, II, III, aVL, aVF, and R′ > R in either of leads V_1, V_2. (Code as 3-2 in addition if those criteria are met. 7-3 suppresses code 1-2-8.)
- 7-4 Intraventricular block. QRS duration ≥0.12 s in a majority of beats in any of leads I, II, III, aVL. (7-4 suppresses all 2, 3, 4, 5, 9-2, 9-4, 9-5 codes.)
- 7-5 R-R′ pattern in either of leads V_1, V_2 with R′ amplitude ≤ R.
- 7-6 Incomplete LBBB. (Do not code in the presence of any codable Q or QS wave.) QRS duration ≥0.10 and <0.12 s in the majority of beats of each of leads I, aVL, and V_5 or V_6.
- 7-7 Left anterior hemiblock (LAH). QRS duration <0.12 s in the majority of beats in leads I, II, III, aVL, aVF, ***plus*** Q-wave amplitude ≥0.25 mm and <0.03 s duration in lead I, ***plus*** left axis deviation of $-45°$ or more negative. (In presence of 7-2, code 7-8 if axis is $<-45°$ and the Q wave in lead I meets the above criteria).
- 7-8 Combination of 7-7 and 7-2.
- 7-9-1[2] Type 1 Brugada pattern convex (coved) ST segment elevation ≥ 2 mm ***plus*** T-wave negative with little or no isoelectric (baseline) separation in at least 2 leads of $V_1 - V_3$.
- 7-9-2[2] Type 2 Brugada pattern ST segment elevation ≥ 2 mm ***plus*** T-wave positive or diphasic that results in a "saddle-back" shape in at least 2 leads of $V_1 - V_3$.
- 7-9-3[2] Type 3 Brugada pattern. 7-2-1 ***plus*** ST segment elevation ≥ 1 mm ***plus*** a "saddle-back" configuration in at least 2 leads of $V_1 - V_3$.
- 7-10[2] Fragmented QRS.

A4.1.1.8 Arrhythmias

- 8-1-1 Presence of frequent atrial or junctional premature beats (10% or more of recorded complexes).
- 8-1-2 Presence of frequent ventricular premature beats (10% or more of recorded complexes).

- 8-1-3 Presence of both atrial and/or junctional premature beats and ventricular premature beats (so that individual frequencies are <10% but *combined* premature beats are ≥10% of complexes).
- 8-1-4 Wandering atrial pacemaker.
- 8-1-5 Presence of 8-1-2 and 8-1-4.
- 8-2-1 Ventricular fibrillation or ventricular asystole.
- 8-2-2 Persistent ventricular (idioventricular) rhythm.
- 8-2-3 Intermittent ventricular tachycardia. Three or more consecutive ventricular premature beats occurring at a rate ≥100. This includes more persistent ventricular tachycardia.
- 8-2-4 Ventricular parasystole (should not be coded in presence of 8-3-1).
- 8-3-1 Atrial fibrillation (persistent).
- 8-3-2 Atrial flutter (persistent).
- 8-3-3 Intermittent atrial fibrillation (code if 3 or more clear-cut, consecutive sinus beats are present in any lead).
- 8-3-4 Intermittent atrial flutter (code if 3 or more clear-cut, consecutive sinus beats are present in any lead).
- 8-4-1 Supraventricular rhythm persistent. QRS duration <0.12 s; and absent P waves or presence of abnormal P waves (inverted or flat in aVF); and regular rhythm.
- 8-4-2 Supraventricular tachycardia intermittent. Three consecutive atrial or junctional premature beats occurring at a rate ≥100 min^{-1}.
- 8-5-1 Sinoatrial arrest. Unexpected absence of P, QRS and T, plus a R-R interval at a fixed multiple of the normal interval, ±10%.
- 8-5-2 Sinoatrial block. Unexpected absence of P, QRS and T, preceded by progressive shortening of P-P intervals, (R-R interval at a fixed multiple of the normal interval, ±10%).
- 8-6-1 AV dissociation with ventricular pacemaker (without capture). Requires: P-P and R-R to occur at variable rates with ventricular rate as fast as or faster than the atrial rate plus variable PR intervals, plus no capture beats.
- 8-6-2 AV dissociation with ventricular pacemaker (with capture).
- 8-6-3 AV dissociation with atrial pacemaker (without capture).
- 8-6-4 AV dissociation with atrial pacemaker (with capture).
- 8-7 Sinus tachycardia (≥ 100 min^{-1}).
- 8-8 Sinus bradycardia (≤ 50 min^{-1}).
- 8-9 Other arrhythmias. Heart rate may be recorded as a continuous variable.

A4.1.1.9 ST-Segment Elevation

Do not code in the presence of codes 6-4-1, 6-8, 7-1-1, 7-2-1, 7-4, or 7-8.

Anterolateral site (leads I, aVL, V_6)

- 9-2 ST-segment elevation ≥1.0 mm in any of leads I, aVL, V_6.

Posterior (inferior) site (leads II, III, aVF)

- 9 2 ST-segment elevation ≥1.0 mm in any of leads II, III, aVF.

Anterior site (leads V_1, V_2, V_3, V_4, V_5)

- 9-2 ST-segment elevation ≥1.0 mm in lead V_5 or ST-segment elevation ≥2.0 mm in any of leads V_1, V_2, V_3, V_4.

A4.1.1.10 Miscellaneous Items

- 9-1 Low QRS amplitude. QRS peak-to-peak amplitude <5 mm in all beats in each of leads I, II, III, or <10 mm in all beats in each of leads V_1, V_2, V_3, V_4, V_5, V_6. (Check calibration before coding).
- 9-3 P-wave amplitude ≥2.5 mm in any of leads II, III, aVF, in a majority of beats.
- 9-4-1 QRS transition zone at V_3 or to the right of V_3 on the chest. (Do not code in the presence of 6-4-1, 6-8, 7-1-1, 7-2-1, 7-4, or 7-8.)
- 9-4-2 QRS transition zone at V_4 or to the left of V_4 on the chest. (Do not code in the presence of 6-4-1, 6-8, 7-1-1, 7-2-1, 7-4, or 7-8.)
- 9-5 T-Wave amplitude >12 mm in any of leads I, II, III, aVL, aVF, V_1, V_2, V_3, V_4, V_5, V_6. (Do not code in the presence of 6-4-1, 6-8, 7-1-1, 7-2-1, 7-4, or 7-8.
- 9-6[2] Notched and widened P wave (duration ≥ 0.12 s) in frontal plane (usually lead II), and/or deep negative component to the P wave in lead V_1 duration ≥ 0.04 s and depth ≥ 1 mm.
- 9-7-1[2] Definite Early Repolarization. STJ elevation ≥ 1 mm in the majority of beats, T wave amplitude ≥ 5 mm prominent J point, upward concavity of the ST segment, and a distinct notch or slur on the down-stroke of the R wave in any of $V_3 - V_6$, OR STJ elevation ≥ 2 mm in the majority of beats and T wave amplitude ≥ 5 mm prominent J point, and upward concavity of the ST segment in any of $V_3 - V_6$.
- 9-7-2[2] Probable Early Repolarization. STJ elevation ≥ 1 mm in the majority of beats, prominent J point, and upward concavity of the ST segment in any of $V_3 - V_6$ and T wave amplitude ≥ 8 mm in any of the leads $V_3 - V_6$.
- 9-8-1[2] Uncorrectable lead reversal.
- 9-8-2[3] Poor Quality/Technical problems which interfere with coding.
- 9-8-3[2] Correctable lead reversal.
  - i. Correctable limb lead connection error.
  - ii. Correctable chest lead connection error in $V_1 - V_3$.
  - iii. Correctable chest lead connection error in $V_4 - V_6$.
  - iv. Correctable other chest lead connection error.
- 9-8-4[3] Technical problems that do not interfere with coding.

A4.1.1.11 Incompatible Codes

Table A4.1 gives a list of incompatible codes. The codes in the left-hand column suppress the codes in the right-hand column.

A4.1.1.12 ECG Criteria for Significant Serial ECG Change

A detailed explanation of criteria for serial change can be found in Chapter 15 of the recently published Minnesota Code Manual of Electrocardiographic Findings (RJ Prineas, RS Crow, Z-M Zhang, Springer, 2009). An extract is given here.

Evolving Q-wave

Q1. No Q-code in reference ECG followed by a record with a diagnostic Q-code (MC 1-1-1 through 1-2-7) ***OR*** an Equivocal Q-code (1-3-x) in reference ECG followed by record with any code 1-1-x Q-code.

---

[1] 1-3-8 was previously 1-2-8

[2] New code from first edition

[3] 9-8-2 in the first edition was 9-8-1, and 9-8-4 was 9-8-2 in the first edition.

Table A4.1
Incompatible codes

Code	Suppresses this code(s)
All Q, QS codes	7-6
Q ≥ 0.03 in Lead I	7-7
3-1	1-3-2
3-2	1-3-8, 7-3
6-1	All other codes except 8-2
6-4-1	All other codes
6-8	All other codes
7-1-1	1-2-3, 1-2-7, 1-3-2, 1-3-6, 1-3-7, 1-3-8, all 2-, 3-, 4-, and 5-codes, 7-7, 7-8, 7-9, 7-10, 9-2, 9-4, 9-5, 9-7-1, 9-7-2
7-2-1	1-3-8, all 2-, 3-, 4-, and 5-codes, 9-2, 9-4, 9-5, 9-7-1, 9-7-2
7-3	1-3-8
7-4	All 2-, 3-, 4-, and 5-codes, 9-2, 9-4, 9-5
7-8	1-3-8, all 2-, 3-, 4-, and 5-codes, 9-2, 9-4, 9-5, 9-7-1, 9-7-2
8-1-2	8-2-4
8-1-4	8-1-1, 9-3
8-2-1	All other codes
8-2-2	All other codes
8-2-3	8-1-2
8-3-1	8-1-1, 8-1-2
8-3-2	6-2-2, 8-1-1, 8-1-2
8-3-3	8-1-1, 8-1-2
8-3-4	6-2-2
8-4-1	6-5
8-4-1 + heart rate ≥140 bpm	All other codes except 7-4 or 6-2
Heart rate >100 bpm	6-5
8-4-2	8-1-1
9-1	All 2-codes

Q2. An Equivocal Q-code (any MC 1-3 x code) and no major ST-segment depression (MC 4-0, 4-4, 4-3) in reference ECG followed by a record with a diagnostic Q-code (MC 1-2-1 – 1-2-7) ***Plus*** a major ST-segment depression (MC 4-1-x or 4-2).

Q3. An Equivocal Q-code (any MC 1-3-x) and no major T-wave inversion (MC 5–4, 5-3 or 5-0) in reference ECG followed by a record with a diagnostic Q-code (MC1-2-1 through 1-2-7) ***Plus*** a major T-wave inversion (MC 5-1 or 5-2).

Q4. An Equivocal Q-code (any MC 1-3-x) and Q-code (MC 1-2-1 through 1-2-7) ***Plus*** and ST-segment elevation (MC 9-2).

Q5. No Q-code and no MC 4-1-x or 4-2 in reference ECG followed by a record with an Equivocal Q-code (any MC 1-3-x) ***Plus*** MC 4-1-x or 4-2.

Q6. No Q-code and no MC 5-1 or 5-2 in reference ECG followed by a record with an Equivocal Q-code (any MC 1-3-x) ***Plus*** a MC 5-1 or 5-2.

Q7. No Q-code and no MC 9-2 in reference ECG followed by a record with an Equivocal Q-code (any MC 1-3-x) ***Plus*** a MC 9-2.

Evolving ST-Elevation

STE-1 MC 9-0 in reference ECG followed by a record with MC 9-2 in at least 2 leads and > 100% increase ST elevation in both leads.

ŞTE-2 MC 9-2 in reference ECG followed by a record with MC 9-2 in at least 2 leads and > 100% increase in ST elevation in both leads.

STE-3 MC 9-2 and no MC 5-1 or 5-2 in reference ECG followed by a record appearance of MC 5-1 or 5-2 with 100% increase in T wave inversion in at least 2 leads.

STE-4 Reversal of evolving STE-1 (within the hospital ECG only).

STE-5 Reversal of evolving STE-2 (within the hospital ECG only).

Evolving ST-Depression/T Wave Inversion

ST-T1 Either MC 4-0 (no 4-code), 4–4 or 4–3 in reference ECG followed by a record with MC 4–2 or 4–1-2 or 4-1-1 and > 100% increase in ST segment depression.

ST-T2 Either MC 4-2 4–1–2 in reference ECG followed by a record with MC 4–1–1 and > 100% increase in ST segment depression.

ST-T3 Either MC 5-0, 5-4 or 5-3 in reference ECG followed by a record with MC 5-2 or 5-1 and > 100% increase in T-wave inversion.

ST-T4 MC 5-2 in reference ECG followed by a record with MC 5-1 and > 100% in T-wave inversion.

ST-T5 MC 4–1-1 in reference ECG followed by a record with MC 4–1–1 and > 100% increase in ST depression.

ST-T6 MC 5-1 in reference ECG followed by a record with MC 5-1 and > 100% increase in T-wave inversion

ST-T7 MC 5-2 in reference ECG followed by a record with MC 5-2 and > 100% increase in T-wave inversion.

ST-T1R Reverse of ST-T1[4]

ST-T2R Reverse of ST-T2[4]

ST-T3R Reverse of ST-T3[4]

ST-T4R Reverse of ST-T4[4]

ST-T5R Reverse of ST-T5[4]

ST-T6R Reverse of ST-T6[4]

ST-T7R Reverse of ST-T7[4]

Evolving Bundle Branch Block

E-BBB1 No MC 7-1 in the reference ECG followed by an ECG with MC 7-1-1 in follow-up ECG ***and*** QRS duration increase by > 0.02 s.

E-BBB2 No MC 7-2 in the reference ECG followed by an ECG with MC 7-2-1 in follow-up ECG ***and*** QRS duration increase by > 0.02 s.

E-BBB3 No MC 7-4 in the reference ECG followed by an ECG with MC 7–4 in follow-up ECG ***and*** QRS duration increase by > 0.02 s.

Evolving ECG – LVH

E-LVH 1 MC 3-0 in reference ECG flowed by an ECG with a MC 3-1 in the follow-up ECG, confirmed as a significant increase.

[4]Requires > 100% decrease in ST depression or T-wave inversion of follow-up record compared to reference ECG, and code changes must occur in the same lead groups.

E-LVH 2 MC 3-0 in reference ECG flowed by an ECG with a MC 3-3 in the follow-up ECG, confirmed as a significant increase.

E-LVH 3 MC 3-1 in reference ECG flowed by an ECG with a MC 3-0 in the follow-up ECG, confirmed as a significant decrease.

E-LVH 4 MC 3-3 in reference ECG flowed by an ECG with a MC 3-0 in the follow-up ECG, confirmed as a significant decrease.

E-LVH 5 MC 3-1 in reference ECG flowed by an ECG with a MC 3-1 in the follow-up ECG, confirmed by a significant increase or a significant decrease.

E-LVH 6 MC 3-3 in reference ECG flowed by an ECG with a MC 3-3 in the follow-up ECG, confirmed by a significant increase or a significant decrease.

A4.2 The Punsar Code

One of the authors of the original publication of the Minnesota code, Punsar, developed an alternative scheme, in collaboration with others, for classifying the ST-T segment. The code is described simply in ➔ Table A4.2 and ➔ Fig. A4.1 which is reproduced from: Punsar S, Pyorala K, Siltanen P. Classification of electrocardiographic ST segment changes in epidemiological studies of coronary heart disease. Ann. Med. Intern. Fenn. 1968; **57**:53–63, with the permission of Annales Medicinae Internae Fenniae, Helsinki.
The authors tested their code in a 5-year follow-up of 1,534 men aged 40–59. The incidence of events including death or myocardial infarction was highest in the ischemic group and decreased in the remaining groups in a progressive fashion. It was also noted that with each group, the prognosis varied according to the amount of ST depression present.

Table A4.2
Correspondence of the items in the new, modified code to those in the Minnesota code

Modified code		Minnesota code			
		At rest		Postexercise[a]	
I	1	IV	1	XI	1
	2		1		1
	3		2		2
	4		3		3
	5				
S	1	IV	4	XI	4
	2		4		4
	3				
	4				
R	1	IV	4		4
	2		4	XI	4
	3				

[a]The original 1960 Minnesota code had postexercise classifications, coded X to XVI

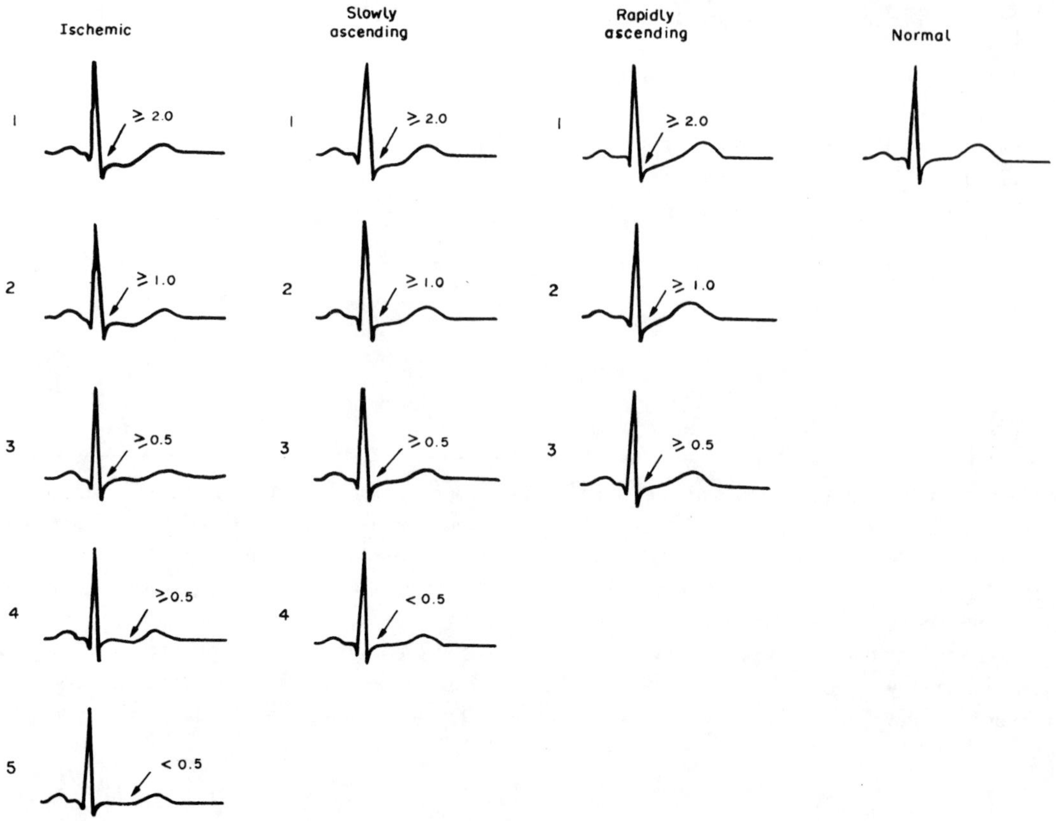

■ Figure A4.1
Categories of ST-segment changes in the new, modified classification (the Punsar code). Note that the figures beside each arrow indicate the amount of ST depression in millimeters.

Appendix 5: Normal Limits of the 12 Lead Vectorcardiogram

P. W. Macfarlane, A. van Oosterom, O. Pahlm, P. Kligfield, M. Janse, J. Camm (eds.), *Comprehensive Electrocardiology*, DOI 10.1007/978-1-84882-046-3_A5

Appendix 5A Normal Limits of Adult 12-Lead Vectorcardiogram

The tables of amplitudes and durations in this Appendix have been obtained from the X, Y, Z leads derived from the 12-lead ECG according to the methods described in Chap. 11, Sect. 11.6 and using the coefficients presented in Table 11.6. Data have been derived from 1,555 adult ECGs whose age distribution is shown in the lower part of Table 43.1. The 215 males over 50 have been subdivided, with 38 men being aged 60 and over, while similarly 139 females include 19 aged 60 and over.

Data are expressed as mean ± standard deviation below which is given the 96 percentile ranges, i.e., 2% of values are excluded at either end of the distribution except in the case of the groups aged 60 and over where the 100% range is presented. Angular data are presented with respect to the reference frames illustrated in Figure 43.9.

A5.1 Scalar Measurements from the Leads X, Y, and Z

A5.1.1 P Wave Amplitude and Duration

Table A5.1
P wave amplitudes (mV) in males

Age (years)	X	Y	Z
18–29	0.07 ± 0.02	0.11 ± 0.05	0.05 ± 0.02
	0.03 → 0.12	0.02 → 0.25	0.02 → 0.10
30–39	0.07 ± 0.02	0.11 ± 0.04	0.04 ± 0.02
	0.03 → 0.12	0.01 → 0.22	0.01 → 0.11
40–49	0.07 ± 0.02	0.11 ± 0.04	0.04 ± 0.02
	0.03 → 0.14	0.03 → 0.22	0.00 → 0.09
50–59	0.07 ± 0.02	0.11 ± 0.04	0.04 ± 0.02
	0.04 → 0.13	0.04 → 0.20	0.01 → 0.08
60–	0.07 ± 0.03	0.11 ± 0.04	0.04 ± 0.03
	0.03 → 0.11	0.04 → 0.18	0.00 → 0.08

Table A5.2
P wave amplitudes (mV) in females

Age (years)	X	Y	Z
18–29	0.07 ± 0.02	0.11 ± 0.05	0.04 ± 0.02
	0.03 → 0.13	0.02 → 0.26	0.01 → 0.08
30–39	0.08 ± 0.02	0.12 ± 0.05	0.04 ± 0.02
	0.04 → 0.13	0.03 → 0.24	0.01 → 0.09
40–49	0.08 ± 0.02	0.11 ± 0.04	0.04 ± 0.02
	0.04 → 0.12	0.03 → 0.24	0.01 → 0.08
50–59	0.08 ± 0.02	0.11 ± 0.04	0.04 ± 0.02
	0.04 → 0.13	0.03 → 0.22	0.00 → 0.07
60–	0.08 ± 0.03	0.12 ± 0.06	0.04 ± 0.01
	0.03 → 0.14	0.03 → 0.23	0.01 → 0.06

Table A5.3
P wave durations (ms) in males

Age (years)	X	Y	Z
18–29	93 ± 14	98 ± 14	98 ± 14
	66 → 122	66 → 122	66 → 122
30–39	103 ± 11	103 ± 11	103 ± 11
	78 → 124	78 → 124	78 → 124
40–49	107 ± 11	107 ± 11	107 ± 11
	86 → 128	86 → 128	86 → 128
50–59	108 ± 11	108 ± 11	108 ± 11
	80 → 128	80 → 128	80 → 128
60–	110 ± 12	110 ± 12	110 ± 12
	90 → 130	90 → 130	90 → 130

Table A5.4
P wave durations (ms) in females

Age (years)	X	Y	Z
18–29	97 ± 11	97 ± 11	97 ± 11
	72 → 116	72 → 116	72 → 116
30–39	100 ± 10	100 ± 10	100 ± 10
	78 → 116	78 → 116	78 → 116
40–49	104 ± 10	104 ± 10	104 ± 10
	82 → 124	82 → 124	82 → 124
50–59	104 ± 13	104 ± 13	104 ± 13
	64 → 128	64 → 128	64 → 128
60–	104 ± 12	104 ± 12	104 ± 12
	88 → 130	88 → 130	88 → 130

A5.1.2 Q Wave Amplitude and Duration

Table A5.5
Q wave amplitude (mV) in males

Age (years)	X	Y
18–29	−0.11 ± 0.09	−0.10 ± 0.06
	−0.37 → −0.02	−0.29 → −0.02
30–39	−0.09 ± 0.08	−0.10 ± 0.06
	−0.33 → −0.02	−0.30 → −0.02
40–49	−0.08 ± 0.05	−0.07 ± 0.04
	−0.21 → −0.02	−0.17 → −0.02
50–59	−0.09 ± 0.06	−0.07 ± 0.05
	−0.26 → −0.02	−0.21 → −0.02
60–	−0.07 ± 0.04	−0.08 ± 0.05
	−0.14 → −0.02	−0.21 → −0.02

Table A5.6

Q wave amplitude (mV) in females

Age (years)	X	Y
18–29	−0.08 ± 0.06	−0.10 ± 0.06
	−0.25 → −0.02	−0.29 → −0.02
30–39	−0.09 ± 0.06	−0.08 ± 0.05
	−0.26 → −0.02	−0.20 → −0.02
40–49	−0.07 ± 0.05	−0.06 ± 0.03
	−0.22 → −0.02	−0.13 → −0.02
50–59	−0.06 ± 0.04	−0.06 ± 0.03
	−0.17 → −0.02	−0.16 → −0.02
60–	−0.07 ± 0.04	−0.05 ± 0.03
	−0.14 → −0.02	−0.10 → −0.03

Table A5.7

Q wave durations (ms) in males

Age (years)	X	Y
18–29	17 ± 5	19 ± 7
	8 → 28	8 → 27
30–39	16 ± 6	19 ± 5
	8 → 27	7 → 31
40–49	16 ± 4	17 ± 5
	7 → 23	6 → 25
50–59	16 ± 4	18 ± 5
	10 → 24	6 → 27
60–	15 ± 3	19 ± 5
	10 → 23	5 → 29

Table A5.8

Q wave durations (ms) in females

Age (years)	X	Y
18–29	15 ± 4	17 ± 5
	7 → 25	7 → 26
30–39	16 ± 4	16 ± 4
	7 → 24	6 → 24
40–49	15 ± 3	14 ± 4
	9 → 21	8 → 24
50–59	14 ± 4	16 ± 4
	7 → 22	8 → 24
60–	15 ± 4	13 ± 6
	9 → 24	5 → 21

A5.1.3 R Wave Amplitude and Duration

Table A5.9
R wave amplitudes (mV) in males

Age (years)	X	Y	Z
18–29	1.66 ± 0.46	1.16 ± 0.46	0.52 ± 0.23
	0.08 → 2.86	0.31 → 2.23	0.13 → 1.02
30–39	1.53 ± 0.43	0.91 ± 0.46	0.42 ± 0.24
	0.69 → 2.44	0.14 → 2.13	0.09 → 1.07
40–49	1.40 ± 43	0.69 ± 0.39	0.37 ± 0.19
	0.66 → 2.41	0.04 → 1.51	0.04 → 0.87
50–59	1.38 ± 0.38	0.58 ± 0.34	0.36 ± 0.17
	0.72 → 2.01	0.11 → 1.46	0.10 → 0.76
60–	1.20 ± 0.36	0.65 ± 0.39	0.30 ± 0.19
	0.75 → 1.93	0.08 → 1.50	0.05 → 0.67

Table A5.10
R wave amplitudes (mV) in females

Age (years)	X	Y	Z
18–29	1.23 ± 0.34	0.91 ± 0.34	0.33 ± 0.15
	0.56 → 1.96	0.32 → 1.75	0.07 → 0.70
30–39	1.23 ± 0.37	0.87 ± 0.32	0.32 ± 0.15
	0.58 → 1.97	0.29 → 1.48	0.10 → 0.67
40–49	1.11 ± 0.32	0.66 ± 0.26	0.25 ± 0.12
	0.62 → 1.75	0.24 → 1.16	0.04 → 0.49
50–59	1.11 ± 0.29	0.61 ± 0.31	0.26 ± 0.14
	0.60 → 1.76	0.14 → 1.31	0.04 → 0.73
60–	1.28 ± 0.46	0.55 ± 0.21	0.32 ± 0.14
	0.54 → 2.23	0.25 → 1.16	0.04 → 0.56

Table A5.11
R wave durations (ms) in males

Age (years)	X	Y	Z
18–29	45 ± 15	55 ± 17	32 ±9
	27 → 79	28 → 86	20 → 45
30–39	46 ± 14	57 ± 16	31 ± 9
	28 → 84	26 → 86	17 → 44
40–49	47 ± 14	55 ± 18	32 ± 8
	31 → 82	15 → 90	15 → 48
50–59	46 ± 14	55 ± 17	33 ± 8
	30 → 79	22 → 90	21 → 52
60–	43 ± 11	54 ± 15	29 ± 8
	29 → 62	34 → 86	14 → 40

Table A5.12
R wave durations (ms) in females

Age (years)	X	Y	Z
18–29	44 ± 12	47 ± 13	28 ± 6
	27 → 76	24 → 78	15 → 41
30–39	44 ± 12	48 ± 12	28 ± 5
	31 → 67	31 → 72	17 → 38
40–49	45 ± 11	52 ± 14	27 ± 7
	31 → 71	28 → 86	10 → 41
50–59	44 ± 11	52 ± 15	27 ± 7
	26 → 86	20 → 77	11 → 43
60–	44 ± 11	51 ± 13	32 ± 5
	28 → 64	33 → 72	24 → 46

A5.1.4 S Wave Amplitude and Duration

Table A5.13
S wave amplitudes (mV) in males

Age (years)	X	Y	Z
18–29	−0.27 ± 0.18	−0.16 ± 0.09	−1.38 ± 0.47
	−0.68 → −0.05	−0.41 → −0.03	−2.50 → −0.59
30–39	−0.26 ± 0.20	−0.18 ± 0.16	−1.05 ± 0.37
	−0.71 → −0.05	−0.44 → −0.04	−1.70 → −0.49
40–49	−0.29 ± 0.18	−0.17 ± 0.13	0.37 ± 0.19
	−0.73 → −0.06	−0.44 → −0.03	0.04 → 0.87
50–59	−0.25 ± 0.17	−0.17 ± 0.16	0.36 ± 0.17
	−0.68 → 0.04	−0.60 → −0.05	0.10 → 0.76
60–	−0.35 ± 0.18	−0.18 ± 0.14	−0.81 ± 0.31
	−0.54 → −0.17	−0.28 → −0.06	−1.04 → −0.48

Table A5.14
S wave amplitudes (mV) in females

Age (years)	X	Y	Z
18–29	−0.17 ± 0.10	−0.15 ± 0.09	−0.92 ± 0.34
	−0.52 → −0.05	0.42 → 0.04	−1.65 → −0.35
30–39	−0.20 ± 0.11	−0.14 ± 0.12	−0.92 ± 0.33
	−0.39 → −0.06	−0.36 → −0.05	−1.65 → −0.35
40–49	−0.18 ± 0.14	−0.14 ± 0.11	−0.99 ± 0.39
	−0.34 → −0.04	−0.25 → −0.04	−1.54 → −0.46
50–59	−0.20 ± 0.14	−0.16 ± 0.09	−0.74 ± 0.30
	−0.42 → −0.06	−0.29 → −0.06	−1.32 → 0.34
60–	−0.16 ± 0.10	0.18 ± 0.10	−0.66 ± 0.25
	−0.19 → −0.11	−0.17 → −0.12	−0.79 → −0.50

Table A5.15
S wave durations (ms) in males

Age (years)	X	Y	Z
18–29	26 ± 12	26 ± 10	53 ± 8
	6 → 48	7 → 48	35 → 71
30–39	28 ± 13	29 ± 14	54 ± 10
	7 → 57	8 → 60	31 → 71
40–49	31 ± 12	32 ± 15	52 ± 11
	10 → 55	7 → 68	25 → 70
50–59	32 ± 12	30 ± 15	50 ± 11
	11 → 53	8 → 67	27 → 73
60–	33 ± 12	28 ± 11	53 ± 9
	12 → 56	11 → 50	31 → 67

Table A5.16
S wave durations (ms) in females

Age (years)	X	Y	Z
18–29	24 ± 9	25 ± 9	51 ±9
	7 → 41	10 → 45	30 → 67
30–39	28 ± 9	23 ± 10	52 ± 8
	8 → 43	8 → 43	35 → 64
40–49	26 ± 10	26 ± 10	52 ± 8
	12 → 41	12 → 50	27 → 75
50–59	27 ± 10	27 ± 11	51 ± 9
	12 → 47	11 → 58	30 → 64
60–	29 ± 11	33 ± 15	47 ± 10
	13 → 53	12 → 50	20 → 59

A5.1.5 T Wave Amplitude

Table A5.17
T wave amplitudes (mV) in males

Age (years)	X	Y	Z
18–29	0.46 ± 0.19	0.25 ± 0.11	0.47 ± 0.18
	0.10 → 0.95	0.02 → 0.55	0.14 → 0.86
30–39	0.42 ± 0.16	0.21 ± 0.10	0.42 ± 0.16
	0.12 → 0.80	0.03 → 0.48	0.08 → 0.80
40–49	0.38 ± 0.17	0.19 ± 0.10	0.39 ± 0.15
	0.08 → 0.82	0.03 → 0.43	0.12 → 0.72
50–59	0.35 ± 0.16	0.17 ± 0.08	0.38 ± 0.16
	0.09 → 0.72	0.04 → 0.38	0.07 → 0.74
60–	0.36 ± 0.18	0.21 ± 0.12	0.33 ± 0.17
	0.06 → 0.66	0.08 → 0.38	0.12 → 0.73

Table A5.18
T wave amplitudes (mV) in females

Age (years)	X	Y	Z
18–29	0.34 ± 0.12	0.20 ± 0.08	0.23 ± 0.11
	0.14 → 0.60	0.04 → 0.46	0.04 → 0.56
30–39	0.33 ± 0.13	0.18 ± 0.08	0.22 ± 0.11
	0.12 → 0.64	0.05 → 0.37	0.02 → 0.43
40–49	0.28 ± 0.11	0.16 ± 0.07	0.20 ± 0.09
	0.05 → 0.51	0.05 → 0.33	0.06 → 0.43
50–59	0.27 ± 0.11	0.17 ± 0.07	0.20 ± 0.09
	0.07 → 0.56	0.03 → 0.32	0.03 → 0.36
60–	0.26 ± 0.14	0.17 ± 0.08	0.21 ± 0.08
	0.03 → 0.62	0.10 → 0.30	0.04 → 0.37

A5.2 Planar and Spatial Measurements

A5.2.1 Direction of Inscription in the QRS Vector Loop

Table A5.19
Direction of inscription of the QRS vector loop in males (%)

	Frontal	Left sagittal	Transverse
Counterclockwise	22.9	88.0	98.1
Figure of 8	19.1	8.4	0.9
Clockwise	58.0	3.6	1.0

Table A5.20
Direction of inscription of the QRS vector loop in females (%)

	Frontal	Left sagittal	Transverse
Counterclockwise	21.8	94.5	97.9
Figure of 8	25.3	3.9	1.2
Clockwise	52.9	1.6	0.9

A5.2.2 Magnitude of Maximal Spatial QRS Vector

Table A5.21
Magnitude of maximal spatial QRS vector (mV)

Age (years)	Males	Females
18–29	2.39 ± 0.62	1.76 ± 0.47
	1.07 → 3.97	0.75 → 3.06
30–39	2.07 ± 0.58	1.74 ± 0.46
	1.00 → 3.61	0.92 → 2.77

Table A5.21 (Continued)

Age (years)	Males	Females
40–49	1.79 ± 0.49	1.46 ± 0.41
	0.85 → 2.94	0.79 → 2.26
50–59	1.65 ± 0.45	1.46 ± 0.37
	0.86 → 2.91	0.78 → 2.22
60–	1.57 ± 0.42	1.37 ± 0.51
	0.99 → 2.20	1.00 → 1.90

A5.2.3 Magnitude of Maximal Planar QRS Vector

Table A5.22
Magnitude of the maximal QRS vector in frontal, sagittal, and transverse planes (mV) in males

Age (years)	Frontal	Sagittal	Transverse
18–29	2.04 ± 0.52	1.75 ± 0.60	2.09 ± 0.52
	0.99 → 3.26	0.70 → 3.41	1.02 → 3.39
30–39	1.80 ± 0.49	1.48 ± 0.58	1.85 ± 0.48
	0.82 → 3.03	0.52 → 3.11	0.92 → 2.95
40–49	1.59 ± 0.46	1.23 ± 0.45	1.64 ± 0.43
	0.69 → 2.82	0.45 → 2.26	0.83 → 2.78
50–59	1.50 ± 0.42	1.07 ± 0.40	1.54 ± 0.39
	0.77 → 2.70	0.44 → 2.05	0.84 → 2.76
60–	1.40 ± 0.37	1.12 ± 0.41	1.42 ± 0.36
	0.92 → 2.03	0.64 → 1.91	0.89 → 2.08

Table A5.23
Magnitude of the maximal QRS vector in frontal, sagittal, and transverse planes (mV) in females

Age (years)	Frontal	Sagittal	Transverse
18–29	1.54 ± 0.40	1.28 ± 0.44	1.51 ± 0.39
	0.67 → 2.56	0.49 → 2.48	0.67 → 2.58
30–39	1.51 ± 0.42	1.25 ± 0.43	1.50 ± 0.40
	0.62 → 2.42	0.60 → 2.25	0.88 → 2.51
40–49	1.29 ± 0.33	1.02 ± 0.35	1.31 ± 0.36
	0.68 → 1.96	0.40 → 1.78	0.84 → 2.76
50–59	1.29 ± 0.33	1.02 ± 0.36	1.31 ± 0.31
	0.68 → 1.96	0.42 → 1.86	0.69 → 2.01
60–	1.27 ± 0.52	0.82 ± 0.30	1.27 ± 0.48
	0.99 → 1.88	0.60 → 1.18	0.93 → 1.77

A5.2.4 Maximal Planar QRS Vector Angle

Table A5.24

96-Percentile ranges of maximal QRS vector angle (degrees) in males

Age (years)	Frontal	Sagittal	Transverse
18–29	34 ± 14	40 ± 24	−37 ± 26
	12 → 62	0 → 96	−105 → 14
30–39	28 ± 14	38 ± 36	−32 ± 28
	3 → 54	−12 → 144	−99 → 19
40–49	23 ± 19	33 ± 34	−28 ± 31
	−2 → 53	−13 → 140	−103 → 27
50–59	20 ± 13	31 ± 41	−25 ± 36
	−5 → 43	−23 → 150	−117 → 25
60–	22 ± 35	32 ± 33	−32 ± 0.35
	0 → 58	0 → 84	−103 → 25

Table A5.25

96-Percentile ranges of maximal QRS vector angle (degrees) in females

Age (years)	Frontal	Sagittal	Transverse
18–29	34 ± 15	40 ± 22	−35 ± 19
	13 → 59	−9 → 89	−85 → 3
30–39	35 ± 11	44 ± 20	−36 ± 22
	14 → 55	−13 → 79	−97 → −2
40–49	27 ± 19	37 ± 31	−33 ± 25
	6 → 48	−20 → 99	−109 → 9
50–59	27 ± 12	34 ± 28	−31 ± 27
	6 → 23	−15 → 142	−108 → 16
60–	21 ± 10	42 ± 54	−16 ± 31
	0 → 47	−41 → 180	−107 → 27

A5.2.5 Maximal T Vector Angle

Table A5.26

96-Percentile ranges of maximal T vector angle (degrees) in males

Age (years)	Frontal	Sagittal	Transverse
18–29	28 ± 15	154 ± 18	45 ± 17
	−6 → 60	110 → 184	13 → 81
30–39	25 ± 17	154 ± 21	44 ± 17
	6 → 48	96 → 180	5 → 78
40–49	27 ± 16	155 ± 18	46 ± 17
	5 → 56	113 → 180	9 → 76
50–59	25 ± 14	156 ± 21	46 ± 20
	3 → 51	98 → 181	2 → 78
60–	32 ± 18	147 ± 26	38 ± 25
	10 → 81	95 → 176	2 → 65

Table A5.27
96-Percentile ranges of maximal T vector angle (degrees) in females

Age (years)	Frontal	Sagittal	Transverse
18–29	30 ± 19	134 ± 26	28 ± 18
	8 → 50	61 → 178	−15 → 62
30–39	29 ± 10	137 ± 26	−30 ± 19
	9 → 48	80 → 183	−15 → 68
40–49	30 ± 16	140 ± 21	34 ± 17
	9 → 56	98 → 185	10 → 77
50–59	29 ± 26	140 ± 26	33 ± 29
	8 → 53	82 → 187	−13 → 76
60–	38 ± 30	144 ± 22	40 ± 24
	0 → 153	78 → 180	−6 → 105

Appendix 5B Normal Limits of Paediatric 12-Lead Vectorcardiogram

The vector data presented in this Appendix have been obtained from X, Y, Z leads derived from the 12-lead ECG according to the methods described in 11.6 and using the coefficients presented in ➋ Table 11.9. Data have been derived from 1,782 neonates, infants, and children, whose age distribution is shown in the upper part of ➋ Table 43.1.

Data are expressed as mean ± standard deviation below which is given the 96 percentile ranges, i.e., 2% of values are excluded at either end of the distribution except when the total number in the age group is small, in which case the 100% range is presented. Angular data are presented with respect to the reference frames illustrated in ➋ Figure 43.9.

B5.1 Scalar Measurements from the Leads X, Y and Z

Table B5.1
Maximal spatial P, QRS, and T magnitudes (mV)

Age	Max P	Max QRS	Max T
< 24 h	0.16 ± 0.05	1.83 ± 0.48	0.25 ± 0.08
	0.09 → 0.25	0.95 → 2.76	0.08 → 0.41
1 day	0.17 ± 0.05	1.71 ± 0.43	0.23 ± 0.07
	0.09 → 0.25	1.03 → 2.48	0.11 → 0.38
2 days	0.17 ± 0.05	1.62 ± 0.41	0.25 ± 0.08
	0.10 → 0.27	0.93 → 2.51	0.12 → 0.43
3 days	0.17 ± 0.04	1.60 ± 0.41	0.28 ± 0.08
	0.10 → 0.25	1.00 → 2.48	0.14 → 0.43
≤ 1 week	0.18 ± 0.05	1.53 ± 0.35	0.34 ± 0.10
	0.10 → 0.31	0.99 → 2.36	0.16 → 0.55
≤ 1 month	0.18 ± 0.06	1.38 ± 0.40	0.37 ± 0.10
	0.08 → 0.26	0.76 → 2.04	0.19 → 0.60
≤ 3 months	0.16 ± 0.04	1.87 ± 0.45	0.43 ± 0.09
	0.07 → 0.24	1.17 → 2.95	0.28 → 0.60
≤ 6 months	0.15 ± 0.04	1.70 ± 0.38	0.44 ± 0.12
	0.09 → 0.27	1.02 → 2.35	0.27 → 0.67

Table B5.1 (Continued)

Age	Max P	Max QRS	Max T
≤ 1 year	0.17 ± 0.05	1.81 ± 0.48	0.49 ± 0.11
	0.11 → 0.25	1.09 → 2.80	0.30 → 0.69
1–2 years	0.15 ± 0.04	1.86 ± 0.49	0.47 ± 0.14
	0.09 → 0.24	1.03 → 2.82	0.22 → 0.70
3–4 years	0.15 ± 0.04	2.20 ± 0.53	0.53 ± 0.15
	0.09 → 0.24	1.25 → 3.21	0.21 → 0.84
5–6 years	0.14 ± 0.04	2.46 ± 0.55	0.58 ± 0.15
	0.08 → 0.24	1.56 → 3.61	0.31 → 0.87
7–8 years	0.14 ± 0.04	2.43 ± 0.48	0.61 ± 0.15
	0.08 → 0.23	1.56 → 3.38	0.36 → 0.90
9–10 years	0.14 ± 0.04	2.51 ± 0.53	0.65 ± 0.18
	0.07 → 0.23	1.55 → 3.33	0.34 → 1.00
11–12 years	0.14 ± 0.05	2.43 ± 0.52	0.60 ± 0.19
	0.07 → 0.24	1.45 → 3.51	0.30 → 0.92
13–14 years	0.14 ± 0.05	2.29 ± 0.59	0.58 ± 0.20
	0.06 → 0.25	1.25 → 3.60	0.26 → 1.00
15–16 years	0.14 ± 0.05	2.17 ± 0.61	0.55 ± 0.20
	0.07 → 0.27	1.20 → 3.23	0.24 → 0.95

Table B5.2

Maximal P, QRS, and T vector angles in transverse plane (degrees)

Age	Max P	Max QRS	Max T
< 24 h	22 ± 38	−6 ± 108	23 ± 69
	−42 → 70	−131 → 181	−76 → 136
1 day	28 ± 35	−10 ± 113	−4 ± 68
	−47 → 78	−139 → 216	−105 → 132
2 days	25 ± 35	26 ± 111	−32 ± 49
	−49 → 83	−135 → 217	−92 → 92
3 days	26 ± 31	10 ± 99	−43 ± 42
	−45 → 73	−138 → 206	−100 → 66
≤ 1 week	24 ± 33	34 ± 88	−44 ± 31
	−46 → 73	−130 → 210	−92 → 20
≤ 1 month	15 ± 36	29 ± 89	−25 ± 33
	−63 → 66	−126 → 185	−80 → 10
≤ 3 months	8 ± 37	14 ± 37	−31 ± 26
	−67 → 62	−111 → 62	−93 → 31
≤ 6 months	6 ± 33	10 ± 44	−38 ± 22
	−41 → 59	−107 → 74	−72 → 14
≤ 1 year	11 ± 34	13 ± 38	−45 ± 19
	−62 → 67	−45 → 83	−87 → −13
1–2 years	23 ± 36	−19 ± 50	−38 ± 25
	−50 → 69	−110 → 89	−83 → 5
3–4 years	20 ± 36	−28 ± 28	−22 ± 21
	−48 → 76	−109 → 27	−64 → 16
5–6 years	28 ± 37	−29 ± 21	−10 ± 17
	−43 → 87	−96 → 7	−45 → −23

Table B5.2 (Continued)

Age	Max P	Max QRS	Max T
7–8 years	19 ± 39	−30 ± 21	−4 ± 17
	−46 → 82	−63 → 16	−40 → 35
9–10 years	17 ± 37	−31 ± 30	3 ± 18
	−50 → 87	62 → 10	−31 → 43
11–12 years	14 ± 39	−33 ± 19	9 ± 17
	−53 → 82	−78 → 4	−21 → 51
13–14 years	10 ± 44	−32 ± 27	9 ± 17
	−62 → 90	−71 → 9	−20 → 41
15–16 years	4 ± 43	−29 ± 50	20 ± 20
	−57 → 81	−93 → 204	−19 → 56

Table B5.3
Maximal P, QRS, and T vector angles in frontal plane (degrees)

Age	Max P	Max QRS	Max T
< 24 h	54 ± 14	79 ± 87	76 ± 69
	27 → 74	−90 → 208	25 → 151
1 day	58 ± 16	88 ± 80	67 ± 39
	18 → 78	−106 → 190	18 → 135
2 days	58 ± 21	77 ± 81	62 ± 43
	12 → 83	−71 → 191	23 → 102
3 days	52 ± 23	67 ± 79	54 ± 22
	−41 → 72	−68 → 194	16 → 106
≤ 1 week	57 ± 14	62 ± 65	51 ± 22
	22 → 77	−46 → 186	19 → 94
≤ 1 month	54 ± 14	61 ± 68	42 ± 19
	30 → 81	−19 → 190	18 → 82
≤ 3 months	56 ± 12	36 ± 25	38 ± 22
	25 → 79	4 → 135	18 → 94
≤ 6 months	56 ± 13	35 ± 27	36 ± 13
	26 → 77	−7 → 105	13 → 56
≤ 1 year	56 ± 14	31 ± 35	41 ± 16
	18 → 76	−18 → 97	21 → 75
1–2 years	57 ± 14	41 ± 38	38 ± 17
	5 → 73	−73 → 151	16 → 72
3–4 years	58 ± 16	39 ± 23	34 ± 10
	23 → 82	12 → 127	18 → 55
5–6 years	52 ± 28	37 ± 20	29 ± 9
	−36 → 82	13 → 101	11 → 46
7–8 years	54 ± 22	38 ± 18	29 ± 8
	3 → 75	13 → 63	12 → 46
9–10 years	50 ± 25	41 ± 19	26 ± 8
	−21 → 80	15 → 65	12 → 45
11–12 years	53 ± 22	39 ± 17	28 ± 8
	−26 → 85	15 → 72	9 → 42
13–14 years	44 ± 41	41 ± 17	28 ± 8
	−91 → 81	13 → 74	12 → 46
15–16 years	50 ± 34	47 ± 32	28 ± 9
	−90 → 82	11 → 181	11 → 47

Index

Note: The page numbers that appear in bold type indicates a substantive discussion of the topic.

A

C

D

E

F

G

I

J

K

L

N

Q

R

S

T

U

V

W